D0913078

McGraw-Hill Yearbook of Science and Technology 1980 REVIEW

1981 PREVIEW

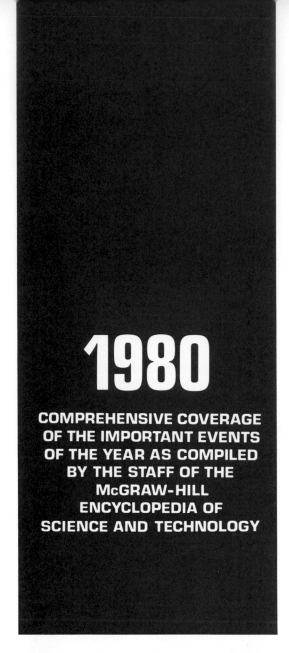

1980

**COMPREHENSIVE COVERAGE
OF THE IMPORTANT EVENTS
OF THE YEAR AS COMPILED
BY THE STAFF OF THE
McGRAW-HILL
ENCYCLOPEDIA OF
SCIENCE AND TECHNOLOGY**

McGraw-Hill **Yearbook of**

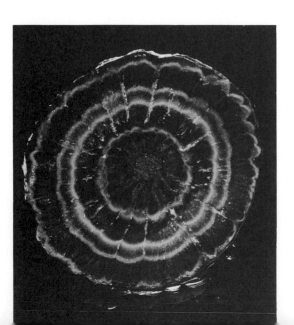

Science and Technology

McGRAW-HILL BOOK COMPANY

NEW YORK   ST. LOUIS   SAN FRANCISCO
AUCKLAND   MONTREAL
BOGOTA   NEW DELHI
GUATEMALA   PANAMA
HAMBURG   PARIS
JOHANNESBURG   SAN JUAN
LISBON   SAO PAULO
LONDON   SINGAPORE
MADRID   SYDNEY
MEXICO   TOKYO
              TORONTO

On preceding pages:

*Cross sections of large hailstones showing the
structure of alternating rings of clear and white ice.
(Alberta Research Council, Edmonton)*

Library of Congress Catalog Card Number: 62-12028

ISBN 0-07-045488-4

Printed in the United States of America
1234567890   DODO   8654321

The Library of Congress cataloged the original printing of this title as follows:

McGraw-Hill yearbook of science and technology. 1962 –
   New York, McGraw-Hill Book Co.

      v.   illus.   26 cm.
      Vols. for 1962 –   compiled by the staff of the McGraw-Hill encyclopedia
   of science and technology.

      1. Science – Yearbooks.   2. Technology – Yearbooks.   I. McGraw-Hill
   encyclopedia of science and technology.
   Q1.M13   505.8   62-12028
   Library of Congress   (10)

Table of Contents

Editorial Advisory Board

Editorial Staff

Consulting Editors

Consulting Editors (continued)

Contributors

A list of contributors, their affiliations, and the articles they wrote will be found on page 415.

Preface

The 1981 *McGraw-Hill Yearbook of Science and Technology,* continuing in the tradition of its 19 predecessors, presents the outstanding achievements of 1980 in science and technology. Thus it serves as an annual review and also as a supplement to the *McGraw-Hill Encyclopedia of Science and Technology,* updating the basic information in the fourth edition (1977) of the Encyclopedia.

The Yearbook contains articles reporting on those topics that were judged by the 56 consulting editors and the editorial staff as being among the most significant developments of 1980. Each article is written by one or more authorities who are actively pursuing research or are specialists on the subject being discussed.

The Yearbook is organized in three independent sections. The first section, a preview of 1981, includes six feature articles, providing comprehensive, expanded coverage of subjects that have broad current interest and possible future significance. The second section, photographic highlights, presents photographs considered noteworthy because of the unusual nature of the subject or the advanced photographic technique. The third section comprises 140 alphabetically arranged articles on such topics as optical communications, computerized tomography, space probes, artificial blood, people movers, military aircraft technology, and memory.

The *McGraw-Hill Yearbook of Science and Technology* provides librarians, the general public, students, teachers, and the scientific community with information needed to keep pace with scientific and technological progress throughout the world. The Yearbook has successfully served this need for the past 20 years through the ideas and efforts of the consulting editors and the contributions of eminent international specialists.

SYBIL P. PARKER
Editor in Chief

McGraw-Hill Yearbook of Science and Technology 1980 REVIEW

1981 PREVIEW

Biotechnology

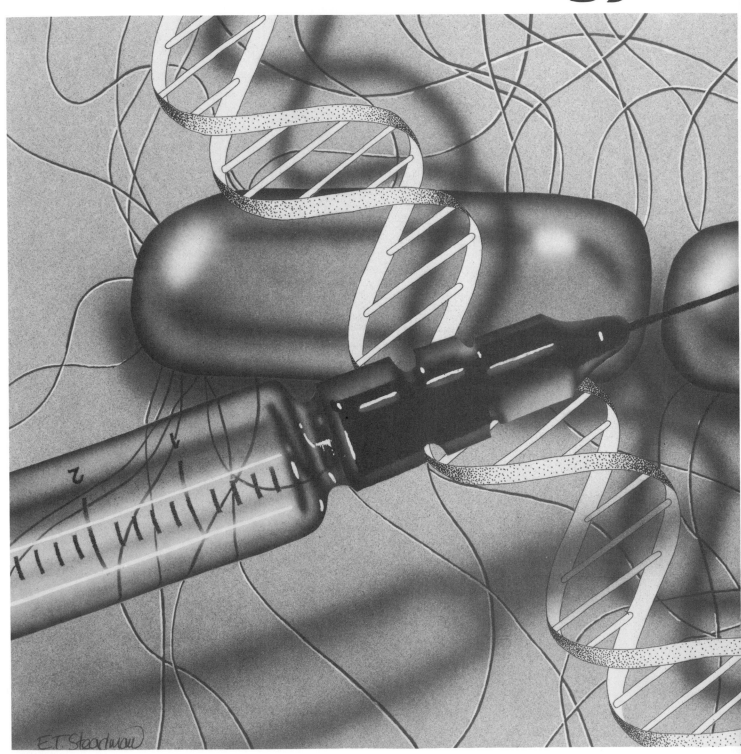

Jonathan King

Cristián Orrego

Jonathan King is professor of biology at the Massachusetts Institute of Technology and director of the biology electron microscopy facility. He has a Ph.D. in genetics from the California Institute of Technology. He has long been interested in the social aspects of genetic technology, and his research has focused on the genetic control of virus structure and assembly.

Cristián Orrego is a postdoctoral associate in bacterial genetics in the biology department at MIT. He has a Ph.D. in biochemistry from Brandeis University and is a consultant to international development agencies in the area of biotechnology.

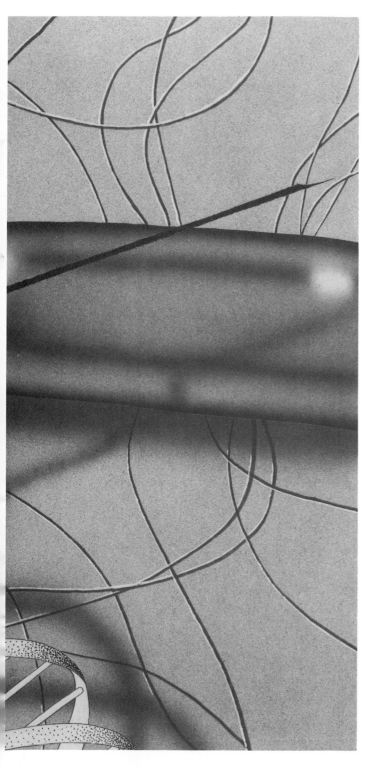

Biotechnology may be defined as the application of living organisms, or their biological systems or processes, to the manufacture of useful products (Table 1). Microbial processes have been utilized in the production of fermented food and beverages for at least 5000 to 6000 years. Little was known about the microscopic agents responsible for these processes until Louis Pasteur demonstrated that the fermentation of alcohol from sugar during wine production was caused by single-cell yeast organisms. This discovery eventually led to the understanding that specific microorganisms are responsible for particular chemical changes in food.

During the past 30 years there have been extraordinary advances in knowledge of fundamental biological processes, particularly at the molecular and cellular level. These advances have derived primarily from the major investment of public funds in the training of biomedical scientists and support for biomedical research conducted by the governments of the industrialized countries since World War II. The 1980 budget for biomedical research in the United States was approximately $3,000,000,000, a thousand times the Federal expenditure for such research in 1948. The U.S. agencies with primary responsibility for dispersing public funds for biomedical research have been the National Institutes of

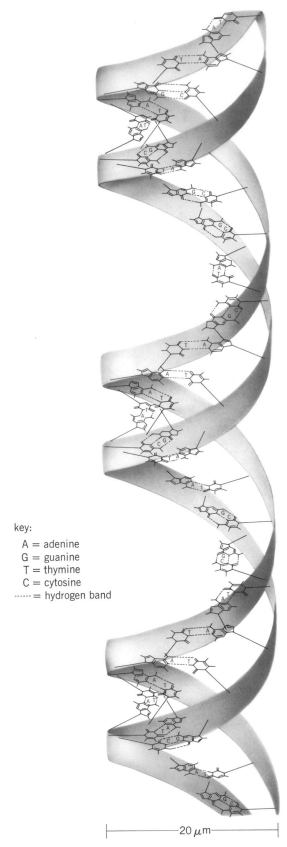

key:

A = adenine
G = guanine
T = thymine
C = cytosine
------ = hydrogen band

├─────── 20 μm ───────┤

Fig. 1. Double-helix structure of the DNA molecule. *(From C. Grobstein, The recombinant DNA debate, in Recombinant DNA, with introductions by D. Freifelder, copyright © 1979 by Scientific American, Inc.; used with permission)*

Table 1. Production capacities for fermentation products in North America*

Fermentation products	Annual production capacity range of each product, metric tons/year
Enzymes, antibiotics, vitamins	100–5000
Xantham gum, monosodium glutamate	5,000–20,000
Ethanol, single-cell protein	10,000–60,000

*From G. E. Tong, Industrial chemicals from fermentation, *Enzyme Microb. Technol.*, 1:173, 1979.

Health and the National Science Foundation. They have supported scientists working primarily in university, hospital, and government laboratories. Most of this research has focused on increasing biological knowledge in order to understand and alleviate disease, and to improve human health in general.

With the development of increasing knowledge of how organisms grow and reproduce has come technologies to selectively modify these organisms. The most dramatic of these procedures, recombinant DNA technology, permits major modification of the DNA molecule (Fig. 1), the blueprint of living organisms. These artificially induced changes are then inherited by subsequent generations.

An important area of biotechnological development will be in manipulating microorganisms to increase the efficiency of production of some desired product per unit of input nutrients provided to the microorganism for its growth. Achieving this depends on genetic modification of the microorganism so that it gains the desired property in a permanent and inheritable manner. Relatively crude techniques have already allowed the industrial microbiologist to genetically modify antibiotic-producing fungi in order to increase the production of penicillin, for example, by factors of more than a thousand from the productive capacity displayed by the fungus when it was first isolated.

Conventional genetic manipulation is limited by the fact that organisms generally exchange genetic material only with members of the same or related species. The biological significance of the mating and sexual exchange of related segments of genetic material is to create combinations which may be slightly better for adaptation to a changing environment. Recombinant DNA technology—genetic engineering—allows scientists to incorporate segments of DNA from one organism into the cells of another organism. The donor and recipient may be closely related, for example, two strains of bacteria, or they may be very different, for example, a mouse and a bacterium. Such genetic exchange between unrelated organisms is extremely rare in nature.

The technology for splicing genes—segments of DNA molecules—into unrelated DNA molecules and introducing them back into a living cell is described in outline in Fig. 2. As the host reproduces, it also replicates the foreign DNA. The process is referred to as cloning a gene. If the foreign DNA expresses its information in the recipient cell as it is translated into its protein product, the hybrid cell will have properties not found in the parent cell.

Cells of different species can also be fused together by appropriate techniques (protoplast fusion) through removal of their cell wall and subsequent regeneration of individual hybrid cells containing genetic material from both parents. This offers another means of obtaining genetic combinations that otherwise would rarely occur in nature.

These technologies of genetic manipulation have the potential to alter the character of biological species, including higher organisms, at rates far more rapid than have occurred in the history of the ecosystem. Further, the same technologies developed to produce a more efficient yeast strain for alcohol production, for example, can also be used for human genetic modification. Application of the technologies of genetic manipulation whether in the environmental, agricultural, or biomedical spheres, generates difficult social, ethical, and economic questions.

Substantial debate over the applications of recombinant DNA technology has been focused on the question of whether new occupational or environmental health hazards might be generated from the introduction, in the laboratory, of foreign DNA into microorganisms. On the assumption that genetically engineered strains of microorganisms could escape into the environment and establish themselves in some niche with potentially unforeseen and eventually undesirable consequences, guidelines have been adopted requiring that certain recombinant DNA experiments be performed with weakened strains of bacteria unlikely to survive outside the laboratory. Physical containment procedures also diminish the likelihood that such strains will escape. However, in certain countries, including the United States, these guidelines apply only to academic research laboratories and not to industrial research and production. The issue of regulation will be discussed below.

Recombinant DNA and other modern technologies of biological manipulation are also being applied in the biomedical sector. However, the scope of this article does not allow for discussion of the applications, nor the social issues associated with them.

In nature, large numbers of different kinds of organisms interact with each other, creating systems which support the growth of all of them. Biological productivity of the ecosystem depends on the recycling and efficient use of nutrients, habitats, and energy sources. Biotechnology, by vitrue of its being based on the elements responsible for maintenance of biological systems in nature, offers increased prospects of designing sustainable production processes which will lessen the dependence on nonrenewable resources such as petroleum.

BASIC FEATURES OF MICROBIAL TECHNOLOGIES

Bacteria, algae, fungi, protozoa, and viruses are fast-growing microscopic organisms that make up about 0.2–2% of the biological carbon of the pla-

net. Only a very small fraction of them are implicated in human and animal diseases. Microbes display great biochemical diversity which, in chemical terms, is a reflection of the enzymes they produce. Enzymes are large protein molecules capable of catalyzing (speeding up) chemical reactions, by factors of many millions, generally at normal temperatures and pressure, and leading to a limited number of products. By contrast, to make

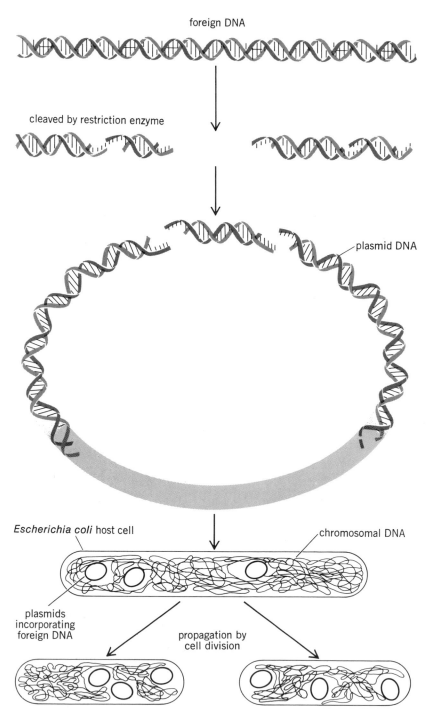

Fig. 2. Artificial splicing of genes using the recombinant DNA technique. (*From C. Grobstein, The recombinant DNA debate, in Recombinant DNA, with introductions by D. Freifelder, copyright © 1979 by Scientific American, Inc.; used with permission*)

the same chemical reaction work in the absence of the particular enzyme, very high temperatures and pressures would be required and a less controlled variety of different products would be obtained. Enzymes catalyze the same chemical change millions of times without themselves undergoing permanent chemical change. Some microorganisms are the only living forms known to have certain enzymes capable of degrading cellulose to its constituent sugars. Another discrete group of bacteria has the capacity of "fixing" the nitrogen gas from the atmosphere into amino acids inside the cell. These in turn become the source of food nitrogen for all higher organisms devoid of nitrogen-fixing enzymes.

Most contemporary biotechnologies require the addition (inoculation) of a particular microbe to a solution containing substances (substrates) whose conversion by the microorganism results in a useful product (product fermentation). In certain cases the microorganism itself is the useful product. Other microbial technologies consist in the more or less controlled management of mutually beneficial associations between plants and specific microbes, such as the nitrogen-fixing bacteria.

Microorganisms derive their material energy for growth and maintenance by utilizing external nutrients and breaking them down through a sequence of metabolic steps catalyzed by enzymes. This transforms the energy contained in the nutrients into biologically useful molecules (Fig. 3). The dependence on or independence from oxygen for growth determines the type of metabolic pathways utilized for energy conversion. Aerobes require oxygen for energy conversion. Anaerobes possess pathways that do not require free oxygen for their multiplication. Facultative aerobes can use either metabolic pathway, depending on the presence or absence of oxygen in the medium.

Water is the single most important component for microbial growth, followed by carbon. Heterotrophic microorganisms can utilize a wide variety of organic carbon materials (mainly sugars) as a source of carbon for building the different constituents of the cell, and as a source of energy. Autotrophic organisms are able to utilize the very simple molecule of carbon dioxide (CO_2) from the atmosphere as their major source of carbon for synthesis, while obtaining energy either from sunlight (through photosynthesis) or from the metabolism of inorganic compounds rich in chemical energy such as ammonia (NH_3), hydrogen (H_2), and a variety of minerals. Microorganisms also require nitrogen, sulfur, phosphate, inorganic salts (such as sodium chloride), and trace metals for growth. Most microbes grow at a temperature of around 30°C, though psychrophilic (cold-loving) organisms thrive best between 0 and 30°C. Organisms such as the latter are frequently associated with spoilage of refrigerated foods. Thermophiles (heat-loving) grow optimally between 40 and 70°C. Most microorganisms grow best at neutral conditions of acidity and alkalinity.

CONTRIBUTIONS OF MICROBIAL PROCESSES

One of the concerns at the center of the renewed interest in biotechnology is the prospect of the eventual unavailability of petroleum and natural gas, not only as a source of fuel, but as a source of organic chemicals which are the starting materials

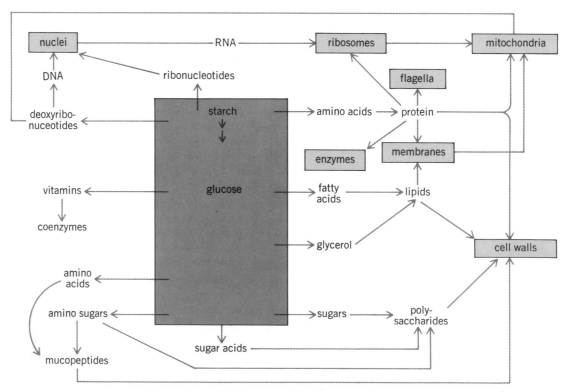

Fig. 3. Steps in microbial metabolism of starch for biosynthesis. (Adapted from D. I. C. Wang et al., Fermentation and Enzyme Technology, copyright © 1979 by John Wiley and Sons; used with permission)

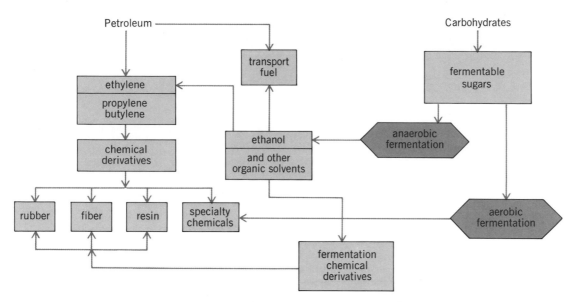

Fig. 4. Alternative routes to organic chemicals of industrial use. (*Adapted from G. E. Tong, Industrial chemicals from fermentation, Enzyme Microb. Technol., 1:175, 1979*)

for many synthetic processes. The diversity of microorganisms in general, and of their enzymes in particular, coupled with recent developments in fermentation plant design offers the opportunity of developing alternative modes of production for some of the chemicals presently derived from non-renewable fossil resources (Fig. 4). Another very important feature of biotechnology will be the utilization of agricultural, industrial, and urban wastes as raw materials in efforts to establish a self-sustaining mode of production closely in touch with natural mechanisms of nutrient recycling in the biosphere. An incomplete list of areas of production where biotechnology already is involved, or soon will be, is as follows:

1. Ethanol production made by microbial fermentation of sugars from plants and agricultural residues. Ethanol is used as fuel for the internal combustion engine and as a chemical precursor for a variety of compounds presently obtained from petroleum.

2. Maintenance of soil fertility by direct application of the sludge that remains after microbial biogas generation of human, animal, and agricultural wastes. The methane gas so generated can be burned for heating and lighting. This process also results in the inactivation of human pathogens and parasites present in human wastes.

3. Replacement of synthetic nitrogen fertilizers by appropiate management of microorganisms which use nitrogen gas from the atmosphere delivering biologically useful nitrogen compounds to crop plants.

4. Replacement of toxic and persistent chemicals used to control plant-damaging insect pests, by microbial preparations pathogenic only to the target insect.

5. Utilization of microorganisms as a protein source in animal foods.

6. Enzyme technology: the utilization of immobilized enzymes or cells in continuous reactors to carry out specific chemical conversions in the petrochemical and pharmaceutical industries.

7. The production of medically important animal proteins by genetically engineered microorganisms.

Conversion of biomass to fuels. Plants capture solar energy in order to fix atmospheric carbon dioxide into sugar carbohydrates by the process of photosynthesis. Most of the energy is actually stored in the sugars themselves. These can be the simple sugar glucose ($C_6H_{12}O_6$) or polymers of glucose starch and cellulose. Simple carbohydrates can be fermented into alcohol (ethanol) by yeasts and certain bacteria. Starch from a variety of roots and grains, and cellulose from trees are enormously abundant polymers which can be degraded back into glucose, usually by means of the appropriate partially purified enzymes. However, three technical obstacles currently prevent the utilization of the vast reservoir of cellulose for purposes of bioconversion. First is the need to dislodge and separate the cellulose fibers from the polymer lignin, a material that tightly glues the fibers together. Second, the cellulose has to be degraded to glucose, which can be obtained either by acid or microbial enzymes. A third limitation common to all biological processes of alcohol generation consists in the relatively large energy requirements posed by the distillation step in order to concentrate the alcohol from the fermentation container. These areas are presently the subject of intense research and development efforts.

Sucrose (glucose linked to fructose) from sugarcane molasses and starch from corn grain are the two sources of biomass in current use for ethanol fermentation.

Starch is a versatile feedstock for many glucose-based fermentations, since degrading it to glucose is much more efficient than with cellulose due to the absence of lignin from plant starch materials. The widely cultivated and starch-rich tropical root cassava, grown in more than 80 countries, promises to be an important source for alcohol production in the future.

The traditional alcohol fermentation procedure

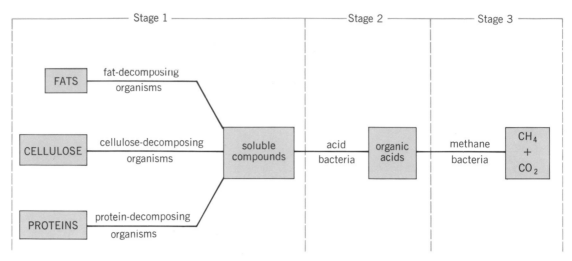

Fig. 5. Sequence of microbial fermentation of organic materials resulting in biogas generation. (*Methane Generation from Human, Animal, and Agricultural Wastes, National Academy of Sciences, 1977*)

utilizes the yeast *Saccharomyces cerevisiae*, capable of breaking down sucrose and subsequently fermenting the constituent sugars to ethanol at 30°C. Two other microorganisms presently under study could play a role in alcohol production in the future. The fungus *Trichoderma reeseii* is currently being used as a source of enzymes to degrade cellulose to glucose. A variety of thermophilic *Clostridium* species are capable of simultaneously degrading cellulose to glucose and converting the latter to ethanol. A major problem not yet explicitly dealt with in the national alcohol programs undertaken by various countries is the treatment of solid residues (stillage) resulting from biomass fermentation. Stillage is generated at a rate 12 to 13 times greater than alcohol. On the average, the pollution potential of 2 liters of untreated stillage equals the sewage produced by one person per day. Microbial conversion of the stillage to methane fuel and subsequent utilization of the stillage thereof as fertilizer offers a promising manner in which to return nutrients to the soil initially used for biomass production.

Microbial biogas production. Methane gas production for cooking and lighting by microbial fermentation of agricultural, animal, and human wastes is being extensively used or introduced in many countries in Asia, particularly China, Africa, and Latin America. A further valuable product, and indeed the most sought-after product resulting from biogas generation, is the residual sludge. This can be used as fertilizer by direct application to crops, regenerating the phosphorus, nitrogen, and potassium of local soils. Biogas production occurs in the absence of oxygen and is the result of bacterial action on organic matter containing cellulose, starch, proteins, fats, and minerals (Fig. 5). The bacteria for this process can be initially obtained by inoculation of municipal and agricultural wastes with animal dung or a sample from a septic tank.

Biogas is a mixture of methane (60%) and carbon dioxide (40%), of which the former can be burned into carbon dioxide. The main type of biogas-producing units are of a size ($3-10$ m³ of gas a day) to satisfy the energy requirements of small families (Fig. 6), although biogas production plants which partially satisfy the energy requirements of small towns are in existence.

Management of nitrogen-fixing microorganisms. Production of nitrogen fertilizer in the form of ammonia (NH_3) requires natural gas, from which hydrogen (H_2) is obtained. The price of ammonia increased ninefold between 1972 and 1974, and prices since then have increased further, putting them far out of reach for many countries. A significant opportunity for diminishing dependence on chemically fixed nitrogen lies in the utilization of biologically fixed nitrogen by the *Rhizobium* bacterium, which lives symbiotically in nodules on the

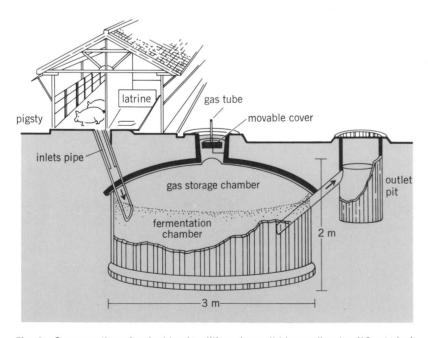

Fig. 6. Cross section of a double-pipe "three-in-one" biogas digester (10 m³/day) constructed in China. The wall of the digester is 4 cm thick. It is initially filled to 60 cm from the top. (*From E. Ariane Van Buren, Biogas beyond China: First International Training Program for Developing Countries, Ambio, vol. 9, no. 1, 1980*)

Table 2. Nitrogen fixed by various Rhizobium-legume associations

Legume plant	Nitrogen fixed (kg/ha/yr)	Value as fertilizer of nitrogen fixed (U.S.$/ha)*
Alfalfa	128–300	38–90
Clover	118–155	35–47
Cowpea	84	25
Faba bean	240–325	72–98
Lentils	103	31
Peanuts	47	14
Soybeans	60–80	18–24
Pasture legumes	118–400	36–120

*The nitrogen fixed is valued at a 1975 "farm gate" cost of U.S.$0.30 per kilogram of nitrogen fertilizer; ha = hectare.

roots of legumes such as soybeans and peanuts (Table 2). Efficient management of the *Rhizobium*-legume association, coupled with a reversal of the trend which has resulted in diminishing production for human consumption of legumes rich in protein compared to low-protein cereal grains, would stabilize nitrogen fertilizer requirements in the coming decades. A number of scientists have proposed designing *Rhizobium* or other microorganisms which would infect the roots of cereals and establish synthetic associations. The complexity of these associations makes it unlikely that this endeavor will be rapidly achieved.

There are, however, three applications of biotechnology to nitrogen economy in agriculture which are currently succeeding. Artificial coating of legume seeds with specific *Rhizobium* strains previously propagated in pure culture has been practiced for the last 70 years. Subsequent optimum *Rhizobium* nodulation provides up to 25% of the total nitrogen required by the plant. Attainment of this optimum is dependent on a number of local variables, and successful utilization of this technology requires local personnel trained to identify the significant problems for proper application of bacterial preparations.

Biological nitrogen fixation by microbes in flooded and nonflooded soils is known to contribute about 70 kg/ha/year to the total nitrogen economy of rice. Rice forms part of the diet of more than half of the world's population, and much of it is grown in small plots by farmers with no access to commercial fertilizers. Artificial inoculation of rice paddies with blue-green algae, organisms which are simultaneously capable of fixing carbon dioxide and nitrogen from the atmosphere, reduces nitrogen fertilizer requirements by one-third. Another contribution to the nitrogen economy of rice is provided by what has been called a natural "floating nitrogen factory." The blue-green algae *Anabaena azollae*, which forms a symbiotic association with the floating fern *Azolla*, has been used for centuries as green manure in the Orient. The *Azolla*-algae complex doubles in mass every 5 days and decomposes rapidly in soil, releasing its nitrogen in a few weeks. This nitrogen can be subsequently assimilated by the rice plant.

Biological control of plant insects. Current agricultural policies in many nations emphasize

short-term food and fiber production, often to maximize financial return on corporate and individual investment, to the detriment of cultivation practices consistent with maintenance of long-term productivity.

The large-scale planting of single varieties renders crops highly susceptible to infestation and the spread of individual pest species. This has led to the increasing use of enormous quantities of chemical pesticides (Fig. 7) to maximize short-term output of crops. Such policies lead in many cases to the decimation of nontarget beneficial insect species, and to the introduction of highly toxic, persistent chemical compounds into the environment which eventually concentrate in human and animal tissues and damage the soil microflora.

Coupled with these problems is the increasing resistance of the target insects to pesticides (Fig. 7).

Though not addressing the fundamental issue of the design of agricultural production, biological agents have been proposed as an alternative to the chemical pesticide strategy. A number of microbial insect pathogens such as viruses, bacteria, and fungi are unique control agents due to the specificity with which they affect insect pests and

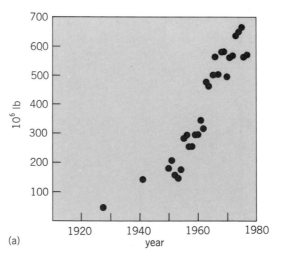

(a)

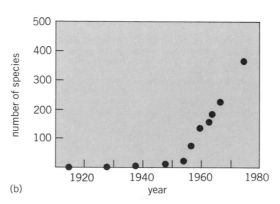

(b)

Fig. 7. Increased use of insecticides is reflected in (a) production levels in the United States. This is coupled with (b) an increase in numbers of insect pests resistant to insecticides. (*From R. L. Metcalf. Changing role of insecticides in crop protection, Ann. Rev. Entomol., vol. 25, 1980*)

their lack of toxicity toward beneficial insect species. The microbe *Bacillus thuringiensis*, pathogenic to more than 100 species of lepidoptera (butterflies and moths), is the bioinsecticide of widest application. It is being used against insect defoliators of cotton, soybeans, forests, grapes, tobacco, fruits, and vegetables. The bacterium also shows promise in the control of several pests of stored flour, cornmeal, almonds, and tobacco. *Bacillus popilliae*, a pathogen of Japanese beetles (which collectively feed on more than 200 species of plants, including turf, pasture, and fruits) is also commercially available. Both of these bacilli are used in spore form, that is, as dormant bacteria which are resistant to desiccation and heat and can be stored for long periods of time. The insecticidal activity of these insect pathogens derives from a protein toxin that is located within the spore and released in the gut of the insect upon its ingestion. The insect virus *Baculovirus heliothis* has been approved in the United States for use against the destructive cotton pest the cotton bollworm and also the tobacco budworm. Other viruses are being tested to control two timber-defoliating insects, the gypsy moth and the tussock moth. Fungi have been used to control a number of pests of soybeans, coconut, sugarcane, and pasturelands.

In general, technology for production of bioinsecticides is relatively simple and utilizes inexpensive media to grow the microorganism. No documented example of the acquisition of resistance to a microbial or viral pathogen by an insect is available. Nevertheless, it is likely that sole reliance on biological insecticides will slowly select for partially resistant insect pests, neutralizing the favorable aspects of this pest control strategy. Foresight would demand that policies be formulated which stimulate the introduction of agricultural methods to assure adequate crop output while maintaining long-term agricultural systems.

Single-cell protein. Large resources have been dedicated in the last 20 years to the development of processes resulting in the growth of microorganisms on alcohols and several petroleum components as carbon sources. Large-scale production of single-cell protein to substitute for soymeal or fish meal in enriching animal feeds is being pursued in a number of industrialized countries. Currently, the most promising single-cell protein process consists in the growth of the bacteria *Methylomonas clara* or *M. methylotrophus* in methanol, CH_3OH, obtained by the chemical oxidation of natural gas, CH_4 (Fig. 8). The final product is a powder containing up to 80% protein.

Agricultural residues constitute approximately two-thirds of total crop production. The conversion of these residues, as well as the conversion of wastes from agroindustrial plants, into microbial biomass is receiving increasing attention. Some of the agricultural and agroindustrial by-products useful as substrates in microbial processes are:

Sugarcane molasses
Maize stover
Straw
Bran
Coffee hulls
Cocoa hulls
Coconut hulls
Fruit peels
Fruit leaves
Bagasses from sugarcane
Oilseed cakes
Cotton wastes
Tea wastes
Bark
Sawdust
Paper mill effluents
Cannery effluents
Milk-processing effluents

The most established bioconversion process of this type to date is the growth of the yeast *Candida utilis* in molasses, whey or sulfite waste liquor, by-products of the sugarcane, cheese, and paper industries, respectively, and sources of simple sugars. The yeast is extensively used in feed and food supplements. The utilization of cellulose and starch-containing crop residues for microbial biomass generation has yet to be developed at a practical level.

It is important to realize that not all of these resources should be dedicated for microbial biomass

Table 3. Industrial applications of microbial enzymes

Enzyme	Industry	Use	Type of reaction and substrate of the enzyme
Amylase	Baking	Reduction of dough viscosity	Degradation of starch
	Brewing	Liquefaction of cereals	
	Paper	Production of sizes	
	Textile	Removal of starch sizes	
Proteases	Leather	Leather bating and dehairing	Degradation of proteins
	Food	Production of fish solubles	
		Meat tenderization	
		Cheese manufacture	
Pectinases	Food	Clarification of fruit juices	Degradation of the sugar polymer pectin
Glucoamylase	Sugar	Production of glucose syrups	Degradation of starch to glucose
Glucoseisomerase	Sugar	Production of fructose syrups	Conversion of glucose to fructose

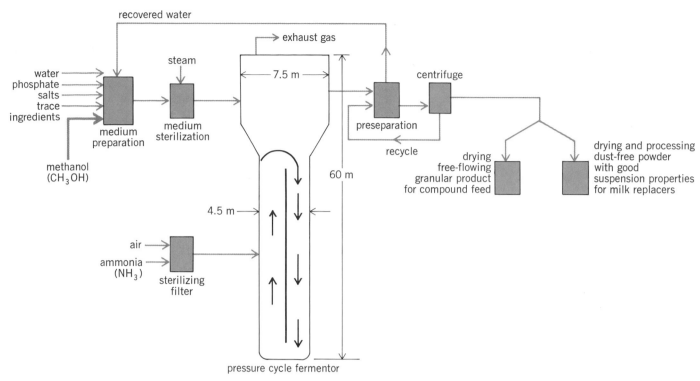

Fig. 8. Process for the production of bacterial single-cell protein using methanol as carbon source. (*Imperial Chemical Industries, Ltd.*)

generation, as most crop residues have to be returned to the soil to replenish nutrients and prevent soil erosion.

Enzyme technology. The enzyme-catalyzed degradation of corn and wheat starch into glucose followed by enzymatic conversion to fructose is now a major industry. The "high-fructose" corn syrups provide a sugar substitute which is twice as sweet as cane sugar. This process is one example of the development and application of partially purified enzymes for industrial production (Table 3). Enzyme technology has recently evolved to the use of immobilized enzymes and immobilized cells in continuous reactors where the stability of the enzyme catalyst is considerably increased over its use free in solution. A variety of stable polymers serve to entrap cells and to absorb or chemically link enzyme molecules to them (Fig. 9). Both then are packed in columns called bioreactors. New uses for enzymes are envisaged for carrying out specific conversions in the petrochemical industries. The utilization of immobilized cells for oxidation of hydrocarbons at mild temperatures and pressures is expected soon to provide an alternative to the conventional industrial process performed at high temperature and high pressures. For example, immobilized cells of the bacterium *Methylococcus capsulatus* (genetically engineered so as to increase the synthesis of the appropriate enzymes) should allow the oxidation of ethylene and propylene to the corresponding oxides, both large-volume chemical feedstocks of the petroleum industry.

Production of animal proteins. As noted above, recombinant DNA technology permits the transfer of DNA from higher organisms into bacteria and other microbes. Use of this technology to produce human proteins, such as insulin, for therapeutic use has been heavily publicized in the last several years. A number of corporations have been formed specifically for the commercial exploitation of these technologies. In some cases it has been possible to obtain expression of the foreign gene in the microbial cell. The protein hormone somatostatin and the human growth hormone have now been obtained from *Escherichia coli* after introduction of the particular gene into the bacterium. Insulin has also been obtained in this manner. Several laboratories, in collaboration with a number of the new genetic engineering corporations, have reported the cloning and expression of human interferon in *E. coli*. This protein has the potential to

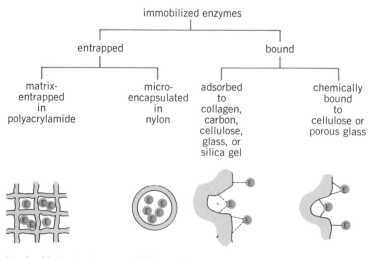

Fig. 9. Methods of enzyme (E) immobilization

become a broad-spectrum antiviral drug. Work suggesting possible use of interferon in cancer therapy is at present inconclusive. The opiate peptide beta-endorphin from the mouse has also been inserted into this bacterium and the production of a biologically active endorphin protein demonstrated. These animal proteins, which have heretofore been extracted from animal tissues in very small amounts and at considerable expense, become economically attractive to produce by using the more efficient microbe-supported synthesis methodology.

Another valuable prospect for recombinant DNA procedures will be the production of vaccines for viral diseases of humans and animals, including some, such as hepatitis, not yet effectively controlled. Much progress has already been made by such means toward the production of a protein component of the hepatitis virus B, which could be used for human immunization. However, therapeutic use in humans, especially on a continuous basis, of some of the above proteins awaits demonstration that products from bacterial cultures approach absolute purity. Contaminating bacterial proteins would be intensely antigenic (rejected by the human body) and fever-inducing. These considerations have inspired the development of alternative modes of production of animal proteins. Very large cultures of animal cells are now possible and provide ways of manipulating cells in culture for the production of human proteins, such as interferon. The protein is secreted into the culture medium, while antigenicity and fever-inducing properties by contaminating substances should be much reduced.

BIOHAZARDS

The introduction of major new technologies of commodity production has generally brought with it side effects causing new forms of damage to human health and the environment. The mechanization of textile production generated fine cotton dusts which caused byssinosis (brown lung disease) in the workers. The development of the aniline dyes industry for coloring cotton textiles led to an epidemic of bladder cancer among the workers due to exposure to the chemicals.

Inadvertent effects. The introduction of recombinant DNA technology into production processes poses a variety of new dangers. The extent and character of these hazards have been hotly debated among molecular biologists involved in the development of the technology.

Two features of recombinant DNA technology in its present form differentiate it from previous technologies such as synthetic organic chemistry, which also introduced new agents into the human environment. First, in biological technologies, the key agents are self-reproducing. If by ignorance, negligence, or accident, or by design, they are released into the environment and establish themselves in some ecological niche, they cannot then be removed. When an area has oil or chemical pollution, eliminating the source eventually eliminates the pollution. When live organisms are once established in the ecosystem, stopping the leak does not stop their further proliferation.

A second area of concern has been the use of recombinant DNA technology to introduce foreign DNA into the bacterium *E. coli*. The laboratory strain of *E. coli* is a relative of the naturally occurring strains and can mate with many of these strains and exchange genetic material. Though *E. coli* strains are normal residents of the intestinal flora of humans and many other animals, certain strains are the leading agents of bacterial infection in the United States, causing urinary tract infections, travelers diarrhea, infantile meningitis, as well as wound, surgical, and bloodstream infections in hospitals.

The pathogenic character of many of these strains is associated with the small DNA molecules they carry (the plasmids shown in Fig. 2). Plasmids are used as carriers of foreign DNA molecules in recombinant DNA technology. In the bacterial hosts the naturally occurring plasmids synthesize proteins which confer special properties on the host: resistance to antibiotics and toxin production (for example, plasmid-specified enterotoxin, acting on the cells lining the large intestine. is responsible for diarrhea induced by certain strains of *E. coli*). These plasmids can often be transferred from one strain of bacteria to another by contact, a feature partially responsible for the rapid spread of resistance to certain therapeutically useful antibiotics among bacteria.

If such strains, already a medical problem, were to acquire the capacity to synthesize human or other potent mammalian proteins, such as certain hormones, they might well cause new kinds of pathology. Even if the proteins are not biologically active, their presentation to the immune system in association with bacteria might induce an immune response, which would then act against the same protein in its normal cellular location. As a result of these concerns and to minimize such potential hazards, laboratory procedures have been designed to prevent the spread of these plasmids from laboratory strains into natural strains. The consistent universal application of these safety procedures has not yet been adopted.

The debates over the potential hazards of recombinant DNA technology led to the convening by the U.S. National Institutes of Health (NIH) of a National Recombinant DNA Advisory Committee. The committee is composed primarily of scientists in disciplines related to the use of recombinant DNA technology, but does have representation from broader public constituencies, such as the environmentalists and organized labor. The committee administers and modifies the NIH Guidelines for recombinant DNA research which define the conditions under which recombinant DNA technology can be applied by research scientists receiving government funds. The guidelines were developed for university and medical laboratory research. They do not apply to the private sector, which remains unregulated. In countries such as Great Britain, the guidelines apply to the whole country and have the force of law. There. the unions representing laboratory workers played an important role in the development of more stringent guidelines.

There are expected dangers in the large-scale use of microbial cultures where process control of microorganisms in 1000–30,000-liter fermentors is difficult. With the advent of production of animal proteins by microbial synthesis, factory workers will be exposed to these organisms and their products, unless stringent precautions are taken during recovery of the products, packaging, and handling, and release of the spent culture medium. Some of the products are of interest precisely because of their powerful biological activity on the human organism, which normally produces them only in carefully regulated quantities. Occupational diseases associated with other large-scale production processes have often only appeared many years after the initial exposure. Ensuring proper monitoring of these populations to catch problems at the earliest time is a problem that has not yet been addressed.

Known hazards. The above discussion concerns hazards that are inadvertent effects of production processes. Another set of hazards derives from the known properties of certain microorganisms in current or future industrial use. For example, valuable microbial products are being presently obtained from large-scale cultivation of plant pathogens as in the case of the antileukemia enzyme asparaginase from *Erwinia chrysanthemi*, pathogenic for chrysanthemums, and xantham gum (used in petroleum recovery operations) from *Xanthomonas campestris*, pathogenic to brassicas. Of concern is the potential airborne dissemination of such organisms to countries where they do not now exist.

Similarly, one strain of *Pseudomonas* (patented by the General Electric Corporation) is claimed to break down thin films of petroleum in the natural environment in a manner not found in naturally occurring strains. Thin films of oil, though unfortunate when part of a petroleum spill in the ocean, are critical to a wide variety of industrial processes. Should this strain establish itself in some different oil-rich niche than that intended, it might well create more problems than it solves.

An ominous aspect for any technology is its potential use in warfare. The new genetic technologies sharply increase the possibilities of developing microorganisms for bacteriological warfare directed against humans and crop plants.

PATENTS AND THE SOCIAL CONTROL

Though the basic technology of genetic engineering was developed entirely through public funding in the United States, private corporations have moved rapidly to exploit it. However, their ability to protect their investment was limited by the traditional exclusion of living organisms from the realm of patent ownership. In June 1980 the U.S. Supreme Court in a narrow decision decided that microorganisms which are genetically modified may be patented. The case was pursuant to a request from the General Electric Corporation to patent a strain of *Pseudomonas* bacteria with a novel capability for degrading oil.

The Court's decision may have set a precedent with respect to patenting any organism which has been genetically modified. The biggest impact of the decision, if the U.S. Congress does not pass new legislation, is in the area of food plants. As a result of the Plant Variety Act, passed at the behest of agribusiness interests in 1970, patent protection was extended to many varieties of food plants. Now, by genetically modifying plant tissues, the agricultural industry will be able to patent a much broader range of crop plants. Large pharmaceutical and petrochemical companies have already acquired a number of international seed companies. This, coupled with their emerging involvement in biotechnology, much of which uses plants as a primary substrate for microbial conversion, points to a pattern of control of the entire sequence from biomass source to microbial transformation to distribution of the final fermentation product, by a few multinationals. It is already evident that the new role for agriculture as a materials source for microbial conversion to liquid fuels, for example, poses the undesirable prospect of substituting land presently dedicated to food and fiber production for "energy farming." Transnational corporate control of land is already evident in many countries and will likely result, especially in countries with low national incomes, in decisions being made which will favor utilization of land for the eventual production of biologically derived fuels, as petroleum becomes more and more scarce. This will result in diminishing food supplies, higher food prices, and consequent intensification of hunger among disadvantaged sectors of the world population. Present trends toward patent protection of genetically modified living cells will, if continued, undermine efforts made to assure the equitable distribution of possible benefits from biotechnology either nationally or internationally.

[JONATHAN KING; CRISTIÁN ORREGO]

Advanced Electric Power Transmission

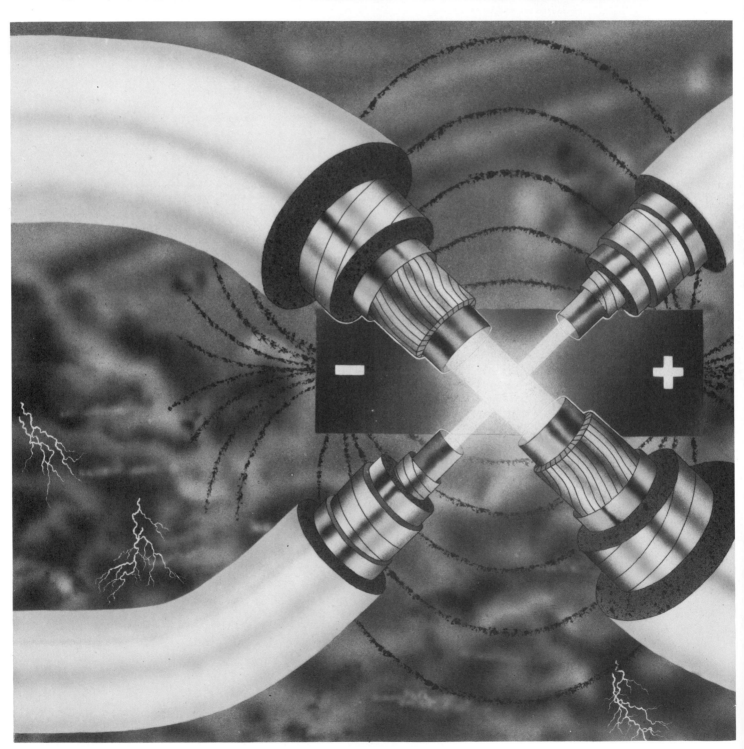

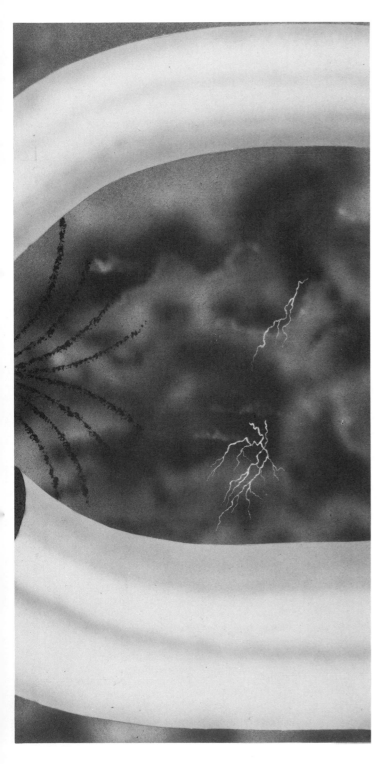

Mario Rabinowitz

Mario Rabinowitz received a doctorate in physics from Washington State University in 1963. From 1963 to 1966 he was a senior physicist at the Westinghouse Research Center. He then spent 7 years at Stanford University doing wide-ranging research in superconductivity, cryogenics, surface physics, radiation effects, and physical electronics. He has also been an adjunct professor at Boston University and Case Western Reserve University. Currently he is senior scientist at the Electric Power Research Institute in charge of advanced projects such as superconducting technology.

Electrical energy represents about 30% of today's world energy consumption. It is anticipated to rise to over 50% of all consumer-used energy sources within the next 50 years because of its versatility, convenience, safety, and the general increase in the world's standard of living. Over the last few decades, the use of electricity both in the United States and in western Europe has nearly doubled every 10 years, with a growth rate of about 7% per year. However, the present energy crisis, beginning in 1974, has slowed this rate down to 3% in the United States. In the 1960s the capacity of the United States electric power industry almost doubled, growing from 175 to 325 GW; in 1974 it was 474 GW; in 1980 it reached 600 GW; and it is expected to exceed 800 GW in 1988. Population growth accounts for only a small fraction of the increase in total electric power demand.

The current arguments in favor of electric power transmission are not much different from those of the past, as can be seen by the following statement: "The transmission of power by electricity both for short and long distances . . . [presents advantages] so enormous that the new system is sure to work great changes in all branches of industrial affairs. . . . [Practical work will be done by] the subtle energy conveyed by wires from central sources of energy — huge furnaces . . ., out of the way waterfalls, tidal currents, even the sun itself. And doubtless this cleanly and trusty servant will serve humanity in ways we are not able to dream of now, and at a cost that will be, by comparison with the present cost of light and heat and working energy, almost nominal." Except for a couple of hints that this statement comes from the distant past, it could well be a summary of current objectives, but it is, in fact, over 100 years old, coming from the January 1880 issue of *Scientific American* during the advent of electric power in the United States.

The world needs reliable, efficient, compact, and inexpensive means of transmitting electricity, particularly because of the present energy shortage. To the 1880 compendium of out-of-the-way energy sources such as waterfalls, tidal currents, fossil-fueled plants, and the Sun, geothermal and nuclear power sources can be added. Either because the energy source is inherently remote, or because of safety and environmental reasons, the generating site may frequently be far from its end use, usually a densely populated metropolitan area. High-density power transmission will be needed to serve densely populated load areas, where transmission line corridors are at a premium and demand for electric power is accelerating.

GROWTH OF TRANSMISSION

The growth of transmission lines has been directly coupled to the growth of turbogenerator and power plant sizes. Due to economy of scale, and increased demand, turbogenerators increased in size from approximately 1 MVA and 10 kV in the early 1900s to the present 1500 MVA and 25 kV without a substantial change in technology. As the generating units and the power plants increased in size, so too did the size of the transmission lines. This is not surprising as it is the main object of the transmission line to deliver power from the generators to the consumers. As more power was carried by a given line, it became desirable to deliver the power at ever higher voltages in order to reduce power losses in the process of transmission. (There are stability considerations that also favor high voltages.) So high-voltage transformers were inserted between the generator and transmission line and again at the distribution end. Thus transmission voltages were increased from 10 to 765 kV, and transmission power levels from 1 to 2000 MVA in less than a century. However, the present energy crisis coupled with technological limits being approached in present turbogenerators, as well as economic obstacles to expansion, will probably limit power line capabilities in the next 10–20 years. Hopefully, if present research is successful, there should be breakthroughs in new technologies for both generating and transmitting electricity. Before considering the possibilities for new means of transmitting electric power, the past and present methods of transmission will be briefly reviewed.

PAST AND PRESENT TRANSMISSION

Transmission of electricity from its earliest days to the present has utilized a simple good conductor such as copper or aluminum. (Sodium has been used successfully on a small scale, but has not received wide acceptance because it burns when exposed to air.) Overhead lines have been the dominant form of transmission from the earliest days on, because of their simplicity, lower cost, and ease of repair. However, underground lines were established almost as early in New York City due to the environmental pressures for undergrounding there. One of the earliest underground transmission lines in the United States was built in 1938 as part of what is now the Northeast Utilities System. Two 69-kV oil-filled pipe-type cables were installed underground from the generating station to connect with overhead lines adjacent to the New Haven Railroad. This system was retired in 1972 after 34 years of service.

The cost of overhead lines has been, and probably will continue to be, less than underground lines for the power levels carried in the past, and even for power levels to be carried in the foreseeable future. However, overhead lines are not a form of high-density power transmission, and when cost of real estate is included overhead lines can be as expensive as underground lines in highly populated areas.

Any kind of high-density power transmission, especially underground lines, requires cooling which adds to the cost. An overhead transmission line is automatically cooled by the large air space surrounding it, which it requires for dielectric strength (electrical insulation), and to keep the electric field at a low level near the ground. Despite the lower cost of overhead lines, it is likely that a decreasing portion of power will be transmitted overhead because of ecological, practical, and esthetic considerations.

Although conventional self-cooled or forced-cooled, underground high-density power transmission (Fig. 1) does not suffer from these drawbacks,

Fig. 1. Three-phase 500-kV high-pressure oil-filled pipe-type cable, a form of conventional underground high-density power transmission.

it has its own disadvantages. It is expensive to manufacture, install, and operate. The high capital cost results mainly from the technical complexity of high-voltage insulation technology and its concomitant cooling requirements. The high operating cost results from high charging currents associated with high voltage and high capacitance, and cool-

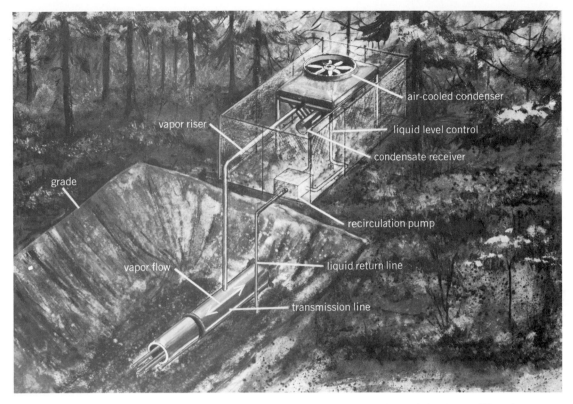

Fig. 2. Vapor-cooled underground transmission line. Heat removal stations are located about every kilometer.

ing (refrigeration) inefficiency in the case of forced-cooled lines. Excavation of large trenches, special accessory equipment, and the introduction of high-thermal-conductivity materials to prevent thermal runaway often make the installation cost as high as the cost for the cable alone.

In addition to the large reactive power outputs required by the generator due to the charging currents and their associated power loss, ac underground coaxial cables are also subject to dielectric power loss, and are more sensitive to ohmic heating. The bare, widely separated conductors of an overhead line are not plagued by any of these problems. The question of relative reliability is not as easily answered and depends upon specifics. Though faults may occur more frequently on overhead lines, they are more easily and quickly repaired. As yet, there are not enough MVA-miles of underground transmission lines to make a meaningful comparison of fault frequency, duration, and cost between overhead and underground transmission.

Because of the additional losses that an underground line incurs, it requires as much as five times the amount of conductor as an overhead line. Thus many underground lines have lower total losses than overhead lines. A typical loss for normal loading of an ac overhead line is 4.4% per 100 mi (2.7% per 100 km) at 345 kV. Comparable ac underground lines have losses of about 3.5% per 100 mi (2.2% per 100 km). However, at 500 kV the ac overhead losses are down to 2.5% per 100 mi (1.6% per 100 km). The losses are even lower than 1% per 100 mi (0.6% per 100 km) for 400 kV dc overhead.

FUTURE TRANSMISSION

At times a new technology may appear to hold great promise precisely because it is remote, and only as it is developed do its drawbacks become evident. On occasion, the remoteness of a technology leads to pessimistic conclusions about its future, as when Ernest Rutherford concluded that nuclear energy would have little or no practical import. Sometimes, even after a technology has been demonstrated, a leading scientist in the field may have doubts. This is exemplified by the 1928 opinion of Lee DeForest, inventor of the triode, who even after television was demonstrated said, "I think that television will never be practical in the home . . ." It became practical after new developments were achieved which DeForest realized were needed but could not foresee. So with any evaluation of future transmission technologies, new developments may well alter presently sound conclusions.

Evaporative cooling. Within the context of conventional technology, evaporative cooled transmission offers attractive possibilities (Fig. 2). The main objective is to increase the cooling rate at the inner coaxial conductor, where it is most needed. Another objective is to cool the conductor by a method that requires the least flow rate, viscous loss, and pressure drop of the coolant. The potential for this lies in the use of liquids having a high heat of evaporation. This would allow large cooling rates with relatively small mass flow of coolant, and hence relatively small hydrodynamic losses.

Extruded cables. Although found extensively in distribution systems (low voltage and low power),

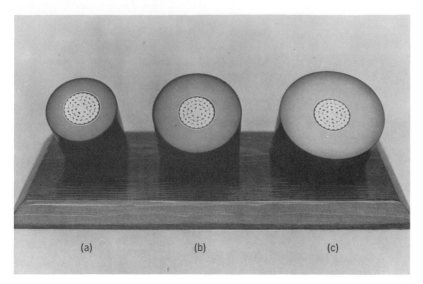

Fig. 3. Extruded cross-linked polyethylene cables (without coaxial sheath), rated for (a) 138 kV, (b) 230 kV, and (c) 345 kV.

extruded solid dielectric cables using aluminum or copper conductors are only now coming into use as transmission lines. Such cables have already been developed at 138 and 230 kV, and research is progressing to increase the ratings to 345 kV and approximately 600 MVA. Compared with low- (LPOF) or high-pressure oil-filled (HPOF) cables using oil-impregnated cellulose tape insulation, extruded cables offer lower cost and are easier to install, splice, and maintain.

Of the three dielectrics that have been used, cross-linked polyethylene (Fig. 3) is preferred over ethylene propylene rubber because of its higher dielectric strength, and preferred over polyethylene (PE) because of PE's low melting point.

Compressed-gas-insulated transmission. Although first introduced into service in 1971, compressed-gas-insulated (CGI) transmission lines remain a system of the future. To date, relatively few compressed-gas-insulated systems have been installed, and these have been mainly in substation getaways. The main insulation is sulfur hexafluoride (SF_6) gas, at about 4 atm (400 kPa) pressure. A hollow aluminum inner conductor is periodically

Fig. 4. Flexible, compressed-gas-insulated cable.

supported by solid insulators from a concentric aluminum sheath.

Cleanliness and absence of contamination are very important in maintaining high dielectric strength in compressed-gas-insulated systems. In addition to highly polishing the conductor, and factory assembly under very clean conditions, electrostatic particle traps are provided near the solid insulators to catch and hold any particulate matter which may be left over or form in the lines.

The majority of existing compressed-gas-insulated lines have been under a third of a mile (0.5 km) in length, at voltages of 145 to 550 kV. An 800-kV prototype system has been developed, and work is being done on a 1200-kV compressed-gas-insulated system. Westinghouse has pursued a rigid system, and ITE-Gould offers either a rigid or flexible compressed-gas-insulated system using corrugated aluminum (Fig. 4).

Cryoresistive transmission. Electrical conductivity is proportional to the electron mean free path in the conductor which increases as the lattice vibrational scattering of the electrons decreases with decreasing temperature. In going from 300 K

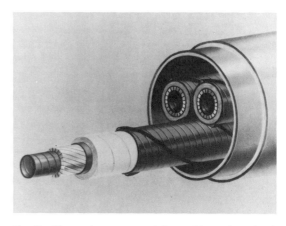

Fig. 5. Three-phase cryoresistive cable, using aluminum conductor and lapped tape insulation, on which work was undertaken at General Electric Company.

to about 80 K, the electron mean free path (and hence the conductivity) increases by approximately a factor of 10, fairly independent of impurities and other lattice defects. Liquid nitrogen, which boils at 77 K at atmospheric pressure (101 kPa), is the coolant for ac resistive cryogenic cables.

The increase in conductivity by a factor of 10 permits a tenfold increase in the power carried, but not without increase in power loss. The reason for the increased overall power loss (despite the increased conductivity) relates to power loss in the refrigeration system which requires about 10 W of refrigerator power for every watt dissipated.

In going from 300 K to below 4.2 K (the boiling point of liquid helium at 1 atm or 101 kPa), the conductivity of most metals can increase by as much as a factor of 10^2 to 10^4, depending on purity and degree of lattice perfection. However, an increase in conductivity by a factor of 10^3 or more would result in a corresponding increase in deliverable power only for dc transmission. This increase would not be as readily achieved in the ac case because the skin depth is inversely proportional to the square root of the conductivity. Thus, the effective gain increases only as the square root of the resistivity ratio unless very thin transposed wires are used. Another problem is that the degree of lattice perfection required to yield a large gain in low-temperature conductivity results in a material that is extremely weak. In addition to the greater expense of making it, there would be an additional expense of mechanically reinforcing it.

All cryoresistive projects to date have been ac at liquid nitrogen temperature. The most active work has been in Japan. In the United States, the General Electric Company (Fig. 5) and Underground Power Corporation have worked in this area, but these projects have been abandoned as they turned out not to be too attractive. A similar fate seems to have befallen such projects worldwide.

Superconducting transmission. Electric current encounters resistance in an ordinary conductor, just as water does when it flows in a pipe. To keep water flowing in a continuous loop, a pump must be placed in the circuit—just as a battery (or other electromotive force) is needed to keep the current flowing in a normal electrical conductor. In either case, even if the pump or the battery is shunted out and removed from the circuit, the flow will not stop immediately. The momentum of the water, and likewise the momentum of the electrons, will still keep them going for a short time until friction or resistance dissipates their kinetic energy. Additionally, electrons receive an added boost as energy stored in the magnetic field is converted into kinetic energy of the electrons to prolong the decay of the current.

As the frictional forces become smaller, these currents will continue to flow longer and longer, until in the limit it can be imagined that they would continue to flow forever without a driving force. Superconductors are supposed to be in this state of infinite conductivity (zero resistivity) at very low temperatures, and hence suffer no power loss in carrying current. If the resistivity is not exactly zero, it is less than can be measured by the most sensitive methods to date. However, above a critical level of current density (on the order of 10^5 to 10^7 A/cm² for dc, and at all levels for ac, superconductors do have a small electrical resistivity, which puts limits on practical applications.

Superconducting materials. Niobium was chosen as the superconductor for some of the early superconducting transmission line development projects in the late 1960s. Niobium seemed to be a reasonable choice as it was then incorrectly thought that ac losses could be kept reasonably low only by operating below the first critical field H_{c1}. With an H_{c1} of approximately 1000 oersteds (80 kA/m) at 4 K, niobium has the highest H_{c1} of any type II superconductor, and its H_{c1} is greater than the critical field H_c of any type I superconductor.

However, work at the Brookhaven National Laboratory and elsewhere in the early 1970s established that the ac losses for type II superconductors [such as niobium-3-tin (Nb₃Sn), at 8 K,

transition temperature 18.1 K] in the mixed state between H_{c1} and H_{c2} (the second critical field, where superconductivity ceases) can be made even lower than for niobium below H_{c1}, and this at even higher temperatures (Fig. 6).

Because of the low transition temperature of niobium, about 9.5 K, a rather low operating temperature had to be chosen between 4 and 5 K. With developments at the Los Alamos Scientific Laboratory (LASL) in making long lengths of high-quality niobium-3-germanium (Nb₃Ge), superconducting tape, the possibility now exists for a superconducting transmission line operating between 12 and 14 K. With a 7 m length of Nb₃Ge tape, 4 to 6 μm thick and 0.6 cm wide on a copper substrate, current densities up to 2.4×10^6 A/cm² have been achieved at 13.8 K. (By comparison, copper and aluminum are usually operated at a current density of 10^2 A/cm².) This tape has operated at a combination of higher temperature and higher current density than has been possible with any other superconductor, primarily because Nb₃Ge has the highest transition temperature, 22.8 K, of all known superconductors (Fig. 7).

Advantages of high transition temperature. The higher the transition temperature of the superconductor, the higher the operating temperature of the cable, with obvious reductions in refrigeration costs. A fringe benefit is that a higher critical current density and a higher second critical field can result, as is the case with Nb₃Ge. This implies that less superconducting material can be used, and that it is possible for the superconductor to stay superconducting through a fault current (approxi-

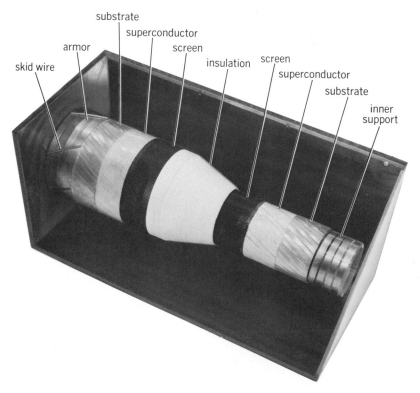

Fig. 6.   One phase of superconducting niobium-3-tin cable proposed by Brookhaven National Laboratory. Operating temperature is 6 to 9 K.

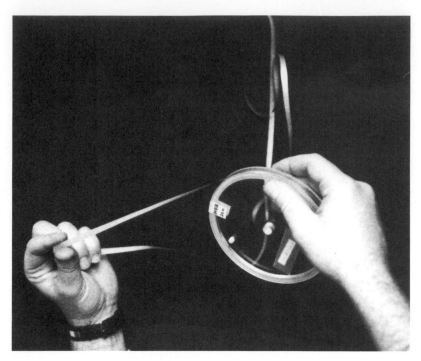

Fig. 7. Chemical-vapor-deposited niobium-3–germanium (Nb₃Ge) tape for use in a superconducting transmission line.

mately 10 times the regular current). Another fringe benefit is that the heat capacity scales approximately as the cube of the temperature; thus, if the operating temperature is increased from 4 to 12 K, the heat capacity is increased by a factor of 27.

An unexpected important benefit from the higher critical current density derives from the fact that a thinner coating of superconductor may be used around the stabilizing normal substrate (usually copper). All the A-15 superconducting alloys, including Nb_3Sn and Nb_3Ge, are quite brittle. In an ac transmission line application, the superconductor must completely surround the copper to shield it from the time-varying self-magnetic field of the cable. If cracks occur in the superconductor, large eddy current losses are set up in the copper which cannot be tolerated. Thinner coatings are more flexible and less likely to develop cracks.

Dielectric. Providing a proper dielectric for superconducting cables has been a difficult problem, which has not yet been satisfactorily solved. Three conditions must be met by the dielectric: high dielectric strength; low dielectric loss; and proper mechanical properties at both high and low temperatures. Deterioration of the properties with time, cycling, strain, and so forth, cannot be tolerated, where an over 30-year operating life is essen-

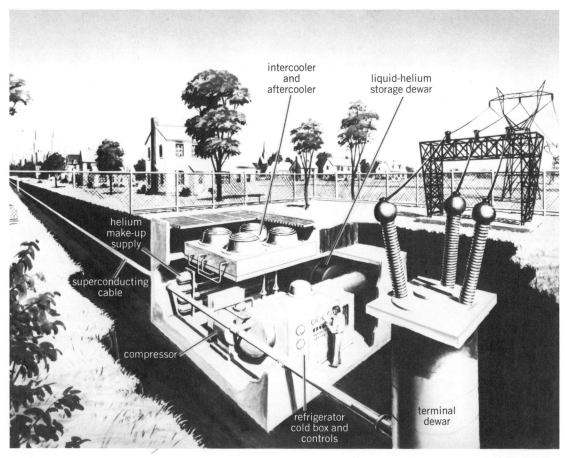

Fig. 8. Superconducting cable terminal and refrigeration station. (*From J. R. Free, Cryogenic power lines: Cool aid for our energy crisis, Pop. Sci. Mag., 201(4):69–71, 130, October 1972*)

tial. In the Brookhaven and other approaches, a tape dielectric is wound around the superconductor. In order to have a flexible cable, the tape cannot be wound tightly, so space is left between tape edges called butt gaps. The cryogen (cooling fluid) gets into the butt gaps. With helium as the cryogen, the electric field across the helium is about three times that across the tape because the dielectric constant of helium is about one-third of the tape's. To further aggravate the situation, helium has lower dielectric strength than the tape.

Hydrogen refrigerant. It may be possible to replace helium with hydrogen as the cryogen in a Nb_3Ge superconducting line. On the one hand, a scarce and relatively expensive resource could be conserved by substituting hydrogen for helium. On the other hand, hydrogen is combustible whereas helium is inert. If hydrogen were simply substituted for helium, the capital cost reduction for a transmission line would be about only 5–10%, as the fractional cost of the refrigerant is roughly no more than this. However, the dielectric strength of hydrogen is significantly higher than that of helium. It is with respect to this feature that substantial savings in both capital and operating costs may be realized. Nevertheless, there are serious public policy safety issues which must be addressed before a hydrogen-cooled transmission line can be a reality. A superconducting cable terminal is illustrated in Fig. 8. Figure 9 shows the proposed dc superconducting cable on which work was undertaken at Los Alamos Scientific Laboratory.

Electron-beam transmission. In 1950 and in 1956 Nicholas Christofilos filed for two patents which form the basis of electron-beam transmission (EBT). This is a method of electric power transmission in which energy is transmitted in kinetic form by means of a magnetically focused electron beam in an evacuated pipe. While the beam is in transit, this is the closest thing to room-temperature superconductivity. However, the return path is in the pipe wall which is just an ordinary conductor, so that only half the line has the advantage of being nearly resistanceless.

The advantage of electron-beam transmission is not in efficiency, as the total power losses may be as large or larger than in a conventional transmission line. There may be a capital cost advantage at very high power levels (10,000 MVA) since high-voltage insulation is not required over most of the line. Basically, magnetic shielding is traded for electrical shielding (insulation).

Electrons are injected from a cathode which is at a large negative potential with respect to a virtual anode at ground potential. They are focused through the virtual anode and drift at a high velocity inside an evacuated spiral quadrupole focused transmission tube. The kinetic energy of the electrons is recovered at the collection end when they enter a retarding electric field. Thus the original accelerating voltage between the cathode and virtual anode is regenerated when the electrons are collected on the true anode (Fig. 10).

One problem with electron-beam transmission is that even a narrow energy spread in the electron beam can translate into large losses at the collector. Maintenance of a high vacuum pressure of less than 10^{-7} torr (10^{-5} Pa) will not be easy. Another problem is that if the electron beam is diverted and strikes the vessel wall, it will burn a hole through not only the tube but any obstacles in its way.

On the other hand, focusing of the electrons is not as difficult as it may seem. In a symmetrical configuration, the electrostatic force of repulsion of the electrons is largely canceled by their self-magnetic field as the velocity of the electron beam approaches the speed of light. In electron-positron storage rings which are similar to electron-beam transmission, the circulating electron bunches have an average power of 200 MW and travel more than 10^9 mi (1.6×10^9 km) in just $1\frac{1}{2}$ hours of storage.

Microwave transmission. In 1897 Nikola Tesla applied for (and was granted) a patent in which he claimed that he had invented, and experimentally demonstrated, an advanced method of wireless electric power transmission through the earth and the atmosphere for collection at a distant point. To this day no one has been able to demonstrate this invention. Although the microwave transmission (MT) of high power through the air sounds a little like Tesla's old dream, the similarity ends with the absence of wires or pipes.

The efficiency of generating, transmitting, and receiving microwave power appears to be significantly less than by conventional means. At frequencies above 10 MHZ there is a problem of beam power absorption by water vapor and oxygen. The sensitivity to water vapor makes microwave transmission vulnerable to rain and snow, and hence quite weather-dependent.

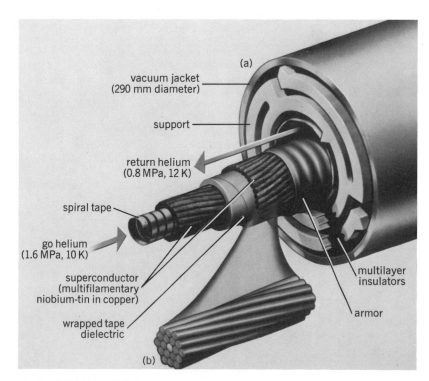

Fig. 9. A 100-kV, 5-GVA dc superconducting cable, on which work was undertaken at Los Alamos Scientific Laboratory. (*a*) Cable in cryogenic enclosure. (*b*) Detail of superconductor.

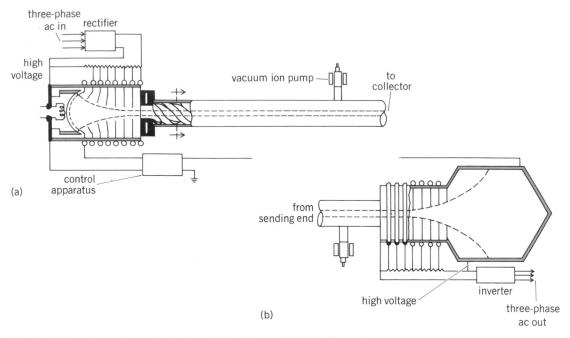

Fig. 10. Electron-beam transmission line. (a) Sending end. (b) Collecting end.

Much interest has been generated in the free-space microwave transmission of power as a means of bringing power to Earth from an orbiting solar satellite. On top of the power losses and high capital and operating costs, such a free-space system has the grave disadvantage that it may be a danger to any life that gets in or near the beam.

Synthetic metal transmission lines. A metal easily conducts electricity because a significant fraction of its electrons move freely from atom to atom with long mean free paths between collisions. A number of organic polymers, in which the outer orbital electron shells overlap extensively, can be made to mimic a metal.

These polymers are essentially quasi-one-dimensional in that they are composed of fine molecular filaments or fibrils, which thread the solid together much like a woven fabric. Along the axes of the threads, the conductivity is high. It is correspondingly low in any nonaxial direction.

With doping, these polymer crystals act like synthetic metals. Because of their reasonably high conductivities, low weight densities, and high tensile strength, the specific conductivity of these materials could exceed that of copper. Polyacetylene, polyparaphenylene sulfide, anthrone polymers, and polypyrrole are such polymers. Many questions regarding their electrical, thermal, mechanical, stability, flammability, toxicity, and other properties must be answered before such esoteric materials can be used for the transmission of electricity.

CONCLUSION

Many, but certainly not all, electric transmission concepts have been reviewed in this article. There are many aspects of transmission lines that have not been discussed, such as transformers, circuit breakers, and fault current limiters. Many new developments will affect these areas as well as the basic transmission line, as this technology has not yet reached full maturity.

Most transmission systems can be made arbitrarily efficient and arbitrarily high in power capacity, given an unlimited capital investment. Regardless of capital, only a few systems can achieve high power density. All factors have to be taken into consideration—capital costs, power, operating costs, power density, reliability, ease of installation, environmental impact, repairability, and security from damage both accidental and intentional.

[MARIO RABINOWITZ]

Bibliography: A. Annestrand, G. A. Parks, and D. E. Perry, *Bonneville Power Administration's 1200 kV Transmission Line Project*, Congrès International des Grands Réseaux Electriques (CIGRE) Pap. 31−09, 1978; G. Bahder et al., *Development of Extruded Cables for EHV Applications in the Range 138−400 kV*, CIGRE Pap. 21−11, 1978; J. W. Bankoske et al., *Biological Effects of ELF Electric Fields*: *Some U.S. Research Results*, CIGRE Pap. 36−05, 1978; A. H. Cookson, T. F. Garrity, and R. W. Samm, *Research and Development in the United States on Three-Conductor and UHV Compressed Gas Insulated Transmission Lines for Heavy Load Transmission*, CIGRE Pap. 21−09, 1978; Electric Power Research Institute, *Workshop Proceedings*: *Public Policy Aspects of High-Capacity Electric Power Transmission*, WS−79−167, September 1979; P. Graneau, *Underground Power Transmission*, 1979; J. R. Kiely and D. E. Hart, *The Role of Electric Energy in the Early 21st Century*, CIGRE Pap. 41−04, 1978; F. Maury and H. Persoz, *Environmental Protection*: *A Factor in the Conception of Power Sys-*

tems, CIGRE Pap. 31–14, 1978; J. H. Moran, *Electrostatics and Its Applications*, pp. 377–424, 1973; E. Occhini et al., *Self Contained Oil Filled Cable Systems for 750 and 1100 kV Design and Tests*, CIGRE Pap. 21–08, 1978; M. Rabinowitz, *Amer. Nucl. Soc. Trans.*, 21:141–142, 1975; M. Rabinowitz, *The Electric Power Research Institute's Role in Applying Superconductivity to Future Utility Systems*, IEEE Trans. Magnet. MAG-11, 1975; M. Rabinowitz, *Materials and Auxiliary Studies Related to Cryogenic Transmission*, IEEE Con. Rec. 76 CH 1119–7–PWR (SUP), 1976; M. Rabinowitz, Prospects for Cryogenic Generation and Transmission, *World Electrotechnical Congress: U.S.S.R. Proc.*, vol. 2, pap. 86; 1977; M. Rabinowitz, *Superconductivity: Worldwide Research*, pts. 1 and 2, EPRI TD–2 and TD–3, 1975; M. Rabinowitz, *Transmiss. Distrib.*, 27(4):64–68, 1975.

Agronomists and the Food Chain

John B. Peterson

John B. Peterson was engaged in
research and teaching in the Iowa
State University agronomy
department from 1928 to 1948,
when he became head of the
agronomy department at Purdue
University. Since 1971, he has
been associate director of the
Purdue Laboratory for
Applications of Remote Sensing.
Author of numerous technical
papers in soil science, he is a
fellow and a past president of the
American Society of Agronomy.

As the world has grown smaller in many
ways, the role of the agronomist as researcher, educator, and practitioner has grown from
responsibility for efficient food and fiber production in localized settings to responsibility
on a worldwide basis. The search for knowledge needed for optimizing crop production
on land in the various climatic patterns has
grown from simple tests of management practices and simple plant breeding to today's
great range of studies aimed at deepening
knowledge of soil, plants, and weather and
the application of this knowledge to expedit-
ing food production over the world.

Early results from studies of soil treatment
and plant nutrition have been spectacular.
Equally spectacular results in genetics and
plant breeding and in plant disease and pest
control have made possible marked increases
in food production wherever the new knowledge has been applied. Agronomists have
been heavily involved in the transfer of these
new advances in technology throughout the
world, and in testing their applicability in
different situations, especially among the
developing nations. In a world where environment, particularly its climatic aspects, varies
widely, and where population growth patterns
also vary greatly, the problem of meeting the
future need for food is the agronomists' greatest challenge.

WORLD SCOPE OF FOOD PROBLEM

Agronomists have always been involved in the food chain, but not until comparatively recently has that involvement been worldwide. They no longer study plant and soil science in isolation. Now they must consider not only yield per unit area, but the worldwide role of the other big unknowns of climate, water, land, and people. The land surface of the world has become their laboratory.

In 1976 the threat of world populations outstripping the food supply seemed imminent. Reflecting their growing interest in the future food supply, the attention of agronomists was directed not only to crop improvement and better cultural practices, but to climate, the worldwide extent of arable land, the world's supply of usable water, and population trends. The moonscape view of the Earth had impressed them with the planet's finiteness. The future of food production and population trends was no longer academic, but on their doorstep. Those problems that research could not solve in time would obviously have to be handled pragmatically.

D. W. Thorne has outlined some of the milestones reached in this century that have increased crop production: (1) doubling, from 10 to 20, the number of elements known to be essential to plant life; (2) fabrication of new and improved fertilizers, especially following the development of the Haber-Bosch process for synthetic nitrogen fixation—fertilizer technology in both production and usage having benefited greatly for more than 50 years from research by the U.S. Department of Agriculture and the Tennessee Valley Authority (TVA); (3) improvement in tillage methods to save energy and reduce erosion; (4) advances in crop genetics and plant breeding which have resulted in greatly increased yields in all staple crops; (5) advances in weed and pest control; and (6) improvements in harvesting and storing.

APPLICATION OF RESEARCH RESULTS

Increased production per hectare resulting from research is comparatively recent. It has increased noticeably in Western Europe and North America only since the 1930s and more recently in other areas. An idea of how recent the improvement in crop production has been is shown by the fact that the yields of corn and wheat in the United States were the same in the 1920s as in the 1870s. Grain yields in England in the early part of the 20th century were no greater than in the mid-19th century. Grain yields in Europe and North America were the same as those in the developing countries during the 1930s, reflecting the high price of fertilizers and the lack of outstanding grain varieties, which research would soon make available. Corn production in the United States corn belt over the last 50 years illustrates how agronomic research can result in increased production per hectare, but also indicates that science has not yet been able to level out the short-term swings in food production caused by climatic variability, such as drought or excessive wetness and unusually warm versus unusually cool seasons. (Fig. 1).

United States corn yields. L. M. Thompson showed that, assuming normal weather, the corn-yield trend in the corn belt has been sharply upward since 1930, reflecting rapid improvement in corn technology. Also, his data show that corn, the major cereal grain in the United States, has increased 50% in yield since 1950, aided by unusually favorable weather in the corn belt. He associates the especially rapid increase in yield since 1960 with the expanding use of less expensive and plen-

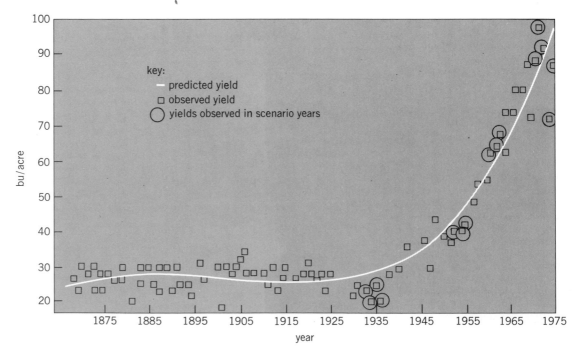

Fig. 1. United States corn yields, 1866–1975. (From The Institute of Ecology, Impact of Climatic Fluctuation on Major North American Food Crops, sponsored by the Charles F. Kettering Foundation, Dayton, OH, July 1976)

tiful nitrogen fertilizer. Nitrogen probably played the major role, but only because of improvements in hybrids, ample and deep application of other nutrients as needed, especially phosphorus and potassium, and high planting rates. Although the value of high rates of nitrogen and high planting populations on corn yields had been known since the early 1940s, there was slow acceptance of research results in the early period by corn belt farmers.

It took a while to disperse the conservatism engendered by the drought and depression years of the 1930s. An evolution in cultural practices has made possible today's high yields. Research results indicated the possibilities of high corn yields from high rates of fertilization and high populations, even without leguminous green manures. Such studies caused a dramatic change in the management of grain crops in level, nonerodible land, resulting in much less forage being grown, a change which caused no problem because it coincided with a rapid reduction in the population of work horses and mules.

Thompson's data show that average corn yields for Iowa may be leveling off at about 120 bushels per acre (1536 kg/ha). There is evidence showing the Iowa average yields to be approaching those of the Iowa Experiment Station's experimental farms, where all known technology is being applied. In other words, Iowa farmers may be capitalizing on new technology as fast as it comes on line.

The increases in grain yields during the last 3 decades resulting from the application of scientific, technical, and managerial skills are without precedent. However, as indicated by the Iowa data, the upward trend in yields has slowed. This raises the important questions of whether this trend, if more than an aberration, is due to unfavorable weather or to a ceiling for agricultural technology.

World grain yields. D. G. Johnson estimates that since 1960 most of the increase in grain production in the developing countries has resulted from greater yields per hectare. American agronomists and their counterparts abroad have contributed greatly to the improvement of crop varieties, cropping patterns, and cultural practices for such countries. Of particular significance was the initiation of cooperative research programs between certain American foundations and the governments of the developing nations. A classic example of such activity is the creation of the International Maize and Wheat Improvement Center (CIMMYT) in Mexico. CIMMYT was successful in producing high-yielding maize and wheat. Within 25 years of its start, corn production in Mexico had trebled, with average yields going from 500 to 900 kg/ha. Even higher yields were achieved for wheat. From 1943 to 1963 the average yield rose from 770 to 3300 to 5900 kg/ha in Mexico's major production areas. During this time Mexico shifted from being a wheat-importing to a wheat-exporting nation.

Another early and highly successful institute is the International Rice Research Institute (IRRI) in the Philippines. It was initiated in 1960 and became operative in 1962. Research in rice and wheat variety improvement by CIMMYT and IRRI resulted in the well-known "green wave." The

successful contributions of these centers stimulated an approach to world food problems which has led to establishment of similar institutions throughout the Third World and to a gradual widening of their base of support. Currently the institutions are under the sponsorship and coordinating influence of the Consultative Group in International Relations.

CLIMATE AND WORLD FOOD PRODUCTION

The ebb and flow of foreign demand for American grain crops emphasizes the dependency of worldwide crop production on climate, and that the swings in climate from year to year can be wide. Grain reserves in the United States were at a 20-year low in 1974. For example, farmers were caught with record numbers of hogs, cattle, and poultry. High prices for grain resulting from the unanticipated short crop resulted in large financial losses to livestock producers. In early 1980, following two good crop years, surpluses existed and market prices were depressed. Farm subsidies were back in the picture.

Climatic changes. As food supply and distribution become more critical, the need grows for the capability to predict climatic trends. Down through the ages climate has continuously changed. The diurnal and annual cycles are known to all. Beyond these, the only known astronomical forcing functions are changes in solar output and the gravitational influences of other planets. Variations in solar activity observed for more than 200 years are becoming more acceptable as a basis for periodicity in climatic patterns, although still questioned by some. J. E. Newman shows evidence of the relation to climate of the so-called Hale's double sunspot cycle, known to vary from 21 to 24 years in length and to consist of two 11-year basic cycles. For example, the dry, hot Midwestern summers of 1932 to 1936 occurred during a major sunspot minimum, whereas the following wet years of 1937 to 1939 were at a sunspot maximum.

Evidence of a 22-year weather cycle is quite convincing. Close agreement with the sunspot period indicates a significant relationship. Cause is still a question. There is evidence that the 22-year variation in solar luminosity may explain the periodicity in the Earth's climate.

Long-time climatic cycles are being studied through paleontology and astronomy. Changes in the shapes of ocean basins and the distribution of continents through continental drift have been factors. There is evidence that the pattern of climatic change during the Pleistocene has been influenced by variations in the Earth's orbit. Changes occur in the composition of the atmosphere with respect to the natural evolution of nitrogen, oxygen, and carbon dioxide content in response to geological and biological processes, as well as from volcanic dust. Whether looking at short- or long-term weather history, climatologists no longer believe climatic variation to be a random process. All evidence indicates that climate is not static but subject to slow, marked changes over long periods of time, as well as to the more apparent short changes in current time and space.

Human impact on climates. In the future, deciphering possible periodicity in climates will be compounded by human impact on them. Concern is growing in some quarters over actions which change the atmospheric composition and interfere with the global heat balance. Land-use trends may be important, such as the present rapid depletion of tropical forests. Carbon dioxide, aerosols, and particulate matter are increasing in the atmosphere. They are not only of consequence to external parameters of the climatic system, but may also be internal variables. For example, there is a changing capacity of the surface layers of the oceans to absorb CO_2, and variations in the amount of windblown dust and the interaction of CO_2 with the biosphere. It is estimated that of the CO_2 going into the atmosphere from the burning of fossil fuels, 50–75% stays in the atmosphere and the remainder in the oceans and biosphere. The CO_2 excess is estimated to increase by 32% by the year 2000. This is calculated to result in a 0.5°C warming of the Earth by the end of the century. The historical growth rate in the release of industrially produced CO_2 is about 4% annually. A worldwide atmospheric CO_2-monitoring network has been developed which will provide much more definitive answers to the CO_2 problem in the future. The effects of particulate load and aerosols on mean atmosphere temperatures cannot be reliably predicted from present information. Whatever the effects, if they should be irreversible, it will be too late to remedy any undesirable changes.

Regional variations. Of significance to the logistics of spreading the supply of food over the world as needed is the growing evidence that climate, and consequently food production, does not vary as greatly on a global basis as it does regionally. In fact, the smaller the area, the more likely is the year-to-year variation to be greater. Furthermore, there are large crop-producing regions of the world in which the variations in climate and in crop yield

are greater than in others. Coefficients of variability for total grain production are greater for the Soviet Union than for Canada, and in turn are greater for Canada than for the United States. The coefficient for world production is less than any of these. This is illustrated by Figs. 2 and 3, which show a much wider year-to-year range in wheat yield in Canada as influenced by climate than the range for wheat yields in the United States. In much of the Soviet Union the climate is subject to even greater variation than in Canada. The difference for wheat between the United States and Canada is also illustrated in the table.

Agronomists must understand enough meteorology and climatology to communicate with specialists in these fields, and to handle the problems of worldwide food production under the constraints of climate which will exist.

Water requirements. Irrigation is a means of extending the world's area of productive agriculture. Where the climate is consistent, the managerial problems are straightforward and depend on knowledge of water requirements of the crops to be grown, the supply and quality of water, and the mechanics of moving the water to the crops. Where climate is variable, the problem becomes complex. Almost every year drought occurs in some part of the United States. Then stream flow decreases, groundwater levels drop, reservoir storage is depleted, and water quality is degraded. In parts of the Far West well drilling and water pumping have increased. The shortages caused by the dry months of 1976 and 1977 in large areas of the United States, which resulted in below-normal stream flow in all parts of every state except for Florida and Louisiana, have sharpened awareness of the critical need for fresh water to support populations. Food production is currently drawing on nonrecurring resources of fossil water in many

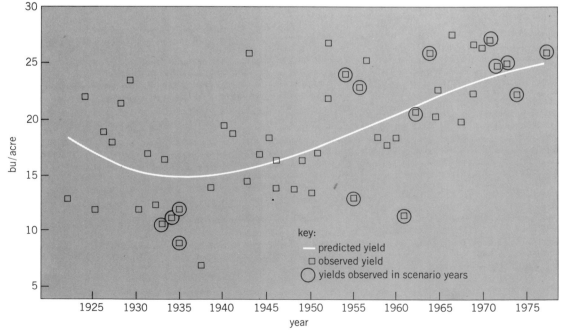

Fig. 2. Canada wheat yields, 1921–1975. (*From The Institute of Ecology, Impact of Climatic Fluctuation on Major North American Food Crops, sponsored by the Charles F. Kettering Foundation, Dayton, OH, July 1976*)

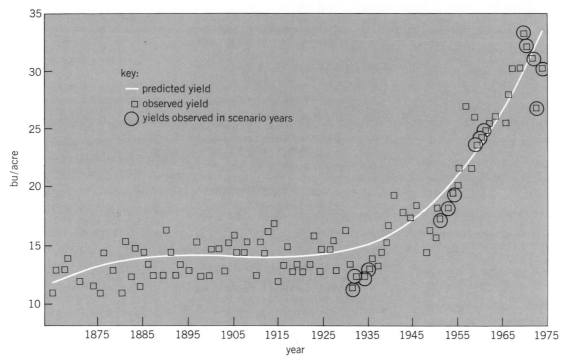

Fig. 3. United States wheat yields, 1866–1975. (From The Institute of Ecology, Impact of Climatic Fluctuation on Major North American Food Crops, sponsored by the Charles F. Kettering Foundation, Dayton, OH, July 1976)

areas of the world, and almost universally on fossil petroleum as a source of hydrogen for making nitrogen fertilizer and fuel for farm machinery.

For example, the United States/Saudi Arabia Joint Commission on Economic Cooperation is studying ways to expand agricultural production in Saudi Arabia. In that climate all agriculture must depend on irrigation. Unfortunately, mainly fossil water has been found. Predictions indicate only a short-time supply. Worldwide food production in the future depends on the future worldwide water budget. Questions on the horizon are the feasibility of freshening sea water and of utilizing icebergs.

UNDEVELOPED AND UNDERDEVELOPED AGRICULTURAL LANDS

Prospects for bringing more land into production depend on many things: adequate rainfall or available water for irrigation, manageable terrain, tillable soil, availability of needed fertilizer, knowledge, and all the items of infrastructure now known to be essential to expanding food production in developing areas and moving the food to those who need it.

Until recently most of the expansion in the world's food supply during the last 2 centuries resulted from bringing new land into cultivation. For example, Johnson estimates that from 1935 to 1960 approximately 75% of the increased grain output in the developing countries resulted in a large part from expansion of the grain area. Areas presenting special difficulties to reclamation were often bypassed. The time has come, however, when it is necessary to claim all potentially arable areas.

Tropical lands are among the most promising areas for expanding food production. Wherever water is available, the climate of the tropics can support long growing seasons and, in much of the area, year-round agriculture. Yet only a small fraction of the world's food is produced there.

The first international center to be primarily committed to a program to improve the productiveness of soils in developing nations, particularly in the tropics, was initiated in 1974 with the incorporation of the International Fertilizer Development Center (IFDC). Stimulus for establishing such a center came from awareness of the need to increase the productiveness of the vast areas of little-developed tropical lands. These have been estimated at more than 50% of the potentially arable lands of the world, equal to 4×10^9 acres (1.6×10^9 hectares). Another 4×10^9 acres of underdeveloped grazable land lies in the tropics.

Successful results on tropical soils with natural

Annual crop yield variability

Crop	Crop years	Mean yield, bu/acre*	Standard deviation, bu/acre*	Coefficient of variability, %
United States wheat	1866–1975	16.5	1.70	10.3
Canadian wheat	1922–1975	18.6	4.30	23.1

*1 bu/acre = 87.1 liters/hectare.

phosphates, slow-release nitrogen fertilizers such as sulfur-coated urea, and other fertilizer materials indicate that great improvements are possible in crop-producing capabilities of such soils. IFDC is rapidly developing a research and outreach program which includes the development of modest indigenous fertilizer industries where needed and feasible, as well as good agronomic practices. In a recent IFDC report it is stated that developing countries will need to increase their production, distribution, and use of fertilizer fivefold by the year 2000 in order to provide adequate food supplies for growing populations. A goal of a fivefold increase in fertilizer use in this time span is deemed obtainable. There is a great need for efficient and rapid inventorying of land and water resources and of land use. In many of the developing countries, adequate statistics and maps are lacking.

POPULATION GROWTH

In 1976 J. B. Kendrich, Jr., stated that the statistics were not comforting as he pointed out that by the year 2000, according to some estimates, food consumption will double or triple; water withdrawals also may increase two to three times; and fuel consumption will be up three times, iron and steel more than two times, fertilizers four to five times, and lumber more than three times. A 1976 report

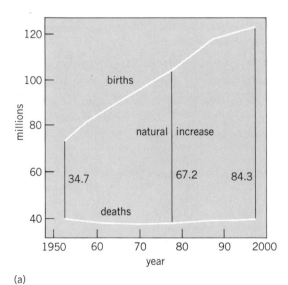

(a)

(b)

Fig. 4. Differences in population growth rates between (a) the developing world and (b) the developed world. (*From Rockefeller Foundation, Population Sciences: The State of the Art, 4(4):11–12, April 1980*)

by the National Research Council estimates the present 4×10^9 world population will grow to 6×10^9 by the year 2000. It further states that although the agricultural output of the developed countries has doubled in the last 40 years, it has increased by only about 15% in the developing countries.

One of the fastest-growing areas in population is the Caribbean. Central America has one of the world's highest population growth rates, with projections showing that at the present approximately 3% annual rate it will double in 25 years. Aiding in this rapid growth is a declining death rate, producing the classic pattern of the population explosion. In 1920 El Salvador's annual birth rate was 47 per 1000 and death rate 33 per 1000. By the late 1950s the death rate had fallen to 15–24 per 1000. Even if family planning brought the birth rate down to 2 per family, the region's population is calculated to fall to only 34×10^6, compared with the forecast 39×10^6 by the year 2000. Mexico's population, now at 64×10^6, may reach 132×10^6 by the end of the century. It is estimated that to nourish properly the people of the world the equivalent of $2\frac{1}{2}$ times the United States production of food will be needed by the year 2000 in addition to current world supplies.

The need for a slowdown in population growth is clear. By the close of 1975, the U.S. Agency for International Development had provided $732,000,000 for population program assistance. Where funds were available to support the programs over the last decade, the world population situation has greatly changed. More than 50 countries now have national family-planning programs. Such programs have already accomplished their purpose in Singapore, Hong Kong, and Taiwan, where birth rates have fallen to approximately 20 per 1000. Noticeable results are apparent when 12 countries, mostly small, with vigorous family-planning programs during the 1960s are compared with 12 without such programs. Family-planning programs are under way in the Phillippines, Indonesia, India, Pakistan, Afghanistan, Nepal, Bangladesh, and a number of Latin American and most African countries.

From 1950 to 1975 the world's population grew by 60%, from approximately 2.5×10^9 to 4×10^9 people. Most of this growth was due to increased longevity (Fig. 4). In that period the worldwide life expectancy rose from about 45 to 60 years. It is estimated that the expectancy rose from 65 to 73 in the "rich" world, whereas in the "poor" world it rose from under 40 to 55 years. Furthermore, a less noted change in demography has been taking place since about 1960. A widespread decline in fertility was beginning in parts of the underdeveloped world. This decline has not affected all countries uniformly. South Korea has brought its fertility rate down to the level of Ireland's. Other countries, including almost all in black Africa and most in the North African–Middle Eastern Muslim expanse, have shown little change.

INVENTORYING NATURAL RESOURCES

As the ratio between population needs and resources narrows, the need to improve the quality of resource information systems becomes more critical. In additional to knowing the volume and potential of natural resources, there is a growing belief that the world would be better served in terms of socioeconomic benefits by having the capability to appraise and even predict the world production of certain major crops on as accurate and timely a basis as possible.

Planners have been greatly handicapped by their limited capability to predict climatic trends and weather and their lack of accurate inventories of land and water areas over much of the world. Fortunately, new technology for recording and tracking climatic changes and for inventorying natural resources has been developing rapidly over the last few years. A capability now exists to acquire great quantities of spectral data and to process, analyze, and interpret the data rapidly. The ability to sense and interpret the radiance of rangelands, forests, crops, soils, and water bodies is now at an operative stage and is constantly being improved. From such data, much information, especially of an agricultural nature, can be obtained with a minimum of ground sampling. Few countries have reliable methods for gathering the data necessary for good planning, much less for forecasting and estimating crop production. Remote-sensing techniques have come on line in time to provide the planners with the capability of securing the essential data.

The capability to secure worldwide data was advanced with the launching of *Landsat 3* in March 1978. Even greater improvement in capability is expected when *Landsat D*, with seven spectral bands and 30 m spatial resolution compared to four bands and 80 m spatial resolution of the current Landsat sensors, is launched in 1982.

Crop inventorying by remote sensing, although new, has rapidly improved and, in view of the encouraging results from current research efforts, promises to become an effective means of obtaining not only areas of, but also production estimates of, certain major crops on a large-area basis, even worldwide.

The need for agricultural information on a global scale is growing. Accomplishment is in reach through advances in computer-implemented methods for such systems and through possibilities for inventorying vast areas of the world with remote-sensing techniques with less than 1% error on a geographic basis. Also from the standpoint of resource development, there is a growing awareness of the need to implement a broad, interdisciplinary, in-depth examination of the total flow of information essential to planning. The report of the World Food and Nutrition Study includes recommendations for high-priority research on information needs of producers, crop-monitoring systems, international data bases for land and nutrition, and a total information systems design.

The new techniques of remote sensing, combined with advances in computer-implemented information, now make possible coordinated, continuously updated information systems which are capable of functioning on a worldwide basis. Scientists are already considering the possibilities of such a system for agriculture. The need for an in-depth information system is critical to the worldwide food problem.

It is obvious that future agronomists will need to know more than the traditional background courses. They will also need understanding of climatology and meteorology, remote-sensing techniques, and computer-implemented procedures for data processing, analysis, and interpretation to be able to function in the worldwide scene.

CONCLUSION

The preceding glimpses of the world situation emphasize the awesome role of agronomists in the world's food chain. Because of the many uncertainties, predictions can only be approximate. Fortunately, agronomists have had, for a long time, a strong, sincere interest in the welfare of the world's people. They have been active in developing the organizational structures, both public and private, to expedite such programs and in helping develop ways to fund them. They are heavily engaged in helping the developing countries optimize their yet undeveloped resources of people, climate, and land. Through advances in the technology of remote sensing, techniques for inventorying resources on a global scale have been rapidly improving. If political considerations at home and abroad are favorable and if population growth slows down in time, the efforts of agronomists in research and education will play a major role in averting worldwide suffering.

[JOHN B. PETERSON]

Bibliography: Committee on Climate and Weather Fluctuations and Agricultural Production, *Climate and Food*: *Climatic Fluctuation and U.S. Agricultural Production*, National Academy of Sciences, 1976; D. G. Johnson, *Amer. Statist.*, 28:89–93, 1974; Rockefeller Foundation, *R. F. Illustr.*, 4(4):11–12, April 1980; L. M. Thompson, in *Food*: *Politics, Economics, Nutrition and Research*, pp. 43–49, American Association for the Advancement of Science, 1975; D. W. Thorne, *Agronomists and Food*: *Contributions and Challenges*, Amer. Soc. Agron. Spec. Publ. no. 30, 1977.

Nuclear
Reactor Safety

Edwin L. Zebroski

Edwin L. Zebroski is director of the Nuclear Safety Analysis Center at the Electric Power Research Institute in Palo Alto, CA. He received his doctorate in physical chemistry from the University of California. He has published in several areas of nuclear energy research, development, and design in previous associations with the Manhattan Project, General Electric Company, Stanford Research Institute, and EPRI.

The legacy of the accident at Three Mile Island, PA, is an intensified concern over three of the main issues of nuclear safety. Two of these issues are: the effects of low-level radiation from routine operation and from minor accidents; and the possibility of major releases of radioactivity from accidents which breach the three levels of protective barriers, the fuel cladding, the reactor vessel and piping, and the containment building. The clean up of the contaminated reactor building involves the third issue, the processing and storage of radioactive waste. The regulatory and legal aspects of such operations for civilian purposes are still in disarray, and subject to public concerns and many procedural delays in the United States. (The military program and overseas nuclear programs have handled and stored larger amounts of similar kinds of radioactivity routinely for over 30 years, but this experience is largely unknown to the public.)

The most basic issues of reactor safety involve questions such as: What is the likelihood of serious accidents which spread large amounts of radioactivity? What are the human and environmental consequences of such accidents? Are the means for preventing or limiting such accidents well in hand? Are small releases of radioactivity a serious hazard to health, even if much smaller than natural background radiation? Would the risks associated with energy production be increased or reduced by limiting the use of nuclear energy?

The idea of limiting nuclear science and technology to peaceful uses was the basis of the "Atoms for Peace" initiatives of the Eisenhower era (1954), which employed the analogy of "swords into plowshares." A large investment was made by the United States in building the diffusion plants which originally made concentrated uranium-235 (about 90%) for nuclear weapons. This investment was converted to a civilian use by modifying the plants to make slightly enriched uranium-235 (about 4%), which is the fuel for power reactors.

This association with the source of nuclear weapons raised the basic safety concern that a power reactor might explode like a nuclear bomb. This is physically impossible. The fuel used for power can be made critical only with thermal neutrons (slow neutrons). The rapid chain reaction with fast neutrons which make a bomb can be made to occur only by using a sufficient amount of concentrated fissionable material.

The operation of a large power reactor produces large amounts of fission products, some of which are radioactive. The possibility of major leakage or accidental dispersion of fission products constitutes the main hazard resulting from reactor operation. The possible threat to health and environment from fission products is dramatized by the effect of the worldwide fallout from testing nuclear weapons. For example, a nuclear test in China resulted in fallout which produced about 50 times the minimum detectable amount of radioactive iodine in milk measured in Pennsylvania. (The iodine level resulting from this fallout was also about 25 to 50 times higher than the radioactive iodine measured in milk after the Three Mile Island accident.)

LEAKAGES OF RADIOACTIVITY

The prevention of conditions which might lead to the spread of radioactivity either by routine small leaks or by larger accidents is the main aim of reactor design, operation, and the regulatory and inspection work of goverment and industry.

Local variations in radiation from nature range from about 50 to over 200 millirem (0.5 to over 2.0 millisieverts) per year in the United States and from about 50 to 150 millirem (0.5 to 1.5 mSv) in the southeast corner of Pennsylvania (Fig. 1). The leakage of small amounts of radioactivity—resulting in possible human exposures to radiation (maximum of 5 millirem or 50 μSv per year) which are small compared with the local variations in radiation from nature—was originally considered an acceptable by-product of energy production. Small releases of radiation have occurred over the 25 years during which civilian reactors have been operated. The radiation is measured at the plant near the point of leakage. Prior to the accident at Three Mile Island, these leakages have been small compared with the permissible amounts and very small compared with the variations in natural background of radiation.

The accident of March 28, 1979, at the reactor at Three Mile Island near Harrisburg, PA, resulted in the release of relatively large amounts of inert-gas

radioactivity (a few million curies, 1 Ci = 3.7×10^{10} Bq) and small amounts of radioiodine (a few curies). The large amount of radioactivity nevertheless resulted only in low-level radiation exposures because inert-gas radioactivity does not react chemically, leaves no residue, is effectively diluted by mixing with air, and is carried away by the wind. Most of it decays in a few weeks. Traces of radioiodine can be absorbed by living things such as grass, which can then enter the food chain via cows and milk. Hundreds of samples of grass and milk were measured after the Three Mile Island accident. Traces of iodine were detected (up to 40 picocuries or 1.5 Bq per liter) in a few percent of the samples. Such levels were almost twice natural background radioactivity and about 3/1000 of the level considered by health authorities to be hazardous for continuous human consumption (12,000 pCi or 4.4×10^2 Bq per liter).

Despite the actual low levels of radiation exposure at the time of the accident, the measurements made by sampling from helicopters in the plume near the building vent gave momentary values which were well above the limits which would be considered safe for prolonged exposure (1500 millirem or 15 Sv per hour). The high measurements in the plume apparently were confused with possible high doses at ground level. A similiar situation would be encountered if the measurement of the substances in the gas plume from a conventional smokestack or even a home fireplace (which can be at lethal levels) would be mistaken for a measurement at ground level.

The actual upper limit of the exposure to the public was summarized by the two government commissions to be an average of 1.4 millirem (14 μSv) to 2,000,000 people. This is expected to produce zero to 1.5 cancers (fatal plus nonfatal) over the next 30 years, in addition to the 325,000 cancers normally expected from other sources (see table).

The maximum exposure that any individual might have received was less than 100 millirem (1 mSv). In order to be exposed to this amount, a person would have had to stand naked for 6 days and nights at the north gate of the plant. One fisherman was thought to have stayed on an island next to the plant. It is estimated that he might have been exposed to 40 millirem (0.4 mSv).

ACCIDENT AT THREE MILE ISLAND

The damaging accident at Three Mile Island Unit 2 resulted from a prolonged period of water

Cancer risks for 2,100,000 people near Three Mile Island

Cancer source	Lifetime dose, 10^6 person-rem*	Deaths from cancer
All causes	—	325,000
Natural and medical radiation	10–20	20–40
Three Mile Island accident	0.002	Less than 1% chance of one death†

*1 rem = 10^{-2} Sv.

†Genetic effects are estimated at .01 to 0.2% chance of one ill-health effect.

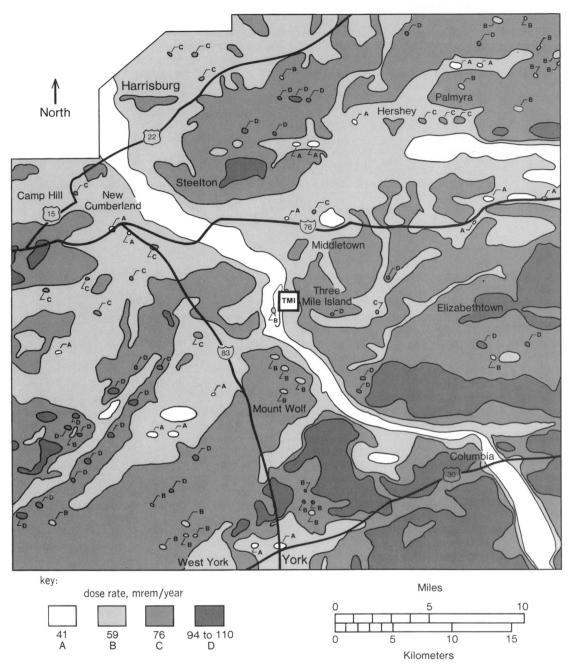

key:

dose rate, mrem/year

41	59	76	94 to 110
A	B	C	D

Miles

Kilometers

Fig. 1. Local variations in annual background radiation from nature (largely from radon and its daughter products) surveyed prior to operation of reactors at Three Mile Island. 1 rem = 10⁻² Sv. (From EG&G, EGG-1183-1710, March 1977)

loss from a jammed open relief valve. The valve was open because of a temporary loss of feedwater supply to the steam boilers (steam generators). While this was nominally the initiating event, the temporary loss of feedwater occurs many times each year in both nuclear and fossil boilers, normally with no damaging consequences. In this case, the operators became confused by seemingly conflicting indications from their instruments, and conflicting interpretations of procedures and regulatory requirements. For several hours they failed to close the block valve which is provided to stop the loss of water from the reactor in such an event.

They also cut back the flow of the automatic safety system which started to inject water and maintained a letdown flow of water to another building which increased the loss of water from the system. Before they recognized that the reactor was losing water, between 2 and 3 h after the start of the accident, the reactor core was partly uncovered.

Although the reactor shut down automatically, residual heat continued to be produced—a few percent of full power immediately after shutdown—which then gradually decayed away. The residual heat was sufficient to cause the uncovered part of the core to get hot enough to damage the

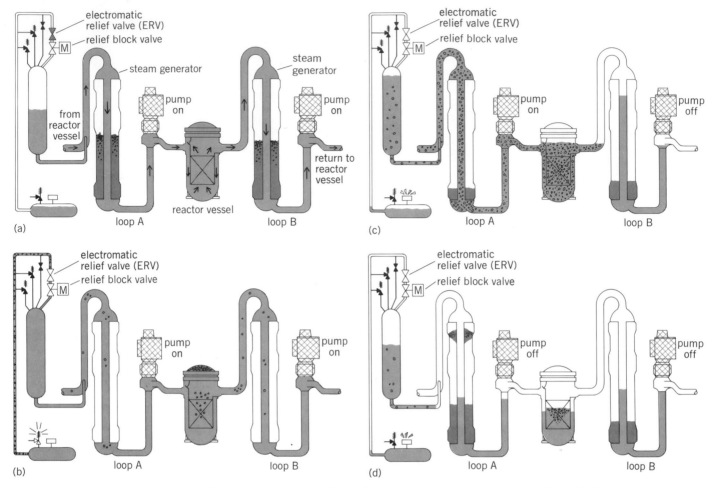

Fig. 2.   Conditions in the reactor at Three Mile Island at various stages of the accident. (*a*) Time *t* = 0. (*b*) *t* = 8 min. (*c*) *t* = 1.5 h. (*d*) *t* = 2 h. (*e*) *t* = 2.8 h. (*f*) *t* = 3.5 h. (*g*) *t* = 16 h.

normally gas-tight zirconium metal cladding tubes which enclose the uranium oxide fuel. The leakage or rupture of these tubes released a large part of the fission gases to the reactor system. Some of the gases then leaked out of the reactor system along with water and steam. The cladding of the overheated fuel reacted with steam to make zirconium oxide and produced hydrogen gas which accumulated in the reactor vessel top, and in the containment building. The reaction with steam is analogous to the rusting of iron: Zr (metal) $+ 2H_2O$ (steam) $\rightarrow ZrO_2 + 2H_2$ (gas).

In addition to the concerns over the known leakages of radiation, there was a concern that the reactor vessel and containment building might be ruptured by burning or explosion of the hydrogen gas which was known to be present. If this occurred, it might lead to the eventual dispersion of much larger quantities of radioactive elements, including the more dangerous fission products such as strontium and cesium. These can leave long-lasting residual radiation in the environment if they are dispersed. There actually was a low-pressure "burn" or explosion of some of the hydrogen which leaked from the reactor into the containment building, but it caused no apparent damage to the containment building. It was later concluded that the apprehension over a more serious explosion in the reactor vessel was never technically

justified, since there was no source for the amount of oxygen which would have been required to produce an explosive mixture. Hydrogen is normally added to reactor coolant water in order to scavenge (to remove by radiation-induced reaction) traces of oxygen.

The reactor core was covered with water again after about 3.3 h, which stopped further damage. However, the containment building basement was heavily contaminated with fission products carried by the water that escaped during the time the relief valve was open, along with the gases xenon and krypton, and small amounts of radioiodine. Some of this radioactivity also entered the auxiliary building with the letdown water. The xenon decayed away naturally, and the krypton was vented from the containment building (July 1980). The maximum exposure experienced by any member of the public from venting was less than 0.05 millirem (0.5 μSv).

Schematic diagrams of the conditions at various stages of the accident are shown in Fig. 2. At the start of the accident (Fig. 2*a*), the plant is in normal operation. Primary water is full, and secondary water in the steam generator is normal, producing steam for the turbine (not shown). At this point the reactor shuts down. At a time *t* after the start of the accident of 8 min (Fig. 2*b*), the primary water is still full, but feedwater is lost to the secondary sys-

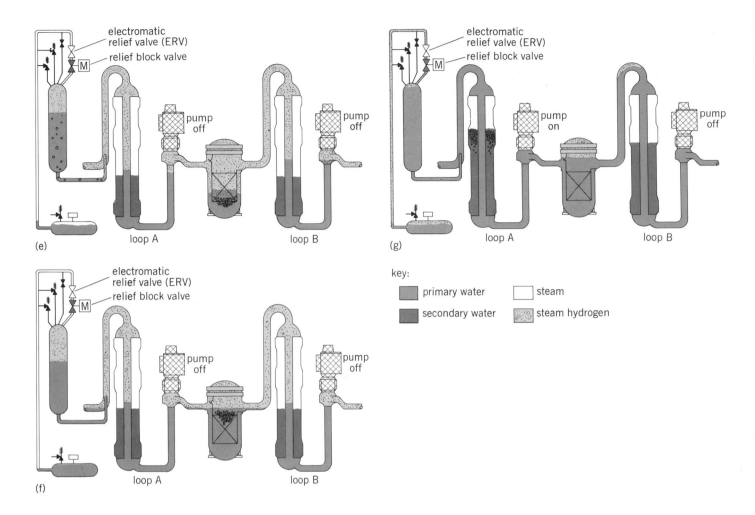

key:

▨	primary water	☐	steam
▨	secondary water	▨	steam hydrogen

tem, which boils dry. Thermal expansion lifts the electromatic relief valve (ERV), which sticks open. At $t = 1.5$ h (Fig. 2c), primary water continues to escape from the open relief valve. Core residual heat is cooled by circulating a mixture of water and steam. Operators, misled by the high water level indicated in the pressurizer, fail to close the relief block valve, and manually reduce the flow of the automatic water injection system. At $t = 2$ h (Fig. 2d), the main pumps are shut off because of excessive vibration due to the mixture of water and steam. The primary water level falls so as to uncover the upper half of the core. At $t = 2.8$ h (Fig. 2e), the uncovered part of the core, now cooled only by steam, overheats and zirconium reacts to produce hydrogen. Fuel tubes corrode and rupture, releasing mostly xenon and krypton. At this point, the block valve is finally closed. At $t = 3.5$ h (Fig. 2f), the closure of the block valve and continuation of water injection result in covering of the core, terminating further core damage. But hydrogen in the top of the vessel and in the "candy cane" pipe regions prevents normal cooldown by natural circulation of primary water to the steam generator. At $t = 16$ h (Fig. 2g), after various attempts to attain natural circulation, one main pump is turned on and stable cooldown attained.

Cleanup of the Three Mile Island buildings is expected to take several years largely because of regulatory restraints on the handling of the concentrated waste from the cleanup of the contaminated water.

LESSONS OF THE ACCIDENT

The lessons learned are different for the general public, utilities, manufacturers, and the government.

General public. The accident and the studies by government commissions appeared to reinforce concerns about the safety of nuclear energy and about the adequacy of the Nuclear Regulatory Commission (NRC), the Federal agency which regulates nuclear energy. For those who believe that an additional domestic energy supply not based on oil is needed and requires acceptance of some degree of risk, the accident could be regarded as confirming the adequacy of public protection. No member of the public was hurt, except for mental anguish. The accident reinforced the concerns of those who doubt the need for additional energy sources, or the need to accept any risk for production of energy (or who believe that alternate sources such as solar energy are immediately practical and economic). A high level of public apprehension was generated by the continuing assertions that a much more damaging catastrophe was narrowly averted. From both viewpoints the information provided by the Federal government or the utilities, or the treatment by some of the media were clearly not reassuring. Public needs appear to include: regulatory agencies which exhibit more competence and inspire more confidence; utilities which are more reliable in avoiding periodic mishaps which arouse concerns as well as serious accidents; and more organized and credible practices for public information and emergency planning by both utilities and state and federal agencies.

Utilities. From the standpoint of the utility operating the reactors at Three Mile Island, the lessons learned include the need for a greater degree of operator training and education than that provided by the reactor designer or required by the NRC. More generally, the operating utility recognizes the need to take a larger degree of responsibility for the technical basis of operation as well as the operation itself. Reliance on regulations or supplier advice may not always be adequate. There is a documented need to take full and detailed cognizance of operating experience from other reactors. The utility cannot depend primarily on Federal or manufacturer coverage of these subjects at the level of detail required for good operation of its own power plants. The financial consequences of a damaging event require greater insurance coverage, both for the added costs of replacement power and for the costs of cleanup and recovery. The increased worth of plant investment (more than tripled since 1974) requires increases in insurance coverage to cope with the consequences of outages. Proportionate increases in staff capability and effort for preventing damaging events is also prudent.

From the standpoint of the utility industry overall, the direct financial impact of the Three Mile Island accident has been delay in the licensing and opening of many plants. The cumulative effect as of mid-1980 was estimated to be a loss of domestic energy production (spread over several years) which would total the equivalent to 4,000,000 barrels (6.4×10^5 m³) of oil per day for a year. This is worth about $30,000,000,000 in added fuel costs to rate-payers for replacement power from more expensive sources, mostly coal and oil. There is also an increase in bond rates equivalent to increased annual costs of about $1,000,000,000 per year, which are also costs to rate-payers.

The lead organizations in the utility industry, both public and private, jointly decided that new industrial institutions were necessary to prevent future accidents of this magnitude and to implement the "lessons learned" in the industry as a whole. The actions which were taken included the following:

1. The Nuclear Safety Analysis Center (NSAC) in Palo Alto was formed in May 1979. This is an industry-supported technical group which analyzes operating experience and provides technical recommendations for refinements in design and operation which can reduce the likelihood of serious malfunctions or accidents. It also studies the "generic safety issues," which are seemingly unresolved questions on safety principles or methods.

2. The Institute for Nuclear Power Operations (INPO) in Atlanta was formed in the fall of 1979 to assist in establishing higher levels of training and better criteria and curricula for plant operators, as well as for line managers of utilities. In addition, there are periodic evaluations of the quality of operations, and advice on the need for additional training or technical support staff is provided.

3. Nuclear Electric Insurance Limited (NEIL), a mutual insurance pool, was formed to offset part of the extra cost for fuel which the utilities sustain when a nuclear plant is shut down for a prolonged period. The direct extra fuel costs per kilowatt-hour are about 100% higher if coal is the replacement fuel and 500–1000% higher if oil is the replacement fuel.

The United States utility industry is hampered by the variablity of requirements by both the state regulatory commissions and the Federal power and regulatory agencies. Their diversity has tended to prevent effective national long-term cooperative efforts between groups of utilities. The establishment of the Electric Power Research Institute (EPRI) in 1972 provided the first cooperative large-scale, long-term basic research effort on methods of power production, distribution, and efficient use. EPRI is now the principal research arm for both private and public utilities, including regional public agencies such as Tennessee Valley Authority (TVA) and Bonneville Power, and municipal utilities and rural electric co-ops. The success of EPRI in providing a coherent national focus and development efforts in many areas of energy technology and use has helped to establish the cooperative pattern for the new institutions (NSAC, INPO, and NEIL). For the whole utility industry, increases in insurance for consequences, and proportionate increases in staff capability for operation, management, and technical support to prevent accidents and outages, are desirable. A key objective for utility managements and public utility commissions is to provide the resources for such added insurance.

Reactor manufacturers. The need to maintain cognizance of the entire plant in both design and operation is evident. The conventional equipment outside the nuclear steam supply system is often provided by other manufacturers and assembled in the field by still other companies. However, there is an obvious need to take continuing detailed account of such equipment since it can contribute to the possibility of outages, which reflect on the reactor manufacturer, as well as being the possible source of damaging accidents. The traditional patterns of commercial secrecy, proprietary restraints on safety-related design and analysis, and limited views of continuing responsibility for adequacy of design need to be modified.

Federal regulatory agency. One evident lesson is that the regulatory emphasis on hypothetical extreme cases and on highly improbable accidents has led to the neglect of less dramatic but more probable accidents. This emphasis has dominated NRC research and regulation. The overuse of formal due process—using protracted licensing hearings—does not necessarily lead to good decisions with respect to design, operation, or regulation. It does involve a large amount of both government and industry worker-power in primary "paper-chase" activities. The regulatory organizations' failure to coordinate their various branches contributes to this inefficiency. Regulations have sometimes vacillated on the directives given to industry; these may be at cross purposes with prior regulations. Minor regulations often overshadow those bearing on potentially serious safety issues. There is a tendency for regulatory agencies to provide overly prescriptive regulations which mandate minutiae of design or operation and imply an impractical level of detailed technical awareness of many dozens of plants by the NRC. General regulations are more practical and productive, analogous to the practice of air-worthiness certification of commercial aircraft.

REMEDIES IMPLEMENTED

Nearly all domestic utilities have joined together in the new institutions and practices reflecting the larger degree of acceptance of detailed responsibility for technical content of operation, for training, and for equipment.

Training. The Institute for Nuclear Power Operations has developed criteria for instructors and training curricula for reactor operators and supervisors. INPO also evaluates operators and management of utility organizations, suggesting where remedial training or additional staff strength is needed.

Analysis of operation experience. The Nuclear Safety Analysis Center provides a comprehensive compilation, documentation, review, and technical analysis of operation experience from all nuclear power plants. Jointly with INPO, recommendations for improved procedures, maintenance, and, in some cases, design are provided.

Procedures. The Three Mile Island accident highlighted the fact that not all possible circumstances are or can be covered by written procedures and rules. A more specific basis for conditions must be established under which operator judgment is appropriate. This had led to a more extensive analysis of the less severe but more probable types of accidents (for example, a small break loss of coolant accidents) and to development of procedures for recognizing and coping with such situations, when they are not covered by existing procedures.

Equipment. A considerable number of equipment changes have been implemented, most of them also mandated by the NRC. These include positive indications of safety- and relief-valve positions, and instruments which directly read the temperature margin to boiling in the reactor cooling water. Several hundred minor equipment and procedure refinements have also been directed by the Nuclear Regulatory Commission that are responsive to various facets of both the Three Mile Island accident and other less damaging incidents or postulated concerns which have occurred over the years. The apparent need for more major changes in design should be tempered by the observation that the Three Mile Island plant was capable of protecting itself against damage automatically, had the operators not intervened. No core damage occurred for nearly 2 h. The accident could have been rapidly and reliably terminated at any point, once the operators understood the situation. This observation applies despite the presence of many less-than-ideal conditions in the plant design, maintenance, and operation practices.

Instrumentation and display. Considering the key role of clear information to the operator, refinement of the area called human factors design is perhaps the paramount improvement needed to prevent future damaging accidents. Of the several thousand indications which are provided in the control room, only a few dozen are essential indicators of the safety state of the system. Some of these key safety parameters were partially lost or temporarily unavailable to the operator in the accident at Three Mile Island. This contributed to the period of confusion which led to severe damage of the reactor core. The provision of a simple and compact display of the key safety parameters, independent of the control and operating functions of the control room, is now under way and is expected to be the major step in improving the human factors aspect of control room design.

Emergency response. The deficiencies in emergency response at Three Mile Island, and in some other less serious incidents, have led to major efforts in improving emergency response planning by both the operating utilities and by the state, regional, and national agencies. The formal organizational requirements have been specified and put in place. Federal requirements for notification and potential evacuation have been developed. Emergency drills are now being routinely conducted by all operating utilities. Arrangement for nearly instantaneous transmission of the key safety parameters of the plant to the emergency response centers, to technical support groups, and to Federal regulators is scheduled. This should help prevent the chaotic situation which has prevailed in some emergencies, and discourage the media from focusing on the worst "what-if" speculations which occur in the absence of solid, reliable information.

REACTOR MELTDOWN

A common feature of some reports of the Three Mile Island accident, even many months after the accident, has been the assertion that meltdown of the core was only moments away and this would have resulted in much more severe consequences to the environment. The studies of the President's Commission on Three Mile Island (Kemeny Commission) and of the Rogovin Report Panel, as well as independent studies by industry and national laboratories, generally support the following conclusions: (1) Even if a core meltdown had occurred at Three Mile Island, there would still remain an extremely small likelihood that the integrity of the containment building would be threatened or that massive amounts of persistent radioactivity would be released to the environment. (2) Various hypothetical extensions of the Three Mile Island accident have been postulated, including scenarios which lead to melting of the core, melting of the reactor vessel, and dumping of the molten core on the floor of the reactor building. Even for such a succession of events there are many easily observable characteristics. A large number of installed systems (all of which were operable in the case of the Three Mile Island accident) can add water and remove heat at any point to terminate the progression of damage. For a typical reactor (Fig. 3) there are eight separate pipes and 14 independent sources (pumps and tanks) for adding water to the reactor vessel. Any one of these can prevent core

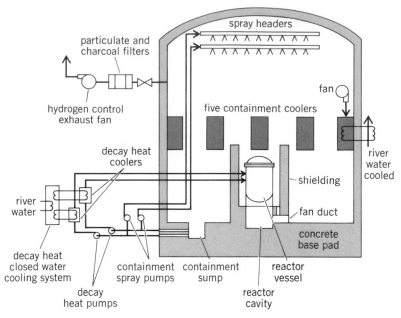

Fig. 3. Systems for heat removal and pressure reduction in containment building of a nuclear reactor.

melting, or terminate a postulated meltdown before it could threaten the integrity of the containment building. (3) Even if all normal power supplies and water supplies are lost, the progression of damage is sufficiently slow that there is ample time to bring up portable water supplies—for example, a fire engine can pump ample water to terminate such an accident even if all other systems are unavailable.

It is always possible to postulate a series of circumstances which might lead to eventual release of massive amounts of long-lived radioactivity to the environment. The Reactor Safety Study (Document WASH-1400) examined the possible chains of events which might lead to this result, and estimated the probabilities and consequences. For the Three Mile Island accident this requires the assumption of a considerable number of conditions which were not acutally present. The common perception that more massive releases of radioactivity were narrowly averted appears unfounded.

PUBLIC RISK PERCEPTION AND MANAGEMENT

The regulatory process has lacked a well-defined goal for reactor safety. Recent legislation has proposed that regulatory activities, nonnuclear as well as nuclear, take account of relative and alternate risks, rather than attempting absolute statements of risk, which inherently have large uncertainties. The bioethics community has consistently observed that the indefinite reduction of any given risk to society often brings with it an increase in other risks. At present, regulators have no authority or incentive for taking account of the possibility that in reducing a particular risk they may be forcing the acceptance of much larger and more probable risks. For example, the delay, limitation of future expansion, or reduction in use of domestically-based energy sources, such as coal or nuclear, reduce the direct risks from these sources, but increase other risks. Some of the comparisons of

alternate risks which have not been seriously investigated include:

1. The social and economic impacts of sharply increased energy costs to two or three times their cost during the 1980s. Some industries die, many items become more expensive or unavailable to some, and certain personal freedoms become more restricted.

2. The possibility of abrupt and continuing interruptions of a large part of imported oil supply. Major disruptions have large costs in human misery and in money—for example, the half-day blackout of New York City was estimated to cause economic losses of over $200,000,000.

3. The potential for severely damaging social chaos resulting from growing dissatisfaction with reduced living standards and perception of continuing reduced prospects for the future.

4. The possibility and consequences of regional wars or world war. These can be triggered in part by excessively procedural regulation leading to inadequate domestic energy supply in the United States. This has led to the need to protect, or threaten to protect, off shore oil sources. The risks inherent in the United States military buildup in the Mideast may overshadow all of the real or hypothetical risks of energy production from all conceivable domestic sources.

REALISTIC SAFETY GOALS

The key elements of a workable safety goal include criteria for the expected frequency of damaging events and their probable consequences and risk relative to other sources of energy the effects of conservation, or energy deprivation. It is essential that any safety goal include a workable practical method for guiding design and operation on the one hand, and regulatory activities on the other. Another essential attribute of the safety goal is that it be understood and accepted by a majority of the public and the responsible public officials. Many different proposals are being considered. One possible formulation of a safety goal includes the following points.

1. The frequency of accidents involving significant damage to reactor cores shall be less than one per so many years for the entire population of operating reactors in the United States. (A 30- to 50-year interval appears determinable.)

2. The design of the containment system shall be such that even if core melt occurs, there is at least 99% assurance that containment integrity will be maintained so that no member of the public can receive a dose of radiation as high as 5 rems (50 mSv).

3. Plans for warning and evacuating the nearby public should be maintained such that even if containment were eventually to fail, there is a 90% or better chance that the dose to any member of the public does not exceed 25 rem (0.25 Sv); and the collective dose to the population does not exceed 25,000 person-rem (250 person-Sv). [In determining collective dose, contributions from added exposures to individual members which do not exceed one-half of the average natural background (approximately 50 millirem or 0.5 mSv), should be excluded.]

4. The relative risks of widely available economic sources of electricity should be periodically

evaluated and compared with the risks estimated for nuclear power plants. To allow for uncertainties in such comparison, the cumulative risk goals for nuclear plants should be maintained in the region between one-half and one-fifth of the cumulative risk of alternate current practical energy sources. Measures which are aimed at reducing cumulative risk from nuclear plants to less than one-fifth of the cumulative risks from alternate energy sources should be subject to rigorous cost-benefit analysis and implemented only if they can be achieved in a manner which has negligible effect on the cost or availability of energy. Risks and social costs associated with potential deprivation, excessive increases in energy costs, or disruptions in supply are to be taken into account.

5. Administration of safety regulations should be established with the highest possible levels of professional, technical, and administrative competence, in order to warrant public confidence.

Fulfilling of the first safety goal above would require the objective determination of about a five-fold reduction in the expected frequency of core-damaging accidents relative to about 1000 operating years of commercial reactor operating experience (world wide) prior to the Three Mile Island accident. Given the prior operating experience as a point of reference, analyses using conventional engineering trade-offs (supplemented by the discipline of probabilistic risk assessment) provide an objective method for determining the effectiveness of improvements in design, operation, and inspection.

[EDWIN L. ZEBROSKI]

Bibliography: Committee on the Biological Effects of Ionizing Radiations, *The Effects on Populations of Exposure to Low Levels of Ionizing Radiation*, National Research Council/National Academy of Sciences, 1980; Interagency Report, *Population Dose and Health Impact of the Accident at TMI*, USGPO no.01701100408-1, May 1979; E. Kemeny et al., *Report of the President's Commission on the Accident at Three Mile Island*, USGPO, October 1979; M. Rogovin, *Report to the (NRC) Commissioners and to the Public*, USGPO, January 1980; E. L. Zebroski et al., *Analysis of the Accident at Three Mile Island*, Rep. NSAC-1, July 1979, revised (NSAC 80-1) March 1980.

Science and Technology for the Developing World

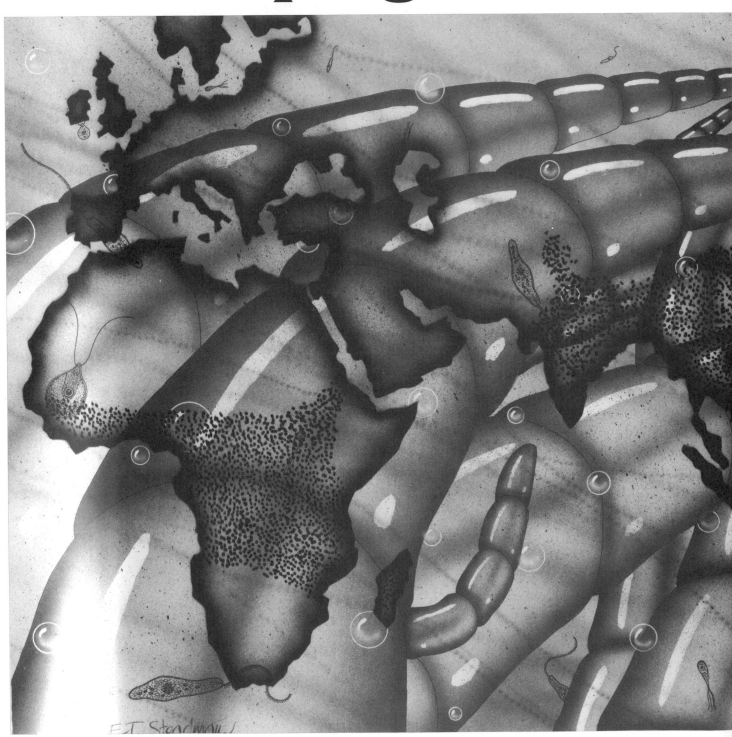

Louis Berlinguet

Louis Berlinguet received a
doctorate in biochemistry from
Laval University in 1950. Until
recently he was senior vice-
president of the International
Development Research Centre of
Canada. During 1976–1979 he
was vice-chairman of the UN
Advisory Committee on Science
and Technology.

During the last 40 years a wealth of studies, documents, and books have been written on the interrelations between science, technology, and development. Although there are marked differences between the types of development in countries from the industrialized north, countries with centrally planned economy, and countries from the south, there is at least one similarity: science and technology provide the tools by which countries can accelerate and improve the course of their economic, social, and cultural development. Scientific and technological factors also determine to a considerable extent the patterns of international comparative advantage, investment, production, and trade.

The potential of science and technology is enormous, yet this potential coexists with desperate poverty, underdevelopment, unemployment, hunger, illiteracy, and illness throughout the developing world, and only a tiny fraction of it is systematically applied to the problems of developing countries. Although there is wide concern about how science and technology may be judiciously applied to these problems, the coexistence of enormous scientific and technological potential in the developed countries with vast and growing unmet needs in the developing countries is a contradiction which cannot be allowed to continue.

The growing interdependence of the world community, the increasing pressures on resources of all kinds, and the scale of present and emerging problems all imply an urgent demand on the world community to reach agreement on how the potential of science and technology can be brought to bear on development problems. Some years ago, this would have been viewed largely as a question of how to transfer the necessary solutions from north to south.

Attitudes have changed greatly over the respective roles of science and technology in the developmental process. The United Nations Conference on Science and Technology for Development (UNCSTD), held in Vienna in August 1979, has produced several hundreds of studies and documents giving on the subject the views of most countries of the world and of the various specialized agencies of the UN family.

Because the attitudes toward science and technology in the development process are evolving very rapidly, not only in the developed countries but even more so in the developing world, it is appropriate to look at some of the important aspects of the current thinking on the subject.

CONCEPTS ON THE ROLE OF SCIENCE AND TECHNOLOGY

In an effort to find answers to the question of how science and technology can best be employed to serve the interests of developing countries, the International Development Research Centre of Canada (IDRC) financed more than 50 developing-country teams in 35 countries to carry out research in science and technology policies. The results of

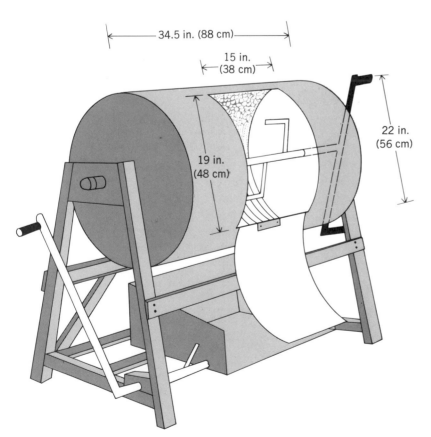

Fig. 1. A tumbler for semisolid fermentation, made of a discarded oil drum.

this important research done by a network of local researchers in developing countries were produced in time for the UN Conference on Science and Technology for Development.

The first main conclusion of the research was that the transferred technology is actually sold, and not given. Furthermore, most technologies can be "unpacked," that is, divided into several component technologies which can be sold and bought individually. From an exchange-value point of view, therefore, technology can be considered as a privately appropriated asset conveying market power. Proprietary knowledge, irrespective of its usefulness, has a market price. Therefore, the access to a technique, and hence its use value to those who need it, is conditioned by its exchange value and the associated market price, which in turn is determined by the relative bargaining power of the buyer and the seller. Finally, the relative bargaining power (and hence the market price associated with the exchange value of a technology) will be determined by the structure of productive forces and the overall form of social organization.

The second finding was that each country had to decide which technology to select and then try to adapt it to the environment. The old idea that all technologies could be easily transferred to all countries has given way to concepts of selection and adaptation.

The third concept which has gradually changed over the years is about the very nature of the technologies. In the past, obsolete technology in the industrialized world was given to developing countries. Now, rightly or wrongly, some technologies are regarded by the developing countries as second-class technologies and are rejected out of hand. The countries insist on having access to the best technology available. Some Third World countries are even prepared to pay a high price. According to prevalent thinking, each country has the right to select the technology which best fits its needs at a given time.

For any developing country to carry on a healthy, sustained process of self-reliant development, it must have an adequate capability to create, choose, adapt, reproduce, and acquire the technologies that are needed in every sector of its economy. But it must also have a healthy base of scientific activities and expertise, to create new knowledge to nourish technological creativity, and to discriminate in the choice, acquisition, and adaptation of technology.

The question of what is appropriate technology has to be considered carefully, case by case, which may lead to surprising results. A large country with a dispersed population might find the latest satellite in some ways a more appropriate technology than a telephone system which requires burying thousands of miles of old-fashioned cables in the ground. India sold Tanzania an "appropriate" system for converting waste into usable gas; it turned out that Tanzanian wages were too high to justify the labor-intensive system the Indians had developed.

EXAMPLES OF TECHNOLOGIES

The remainder of this article will consider some examples of technologies which are now being studied and adapted to local conditions in different parts of the world.

Microbial protein production. Applied microbiology is becoming a very important field of activity in developing countries. This science is being used for detoxification of wastes, purification of polluted waters, production of vaccines, fermentation of local food products, microbial fixation of nitrogen, production of methane fuel from garbage and manure, and microbial capture and conversion of solar energy.

A good example is the fermentation of agricultural wastes to produce protein for animal feeds. This relatively simple technology is being adapted to local conditions in Quezon City, Republic of the Philippines. In this project, ripe and unripe banana crops which do not meet export standards are being used as substrate. They are mixed in a slurry form (1:3 banana-to-water ratio) to which strains of fungi and yeast are added at the proper pH with a low concentration of sugars. Because of the high protein content of the resulting product, which is a dry, grainy cake, it can be used for feeding animals. The protein content is comparable to that of the current protein sources such as plant proteins and the commercial animal feeds (such as soymeal 45–50%, fish meal 60–65%). The fermentation takes between 24 and 48 h and is done in a batchwise manner in a tumbler made of discarded oil drums which can be rotated by human power or by any kind of mechanical driving force (Fig. 1). The process is suitable for villages because no high technology is involved.

Hand pump in rural areas. The World Health Organization estimated in 1975 that over 2,000,000,000 rural residents did not have reasonable access to safe water. According to a World Bank study published in 1976, only 14% of the rural population in developing countries have reasonable access to safe water.

Rural domestic water supplies in the Third World usually come from surface sources, such as streams or open wells. But these are susceptible to contamination, are often relatively far from dwellings (in an extreme case in Ethiopia, water trips in the dry season extend over 15 km) and in the dry season are mostly unreliable. The net result is a high mortality rate, a string of enervating diseases, malnutrition, and abject poverty, with a great time loss, particularly for women and children who could be much more usefully occupied.

To compound the problems, existing pumping installations in developing countries are notoriously unreliable. According to UNICEF in 1975, as much as 70–80% of the hand pumps used in India are out of order at any time. The estimated life of these pumps is only about 2 to 2½ years. The original hardware is often grossly inappropriate and difficult to maintain. Spare parts and tools are either missing or too costly, and there is a lack of a local institution or business willing to service or capable of servicing the pumps. A further problem is cost. Power pump installations cost in excess of $1300 and require sophisticated support services.

One main reason for the water underdevelopment prevailing in the Third World is that until recently very little engineering attention has been focused on the particular needs of rural water sys-

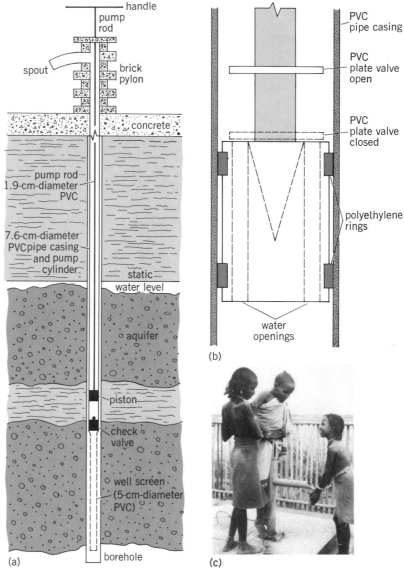

Fig. 2. Agricola pump technology. (*a*) Schematic of a well. (*b*) Cross-section view of a pump piston. (*c*) Pump site in Ethiopia. (*IDRC photo*)

tems. In fact, villagers in these countries have so far had to make do with pumps designed in industrialized countries a century ago!

Among the particular water supply problems of rural areas in developing countries, four recurrent themes are most noticeable: manually pumped tube wells should be used wherever possible, to make groundwater available at lowest cost and without a power supply; local people should be able to maintain the systems themselves; designs should incorporate materials for replaceable components that lend themselves to local fabrication; the pump should be low-priced.

The Agricola hand pump (Fig. 2) is a technological breakthrough incorporating the latest developments in engineering materials and production design. This pump meets all the conditions listed above. One of its major features is that it substitutes plastic pipe for the traditional and expensive steel or cast-iron pipe. The simplified design and use of much better materials significantly reduce the cost and improve the durability of the pump. In addition, all parts of the pump can be made easily in the developing countries, and replaceable components can be improvised from local materials if necessary.

One of the key features of the design of this pump is that the piston and check valve are identical; they can be used interchangeably by simply removing or adding minor components. Another key feature is that the base of the piston rod is embedded solidly into the piston, thus providing greater strength and reducing the risks of breakage.

The simple operating principle of the Agricola pump is as follows: On the downstroke, water pressure forces the polyvinyl chloride plate valve in the piston to open and the identical plate valve in the check valve (foot valve) to close. On the upstroke, this configuration is reversed—the piston valve closes, and water is drawn into the pipe casing through the open check valve and is lifted a distance equal to the length of the stroke. With repeated strokes, water eventually reaches the collar of the well, where it remains thereafter for easy hand pumping. Thus, the piston and check valve remain below the original surface of water in the well. This means not only that the piston rings are always water-lubricated, but also that the pump is always ready for use and need never be primed.

Remote sensing. In early 1971 a project was launched to study the possible utilization of remote sensing for resource surveying, mapping, planning, and development in the Sudan. This is a vast and expansive country with limited technical and financial resources, but with many climatic and ecological zones, extending from the desert in the north to the tropical forests in the extreme southwest. Thus, it would be impractical from technical, economic, and time considerations to rely on ordinary aerial photography for resource surveys of the country's more than 2,500,000 km².

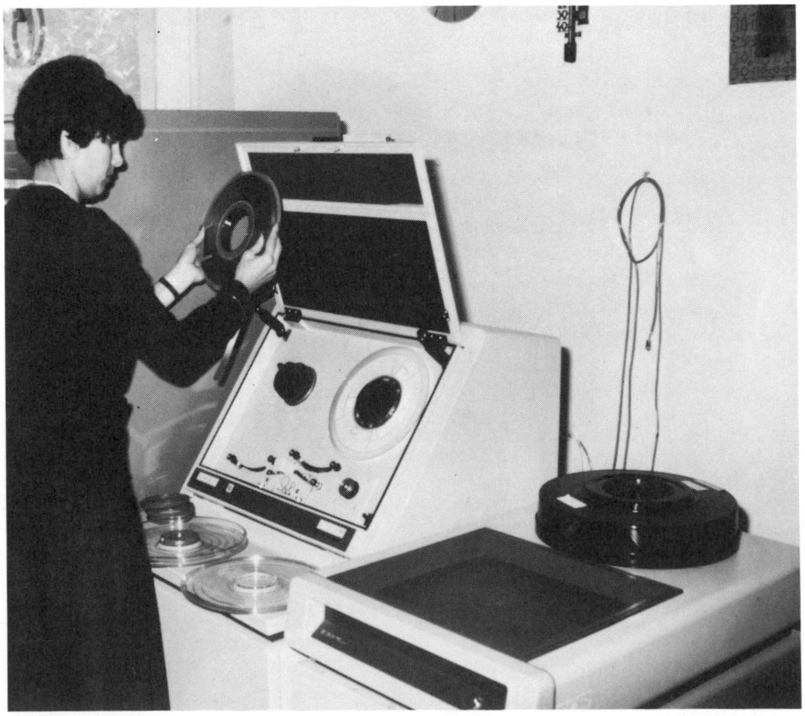

The objective of the project was: to transfer new technological information to members of the Sudan Remote Sensing Unit through training and research; to provide a high-quality data base (Landsat imagery) for present uses and future reference; and to produce maps of natural resources on a selected area of Sudan's savanna belt by using visual interpretation of Landsat false-color imagery and digital analysis of Landsat multispectral scanner data.

This experiment has shown that it is relatively easy to identify and map gross features such as land systems, large forested areas, sand dunes, and water bodies. Major difficulties were encountered only in the identification of small areas of spectrally separable features for which little or no precise ground observation data were available. This is a case where high and sophisticated technology can be of great help to developing countries in exploring and exploiting their natural resources.

Information retrieval and library management. In the last 10 years the operating speed and reliability of computers have increased markedly, while the physical size and the cost of storage have been reduced by factors ranging from 100 to 1000. Many sophisticated hardware capabilities are now being offered as standard items on low-cost minicomputers. Coupled with this, minicomputer manufacturers are now providing fairly elaborate operating system software responding to the diverse needs of the users.

Thus, when new applications requiring automation are being considered, users are turning more and more toward the minicomputer. Governments, institutions, and industries with limited budgets have opted for minicomputers to meet their specific needs. Some of these users not only have proved that minicomputer-based systems are cost-effective, but have shown a high level of sophistication in the use of the equipment. Many developing countries have expressed interest in acquiring a low-cost computer system to allow them to participate in decentralized international information systems. The International Development Research Centre has developed the software of an Interactive Minicomputer System (MINISIS) which has been installed in several developing countries (Fig. 3). International and local banks of data are now available, and many technicians and professionals from these countries are being trained in this high technology.

Wastewater treatment and resource recovery. Much attention is being focused on the problem of water quality, giving rise to a search for low-cost methods of wastewater treatment. Domestic sewage is extremely dilute (99% water) and therefore difficult to treat. The Western world has, with few exceptions, chosen to regard such wastes as nonrecoverable and has managed them at considerable cost. There are, however, techniques for treating wastewater while reclaiming wasted nutrients in the form of reusable by-products.

One of these techniques, the high-rate algae

Fig. 3. A Tunisian technician working on the computer used for the MINISIS system. (*IDRC photo*)

Fig. 4. Wastewater reclamation in high-rate ponds for treating piggery wastes in Singapore. (*IDRC photo*)

pond, intensifies the natural process of microalgae growth in open ponds. The algae use waste nutrients and solar energy to grow and, when harvested, can be used as a protein-rich component of livestock feeds. The pond also provides the possibility of reusing the treated water for irrigation or other uses. The protein content of algae is 50%, and annual yields of algae on the basis of a kilogram of protein per hectare are far higher than yields of conventional crops such as rice, corn, or soybean.

Most of the early experimental work with algal systems involved algae growth on media fertilized with inorganic chemicals; however, it is more economical to grow algae on wastes. The addition of costly fertilizers is not required, as municipal and agricultural wastes contain all or most macro- and micronutrients required to maintain peak algal growth. Stabilization ponds that are 20–40-cm deep maximize algae production because sunlight can penetrate throughout. The conversion of nutrients to algae is extremely rapid under these conditions, taking only 3 to 4 days. This growth/treatment pond is usually referred to as a high-rate algae pond (Fig. 4).

The process (Fig. 5) begins with influent wastewater, which may be domestic or agricultural. Settleable solids are removed from the influent by sedimentation and are treated by anaerobic digestion. The digestion by-product, methane, may be used as an energy source for sterilization and drying of the final algal product.

Clarified influent is added continuously to the pond, which is mixed daily so that settleable solids are kept in suspension. In terms of biochemical oxygen demand (BOD), the pond may be loaded as heavily as 350 kg/ha-day in the tropics. At this loading rate, one may expect the effluent to have a filtered BOD of less than 20 mg/liter when treating sewage.

Biomass yields from high-rate ponds may reach 40 g/m²-day under very favorable climatic conditions. However, in practical terms, biomass productivity of 30 g/m²-day is more realistic considering occasional unfavorable weather conditions, especially in tropical monsoon regions. Thus, it is possible that biomass yields could average 109 metric tons/ha-year or 49 tons/ha-year of protein. This is a 37-fold better yield of protein per unit area than the peak yield attained by soybean in the United States.

The treatment and recycling of wastewater achieves several purposes: Primary among them is sanitary disposal of wastewater. Also there is recovery of valuable resources in the form of protein and water. The water discharged from high-rate algae ponds is suitable for irrigation, fish culture, or cleaning and maintenance of animal shelters. This last use would reduce the farmers' need for additional supplies of costly water. As well, the settled solids, after biogas digestion, produce sludge suitable for use as agricultural manure.

The provision of feed for fish and livestock constitutes more than 50% of total operating costs of a farm. Harvested algae provide a valuable protein source for animal feed. Production of algae protein can therefore contribute markedly to the livestock economy, particularly in areas that are dependent on imports of soybean products for animal feed. If agricultural and aquacultural practices are to be rapidly expanded in developing countries to provide supplies of much-needed animal protein, then the above process is an attractive proposition, provided an economical, reliable system can be achieved.

Biogas technology. Biogas technology represents one of a number of village-scale technologies that are currently enjoying a certain vogue among governments and aid agencies and that offer the technical possibility of more decentralized approaches to development. Biogas technology is based on the phenomenon that when organic matter containing cellulose is fermented in the absence of air (anaerobically) a combustible gas (methane) is formed. However, the technical and

economic evaluation of these technologies has often been rudimentary. Therefore, there is a real danger that attempts are being made at wide-scale introduction of these techniques in the rural area of the Third World before it is known whether they are in any sense appropriate to the problems of rural peoples.

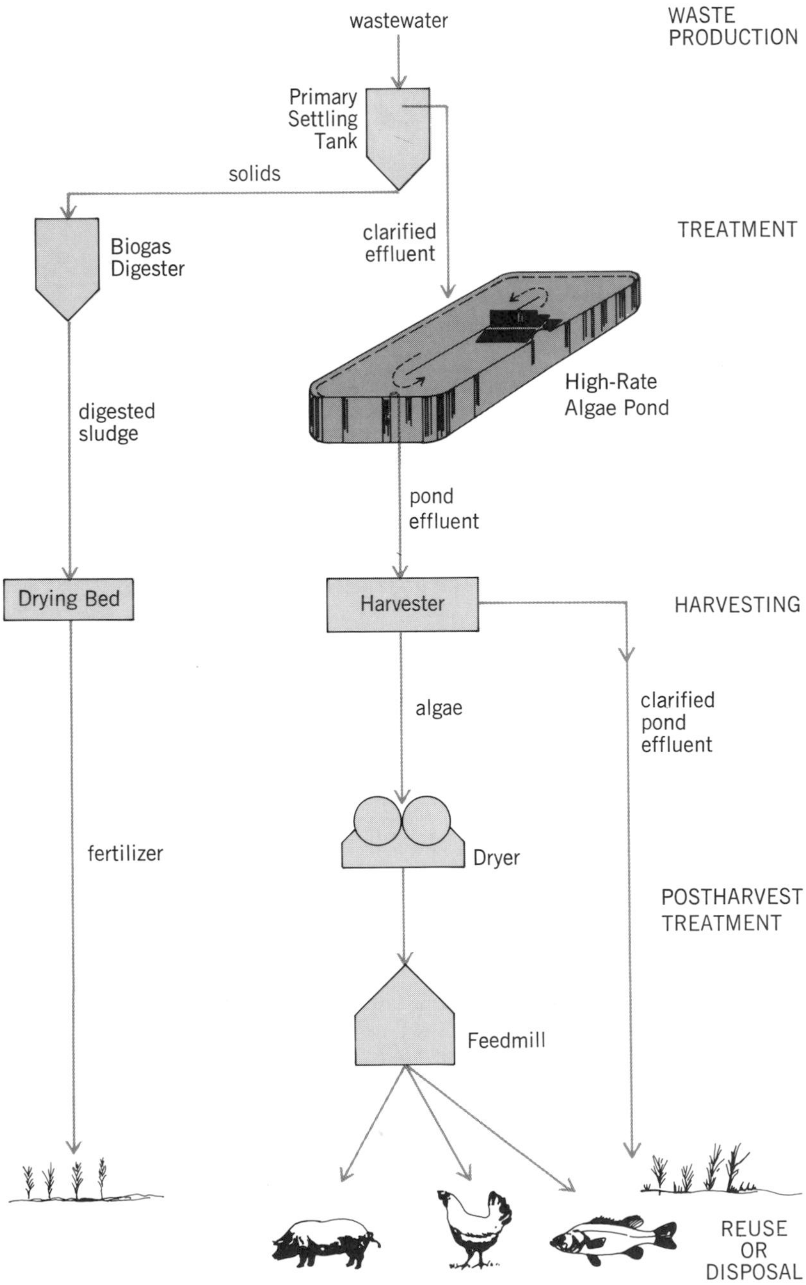

Many of the technical and economic evaluations that have been carried out so far have been applied to only a limited set of the known techniques, and comparisons have been made between biogas and other systems at the "high" end of the technology spectrum. Given the current bias in the distribution of the world's research and development effort, it is hardly surprising that in these comparisons the underdeveloped small-scale techniques sometimes appear to be inferior. With the fluid state of biogas technology and the unusual interest currently being shown in it, it would seem to be relatively easy to design and build biogas plants that could be operated in rural situations to meet certain social objectives and yet still compete with higher technologies even in conventional terms of profit and capital required per unit of output.

The viability of a particular biogas plant design depends on the particular environment in which it operates. Therefore the research problem becomes one of providing a structure in which technologists, economists, and users of the technology can combine to produce both the appropriate hardware for various situations and the infrastructure that is necessary to ensure that the hardware is widely used.

The major factors in considering the importance of biogas to a rural or village environment include its contribution to supplies of fuel, energy, and fertilizer, and to waste treatment, public health control, and sanitation, as well as its use of local resources (material and human).

Alternative energy and fuel sources. The main sources of energy that could be provided to rural areas are listed in the table. This gives a qualitative picture of the substitutability of the different energy sources for household, agricultural, or (small) industrial use, but gives no indication of the costs or likely appropriateness of the energy sources.

Feed materials can be fermented anaerobically (at 30–60°C) to produce biogas. Three methods are outlined in Fig. 6. Whether the wastes are animal or vegetable products, they can be considered to have their basic energy supply in the capture of solar energy (Fig. 7).

Biogas plants. Perhaps the most obvious initial biotechnology to investigate, with probably the most potential for raising both energy supply and crop yields, is the biogas plant. Over 75,000 of these have been set up throughout India over the past 5 years, and the Indian government hopes to establish a further half million by 1983. There are many more in China and the Far East already. Experience has shown that only the wealthiest 10–12% of rural households in India can actually benefit from the small family-size units, while the physical output and economic advantages of larger community plants are greater and can be shared by all. Thus, virtually all the biogas plants in India are family-size units capable of receiving the dung

Fig. 5. Wastewater treatment and resource recovery through algae production in a high-rate pond.

Main energy sources that could possibly be provided to rural areas*

Energy source	Household			Agriculture/industry		
	Cooking	Lighting	Heating	Power†	Transport	Heat energy‡
Electricity	(X)	X	....	X	....	....
Coke, coal	X	....	....	....	....	X
Kerosine	X	X	X	X	....	X
Diesel	....	....	X	X	X	X
Gas	X	X	X	X	X	X
Wood	X	....	X	....	....	X
Straw, vegetable wastes, crop residues	X	....	X	....	....	X
Dung	X	....	X	....	....	....
Solar energy	(X)	....	....	(X)	....	X
Hydro	....	....	....	X	....	....
Wind	....	....	....	X	....	....
Alcohol	....	....	....	X	X	....

*An X in parentheses represents methods that are likely to be very expensive, be of limited application, or need further development.

†Includes, for example, pump sets.

‡Includes steam.

from three to five cattle. These have been installed under the Khadi and Village Industries Commission's (KVIC) biogas program. However, an improved Chinese design has recently been introduced which is half the cost of the KVIC model and combines the fermentation pit and gas holder into one brick structure, as opposed to the mild steel gas holder floating over the pit which is a characteristic of the KVIC design. The KVIC is now experimenting with ferrocement structures, about 20% cheaper than their previous type, while the Murugappa Chettiar Research Centre (MCRC) of Madras has installed in Injambakkan village a modified KVIC-designed plant with an underground digester of brick wall and concrete floor and a gas holder consisting of a hemi-geodesic dome of wood to which is secured a vinyl balloon. Simplicity and cheapness are two of the basic criteria for these family-size units; the MCRC model costs about 50% of the price of a standard KVIC plant. Two cubic meters of gas (60–67% CH_4, the remainder mostly CO_2) can be produced per day from the dung of four cows, sufficient energy to satisfy the cooking and lighting demands of four people.

Digesters can be made from relatively cheap local materials (stone, mortar, cement), but must be constructed extremely carefully to avoid leakage. One ingenious possibility is a digester design of Chinese origin presently being promoted in Pakistan (Fig. 8). The construction is entirely brick and cement and incorporates no moving parts. It is possible to maintain an approximately constant gas pressure because increasing gas volume in the storage chamber expels some of the liquid content of the digester. If the cross-sectional area is large, the change in liquid height and thus gas pressure is small. The only possible drawback lies in exposure of the fermentor contents to the air, but diffusion of oxygen into the digester slurry is usually negligible.

Likely developments. A major characteristic of anaerobic digestion in practice is that the process depends on an acclimatized mixed culture of bacteria. Very little is known of the population dynamics or ecology of these cultures, and still less of ways in which particular strains might be encouraged or suppressed if this were useful in improving process efficiency. The room for improvement in speeding up the fermentation process and for improving its robustness or stability is enormous, and it may well be that unless such improvements can be made, biogas fermentation will always be at best a marginal contributor to rural and industrial development. Nutrient requirements and the ways in which different species compete for limited nutrients are little understood; at a practical level

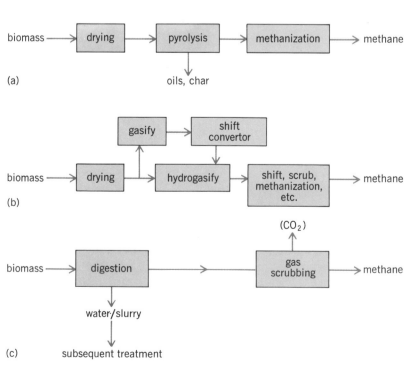

Fig. 6. Three principal methods for anaerobic conversion of feed materials to biogas: (a) pyrolysis, (b) gasification, and (c) anaerobic fermentation.

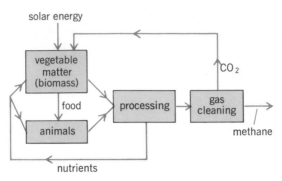

Fig. 7. Wastes have their basic energy supply in the capture (by photosynthesis) of solar energy.

there is little information on nutrient requirements and the returns on them.

Doubtless, as the role of the different bacterial strains becomes clearer, it will become possible to devise methods or to create environments to give improvements in efficiency.

Aquaculture. Despite its relatively ancient origins, it is only comparatively recently that any systematic research has been devoted to aquaculture. Although the practice is widespread throughout much of Asia, what little research has been undertaken has usually been capital-intensive, aimed at the lucrative markets of the industrialized nations—the mass production of trout or shrimp, for example.

There is little, if any, research in aquaculture devoted to the needs of the protein-poor Third World countries.

First and foremost among numerous reasons for aquaculture research is the need to increase food production in the Third World. Fish is high-protein food, but the world's oceans are no longer the inex-

haustible source of fish they were once thought to be. Some experts believe that the maximum sustained yield of the oceans—the point beyond which total fish stocks actually start declining—is now being approached. Most agree that many important species will be "fished out" by the end of this century. In addition, the technology for marine capture fisheries is becoming increasingly sophisticated and energy-consuming, and the capital cost of establishing and maintaining a deep-water fishing fleet is overwhelming.

By contrast, the potential for increasing fish production through aquaculture is considerable, and much of the technology is relatively simple, inexpensive, and consumes a minimal amount of energy. There are vast underdeveloped, underutilized areas of water throughout the world—natural and artificial lakes and ponds, rivers, irrigation canals, estuaries, and coastal waters—that are suitable for aquaculture in one form or another. In India alone, for example, there are an estimated 4,000,000 ha of village ponds, but less than a half million hectares of these are used for aquaculture.

Then there is the nature of the product itself. Fish are cold-blooded animals, and adapt to the temperature of the water that surrounds them rather than wasting energy keeping warm. Consequently they are much more efficient feed converters than are most land animals: fish are about 50% more efficient than hogs, and perhaps three times more efficient than cattle. Equally important, particularly from the point of view of the developing countries, is the fact that fish grow faster in warmer waters. There is a biochemical rule of thumb that for every 10° rise in water temperature the reaction rate doubles. The implied potential for fish production in warm tropical waters is obvious.

Fish production offers another advantage over both animals and plants: the fish occupy a three-dimensional space. In a polyculture system, for example, different species of fish that habitually feed at different water levels are raised together in one pond. Thus a pond with a surface area of 1 ha, when stocked with three such compatible species of fish, becomes in effect a much larger pond.

Fish also require little space. Experiments using advanced technology to recycle the water around the fish have shown that trout can grow to their maximum size in a body of water no greater than their own volume. In coastal waters off Singapore, mussel culture experiments have produced yields as high as 250 kg/m². Such yields theoretically could produce as much as 100 metric tons of protein from a 1-ha sea surface area of mussels. One hectare of land planted with protein-rich soybean will produce perhaps 1 ton of protein.

Finally there is the fact that aquaculture is much closer to husbandry than it is to hunting, and it is therefore more compatible with farming than with conventional fishing. The problems to be overcome to improve aquaculture production are problems that the farmer can relate to. Even the terminology—stocking and breeding, feed and fertilizer—is familiar. A project to more fully utilize village ponds in several states of India has shown that farmers are both able and willing to adapt to new systems, and to increase yields by as much as 20 times (Fig. 9).

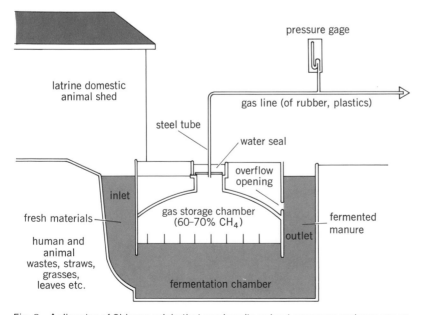

Fig. 8. A digester of Chinese origin that can handle animal manures and some vegetable wastes.

Fig. 9. SEAFDEC fish pond in the Philippines. (*IDRC photo*)

There is a remarkable potential benefit to be gained from the development of tropical aquaculture. But there are vast gaps in knowledge related to aquaculture in general. Research needs to be done in order to bring economical aquaculture within reach of the rural poor. To better understand the magnitude of the problem, consider that there are only about 10 farm animals of economic consequence. By comparison, there are literally thousands of different fish which could possibly be cultured, and about which very little is known. Information is needed to determine which species are most suitable for aquaculture, and to learn more about reproductive life cycles, nutritional requirements, their relationship with the environment, and a score of other factors that are essential knowledge before any fish species can be effectively farmed.

CONCLUSIONS

The technologies discussed above show that advanced technologies like remote sensing can play as important a role in the development process of a country as older technologies like aquaculture. Advances in the various technologies do require the close collaboration of the scientists and the technologists from the developed and the developing countries. The scientific community of the developed countries will have to pay more attention to the scientific and technological problems confronting the people in the rural areas of the developing countries. At the same time, the governments of the developing countries will have to increase the scientific capabilities of their societies. It is only then that they will be able to select and adapt the technologies which have been produced and worked out by the scientific community of the north. Given time, they will be able themselves to invent or to modify the technologies which are needed for their development.

[LOUIS BERLINGUET]

Bibliography: *Biogas Technology in the Third World: A Multidisciplinary Review*, IDRC-103e, 1978; *Fish Farming: An Account of the Aquaculture Research Program*, IDRC-120e, 1978; *Information Retrieval and Library Management: An Interactive Minicomputer System*, IDRC-TS14e, 1978; *Murugappa Chettiar Research Centre*, Periodical Tech. Notes no. 1, MCRC, Madras, 1977; *Remote Sensing in the Sudan*, IDRC-TS9e, 1978; *Science and Technology for Development—STPI Module 1: A Review of Schools of Thought on Science, Technology, Development, and Technical Change*, IDRC, 1980; *Science, Technology and World Development: The Views of UNDP*, Doc. A/CONF.81/BP/UNDP presented to the UN Conference on Science and Technology for Development, July 1979; A simple technology to augment fish production, *Technology*, vol. 2, no. 1, Philippine Council for Agriculture and Resources Research, Los Banos, 1980; *Wastewater Treatment and Resource Recovery: Report of a Workshop on High Rate Algae Ponds, Singapore, 27–29 February 1980*, IDRC-1543, 1980.

Nonparasitic Plant Pathogens

Shreve S. Woltz

Shreve S. Woltz is professor of plant physiology at the University of Florida. He received his B.S. degree at the Virginia Polytechnic Institute in 1943 and his Ph.D. in soils and plant physiology from Rutgers University in 1951. His research has been in the area of the physiology of plant stress and disease, with his current work focusing on air-pollution physiology.

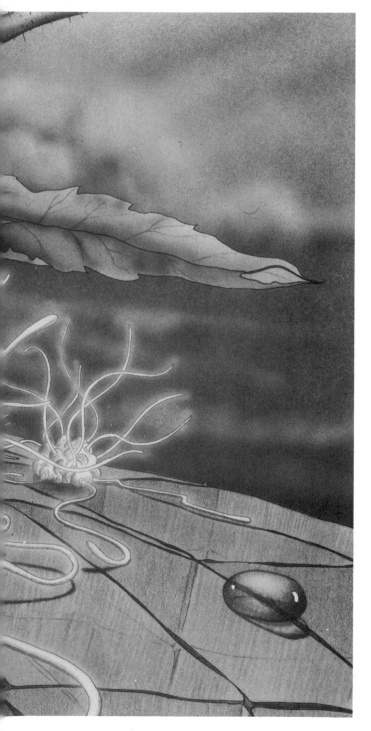

An unusual group of microorganisms (bacteria, fungi, actinomycetes, and algae) cause disease in higher plants, although they do not live within the plant tissues. The concept, unique and somewhat new to plant pathology, is that an environmental microorganism living in close but noninvasive contact with a plant can produce prominent, disabling symptoms of disease at a distance. This action is possible through an "arrow" of a toxin possessing some degree of suscept specificity or by some physical effect that produces abnormal growth but less of a classical "disease." The concept and terminology of a nonparasitic disease seem somewhat contradictory at first glance. This is because of the development in phytopathology of disciplinary criteria based on the classical host-parasite relationship. S. S. Woltz suggests the use of the terms exopathogen (an external, living plant pathogen) and exopathogenesis (external incitement of disease) for this small area of phytopathology. Toxins are the base of most exopathogenic disease, whereas in endopathogenic disease toxins are important, but tissue maceration and plugging of conductive vessels are also important.

TERMINOLOGY

In nonparasitic disease neither parasite nor host is deemed an appropriate term when using a definition of parasitism that involves invasive action. (External parasitism occurs, but for the purposes of this article it is necessary to restrict terminology to the narrower definition.) Preferred terminology lists the affected plant as the suscept, and the externally operating pathogen as an exopathogen. The pathogen obtains its nutrition outside the plant, although the nutrition may be supplemented by plant secretions which may even be enhanced by the exopathogen; the external microorganisms cause continuous irritation to the suscept plant, which fact justifies the description of them as pathogens.

Toxins produced by exopathogens are quite specific in their action on the affected higher plants, producing yield reduction and morphogenetic changes. A susceptible plant reacts by developing visible symptoms such as chlorosis (yellowing because of chlorophyll destruction), malformation, and necrosis. The symptoms of diseases produced by an exopathogen are quite specific with regard to species and cultivar of plant affected. Nonparasitic pathogens of plants have been able to develop an ecological niche relative to competing microorganisms and higher plants with which they establish an external relationship. The development of the ecological niche may depend on the presence of the suscept. The rarity of such disease results from the infrequency of simultaneous occurrence of the requisite environmental conditions when the suscept and exopathogen are in close contact under appropriate environmental conditions for the production and absorption of toxin.

The diseases which are known or suspected as being caused by nonparasitic pathogens of exopathogens are few in number, poorly categorized, and have incomplete descriptions of etiology. Exopathogenesis occurring in the absence of parasitism is, because of a lack of a firm association, difficult to link as pathogen-related and suscept- or victim-related. The concept of exopathogenesis brings forth a few changes in the classical terminology of plant pathology. Disease remains the same, a condition which is a result of altered physiological or morphological development, leading to visible symptoms of dysfunction. A saprophyte subsists on nonliving materials; a saprophyte is commonly thought of as the opposite of a pathogen but may, in fact, be an exopathogen as well as a saprophyte. A saprophyte that secretes a disease-inducing toxin is properly termed a pathogen. Suscept is the definitive term for a plant that is susceptible to disease caused by either parasitic plant pathogens or nonparasitic plant pathogens. Exopathogen is equivalent to nonparasitic plant pathogen, and is a more convenient term for reference and information retrieval. Both saprophytes and exopathogens are organisms that live outside their suscepts and induce disease by liberating a toxin that can be absorbed by some part of the plant, producing abnormal metabolism, chlorosis, necrosis, or growth derangement. In the reference of these definitions, specific examples of established, probable, or potential nonparasitic diseases can be considered.

FRENCHING OF TOBACCO

By far the oldest example of a probable nonparasitic disease caused by a living exopathogen is frenching of tobacco, described in a letter to the Royal Society of London in 1688. The disease has been very troublesome to plant scientists because the etiology has been elusive, ephemeral, and incompletely explained. Control has been accomplished by sterile or gnotobiotic culture. The disease occurs in a sporadic manner, occurrences being somewhat unpredictable. The disease was attributed to a number of causes, but appears most likely to be caused by one or more amino acid analogs synthesized by microorganisms and released into the soil milieu. These compounds structurally similar to leucine are likely resistant to rapid microbial decomposition in the soil and capable of uptake by plants. They are available for plant uptake and cause morphological derangement and stunting in the tobacco plants at very low concentrations, apparently as an antimetabolite of the amino acid leucine. Frenching soils can be used to inoculate healthy soils and thereby produce frenching of tobacco under a continuously warm, moist soil environment. Frenching is characterized by a green-netting pattern on a chlorotic background of leaf lamina of fine veins in affected plants. Plants have a marked reduction in internode length, have rosetting of small strap-shaped leaves, and exhibit severe stunting (Fig. 1).

The fact that applications of inorganic nitrogen fertilizer to tobacco reverse trends toward frenching may be interpreted as being related to increased endogenous production of leucine or other affected natural amino acids which overcomes a certain level of amino acid antimetabolite pro-

Fig. 1. Tobacco with frenching syndrome induced by injection of 1-amino-2-nitrocyclopentane-1-carboxlyic acid (ANCPA). (a) Plant injected with 10 μg ANCPA. (b) Control plant. (Photo by P. W. Brian and D. Broadbent)

duced in the soil. Also, organic nitrogen, in contrast, favors soil microorganisms by providing assorted organic metabolic intermediates, as well as an energy or carbon source. Pateurizing soil with heat prevents frenching. Reinfesting with "sick" soil in a very small proportion (1 part sick soil to 2000 parts pasteurized soil) restores the disease-producing capacity of the soil.

Practices which disturb the soil and root zone such as cultivation or lifting and resetting plants frequently cause tobacco plants to "grow out" of the diseased condition. Also, pulling on plants to partially break and disturb the root system causes an amelioration of the disease. These observations apparently relate to disease control by improving soil aeration and perhaps by root pruning which will result in a decrease of uptake of toxin by roots from the closely associated microflora surrounding each root.

Low soil pH, low soil temperature, and growing plants with low-soil-moisture content reduced the tendency for frenching or prevented the disease entirely. Frenching soils also produced the respective types of frenching in ragweed, sorrel, tomato, and squash, but not in pepper. Family and genus were apparently not as important in relation to susceptibility as biochemical or metabolic characteristics.

Soil features favoring frenching include poor aeration, relatively high temperatures, soil pH in the range 5.8–7.5, and higher soil temperature associated with summertime in warm geographical areas. The chemical disease-producing factor (toxin) can be generated in one container of sick soil, collected by decoction, and then applied to tobacco plants growing in sand culture for successful production of frenching symptoms. The best theory from this observation is that frenching is a toxicity disease rather than a deficiency disease, even though addition of inorganic nitrogen frequently overcomes the diseased condition.

YELLOW STRAPLEAF OF CHRYSANTHEMUM

The yellow strapleaf disease has been observed in all major chrysanthemum growing areas of Florida from time to time for the past 25 years, as well as in other areas of the United States. In a survey of the commercial plantings in Florida in 1958, the disease was found in almost every large commercial planting. Less than 10% of the plants in any single planting were affected, and in most cases fewer than 0.1 percent were found. Incidence of the disease has generally declined over the years, apparently associated with adoption of preventive cultural techniques by growers.

Symptoms. The name yellow strapleaf was selected because narrow pale-yellow leaves are noticeable symptoms of the disease. The disease commonly appears 3–4 weeks after planting and most often after pinching if the crop is pinched. First symptoms are usually seen in young developing leaves in axillary shoots (pinched crop) or terminal shoots. These leaves fail to expand normally, becoming claw-shaped with upward rolling of margins and incurved tips (Fig. 2). Pinching seems to cause the symptoms to develop more quickly and uniformly. As the narrow pale-yellow leaves expand, they begin to flatten out and assume a dull

Fig. 2. Chrysanthemum shoots. (a) Shoots exhibiting yellow strapleaf disease symptoms. (b) Normal shoots.

yellow or ivory color. At this stage, leaves frequently remain narrow, are very much elongated, and slightly brittle with small lobes, or entire margins. A characteristic symptom is the retardation of new growth after pinching. Axillary buds turn yellow and may be swollen. New leaves open slowly and internodes are short. The upper portion of the stems of affected plants are often larger in diameter, and the entire stem becomes abnormally hard and brittle. Symptoms are similar for various commercial cultivars. The disease has been noted in 20 or more cultivars, but is seen most often in Iceberg, Blue Chip, and Shasta. The disease may occur in plants at later stages, including plants on which flower buds are visible. The symptoms are sufficiently different from those of other diseases to simplify identification.

In severe cases, diseased plants remain stunted and yellow for 6–8 weeks, during which time they grow only 4–6 in. (10–15 cm). Root development is normal, but may be slightly greater in diseased plants. Stunted plants often resume normal growth after passage of time. Leaves which are already strap-shaped do not develop normal shapes, but yellow tissues turn green and new growth may be normal except for a residual stunting. Flowers produced on recovering plants are frequently small and distorted, with few expanded petals. The centers of heads are abnormally compact and green, partly due to excessive growth of bracts around the disk flowers.

Soil conditions. In Florida a yellow strapleaf appearance is usually associated with warm weather, roughly from April to November. High rainfall and ample irrigation, together with soil conditions favoring retention of soil moisture and reduced aeration, all contribute to development of the disease, which is usually found in the lower parts of a planting. The difference in elevation may be very slight, but the soil surface condition will indicate higher moisture conditions in terms of darkness of color, cohesiveness of soil, and presence of algae or fungi growing on the soil surface. The maintenance of seep-water level that is optimum or slightly high for most beds may saturate the root zone soil in low areas. In potted greenhouse plants growing in soil from around yellow strapleaf plants, the disease developed much more readily in plastic pots kept wet than in clay pots. The difference appears to be less aeration and greater moisture retention in plastic pots.

Disease occurrence has been observed to be less with steaming of soil compared with various methods of fumigation. When entire plantings are steamed, there is usually a low incidence; occasional plants may have yellow strapleaf, however, when nonsterilized subsoil is brought up to the soil surface in digging post holes. Another indication of the effects of contamination of soil microorganisms is the occasional observation that surface flooding by heavy rains may cause a widespread and randomly distributed occurrence apparently associated with an influx of soil microorganisms under conditions favoring disease development.

Occurrence. In addition to the observed occurrences of yellow strapleaf on both the east and west coasts of central to southern Florida, it was found in plantings in St. Johns County for a number of consecutive years. Reports from Georgia, Massachusetts, New York, North Carolina, and the Republic of Panama indicate the occurrence of yellow strapleaf in those areas, verified by persons familiar with the symptoms. The report from New York (Long Island) was of special interest, since it constituted the single observance in commercial pot plant production, involving about 1000 pots. While yellow strapleaf can be induced in potted chrysanthemums and pot varieties, the cultural procedures used in growing potted chrysanthemums do not favor the disease; the common pot varieties which do not grow to a large size appear less susceptible than other varieties.

Cause. The specific cause of yellow strapleaf has not been determined; the general nature of contributing factors, however, has been delineated. Accumulated information strongly suggests that yellow strapleaf is caused by an amino acid toxin released in the soil by one or more soil microorganisms. The toxin disrupts amino acid metabolism within the plant after being accumulated from the soil. While there is no indication that the organism directly invades chrysanthemum plants in producing disease, there is indication that disease may be prevented by controlling the population of soil microorganisms.

Exploratory research was carried out with the object of developing leads to the cause and control. Herbicides and viruses were studied as possible causes and ruled out. Simple inoculation and grafting procedures indicated that the disease was not directly transmissible in terms of an infectious disease. Thallium toxicity, studied with the similar disease frenching of tobacco, did not produce yellow strapleaf in chrysanthemums.

Complete soil and plant chemical analyses did not reveal any abnormal nutritional situation in paired yellow strapleaf – no yellow strapleaf samples from various areas of Florida. Complete nutrient solutions of known inorganic nutrients did not correct severe disease in the field when used repeatedly as combination spray-drench treatments. Individual micronutrients (boron, copper, iron, manganese, molybdenum, and zinc) were tested, similarly without evidence of marked improvement. Although deficiencies of nutrients do not appear to be the basic cause, the development of the disease is subject to the effects of many soil-nutritional and environmental factors. This is not unexpected in view of the complexity of the disease situation with soil-plant-microorganism interactions. There are a great many similarities in the etiology of yellow strapleaf and frenching disease.

Bacillus cereus. Sophisticated research was carried out with the tobacco disease, and it was found that certain forms of the amino acid protein-building unit isoleucine would reproduce the natural disease syndrome. This was true for tobacco grown in germ-free culture when small amounts of isoleucine were fed to the plants by way of the roots. In soil, however, "very large" amounts of isoleucine were required to produce frenching symptoms. It was suspected that *Bacillus cereus* found in the root zone of tobacco plants might be producing a toxin—perhaps an amino acid—that would produce the same effects as isoleucine compounds. Tobacco was therefore grown in closed flasks in the absence of contaminating microorganisms, and then physically separated chambers were inoculated with an isolate of *B. cereus.* Under this aseptic (gnotobiotic) microculture, symptoms of frenching were obtained due to the diffusion of a toxin from *B. cereus* into very small plants. This aspect of the work has not been extended to soil culture, except to the extent that *B. cereus* has been found commonly in tobacco soils and in especially large numbers in the root zones of frenched tobacco plants.

Isoleucine. Frenching of tobacco and yellow strapleaf of chrysanthemum are apparently caused by similar soil conditions and result in similar syndromes. Both disorders are favored by relatively high soil pH, periods of high soil-moisture levels, and high soil temperatures. The symptoms displayed with the diseases have three common features: green netting or reticular chlorosis of leaves; narrow, strap-shaped leaves; and growth retardation. Tobacco and chrysanthemum plants growing together in containers of soil that had produced yellow strapleaf in the field developed symptoms of the respective diseases. Following the information that isoleucine and certain other amino acids were effective in producing frenching, DL-isoleucine with DL-alloisoleucine was applied to the root zone of chrysanthemum plants. This amino acid treatment uniformly and quickly produced yellow strapleaf symptoms with plants growing in solution culture, steamed soil, methyl bromide--treated soil, untreated soil, quartz sand, and expanded volcanic glass (perlite). Six chrysanthemum varieties of varying degrees of susceptibility were treated with isoleucine in this experiment. Symptoms developed in all varieties, with severity approximately in the order of observed field susceptibility.

Aspergillus wentii. A new surge of work and interest in the subject of yellow strapleaf developed following reports from England that the fungus *Aspergillus wentii* synthesizes an analog of leucine, 1-amino-2-nitrocyclopentane-1-carboxylic acid (ANCPA), which might also be called 2-nitrocycloleucine. ANCPA was shown to be a potent antagonist of leucine, affecting the growth and development of the pea (*Pisum sativum*) (Figs. 3 and 4). P. W. Brian and coworkers reported on the physiological effects of ANCPA on various plants, especially pea and tobacco. The syndromes were quite similar to those of frenching of tobacco and yellow strapleaf of chrysanthemum. They did not explore the effects of *A. wentii* in soils.

Since ANCPA is a naturally occurring antagonist of leucine, experiments were undertaken with *A. wentii* and ANCPA to establish whether *A. wentii* might be a natural causal agent. Pure ANCPA and ANCPA contained in filtrates from cultures of *A. wentii* readily produced the characteristic, easily identified yellow strapleaf syndrome. ANCPA produced symptoms equally well whether sprayed onto plants, injected into them, or poured onto root zones of plants growing in agricultural soils or autoclaved media. *Aspergillus wentii* spore suspension was poured into germ-free vermiculite in which *Chrysanthemum morifolium* seedlings were growing. These inoculated seedlings developed severe yellow strapleaf, while control seedlings grew normally. In recent experiments, non-aseptic infestation of soils with *A. wentii* resulted in the development of severe, lasting disease symptoms, indicating that *A. wentii* might cause a problem in field culture. Research is continuing with the objective of identifying specific soil microorganisms and their toxins that cause natural yellow strapleaf in the field. The possibility exists that yellow strapleaf may be caused by a bacterium (as by *B. cereus*) or by a fungus (*A. wentii*) and—more directly—by their individual toxins, which may not be the same compound. Procedures suitable for control of a soil-borne bacterium would differ from those for a fungus in many respects, since the physiology and adaptations of bacteria and fungi represent two extremes for soil microflora.

While it has not been established that *A. wentii* is the causal agent in naturally occurring plant diseases such as yellow strapleaf, its potential requires evaluation.

Aspergillus wentii is cosmopolitan in its distribution, occurring in all parts of the world and upon a variety of organic substrates, such as moist grains and decaying vegetation; *A. wentii* possesses antibacterial activity which gives it a strong competitive advantage in soil. This fungus species is capable of carrying out a great many enzymatic reactions, utilizing chemicals and materials in the environment as growth substrates. Because of the extremely large numbers and kinds of microorganisms growing competitively in soil, the likelihood of the rise or decline of a given species is difficult to predict. *Aspergillus wentii* is commonly found associated with the roots of pineapple and the drug plant *Coptis japonicus*, and grows readily at a low relative humidity on organic debris of plant or animal origin. Populations of this fungus have reportedly increased in soil between forest strips planted as windbreaks. Work is under way to determine the frequency of occurrence of *A. wentii* in yellow strapleaf versus nonyellow strapleaf soils. It is frequently difficult to recover the fungus from soils because of competition of other microorganisms during the process of isolation. The nature of the problem can be visualized when one considers that there are frequently millions of cells of mi-

Fig. 3. Comparison of (a) normal (control) Pilot pea plant and (b) plant sprayed with autoclaved *Aspergillus wentii* culture filtrate. (*Photo by D. Broadbent and P. W. Brian*)

56 NONPARASITIC PLANT PATHOGENS

Fig. 4. Close-up of Pilot pea plant which had been sprayed with autoclaved *Aspergillus wentii* culture filtrate. Note narrow, strap-shaped leaves, stunting, and deformity of tendrils. *(Photo by D. Broadbent and P. W. Brian)*

croorganisms per gram of soil, and commonly millions of *Aspergillus* spp. cells per gram of slightly decomposed hay (as an example of organic soil debris). Also, the air in comparatively "clean" environments, such as hospital wards, may have as many as 2000 *Aspergillus-* type spores per cubic meter. In agricultural buildings where hay is handled, the count may range from 12,000,000 to 20,000,000 spores per cubic meter.

The distribution of *A. wentii* and other strongly competitive microorganisms is dependent on their ability to survive and grow under adverse conditions, as well as their capacity to wage biochemical warfare. *Aspergillus wentii* has been reported to produce an unidentified antibiotic active against bacteria, β-nitropropionic acid active as a toxin, and the potent leucine antagonist ANCPA. Experimental evidence indicates that *A. wentii* steadily secretes ANCPA into the aqueous phase of the environment. ANCPA produces the yellow strapleaf syndrome when sprayed on chrysanthemums at the rate of 1 oz per 100 gal or 8 ml per 100 liters. Much lower levels are effective when completely absorbed within the plant, that is, injected; 1 g is capable of causing severe yellow strapleaf symptoms in 60,000 plants (1 acre or 0.4 hectare). ANCPA is also inhibitory to seed germination and certain microorganisms.

Aspergillus wentii is capable of growth and sporulation in environments unfavorable to most organisms; under favorable nutrient, temperature,

and moisture conditions, growth can be very rapid. However, the competitive capability of an individual microorganism species such as *A. wentii* is usually surpassed by the combined effect of microorganisms in the soil. Therefore, to obtain individual effects of a soil microorganism on host plants, such as yellow strapleaf in chrysanthemums, highly specific soil environmental conditions must prevail.

Control. The effect of ANCPA or alloisoleucine in producing the yellow strapleaf syndrome is primarily that of a biochemical similar in structure to leucine (and in the case of alloisoleucine similar to valine) entering the chrysanthemum plant and blocking reactions by substituting physically for the normal amino acids. The effect is not that of a corrosive chemical or acute poison to the plant, but rather a metabolic inhibitor that limits new growth and synthesis of proteins and chlorophyll. Since leucine and sometimes valine are "antidotes" for the yellow strapleaf chemically induced syndrome, natural yellow strapleaf plants were sprayed with leucine (the DL-2-methyl form was used because of its higher solubility) and valine. Leucine resulted in new growth and greening, but valine did not on replicated plots in chrysanthemum plantings on both the east and west coasts of Florida. This constitutes preliminary evidence which may be of value in identifying the microorganism pathogen and the toxin that it apparently produces in agricultual soil. Natural amino acid antidotes are exported by mature chrysanthemum leaves, especially when well supplied with nitrogen and other elements that permit export of manufactured amino acids to immature plant tissue.

In yellow strapleaf new growth is severely limited, there is an accumulation of sugars and amino acids in the plant, and older leaves become thickened with a downward roll. These are symptoms of excess carbohydrate. Yellow strapleaf, which frequently appears after a pinch of plants, is probably brought on by pinching, which temporarily retards vegetative growth. Pinching is likely to cause an accumulation of sugars and amino acids, which would then be lost by the roots. The first stages are probably autocatalytic in that free amino acids, present in affected plants in large amounts due to the disease, can diffuse into the soil and support growth of microorganisms and further production of an amino acid inhibitor in the soil rhizosphere for uptake by the plant. Large amounts of ANCPA or alloisoleucine introduced into chrysanthemum plants have long-lasting effects associated with a low rate of destruction or alteration of these chemicals by the plants; acute naturally induced yellow strapleaf may be very long-lasting.

Susceptible cultivars. Some chrysanthemum cultivars are more susceptible than others. Two factors that would influence this reaction would be: some cultivars have greater native capacity for synthesis of leucine; and faster-growing cultivars would likely be more susceptible due to a depletion of free-leucine supply within the plant which would render the plant more susceptible to an induced leucine deficiency, the apparent cause of yellow strapleaf expressed in terms other than those customarily used involving the toxin.

ANCPA EFFECTS ON VEGETABLE PLANTS

Twenty vegetable species were divided into three groups according to response to a foliar spray application of ANCPA. Six species were affected severely, with long-lasting effects (2–4 weeks). These plants eventually resumed normal growth, but many of the growth deformities in the older leaves were irreversible. Seven species showed no response to the ANCPA spray application.

The symptoms produced in the affected species included chlorosis of expanding leaves, a failure of the growing leaves to expand laterally, thereby causing a narrowed appearance, and sometimes a green netting produced by green veinlets surrounding areas of chlorotic leaf lamina. It is only in developing plant organs or in those rapidly synthesizing chlorophyll that the response is readily apparent. In addition to the effect on leaves, there was a retardation in stem elongation closely associated with the development of other symptoms. A longitudinal stem splitting was observed in watermelon sprayed with ANCPA. Small tomato fruits (1–2 cm in diameter) developed severe catface in about 50% of the fruits that were sprayed. Injection of 10–25 μg of ANCPA into the peduncle of fruit hands bearing similar fruit caused catface development in all of the fruits on each hand injected. The catface was moderate to severe. Uninjected fruit hands had very little catface. Naturally occurring catface in control fruit hands was not severe. Flower development was noticeably altered in one of the nonvegetable test species, morning glory. The affected flowers had narrowed and twisted petals, and were polypetalous instead of gamopetalous. In cases of less severe response the petal margins of the trumpet-shaped flowers had indentations.

ANCPA was very effective in producing growth retardation when injected into the stems of seedlings of five plant species. For bean, tomato, chrysanthemum, sunflower, and tobacco, ANCPA probably acts as an antimetabolite of L-leucine. It was observed that L-leucine also prevented or reduced the severity of the typical ANCPA effects of narrowing and yellowing leaves. ANCPA injected into seedlings of the plant caused the development of very defective leaves. The leaves were narrowed to the extent of appearing stringlike with the higher amounts of ANCPA (15 and 25 μg per seedling).

ANCPA is a very potent toxin capable of disrupting the growth and normal morphological development of many plant species. Since it is of natural occurrence, being produced by the fungus *A. wentii*, it could under unusual soil and climatic conditions cause disorders of the type described as frenching of tobacco or yellow strapleaf of chrysanthemum, as well as the similar disorders in tomato, squash, ragweed, and sorrel. There is reason to suspect other microorganisms such as *B. cereus* as being possibly implicated in natural disorders of the type under discussion.

The observation that catface symptoms in tomato fruit are reproduced by ANCPA is of considerable interest. While there is no demonstrated link between the toxin and naturally occurring catface, there is still the strong possibility that the mode of action of ANCPA may lead to an understanding of the nature of the processes leading to catface development, and thereby to some control measures.

NONPARASITIC STUNTING OF TOBACCO

A nonparasitic stunting of tobacco plants has been described; it is caused by *Phytophthora cryptogea*. Plants grown in autoclaved *P. cryptogea*–infested soil were stunted as severely as plants that were grown in soil containing the living organism. In addition to stunting, transplanted seedlings displayed chlorosis and laminar and veinal necrosis in infested soil autoclaved or nonautoclaved. Leaves developing later were normal but smaller than control leaves. Methyl bromide fumigation did not prevent stunting, chlorosis, or necrosis associated with infestation with *P. cryptogea*. Tobacco plants were stunted by the inclusion of nonviable mycelium of *P. cryptogea* in soil cultures. This showed that the toxicity of *P. cryptogea* was independent of the viability of the fungus. It was concluded that reduction in growth was not due to parasitism, but resulted from the phytotoxicity of the mycelium. The problem of irregular plant growth of tobacco appears to be associated with actions of *P. cryptogea* as an exopathogen, as well as the toxicity of substances produced by the fungus in soil.

Phytophthora cryptogea was compared with *P. parasitica*. The former is a documented nonparasite of tobacco and the latter is a strong parasite, causing the black shank disease of tobacco. Extracts of *P. cryptogea* oat cultures and mycelial material were very toxic to tobacco, causing veinal and laminar leaf necrosis, stunting, and death. The severity of effects was dependent on the extent of exposure to extracts. Seedlings transplanted into soil infested with *P. cryptogea* were severely stunted. Minor veinal necrosis and laminar collapse resulted, but the plants did not have stem lesions or visible root necrosis. The black shank pathogen, *P. parasitica* var. *nicotianae*, reduced growth and killed plants transplanted into infested soil, but the extracts of this fungus were not toxic. It was concluded that growth inhibition by these fungi is caused by different mechanisms. *Phytophthora cryptogea* operates possibly as a nonparasitic producer of a potent toxin, and *P. parasitica* as a potent parasite but not a strong producer of toxins in laboratory culture. A mycelial toxin can produce all symptoms produced by living cultures of *P. cryptogea*. Nothing is known of the saprophytic capabilities of *P. cryptogea*. To support the hypothesis that *P. cryptogea* causes nonparasitic stunting of tobacco plants in the field, it will be necessary to demonstrate the presence of the organism and adequate toxin production in association with stunting.

OTHER EXOPATHOGENIC DISEASES

The general concept that a living pathogen can incite disease even though it is located externally to the affected plant is additionally supported by four examples: seedling diseases caused by pathogens inhabiting the nonliving seed coat and exporting toxins into living tissue; the almond hull-twig disease, caused by a class of pathogens inhabiting dead tissue and producing toxins that diffuse into living tissue and cause disease; the milo disease, in which a potent, stable toxin is produced by *Periconia circinata* on milo tissue with endopathogenesis and in some cases exopathogenesis; and artificial exopathogenesis where a dialysis membrane pod holds a culture of, for example, *Rhizoctonia solani* that then produces a toxin that diffuses outward and causes disease through absorption by soybean roots growing in soil around the pod containing *R. solani.*

SUMMARY

Nonparasitic pathogens of plants (exopathogens) should be included in the list of pathogens to be considered in plant disease studies. If a disease does not fit the classic pattern of host and parasite relationships, one should explore the possibility of external pathogenesis by a living microorganism. If the existence of a "messenger" toxin that is suscept-specific can be established, the battle is partly completed. Following this, a complete delineation of etiology should be undertaken.

The establishment of the occurrence of an exopathogenic disease could well include a sequential development of the case as follows: ascertain that the disease is not endopathogenic, anthropogenic, or physiological (unrelated to pathogens); develop a list of microbial associates of the disease; search for a potential toxin that fits the biological pattern represented; test potential biochemical compounds for toxin roles as leads; test microorganism associates for production of toxins by using the suscept plant as a bioindicator, along with other appropriate methodology; obtain cultures of potential exopathogens, culture them, and test for toxin production; employ gnotobiotic and living-plant infestation procedures to identify the exopathogen, and then produce the disease gnotobiotically in laboratory culture; from the above, reconstruct the potential disease situation and produce the disease in the living plant; identify the environmental conditions necessary for the exopathogenic disease; and develop disease control procedures by adjustment of environmental conditions or the use of specific antitoxin compounds, such as the metabolite that is adversely affected, to overcome the toxicity of the externally produced toxin.

[SHREVE S. WOLTZ]

Bibliography: A. Csinos and J. W. Hendrix, *Phytophthora* species producing toxin active on tobacco, *Soil Biol. Biochem.*, 10:475–81, 1977; M. Mandryk, Frenching of tobacco in Australian soils and in soil leachates, *Aust. J. Agric. Res.*, 1969:709–717, 1969; R. A. Steinberg and T. C. Tso, Physiology of the tobacco plant, *Annu. Rev. Plant Physiol.*, 9:151–174, 1958; S. S. Woltz and R. H. Littrell, Production of yellow strapleaf of chrysanthemum and similar diseases with an antimetabolite produced by *Aspergillus wentii*, *Phytopathology*, 58:1476–1480, 1968.

Photographic Highlights

These photographs have been chosen for their scientific value and current relevance. Many result from advances in photographic and optical techniques as humans extend their sensory awareness with the aid of the machine, and others are records of important natural phenomena and recent scientific discoveries.

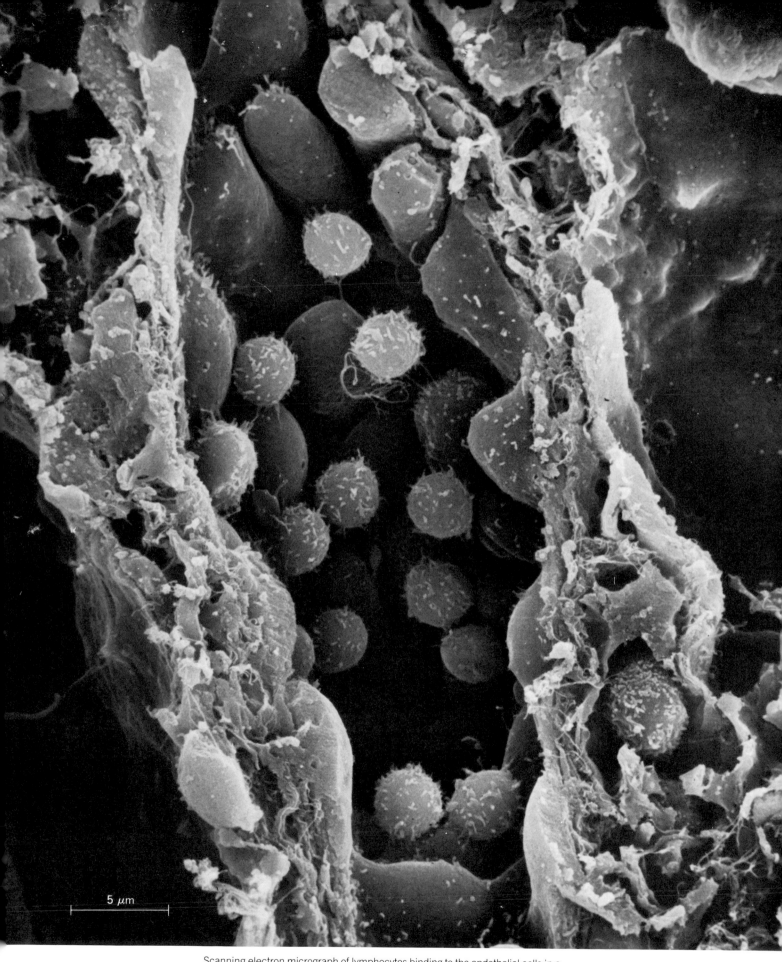

Scanning electron micrograph of lymphocytes binding to the endothelial cells in a specialized lymph node venule in the mouse. *(From E. Butcher, R. Scollay, and I. Weissman, Nature cover, vol. 280, no. 5722, August 9, 1979)*

5 μm

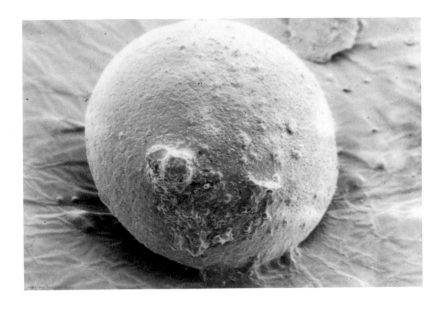

Top: Scanning electron micrograph of sulfur sphere from Poás volcano, Costa Rica.
Bottom: Sulfur encrusting the face of an ejected block from Poás volcano. *(From
P. W. Francis et al., Pyroclastic sulfur eruption and Poás volcano, Costa Rica,
Nature, 283(5749):754-756, February 21, 1980)*

Scanning electron micrographs of carbon particles extracted from Lake Michigan sediments and containing (top to bottom) oil, coal, and wood. For each cenosphere (left) there is a close-up of the surface area inside the outlined rectangle. *(From J. J. Griffin and E. D. Goldberg, Morphologies and origin of elemental carbon in the environment, Science, 206:563-565, November 2, 1979; copyright © 1979 by the American Association for the Advancement of Science)*

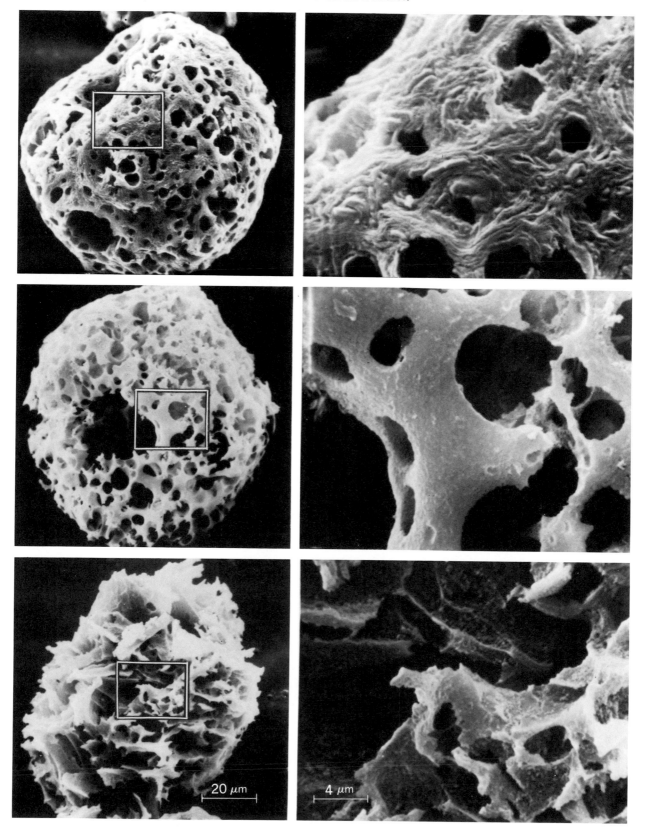

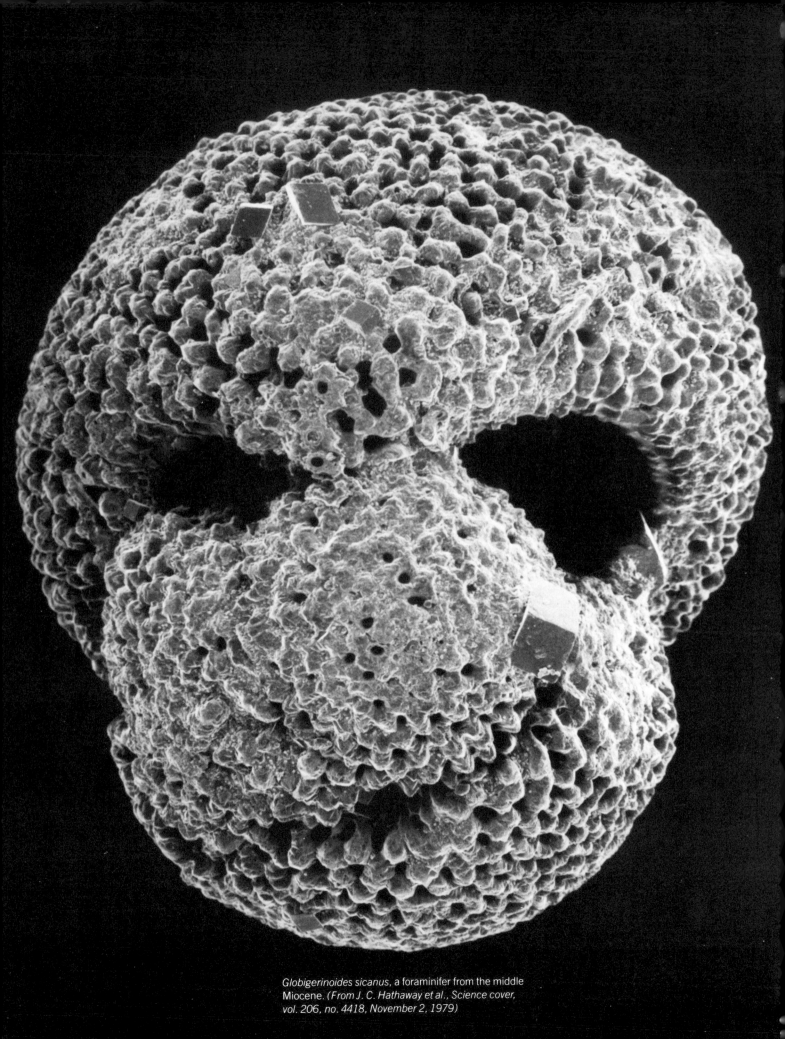

Globigerinoides sicanus, a foraminifer from the middle
Miocene. *(From J. C. Hathaway et al., Science cover,
vol. 206, no. 4418, November 2, 1979)*

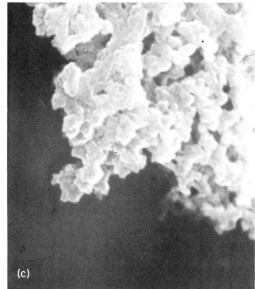

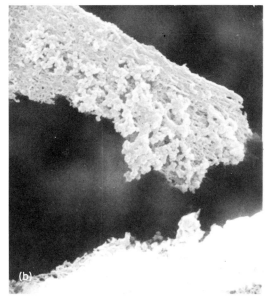

Scanning electron micrographs of *(a–c)* thin polyacetylene film polymerized directly onto a gold grid and *(d, e)* thick film polymerized on the glass wall of a reactor. *(From F. E. Karasz et al., Nascent morphology of polyacetylene, Nature, 282:286-288, 1979)*

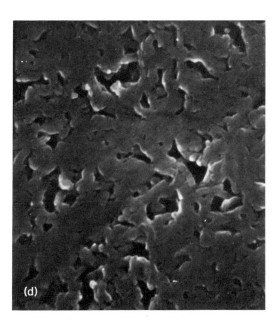

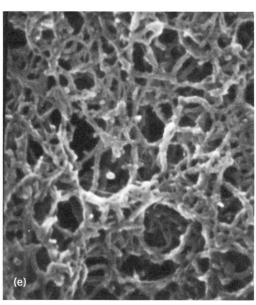

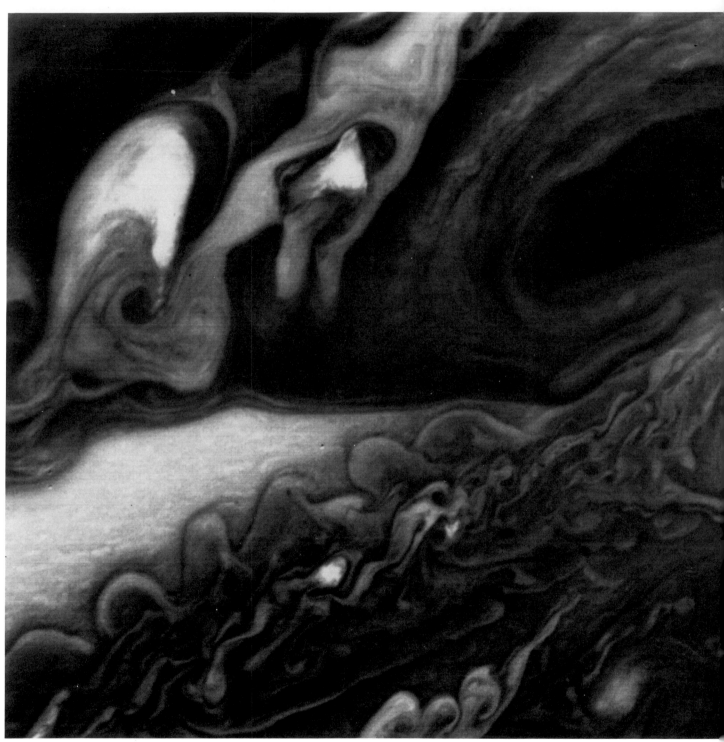

Jupiter viewed from *Voyager 1,* on March 1, 1979, at a distance
of 1,000,000 km. The Great Red Spot is seen at the upper right,
and the turbulent region immediately to the west. *(NASA)*

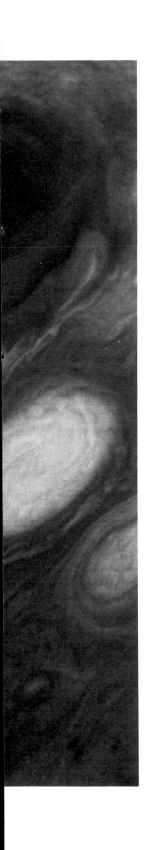

Callisto, Jupiter's outer large satellite, viewed from
Voyager 1, on March 6, 1979, at a distance of
350,000 km. *(NASA)*

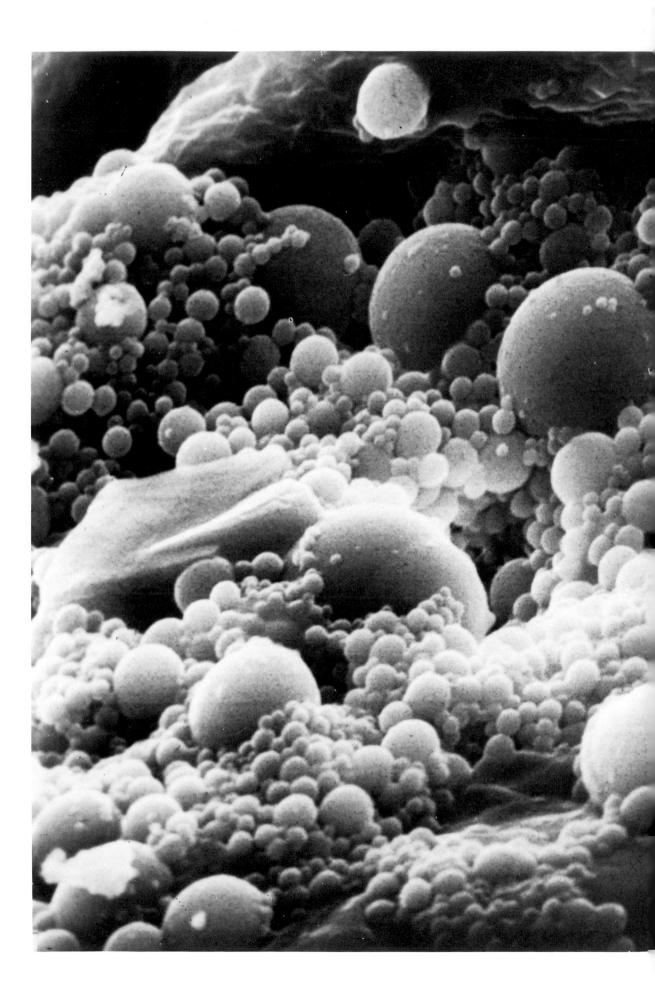

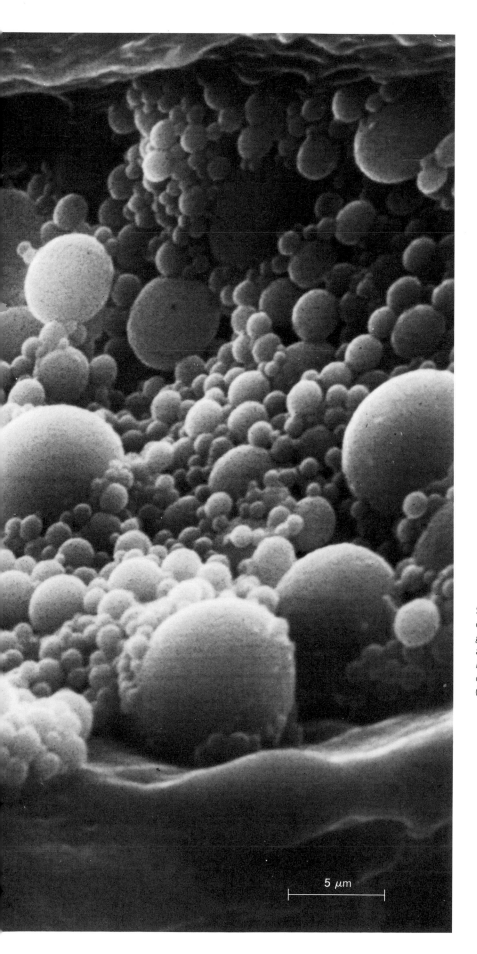

5 μm

Scanning electron micrograph of the edge of an emptied spermatophore from *Blatella germanica*, the German cockroach, with adhering urate spherules. *(From D. E. Mullins and C. B. Keil, Paternal investment of urates in cockroaches, Nature, 283 (5747):567-569, 1980)*

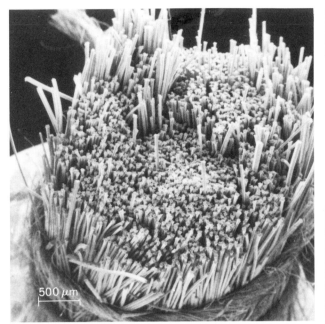

Freeze-formed silica fibers. *(From W. Mahler and M. F. Bechtold, Freeze-formed silica fibers, Nature, 285(5759):27-28, 1980)*

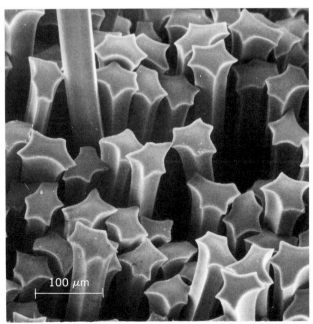

Specimens used for metallographic evaluation of defective metal products. *(a)* Profile of a burr on a low-carbon-steel toy stamping. *(b)* Defects in a welded steel bicycle wheel rim. *(c)* Case depth profile on a critical wear surface of a precision instrument part. *(d)* Weld in a heart pacemaker battery. *(e)* Multiple plated layers on an electrical contact. *(Courtesy of Metal Progress)*

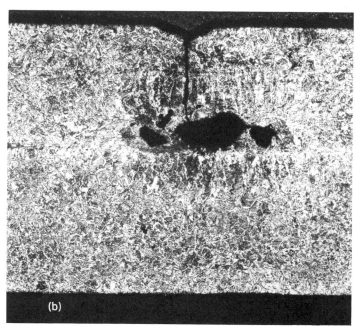

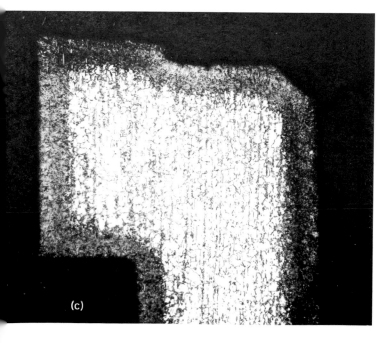

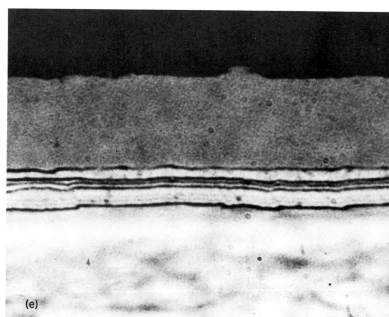

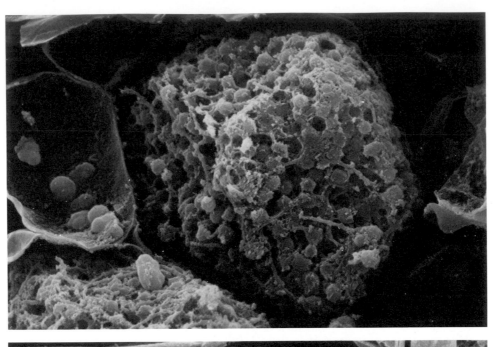

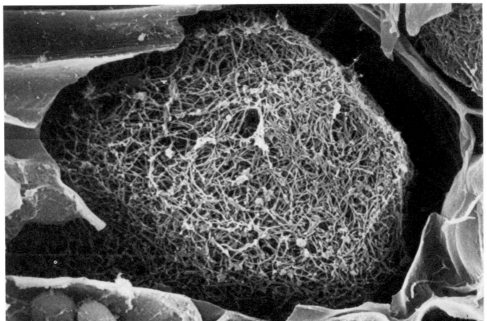

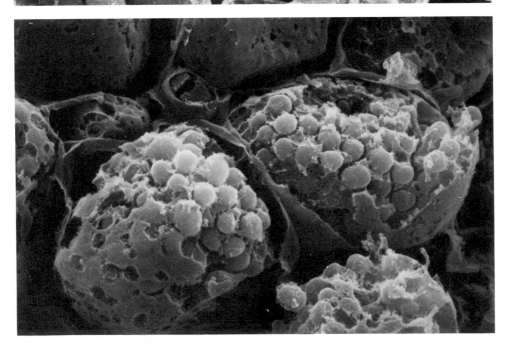

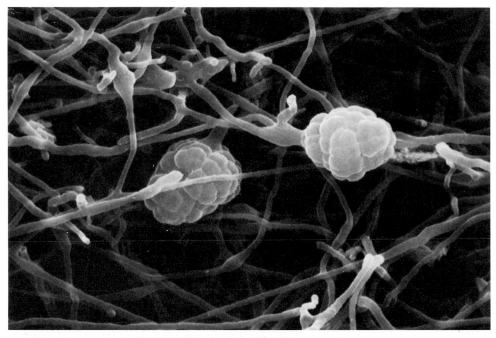

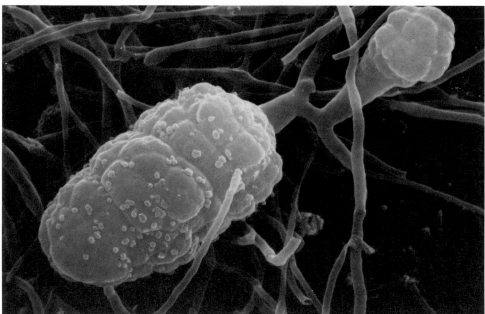

Root nodule actinomycetes (opposite page) in host plant cells and (above) cultured after separation by sucrose-density sedimentation. *(From D. Baker, J. G. Torrey, and G. H. Kidd, Isolation by sucrose-density fractionation and cultivation in vitro of actinomycetes from nitrogen-fixing root nodules, Nature, 281:76-78, 1979)*

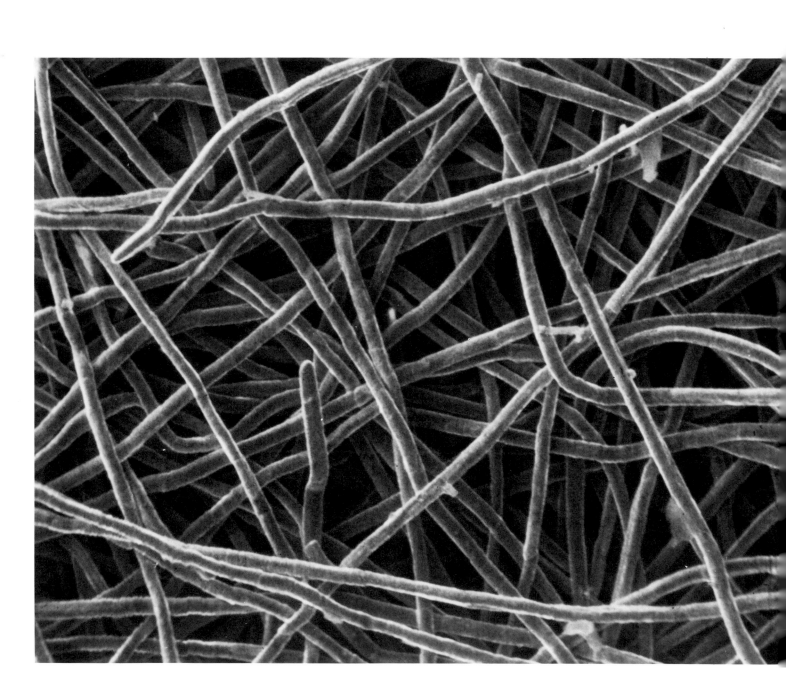

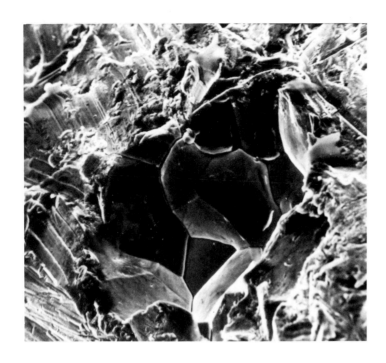

Opposite page: Scanning electron micrograph of a new filter medium produced by
sintering metal fibers. Above: Cracking
along prior austenite grain boundaries
leads to spalling in the worn surface.
(Courtesy of Metal Progress)

Sawtooth wetting line, typical of high-speed movement, formed when a piece of tape is slowly withdrawn from a liquid at a steady speed greater than the maximum speed of dewetting. *(From T. D. Blake and K. J. Ruschak, A maximum speed of wetting, Nature, 282(5738):489-491, 1979)*

A-Z

Aegyptopithecus and Propliopithecus

Aegyptopithecus and *Propliopithecus*, the two oldest known kinds of apes, are sometimes referred to as the "dawn apes." Today they provide the best evidence as to the origin and earliest characteristics of the fossil group of the hominoids, close relatives and ancestors of humans and consisting of apes, humans, and prehumans. These two genera may be ranked in a primitive taxonomic family, the Propliopithecidae, but in tooth shape and construction they are not very different from a group of Miocene apes 10,000,000 years younger, the dryopithecine pongids. *Aegyptopithecus* particularly resembles the earliest of the dryopithecines or "oak apes," a later group recovered from the Miocene of eastern Africa and Eurasia. All known propliopithecids come from Oligocene rocks, about 29,000,000–35,000,000 years old, in the Fayum Province, Egypt.

History of study. The first specimen of *Propliopithecus* was found in the Fayum badlands by Richard Markgraf. However, he did not record anything about where or at what level he recovered it. The jaw was sold to the Stuttgart Museum, and the type species based on it, *Propliopithecus haeckeli*, was named by Max Schlosser in 1911 in honor of the renowned biologist Ernst Haeckel.

Fifty years after the naming of *Propliopithecus*, during an expedition led by Elwyn Simons in the fall of 1961, an isolated tooth of *Propliopithecus* turned up at Quarry G, north of the Fayum lake, Birket Qarun. Throughout the first part of this century, no one ever doubted that *Propliopithecus* was related to apes and to humans, but whether it was closer to the ancestry of the lesser apes or gibbons or to the forerunners of humans remained uncertain. This uncertain placement was because some female fossil gibbons *(Pliopithecus)* and most humans and human relatives do not show distinct enlargement or elongation of the front lower premolar (P_3). A short front premolar characterizes *Propliopithecus haeckeli* from the Fayum. Although details of the crown cusp anatomy differ between the P_3 and the tooth behind it, P_4, in the type specimen of *Propliopithecus* both these teeth are about the same size and shape. When these two premolar teeth (bicuspids) are similar-sized, they are called homomorphic premolars, in contrast to heteromorphic premolars, seen in Old World monkeys and apes, where the front premolar is elongate, bladelike, and distinctly larger than the back premolar. Besides *Propliopithecus*, homomorphic premolars also characterize female *Pliopithecus*. The latter French Miocene ape is the first fossil ape ever to have been found, and was named by E. Lartet in 1836. When Schlosser defined *Propliopithecus*, meaning "before – more of – an ape," presumably he had in mind the suggestion that it was the forerunner of *Pliopithecus*. The unique specimen of *Propliopithecus* described in 1911 posed a problem that was not solved for nearly 70 years: were the small canine and homomorphic premolars (both resemblances to humans) due to its being from a female or were these characteristics present in both sexes? In the former case the resemblance would be to a small, ancient ape, *Pliopithecus*, but in the latter case the similarity would be with the specific group of humans and their close allies. While only the single type specimen was known, this problem could not be resolved.

Hoping to clarify this and other points about the dawn apes, Simons directed a series of expeditions to the Oligocene badlands of Egypt in search of more ape fossils like *Propliopithecus*. This program was in two phases. There were six field seasons in the early and middle 1960s and three in the late 1970s. Each phase of this field work produced important finds about the earliest apes. Although initially finds of *Propliopithecus* were scarce and limited to isolated teeth, another larger and quite different ape turned up in abundance in the uppermost part of the Egyptian Oligocene deposits. The discovery of this new kind of ape led to the naming of *Aegyptopithecus zeuxis*, the "yolking-Egyptian-ape," by Simons in 1965. At the same time he named another ape, *Aeolopithecus chirobates*, meaning the "hand-walking wind-ape." In 1967 an ulna of *Aegyptopithecus* was found, and at other times tail bones, a toe bone, and some finger bones of this ape were recovered; during the 1970s these postcranial bones were described. A well-preserved skull of *Aegyptopithecus* was also found in 1967. This is the smallest mammalian skull ever found in the Egyptian Oligocene deposits. It is the only Old World primate skull from a great period of time running from about 38 MY BP (million years before present) up to 18 MY BP. Its age is estimated at about 28 MY. This cranium shows primitive features, such as a long snout and large upper parts of the premaxillary bone which houses the upper incisor teeth (Fig. 1); these details resemble earlier Eocene primates like the North American prosimian (meaning premonkey or submonkey) *Notharctus* or the European Eocene premonkey *Adapis*. In addition to such primitive features, the skull of *Aegyptopithecus* shows progressive characteristics which combine elsewhere only in later and more modern forms. These include fusion of the two separate frontal bones of the forehead into one bone, and closing in of the eye area with plates of bone to make up an eye socket. Some structures of this skull are intermediate in character; for instance, the tympanic ring of the ear is more advanced than in premonkeys like *Notharctus* and less advanced than in later apes. This complex blending of primitive, intermediate, and advanced features places *Aegyptopithecus* as a true connecting link between ancient and modern primates. Such a construction can be explained only as an example of the evolutionary process.

Significance of recent discoveries. After a delay of nearly 10 years, three recent seasons (1977, 1978, 1979) in the Fayum initiated an important second phase of research. Many new jaws of *Aegyptopithecus* were recovered, as well as four upper-arm bones (or humeri) and some other postcranial bones. These jaws showed that *Aegyptopithecus zeuxis* had canine dimorphism of the sort that in modern monkeys and apes correlates with a large social group of females with several males who vie with each other for dominance (Fig. 2).

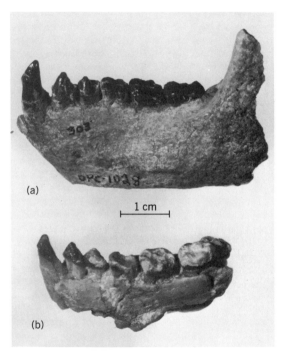

Fig. 2. *Aegyptopithecus* jaws: (*a*) male and (*b*) female.

This sexual size difference thus indicates a social organization more advanced than most premonkeys. The upper-arm bones, like the ulna found earlier, indicate that Oligocene apes, particularly *Aegyptopithecus*, were probably generalized arboreal quadrupeds somewhat resembling the modern South American howling monkey.

Perhaps the most important new finding, based on material found in December 1979, is that *Propliopithecus* has an advanced species, from the uppermost beds in the Fayum. It is contemporary with *Aegyptopithecus* but much rarer. The several excellent specimens recently found also make clear that *Propliopithecus*, like *Aegyptopithecus*, had sexes with very different-sized canines. Moreover, the amount of variation in this late species of *Propliopithecus* is great enough to encompass in it the type specimen of *Aeolopithecus chirobates*. This placement means that the species, as now understood, should be called *Propliopithecus chirobates*; *Aeolopithecus* should no longer be used. The most important thing about these new discoveries is that the homomorphic premolars of

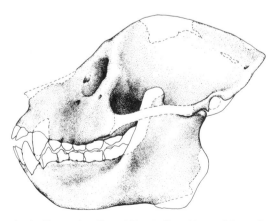

Fig. 1. Reconstruction of the skull and jaws of *Aegyptopithecus zeuxis*.

the type and only known specimen of *Propliopithecus haeckeli* can be understood to be due to its being from a female. Premolar homomorphy need not now be seen as a uniform feature of *Propliopithecus*. From this, it is possible to deduce further that the presence in the type species of homomorphic premolars is coincidental with the similar condition in humans and their allies. Thus *Propliopithecus* is removed from consideration as a direct human ancestor, a ranking which some students had given it.

Relationships to other forms. The first thing to consider about relationships of the dawn apes is their relation to each other. *Aegyptopithecus* is about 20% larger than *Propliopithecus*, and differs in a number of dental details that are important in understanding its affinities. The lower front teeth (incisors) relative to the back teeth are larger in *Propliopithecus* than in *Aegyptopithecus*; the bicuspids of *Propliopithecus* are more similar-sized, and distinctly larger relative to the first molars than in *Aegyptopithecus*. Their lower molars also differ in two principal features: *Propliopithecus* has molars of similar length throughout and small, more laterally placed cusps upon them, while *Aegyptopithecus* has larger, more rounded cusps and molars that increase in size markedly between the first and second and less so between the second and third. The upper cheek teeth are broader compared to length than in *Aegyptopithecus*. It appears that shelves on the molars with many small cusps (beaded cingula) make *Aegyptopithecus* more closely resemble *Proconsul* from the succeeding Miocene deposits of eastern Africa. In a similar manner, the small, laterally placed molar cusps and similar molar lengths from front to back of *Propliopithecus* resemble *Micropithecus*, *Dendropithecus*, and *Limnopithecus* from the east African Miocene. Some of the latter three Miocene genera, as well as *Pliopithecus*, are regarded as primitive gibbons, whereas *Proconsul* is widely recognized as the basal ancestor of the great apes and humans. This would make it appear that the two main kinds of apes had already separated at about 28 MY BP.

Of course, balanced against this possibility, the primitiveness of the skull and limb bones of the propliopithecids could mean that they were near the base of ancestry to both groups, lesser and greater apes. In this latter case a more recent split might be implied. Advocates of molecular clocks would find an ancient split between these groups difficult to reconcile with new views. The molecular clock data, derived from the degrees of chemical difference between sets of living species, would place the split time of these groups more recently. They believe that only about 15 MY have passed since greater and lesser apes had a common ancestry. Even if the propliopithecids prove to be the common ancestors of both gibbons and great apes, early Miocene forms (18–20 MY old) do document that the split had then occurred, and this evidence argues against the validity of a molecular clock here.

Another scheme which has been suggested by the molecular clock advocates is that *Aegyptopithecus* stands even before the split between the ape-human group and that of the catarrhine monkeys. There is little present evidence to suggest that this could be so, for the monkeys from the Fayum provide a much better candidacy than does *Aegyptopithecus zeuxis* for the approximate ancestry of Old World monkeys. *See* APIDIUM AND PARAPITHECUS.

Adaptations of the earliest apes. *Aegyptopithecus* is the largest primate from the Fayum and one of the largest primates that had existed up to that time, although it was still comparatively small by modern catarrhine standards. It was about the size of a house cat, about 12–14 lb (5.4–6.3 kg). *Propliopithecus* species are smaller; they weighed perhaps about 6–8 lb (2.7–3.6 kg).

There is little evidence about the limb skeleton of these species on which to judge their probable mode of locomotion. The sediments in which these animals have been found were not deposited in environments conducive to preservation of complete skeletons. However, from those fragments of fore and hind limb bones that have been recovered it is possible to piece together a picture of what these animals must have looked like. The limb bones of these early apes resemble those of deliberate, slow-moving arboreal quadrupeds, like the bearded saki and howling monkey of Central and South America. Species of both these monkeys move about in the forest canopy predominantly by quadrupedal walking and running on large branches and boughs. Leaping and climbing are infrequently used by sakis, either when traveling or feeding: howling monkeys climb considerably more than sakis during travel and especially while feeding. Thus apparently the earliest apes had not then attained any of the arm-swinging acrobatic abilities of their similar-sized living descendants, the gibbons.

Study of the anatomy of the cheek teeth of these apes provides information about their dietary patterns. Both *Aegyptopithecus* and *Propliopithecus* have relatively low-crowned, bunodont molars resembling greatly those of the New World spider monkeys and Asian gibbons. By analogy with the living species, this indicates that the diets of the earliest apes consisted of nuts, forest fruits, and other herbaceous foods. Both kinds of these apes had spatulate incisors similar in shape to those found among living apes and monkeys. There are differences in tooth proportions between *Aegyptopithecus* and *Propliopithecus*. *Aegyptopithecus* had relatively smaller incisors compared to molars and a considerable molar-size increase toward the back of the jaw; *Propliopithecus* had large incisors and uniformly sized molars. These differences suggest that *Aegyptopithecus* was using its incisors less than was *Propliopithecus* for acquiring and manipulating food items. *Aegyptopithecus*, with its large back molars and relatively more massive jaws, may have been eating tougher food items.

There is practically no information about the skull of *Propliopithecus*. However, a nearly complete skull of *Aegyptopithecus* reveals certain insights about the adaptations of this animal. *Aegyptopithecus* had a comparatively small brain by modern anthropoid standards, but shows a precocious advance toward the modern anthropoideans by the large size of the visual cortex and the substantial reduction of olfactory lobes. These features, together with the small eye sockets, suggest a daytime-active animal that utilized vision over smell as a principal means of interacting with its environment, just as do the modern anthropoids.

As a result of the greatly increased number of jaws and teeth recently recovered from the Fayum, it has been possible to demonstrate for the first time that the earliest apes were sexually dimorphic in body size and canine size. These related features are advanced anthropoid characteristics. Among living anthropoids, the degree of dimorphism is correlated to some extent with the typical social organization of the species. Solitary and monogamous species show very little dimorphism, while species which live in more complex polygynous groups show considerable dimorphism. This suggests that the dawn apes were probably living in complex social groups.

For background information see FOSSIL MAN; PRIMATES in the McGraw-Hill Encyclopedia of Science and Technology.

[ELWYN L. SIMONS; RICHARD F. KAY]

Bibliography: R. F. Kay and E. L. Simons, Int. J. Primatology, 1:21–37, 1980; E. L. Simons, Sci. Amer., 217(6):28–35, 1967; E. L. Simons. Primate Evolution, pp. 210–222, 1972; E. L. Simons, P. Andrews, and D. R. Pilbeam, Cenozoic apes, in V. J. Maglio and H. B. S. Cooke (eds.), Evolution of African Mammals, pp. 120–146, 1978; E. L. Simons and J. G. Fleagle, Nature, 276:705–707, 1978; E. L. Simons, J. G. Fleagle, and G. C. Conroy, Science, 189:135–137, 1975.

Agricultural engineering

In this article, agrotechnology transfer refers to the taking of an agricultural innovation from one location to another location where the innovation is likely to succeed. Success is generally measured in terms of adoption rate and performances obtained by the adopter. An immense quantity of agrotechnology is currently being developed by a large number of national and international agricultural research centers for the resource- and technology-poor farmers of the tropics and subtropics. The principal aim of these centers is to prevent food shortages and world unrest by providing resource-poor farms with the means of keeping pace with their own rising expectations.

The transfer of agrotechnology from research centers to farm fields has a horizontal and a vertical component. A long-distance transfer of biologically and physically sound agrotechnology from one research center to another is called horizontal transfer. The rendering of scientifically correct agrotechnology appropriate to local conditions enables an innovation to be transferred vertically from research centers to farm fields. A high adoption rate is obtained when the degree of mismatch between the technological requirements of the innovation and the resource characteristics of the farmer is minimized. Keeping the number and magnitude of mismatches to a minimum is achieved through analog and matching transfers. Technological and socioeconomic mismatches result in lower and extremely variable performances of the innovation in the field. Gap analysis, a systematic study of the performance difference between the researcher's plot and the farmer's fields, promises to accelerate the flow of agrotechnology from research centers to farms.

Horizontal transfer. A highly effective and transferable agricultural innovation is a new high-performance cultivar of a food or fiber crop. In this instance, researchers are able to condense their work into a compact and easily transferable seed. But unlike a product of industrial technology, the plant that emerges from a seed is sensitive to agroenvironments. Since the resource- and technology-poor farmer operates in agroenvironments ranging from deserts to marshes and from steaming jungles to freezing mountain slopes, most research centers concentrate on a few major food crops adapted to a given agroclimatic zone. A good example of such a center is the International Crop Research Institute for the Semi-Arid Tropics (ICRISAT), located in Hyderabad, India. ICRISAT concentrates on sorghum, pearl millet, pigeon pea, chick-pea, and ground nut, all of which are adapted to, and are important food crops of the resource-poor farmers of, the seasonally dry semiarid tropics. Thus, when ICRISAT scientists succeed in improving the genetic potential for the grain yield and the nutritional quality of these crops, the new technology is applicable not only to the Indian subcontinent, but to the whole of the seasonally dry semiarid tropics.

The transfer of agrotechnology from its site of origin to distant locations around the world becomes crucial to the mission of all international research centers. Very frequently, the technology is transferred directly from the international center to national centers for local testing and modification. This long-distance transfer between research centers is only the first step and constitutes the horizontal component of agrotechnology transfer. Such a transfer is designed to ensure that serious mismatches between the environmental requirements of the agrotechnology and the agroclimatic characteristic of the new location are avoided. By minimizing these mismatches, the scientist is assured that uncontrollable climatic variables will not adversely affect the performance of the new product. Horizontal transfer is necessary but not sufficient to ensure adoption of the new technology by the farmer. The farmer's objective is first to minimize risk and second to maximize profit, and rarely to maximize yield so that a mismatch between the requirements of the new technology and the socioeconomic characteristic of the farmer must also be considered.

Vertical transfer. This involves making technically sound innovation economically feasible and socially acceptable so that the farmer willingly adopts the innovation. When vertical agrotechnology transfer fails, it frequently happens that low-yielding, traditional varieties are cultivated by the farmer alongside high-performance cultivars of the adjoining research station. Too often, the factors which render adoption profitable fall outside the control of the farmer or researcher. One such factor is the ceiling placed on farm prices by governments. The farmer knows that more can be produced by adopting a particular innovation but is unwilling to do so unless convinced that the change will yield benefits. For this reason, agrotechnology transfer often can be only as successful as government policy makers wish it to be. Thus, scientists and others concerned with the development and dissemination of agrotechnology must strive to separate those aspects of technology transfer that are subject to biological and socioeconomic analyses from those that are not.

Since successful agrotechnology transfer is heavily dependent on minimizing mismatches between the requirements of the agrotechnology and the resource characteristics of the farmer, the adoption rate is high when the technology is simple, effective, and stripped of complex technology. A good example of an effective technology is the use of hybrid corn seed by farmers of the industrialized countries. The complex technology of producing hybrid seed is outside the province of farmer responsibility. In contrast, the resource-poor farmer who grows open-pollinated corn is still burdened with the technology of selecting and storing seed. Although high-technology farming is the phrase used in industrialized countries, the technology, for the most part, has been taken from the farm and placed in the hands of highly specialized industries. The success of modern agriculture has largely come about by freeing the farmer of technology and permitting concentration on production. Thus, agrotechnology transfer relieves the farmer of the burden of using complex and costly technology. To achieve this, agrotechnology transfer must be based on minimizing mismatches between requirements of the agrotechnology and the farmer's resources because it is in the rectification of mismatches that costly and complex technology becomes necessary. Two approaches for minimizing mismatches are described below.

Analog transfer. The transfer of agrotechnology between sites with similar soils and climate is called analog transfer. It is the simplest and most straightforward type of agrotechnology transfer.

If an innovation is found to be successful in one location, it is reasonable to assume that it will be equally successful from a biological and physical standpoint in other locations with similar soils and climate. In 1975 the Soil Conservation Service of the U.S. Department of Agriculture issued a handbook, *Soil Taxonomy*, describing a system of soil classification for making and interpreting soil surveys. Soil taxonomy is a multicategorical system of soil classification that employs soil and climatic parameters to stratify soils of the world into groups which behave and perform alike. In this system, those characteristics that are useful in interpreting soils for agricultural use are provided in a category called the soil family. Soils are grouped so that responses to human manipulation are nearly alike for comparable phases of all soils in a family.

The Benchmark Soils Project of the Universities of Hawaii and Puerto Rico is currently testing the hypothesis that agrotechnology can be transferred from its site of origin to distant locations on the basis of the soil family. To test this hypothesis, the project has established an international network of soil families with sites in Brazil, Cameroon, Hawaii, Indonesia, the Philippines, and Puerto Rico. Standard experiments are conducted on each site to discover whether crops respond alike to similar management practices on similar soil families. A statistical model to test the hypothesis has been developed by the project, and the preliminary test of the hypothesis using maize as the test crop shows promising results. By stratifying land according to soil and climate, the soil family also stratifies the occurrences of insects, weeds, and diseases, as well as the potential of a crop to perform in that agroenvironment.

Matching transfer. A crop and its accompanying agrotechnology will perform equally well on soils from a large number of families if the soils have in common a set of characteristics that match the requirements of the crop and related management practices. Unlike analog transfer, which requires near-complete similarity of soil and climate, matching transfer depends on selecting only those characteristics crucial to the performance of a particular crop and associated farming practices. In order to select these characteristics, it is necessary to know the cause-and-effect relationship between land characteristics and the requirements of the new agrotechnology. Although these relationships are known in a qualitative way, they are not documented in sufficiently quantitative terms to enable low-risk transfers to be made. The formulation of standard procedures for matching agrotechnology to land type has been slow because it requires coordinated multidisciplinary efforts on an international scale. The Food and Agricultural Organization of the United Nations has made a first attempt to develop a framework for the matching procedure.

Gap analysis. Mismatches between the requirements of an innovation and a farmer's characteristics and resources result in gaps between potential and actual performance in the user's field. If horizontal agrotechnology transfer has been properly made, differences in potential performances among sites should be negligible. While it is rarely profitable or possible to achieve this potential completely, the farmer must have a good understanding of what inputs and practices are required to achieve it and of the cause-and-effect relationships between performance and farming practices. This knowledge enables limited resources to be used effectively and innovations to be exploited to the fullest. But it is unfair to expect the technology- and resource-poor farmer to know these relationships. Farming is an art as well as a science. Mismatches such as the effect of day length or soil and air temperature on crop performance that are not rectifiable by farming practices can be avoided or minimized by proper horizontal agrotechnology transfer. Mismatches that can be corrected by inputs and practices such as soil amendments, fertilizers, chemicals for disease, insect and weed control, soil preparation, water management, time of planting, planting density, and cropping sequence are those that affect vertical agrotechnology transfer and also are amenable to gap analysis.

Success in correcting mismatches is farmer-dependent. A rich farmer can afford to take risks and has the resources to make corrections, but it is virtually impossible for a resource-poor farmer with the same number of mismatches to exploit an innovation to the same degree. For this reason a great deal of time and effort is devoted by the international research centers to tailor agrotechnology for the small farmer. The analysis of performance gaps caused by mismatches between the requirements of a new agrotechnology and the resource characteristics of the farmer is receiving increasing attention by researchers in the international centers. Gap analysis also serves as a feedback mechanism to alert researchers of farmer needs and capabilities, and promises to strengthen the links between development of agrotechnology and its utilization.

For background information *see* AGRICULTURAL ENGINEERING; AGRONOMY; SOIL in the McGraw-Hill Encyclopedia of Science and Technology.

[GORO UEHARA]

Bibliography: Benchmark Soils Project, University of Hawaii, *Progress Report no. 2*, 1979; Food and Agricultural Organization of the United Nations, *A Framework for Land Evaluation*, Soils Bull. no. 32., 1976; International Rice Research Institute, *Farm Level Constraints to High Rise Yields in Asia: 1974–77*, Los Banos, Philippines, 1979; H. A. Nix, The assessment of biological productivity, in G. A. Stewart (ed.), *Land Evaluation*, 1968; L. D. Swindale, Problems and concepts of agrotechnology transfers within the tropics, *International Symposium on Development and Transfer of Technology for Rainfed Agriculture and the SAT Farmer*, ICRISAT, Patnacheru, India, 1979; L. D. Swindale, *Soil-Resource Data for Agricultural Development*, Hawaii Agricultural Experiment Station, College of Tropical Agriculture, University of Hawaii, 1978.

Agricultural systems

The problem of feeding the world's population is not solely a technical and biological issue. There is little doubt that much more food can be produced with current levels of technology. The biological and technical aspects of the problem simply represent one subsystem of a much larger and more complex system which ultimately involves the political, economic, and scientific arenas of all nations. The complexity of the system is such that scant progress can be made by any one discipline or by working on a particular subsystem in isolation. Too often the disciplinary nature of training and the professions militates against a real understanding of the problem and tends to overemphasize microproblems which are of great interest at a local level but which may have minimal impact on the operation of the system as a whole.

The drawbacks of the disciplinary approach in solving the world's food problems suggest the need for an alternative approach. The systems approach is important in this respect for it emphasizes: the nondisciplinary nature of major systems; the hierarchical structure of systems; the interactions among subsystems; and the driving forces on a system that may be controlled by humans. Interrelationships within the world food supply system occur at many levels, and the whole is tremendously complex. This accentuates rather than excuses the need to examine the whole before developing policies concerning the parts. In this respect the holistic systems approach appears to have much to recommend it.

Simulation modeling. Recently the conceptual framework of the systems approach has been applied by computer-based simulation techniques. The concept is simple: a computer model is constructed of the system under study, and by working with the model the sensitivities of the system can be quantified and appropriate alternative management policies can be explored as if this experimentation were proceeding on the real system. Substantial advantages in this method are: real systems are not themselves disturbed, at least until new management strategies are designed and implemented—a significant point when delicate biological systems or financially sensitive economic or business systems are involved; and the time required to explore the system is much reduced, since in a computer-simulated system real time can be greatly compressed and quantitative measurements are often much easier to take. Against such advantages must be stacked the problems related to designing, building, and validating the computer models, and it must be accepted that while a number of attempts have been made to develop models for world food supply and demand and for international economic activity, the sheer magnitude of these problems is daunting. For detailed models, the data requirements are staggering, both for construction and for validation, and the computer requirements are large; for more generally phrased models, the value of the results are frequently questionable. As always in computer simulation, some compromise is essential between these two extremes: while compromise to develop valuable decision-support models has been achieved in many fields, including agriculture, it is intellectually difficult and operationally awkward to find the appropriate compromise when the defined system relates to world food supply. Relatively complex models concerned largely with the primary sector at national level have been developed for relatively simple third-world economies, and relatively simple national models have been designed for relatively complex developed economies.

While this activity means progress in terms of data acquisition and computer-modeling capability, operationally useful systems models of world food supply have yet to be constructed. However, it is important that the general format for such a model be understood by the spectrum of those concerned with all levels in this system: cellular biologists, applied scientists, production economists, rural sociologists, trade economists, and political economists. A general framework of this system is most valuable at this point in time, and the illustration attempts to present the juxtaposition of the major subsystems involved with emphasis on the interrelationships between the developed economies and the third world.

Framework of world food supply system. The framework presented here recognizes that while vast discrepancies exist within individual third-world economies, there is still great dependence in most parts on subsistence farming, with a concomitant strongly skewed distribution of income. Socioeconomic policies which govern income distribution within countries are the touchstone for improvement in food production and the satisfaction of nutritional expectancy; the world's problem is poverty, not food production. Hence socioeconomic policy in third-world countries,

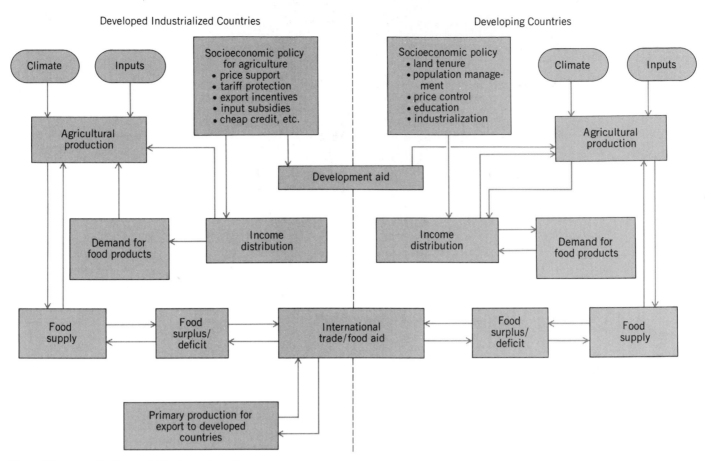

General framework of the world food supply system.

particularly related to land tenure, population management, and food-price control, is seen in the illustration to influence income distribution. This factor affects food production perhaps more than others shown in the illustration relating to climate, management of farming inputs, production policies, and aid. Production of a marketable (and possibly exportable) surplus creates the opportunity for improving the level of farm inputs, and hence subsequent production, only if price and other conditions make marketing worthwhile. In a market economy the price perceived by farmers is a function of domestic price policy and local and international supply and demand. In a subsistence economy, however, the relationships between supply and demand can be tenuous. For example, if the harvest is good, the real income of farmers increases and they can afford to eat more themselves. As a result, the quantities brought to the market may not be as large as expected. Likewise, rising food prices may not always induce farmers to sell more on the market, because they get relatively more for a small quantity. Consequently, market supply is residual and variable, and not particularly responsive to price.

Deficiency in food supply is partly met by trading and by aid from developed economies; the remaining gap constitutes the poverty problem, particularly among the urban poor, but also among the rural poor involved in low-productivity farming systems. A surplus of production, on the other hand, may provide a source of valuable foreign exchange and a stimulus for development.

International trade/food aid subsystem. International trade and aid provide the potential for filling the gaps in food production and for earning foreign exchange from exportable surpluses. Unfortunately, this potential is only partially realized because international markets for primary commodities are far from perfect. In particular, they are adversely affected by socioeconomic policies of many industrialized developed countries such as the United States and European Economic Community members; these policies are primarily designed (for political reasons) to support farm incomes in the respective countries. Once again, distribution-of-income considerations are the main beacons of policy. Such policies involve varying measures of protection and subsidization for the farming community, including concessionary credit, minimum price fixing for primary commodities, tariffs, export subsidies and incentives, and investment in agricultural support such as research and extension services.

Distortions in trade and price are created by such protection policies, and lead to an uneconomic expansion of farm output in developed countries. This disrupts the normal forces of world trade and comparative advantage to the detriment of "efficient" resource use in world food supply. For third-world countries, the availability of cheap food imports and food aid from developed countries may disrupt local markets and discourage the improvement of local production and marketing systems. Undoubtedly in times of food shortage third-world countries benefit from food aid; however, at other times there is a danger that they may be put under pressure to accept food aid in a form that may unnecessarily disrupt local markets. If subsistence farmers are to become increasingly responsive to price incentives, it is important that food aid does not degenerate into surplus disposal.

Protection of agriculture in developed industrialized countries also erodes export opportunities for food-exporting developed economies and for third-world countries. Surplus production in developed countries substitutes for imports and may be traded (effectively with an export subsidy), or provided as aid, to other countries. This depresses and destabilizes international markets, and reduces the potential for developing countries to trade profitably. Lack of export opportunities for the third world can burden not only agricultural development but also tertiary development, and therefore can aggravate urban employment problems. In turn, resultant urban unemployment fosters poverty and dampens effective demand for primary produce.

Conclusion. The world food supply system has massive inconsistencies if examined from the entirely rational viewpoint of efficient use of limited resources. From any viewpoint, however, it is clear that the problems do not lie only within the technical subsystem of farming. Much research is directed to this subsystem throughout the world, and improvements in managing the biology, chemistry, and physics of production have resulted; it is important that research be maintained at some level. But what of its benefit/cost ratio? The application of the technical subsystem and therefore the benefits stemming from it depend on millions of farmers making individual decisions. These decisions are not technologically oriented, but related to improving their expected well-being or minimizing their expected misery. The world is so linked in food issues that policy decisions of governments must be coordinated and must be as rational within the world framework as the political realities of individual countries will allow. Ultimately politics is the driving variable of the world food supply system, and so perhaps the most crucial elements for research and development lie in the assessment of economic alternatives: in developed economies, to provide conditions for less restricted international trading postures in primary commodities; and in third-world countries, to stimulate economic development. Perhaps this understanding will emerge as more operationally effective simulation models of the world food supply system are constructed. Meanwhile the overview created by systems thinking may be helpful in establishing some immediate broad priorities.

[J. B. DENT; A. C. BECK]

Bibliography: J. B. Dent and M. J. Blackie, *Systems Simulation in Agriculture*, 1979; J. L. Dillon, *Agr. Syst.*, 1:5–22, January 1976; J. W. Forrester, *World Dynamics*, 1971; D. G. Johnson, *World Agriculture in Disarray*, 1973; L. Joy, *J. Agr. Econ.*, 24: 165–192, January 1973.

Alcohol

The cellulosic fraction of agricultural, forestry, and municipal residues is gaining increased attention for its material and energy values. These residues have the potential of producing 5×10^9 gal (19 $\times 10^9$ liters) of ethanol per year by 1985, 25×10^9 gal (95×10^9 liters) by 1990, and more than 40×10^9 gal (150×10^9 liters) by 2000. Typically these residues would first be subjected to hydrolysis to form simple sugars; subsequent fermentation of these sugars to ethanol, followed by distillation to remove water, would produce anhydrous ethanol for use in "gasohol" as a 10% blend with gasoline.

It is inevitable that as the prices of nonrenewable energy forms increase, large sums will be spent to develop alternatives. Cellulosic wastes, although diffuse in nature, are generally low in cost and represent a prime candidate for conversion, especially since these wastes can be processed to a liquid fuel which can effectively replace a fraction of oil for transportation. The key to a viable approach is the development of an economical, energy-efficient hydrolysis process.

Over the past 15 years, a significant effort has been made to develop a competitive hydrolysis process. There are two basic approaches, namely, acid and enzyme hydrolysis, which can break down the cellulose to simple sugars. These sugars, primarily glucose, but with some xylose or mannose, can be transformed by chemical or biological processes to chemicals, protein, or fuels such as ethanol or methane (Fig. 1). Figure 2 illustrates possible routes to petrochemicals from cellulosic wastes. This article will discuss new developments in processes for the manufacture of ethanol from waste cellulose.

Enzyme hydrolysis. A number of organisms, including bacteria and fungi, produce enzymes called cellulases, which degrade cellulose to glucose. The principal advantages in adapting these to industrial use lie in the relatively mild "biolog-

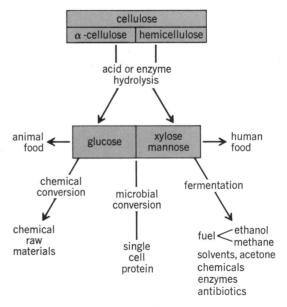

Fig. 1. Waste cellulose utilization routes.

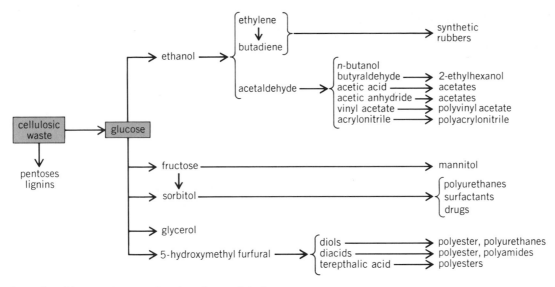

Fig. 2. Possible routes to petrochemicals from cellulosic wastes.

ical" conditions — modest temperatures, pressures, and acidity — under which they function, so that the plant design may be relatively simple. The principal disadvantages of enzyme hydrolysis include the sensitivity to contamination and the major costs for cell growth and enzyme production. Also, the hydrolysis reactions have been found to have long time scales (on the order of 40 h), thus requiring large reaction volumes and associated capital costs. In order to make the cellulose more accessible to enzymes, and thus to achieve reasonable conversion times, special pretreatments are required. Ball milling to a small particle size, steam explosion, ultrasonic energy, and many other techniques have been tried.

Enzyme hydrolysis has been under investigation for several years at a number of facilities. At the University of California, Berkeley, extensive design and economic studies have been accomplished on conversion of waste newsprint and corn stover. Gulf Oil Corporation has developed simultaneous hydrolysis of cellulose and fermentation of glucose. At the Massachusetts Institute of Technology, improved strains of bacteria have been produced which are tolerant of unusual environments, including high temperatures. Scientists at the U.S. Army Laboratories, Natick, MA, have studied pretreatments for cellulose fibers and the development of improved cellulases from fungi. With the advent of genetic engineering, it is conceivable that in the near future, researchers will develop faster acting, more tolerant enzyme strains with a high cellulose conversion efficiency. Under such conditions, enzyme hydrolysis may become an extremely attractive method.

A different approach has been developed at Purdue University. In this process, the hemicellulose portion of the cellulosic waste is hydrolyzed first under relatively mild acidic conditions. The subsequent step to improve the α-cellulose accessibility utilizes a cellulose solvent, cadoxen (cadmium oxide/ethylene diamine) or concentrated sulfuric acid, which acts to reduce cellulose crystallinity and break the lignin bonds. Hydrolysis can then be easily completed by using either enzymes or acid. The economic viability of this process depends greatly on efficient recycling of the cellulose solvent.

Acid hydrolysis. H. Bracconot in 1819 first demonstrated acid-catalyzed breakdown of cellulose to glucose; essentially this can be done either with concentrated acid at room temperature or with dilute acid at high temperature. Chemically the mechanism of the reaction involves protonation of the glucosidic oxygen, followed by a rate-determining cleavage to the glycosyl carbonium ion, which rapidly takes up a hydroxyl ion to form the product. The kinetics can be approximated by a system of homogeneous first-order consecutive reactions for the formation of the glucose, and its subsequent decomposition to furfural derivatives. Optimum yields of glucose are obtained by allowing the reaction to proceed long enough to convert a high proportion of cellulose. Termination of the reaction is determined by glucose breakdown; high temperatures (around 230°C) coupled with short residence times give the best conversions. This is a principal advantage of acid hydrolysis, and allows for the processing of large amounts of materials in short time periods. The disadvantages of acid hydrolysis include the need for acid-resistant materials of construction and the requirement for accurate process control in the particular case of high-temperature, dilute-acid processes.

History. Concentrated acid processes which have been developed include those of F. Bergius and of J. Schoenemann, using hydrochloric acid in batch treatments, and the Hokkaido process, using sulfuric acid, which was operated in Italy and Japan. While technically feasible, such processes incur high materials costs due to corrosion.

Dilute acid processes include that of H. Scholler, implemented in Germany and Switzerland in the 1930s; this involved percolation of a 1% sulfuric acid solution for 3 h at 130°C. Modifications of the Scholler process led to the Madison process in

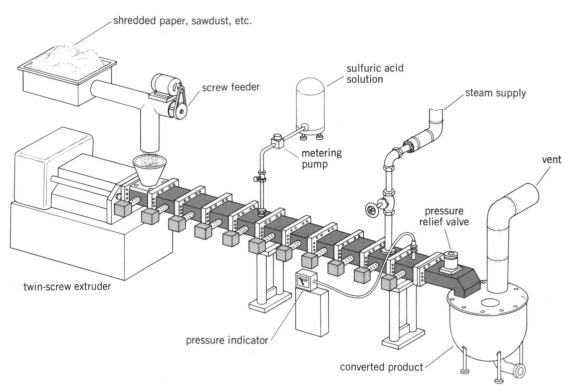

Fig. 3. Continuous acid hydrolysis schematic.

the United States during World War II, although this was discontinued due to unfavorable postwar economics. In the Soviet Union extensive acid hydrolysis operations have been carried out for a considerable time, yielding a wide range of products.

Current research and development. Promising approaches are essentially directed toward the development of a continuous high-temperature process rather than the batchwise ones employed previously. The kinetics of high-temperature, dilute-acid hydrolysis require relatively short residence times, of the order of a few seconds, for optimum conversion. At Dartmouth College an isothermal plug-flow reactor has been extensively studied in theory, and experiments are being conducted giving bench-scale yields of 55% conversion (available α-cellulose to glucose). At the same time, a maximum yield of 70–80% has been established theoretically.

At New York University, a pilot plant with capacity of 1 ton (0.9 metric ton) feedstock per day has continuously converted waste newspaper pulp, sawdust, and so forth to glucose at yields of 55–60%. This process illustrates various features common to all such operations, and is near the scale-up to demonstration stage. It employs a commercially available twin-screw extruder, used in the plastics industry, with continuous feed and discharge, intensive mixing, and accurate process control (Fig. 3). Operating conditions are around 230°C, 400–500 psi (2.8–3.4 MPa), and approximately 1% sulfuric acid, with residence times of several seconds. The machine design includes acid- and wear-resistant materials at appropriate points. The process has the advantage of handling a diverse range of feedstocks at low moisture content, minimizing the energy requirements.

Preliminary cost analysis. A preliminary cost analysis has been undertaken for a plant processing 2000 tons (1800 metric tons) per day of hardwood sawdust utilizing state-of-the-art hydrolysis technology. The plant produces 43,000,000 gal (163 × 10⁶ liters) per year of fuel-grade ethanol, and had a capital cost of $104,000,000 in 1980. The plant assumes the advantages of continuous fermentation and energy-efficient distillation. The lignocellulosic residue is utilized to generate process steam. A by-product credit of 2.5¢ per pound (5.5¢ per kilogram) of CO_2 is taken. Operating at 90% capacity and using sawdust costing $30 per ton ($33 per metric ton), the plant can manufacture alcohol at 98¢ per gallon (26¢ per liter). If an 80:20 debt/equity ratio is assumed, with 15% interest on debt over 20 years and 15% return on investment after taxes, an additional 65¢ per gallon (17¢ per liter) must be added. This would result in a sale price at the plant of $1.63 per gallon (43¢ per liter).

Three factors must be considered for a legitimate cost comparison of alcohol to be used in gasohol and gasoline: (1) the current (1980) wholesale price for gasoline at the refinery is 90¢ per gallon (24¢ per liter); (2) an allowable premium of 3¢ per gallon (0.8¢ per liter) due to 3 octane points increase when using 10% ethanol is equivalent to a subsidy of 30¢ per gallon (8¢ per liter); and (3) a Federal tax elimination on gasohol of 4¢ per gallon (1.1¢ per liter) granted through 1992 is equivalent to a subsidy of 40¢ per gallon (11¢ per liter). Therefore it should be possible to sell ethanol for gasohol at $1.60 per gallon (42¢ per liter), while competing with gasoline at 90¢ (24¢ per liter). The price of $1.63 per gallon (43¢ per liter) for ethanol appears competitive under these circumstances.

Thus if one can meet these assumed capital investment cost and other associated projections, low-cost alcohol from waste cellulose can be a reality in the near future. One additional incentive is the availability of local state subsidies equivalent to as much as 80¢ per gallon (21¢ per liter) in states such as Nebraska and Florida; numerous other locales are now considering such incentives.

For background information *see* ALCOHOL in the McGraw-Hill Encyclopedia of Science and Technology. [BARRY RUGG]

Bibliography: J. F. Harris, Acid hydrolysis and dehydration reactions for utilizing plant carbohydrates, *Applied Polymer Symposium No. 28*, pp. 131–144, 1975; *Report of the Alcohol Fuels Policy Committee*, U.S. Department of Energy, DOE PE OO 12 June, 1979.

Alkali metals

The alkali metals, Li, Na, K, Rb, Cs, and Fr, are the most powerful reducing agents known. They tend to donate one electron per atom to a reaction partner, forming the monopositive ion, M^+, in the process. Until recently, chemists have assumed that only the +1 oxidation state of the alkali metals can exist in compounds. It is now known, however, that under certain conditions an alkali metal atom can accept an electron to form the alkali metal anion. A number of crystalline salts of this class (alkalides) have been synthesized. All of them owe their stability to the complexation of an alkali cation by a crown ether or cryptand, examples of which are given in Fig. 1. The nomenclature 18-crown-6 (18-C-6) refers to an 18-membered ring

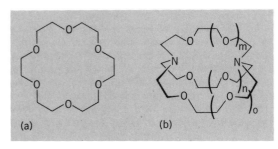

Fig. 1. Complexants for alkali metal cations. (*a*) A representative crown ether. (*b*) Structure of macrobicyclic cryptands.

with six ether oxygen atoms (Fig. 1*a*). Cryptands have Cmno abbreviations; for example, cryptand (2·2·2), C222, has m=n=o=1. Stabilization of the trapped cation prevents it from accepting an electron from M^- to form the metal. X-ray crystallography has shown that the representative alkalide Na^+ C222·Na^- can be most simply described as an ionic solid in which the large cryptated sodium cations form hexagonally close-packed layers and the anions, Na^-, occupy the octahedral holes formed by these layers.

Solution properties. Alkali metals dissolve readily in liquid ammonia to form deep blue solutions. (Concentrated solutions are metallic bronze in color.) Evaporation of ammonia leaves finely divided pure metal. When the metal dissolves, it yields solvated alkali cations, M^+, and solvated electrons, e^-_{solv}. In very concentrated solutions,

most of the solvent is coordinated to the cation and the electrons occupy a metallic conduction band. Some metals, notably lithium and the alkaline earth metals Ca, Sr, Ba form solid metal ammoniates such as $Li(NH_3)_4$ and $Ca(NH_3)_6$. It is generally recognized that these so-called expanded metals contain ammoniated cations such as $Li(NH_3)_4^+$ or $Ca(NH_3)_6^{++}$ with the released electrons in a conduction band. All of these compounds dissociate in vacuum to form the metal and ammonia vapor.

Except for Li in methylamine, metal solubilities in other nonreactive solvents are very low. In 1969 the presence of genuine alkali metal anions in such solutions in which M^- has two electrons in its outer *s*-orbital compared to no electrons in the case of M^+ and one electron for the neutral atom was suggested. Figure 2 shows these configurations for sodium. Over the next few years this postulate was verified in a number of laboratories.

The key to solving the twin problems of low metal solubility and reversion to the metal upon removal of the solvent came in 1970, when it was shown that the crown ethers acting as cation complexing agents greatly enhanced metal solubilities in amines and ethers. A year later it was shown that the cryptands are even more effective. The chemistry associated with such processes can be summarized by reactions (1)–(3), in which

$$2M(s) \rightleftarrows M^+ + M^- \quad (1)$$
$$M^- \rightleftarrows M^+ + 2e^-_{solv} \quad (2)$$
$$M^+ + C \rightleftarrows M^+C \quad (3)$$

C is a crown ether or cryptand. By the law of mass action, the removal of free M^+ by C shifts the equilibrium shown in reaction (1) to the right and this can enhance the solubility of the metal by many orders of magnitude. By using 2 moles of metal per mole of complexant in solvents in which reaction (2) lies to the left, solutions containing primarily M^+C and M^- can be prepared. On the other hand, when equimolar amounts of metal and complexant are used in a solvent which permits reaction (2), the solution contains largely M^+C and e^-_{solv}. It should be noted that both M^- and e^-_{solv} are very reactive and require special vacuum techniques, exceptionally clean glassware, and purified solvents for their preparation.

Crystallization of alkalide salts. With the ability to prepare 0.05–0.4 molar solutions of M^+C·M^-, it was possible to form saturated solutions from which crystals could be grown by slow cooling. Rapid cooling causes precipitation of small crystals which can be separated from the solution by decantation, washed with cold diethylether, and transferred to glass or quartz tubes which are sealed for storage. Analysis of the crystalline solid verifies the proposed stoichiometry. To carry out this analysis, a sample of $M^+C·M^-$(s) is decomposed with water vapor in a closed preevacuated system, and the amount of hydrogen evolved according to reaction (4) is measured by determin-

$$M^+C·M^- + 2H_2O \rightarrow M^+C + M^+ + 2OH^- + H_2 \quad (4)$$

ing its volume, temperature, and pressure. Titration of M^+C and OH^- with standard acid is followed by flame emission analysis for total metal. Finally, the amount of complexant is measured by proton magnetic resonance techniques. This also verifies the absence of decomposition of the crown

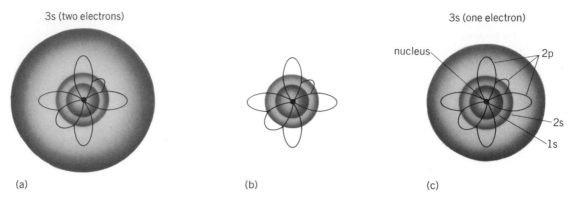

ether or cryptand.

Mixed metal systems can also be formed in which one metal provides the cation and another the anion. Which metal goes where depends on the relative affinities of the complexant for the two cations and the relative stabilities of the respective alkali metal anions. With equimolar K, Na and C222, for example, the crystals which form are K⁺ C222·Na⁻. Other cations can also be used. For example, crystals of Ba⁺⁺C222·Na₂⁻ have recently been prepared. The homonuclear compounds M⁺C222·M⁻ have been formed with M = Na, K, Rb, and Cs, while salts of Na⁻ (sodides) have been formed with K⁺C222, Rb⁺C222, Cs⁺C222, Cs⁺C322, and Cs⁺18C6. No salts of Li⁻ have yet been prepared. The crystal structure of Na⁺C222·

Na⁻ has been determined. It closely resembles the structure of the iodide salt Na⁺C222·I⁻.

Transmission spectra of thin films. Crystals of M⁺ C·M⁻ are remarkably metallike in appearance, varying from gold-colored to dark bronze-colored. Powder conductivities of all alkalides tested so far show semiconductor behavior, not metallic conduction. Thin films prepared by solvent evaporation have optical absorption bands of M⁻ similar to but generally narrower than the corresponding absorption maxima of M⁻ in solution (Fig. 3). Such spectra provide valuable clues for synthesis. For example, films formed by evaporating equimolar mixtures of K, Na, and C222 show only the absorption peak of Na⁻, not that of K⁻. This indicates that K⁺C222·Na⁻ is formed rather than Na⁺C222·K⁻. The result is in agreement with the greater stability of Na⁻ and the more effective complexation of K⁺ by C222. Similarly, evaporation of an ammonia solution which contains Ba, C222, and Na in the molar ratios 1:1:2 shows only the absorption band of Na⁻ in agreement with the formation of Ba⁺⁺ C222·(Na⁻)₂. In general, there is complete agreement so far between the predictions of the transmission spectra of thin films and the subsequent analyses of crystals.

Properties of alkalide salts. The study of alkalide salts is hampered by high reactivity toward air, moisture, and other reducible substances. Even in a high-vacuum system, the compounds can self-destruct by reaction of M⁻ with the cryptand. This reaction, which results in rupture of the ring, appears to be accelerated by light and is more rapid at higher temperatures. The nature of the compound is also important. Thus, crystals of Na⁺C222·Na⁻ have been stored in sealed tubes in the dark at −10°C for over 3 years without apparent change, while K⁺C222·Na⁻ must be stored at −30°C or below to ensure long-term stability. The crystals can be readily dissolved in amine and ether solvents and recrystallized if desired. Most of the alkalides tend to form hexagonal crystals, but several grow as long dendritic needles. The most useful solvents for growing crystals are ethylamine and isopropylamine, sometimes with diethyl ether added.

Electrides. Of potentially great interest are solvent-free solids of stoichiometry M⁺C·*e*⁻ prepared

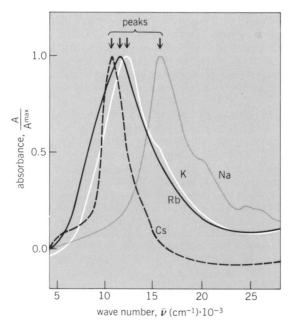

Fig. 3. Absorption spectra of thin solvent-free films of M⁺C222·M⁻. The arrows indicate the positions of the corresponding peaks for alkali metal anions in liquid ethylenediamine solutions. (*From J. L. Dye et al., Optical spectra of alkali metal anion and electride films, J. Chem. Phys., 68:1665–1670, 1978*)

from solutions which contain M+C and e⁻ₛₒₗᵥ. All attempts to grow crystals from solutions of this stoichiometry have failed so far and the powdered samples which have been studied have all been prepared by solvent evaporation. Thin films show optical absorptions in the infrared with the maximum absorption at 1300–1800 nm depending on the metal and complexant used. Often, multiple features such as shoulders or multiple peaks are present, probably because of the precipitation of mixed phases. Electron paramagnetic resonance (EPR) studies of solid electrides indicate that the electrons do not interact strongly with the complexed cations, and suggest that the electrons may be trapped in the vacancies between the cations. If this proves to be the case, electrides will represent a new class of ionic compounds with the simplest possible anion—a trapped electron. Unfortunately, electrides are even more subject to decomposition than alkalides.

Expanded metals. Electrides contain one electron per cationic center. Because of the large size of the cryptated cation, this results in a number density of about 10^{21} electrons per cubic centimeter. This is near the predicted critical concentration for the onset of metallic behavior. Indeed, some electrides, most notably those prepared with Li^+C211 as the cation, have transmission spectra, EPR spectra, and microwave conductivities which suggest metallic character or at least very weak traps for the electrons. The factors which lead to electron delocalization are not clear at this time. Understanding of these systems would be greatly improved if homogeneous crystalline samples could be prepared. Thus, crystallization of compounds with the stoichiometry $M^+C \cdot e^-$ is a major research goal.

For background information *see* ALKALI METALS in the McGraw Hill Encyclopedia of Science and Technology.

[J. L. DYE]

Bibliography: J. L. Dye, *Angew. Chem. Int. Ed. Engl.*, 18:587–598, 1979; J. L. Dye, *J. Chem. Educ.* 54:332–338, 1977; J. L. Dye et al., *J. Phys. Chem.*, 84:1096–1103, 1980; J. C. Thompson, *Electrons in Liquid Ammonia*, 1976.

Alpine landforms

Mass movement of rock debris occurs on mountain slopes above the tree line all over the world, primarily because of the rigorous climate that exists at such high altitudes. Research on alpine landforms indicates that much can be learned from these relatively unstudied features to estimate past and present climates in mountainous regions. Movement takes place due to some combination of factors, such as sudden falling or sliding of rocks because of gravity, loosening of rocks due to the wedging effect of water freezing in cracks in the rocks, gradual downhill movement induced by repeated freezing and thawing of water in pore spaces in the ground, rainwash due to deluges of water from intensive rainstorms, or avalanching of windblown snow and ice. Alpine forms of mass movement occur wherever large rocks fall, or smaller rocks and fine debris wash from cliffs and steep slopes, or debris is carried downslope by slush flows or snow avalanches. Bouldery debris

creeps downhill as tongue-shaped masses in the centers of valleys or as lobes moving out from valley sides. Coarse blocks either cover high mountain slopes as broad blankets or fill narrow ravines; some collect as ramparts beyond the toe of snowbanks.

Rockfall talus. This is the collection of angular rocks of all sizes below cliffs or steep rocky slopes, derived mainly by falling, then rolling, bouncing, or sliding to the slope bottom (Fig. 1). Other names

Fig. 1. Rockfall talus on a south-facing slope in the Colorado Front Range. Two small avalanche taluses are at the far right.

for rockfall talus are scree, talus cone, and talus slope (incorrectly used, since talus is French for slope). Due to earlier rock removal by erosion and formation of valleys, and thus, the release of confining pressure from the weight of overlying rock, adjacent rock that is deeper within a valley wall will split away as thick sheets more or less parallel to the hillside. The sheets of rock produced are again broken in other directions due to tension. Then, because of gravity, and pressure also exerted by the wedging of water changing into ice in the previously formed cracks, large blocks of rock are gradually worked into unstable positions and fall. Rainwater in the cracks also may reduce friction between corners and the edges of blocks and aid in their instability. After falling, the rocks shatter into pieces. The smaller pieces stop high on the talus, and the larger ones go all the way to the bottom. Distribution of smaller sizes at the top and larger at the bottom helps identify this talus. Fine sizes are in the minority, so little or no vegetation grows on this talus. The larger the blocks, the steeper the slope. Rockfall taluses are much steeper than 32° and can be up to 43°, with a straight profile that is slightly concave-up only at the bottom. Some rockfall talus may be 130 m high (Fig. 1).

The orientation of cliffs above the talus relative to the direction of the morning sun, the small area or steepness of the mountainside above the talus hindering winter snow accumulation and later snow meltwater or snow avalanches, and the availability of jointed or fractured rock all determine occurrence of rockfall talus. Some rock debris, however, may be brought down by rain, snow melt-water, or avalanches. Hence, some rockfall talus-es may have alluvial or avalanched debris mixed with rockfall debris (Fig. 1).

Alluvial talus. This is the accumulation of rocks of any size or shape, derived during and after heavy rainstorms or by snow meltwater, passing through a gully or couloir in a cliff face to rest against a valley wall (Fig. 2). Other terms for such talus are scree, talus cone, and alluvial cone. Any loose rock or partly weathered rock debris may be moved by the rain. Larger blocks cannot be carried far by water so they collect at the talus top, whereas the finer sizes are carried to the bottom. During storm flow or after a period of prolonged snow melting resulting in slush flows, long narrow debris flows (not mudflows, since the majority of particles being moved are too large) carry the rock debris to the bottom. Parallel ridges or levees along the debris-flow sides on the talus surface reveal this method of transport and help identify alluvial talus. Upper talus slopes are 38 to 35°, whereas the talus toe is less than 28°, with a curved concave-up profile from top to bottom. Since fine sizes fill the spaces between larger rocks, alluvial taluses may support vegetation except where plowed out or covered by debris flows.

Movement of rockfall talus and of alluvial-and-rockfall talus combined takes place in several ways. From information available of actual measurements, it is known that movement of talus, referred to as talus shift, occurs rapidly when actually in motion, and in an extremely variable manner from sector to sector and through time, as creep or as sliding, and even as overturning. Talus shift varies from 5 to 450 cm/year in some

Fig. 3. Avalanche talus in foreground, below the Continental Divide, in the Colorado Front Range. Light snow reveals boulder-protected debris tails in a zone of avalanche scour. Zone of deposition is in lower right.

mountains (northern Sweden), 6 to 110 cm/year (Canadian Rocky Mountains), to no shift at all for as long as 8 years, for example, and then sudden shift of 8 to 10 cm/year for the next 3 years (Colorado Front Range). Movement of individual blocks or units of talus also is by direct impact of falling rocks, nudging or tipping large blocks 1 to 10 cm from stable positions, or scattering fragments 1 to 10 m down the talus (Colorado Front Range).

Alluvial talus below signifies wide mountain slopes above on which snow collects in winter or rain concentrates in summer, or both. This in turn implies a certain ridge orientation and broad, windward upland surfaces across which winter winds can blow the snow. Some alluvial taluses also may have rockfall debris or avalanched rocks on them, and such taluses may rise 260 m above the valley floor (Fig. 2).

Avalanche talus. This is the assemblage of rocks of any size, usually angular in shape, derived from avalanched snow and ice mixed with rocks carried from cliffs or steep rocky slopes below ridge crests (Fig. 3). Avalanche taluses occur on the downwind lee sides of ridges where wind-packed snow builds as overhanging ledges or cornices. When the cornices collapse, as in midwinter and early spring, down come snow, ice, and rocks. The talus slope usually is less than 26°, with a concave-up profile. Some avalanche taluses may be 800 m in length and ascend 500 m. The upper and midportions of these taluses are zones of erosion, since previously deposited debris is scoured by later avalanches (Fig. 3). Large blocks or boulders

Fig. 2. Alluvial talus, more than 260 m high, on south-facing slope in the Colorado Front Range.

rising above the surface protect the talus down-slope from being eroded by later avalanches, and hence the protected talus stands slightly higher as a tail of boulder-protected debris extending a short distance downslope (Fig. 3). These avalanche-debris tails are distinctive and typical of this talus. Downslope from the scour zone of erosion is a rough-appearing area of deposition where most of the rock debris remains. Small rocks rest in hap-hazard positions on large ones, and larger blocks may be crushing small ones, divulging the delicate manner by which the avalanched debris is gently lowered by the melting snow. Vegetation may grow in the fine debris of the zone of erosion, and occurs in long stripes where not scoured away. No move-ment measurements are known for avalanche talus.

Avlanche talus indicates the direction of winter winds as well as the positions where cornices form and collapse, information not determined if a mountain range is visited only in mid or late sum-mer. Accumulation may occur by rockfall or debris flow on some avalanche taluses.

Avalanche boulder tongues are long, narrow forms of avalanched debris, with flat tops and steep embankment edges rising sharply above the surrounding slopes (Fig. 4). Avalanche boulder tongues occur on mountainsides where avalanche tracks start high on the slopes above the ava-lanched debris. The location of the upper ava-lanche tracks may be controlled by long chutes or couloirs above the cliffs. The location of the lower avalanche tracks is controlled by thick winter snow on each side of the track, so that avalanches are funneled down the centers of the tracks. Steep embankment edges appear after the snow on each side melts during summer. The avalanched debris may resemble a road or highway embankment. Some avalanche boulder tongues are 1000 m in length and extend 600 m high, as in northern Swe-den (Fig. 4). They have a concave-up profile, with a slope at the bottom of less than 25°. All have an upper scour zone with avalanche-debris tails and an avalanche runout zone at the bottom. If thick winter snowbanks do not control the location of the lower end, they become broad avalanche fan tongues.

Avalanche boulder tongues provide evidence of avalanche erosion as a separate and, in mountains where they occur in great numbers, an important type of slope reduction. If avalanches reach only to the tree line today, then older vegetated avalanche boulder tongues below the tree line give evidence of the previous kind and amount of precipitation and of wind directions.

Rock glacier. This is an accumulation of unsort-ed, coarse-to-fine rock debris occurring in moun-tain valleys and having a glacier shape (tongue shape) spreading downvalley or having a lobate form extending out from a valley wall. Tongue-shaped rock glaciers that move beyond the narrow confines of valley walls spread laterally and be-come spatulate rock glaceirs. Active rock glaciers may move because of the continuous slow creep of a buried glacier remnant inside (ice-cored), or by recrystallization and creep of tiny ice grains filling the spaces (interstitial ice) within the fine debris inside the rock glacier (ice-cemented). Tongue-

Fig. 4. Avalanche boulder tongues in Tarfala Valley, northern Sweden.

shaped rock glaciers may or may not be ice-cored, but lobate rock glaciers against valley walls can be only ice-cemented. Active rock glaciers imply by their presence that the climate of the mountains is cold enough to have glacier ice remain inside them or to maintain interstitial ice.

Tongue-shaped. This type is an elongate mass of rock debris, longer than broad and of glacier shape, moving downvalley, usually in a cirque or moving out of a cirque (Fig. 5). Other terms for these are rock stream, block glacier, debris-covered glacier, and cirque-floor rock glacier. They may be 150 to 1500 m long, 50 to 750 m wide, and 20 to 50 m thick, with a steep front of 38 to 50°. Many show longitudinal ridges along their sides, parallel to movement, that wrap around the front as transverse ridges, giving the appearance of a glacier in motion (Fig. 5). After years of summer rainfall, the fine debris will be washed down in-side, leaving an upper layer of coarse blocks. Tongue-shaped rock glaciers with a glacier inside may have a saucer-shaped depression between the base of the cirque headwall and the head of the rock glacier, indicating the buried glacier has par-

Fig. 5. Tongue-shaped rock glacier moving 5 cm/year from left to right, in the Colora-do Front Range.

Fig. 6.   Lobate rock glacier moving 1.4 cm/year on a north-facing slope in the Colorado Front Range.

tially melted. Others may be fed by avalanched snow and ice mixed with rock debris, or their source might be the till of a large moraine containing interstitial ice. Small rock glaciers or those on gentle slopes may move 5 to 10 cm/year (Fig. 5), but large rock glaciers or those on steep slopes may move 50 to 150 cm/year.

Almost all tongue-shaped rock glaciers occupy cirques where ice glaciers once flourished. Some even grade downvalley from ice glacier to buried glacier to rock glacier. All imply some retreat or total disappearance of former glaciers and a return to colder, more severe climate, with regrowth or rebirth of glaciers in some valleys and growth of rock glaciers in other valleys. In those mountains with many active rock glaciers, that severe climate is continuing; in those mountains with most if not all rock glaciers inactive and with buried glacier or interstitial ice melted, there has been a change to a warmer, more tolerant climate.

Lobate. This type is a lobe-shaped pile of rock rubble, broader than long, occurring singly or in groups, moving away from a valley wall, usually as an extension of talus (Fig. 6). Other names once

Fig. 7.   Block slopes, extending to ridge crest, in the Colorado Front Range. Block field is in the foreground.

applied to this alpine feature are talus flow, talus glacier, protalus lobe, protalus rock glacier, and valley-wall rock glacier. Most occur on shaded sides of valleys, below alluvial talus which provides the fine debris permitting interstitial ice to form. Single lobate rock glaciers may be 100 to 200 m broad and long, 10 to 30 m thick, with steep fronts of 40 to 55°. Multiple ones may extend along valley floors 500 to 1500 m below steep cliffs. Most have transverse ridges and furrows, with blocky rubble on top, and if they are active, interstitial ice inside. Movement may be only 1 to 6 cm/year (Fig. 6).

Lobate rock glaciers imply a climate severe enough to allow snow meltwater and rain to freeze and form interstitial ice which deforms under pressure, providing movement. The weight of the talus behind some lobate rock glaciers may be important in their movement.

Block field and block slope. A block field is a thin blanket of rocks, with no fine sizes visible, that is located over solid or weathered bedrock or over colluvium, with no cliff or ledge rock above as a source. It is found in mountains, usually above the tree line, on slopes of less than 10°. The more

Fig. 8.   Block stream in Hickory Run, PA. (*Photo by Tim Frable*)

common block slope is similar to a block field but occurs on slopes of more than 10° (Fig. 7). Block fields are rare, with only a few at low divides and ridge crests. Other names formerly used for these features are block glacis, block sea, debris mantle, felsenmeer, frost-moved rubble, frost rubble sheet, and mountaintop detritus. Block fields and block slopes are more extensive along slopes parallel to the contour. Since no visible rock source is at their head, they may reach all the way to the ridge crest. The blocks may be subangular to subrounded, but block slopes higher on mountain sides and peaks more likely come from the underlying bedrock, and their blocks are more angular. Blocks are smaller below the surface blocks and are interlocked together. Their thickness may vary from a few blocks to tens of meters. The long axes of blocks may be aligned parallel to slope direction, suggestive of creeping or sliding.

The few measurements ever made indicate movement of 1 to 2 cm/year.

Block stream. A block stream is an assemblage of boulders or angular blocks, with no fine sizes visible, over solid or weathered rock, colluvium, or alluvium as a narrow body on mountain sides or in heads of ravines above or below the tree line (Fig. 8). Other names once applied to block streams are boulder field, boulder stream, felsenmeer, rock stream, rubble stream, stone run, and incorrectly, rock glacier and block field. Block streams are more extensive in a downslope direction than along a slope. They may go down into forests from above the tree line and cover a valley floor. They exist on hillsides of any angle. The blocks may be angular to round depending on origin, and, as with block fields, larger blocks are near or at the surface (Fig. 8) and are interlocked together; also, no fine sizes are in their uppermost part. They vary from a few to tens of meters thick. The long axes of blocks usually are oriented in a downslope direction. Movement may be less than 1 cm/year.

Any hypothesis for the origin of block fields, block slopes, and block streams must consider conditions existing on mountain slopes and peaks. Rocks first must be jointed or fractured into sheets, as explained for rockfall talus. The rock then weathers along the sheeting joints, producing angular to round blocks surrounded by partly weathered rock detritus. The detritus in turn is further reduced in size by hydration shattering and chemical reactions. If weathering proceeds to its expected conclusion, a rocky mountain soil may be formed. But if the process is slowed or arrested by change to a cold freeze-thaw regime before the large blocks are too reduced in size, then the stage is set for the creation of block fields, slopes, and streams. With rain and snow meltwater then forming interstitial ice, the whole assemblage of weathered detritus and blocks moves downslope due to gravity, the wedging effect of ice, and interstitial ice creep. The downslope orientation of long axes of blocks is taken at this time. Flowage of block fields, slopes, and streams to their present positions occurred in the past when pore spaces between the blocks were filled with fine sediment. After a long period of time, however, the fine sediment is removed by rainwash and water flowing through the deposit, and the larger blocks remain behind. Motion of the whole mass slows as more and more fine grains and sediment are washed away, and the interstitial ice melts. The deposit eventually consists of large round to angular blocks interlocked together in an almost inert or stable position. Accordingly, most block fields, block slopes, and block streams are relic, either created during a past cold climate or formed during the vigorous climate that exists in the mountains today.

Protalus rampart. This is a ridge or rampart of angular blocks, derived by single rocks falling from cliffs or steep rock slopes, that slid across a snowbank and mark the downslope edge of that snowbank (Fig. 9). Other names once used are winter protalus ridge and, incorrectly, moraine. When a protalus rampart is forming, rocks roll or slide to the bottom of the snowbank, but no fine detritus can or does reach its lower edge. After the snow

Fig. 9. Protalus rampart, in left foreground, in Colorado Front Range, Rockfall talus is at right; snowbank in center.

melts in summer, the rampart stands some distance away from any talus directly below the cliff or valley wall. Protalus ramparts may be 10 to greater than 100 m long, often arcuate if in cirques, but straight and extensive if built below a long valley wall. The front may rise abruptly at angles of 40 to 50°, but the upslope edge is gentle and may merge with the talus toe or be hidden under snow (Fig. 9). Thickness varies up to several tens of meters. The blocks of rock rest in haphazard positions inherited from their travel across snow, and although some blocks may be unstable, movement of the rampart itself is not expected. Protalus ramparts most likely occur in front of rockfall talus. Those in the middle of cirques or far away from valley walls, by their mode of origin, indicate larger snowbanks and thus greater snowfall in the past, and are relics of a colder climate. Several nested inside each other imply abrupt changes from cold to less cold to the present climate.

For background information *see* CIRQUE; GLACIATED TERRAIN in the McGraw-Hill Encyclopedia of Science and Technology.

[SIDNEY E. WHITE]

Bibliography: J. S. Gardner, *The Nature of Talus Shift on Alpine Talus Slopes: An example from the Canadian Rocky Mountains*, Research in Polar and Alpine Geomorphology, GeoAbstracts Ltd., Norwich, England, pp. 95–106, 1973; A. Rapp, *Geogr. Ann.*, 41:34–48, 1959; A. L. Washburn, *Periglacial Processes and Environments*, 1973; S. E. White, *Quaternary Res.*, 6:77–97, 1976.

Animal communication

A recent experiment on symbolic communication between pigeons is part of an ongoing project being conducted at Harvard University by behaviorist B. F. Skinner and Robert Epstein. Called the Columban Simulation Project (after *Columba livia*, the taxonomic name for pigeon), the project is attempting to simulate complex human behaviors, normally attributed to "cognitive" processes, with pigeons. Language is only one of many human phenomena being simulated; others are competition and sharing, insight and problem solving, self-awareness, tool use, and so on.

The project has two major goals: to provide plausible accounts of the origins of complex human behavior in terms of specifiable environmental histories, and to provide a data-based commentary on current nonbehavioristic psychology.

Behaviorists such as Skinner are strict determinists. They believe that human behavior—a category in which they include thoughts and feelings—can, in theory, be completely accounted for by a person's genetic endowment, environmental history, and the current circumstances. The role of genes is not yet well understood and is relatively difficult to study experimentally. Experimental psychologists concentrate on understanding the role of the latter two variables. They reject explanations of human behavior based on traits, mental states, or feelings, which are often proffered by cognitive psychologists, popular psychologists, and laypersons. Thus, an insightful performance is not explained in terms of "insight" or other mental operations, but in terms of experience; behavior said to show that one is aware of one's self is not then accounted for by one's "self-concept," but, again, by past events; and language is not explained by "capacities," "traits," "knowledge," or "mental structures," but by a genetic endowment, an environmental history, and current circumstances. The communication experiment was an attempt to show the possible contribution of an environmental history in determining certain languagelike behavior.

The details of the procedure were prompted by recent language research done with chimpanzees. "Symbolic communication" was achieved as follows: One chimp watched a trainer hide some food and then, in the presence of a second chimp, was asked by the trainer to indicate the symbolic name for that food by pressing (and thus illuminating) it on a keyboard. If the second chimp then asked for the food by using its symbolic name (again, by pressing keys), both chimps were fed.

Communication sequence. A variation of this performance was achieved with two pigeons as follows: The pigeons, named Jack and Jill, were placed in a Plexiglas chamber with a Plexiglas partition between them. Each could peck (and thus illuminate) various keys embossed with colors or letters, and each could see the other's keyboard through the partition. Jack's task was to name a color to which only Jill had access. Jack would initiate a conversation by pecking a key labeled WHAT COLOR? Jill would then poke her head behind a curtain where one of three colors (red, green, or yellow) was illuminated. Having seen the color, she would peck (and illuminate) a corresponding black-on-white letter (R, G, or Y) on her keyboard. Jack would then peck a key marked THANK YOU on his keyboard, thus operating a feeder for a few seconds on Jill's side of the partition. Finally, Jack would double-check the illuminated letter on Jill's keyboard and peck the corresponding color on his own keyboard. If his selection was correct, the equipment would automatically operate his feeder. Jack invariably then asked for an-

other color. (Hidden colors appeared in a random sequence.) The pigeons could engage in this exchange with 90% accuracy for sustained periods of time. If they had been responding at random, overall accuracy would have been about 11%.

Although the exchange has been described in terms of "meaning," "information," "knowledge," and "purpose," the animals in fact behaved as they did because of a complicated "history of reinforcement." The experimenters suggest that "a similar account may be given of . . . comparable human language." Consistent with the goals of the project, the adequacy of current popular accounts of language was questioned and a plausible account of certain languagelike behavior constructed based on a specifiable environmental history.

Training. The communication sequence was established after about 5 weeks of training, the major steps of which were as follows:

1. Adaptation: The birds were placed in the experimental chamber, one at a time, and housed there for a short time until they showed no signs of distress.

2. Hopper training: The food hopper (feeder) was operated repeatedly, giving them access to grain for several seconds with each operation, until they approached and ate from it readily. Since the birds were kept slightly hungry at all times, the operation of the hopper would now serve to strengthen ("reinforce") whatever behavior it followed.

3. Key pecking: Pecking any key was reinforced (that is, was followed immediately by a hopper operation).

Jill's training now proceeded in Jack's absence as follows:

4. Chaining: What would eventually be the hidden color was at first flush with the panel and not covered by a curtain. A peck at the color (red, green, or yellow) followed by a peck at any of the symbol keys (R, G, or Y) was reinforced.

5. Matching: The second peck was reinforced only if the symbol corresponded to the illuminated color. Jill was thus taught to "name" colors.

6. Shaping: The color was gradually moved into a recess in the panel and then gradually covered with a curtain.

7. Discrimination training: The matching sequence was reinforced only when the WHAT COLOR? sign was illuminated.

Jack was trained in Jill's absence as follows:

8. Chaining: The center partition was removed. A peck at an illuminated symbol key (on Jill's side of the keyboard) followed by a peck at a color key (on Jack's side) was reinforced.

9. Matching: The second peck was reinforced only if the color corresponded to the illuminated symbol. Jack was thus taught to select the right color, given its "name."

10. Shaping: The partition, at first placed directly over the symbol keys, was gradually restored to its proper position. The first response in the chain was now only a "look" rather than a peck.

11. Chaining: A symbol key would be illuminated only after a peck at the WHAT COLOR? key.

12. Chaining: A peck at the THANK YOU key was required before a peck at a color key would be reinforced.

The two birds were next placed in the chamber together, on either side of the partition. They were housed together until they showed no signs of distress in each other's presence ("adaptation"). With chamber lights illuminated, they would now engage in the communication sequence described above. Errors were followed by brief "time-outs" (all chamber lights were extinguished) throughout training.

Jack and Jill later learned each other's roles and acquired still other languagelike performances.

[ROBERT EPSTEIN]

Bibliography: A. C. Catania, *Learning.*, 1979; R. Epstein, R. P. Lanza, and B. F. Skinner, *Science*, 207:543–545, 1980; E. S. Savage-Rumbaugh, D. M. Rumbaugh, and S. Boysen, *Science*, 201:641–644, 1978.

Animal virus

Several viruses have been implicated as causative agents for many congenital defects in humans. The role of a specific virus as a teratogenic agent for humans was first recognized in 1941 by N. M. Gregg, an Australian ophthalmologist who noticed that there was an unusually large number of newborn babies with congenital cataracts. A careful examination of his records indicated that most of the mothers had contracted rubella (German measles) during the first trimester of pregnancy. Upon further investigation, additional congenital defects were discovered, including deafness, cardiac abnormalities, and mental retardation, which resulted from a maternal rubella infection contracted during pregnancy.

Subsequently, it has been shown that other viruses also have the capability of traversing the placental barrier and consequently may produce adverse effects on the fetus. During pregnancy a unique situation exists in that a susceptible mother may contract a virus infection and, following viremia, such infection may proceed to the fetus. In general, the type and degree of fetal malformation are determined primarily by the age of the fetus at the time of infection, with particular susceptibility during the first trimester of pregnancy, although other factors may also contribute to resultant teratogenesis.

Viruses involved. Little is known concerning the pathogenesis of congenital virus infections. Those agents capable of traversing the placental barrier and infecting the human fetus with adverse consequences have been grouped under the acronym TORCH by A. Nahmias (see table). With the exception of *Toxoplasma gondii*, a protozoan, all are viruses. The commonly observed clinical manifestations for each virus are listed, suggesting possible diagnostic value.

Factors determining teratogenesis. The majority of congenital infections resulting in birth defects are a consequence of maternal viremia occurring during pregnancy or, in a small number of cases, a maternal infection before conception. After the maternal infection has become established, the degree of teratogenic consequences is dependent upon a number of factors. One very important factor is the virulence of the virus. Congenital infections with a highly virulent virus such as variola often produce fetal wastage. Less virulent viruses cause mild infections in which the fetus survives, but sufficient damage may occur to result in teratogenesis.

Infections due to rubella and cytomegaloviruses are generally mild, innocuous human diseases except for the fetus. The effect of virus multiplication appears to be greatest in cells which grow most actively. Infections acquired during the first trimester, a time when embryological growth and differentiation are most active, appear to be more teratogenic than those occurring later in development. The absence of a fully developed immunological system in the early fetus no doubt is also an important factor increasing the seriousness of congenital infections.

Clinical features. The results of congenital virus infections may range from inapparent infection to fetal death. In the former case, IgM antibodies specific for the infectious agent may be the only evidence of infection in the infant. While a number of viruses may be responsible for congenital infections, the clinical patterns for these infections are often quite similar. The infant infected with TORCH agents is commonly premature or shows retarded growth, with microcephaly, chorioretinitis, hepatosplenomegaly, and disorders of the circulatory system. Oftentimes, congenital infections may be characterized by target organ involvement depending on the specific virus responsible.

Cytomegalic inclusion disease. This disease of the newborn, acquired by transplacental transmission of cytomegalovirus, is often fatal within the first few weeks of life. It accounts for up to 1% of all infant deaths and is considered to be the most important cause of microcephaly.

The greater number of infants with congenital cytomegalovirus infection show few or no clinical manifestations at birth. Affected infants may experience neurological sequelae that become apparent in the neonatal period or in later life, for example, during the early school years. Abnormalities may include mental retardation, hearing defects, speech defects, behavior disturbances, or a combination of these. The effect of inapparent cytomegalovirus infections on the newborn and on subsequent growth and development is presently an active area of research.

Congenital rubella syndrome. Infection of the fetus with rubella virus during early development may damage any organ system and result in congenital defects. Frequently, a clinically recognizable pattern of defects is encountered, termed the rubella syndrome. Severe defects are most frequent when congenital infection occurs within the first trimester of pregnancy, particularly the first month. Such infections may produce loss of function in hearing and vision, disorders of the cardiovascular and central nervous systems, or mental retardation with microcephaly. Approximately 10–20% of infants infected at birth with rubella die within the first year.

Herpesvirus hominis infections. Infants with neonatal infections due to herpesvirus hominis (Herpes simplex virus) are usually infected with the type 2 or genital strain, but approximately 40% may be infected with the type 1 or "nongenital" virus strain. A high proportion of infants with perinatal infections due to herpes simplex virus show

TORCH* agents and manifestations of infection

Major viruses	Clinical manifestations
Toxoplasma gondii (protozoan)	
Other	
Coxsackie group B	Myocarditis; hepatitis; encephalitis; interstitial pneumonitis; interstitial pancreatitis; focal adrenal cortical necrosis
Hepatitis type B virus	Predominantly asymptomatic infection with no identified sequellae; neonatal hepatitis; neonatal or juvenile cirrhosis
Lymphocytic choriomeningitis	Meningoencephalitis (?)
Mumps	Parotitis; endocardial fibroelastosis (?)
Polioviruses	Meningoencephalitis; aseptic meningitis; myocarditis (?)
Rubeola	Measles
Vaccinia	Fulminating generalized disease; chorioretinitis; hypoplasia or destruction of osseous element
Varicella-zoster	Fulminating generalized infection with miliary areas of coagulative necrosis or in utero embryopathy characterized by hypotrophic limbs and digits, ocular deficits may include: chorioretinitis, microophthalmic cataracts, nystagmus or optic atrophy and neurological abnormalities which may include psychomotor/sensory deficits, paralysis, dysplasia or psychomotor retardation
Variola	Fulminating generalized infection with miliary areas of coagulative necrosis
Western equine encephalitis	Meningoencephalitis (?)

Rubella

Overt disease

Microcephaly; congenital heart disease (especially patent ductus arteriosus ± pulmonary stenosis); central deafness; cataracts; myocarditis; myocardiopathy; hypoplasia of the ocular bulb; inflammation of iris, retina, and ciliary body; iris hypoplasia; incomplete development of the chamber angle; interstitial pneumonitis; interstitial nephritis; hypoplasia of the extrahepatic biliary system; hepatitis; osteomyelitis; meningoencephalitis; disseminated intravascular coagulopathy with thrombocytopenia; hemolytic anemia with marked extramedullary hematopoiesis; panorgan involvement

Cytomegaloviruses

Overt disease

Microcephaly ± intracranial calcifications; chorioretinitis; interstitial pneumonitis; disseminated intravascular coagulopathy with thrombocytopenia; hemolytic anemia with marked extramedullary hematopoiesis; hepatitis with jaundice; interstitial nephritis; panorgan involvement; subtle "forma frusta" of the disease includes hepatosplenomegaly, transient evidence of disseminated intravascular coagulopathy, jaundice

Herpesvirus hominis types I and II

Congenital

Microcephaly with intracranial calcifications; chorioretinitis; hemolytic anemia with extramedullary hematopoiesis; focal miliary necrosis of liver, lung, adrenal, and brain; interstitial pneumonitis; herpetic ulcers of the gastrointestinal tract

Neonatal

Fulminating in disseminated syndrome; widespread organ involvement with preferential replication in brain, liver, adrenals, and lung; if infants survive sequellae common

Limited disseminated syndrome: predominantly manifested by recurrent crops of cutaneous lesions and relatively minor evidence of systemic involvement; prognosis good

*For derivation of the acronym TORCH, see the boldface letters in the table.
SOURCE: G. R. G. Monif, *Infectious Diseases in Obstetrics and Gynecology*, 2d ed., Harper & Row, 1980; used with permission.

recognizable sickness. Transmission of the virus to the fetus may occur across the placenta or, more frequently, during birth via passage through an infected birth canal. Herpetic infections are more prevalent in premature than in full-term infants, and some infections may be asymptomatic. Neurological sequelae have been observed in some patients, and consequently there is concern with regard to complications such as mental retardation arising in later life as mentioned with respect to rubella and cytomegalovirus congenital infections.

Other viral agents. Other viruses have been incriminated as causes of fetal damage resulting from infection during prenatal development. These include some members of the enterovirus group, principally the poliovirus, Coxsackie and ECHO viruses, as well as varicella-zoster, smallpox, and vaccinia viruses, hepatitis viruses, and certain others (see table). The last include viruses which rarely infect the fetus or those whose status as agents of congenital infection is still uncertain.

Prevention. The absence of any chemotherapy specific for most human virus infections, especially when pregnancy must also be taken into consideration, makes prevention the most important method of control. The protection afforded by circulating antibody is a matter of current investigation. In the case of cytomegalovirus, asymptomatic mothers with serological evidence of previous infection may deliver infected infants. A vaccine for rubella has been developed and does afford protection against the congenital rubella syndrome. The immunization of women of childbearing age who show no serologic evidence of previous infection is encouraged. Immunization 2 or more months before pregnancy is recommended as a precaution against possible rubella syndrome by the vaccine virus. However, no evidence exists that any cases have been caused by the vaccine virus strain.

For background information *see* ANIMAL VIRUS; CONGENITAL ANOMALIES; TERATOLOGY in the McGraw-Hill Encyclopedia of Science and Technology.   [NORMAN S. SWACK]

Bibliography: J. A. Dugeon, *Brit. Med. Bull.* 32: 77–83, 1976; J. B. Hanshaw and J. A. Dugeon, *Viral Diseases of the Fetus and Newborn*, vol. 17 in A. J. Schaffer and M. Markowitz (eds.), *Major Problems in Clinical Pediatrics*, 1978; S. Krugman and A. A. Gershon (eds.), *Infections of the Fetus and Newborn: Progress in Clinical and Biological Research*, vol. 3, 1975; G. R. G. Monif, Congenital viral infections, in D. Seligson (ed. in chief), *CRC Handbook Series in Clinical Laboratory Science*, sec. H, *Virology and Rickettsiology*, vol. 1, p. 2, 1978.

Annealing

The use of laser or electron beams for directed energy processing offers a new and promising technique for rapidly heating a spatially localized region in the near surface of a material to high temperatures for a short period of time to improve microelectronics device fabrication, or to investigate the properties of materials, or to develop unique metastable systems which cannot be formed by conventional techniques.

Semiconductor processing. Laser and electron-beam annealing show great promise for applications to semiconductor processing leading to improved device performance, yield, and circuit packing densities compared with conventional furnace annealing. The performance of semiconductor circuits depends strongly on precise control of the introduction, spatial distribution, and crystalline damage from impurity dopant atoms. Ion implantation is now a commonly used process for achieving precisely defined dopant doses and depth profiles needed for advanced solid-state device fabrication. The ion implantation process, however, produces considerable damage in the implanted host crystal (usually silicon) in the form of displaced atoms, dislocations, and other crystalline imperfections. As shown in the illustration, many of the implanted dopant atoms are interstitially located or trapped at defects in the host crystal and therefore do not have the desired electrical activity until a further annealing process is used to move them into substitutional positions. Conventional annealing introduces the entire implanted wafer into a high-temperature furnace for relatively long periods of time. Although this is effective in removing the defective nature of the crystal, it suffers a drawback in that a further redistribution of the dopant atoms also occurs and sets ultimate limits on the precision of dopant

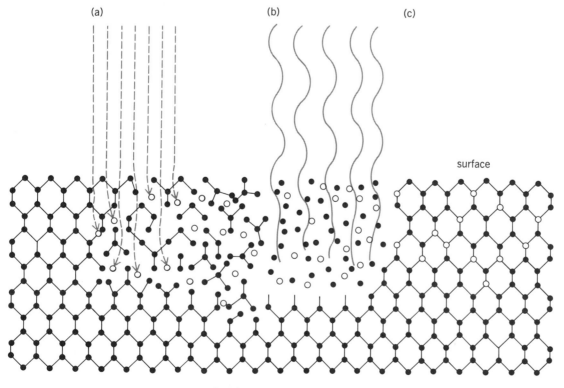

Steps in laser-beam annealing or electron-beam annealing process. (*a*) Implant dopant. (*b*) Laser or electron-beam anneal. (*c*) Epitaxial regrowth.

spatial distributions and hence on device performance and dimensions. Lateral dopant redistribution results in increased parasitic capacitances and thus a reduced operating speed of metal oxide semiconductor (MOS) transistors. Increased spread in the dopant's depth profile limits the maximum frequency of devices which require abrupt junctions and defined drift regions such as impact avalanche transit time (IMPATT) diodes.

Laser annealing. Both pulsed and scanned continuous-wave (CW) laser annealing have been found to restore crystallinity to implanted silicon and produce high electrical activation of implanted dopants. Pulsed laser annealing (involving pulses carrying energies of approximately $1-3$ joules/cm^2 in a period of approximately 10 nanoseconds) provides a somewhat higher degree of electrical activation than does continuous-wave laser annealing, but is accompanied by more dopant redistribution.

Pulsed laser annealing involves melting of the implanted surface to a depth greater than the dopant distribution followed by liqud-phase epitaxial regrowth from the solid-liquid interface out to the semiconductor surface. There is a threshold energy density for the laser pulse above which the implanted layer melts and regrows epitaxially. For slightly lower energy densities in which the melt depth does not penetrate through the disordered layer, the layer regrows as a polycrystalline layer. The time available for regrowth of 0.1 to 0.3 μm depth is determined by heat conduction and is of the order of 0.1 to 1 microsecond. In contrast, continuous-wave laser annealing can be used in a manner which brings about solid-phase epitaxial regrowth of the damaged layer.

Electron-beam annealing. Electron-beam annealing is not yet in as advanced a stage of application as laser annealing, but similar results have been reported for implants into silicon. Electron-beam annealing has the advantage that the energy deposition is not limited by the optical properties of the target, and thus the process is particularly attractive for metals, both for ohmic metal contacts to semiconductors and for investigations of unique implanted alloy surface layers. The absorption of laser energy depends strongly on its wavelength and on the reflectivity and band gap of the semiconductor, whereas the absorption of the electron energy depends primarily on the atomic number of the target and the energy and angle of incidence of the electrons.

Future prospects. The material for which the most promising practical results have been demonstrated to date is silicon, but laser or electron-beam annealing may be of even more potential benefit to compound semiconductors such as gallium arsenide (GaAs), which require controlled ambients or specially deposited surface layers to prevent dissociation of the compound during furnace annealing. However, more work is needed since compound semiconductors have much greater sensitivity to laser-induced defects and damage than does silicon. New promising applications also include the formation of otherwise unobtainable metallurgical phases by pulsed annealing in which rapid heating and cooling occur, for example, $10^9°$C per second for Q-switched lasers and approximately $10^{14}°$C per second for picosecond-pulse lasers.

For background information *see* ION IMPLANTATION; LASER; MICROWAVE SOLID-STATE DEVICES; SEMICONDUCTOR; TRANSISTOR in the McGraw-Hill Encyclopedia of Science and Technology.

[F. L. VOOK]

Bibliography: S. D. Ferris, H. J. Leamy, and J. M. Poate (eds.), *Laser-Solid Interactions and Laser Processing*, 1979; C. W. White and P. S. Peercy (eds.), *Laser and Electron Beam Processing of Materials*, 1980.

Annelida

Vision has evolved by many routes. However, only two fundamentally different ways of resolving images exist. These are seen in the compound eyes of arthropods and the lensed eyes of mollusks and vertebrates. Indeed, all forms studied so far appear to use a single class of visual pigments consisting of a visual protein (an opsin) combined with a color group, 11-cis retinal (a form of vitamin A). The universality of these mechanisms reflects a considerable degree of conservation throughout evolution, and possibly evolutionary convergence. The old but largely unrecognized observation that one family of annelid worms—the alciopids—has highly structured image-resolving eyes offers a special circumstance in which to evaluate aspects of evolutionary dynamics. These alciopid worms apparently utilize the universal photochemistry. Like deep-sea fish and squid, they have accessory retinas, raising the possibility that each group separately evolved these structures to serve in depth perception. Furthermore, deep-sea alciopids reflect a second striking instance of evolutionary convergence in possessing light sensitivities closely matching those of other deep-sea organisms.

Worms with well-developed eyes are confined to a single family of polychaetes, the alciopids. All are marine, living in warm waters. Something has lately been learned of the visual physiology in two genera, that of the surface form *Torrea* and of deep-sea *Vanadis*. *Torrea candida* (Fig. 1) is a long $(12-17$ cm), slender $(1.5-2$ mm wide) worm constantly moving by parapodial beating and bodily undulations. The eyes, which dominate the head, are just under 1 mm in diameter and bright red-orange. The pupils glow about the same color with light reflected from inside the eye, and are surrounded by a silver, iridescent iris. Black podial glands contain the only other pigmentation in the animal and give the body a ladderlike appearance.

Eye structure. The eye of the alciopid *T. candida* (Fig. 2) bears a certain similarity in structure to that of humans. Light enters the eye through the cornea, which in *Torrea* is two-layered, one layer an extension of the skin, the other of the iris and retina, is focused by the lens which is large and spherical, and then passes through the vitreous which consists of two layers of humoral material that fill the inside of the globe, stretching from the lens to the retina. Directly to either side of the lens lie paired accessory retinas containing a few relatively short and stubby light-sensitive cells (photoreceptors) arranged similarly to the cells making up the main retina. The main retina itself is formed by about 10,000 photoreceptors whose individual structure gives the retina a three-layered appearance. Light having passed through the vitreous first strikes the receptive region of the photoreceptor cells, a specialized region marked by extensive folding of the cell membrane into microvilli which contain the visual pigment. This arrangement is quite the opposite in vertebrates, where the retina develops in an inverted orientation so that light must pass through the entire retina before encountering the receptor layer. The visual pigment transforms the light signal into electrical information which is conveyed down the receptor cell through a pigmented layer, past the nuclei of the receptor cells which make up the third layer, and into the optic fibers that in turn connect with the optic ganglion and the rest of the nervous system.

Visual physiology. By measuring the electroretinogram with electrodes placed on the cornea and

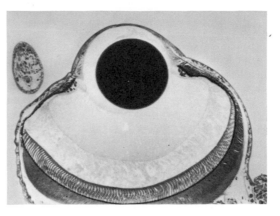

Fig. 2. Cross section of *Torrea candida* eye showing paired accessory retinas to either side of the spherical lens. The main retina, lining the back of the eye, is made up of a single layer of about 10,000 photoreceptor cells. (*From G. Wald and S. Rayport, Vision in annelid worms, Science, 196:1434–1439, copyright 1977 by the American Association for the Advancement of Science*)

its position in the eye, which is such that it responds primarily to light reflected off the main retina.

Photochemistry. The shape of the sensitivity curves is important because it offers a clue to the photochemistry. Many of the known visual pigments have quite similarly shaped absorption spectra when plotted on a frequency basis. The spectral sensitivity curve for the main retina of *T. candida* closely matches that of other well-studied visual pigments, while that of the accessory retina differs considerably. The poor match for the accessory retina means little because its sensitivity is determined both by the properties of its receptors and by the reflective properties of the eye. However, the implication of the good match for the main retina is that it, like vertebrate and inverte-

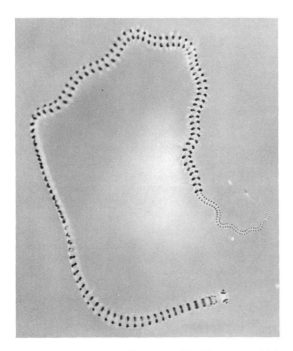

Fig. 1. Free-swimming *Torrea candida*. The eyes, which are bright red-orange, dominate the head. The striped appearance is due to black podial glands, one pair per segment. (*From G. Wald and S. Rayport. Vision in annelid worms, Science. 196:1434–1439, copyright 1977 by the American Association for the Advancement of Science*)

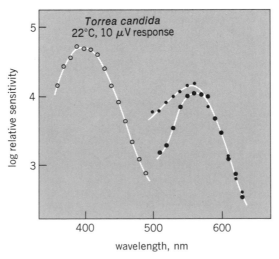

Fig. 3. Spectral sensitivities of the *Torrea candida* eye. The main retina is maximally sensitive at about 400 nm, the accessory retina at about 560 nm. The data from the accessory retina vary, primarily because light strikes it only after reflection off other parts of the eye. (*From G. Wald and S. Rayport, Vision in annelid worms, Science, 196:1434–1439, copyright 1977 by the American Association for the Advancement of Science*)

back of the eye, recordings can be made of the electrical potentials created when light strikes the main and accessory retinas, generating cornea negative and positive signals respectively. Stimulation with light of different wavelengths allows a determination of the spectral sensitivity of the eye. For the best-studied alciopid, *T. candida*, the spectral sensitivity (Fig. 3) of the main retina peaks in the blue at about 400 nanometers, that of the accessory retina in the yellow-green at about 560 nm. The sensitivities overlap considerably; in fact, the main retina retains measurable sensitivity at 560 nm, the peak sensitivity of the accessory retina. The spectral sensitivity of the accessory retina is quite narrow and variable (examples from two worms are shown in Fig. 3), probably as a result of

brate retinas studied before, may share a similar chemistry based on 11-cis retinal (11-cis vitamin A aldehyde) as the color group.

Accessory retina. While shared visual chemistry may primarily reflect conservation of photochemistry through evolution, the presence of accessory retinas in alciopid worms, in deep-sea fish with tubular eyes, and in certain deep-sea cephalopods presents an especially clear demonstration of evolutionary convergence. Other than the shared position and possible role, the characteristics of the accessory retinas are subtly different and stand as evidence of their independent evolution. Those of the cephalopods and fish are an extension of the main retina running up to lie beside the lens, much too close ever to see a focused image, as compared with the worms, where the accessory retina is in a similar location but discrete from the main retina. Furthermore, in the invertebrates (the worms and cephalopods) both retinas face the inside of the eye, whereas in the fish (as in all vertebrates) they face away, and are by comparison inverted. Toward what purpose then should the process of evolution have resulted in accessory retinas, following three presumptively separate routes, in the evolution of eyes in three different phyla?

One possibility is that the role of the accessory retinas takes advantage of the wide discrepancy in wavelength sensitivities of the main and accessory retinas. Long-wavelength light is attenuated by depth in the ocean much more than is blue light (looking up when underwater, one is struck by the blueness of the water). For *Torrea*, with retinal sensitivities at 400 and 560 nm, the ratio of energies of sunlight at these two wavelengths that reach increasing depths in a clear ocean are at the surface, 0.6; at 10 m, 1.4; at 25 m, 3.3; at 50 m, 11; at 75 m, 25; at 100 m, 40; and at 150 m, about 70, rising rapidly with greater depths. As a result, comparisons in the short- and long-wavelength regions could provide a sensitive measure for the determination of depth in the sea. This may be the functional role of accessory retinas not only in *Torrea*, but in other alciopids and perhaps in other animals as well. To test the generality of this conclusion, an exploration of spectral sensitivities of main and accessory retinas in the other phyla is needed.

Deep-sea alciopids. Deep-sea alciopids offer a second instance of evolutionary convergence. Two different species of alciopids in the genus *Vanadis* (collected at 300 m depth) have main retinas with spectral sensitivities maximal at about 460 and 480 nm. It is striking but not surprising that these worms, like deep-sea fish, have maximal sensitivities in the 460–480-nm range, which is the center of the narrow band of sunlight that filters through the ocean to depths beyond 200 m. At greater depths and at night, furthermore, vision depends on light provided by luminescent bacteria generating light primarily in the same wavelength range. So deep-sea alciopids have achieved along with other deep-sea creatures a visual fitness optimal for their limited-light environment.

Summary. Similarity in eye structure and possibly in visual chemistry suggests that alciopid worms may have evolved vision from ancestors shared with arthropods, mollusks and vertebrates. The presence of accessory retinas in alciopids reflects a striking instance of evolutionary convergence; alciopids as well as deep-sea cephalopods and fish may use the accessory retinas in conjunction with the main retina to perceive depth. Finally, deep-sea alciopids share with other creatures of the depths maximal spectral sensitivities tuned to the blue light which alone penetrates the sea to reach their environment. Thus vision in alciopid worms has reached an evolutionary fitness commensurate with that of other organisms in several other phyla that have eyes.

For background information *see* EYE (INVERTEBRATE); PHOTORECEPTION in the McGraw-Hill Encyclopedia of Science and Technology.

[STEPHEN RAYPORT]

Bibliography: C. O. Hermans and R. M. Eakin, *Z. Morphol. Tiere.*, 79:245–267, 1974; R. Menzel, Spectral sensitivity and color vision in invertebrates, in H. Autrum (ed.), *Handbook of Sensory Physiology*, vol. VII/6A, 1979; G. Wald and S. Rayport, *Science*, 196:1434–1439, 1977.

Antibody

Since the discovery that antibodies formed in the blood of an animal are usually a heterogeneous mixture of molecules, immunologists have sought means of generating homogeneous antibodies for both theoretical and practical reasons. For example, it has not been possible to ascertain the molecular structure or size of the antigen-combining site of an antibody owing to the fact that "antibody" is actually a pool of many different antibodies with similar but not identical structure and antigen-binding characteristics.

Tumor proteins. It was equally difficult to ascertain a structural basis for the mechanisms of interaction of "antibody" with other assessory factors, such as complement. From a practical standpoint, antibodies have been useful diagnostically in areas such as blood grouping and blood typing. It was probably fortuitous that blood group antigens are composed of very strong and very different immunodominant chemical groups. Thus a predominant number of the heterogeneous antibodies were directed toward the most immunodominant determinants. However, this circumstance may be related to the more serious dilemma of how antibodies can be produced which will react with, and allow detection of, weaker antigens, such as some hormones or enzymes and most neoplastic tumor antigens.

A milestone was set in immunology when it was discovered that the paraproteins in the serum of animals and humans were homogeneous immunoglobulins. These myeloma proteins were discovered to be secreted by the cells of tumors of the lymphoid system in humans and animals. Moreover, similar plasmacytomas of the lymphoid system could be induced in mice. Several of these tumors were found to secrete immunoglobulins capable of binding specifically with antigenic groups such as dinitrophenyls, α-1→6-dextrans, α-1→3-dextrans, and phosphorylcholine. It was recognized that many of the carbohydrate classes of determinants with which these myeloma and plasmacytoma proteins bind with high avidity are also core-group antigens of many microorganisms. This observation was consistent with the premise that such tumor proteins were somehow generated by natural exposure to the respective antigens. Tests of this hypothesis were disappointing, since it turned out to be virtually impossible to produce plasmacytomas which would secrete immunoglobulins with a desired antigen-binding specificity by simultaneous immunization or by incorporation of the respective antigens into the various oils used to induce plasmacytomas in mice. Current knowledge of the molecular genetics of the immune system suggests that plasmacytomas result from precursors of antibody-secreting cells (B cells) which become leaky with respect to expression of their V-region gene product (active site) and, for yet unknown reasons, become malignant.

Hybridomas. The genesis of malignant plasmacytomas and myelomas remains an elusive and general problem in biology. The phenomenon has been exploited recently in the discovery that the experimental fusion of monoclonal plasmacytoma cells with immune lymphocytes, taken from animals immunized deliberately with antigens of choice, resulted in a monoclonal line of cells with the antibody specificity of the immune lymphocyte partner and the continuous growth characteristic of the malignant cells of the plasmacytoma. The

successful hybrid clones are capable of continuous growth and secretion of monoclonal, antigen-specific immunoglobulins or monoclonal antibodies. This powerful technique, which was successfully achieved by G. Köhler and C. Milstein in the mid-1970s, is revolutionizing the field of immunology.

The general procedure for construction of hybridomas involves the synthesis of two well-known and widely used protocols. An animal is immunized by conventional injection of a suitable antigen. An appropriate lymphoid tissue, usually the spleen, is dispersed into a single-cell suspension. The lymphocytes obtained are then mixed with one of several possible malignant plasmacytoma cell lines having a propensity to fuse with other cells in the presence of a fusion-promoting agent. While Köhler and Milstein used the Sendai virus as their fusing agent, most investigators now use polyethylene glycol (PEG). PEG promotes cell fusion 100- to 300-fold the rate of Sendai-virus fusion, and usually up to 50% of cells exposed are observed to fuse in the presence of PEG. Successful hybridization and ultimate propagation and cloning of interesting hybridomas depend on the technical skill of gradual removal of the PEG, promotion of survival and growth of hypoxanthine-guanine phosphoribosyl transferase positive

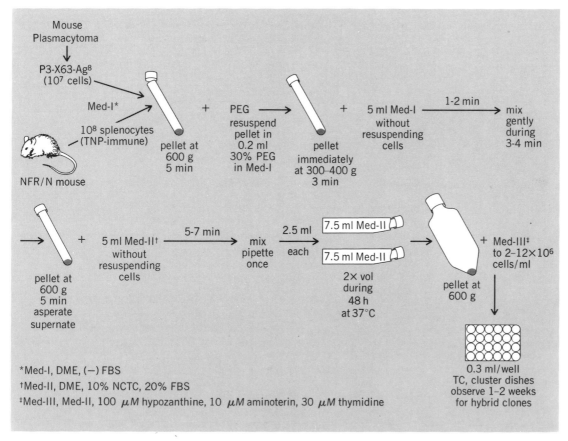

Fig. 1. Technique for the construction of hybridomas which produce monoclonal antibodies. There is no unique requirement for the NFR/N mouse strain. DME = Dulbecco's modified Eagle medium; FBS = fetal bovine serum. (*From E. E. Hanna et al.. Hybridomas: Fusion of lym-* *phocytes and mouse plasmacytomas—an approach to specific monoclonal antibodies. in D. Schlessinger. ed.. Microbiology—1980. American Society for Microbiology. 1980*)

(HPRT⁺) hybrids in HAT (hypoxanthine, aminopterin, and thymidine) medium, and transfer of growing hybrids to new growth conditions.

The method is theoretically adaptable for the construction of hybridomas which can biosynthesize monoclonal antibodies specific for single determinants (finite chemical groups, defined or undefined) present on any immunogen. The achievement of a desired monoclonal antibody seems to be greatly facilitated when the finite antigenic determinant in question is immunogenic in the donor of the normal lymphocytes. However, for theoretical reasons not considered here, this principle may not be an absolute requirement. There are various tumor cell lines available to meet special requirements. For example, P3-X63-Ag8 or simply X63 secretes as immunoglobulin. However, it has no known binding activity. Thus there is no problem when the experimental purpose is to produce monoclonal antibodies reactive with a known antigen. On the other hand, if the experimental purpose is to determine something about the structure (for example, class or allotype), which may not be directly related to antigen binding, problems may arise, since it will be necessary to separate the immunoglobulins coded by the genome of the plasmacytoma from the immunoglobulins coded by the genome of the immune spleen lymphocyte parent,

both of which are usually expressed in the hybrid or hybridoma. Thus it is obviously simpler to use a tumor or mutant plasmacytoma line which does not secrete immunoglobulins. Such cell lines are available; for example, P3-NS1-1-Ag4-1 or simply NS-1 is a nonsecreting mutant subline of X63.

The procedure used by E. E. Hanna is depicted in Fig. 1. It illustrates the protocol for the construction of a hybridoma which will secrete monoclonal antibodies specific for the trinitrophenyl (TNP) chemical group. The procedure involves PEG-promoted fusion of specifically immune mouse lymphocytes (TNP-hapten immunized for TNP-immune splenocytes) which are not capable of continuous growth, with the mouse plasmacytoma cell line P3-X63-Ag8. The normal immune lymphocytes are HPRT⁺, and the cells of the plasmacytoma are HPRT⁻, so that HPRT⁺ antibody-secreting hybrid cells (hybridoma) capable of continuous growth may be selected on the basis of their capacity to grow in HAT medium. As indicated in the final step of the hybridization procedure, initial HAT-resistant hybrid clones may begin to be visualized in the cluster dish wells after 1–2 weeks (Fig. 2a). Proliferating hybrids eventually occupy the entire well (Fig. 2b) Depending upon the initial seeding density, more than one and usually several hybrids develop in a well. At this stage it may be advantageous to screen the wells for clones

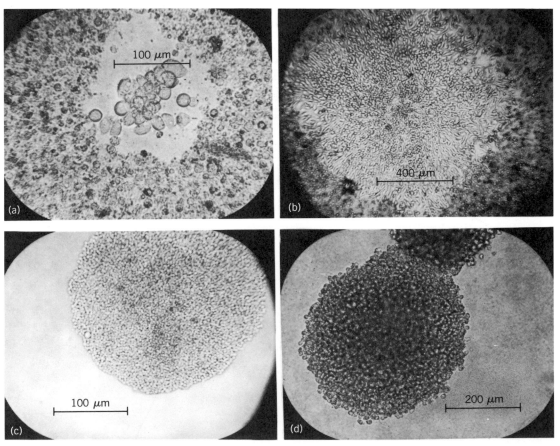

Fig. 2. Photomicrographs of representative hybridomas at various stages of development. Appearance (a) at 6–8 days and (b) at 15–18 days in wells of a cluster dish such as that shown in Fig. 1. Note that the foci of HAT-resistant hybrid cells growing profusely at the centers are surrounded by dying HAT-sensitive cells. (c, d) Typical soft-agar-cloned colonies of hybridomas successfully transferred from the cluster dishes and propagated.

which are secreting the required antibody by an appropriate assay for the antibody of interest. It is sometimes possible to transfer the growing clone from wells showing evidence of only one clone to a larger growth area (dish or flask), and continue to harvest and concentrate or purify the antibody from appropriate supernatant fluids. However, in the usual case it becomes necessary to isolate the individual clones in order to obtain a single pure clone. Several procedures are available to do this. The most common procedure is by limiting dilution of the cells, followed by allowing the single cells to form single isolated colonies in soft agar (soft-agar cloning; Fig. 2c and d). It is then sufficient to transfer the soft-agar colonies once again to liquid media and to grow each selected colony (clone) to sufficient density to harvest the desired monoclonal antibodies. In many instances it is possible to use an appropriate recipient animal host for the hybridoma, and to harvest the monoclonal antibody in large amount from the ascitic fluid of the animal. It is usually necessary in this procedure to separate the antibody from the other host proteins of the ascites. Provided that the cloned hybridoma is not exceedingly fastidious in its growth requirements, it may be advantageous to propagate the hybridoma in mass culture and then to concentrate the antibody from large quantities of supernatant fluid.

Applications for monoclonal antibodies. Monoclonal antibodies are being applied broadly toward ascertaining the identity of histocompatibility antigens, cellular differentiation antigens, and neoplastic tumor antigens. It is likely that new important cell-surface differentiation antigens will be discovered by the elimination of many heretofore puzzling cross reactions by the use of hybridoma-produced antisera. Hybridomas are being constructed for the production of monoclonal antisera applicable in other areas of biology as well; for example, endocrinology and enzymology are being greatly aided by the identification of important hormones and enzymes which were predicted to exist on the basis of biologic activity, but had not been identified or isolated, since heteroclonal antisera produced in animals could not distinguish such molecules because of cross reactions. Protective monoclonal antibodies are now being produced in certain experimental parasitic diseases of animals. In a different approach, still using the hybridoma methodology, but toward the production of antigen rather than antibody, hybrids between unicellular protozoan parasites (such as trypanosomes) and X63 have been constructed. These hybrids express parasite antigens at their cell surfaces.

New avenues of research. There are other more fundamental questions relevant to the cellular and molecular biology of the immune system being approached through hybridoma technology. For example, much information is being acquired as to the size of the antibody gene repertoire (library) encoded within antibody-producing B lymphocytes. Moreover, the phenotypes and functions of the various regulatory lymphocytes, such as the several subpopulations of T lymphocytes and other accessory cells (macrophages), are being ascertained by the use of the hybridoma-mono-

clonal antibody technology. Monoclonal gene products of these regulatory cells, which may be structural or secretory, are being delineated. In this respect several hybridomas have been obtained by the fusion of fractionated mouse spleen T lymphocytes with the AKR mouse T-cell lymphoma cell line BW5147. A number of these hybrids have now been cloned, and at least one of them has the phenotype and function of helper T lymphocytes which are required in the control of antibody biosynthesis to certain immunogens.

It is excitingly apparent that hybridoma technology, with the assistance of molecular gene cloning technology, is initiating a new round of progress and understanding in immunology and cell biology.

For background information *see* CELL MEMBRANES; GENETICS, SOMATIC CELL in the McGraw-Hill Encyclopedia of Science and Technology.

[EDGAR E. HANNA; MICHAEL L. MISFELDT]

Bibliography: M. St. J. Crane and J. A. Dvorak, *Science*, 208:194–196, 1980; E. E. Hanna et al., in D. Schlessinger (ed.), *Microbiology—1980*, 1980; G. Köhler and C. Milstein, *Nature*, 256:495–497, 1975; F. Melchers, M. Potter, and N. L. Warner (eds.), *Current Topics in Microbiology and Immunology*, vol. 81, 1978; N. Yoshida, R. S. Nussenzweig, and M. Aikawa, *Science*, 207:71–73, 1980.

Apidium and Parapithecus

Apidium and *Parapithecus* are members of a family of extinct primates, the Parapithecidae, the oldest and most primitive African monkeys. As such, they are the key to understanding the origins of the Old World anthropoid primates, the group which includes monkeys, apes, and humans. Parapithecids are found only in rocks of Oligocene age in Fayum Province, Egypt, in an area of badlands at the eastern edge of the Sahara Desert. The history of their discovery, the conflicting theories as to their place in higher primate evolution, and the evidence for their way of life are discussed below. Much of this information is based on the collections made between 1977 and 1979 by joint Egyptian government–Duke University expeditions.

History of study. The first recovered parapithecid was *Apidium phiomense*, a name which approximately translates into "little sacred bull of the Fayum." A single jaw of a young *A. phiomense* was found by Richard Markgraf early in 1907 and described by Henry Fairfield Osborn in 1908. Osborn suspected primate affinities or that it was a hoofed mammal, hence the name. Later in 1908 Markgraf collected a more complete adult lower jaw of a second kind of monkey which he sold to Dr. Fraas of the Stuttgart Museum. This find was described by M. Schlosser in 1911 with the name *Parapithecus fraasi*, meaning "next-to-an-ape." No other specimens of *Apidium* or *Parapithecus* were recovered until October 1961.

These two monkeys were difficult to relate to modern primates, so they remained of uncertain evolutionary relationship throughout the 50 years that followed their description. Osborn thought *Apidium* might be an odd sort of pig or a primate, while others later considered it a possible archaic ungulate, a monkey, or an ancestor of the extinct Italian primate *Oreopithecus*. *Parapithecus* was enigmatic partly because of damage to the specimen at the front of the jaw which caused loss of teeth and tooth sockets. This damage, in turn, led to misinterpretation of the numbers and kinds of its teeth, information that would have been useful in judging its affinities. Hence, opinions about its closest relatives ranged from tarsiers to monkeys, apes, or even humans.

Many new finds from the paleontological expeditions led by Elwyn Simons in the Fayum badlands between 1961–1968 and 1977–1979 have considerably clarified the systematics of these animals. Recent discoveries have provided new information about their geological antiquity, paleoecology, anatomy, and adaptations. Simons described two new species of parapithecids, the first, *Apidium moustafai* in 1962 and the second, *P. grangeri*, in 1974. He demonstrated that the two genera are closely related. He assigned *Apidium* to Schlosser's *Parapithecus* family group of Parapithecidae.

Geological studies indicate that Fayum mammals occur between about 27 and about 37 million years (m.y.) before present (BP), so that 29–32 m.y. is a likely age range for all parapithecid fossils. *Apidium moustafai* is geologically older and could be in the direct ancestry of *A. phiomense*. *Parapithecus grangeri* was a contemporary of *A. phiomense*. The age of *P. fraasi* is unknown, but it may be older than *P. grangeri* as it is an ideal structural ancestor for the latter. Further information may confirm the probability that there were two parapithecid lineages, one for *Apidium* and one for *Parapithecus*.

Anthropoid status of Parapithecidae. The finds of the 1960s and 1970s included many relatively complete specimens, including well-preserved jaws, parts of the skull, and many limb bones. These show that parapithecids had reached the anthropoid or "monkey" grade of organization. Parapithecids seem more modern than all primates of the preceding Eocene Epoch (the geological time period between about 58 and about 37 m.y. BP). They resemble anthropoids (apes, humans, and Old and New World monkeys) and not Eocene primates or modern Madagascar lemurs. This is because they have reduced olfactory lobes of the brain, show an anthropoid configuration in bony-ear structure, possess a partition between the eye socket and the space behind it housing jaw muscles (postorbital closure), exhibit closely packed cheek teeth, and have spatulate incisors and projecting canines. Both *Parapithecus* and *Apidium* show sexual size differences. This advanced combination of characteristics has led all authorities to accept their status as among the oldest well-known anthropoids.

Possible New World monkey affinities. A few paleontologists, particularly R. Hoffstetter, have suggested that parapithecids are likely to be ancestors of New World monkeys (=platyrrhines). Parapithecids do show several resemblances to platyrrhines not seen in modern catarrhines (humans, apes, or Old World monkeys). For example, the tympanic ring of the ear resembles that of platyrrhines rather than that of catarrhines. Similarly, the presence in parapithecids of three, not two, premolar teeth, and the incomplete separation between the eye sockets and jaw musculature

(that is, the incomplete postorbital closure) are like New World monkeys. Further, the limb bones of parapithecids show considerable overall similarity with small- to medium-sized platyrrhines such as capuchin and squirrel monkeys. What is important, however, is that all these similarities are most likely holdovers from the last common ancestor of both these two groups and do not indicate ancestral-descendant relationships. Primitive features shared in common do not make special relationships. Put another way, it is reasonable to expect that the ancestors of both catarrhines and platyrrhines should have had common primitive characteristics such as the foregoing.

Thus, parapithecids have no special or advanced similarity to platyrrhines that would place them exclusively in the line of platyrrhine ancestry. Instead they show several characteristics of teeth and limb bones which are seen elsewhere only among catarrhines, and which are not of the sort to be expected in ancestral platyrrhines. Specifically, the quadrate outline of the main cusps of the molar teeth is distinctively catarrhine and nonplatyrrhine, as is the three-surfaced structure of the iliac part of the pelvis.

These endemically catarrhine features of parapithecids go along with zoogeographic evidence. Although at one time in the past the South Atlantic Ocean had not yet formed and South America and Africa were parts of one continent, the breakup of this continent was already well under way by 100 m.y. BP and separation was advanced at 30 m.y. BP. By the latter date, the South Atlantic Ocean would have been no less than 700 km at its minimum width. The possibility of a parapithecid family being rafted from Africa to South America across so wide an ocean appears remote in the extreme.

Parapithecid relations to Old World monkeys. The evidence of parapithecid anatomy is mixed regarding their direct assignment to the taxonomic group of the Old World monkeys, the Cercopithecoidea. Many of the details of parapithecid teeth, especially those of *Parapithecus grangeri*, strongly indicate such a ranking. The lower molar teeth show the beginnings of the cross-cresting or bilophodont condition which is the hallmark of modern cercopithecoids. Furthermore, *P. grangeri*, although still retaining three premolars, has the first of these distinctly reduced, which would be expected in the ancestor of two-premolared Old World monkeys.

Opponents of the proposed common grouping of Old World monkeys and parapithecids point to a few specialized cusps on the teeth which are not seen in the cercopithecids. Nevertheless, since the hallmark of cercopithecid teeth is simplicity, many cusp structures of their ancestors almost certainly must have been lost. Opponents of the relationship also argue that the primitive characteristics of parapithecids would imply that if animals like these are directly ancestral to Old World monkeys, then many of the apparently specialized characteristics shared by Old World monkeys and apes must have evolved in parallel. Such a list would include independent loss of the first of the three premolar sets, separate ossification of a tubelike extension of the bony ring which supports the eardrum, and a few minor features of the limb bones. The balance of the evidence does not support the hypothesis that parapithecids are Old World monkey ancestors, but parallelism is commonplace within the best-known mammalian orders.

Parapithecid adaptations. Parapithecid fossils come from continental sediments which were deposited by rivers, lakes, and streams in an area of low relief. The Fayum climate during the Oligocene is thought to have been humid, subtropical to tropical, and densely forested (along the major streams at least). It is probable that there were more open savannas in interstream areas.

Estimating from tooth size and from the size of skeletal elements, parapithecids ranged in weight from about 700 to about 1800 grams (between the weight of the South American squirrel monkey and that of the capuchin monkey).

Where the two are found together in the same fossil quarry, *Apidium* is about 20% smaller than *Parapithecus*.

Judging from the structure of their limbs and pelvis, *Apidium* and *Parapithecus* were agile, free-leaping quadrupeds. *Apidium* appears to have been a highly arboreal animal. *Parapithecus grangeri*, on the other hand, has very high-crowned cheek teeth, a feature common in living Old World monkeys which spend a significant amount of time on the ground. (Apparently, grit in food found on the ground subjects the cheek teeth to greater wear; higher molar crowns are selectively advantageous for resisting such wear.) Therefore *Parapithecus* may have been semiterrestrial. *Apidium* species have low, rounded cusps on their cheek teeth resembling the teeth of predominantly fruit-eating monkeys and apes. Cheek-tooth enamel of *Apidium* was relatively quite thick. Thick enamel suggests that these animals may have eaten nuts or seeds. *Parapithecus grangeri* has cheek teeth with sharper cutting edges implying a greater leaf component in its diet, judging by analogy with the structure of cheek teeth among living leaf-eating mammals.

The relatively small size of the eye sockets of *Apidium* places it among daytime active (diurnal) mammals, as are the living anthropoids, but distinct from many prosimians with relatively large eyes (and eye sockets) and nocturnal habits.

Probably the closest living ecological parallels to the parapithecid primates may be found in the South American squirrel monkey (for *Apidium*) and the African vervet monkey (for *Parapithecus*).

For background information *see* PRIMATES in the McGraw-Hill Encyclopedia of Science and Technology.

[RICHARD F. KAY; ELWYN L. SIMONS]

Bibliography: J. G. Fleagle and E. L. Simons, Anatomy of the bony pelvis in parapithecid primates, *Folia Primatol.*, 31:176–186, 1979; R. F. Kay and E. L. Simons, The Ecology of Oligocene African anthropoids, *Int. J. Primatol.*, 1:21–37, 1980; E. L. Simons, The deployment and history of Old World monkeys (Cercopithecidae, Primates), in J. F. Napier and P. R. Napier (eds.), *Old World Monkeys*, pp. 99–138, 1970; E. L. Simons and E. Delson, Cercopithecidae and Parapithecidae, in V. J. Maglio and H. B. S. Cooke (eds.), *Evolution of African Mammals*, pp. 100–119, 1978.

Archeology

Soils are the most important resource of civilizations past, present, and future. Recent advances in mapping and classification of soils permit artifactual and anthropic strata affected by humans to be related to pedological strata (soil horizons) formed by natural genetic processes. Each soil (11,000 types in the United States) has defined properties (Fig. 1), and anthropic effects can be readily identified in sequence with the natural soil-forming processes. Distributions of soils from place to place over landscapes as delineated on soil maps also permit ancient settlement and land-use patterns to be studied, and predictions can be made about future human activities. Increasing recognition is being given to the concept of the "tragedy of the commons" in ecological and environmental research related to past landscape abuse and future planning. Peaks of cyclic population increases (human and animal) have devastated landscapes in the past and contributed to collapse of productivity of ecological and agricultural systems. Present population growths and associated environmental abuse appear to be projections and continuations of past cycles in competition for scarce resources, and continuations of environmental abuse are likely to have dire consequences in the future.

Soil survey of Tikal. Much of the recent research on relationships of soils to archeology has been done in Maya areas. At Tikal in the Peten region of Guatemala, soils were described, sampled, and mapped at a scale of 1:2000. The Maya apparently achieved good mastery of the visible soil properties related to engineering use and management through experience; the invisible and less visible degenerative soil processes (soil fertility depletion and erosion), however, caused many insidious long-run problems for the Maya. The soil survey of Tikal is a good example to examine and to generalize from because it is the most detailed (large-scale) and comprehensive soil survey ever conducted specifically for soil and archeological correlations in the Maya areas.

At Tikal, Mollisols, Vertisols, Inceptisols, Entisols, and various other soils were discovered in the central core of the abandoned Maya city. Gilgai (mound-and-pit microrelief) and other soil features were also identified and have had a significant impact on the considerations of archeologists working in Maya areas. Excellent correlations were delineated between Maya architecture and artifacts and the soils, and significant effects of erosion, sedimentation, and nutrient depletions were quantified through descriptions of physical soil conditions and later chemical analyses conducted on soil samples.

Mollisols of the uplands at Tikal are naturally very fertile—but extremely vulnerable to erosion and damage by Maya populations because of the topsoil shallowness to limestone bedrock. The dark rich Mollisol appearance of the thin surface soil accumulation misled the Maya developers, because the topsoil is quickly oxidized and depleted through erosion when the forest is cleared. In areas disturbed by the Maya, including causeways and built-up areas, light-colored Inceptisols and

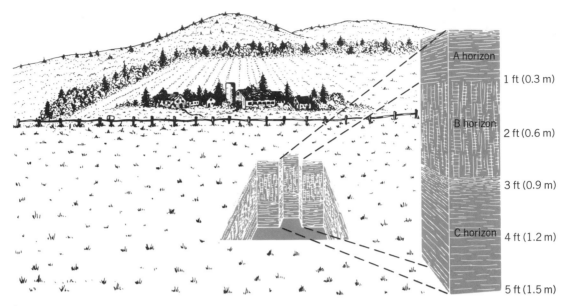

Fig. 1. Soil profile and horizons in a landscape segment.

Entisols were found which were distinctly different from the dark-colored undisturbed Mollisols. Areas in the uplands of Tikal disturbed by the Maya could be mapped in the silt loam places based on soil color alone. Soils at Tikal have not recovered from the Maya occupation even after more than 1000 years of abandonment to the rain forest.

The bajo (lowland) Vertisols at Tikal were strikingly avoided by the Maya for construction sites. These places are wet in the rainy seasons, and the soils with montmorillonite clays are extremely sticky even to pedestrian traffic. The Vertisols were used by the Maya for agriculture and less intensive purposes, however, because sherds were commonly found in the soil survey borings made in the bajos, especially near the edges. The Vertisol soils could be identified by "slickensides" within the soils (as examined in pits) where structural peds rubbed against one another, verifying the churning action of the soils with cracking and expansion upon drying and wetting.

Mapping at Tikal (Fig. 2) identified 11 different soils on six different slopes. The soils and slopes are explained in the table.

Maya yield declines. Maize on the Mollisols at Tikal (in a few patches tilled by shifting cultivation) yields about 1000 kg/ha in the first year after clearing, about 750 kg/ha in the second year, about 500 kg/ha in the third year, and about 250 kg/ha in the fourth year—without any fertilizer additions. For shifting cultivation, the Food and Agriculture Organization of the United Nations recommends a 15-year rotation on these soils—where an area cleared from the forest is cropped for 5 years, and left fallow for 10 years for forest regrowth before clearing the area again for cropping. Obviously, the Maya could not have maintained this ideal rotation as their population increased, and the consequent soil decline had a considerable influence on the culture during the centuries of Maya occupation. The important but gradual soil changes taking place were probably almost unnoticed by the Maya, due to the slow and insidious nature of soil erosion and nutrient depletion.

Sediments eroded from the upland soils at Tikal damaged the lowlands also. Large deposits of eroded sediments were found in drainageways and in Maya reservoirs. Sedimentation from soil erosion had a highly detrimental effect upon the limited Maya water supply in the urban setting.

Raised fields in Belize. Each Maya site is unique, but all show common evidence of soil

Soils and slopes at Tikal (see Fig. 2)

Soil symbol	Brief description
1/	Well-drained clay loam upland soil 45 cm to soft limestone
2/	Well-drained silt loam disturbed alluvial soil 128 cm to soft limestone
4/	Well-drained silt loam upland soil 28 cm to hard limestone
5/	Somewhat poorly drained clay swamp soil 48 cm to soft limestone
6/	Well-drained silt loam disturbed upland soil 142 cm to soft limestone
7/	Moderately well-drained clay upland soil deeper than 150 cm to soft limestone
8/	Well-drained silty clay loam upland soil 40 cm to soft limestone
10/	Poorly drained clay swamp soil deep to soft limestone
11/	Very poorly drained clay swamp (water hole) soil deep to soft limestone
18/	Well-drained silt loam disturbed upland soil formed in soft silt loam limestone material
19/	Well-drained silt loam upland soil 65 cm to moderately hard limestone

Slope	designations
A	0–3% slopes
B	3–8% slopes
C	8–15% slopes
D	15–25% slopes
E	25–35% slopes
F	35–45+% slopes

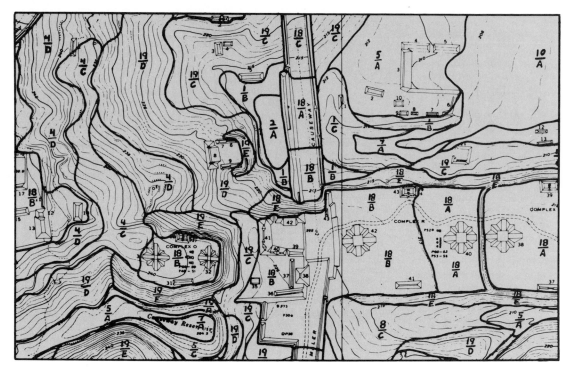

Fig. 2. Small sample of Tikal soil map (see table). In the Tikal ruins, 11 soil map units were found, including low-land Vertisols (marked 10/A) and steep-slope upland Mollisols (4/C, 4/D). *(From G. W. Olson. The bitter harvest of the ancient Maya, Garden Mag., 3(1):26–29, January/February 1979)*

abuse by the Maya. Along the Rio Hondo, in northern Belize, remains of raised fields have been discovered in backwater areas. The raised fields were constructed by the Maya to make an intensive agricultural system utilizing canals. Similar canal systems were used by the Maya at Edzna in Campeche, Mexico, and in the present *chinampas* (floating gardens) constructed in shallow lakes near modern Mexico City. In Belize recent alluvial clay soils without any Maya artifacts cover the artifact-containing soils of the raised fields. The floodings and sedimentations of the raised fields were apparently associated with ancient deforestation of the uplands.

Ancient land abuse in Honduras. Soils in Valle de Naco, Honduras, show similar evidence of ancient land abuse. In stream banks, buried soils were discovered underneath modern alluvial sediments more than a meter thick; the buried soil contained artifacts such as pottery sherds indicating that it had been farmed more than 1000 years ago. Analysis of the buried soil showed accumulations of high levels of phosphorus, a finding common where there has been human activity and human waste. The recent alluvial sediments that covered the buried soil had no such chemical artifacts, another indication that at some point (just before or during severe flooding) farming suddenly ceased in the region.

El Salvador volcanic soils. In El Salvador's Zapotitan Basin, excessive erosion resulting from Maya land abuse seriously depleted volcanic ash deposits of their nutrients. The relatively recent ash deposits, about 1000 and 2000 years old, buried the better older soils which the Maya had originally farmed. Excavations of several meters of volcanic ash layers have recently disclosed even the plant molds from the standing Maya maize crop destroyed by a catastrophic volcanic ash fall. Maya artifacts in the soils formed in the various volcanic ash layers indicate that each volcanic eruption forced migrations from the area, but that after a time there was some resettlement on each successive ash layer. In spite of the persistence of the Maya in reoccupation, the soils were degraded to successively lower levels of agricultural productivity.

Outside the Maya area the soil and archeological correlations hold equally true, but each environment has its own soil relationships to each ancient culture. In New Guinea, forest clearance about 2500 years ago increased soil erosion of the surrounding slopes and caused increased swampiness in ancient farmed areas; the environmental degradation was reflected in pollen analysis and in changing drainage systems where ditches were dug shallower and spaced more closely together over time.

Middle East Crescent of Crisis. Environmental degradation and consequent human suffering are probably best documented in the soils of arid regions. Throughout the Middle East and the present Crescent of Crisis, ancient forests in mountains have been cut, steep areas have been overgrazed, and erosion has depleted the uplands and buried the lowlands under meters of sediments. In Iran alone, more than 250,000 archeological sites mark places where ecological shifts have had devastating effects upon the communities that once existed in those areas.

Future tragedy of the commons. The archeological evidence of the tragedy of the commons in the soils gives cause for future planners to ponder. Wherever people have abused the soils and the environment, their culture and civilization have suffered over the long run. Severe consequences result when each individual of large populations can exploit the ecological system without limits, as is the case where forests are destroyed or areas are overgrazed. Whatever is in store for humans in the future, the soils and the associated archeological record bear witness that better management of natural resources is needed than has been practiced in the past.

For background information *see* ARCHEOLOGY; SOIL; SOIL, SUBORDERS OF; SOIL CONSERVATION in the McGraw-Hill Encyclopedia of Science and Technology.   [GERALD W. OLSON]

Bibliography: E. P. Eckholm, *Losing Ground: Environmental Stress and World Food Prospects*, 1976; G. Hardin, *Science*, 162:1234–48, 1968; P. D. Harrison and B. L. Turner II, *Pre Hispanic Maya Agriculture*, 1978; G. W. Olson, *Garden Mag.*, 3(1): 26–29, January/February 1979; G. W. Olson, *The Soil Survey of Tikal*, Cornell (Univ.) Agron. Mimeo 77–13, 1977; G. W. Olson, *The Strategic Situation in Iran according to the Soils*, Cornell (Univ.) Agron. Mimeo 80–1, 1980; J. P. White and J. Allen, *Science*, 207:728–734, 1980.

Astronomy

Within the past few years, a remarkable stellar object known as SS 433 has become the focus of attention of a large number of observational and theoretical astronomers. This object has properties which may truly be termed unique, even among the myriad of unusual stars cataloged in the Milky Way Galaxy. SS 433 shows evidence for ejection of two narrow streams of cool gas, traveling in oppositely directed beams from a central object, and moving at a velocity of almost one-quarter the speed of light. Furthermore, these beams execute a repeating, rotating pattern about the central object once every 164 days.

Discovery. SS 433 is an example of an astronomical object discovered, forgotten, and rediscovered several times over a period of 2 decades. The initial observation of the object was made in 1959 by C. B. Stephenson and N. Sanduleak, during a survey of the Milky Way for peculiar stars. The Stephenson-Sanduleak (SS) published survey lists this star as the 433d entry, thus the nomenclature SS 433. The SS survey searched for objects with a specific anomaly in their spectrum. Ordinarily, when starlight is dispersed via a prism or grating into its component colors, the resulting spectrum is relatively smooth, with comparable amounts of light reaching Earth at every wavelength (color). However, occasionally stars are found which have intense peaks in their spectra at very specific wavelengths. When seen on a spectrum recorded on a photographic plate, this phenomenon is called an emission line, because the convergence of light at this one particular wavelength causes a linear thickening of the developed emulsion at a specific location. Emission lines in stars are caused by light emitted from atoms in the star which have become excited due to collisions with other atoms or with electromagnetic radiation (light). The relaxation from this excitation is normally accompanied by emission of light at certain specific, unchanging wavelengths characteristic of the chemical element involved. Thus the detection of emission lines is a valuable insight into the nature of the star, since they provide a "fingerprint" enabling identification of chemical elements present in the object. Perhaps 5–10% of all stars show emission lines in their spectra, so this characteristic of SS 433 was not in itself sufficiently unusual to provoke more detailed observations of the object.

During the 1960s and 1970s a variety of astronomers mapping the skies at x-ray and radio wavelengths unknowingly rediscovered SS 433, as they found that intense x-ray and radio emissions were emanating from this region of sky. However, the precision of these observations was insufficient to permit these workers to associate the source of the emissions specifically with SS 433, as opposed to numerous other nearby stars (Fig. 1). Finally in 1978 three independent groups of English and Canadian astronomers recognized that the visible object SS 433 and the previously cataloged sources of radio and x-ray emission were in fact all the same object. This is extraordinary, as only the tiniest fraction of all stars emit detectable amounts of either radio or x radiation. The first modern spectroscopy of SS 433, obtained by D. Clark and P. Murdin of Australia, confirmed the results of the SS catalog by showing the object to have extraordinarily intense emission lines of hydrogen and helium. These workers also pointed out that SS 433 is surrounded on the sky by a large, diffuse glow of radio emission, itself a previously cataloged object,

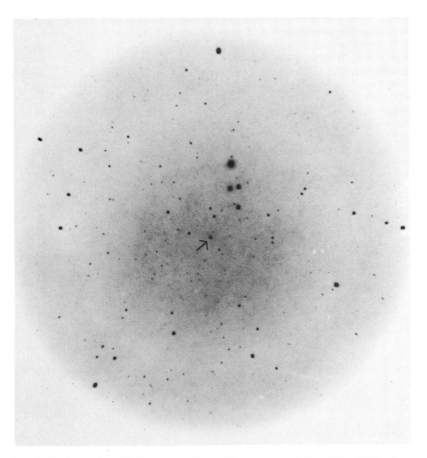

Fig. 1. Photograph of SS 433 made at Lick Observatory with the 36-in. (0.91-m) refractor telescope. The obvious lack of any features which distinguish the object from its numerous neighbors in this dense portion of the Milky Way helps explain why the star was overlooked for so long.

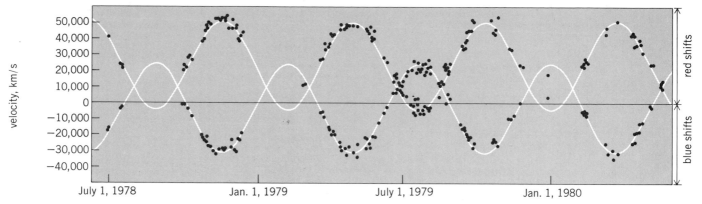

Fig. 2. Graph of the value of red shifts and blue shifts of SS 433 over a period of about 2 years. Large gaps in the data are caused when the object is too close to the Sun for nighttime observation; smaller gaps are due to the proximity of the Moon, making observations difficult. Curves show the predicted behavior if the Doppler-shifted gas is located in two narrow, oppositely directed, rotating beams.

termed W50. The structure of W50 has led most astronomers to conclude that it is the remnant of an ancient exploded star, or supernova, in this case probably occurring more than 100,000 years ago. The central location of SS 433 within W50 leads to the speculation, appealing but unproved, that the two objects are in fact associated.

Jets of matter. The most peculiar characteristics of SS 433 have been revealed by an intensive series of spectroscopic observations made from Lick Observatory by a group of University of California astronomers led by B. Margon. These observations show that the spectrum possesses not only a set of emission lines due to hydrogen and helium, but two further sets of lines, one displaced to longer (redder) wavelengths from the familiar lines, and the second displaced to shorter (bluer) wavelengths. These displacements can be understood in terms of the Doppler effect, a familiar mechanism which lengthens the apparent wavelength of any wave phenomenon (including light or sound) when there is a recessional motion between the source and the observer, and shortens the wavelength if there is approach. Thus the observations imply that in addition to a stationary object, SS 433 possesses some gas (a mixture of hydrogen and helium) approaching the Earth, while some presumably different patch of gas recedes. The remarkable property is the velocity of approach and recession, calculated simply from the magnitude of the observed spectral shifts. The velocity of this gas in the initial observations was found to be up to 50,000 km/s, that is, about 16% of the speed of light. Because the escape velocity from the Milky Way Galaxy is only a few hundred kilometers per second, one never observes stellar objects with velocities in excess of this, since they would rapidly leave the Galaxy.

Further spectral monitoring of SS 433 has revealed spectacular changes in these two sets of Doppler-shifted lines. The wavelengths of the lines change every night in a smoothly progressing pattern, indicating that the velocity of the emitting regions is also changing. Each set of lines proves to cycle in a regular pattern between a recessional velocity of 50,000 km/s and an approach velocity of 30,000 km/s, with the cycle lasting precisely 164 days (Fig. 2). The pattern then begins again. The currently accepted interpretation of this periodic behavior, provided originally by M. Milgrom and then elaborated upon by G. O. Abell and Margon, is that the "moving" emission lines are due to light from two narrow streams or jets of matter ejected from a central object in opposite directions. A slow rotation of the axis of these jets, once every 164 days, is then responsible for the changing velocities observed at Earth. Different velocities are seen on different days because the moving axis of the jets may be more or less directly pointed at Earth at a given time. Interpretation of the observations shown in Fig. 2 using this concept shows that the true velocity of the ejected beam is about 80,000 km/s, about one-quarter of the velocity of light. A velocity this high is never directly observed, since the Earth would have to be fortuitously located exactly in the equatorial plane of the rotation pattern to have the jets point exactly toward and away from it. The tremendous beam velocity inferred implies a huge energy source to accelerate a substantial amount of gas to this speed: the kinetic energy in the beams is approximately 1,000,000 times larger than the total amount of light energy radiated by the Sun every second.

Time dilation. Especially intriguing is the observation (Fig. 2) that on a given night the average velocity of the approaching and receding beams is not zero, but rather a large positive value, about 12,000 km/s, despite the fact that SS 433 is approximately stationary with respect to Earth. This proves to be a direct consequence of Einstein's theory of special relativity. An outside observer perceives a change in measured times and lengths of a system moving at very large velocity. The beam velocity in SS 433 is large enough that special relativity is important, and this "time dilation" effect causes a permanent red shift of the spectral lines of 12,000 km/s; the 164-day rotational component is then superposed on top of this underlying effect.

Star type and matter source. There has been much speculation as to the type of star present in SS 433, with many astronomers now agreeing that the enormous velocities in the beams require a highly collapsed, compact star with a strong gravitational field. Either a neutron star, the same end point of stellar evolution responsible for pulsars, or, possibly, a black hole could satisfy this requirement. The source of matter ejected through the beams is also a problem; it seems probable that this may be supplied by a relatively normal, nearby companion star trapped in an orbit about SS 433. A recent discovery by D. Crampton and colleagues shows that the central set of emission lines in SS 433 cycle through a very-small-amplitude period every 13 days, implying that the orbital motion of this unseen companion may have been detected.

Perhaps the most vexing question of all is why there is only one object like SS 433 known in a galaxy of 10^{11} stars. Only the accidental discovery of a second such star can determine whether this is a possible end point of stellar evolution for a certain class of binary star, or whether SS 433 is truly a unique and obscure stellar accident.

For background information see ASTRONOMICAL SPECTROSCOPY; RADIO ASTRONOMY; RELATIVITY; X-RAY ASTRONOMY in the McGraw-Hill Encyclopedia of Science and Technology.   [BRUCE MARGON]

Bibliography: G. O. Abell and B. Margon, Nature, 279:701–703, 1979; B. Margon et al., Astrophys. J. Lett., 233:L63–L68, 1979; D. Overbye, Sky Telesc., 58:510–516, 1979.

Atmospheric chemistry

Air pollution is generally thought of as anthropogenic changes (not necessarily increases) in the atmospheric gas or particle content. Changes can occur in the concentration level of gases and particles already present in the atmosphere or through the introduction of new gaseous species or particles with different chemical composition and sizes.

Most species' sources and final sinks (a body or process that acts as a disposal mechanism) are the continents and the oceans. The atmosphere is a temporary holding reservoir and transport medium in which chemical or physical transformations occur. These transformations are often referred to as cycles, and include: the water cycle (hydrological cycle), the particle cycle, and the trace gas cycles. The composition of the atmosphere near sea level is reported for three categories: permanent gases, variable gases, and particles (see table).

Trace gases. Atmospheric trace gases form three main cycles: the carbon cycle, the nitrogen cycle, and the sulfur cycle. Sources of these trace gases are both natural and anthropogenic. The natural sources include bacteriological decay (methane, ammonia, nitrogen dioxide, nitrous oxide, hydrogen sulfide, dimethyl sulfide, and so on), lightning (oxides of nitrogen), volcanoes (sulfur dioxide, carbonyl sulfide, and so on), vegetation (nonmethane hydrocarbons, that is, terpenes, isoprenes, and so on), and forest fires (hydrocarbons, carbon monoxide, and so on). Anthropogenic sources include emissions from all types of fossil-fuel burning (coal, oil, wood, gas), yielding oxides of sulfur, oxides of nitrogen, hydrocarbons, carbon monoxide, and so forth, and from processing materials. The overall mass of the variable trace gases in the atmosphere is minute compared with the permanent gases, but their impact on humans and the ecosystem is very profound.

At elevated concentrations, trace gases can induce health effects and cause ecological or material damage. Consequently, their concentrations in the air are monitored, and their anthropogenic emissions into the atmosphere are restricted. If controlled in such a way, the respective trace gas

The composition of the atmosphere near sea level

Constituent	Chemical formula	Parts per million by volume
Permanent gases		
Nitrogen	N_2	780,900
Oxygen	O_2	209,400
Argon	Ar	9,300
Neon	Ne	18.2
Helium	He	5.2
Krypton	Kr	1.1
Hydrogen	H_2	0.5
Nitrous oxide	N_2O	0.3
Xenon	Xe	0.09
Variable gases		
Water vapor	H_2O	
Carbon cycle { Carbon dioxide	CO_2	
Methane	CH_4	~1.5
All hydrocarbons other than methane		~0.05–0.5
Carbon monoxide	CO	~0.1
Ozone	O_3	~0.02–0.5
Nitrogen cycle { Oxides of nitrogen	NO/NO_2	0.001–0.5
Ammonia	NH_3	0.001–0.01
Sulfur cycle { Sulfur dioxide	SO_2	0.001–0.5
Hydrogen sulfide (also carbonyl sulfide, dimethyl sulfide, etc.)	H_2S	0.001–0.01
Particles		
All sizes ranging $0.002–100 \times 10^{-6}$ m diameter; $1–500 \times 10^{-6}$ g/m³		

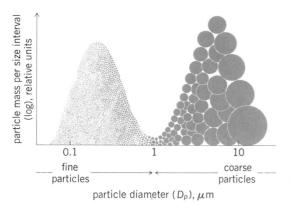

Fig. 1. Mass distribution of the atmospheric aerosol.

is called a criteria pollutant. Currently, sulfur dioxide, oxides of nitrogen, nonmethane hydrocarbons, carbon monoxide, and ozone fall into that category.

While in the atmosphere, the trace gases continuously interact—physically and chemically—with each other, with sunlight, and with water vapor, cloud droplets, and rain. This atmospheric chemistry process yields hundreds of intermediate products that are grouped into so-called secondary reaction products.

Particles. These secondary reaction products can remain in molecular form, or they can undergo a phase transition which is called—in most general terms—gas-to-particle conversion. The products, that is, particles, then become part of the atmospheric aerosol system which defines the size range and relative number concentrations of all at-

mospheric particulates. The fine particles produced by the gas-to-particle phase transition make up the fine-particle mode of the aerosol system (Fig. 1). Fine particles are almost entirely formed by phase-transition processes, whereas coarse particles are emitted into the atmosphere as dust or sea salt from ocean spray and breaking bubbles. These fine particles can penetrate deep into the alveoli of the human lungs and are therefore a potentially serious health hazard. They also can be transported over very great distances, since their gravitational settling is almost negligible—a phenomenon called long-range transport. The fine particles also interact very efficiently with sunlight, causing visibility degradation. The Smoky Mountains are a classic example of natural visibility degradation, while the Los Angeles smog and the haze over many urban areas are typical illustrations of gas-to-particle conversion due to increased anthropogenic emissions of reactive gases. A simplified diagram for gas-to-particle conversion is shown in Fig. 2.

Phase-transition processes. Gas-to-particle conversion processes include: (1) homogeneous, homomolecular nucleation, that is, the formation of a new, stable liquid or solid ultrafine particle from the gas involving one gaseous species only; (2) homogeneous, heteromolecular nucleation, that is, the formation of a new particle involving two or more gaseous species, typically one of these being water; and (3) heterogeneous heteromolecular condensation, that is, the growth of preexisting particles due to deposition of molecules from the gas phase.

There are chemical or physical forces that gov-

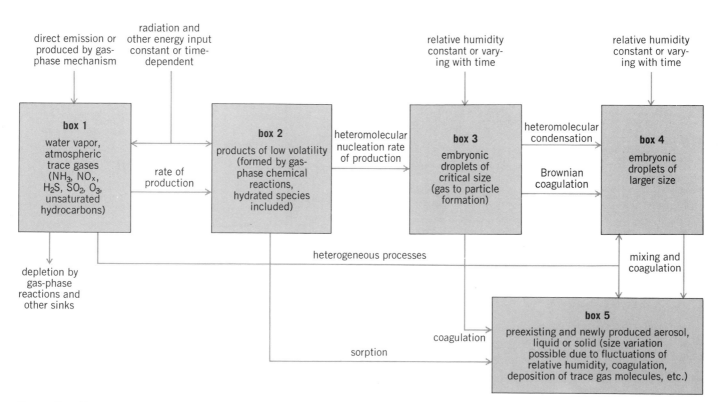

Fig. 2. Box diagram for gas-to-particle conversion. The boxes contain the substance, and the arrows describe the process.

ern these phase-transition processes. Chemical or photochemical reactions involving atmospheric trace gases [box 1 in Fig. 1] produce reaction products of low volatility (possessing a low vapor pressure) [box 2]. As more of these secondary low-volatility products are formed, the atmosphere becomes supersaturated with respect to these molecules. The degree of supersaturation will determine the degree of nucleation (formation of new particles) [box 3] and condensation (deposition on preexisting particles) [boxes 4 and 5]. From a thermodynamic point of view, more energy is required to form a new particle (overcoming the energy barrier for nucleation characterized by the Gibbs free energy) than to enlarge the surface of an already existing particle.

The particles formed by either mechanism in the atmosphere are mostly in the fine-particle size range (Fig. 1). It is quite helpful to compare this gas-to-particle conversion process of trace gases with the well-known water nucleation/condensation process leading to clouds: as the relative humidity (water vapor) in the atmosphere increases and eventually exceeds 100% (becomes supersaturated), the water vapor condenses out on preexisting particles called cloud condensation nuclei. However, if there are no preexisting particles to condense on, or if the supersaturation increases suddenly and rapidly, water vapor will nucleate homogeneously. All clouds in the lower troposphere are formed around cloud condensation nuclei, but most clouds at very high altitudes (that is, cirrus) are formed via homogeneous nucleation. The vapor pressure of water at room temperature is around 20 torr (2.7 kPa); however, the vapor pressure of some of the secondary reaction products is 100 to 100,000 times lower. Therefore, their ability to nucleate and condense at even very low concentrations is dramatically increased.

Transformations. The photochemical and chemical reactions which initially transform high-volatility gases [box 1 Fig. 2; all variable gases in the table] into secondary reaction products, very few of which possess a low enough vapor pressure to participate in and undergo a phase transition, are extremely complex and not yet fully understood. In principle, however, a series of transformations are thought to occur: Sulfur dioxide reacts with hydroxyl radicals eventually to form sulfuric acid molecules [the vapor pressure of sulfuric acid, 2×10^{-5} torr (2.7×10^{-3} Pa), is almost a million times lower than that of water]. Nonmethane hydrocarbons react with ozone or hydroxyl radicals to form aldehydes, alcohols, carboxylic acids and dicarboxylic acids. Most secondary reaction products of nonmethane hydrocarbons react with oxides of nitrogen to form organic nitrites and nitrates.

The volatility sequence for C_n compounds ($n=$ number of carbon atoms in the original carbon molecule) decreases in the following order: alkane > aldehyde > alcohol ≅ nitrate ester > carboxylic acid ≅ dialdehyde > diol ≅ dinitrate ≅ acid aldehyde >> dicarboxylic acid. Vapor pressures as low as 10^{-10} torr (1.33×10^{-11} kPa) have been measured, although they need to be only less than 1 torr (1.33×10^{-1} kPa) or so to be important for gas-to-particle conversion.

Ozone (O_3) and the hydroxyl radicals (HO and HO_2) play a very important role in atmospheric chemistry. They are all directly or indirectly a product of photon absorption, that is, photochemistry. Hence, gas-to-particle conversion processes in the atmosphere typically exhibit a strong diurnal pattern.

It has been estimated that over 100,000,000 tons of fine particles are formed every year. Their lifetime in the atmosphere of 1–3 weeks is governed mainly by the water cycle because they are captured by cloud or precipitation elements and eventually returned to the ground.

For background information see ATMOSPHERIC CHEMISTRY; ATMOSPHERIC POLLUTION; PHOTOCHEMISTRY in the McGraw-Hill Encyclopedia of Science and Technology.   [VOLKER A. MOHNEN]

Automobile

In the last decade, the extremely rapid growth of the electronics industry has made the integrated circuit a ubiquitous part of life. The microprocessor is fast becoming an essential part of many appliances, as well as making computers a commonplace tool. While the quantity of microprocessors in automobiles is relatively minor in terms of the total number of microprocessors being utilized, their use has allowed significant advances in automotive engine control.

Engine control systems. In order to understand the use of microprocessors in automobiles, the engine will first be described in terms of its control systems. This discussion will be restricted to four-stroke spark-ignited gasoline engines, since they represent most of the applications to date. In the gasoline-engine vehicle, the major control variables affecting engine speed and torque are air flow, fuel flow, spark timing, and exhaust-gas recirculation.

Fuel and air flow. The regulation of fuel and air flow is the primary control available for the automotive engine, since this represents the control of the input energy to the engine. In a conventionally carbureted vehicle, the carburetor controls both of these flows. Air flow is adjusted by changing the position of the throttle plate in response to the driver's demand. Fuel is metered into the air stream by one or more venturi tubes whose flow characteristics determine the relationship between air flow and fuel flow. The relationship represents a compromise between good fuel economy, smooth operation, performance, and absence of engine knock. For most of the range of engine operation, a ratio between air flow and fuel flow at or near stoichiometric balance is desirable. This is the ratio at which there are exactly enough oxygen atoms in the air stream to combine with all the carbon and hydrogen atoms in the fuel flow. The ratio will vary somewhat with fuel composition, but for most gasoline blends it is about 14.7:1 by weight. In most engines it is desirable to operate at a somewhat richer ratio at large throttle settings to obtain maximum power output and to avoid engine knock.

Spark timing. Fundamentally an internal combustion engine is a heat engine, subject to the laws of thermodynamics governing the conversion of heat to work. This implies that only a certain fraction of the input energy is converted to work, and the remainder of the energy is rejected as heat. This division of energy into work and rejected heat is most directly affected by spark timing, that is, the time in the engine cycle at which the combustion of the air-fuel mixture is initiated. To achieve the maximum amount of work from the combustion, the spark must be timed so that the maximum pressure rise occurs slightly after the piston has reached the top of its travel, top dead center (TDC). In this manner, the pressure of the expanding gases will produce the greatest downward force on the piston. To achieve this, however, the spark must be fired well before the piston has reached TDC, in a manner which depends on the particular conditions of engine operation. In engine operation it is much more convenient to measure intervals in units of engine crank angle than in units of time; however, spark timing for optimum engine operation is more nearly constant in time than in engine crank angle. This implies that as engine speed increases, the spark must be fired at ever-increasing values of crank angle before TDC. Also, as the engine torque increases, there is a decrease in the spark interval required. A conventional distributor satisfies these requirements by means of a centrifugal governor to give greater spark timing intervals at higher speed, and a vacuum diaphragm to increase the interval at lighter load.

Exhaust-gas recirculation. Exhaust-gas recirculation (EGR) is the process of returning a fraction of the exhaust gases back into the cylinder along with the combustible mixture. As a diluent, the gas has the primary effect of reducing flame temperature during the combustion process. Although originally used as a means of control of engine knock, EGR was reintroduced in the 1970s as an emission-control technique, specifically for control of oxides of nitrogen. Normally, without EGR the engine has a small fraction of burned gases in the cylinder when the combustible mixture is introduced. This is because the piston does not sweep out the entire volume of the cylinder on the exhaust stroke, and because there is overlap in the action of the intake and exhaust valves. The fraction of burned gases left in the cylinder is greatest at lowest speed and power, for example, at idle. Therefore the controllers of exhaust-gas recirculation usually shut off such flow at idle. At high power conditions, however, exhaust-gas recirculation is also inhibited. This is because exhaust gas would displace the air-fuel mixture and limit the maximum power of the engine. It is not immediately obvious, but if the maximum power of an engine is to be reduced, it is much more efficient to do so by building a smaller engine (less displacement) than by limiting the power with exhaust-gas recirculation. Therefore, while the exhaust-gas recirculation must be off at idle and at wide-open throttle, it can be used at intermediate load conditions to satisfy emission requirements. This is usually done with control valves actuated by vacuum signals.

Engine calibration. The scheduling of these variables (spark timing, exhaust-gas recirculation, and air-fuel ratio) is called the engine calibration. Before the advent of emission controls, the principal criteria used in developing an engine calibra-

tion were smooth operation, fuel economy, performance, and absence of knock. With the enactment of the Clean Air Act of 1970 (and amendments), a limitation on the amount of hydrocarbons, carbon monoxide, and oxides of nitrogen also became part of the criteria. To some degree these criteria are incompatible. The simplest control of unburned hydrocarbons is to delay the spark firing so that more of the available energy is rejected as heat. This heat ensures the combustion of unburned hydrocarbons in the exhaust system. However, the vehicle fuel economy suffers. Similarly, the introduction of exhaust gas into the cylinder limits the temperature rise in the cylinder during combustion, thereby reducing the formation of oxides of nitrogen. However, this also reduces the rate of flame propagation, and so leads to higher hydrocarbon emissions, and can interfere with smooth operation. The illustration shows some of the prob-

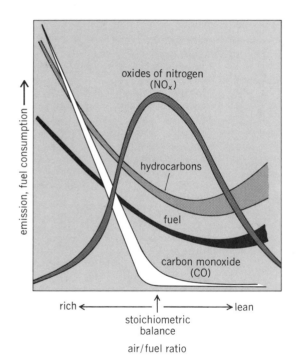

Dependence on air-fuel ratio of fuel consumption in, and emission of various atmospheric pollutants from, an automotive engine.

lems of choosing an appropriate air-fuel ratio. At an air-fuel ratio slightly lean of stoichiometry, the fuel consumption is minimized, carbon monoxide is greatly reduced, unburned hydrocarbons are minimized, but the production of oxides of nitrogen is at a maximum. The introduction of exhaust-gas recirculation controls the oxides of nitrogen formation, but can exacerbate the hydrocarbon fuel economy problem.

The introduction of the catalytic converter improved the situation in that the control of hydrocarbons and carbon monoxide can now be effected outside the engine. The catalyst makes it possible to avoid sacrificing fuel economy to achieve emission control of hydrocarbons. Nevertheless, engine calibrations are still much more complex than prior to the introduction of emission controls. Also, the increased importance of fuel economy has placed a premium on the precision of the control system, and on minimizing the variation of the manufactured product from the design specifications. The microprocessor-based digital control system answers these needs.

Microprocessor control. In practice, microprocessor-based systems have been used to control one or more of these important engine variables, and to perform other minor functions as well. Thus, for example, some vehicles use a computer solely to control spark timing, others to control fuel delivery, in conjunction with an electronically controllable carburetor or some form of fuel injection, and other systems control several of these variables simultaneously. These systems vary from vehicle to vehicle and from one model year to another. Only common features found in most systems will be described here.

Computerized spark timing. To handle the spark-timing function, the computer needs a timing signal which corresponds to engine crank angle position. This is usually done by some form of pulse transducer attached to the engine crankshaft or cam shaft. By differentiating this signal with respect to an internal clock, the computer obtains the engine speed, needed for the spark-advance calculation. An additional input is the engine intake-manifold absolute pressure, which is an approximate indication of engine load. A pressure transducer is usually fitted to the engine intake manifold, and if the transducer provides an analog voltage output, the computer uses an analog-to-digital converter to convert this value to a digital form. An alternative approach is to fit an analog-to-frequency converter to the transducer so that the unit supplies a variable-frequency pulse stream which the computer reads as the value of the pressure. Other signals which may be used to refine the timing calculation include the engine-coolant temperature, the temperature of the air entering the carburetor, the position of the throttle, and the barometric pressure. Using this information, the computer performs a calculation of the proper time to fire the spark plug, and when that time occurs, sends a signal to an electronic ignition module which actually generates the high-voltage spark. The calculation entails looking up appropriate values in a prestored table (the results of extensive engine testing to determine an optimum calibration) and performing interpolations on this table. The tables, as well as the computer programs, are normally contained in read-only memory.

Computerized exhaust-gas recirculation. The control of exhaust-gas recirculation uses many of the same input values for its calculations. Engine crank angle is not necessary for this calculation, but engine speed is usually required. To effect the motion of valves, the computer may control a stepper motor directly, or it may actuate solenoids which control the application of several pressure sources to a control valve. These pressure sources can include a vacuum derived from the intake manifold, a pressure source from an air pump, and the ambient air.

Computerized fuel delivery. To control fuel delivery, fuel-injection systems have been used with electronic control. In such systems solenoid-actuated injectors are used to meter fuel into the air in the intake manifold. The carburetor is replaced by a simple throttle body whose sole function is to control the flow of air into the engine. In many systems there is one fuel injector placed near each cylinder, but recently alternative systems with one or two centrally placed injectors have also been used. The injector receives its fuel supply from a pressure-regulated pump, and the control of fuel flow is done by controlling the length of time that the injector is opened in each cycle. Such systems are different from injectors in a diesel engine in that the diesel injects fuel directly into the cylinder.

Another approach to computer control of fuel delivery is the electronically controlled carburetor. In this system the flow characteristics of the carburetor are altered by controlling the pressure of the fuel delivered to the carburetor. While the gross behavior is that of an ordinary carburetor, the computer is able to make fine adjustments to effect more precision in the air-fuel ratio. Such systems are driven with stepper motors or air-pressure control valves.

The inputs necessary for determining fuel schedules are largely those required for spark timing, with one important addition. To determine the air-fuel ratio actually entering the engine, a zirconia sensor is used in the exhaust stream. Zirconia acts as a galvanic cell at high temperatures, and its voltage output is sensitive to the oxygen content of its environment. If the engine is running on the rich side of stoichiometry, there will be practically no excess oxygen in the exhaust stream. If, on the other hand, the engine is running somewhat lean, there will be a small amount of oxygen present, and the zirconia sensor will indicate this. Operating as a feedback control system, the computer reads the output of the zirconia sensor and alters fuel flow until the air-fuel ratio is at its desired value.

Microprocessor design. The first microprocessors used for automotive engine control have been custom designs rather than the regular commercially available units. Automotive microprocessors have been designed for their real-time numeric capability and for their ability to withstand rugged environments. The engine-control algorithms tend to require more precision in their numeric calculations than can be obtained from 8-bit processors, the most commonly available commercial units. Therefore, the first engine control systems were designed with 10- or 12-bit data words. Subsequent applications have used derivatives of commercial chips, but the environmental and reliability requirements still impose special requirements on these microprocessors.

For background information *see* INTEGRATED CIRCUITS; INTERNAL COMBUSTION ENGINE in the McGraw-Hill Encyclopedia of Science and Technology.

[NEAL L. LAURANCE]

Bibliography: Society of Automotive Engineers, *Automotive Electronics II*, SP-393, 1975, *Automotive Applications of Sensors*, SP-427, 1978, *Sensors for Automotive Systems*, SP-458, 1980; U.S. Department of Transportation, *Automobile Engine Control Symposium*, July 8–9, 1975.

Bacteria

Several species of aquatic bacteria which orient in the Earth's magnetic field and swim along magnetic field lines in a preferred direction (magnetotaxis) have been observed in marine and fresh-water sediments of the Northern Hemisphere. Their orientation is due to one or more intracytoplasmic chains of single-domain magnetite particles. These linearly arranged particles impart a net magnetic dipole moment to the bacterium, parallel to the axis of motility. Northern Hemisphere magnetotactic bacteria with unidirectional motility swim consistently in the direction of the magnetic field, that is, to the geomagnetic north. This implies that their magnetic dipole is systematically oriented with the north-seeking pole forward. Bacteria from aquatic environments in New Zealand and Australia orient in the Earth's magnetic field and, when separated from the substrate, swim along magnetic field lines to the south. This implies that their magnetic dipole is oriented with the south-seeking pole forward. Consequently both Northern and Southern Hemisphere magnetotactic bacteria observed to date migrate downward by swimming along the Earth's inclined magnetic field lines.

Magnetotactic bacteria were first discovered in sediments near Woods Hole, MA, by R. P. Blakemore. It was observed that most of these bacteria orient and swim along the magnetic lines toward the north. Reversal of the ambient magnetic field by Helmholtz coils causes the cells to make U-turns within 1 s and swim in the opposite direction. Killed cells also orient in uniform fields as low as 0.1 gauss (1 microtorr). In these and other respects, the cells behave like single magnetic dipoles.

Magnetite particles. A magnetotactic spirillum designated strain MS-1 has been isolated from a fresh-water swamp and grown in pure culture. These magnetotactic cells each have an intracytoplasmic chain of approximately 22 iron-rich, electron-opaque particles (see illustration). Each particle is roughly octahedral, 50 nm along each major axis. Mössbauer spectroscopy on ^{57}Fe in freeze-dried cells shows conclusively that the particles are primarily composed of magnetite, Fe_3O_4. Another iron-containing material with a room-temperature Mössbauer spectrum similar to that of the iron-storage protein ferritin is also observed. Thus magnetotaxis is associated with intracellular magnetite. Since the bacteria were grown in chemically defined media containing soluble (chelated) iron, the presence of intracellular magnetite implies a process of bacterial synthesis.

The magnetic properties of magnetite particles depend on their size and shape. For a particle of roughly cubic shape with side dimension d, there is a range of d over which the particle will be a single magnetic domain. The magnetic moment of a single domain for temperatures well below the Curie point is the saturation or maximum magnetic moment of the particle and is unchanging in time. Magnetite particles with dimensions d equal to or less than 40 nm are superparamagnetic; that is, thermal activation induces transitions of the magnetic moment between different easy magnetic axis directions with a consequent loss of magnetic "memory." Particles with dimensions d equal to or

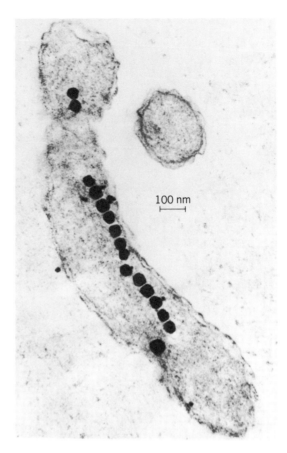

Electron micrograph of a section of a magnetotactic spirillum. The black dots are some of the magnetite particles that constitute the bacterium's biomagnetic compass. (*Courtesy of D. Balkwill, D. Maratea, and R. P. Blakemore*)

less than 80 nm are multidomain and consequently have macroscopic moments which are either nulled or less per unit volume than single-domain particles. With $d = 50$ nm, the magnetite particles in strain MS-1 are within the single-domain size range.

Because of strong interparticle interactions, the preferred orientation of the individual particles is such that their axes of magnetization are parallel, north to south along the chain direction. Thus the entire chain acts as a single magnetic dipole with a moment equal to the sum of the particle moments. For a cell containing the average chain length of 22 particles, the total moment is $M = 1.3 \times 10^{-12}$ erg/G $= 1.3 \times 10^{-15}$ joule/T. In the geomagnetic field of 0.5 G $= 5 \times 10^{-5}$ T, the total magnetic energy of a cell is $MH = 6.6 \times 10^{-13}$ erg $= 6.6 \times 10^{-20}$ J. This value is more than an order of magnitude greater than the thermal energy, $kT = 4.1 \times 10^{-14}$ erg $= 4.1 \times 10^{-21}$ J at 300 K, (where k is Boltzmann's constant and T is the absolute temperature). Thus each bacterium contains a sufficient but not an excessive amount of single-domain-sized magnetite in an appropriate configuration to be oriented in the Earth's magnetic field at ambient temperature. In other words, the cell's chain of magnetite crystals functions as a biomagnetic compass.

Mechanism of magnetotaxis. The simplest hypothesis for the mechanism of magnetotaxis is passive orientation of the bacterium resulting from the torque exerted by the ambient magnetic field on its biomagnetic compass as it swims. Since most magnetotactic bacteria from the Northern Hemisphere are observed to swim northward, the compass in these cells must have a fixed orientation with respect to the flagellum, with the north-seeking pole opposite to the flagellum. This orientation could be preserved in cell division if the compass is partitioned between the two daughter cells. Subsequently, during magnetite biosynthesis, the magnetic moments of nascent magnetite particles at the ends of the preexisting chains would become oriented along the chain direction by interaction with the chain dipole moment.

Downward directed motion. Due to the inclination of the Earth's magnetic field, magnetotactic bacteria which swim to the north in the Northern Hemisphere are directed downward at an angle increasing with latitude. It has been suggested that this downward-directed motion confers a biological advantage by guiding the bacteria, when dislodged, back to the sediments. On the basis of this hypothesis, magnetotactic bacteria of the Southern Hemisphere would be expected to swim south in order to reach the bottom. Recently several morphological types of magnetotactic bacteria have been observed in sediments of Australia and New Zealand. These bacteria indeed swim consistently to the south, and hence downward along the Earth's inclined magnetic field lines, as hypothesized. As revealed by electron microscopy, they contain internal chains of electron-opaque particles similar to those observed in magnetotactic bacteria from the Northern Hemisphere. Like that of their Northern Hemisphere counterparts, their magnetic polarity can be permanently reversed and they cannot be demagnetized.

The prevalence of south-seeking magnetotactic bacteria in Southern Hemisphere sediments and north-seeking magnetotactic bacteria in the Northern Hemisphere verifies the hypothesis that downward-directed motion is advantageous for and upward-directed motion detrimental to the survival of these magnetotactic bacteria with unidirectional motility. Magnetotaxis is a reliable means of keeping these microorganisms in or near the bottom sediments. Since particles in the micrometer size range with densities close to 1.0 tend to remain suspended in water, gravity is virtually inconsequential in determining the vertical distribution of the bacteria.

The Earth's magnetic field provides a global orientation cue to which various organisms, including homing pigeons, are known to respond. In the bacteria, nature has shown the biological feasibility of synthesizing a highly organized ferromagnetic structure equivalent to a magnetic compass needle. Similar structures may provide the basis for a magnetic sense in more complex organisms as well.

For background information *see* MAGNETIC MOMENT; MAGNETIC RECEPTION (BIOLOGY) in the McGraw-Hill Encyclopedia of Science and Technology.

[RICHARD B. FRANKEL]

Bibliography: R. P. Blakemore, *Science*, 190: 377–379, 1975; R. P. Blakemore, R. B. Frankel, and A. J. Kalmijn, *Nature*, 286:384–385, 1980; R. B. Frankel and R. P. Blakemore, *J. Magn. Magn. Mater.*, 15–18:1562–1564, 1980; R. B. Frankel, R. P. Blakemore, and R. S. Wolfe, *Science*, 203:1355–1356, 1979; A. J. Kalmijn and R. P. Blakemore, in K. Schmidt-Koenig and W. T. Keeton (eds.), *Animal Migration, Navigation and Homing*, pp. 344–345, 1978.

Bacteriochlorophyll

There are fundamental as well as practical reasons for studying the redox properties of bacteriochlorophyll (BChl), the major chromophore in most photosynthetic bacteria. From a fundamental viewpoint, values for the oxidation and reduction potentials can be related to theoretical calculations concerning the electronic structure of BChl. These values are important criteria for testing the validity of the theory. The energetics of photoinduced electron transfer involving BChl as an electron donor or acceptor can also be related to its redox properties. The difference between the oxidation potential of the primary electron donor (special pair BChl or $BChl_{SP}$) in the bacterial photoreaction center (RC) and the reduction potential of the intermediate electron acceptor (designated I) gives a measure of the efficiency of the photoinduced charge separation when compared to the energy of a single photon trapped in the RC. The overall efficiency, however, requires that energy-wasteful recombination between the initial charged pair be minimized. This may be accomplished in the reaction center by the rapid electron transfer through one or more intermediate electron acceptors. The rate of electron transfer between two species is dependent upon their relative reduction potentials. Hence these values are important to the mechanism of charge separation as well. From a practical viewpoint, insight into the mechanism of photosynthesis may suggest experimental approaches to the design of efficient solar-energy devices.

The effect of solvent on the one-electron reduction and oxidation potentials of BChl in solution has been examined and found to produce substantial (as much as 200 mV) variations in some cases. The relevance of these findings to photosynthesis is discussed below.

Determination of redox potentials. The reduction or oxidation potential of a molecule is a measure of its tendency to accept or give up one or more electrons. The standard IUPAC convention requires potentials to be expressed in terms of a reduction reaction, as in Eq. (1), where k_f and k_r

$$\text{Ox} + ne^- \underset{k_r}{\overset{k_f}{\rightleftharpoons}} \text{Red} \qquad (1)$$

are the rate constants for the forward and reverse reactions. When the activities of Ox and Red are equal to 1.0 M and the temperature is 25°C, the measured potential for the above reduction is termed the standard formal reduction potential ($E^0_{1/2}$). In most experimental situations the activities or concentrations of Ox and Red are not equal to 1.0 M, and the potential of the solution is described by the Nernst relation, Eq. (2), where R is the universal gas constant, T the temperature, n the num-

$$E_{1/2} = E^0_{1/2} + \frac{RT}{nF} \ln \frac{(\text{Ox})}{(\text{Red})} \qquad (2)$$

versal gas constant, T the temperature, n the number of electrons transferred, and F the faraday. This equation accurately describes the potential when both k_f and k_r are large (that is, the electrochemical reaction is reversible).

Redox potentials are determined by chemical potentiometry or electrochemical techniques. In chemical potentiometry the compound of interest is titrated with an oxidant or reductant of known potential. Electrochemical methods, such as cyclic voltammetry, do not usually require the addition of extraneous redox buffers. The current through the solution containing the unknown redox couple is measured as a function of applied potential. There are several advantages to electrochemical procedures over chemical potentiometry. These include the use of simple, defined solutions devoid of redox buffers, the short time needed for the measurement, the wide potential range accessible in both the anodic (positive) and cathodic (negative) limits, and the large amount of information available in addition to redox potentials.

Experimental aspects of voltammetry techniques. In both cyclic voltammetry and the recently developed technique of cyclic differential pulse voltammetry, solutions of the redox active material are prepared in dry, purified aprotic solvents at millimolar concentrations. An inert electrolyte, such as tetrabutylammonium perchlorate, is added at a concentration 100 times that of the electroactive species to minimize ion migration of the redox material and to reduce the high electrical resistances usually encountered in nonaqueous solvents.

A high-vacuum electrochemical cell used in both voltammetry techniques is shown in Fig. 1.

There are four electrodes in this cell, three of which are used for voltammetry measurements. The working electrode consists of a platinum bead sealed in soft glass. The auxiliary electrode is platinum mesh, and the reference electrode is a platinum wire. The potential of the reference electrode is calibrated by the pilot ion technique, and corrected to values relative to the standard calomel electrode. Ideally the reference electrode does not carry current, but serves only to sense the solution potential. Current through the solution is carried between the working and auxiliary electrodes.

The potentiostat in an electrochemical experiment serves to control the solution potential by means of a feedback circuit incorporating the reference electrode. It is also used to vary the solution potential at a selected rate, by means of a wave-form generator or through computer control. In cyclic voltammetry measurements a staircase potential ramp is applied to the solution, and current is measured as a function of potential. Figure 2A shows the cyclic voltammogram of BChl in CH_2Cl_2. An increase in current is observed at the potential where electron transfer occurs. Each electron transfer step results in a peak on the forward scan, together with a peak of opposite current magnitude on the reverse scan. For electrochemically reversible processes, the separation in the forward and reverse scan peaks is predicted from theory to equal $59/n$ mV, where n equals the number of electrons transferred at the particular potential. In a chemically reversible electron transfer, the height of the forward and reverse scan peaks must be equal, indicating no irre-

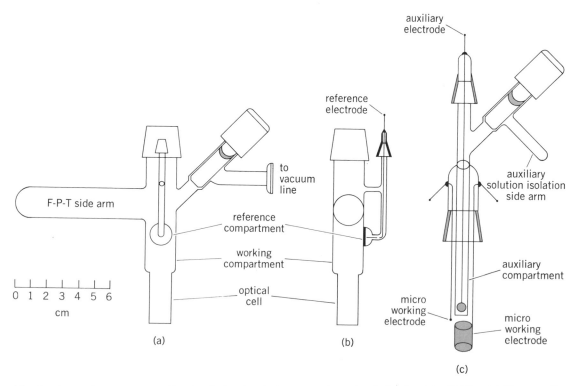

Fig. 1. Low-volume, vacuum spectroelectrochemical cell. (a) Cell bottom, front view and (b) side view. (c) Cell top, side view. (From R. P. Van Duyne et al., Resonance raman spectroelectrochemistry, 6: Ultraviolet laser excitation of the tetracyanoquinodimethane dianion, J. Amer. Chem. Soc., 101:2832–2837, 1979)

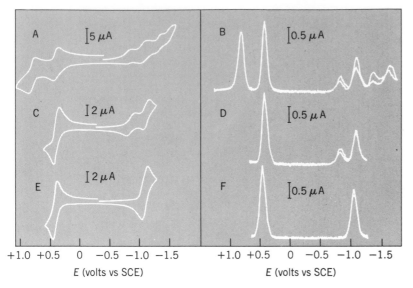

Fig. 2. Cyclic and cyclic differential pulse voltammograms of BChl (A–D) in CH$_2$Cl$_2$ and (E, F) in CH$_2$Cl$_2$ plus tetrahydrofuran (THF). Initial scan direction was cathodic, and starting potential was −0.3 V corrected to the standard calomel electrode. (*From T. M. Cotton and R. P. Van Duyne, An electrochemical investigation of the redox properties of bacteriochlorophyll and bacteriopheophytin in aprotic solvents. J. Amer. Chem. Soc.. 101:7605–7612. 1979*)

versible follow-up reactions. The potential can be determined from the relationship shown in Eq. (3), where E_p is the peak potential on the anodic

$$E_p = E_{1/2} - \frac{0.0285}{n} \qquad (3)$$

scan and n is the number of electrons transferred.

The potentiostat output in the cyclic differential pulse voltammetry experiment also consists of a staircase potential ramp. In addition, however, a short voltage pulse is applied at the beginning of each step. Current is measured immediately before the pulse (i_1) and near the end of the pulse (i_2). Data are displayed as the difference in current ($i_2 - i_1$) versus potential. Figure 2B shows the cyclic differential pulse voltammetry experiment equivalent to that in Fig. 2A. As may be seen, the data display has advantages over that in cyclic voltammetry. The peak potentials can readily be determined, and the electrochemical and chemical reversibility can easily be assessed.

Redox properties in aprotic solvents. BChl is a tetrahydroporphyrin derivative containing Mg as its central metal atom. Bacteriopheophytin (BPheo) is its metal-free analog. Both contain a fifth isocyclic ring, which is common to the class of chlorophylls. Various side chains are present on the periphery of the porphyrin macrocycle. Of particular importance to its coordination and aggregation properties are the central Mg atom, an acetyl group, and a ketone group. It has been established that the Mg atom requires at least one axial ligand, or a coordination number of five. In the presence of limited amounts of weak nucleophiles, this requirement is met by the nucleophile. In neat polar solvents, or in the presence of sufficient quantities of strong nucleophiles (for example, pyridine), two molecules of nucleophile may

coordinate to the Mg atom, resulting in the six-coordinate species. If nucleophiles are rigorously excluded from solutions of BChl in nonpolar solvents (for example, CH$_2$Cl$_2$), the coordination unsaturation of the Mg with coordination number 4 can be satisfied only by the acetyl or keto carbonyl functions on a second BChl molecule. Thus a BChl dimer results. This process of self-aggregation by Mg $\cdots$ O=C interactions can continue and leads ultimately to the formation of trimers and higher oligomers, the size of which depends upon the solvent and the BChl concentration.

Self-aggregation and redox properties. BChl can undergo two successive one-electron oxidation steps, leading to the formation of the monocation and the dication radicals. This process is represented by reaction (4). It should be noted that

$$BChl - e^- \rightleftharpoons BChl^+ - e^- \rightleftharpoons BChl^{++} \qquad (4)$$

according to IUPAC (International Union of Pure and Applied Chemistry) convention and the reported values for the potentials, these reactions should be written in the reverse direction.

Similarly BChl can undergo two successive one-electron reductions as shown in reaction (5). Both

$$BChl + e^- \rightleftharpoons BChl^- + e^- \rightleftharpoons BChl^{--} \qquad (5)$$

the cation and anion radicals of BChl have been found stable in aprotic solvents.

The cyclic voltammetry and cyclic differential pulse voltammetry behavior of BChl in CH$_2$Cl$_2$ was observed to be unusual. Four reduction peaks are present rather than the two previously reported. These are shown in Fig. 2A and B. The peaks are clearly resolved in the cyclic differential pulse voltammetry experiment, and the chemical reversibility of the electron transfer steps is readily evaluated. Both the first and second oxidation peaks are highly reversible. Several possible explanations were considered for the origin of four reduction peaks. These include adsorption of the BChl anion radical and dianion at the electrode surface (adsorption of the product of an electron transfer results in a prewave), the presence of a reversible chemical follow-up reaction for the two reduction steps (such as a protonation), and the presence of BChl aggregates with reduction properties different from those of monomeric BChl. The first two possibilities were discounted by scan-rate studies and experiments with added proton sources. The third explanation appears most probable at present. BChl is known to aggregate in CH$_2$Cl$_2$, and the addition of stoichiometric quantities of tetrahydrofuran (THF) causes disaggregation of the BChl and the formation of the five-coordinate monomer (BChl·THF). Under these conditions, the four reduction peaks collapse to two of enhanced current magnitude. No charges are observed in either the one- or two-electron oxidation potentials. A comparison of the data in Fig. 2C and D with those in E and F illustrates the effect of THF on the reduction potential. The first peak in the aggregate is at −0.83 V corrected to the standard calomel electrode, whereas that of monomeric BChl·THF is at −1.03 V.

Additional evidence favoring BChl aggregates in CH_2Cl_2 as the cause of the four reduction peaks may be had from the electronic transition properties of the reduced species. Exhaustive bulk electrolysis of the BChl solution in CH_2Cl_2 at a potential 90 mV more negative than the first reduction peak produces a species with red-shifted absorption bands in comparison to those of the monomeric BChl anion radical. Shifts of chlorophyll absorption bands to lower energy are typical of aggregates.

The cyclic voltammetry and cyclic differential pulse voltammetry behavior of BPheo in CH_2Cl_2 shows only two reduction peaks. The addition of THF to the solution causes no change in the reduction potential. This would be expected on the basis of the BChl aggregation hypothesis, since BPheo lacks Mg and cannot undergo the same type of self-aggregation as BChl. The one-electron reduction potential of BPheo in CH_2Cl_2 is −0.83 V corrected to the standard calomel electrode, which is equal to that of the first reduction peak of BChl aggregate.

Effect of six-coordination at Mg. BChl dissolved in neat THF contains two THF molecules as axial ligands at the central Mg atom. The increase in ligation number from five to six causes a 100-mV anodic shift in the one-electron oxidation potential. This indicates that six-coordinate BChl is harder to oxidize, or alternatively its cation radical is a more powerful oxidant. A similar observation was made concerning the one-electron oxidation potential of chlorophyll a. It was proposed that six-coordinate chlorophyll a may be present in the reaction center of photosystem II in green plants. The increased oxidation power of the cation radical would make it a more suitable oxidant for the water-splitting reaction.

Conclusions. The results reviewed here indicate that the redox properties of BChl are quite sensitive to solvent, apparently due to its ligation and aggregation behavior. The findings stress the difficulty of assigning a specific value to the one-electron reduction potential of BChl in the reaction center, where its molecular interactions are unknown. There are four BChl molecules in the reaction center, two of which function as the primary electron donor (BChl$_{sp}$) on photoexcitation. The remaining two, referred to as P800 in accordance with their lowest energy-absorption band (800 nm), may function as an extremely short-lived (less than 35 picoseconds) intermediate electron acceptor. This possibility has been suggested by Soviet researchers on the basis of kinetic analysis of optical transients near 800 nm. It is fairly certain that the longer-lived (250-ps) intermediate electron acceptor is one of the two BPheo molecules in the reaction center.

The redox active components of the reaction center are shown relative to an emf scale in Fig. 3. Light energy is used to drive electron transfer from BChl$_{sp}$ to the intermediate electron acceptor, shown as BPheo in the figure. The photoprocess is indicated by the solid arrow. In discussing the energetics of bacterial photosynthesis, a comparison is frequently made between the energy contained in a quantum of light absorbed by the lowest energy-absorption band of the special pair (870 nm or

1.43 eV) and the potential difference generated by transferring an electron from BChl$_{sp}$ to the electron acceptor. The one-electron oxidation potential of BChl$_{sp}$ is known accurately from chemical potentiometry in living cells. The reduction potentials of BPheo and BChl used in these calculations are those determined electrochemically in organic solvents. These values are shown in Fig. 3, together with the newly determined values for BChl aggregate and six-coordinate BChl. Use of the reduction potential of BPheo in the photochemical efficiency calculations results in a value of approximately 70% utilization of 870-nm light. If the reduction potential of monomeric BChl is used, the efficiency is greater than 90%, a value considered too high by researchers. On the other hand, if the reduction potential of aggregated BChl is used, the efficiency is once again near 70%. While this does not constitute evidence that BChl does indeed function as an electron acceptor in the primary photochemistry of photosynthesis, it does make

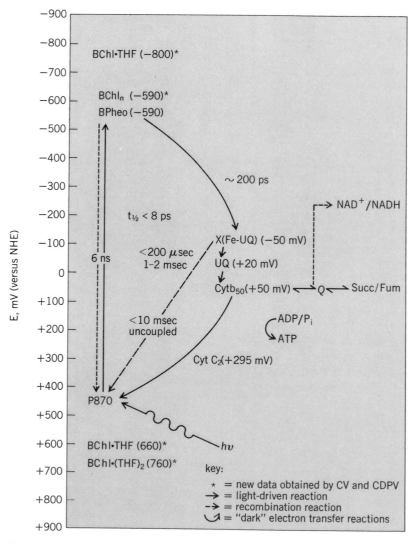

Fig. 3. Scheme for photosynthetic electron-flow pathway in *Rhodopseudomonas spheroides*. The ordinate gives the reduction potentials for the components versus the standard hydrogen electrode (standard calomel electrode +240 mV).

the possibility more plausible. More experimentation is needed to determine the role of BChl P800 in charge separation. The direct determination of the redox potentials of the BChl in the reaction center would be particularly helpful in this respect.

For background information *see* BACTERIAL PHOTOSYNTHESIS in the McGraw-Hill Encyclopedia of Science and Technology.

[THERESE M. COTTON]

Bibliography: T. M. Cotton and R. P. Van Duyne, *J. Amer. Chem. Soc.*, 101:7605–7612, 1979; M. S. Davis, A. Forman, and J. Fajer, *Proc. Nat. Acad. Sci. USA*, 76:4170–4174, 1979; J. J. Katz et al., in D. Dolphin (ed.), *The Porphyrins*, vol. 5, pp. 401–458, 1978; V. A. Shuvalov et al., *FEBS Lett.*, 91:135–139, 1978.

Bioelectromagnetics

The study of the interactions of electromagnetic energy (usually referring to frequencies below those of visible light; Fig. 1) with biological systems. This includes both experimental and theoretical approaches to describing and explaining biological effects. Diagnostic and therapeutic uses of electromagnetic fields are also included in bioelectromagnetics.

Background. The interaction of electromagnetic fields with living organisms has intrigued both physicians and engineers since 1892 when J. A. d'Arsonval, a French physician and physicist, applied an electromagnetic field to himself and found that it produced warmth without muscle contraction. Subsequently, the use of electromagnetic energy to heat tissue became a common therapy, and in 1908 Nagelschmidt introduced the term diathermy to describe this process. During the 1930s "short-wave" diathermy (27 MHz) was in common use by physicians. World War II spurred the development of high-power microwave sources for use in radar systems. Shortly thereafter, concern over the safety of radar was voiced, leading to investigation of the biological effects of microwave radiation. Detailed study of the therapeutic potential of diathermy at microwave frequencies began after World War II as high-power equipment became available for medical and other civil applications.

Rapid growth in the development of electronic systems for industrial, military, public service, and consumer use occurred in the 1970s. Much of this equipment is able to emit significant levels of electromagnetic radiation. The most extensive exposure to radio-frequency energy is from the 22,000,000 transmitters authorized by the Federal Communication Commission (including commercial broadcast stations and 15,000,000 citizens' band stations). The National Institute for Occupational Safety and Health estimates that 21,000,000 Americans are now occupationally exposed to radio-frequency sources mainly from the heating and drying of plastics, textiles, wood products, and other manufactured goods. The use of electromagnetic fields in medicine is increasing as radio-frequency-induced hyperthermia is applied to cancer therapy. To meet the demand for electric energy, the construction of 50- and 60-Hz and dc transmission lines capable of carrying current at a potential of 1 MV is being considered. These transmission lines will produce relatively large, extremely low-frequency electromagnetic fields.

Energy absorption. Electromagnetic energy is not absorbed uniformly across the geometric cross section of an organism. The total quantity of energy absorbed and the sites of maximum energy absorption depend on the frequency and polarization of the electromagnetic field, as well as on the electrical characteristics (the dielectric constant and conductivity—two properties of the tissue that control its interaction with electromagnetic radiation—which differ for different tissues vary with frequency as well), mass, and geometry of the absorbing object. In principle, the distribution of absorbed energy in an animal can be calculated from classical electromagnetic theory. However, the problem of energy distribution has not been solved for an object as complex as an animal. Simple calculations that assume the exposed system is of regular shape (say, spheroidal) and of homogeneous composition allow some generalizations to be made. For an average man maximal absorption (resonance) is predicted at approxi-

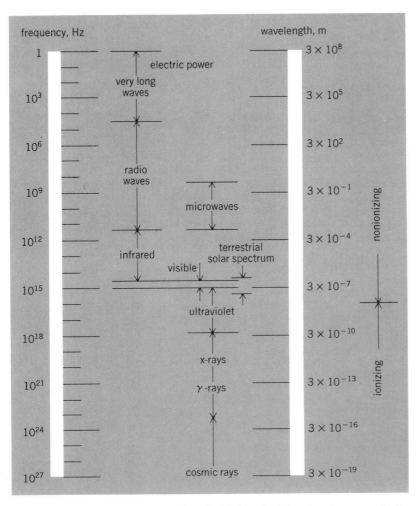

Fig. 1. Electromagnetic spectrum. Speed of light = 3×10^8 m/s = frequency (Hz) × wavelength (m).

mately 70 MHz. When a complex target, such as a human being, is considered, resonant frequencies are also found for anatomically distinct portions of the body, such as head or leg. The internal energy absorption pattern has been calculated for several simple models: semiinfinite slab, homogeneous sphere, multishell sphere, and multiblock model of a human. While none of these models actually simulates a real human being, local regions of higher-than-average levels of energy absorption have been calculated (Fig. 2). Development of instruments that do not interfere with the field has permitted measurement of partial body resonances and the pattern of energy deposition within some experimental systems.

Biological effects. The induction of cataracts is commonly associated with exposure of animals to intense microwave fields. Although heating the lens of the eye with electromagnetic energy can cause cataracts, the threshold for cataract production is very high. Many experimental animals could not survive exposure of their entire bodies at levels required to produce cataracts.

In 1961 it was reported that people can "hear" pulsed microwaves at very low averaged power densities (50 μW/cm^2). It is now generally accepted that the perceived sound is caused by elastic-stress waves that are created by rapid thermal expansion of the tissue that is absorbing microwaves. The temperature elevation occurs in 10 ms, so that the rate of heating is about 1°C/s. However, the temperature increase is only about 0.00005°C.

There are reports that microwave irradiation at very low intensities can affect behavior, the central nervous system, and the immune system, but many of these reports are controversial. Animals exposed to more intense electromagnetic fields that produce increases in body temperature of 1°C or higher (thermal load equal to one to two times the animal's basal metabolic rate) demonstrate modification of trained behaviors and exhibit changes in neuroendocrine levels. Exposure of small animals to weaker fields has been shown to produce some changes in the functioning of the central nervous system and the immune system. While the mechanism is not yet known, a thermal hypothesis cannot be ruled out. In addition, reports of the effect of very low-level, sinusoidally modulated electromagnetic fields on excitable cell and tissue systems have raised fundamental questions about basic understanding of how those systems function.

Exposure of pregnant rodents to intense electromagnetic fields can result in smaller offspring, specific anatomic abnormalities, and an increase in fetal resorption. However, only fields that produce significant heating of the gravid animal have been shown to be teratogenic.

Most biological effects of microwaves can be explained by the response of the animal to the conversion of electromagnetic energy into thermal energy within the animal. However, a few experiments yield results that are not readily explained by changes of temperature.

In an industrial society, the effects of stationary electric and magnetic fields, and of extremely low-frequency fields are important because of the ubiq-

uitous nature of electricity. When the body is in contact with two conductors at different potentials, current flows through it. Typical adult-human thresholds for 50- or 60-Hz currents are:

Reaction	Total body current
Sensation	1 mA
"Let go"	10 mA
Fibrillation	100 mA

The "no contact" case, such as that experienced by an individual or animal under a high-tension

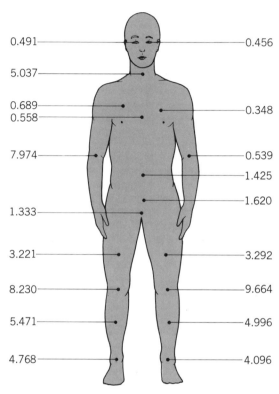

Fig. 2. Distribution of energy deposited in a human being exposed to electromagnetic radiation near the resonant frequency. The numbers provide an indication of the differences in energy absorbed by different parts of the body. *(From O. P. Gandhi. Dosimetry—the absorption properties of man and experimental animals. Ann. N.Y. Acad. Med.. 55(11):999–1020. December 1979)*

transmission line (the field strength directly under a 765-kV line is 4–12 kV/m) is only now being investigated under controlled experimental conditions.

Both strong and weak magnetic fields are being studied—the former to determine if they are hazardous to people working near cyclotrons, magnetohydrodynamic devices, or isotope-separation facilities; the latter to understand how magnetic fields interact with biological systems, and to appreciate the ecological import. Some bacteria swim northward in stationary magnetic fields as weak as 0.1 G (the Earth's magnetic field is about 0.5 G at its surface). These bacteria contain iron

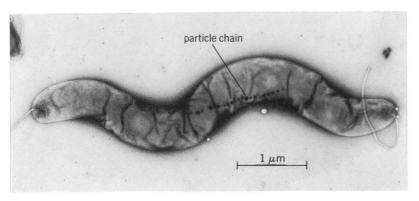

Fig. 3. Transmission electron micrograph of whole cell preparation of a bacterium which orients in a magnetic field. Within the cytoplasm are crystals containing iron which appear as a chain of electron-dense particles. (*From D. L. Balkwell, D. Maratso, and R. R. Blakemore, Ultrastructure of magnetotactic Spirillum, J. Bacteriol., 141:1399–1408, 1980*)

organized into crystals of magnetite (as illustrated in Fig. 3). Magnetite is also found in the brains and harderian glands of birds that can use local variations in the Earth's magnetic field for navigation and orientation.

Mechanisms. The photon energy associated with radio-frequency radiation is orders of magnitude below the ionization potential of biological molecules (see table). Covalent bond disruption, London–van der Waals interactions, and hydrogen-bond disruption, as well as disruption of bound water or reversible conformational changes in macromolecules, all require more energetic photons than that of a single microwave photon. The possibility that absorption of microwaves (or slightly shorter millimeter waves) may produce vibrational or torsional effects in biomacromolecules is being studied. Thus, if radio-frequency electromagnetic fields have a specific action on molecules that alter their biological function, it will be through a route more complicated and less understood than that associated with ionizing radiation.

The most common mechanism by which electromagnetic fields interact with biological systems is by inducing motion in polar molecules. Water and other polar molecules experience a torque when an electric field is applied. In order to minimize potential energy, the dipole attempts to align with the ever-changing electric field direction, resulting in oscillation. Both free and oriented (bound) water undergo dielectric relaxation in the radio-frequency region. The excitation of water, or other polar molecules, in the form of increased rotational energy is manifest as increased kinetic energy (elevation of temperature), but molecular structure is essentially unaltered if elevations are not excessive.

Alternating electromagnetic fields can cause an ordering of suspended particles or microorganisms. This effect, often called pearl-chain formation because the particles line up like a string of pearls, results from a dipole-dipole interaction. Nonspherical particles may also be caused to orient either parallel or perpendicular to an applied field. These effects have been observed only at a very high field strength.

Electromagnetic energy absorbed by biological material can be converted into elastic stress by thermal expansion. This phenomenon is caused by a rapid rise of temperature either deep within or at the surface of the material, and thus creates a time varying thermal expansion that generates elastic-stress waves in the tissue.

Medical applications. The therapeutic heating of tissue, diathermy, has been used by physicians for many years. "Short-wave" diathermy has been assigned the frequencies 13.56, 27.12, and 40.68 MHz, while microwave diathermy has been assigned 915, 2450, 5850 and 18,000 MHz. Short-wave diathermy provides deeper, more uniform heating than does diathermy at higher frequencies.

High-intensity radio-frequency fields have been used to produce hyperthermia in cancer patients. If radiation is focused into the tumor, then the difference between the temperature of the tumor and that of the surrounding tissue can be increased, at times producing tumoricidal temperature elevations of 43–45°C within the tumor while the surrounding tissue is below the critical temperature at

Activation energies of molecular effects in biological systems*

Effect	Activation energy		Radiation parameters	
	kcal/mole	eV	Frequency, GHz	Wavelength, μm
Thermal or Brownian motion (at 30°C)	0.60	0.026	6.3×10^3	47.6
Ionization	230	10	2.4×10^6	0.12
Covalent bond disruption	115	5	1.21×10^6	0.25
London–van der Waals interactions	23	1	2.4×10^5	1.25
Hydrogen bond disruption	1.8–4.6	0.08–0.2	$1.9 \times 10^4 - 4.8 \times 10^4$	15.8–6.25
Proton tunneling	16.1	0.7	1.71×10^5	1.76
Disruption of bound water	12.9	0.56	1.4×10^5	2.14
Rotation of polar protein molecules	0.92–9.2	0.04–0.4	$9.7 \times 10^3 - 9.7 \times 10^4$	30.9–3.1
Reversible conformational changes in protein molecules	9.2	0.4	9.7×10^4	3.1
Charge transfer interaction	138–69	6.3	$1.45 \times 10^6 - 7.25 \times 10^5$	0.2–0.4
Semiconduction	23–69	1–3	$2.4 \times 10^5 - 7.25 \times 10^5$	1.2–0.41
Microwave radiation	2.7×10^{-6}	1.2×10^{-7}		
	0.03	1.2×10^{-3}	0.03–300	$10^7 - 10^3$

*S. F. Cleary, Uncertainties in the evaluation of the biological effects of microwave and radiofrequency radiation, *Health Phys.*, 25:387–404, 1973.

which normal cells are killed. In addition, radio-frequency hyperthermia is often used in combination with x-ray therapy or with chemotherapy. In these cases, the tumor temperature is kept at 41–42°C to enhance the effectiveness of radiation or chemotherapy.

The development of bone tissue (osteogenesis) can be stimulated electrically either with implanted electrodes or by inductive coupling through the skin. The noninvasive technique uses pulsed magnetic fields to induce voltage gradients in the bone. This therapy has been used successfully to join fractures that have not healed by other means.

Electromagnetic fields were first used for medical diagnosis in 1926 when the electrical resistance across the chest cavity was used to diagnose pulmonary edema (water in the lungs). At frequencies below 100 kHz the movement of ions through extracellular spaces provides the major contribution to conductivity through the body. Thus, fluid-filled lungs can be detected by their lower resistance. Another diagnostic use of electromagnetic radiation is based on the fact that the spectrum of radiation emitted by any object depends on its temperature (blackbody radiation). Because a tumor is often at a higher temperature than the surrounding tissue, the spectrum of its emitted radiation is different. A technique of diagnosing tumors by the radiation they emit (radiometry) is now being evaluated.

Internally generated fields associated with nerve activity (EEG) and with muscle activity (ECG, MCG) are used to monitor normal body functions. There may be other uses of electric currents or fields in growth differentiation or development which have not yet been explored.

For background information *see* BIOPOTENTIALS AND ELECTROPHYSIOLOGY; ELECTROMAGNETIC RADIATION in the McGraw-Hill Encyclopedia of Science and Technology.   [ELLIOT POSTOW]

Bibliography: S. F. Cleary, Biological effects of microwave and radio frequency radiation, *CRC Crit. Rev. Environ. Control*, pp. 121–166, June 1977; O. P. Gandhi (ed.), Special issue on biology of radiofrequency radiation, *Proc. IEEE*, vol. 66, no. 1, 1980; A symposium on health aspects of non ionizing radiation, *Bull. N.Y. Acad. Med.*, vol. 44, no. 11, December 1977.

Biomechanics

All of the larger decapod crustaceans (such as crabs, lobsters, and crayfish) characteristically have their first pair of walking legs modified into pincerlike grasping appendages, or chelipeds. These unique appendages have long been of taxonomic interest and value, but only recently have investigators begun to look in detail at their mechanical properties and functional capabilities. Although much remains to be learned, a pattern of remarkable "variation on a common theme" has started to emerge.

General structure. The walking appendages of decapod crustaceans consist uniformly of seven rigid tubular segments joined serially by flexible articular membranes (Fig. 1a). The exoskeleton is composed of a protein/polysaccharide matrix, reinforced and made rigid in each leg segment by the deposition of crystalline calcium carbonate. With-

in the hollow legs, sets of muscles span each joint and insert into the calcified exoskeleton of adjacent segments. The musculature is usually arranged in simple antagonistic pairs, such that each leg segment can be extended or flexed with respect to its neighbors. Formation of a cheliped from this basic appendage plan primarily involves modification of the two terminal segments (Fig. 1b). The penultimate segment (termed the propus) becomes considerably enlarged to accommodate the large flexor (or closer) muscle, and reshaped to include a nonmovable paraxial projection comparable in size to the terminal segment. The movable terminal segment (the finger, or dactyl) is thus capable of occluding—potentially along its entire length—with the opposing surface of the propus. The exoskeleton at the occluding surfaces bears a variety of projections or "teeth," and is always heavily calcified. Articulation of the dactyl is confined to a single plane by peg-and-socket hinges on either side of the propus-dactyl joint.

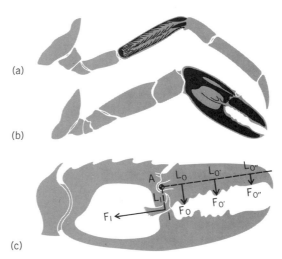

Fig. 1.   Cheliped structure and lever mechanics. (a) Jointed walking appendage. (b) Appendage with two terminal segments in the form of a cheliped. (c) Dactyl lever system.

Lever mechanics. The movable dactyl has been examined as a single lever system; Fig. 1c illustrates the lever action during cheliped closure. Input force (F_I) generated by contraction of the closer muscle is transmitted to the dactyl at the point of tendon insertion (I). This force produces a movement (clockwise) of the dactyl about an axis of rotation (A) passing through the peg-and-socket hinges. The input lever arm L_I is the perpendicular distance between A and I, and the total effective rotational force (or torque) applied to the dactyl lever is $F_I \times L_I$. The comparable output torque applied by the dactyl lever is the product of the force (F_O) exerted at the point of occlusion and the output lever arm (L_O). It is important to note that although the input lever arm is a characteristic constant value for each claw (since the positions of the hinge and tendon insertion remain fixed), the output lever arm can vary (since the point, or points, of force application can be anywhere along

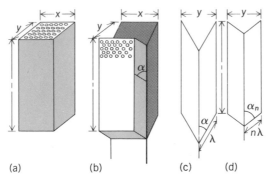

Fig. 2. Muscle geometry. *(a)* Parallel-fibered muscle. *(b)* Pinnately fibered muscle of identical overall dimensions. *(c)* Pinnately fibered muscle in relaxed condition. *(d)* Pinnately fibered muscle in contracted condition. *(From R. McNeill Alexander, Animal Mechanics, Sidgwick and Jackson. 1968)*

the occluding region of the cheliped).

Whenever a crustacean grasps an object in its cheliped, dactyl rotation is stopped and a static equilibrium condition occurs in which input and output torques are equal, that is, $F_I L_I = F_O L_O$. From this it may be seen that, strictly from the viewpoint of lever mechanics, a cheliped exerts the greatest amount of force (for a given level of muscular contraction) at the proximal region of the movable dactyl, where L_O is small (Fig. 1c). Conversely, the least amount of force is delivered at the tip of the dactyl, where L_O is greatest. Reflecting this relationship, blunt chelipeds with short dactyl levers are termed "strong" claws, while slender chelipeds having long dactyl levers are termed "fast" claws.

Musculature. Although the ability of a cheliped to function as an effective grasping appendage is facilitated by the dactyl lever system, it is evident that the dactyl closer muscle is the ultimate source of the delivered force. R. McNeill Alexander has pointed out the unique structural features of this muscle which permit it to function optimally under severe spatial constraints. He considered several aspects of the geometry of muscle contraction in his analysis. First, the force developed by a muscle is dependent upon the total number of contracting fibers, and hence on the total cross-sectional area of the muscle. Second, in typical skeletal muscle the contractile fibers run parallel to the axis of movement. As this type of muscle shortens, it

must increase in diameter, since the total muscle volume remains constant. In the cheliped, however, the rigid exoskeleton would prevent any such diameter increase by a large (that is, inherently strong) parallel-fibered muscle. The alternative—a parallel-fibered muscle with such a small cross-sectional area that the contractile swelling could be accommodated by the space provided—would have relatively few fibers, and thus be inherently weak. The geometrical solution to this dilemma of strength within a confined space is the pinnately fibered muscle (Fig. 2), which is the type actually found in chelipeds. In such a muscle the fibers converge on the central tendon at an angle from either side. As can be seen, there are many more fibers in this type of muscle than are to be found in a parallel-fibered muscle of the same overall dimensions. Moreover, contraction does not alter any of the external dimensions, and thus swelling does not occur. A muscle with pinnate geometry can therefore attain maximal size while retaining complete functional capability for contracting with angular change (Fig. 2c, d). Alexander has calculated that, in the confined volume of a cheliped, a closer muscle of the pinnate type can exert at least twice the force of the largest permissible parallel-fibered muscle.

Functionally, the many varieties of crustacean muscle fibers which exist can be correlated, anatomically, with sarcomere length, and span the range from long-sarcomere fibers capable of slow, strong contraction, to short-sarcomere fibers capable of rapid, relatively weak contraction. It appears that most crustacean limb muscles contain approximately equal amounts of the extreme fiber types, or fibers of intermediate sarcomere length and speed/strength properties. However, in the American lobster the dactyl closer muscle of the strong claw contains a preponderance of long-sarcomere fibers, while the muscle of the fast claw has mostly short-sarcomere fibers. Thus, in this animal at least, the contractile properties of the dactyl closer muscles are closely correlated with the speed/force characteristics of the lever systems.

Occluding surfaces. It has become evident that variation in the geometry and position of the occluding surfaces plays a crucial role in determining the actual functional capabilities of a cheliped. For example, it was found that in representatives from five families of decapod crustaceans, every cheliped type contained from two to four dis-

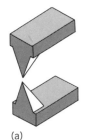 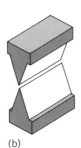 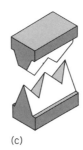 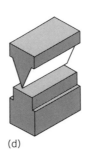 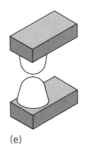 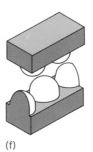 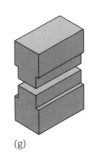

(a)   (b)   (c)   (d)   (e)   (f)   (g)

Fig. 3. Mechanical analogs of cheliped occlusive regions. See text for discussion of *a–g* *(From S. C. Brown, S. R. Cassuto, and R. W. Loos, Biomechanics of chelipeds in some decapod crustaceans, J. Zool. Lond., 188:143–159, 1979)*

tinctive regions of occlusion (Figs. 3 and 4). Cheliped "teeth" were found to have three basic crown shapes: acutely pointed or sharp, broadly rounded, and flat. Acutely crowned occlusive surfaces could be further distinguished on the basis of their edge form (points, ridges, serrations) and linear extent (Fig. 3*a*–*d*). Rounded occlusive surfaces appeared singly (Fig. 3*e*) or in multiples (Fig. 3*f*), while flattened surfaces had only one basic form (Fig. 3*d* and *g*). Further diversity in the pattern of cheliped occlusion could be recognized if the corresponding regions of propus and dactyl were considered together. Figure 3 shows that the majority of occlusive "pairs" were symmetrical (that is, sharp ridge against sharp ridge, and so on). However, at least one asymmetrical occlusive pair (sharp ridge against flat surface, Fig. 3*d*) was found in two of the species studied. Another source of diversity was the degree to which the occlusive surfaces approached one another. For example, more than half of the claws examined contained permanently disjoined regions where the opposing toothed surfaces did not come in contact, even at full dactyl closure. If one adds to the foregoing the possible alternatives of compressive or shear alignment of forces at the occlusive surfaces, it is clear that the structural and geometrical variables provide for great potential diversity in occlusive design. Moreover, since many species of decapod crustaceans are heteromorphic (their right and left claws differ in size and shape), the total diversity of cheliped occlusive regions (both claws) available to an animal may be quite high.

Qualitative and quantitative performance. The presence of structurally specialized regions of occlusion in crustacean chelipeds clearly indicates that such appendages are really multipurpose tools. Common slip-joint pliers—with successive specialized regions for holding flat objects, for holding tubular or irregular objects, and for cutting wire—offer a near-perfect analogy. In addition to obvious holding or crushing regions, the common occurrence and diversity of occlusive regions containing sharply crowned teeth are indicative that the functions subserved by such regions (piercing, pinching, splitting, cutting, shredding, sawing) are of great importance to these animals. Mechanically, sharp-crowned teeth act as force concentrators, delivering the entire output of muscle contraction and lever transmission to an extremely small area. Deliverable pressures (force/area) at such occlusive surfaces should therefore be quite high, particularly in the larger crustaceans. Although common experience indicates that crabs pinch hard, actual quantitative data are few. However, of those crustaceans whose cheliped closing force has actually been measured, the case of the Florida stone crab *(Menippe)* stands out. This crab is reported to be able to fracture the heavily calcified shells of oysters and clams with its chelipeds. Measurements of the pressure exerted by the sharply crowned tooth in the "chisel-and-anvil" region of the fast claw (Fig. 4) show that a 270-g animal can exert sustained pressure of at least 133×10^6 N/m² (19,000 lb/in.²) at the apex of the tooth. It is unlikely that many other biological lever systems will be found to equal this performance.

For background information *see* CRUSTACEA in

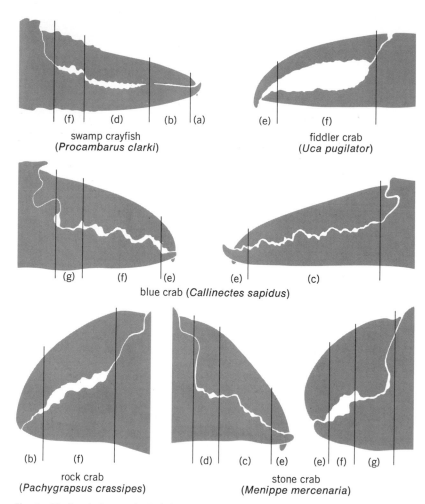

swamp crayfish
(*Procambarus clarki*)

fiddler crab
(*Uca pugilator*)

blue crab (*Callinectes sapidus*)

rock crab
(*Pachygrapsus crassipes*)

stone crab
(*Menippe mercenaria*)

Fig. 4. Chelipeds from selected decapod crustaceans. The labeled occlusive regions correspond to those labeled in Fig. 3. *(From S. C. Brown, S. R. Cassuto, and R. W. Loos, Biomechanics of chelipeds in some decapod crustaceans, J. Zool. Lond., 188:143–159, 1979)*

the McGraw-Hill Encyclopedia of Science and Technology.   [S. C. BROWN]

Bibliography: S. C. Brown, S. R. Cassuto, and R. W. Loos, *J. Zool. Lond.*, 188:143–159, 1979; S. S. Jahromi and H. L. Atwood, *J. Exp. Zool.*, 176: 475–486, 1971; R. McNeill Alexander, *Animal Mechanics*, 1968; G. F. Warner and A. R. Jones, *J. Zool. Lond.*, 180:57–68, 1976.

Blood

Blood is a very complex mixture of salts, proteins, hormones, enzymes, cells, antibodies, and probably thousands of other substances. However, blood can be replaced by a laboratory-concocted mixture and the animal will survive. This has been accomplished by using a mixture of fluorocarbons, detergent, salt, and water. This substitution is possible because the main functions of blood are to carry oxygen and carbon dioxide, sodium and other ions, and high-molecular-weight polymers to keep blood from leaking into the tissues.

The understanding of the essential role of sodium, potassium, and calcium ions in keeping the heart alive was demonstrated by S. Ringer over 50 years ago. The role of plasma proteins in maintaining blood volume has also long been appreciated.

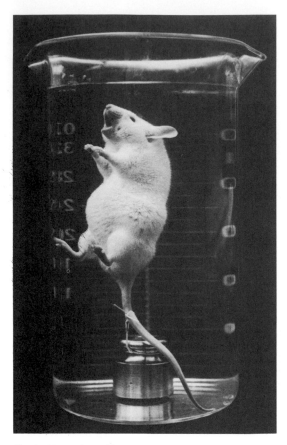

Fig. 1. Mouse breathing oxygen-saturated (by bubbling) fluorocarbon liquid, perfluorobutyltetrahydrofuran.

BLOOD

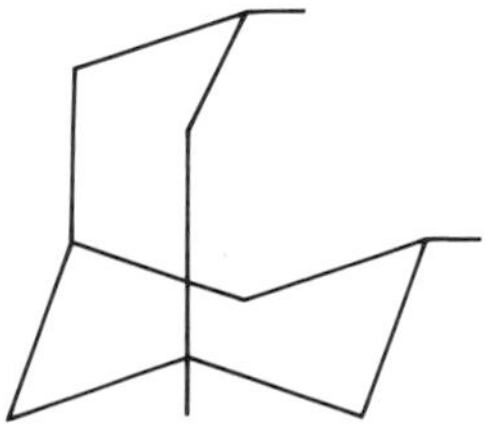

Fig. 2. Structural formula of trimethylbi-cyclononane.

Ringer's solution is used in large quantities for intravenous maintenance of salt and water. Human albumin, and substances such as dextran (a starch polymer), gelatin, and other polymers have been used for many years as plasma substitutes. The oxygen-carrying component in blood, hemoglobin, has defied attempts at practical synthesis, even though its structure is well known. It was not until recently that a synthetic substance was found which could carry oxygen in sufficient quantities to support life.

Artificial blood began with the discovery by L. C. Clark that mice could survive the breathing of inert, highly fluorinated liquids closely related to the fluorocarbon Teflon (Fig. 1), or perhaps it began when J. A. Kylstra found that mice could breathe hyperbarically oxygenated Ringer's solution. The fluorocarbon liquids (F-liquid) used for the first mouse-breathing experiment dissolved large amounts of oxygen, over 20 times that dissolved in salt water, and also large amounts of carbon dioxide. In addition, the F-liquid is so inert that it does not damage the very delicate structure of the lungs after hours of exposure.

Because the F-liquid is insoluble in water, it cannot be given intravenously. Moreover, no glucose or salts are soluble in it, so that it cannot support life. It can be given as an emulsion, a suspension of fine particles in salt water and a detergent (Pluronic F-68, a nonionic surface-active agent). This white, milky liquid was used by R. P. Geyer to totally replace the blood of rats while they

breathed oxygen. Aside from a pale ghostly appearance, they behaved normally.

But the fluorochemicals used for liquid breathing are not perfect for blood because the particles, once deposited in the liver by the macrophages (scavenging cells), tend to stay a long time. Clark then found that another fluorocarbon liquid, F-decalin, worked nearly as well as blood but did not persist in the body. This has started a search for even better fluorocarbons that will make stable fine-particle emulsions. One of the newly discovered fluorocarbons is trimethylbicyclononane (Fig. 2). The F-decalin emulsion has been used in over 150 human subjects.

Because fluorocarbons dissolve oxygen, rather than combine with it as does hemoglobin, they confer the unique ability to deliver large quantities of oxygen at a high pressure, if desired. Therefore, fluorocarbon liquids may be uniquely valuable as blood in hearts or brains with deprived circulation. Synthetic blood does not have blood types or carry disease, and can be stored. Synthetic blood substitutes, like stroma-free hemoglobin, are useful in primates and nonprimates. Synthetic bloods have opened the way to designing special blood for specific purposes.

For background information *see* BLOOD; HEMOGLOBIN in the McGraw-Hill Encyclopedia of Science and Technology.   [LELAND C. CLARK]

Bibliography: L. C. Clark, Jr., et al., *Science*, 181:680–682, 1973; L. C. Clark, Jr., and F. Gollan, *Science*, 152:1755–1756, 1966; R. P. Geyer, R. G. Monroe, and K. Taylor, *Fed. Proc.*, 27:384, 1968; H. Ohyanagi et al., *Clin. Therapeut.*, 2:306–312, 1979.

Buildings

Traditional systems of bracing buildings subjected to seismic loadings have been improved by deliberately introducing bending into certain main members to increase ductility and energy absorption of the structure. This reversal of traditional practice appears to combine the stiffness of a braced frame or shear wall building with the ductility and energy absorption capabilities of the moment frame system.

Bracing a building to resist seismic forces is usually accomplished by one of two basic methods or some combination of them. The first is a method whereby some form of diagonal bracing or the equivalent in shear walls and diaphragms resists the lateral loads (Fig. 1). The second method is by moment frame (Fig. 2), whereby the usually rectangular panel of columns and girders resists forces by making the connections between columns and girders rigid or fixed. When the panel is forced out of square by the lateral loads, the columns and beams are forced to bend into S-shaped curves with the resulting column shears opposing the lateral seismic forces.

Diagonal bracing. Diagonal bracing or walls is the most commonly used method in low buildings where wood, masonry, or concrete walls act as shear walls to transfer the loads from the roof or floors to the foundations. In structural steel buildings with light cladding, diagonal rods or structural shapes are often used. The floors and roofs are usually designed to be capable of acting as horizontal beams or diaphragms to transfer the lateral seismic forces to the walls. Wood, plywood, concrete slabs, or metal decks can all be designed as diaphragms.

In tall buildings, this method can be used with a structural steel or concrete frame with concrete walls or diagonal steel braces in certain locations to transfer loads from floors and roofs to the ground. The walls can have some penetrations for doors, windows, ducts, utilities, and such. Diagonal braces can have any pattern that provides stability, such as X, V, Λ, K, or other more irregular patterns.

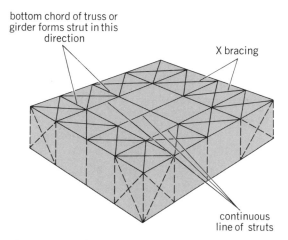

Fig. 1. Diagonal bracing systems in roof and walls. Solid roofs or walls may be designed to resist forces instead of diagonal members.

This first type of bracing has many advantages. It makes economical use of the material and so requires the least steel or concrete, and it is stiff. In many buildings such as homes and low industrial buildings, the same wall that keeps out the weather can also furnish the required bracing with little or no extra cost. Generally, it is the stiffest of all possible systems; consequently it furnishes the best protection to contents in either high wind or earthquake. The system has two disadvantages. First, if interior walls or braces are used, the architectural layout may be somewhat limited because the walls or bracing are permanent—they cannot be shifted around like movable partitions. This disadvantage can be avoided if the braces are placed around elevators, stairs, utility areas, and such, where they do not interfere with the appearance or function of the building. Second, as often detailed, the walls and bracing can easily resist calculated known loads such as gravity or wind loads, but are "brittle" and have little ductility in the postelastic or plastic region of loading which is a necessity for resisting seismic loads. Special detailing can reduce or eliminate this problem in concrete, but it is difficult to prevent buckling of diagonals or tearing of gusset plates in steel.

Moment frame. The use of moment frame action is currently the most popular in tall buildings and often used in low buildings that require an "open appearance." Architects like this system because it does not interfere with their layouts and planning. Partitions and uses can be shifted at will. This system is usually easier to analyze. However, the columns and beams are heavier since they take lateral load bending stresses. Connections are larger and much more costly. The system is very flexible, and experience has shown that it permits a large amount of damage to architectural features and building contents.

Where loads are known, the traditional method of connecting bracing members concentrically with column and beam intersections is the strongest and most economical. However, the system has little energy-absorbing capacity, and unless designed to be very strong, members or connections tend to fail when subjected to strong earthquake loads. Some early designs both in the United States and in Japan avoided this by deliberately introducing bending into various members so that energy-absorbing hinges could form without collapse of the structure.

Braced frame arrangements. C. W. Roder and E. P. Popov have performed analytical studies and laboratory testing on steel members with eccentric connections and on a complete half-size subassembly. Pilot studies were performed to determine the parameters of the hinging mechanism on steel beams both where the plastic hinge is formed by yielding in bending and where it is formed by yielding in shear. It was found that the shear yielding was very efficient, but since tests were performed only on sections meeting the compact criteria, thin webs should have stiffeners to prevent buckling at the web shear ultimate.

Studies and tests have been made to determine the hysteresis stress-strain cycle of bracing members subjected to both compression buckling and tension of bracing members. The subassembly tests by Roeder and Popov have indicated that the eccentrically braced frame is very stiff, approaching the stiffness of the concentric braced frame, and can be designed to provide excellent energy absorption in the inelastic stress range.

The design procedure that has been used for systems with eccentric bracing as illustrated in Fig. 3 is approximately as follows: Preliminary member sizes are determined from an elastic analysis, including the vertical loadings on beams, columns, and braces with the braces both included and omitted. This analysis should include the axial loads, moments, and shears. The effect of the eccentric bending moment and the fixity moment of the beam where it joins the column should be included in the column sizing. Lateral loads for the full code requirements should be added with the bracing included, and it is recommended that 25% of the lateral loads should be considered with the bracing omitted. This gives the minimum sizes required to meet local building code requirements or project criteria.

Since the energy absorption is based on the hinging of the beam or the buckling of the bracing, these must be the elements (one or the other) that reach yield first. If the bracing member is to be enclosed in a wall, it is recommended that the hinge be formed in the beam, but in a project where buckling is not objectional or where fireproofing does not interfere, the critical member could be the bracing member. Where the hinge in

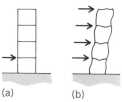

Fig. 2. Moment frame system. Bracing is provided by (a) rigid joints and (b) bending of columns and girders. (*From H. J. Degenkolb and L. A. Wyllie, Jr., in Western Woods Use Book, Western Wood Products Association, Portland, OR, p. 175, 1973*)

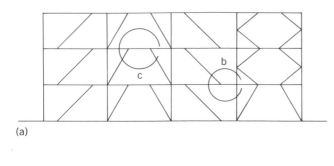

(a)

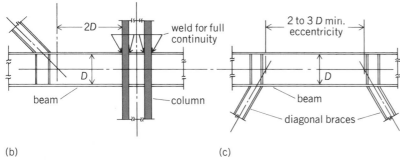

(b)   (c)

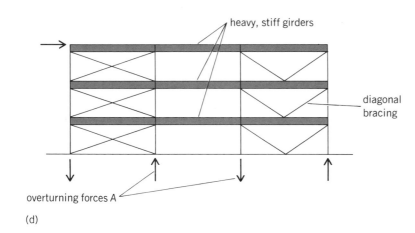

(d)

Fig. 3. Braced frame arrangements using eccentric connections to absorb energy. (a) Elevation of bracing bent showing alternate arrangements of diagonal members. (b, c) Enlargements of areas shown in a. (d) Alternate method of absorbing energy in braced frames using coupling beams. (From H. J. Degenkolb, Practical design (aseismic) of steel structures, Can. J. Civil Eng., 6(2):292–307, 1979)

the beam is controlling, the maximum shear that can be transmitted through the hinge is calculated. That shear is applied at the hinge, and the resulting moment on the column is combined with all other loads on the column to determine the minimum column size. The column must not hinge at the maximum possible load. The hinge shear is also used to determine the maximum load on the brace, and this is resized so that buckling or tension yield does not occur. In this way, all of the inelastic energy absorption is concentrated in the hinge.

In some systems, the designer must be aware of the vulnerability of the bottom flange of the girder to buckling where the hinge is expected to form. The top flange of the girder is usually restrained by the floor system, but the lower flange may be unsupported. If the hinge is near the end of the beam and the girder flange is welded to the column,

there is considerable restraint against buckling, but the extent of the flange bracing must be reviewed for each case.

Another similar method of introducing energy absorption and redundancy into a steel bracing system is through the use of concentric braced frames linked by heavy girders as illustrated in Fig. 3d. This system is useful only in taller structures—probably above 8 or 10 stories. The stiffness of the girder should be sufficient, and its span short enough in relation to the length and stiffness of the bracing system, that hinges form in some of the girders before yield occurs in the bracing system. The rest of the structure should be well into the plastic range before the columns yield. The object of the system is to reduce overturning stresses in the columns and to introduce a continuing plastic deformation in the loading cycle.

In the one 35-story building where this system was tried, the frame was very economical and the architect had adequate freedom to achieve an optimum layout.

For background information *see* BUILDINGS in the McGraw-Hill Encyclopedia of Science and Technology.   [HENRY J. DEGENKOLB]

Bibliography: H. J. Degenkolb, *Can. J. Civ. Eng.*, 6(2):292–307, 1979; A. K. Jain and S. C. Goel, *J. Struct. Div., Proc. ASCE*, 106(ST4):843–860, April 1980; C. W. Roeder and E. P. Popov, *Inelastic Behavior of Eccentrically Braced Steel Frames under Cyclic Loadings*, EERC Rep. 77–18, University of California, Berkeley, August 1977; R. Rush, *Progr. Archit.*, pp. 106–114, February 1980.

Campylobacteriosis

Campylobacter fetus recently has been recognized as an important human pathogen. There are three subspecies: *C. fetus fetus* causes reproductive disease only in cattle; *C. fetus intestinalis* causes reproductive disease in cattle and sheep and a bacteremia and other infections in humans; and *C. fetus jejuni* causes reproductive disease in sheep and bacteremia and enteritis in humans. The last-named subspecies is the one that has caused worldwide interest in its human infection potential. *Campylobacter fetus* is a small, curved, motile gram-negative rod that grows under microaerophilic conditions. Though the organism does require oxygen, the concentration of oxygen normally found in the air (20–21%) is toxic to *C. fetus*. Microaerophilic conditions are required; that is, the oxygen concentration is reduced to 5–6%. *Campylobacter fetus* does not attack carbohydrates, and gets its energy for growth from the tricarboxylic acid cycle. Motility is unique, with a characteristic corkscrewlike motion.

Incidence. *Campylobacter fetus jejuni*, a cause of enteritis and diarrhea in humans, has been recognized as a significant cause of human infections only in the past few years. In 1978 an outbreak of *C. fetus* enteritis infecting about 2000 people occurred in Bennington, VT. There have been reports of enteritis caused by *C. fetus* in many areas of the United States, continental Europe, and Great Britain. Most reports indicate that the incidence of *C. fetus* enteritis is as great as that of *Salmonella* and *Shigella* infections. Various reports have stated that the incidence is between 2 and 30% of patients with enteritis and diarrhea. However, the true incidence of *C. fetus* enteritis will not be known for several years. In addition to enteritis, *C. fetus jejuni* has been implicated in an ileocolitis which is an inflammatory bowel disease.

Symptoms, diagnosis, and treatment. *Campylobacter fetus jejuni* infects people most frequently between the ages of 5 and 34, but can infect people of all ages. Children tend to have a less severe infection than adults. The abdominal pain and tenderness found in patients with campylobacter enteritis may be misdiagnosed as an acute peritonitis or appendicitis.

The incubation period for *C. fetus* enteritis is estimated to be 3–5 days. The onset of the infection is accompanied by central abdominal pain and diarrhea. The stools become liquid and foul-smelling, and in a few days may be tinged with blood and contain pus cells. A bacteremia, infection of the blood, usually precedes the diarrhea. The patient is usually indisposed for 1–3 weeks. In many cases the organism can be isolated from a patient's stools for 2–5 weeks after an attack of enteritis. Death of a patient is rare.

Campylobacter fetus is usually isolated from blood cultures, but *C. fetus jejuni* is best isolated from fecal culture by using special selective media incubated under microaerophilic conditions. In most cases the organism may be seen in stool samples by using phase-contrast microscopy. Treatment of campylobacter enteritis is usually with erythromycin, the tetracyclines, or a nitrofuran.

Transmission. The mode of transmission is not well understood. However, it is suspected that the organism is orally transmitted by the ingestion of contaminated water, raw milk, and foods. Transmission to humans may also be from infected domestic and pet animals. *Campylobacter fetus jejuni* has been isolated from the intestinal tract of cattle, sheep, swine, and poultry, as well as wild birds such as sparrows, starlings, blackbirds, and pigeons. It has also been isolated from the intestinal content of dogs, cats, and pet birds. With the ubiquitous nature of the organism, it is not surprising that the infection is widespread.

Campylobacter fetus intestinalis. This subspecies has also been reported to cause diarrhea. The incidence, however, is much less than that of *C. fetus jejuni*. It is usually associated with bacteremia and other infections such as meningitis and abscesses of various organs of the body. It is also thought to be orally transmitted, but it is not found with great frequency in the intestinal contents of animals or birds. The subspecies has been isolated from the bile of cattle and from vaginal secretions of cattle. Thus the source of this subspecies is more limited than that of *C. fetus jejuni*. *Campylobacter fetus intestinalis* and *jejuni* usually infect the young or the elderly. They are often isolated from patients with some other debilitating disease, such as diabetes or alcoholism. The two subspecies are also frequently found infecting people with cancer or those receiving immunosuppressing therapy. Thus, at any age, people with a compromised immune system can be infected by *C. fetus intestinalis* and *jejuni*.

Pathogenicity. Very little is known about the mechanism of pathogenicity of *C. fetus jejuni*. Various tests to demonstrate an enterotoxin have been unsuccessful. Tests for invasive ability of *C. fetus* were positive. Thus *C. fetus jejuni* is an organism that invades the intestinal mucosa and can invade the blood. Its mode of pathogenicity seems similar to that of *Salmonella*. Very little is known about human immunity to campylobacter infection.

Growth conditions. The microaerophilic nature of *C. fetus* has been studied. The organism is very sensitive to hydrogen peroxide and superoxide anions. These compounds are generated in the culture medium when it is exposed to air and sunlight or room light. Thus, when the organism is inoculated into the medium, it is not able to cope with these compounds and the cells are eventually killed. The organism has intracellular catalase and superoxide dismutase, two enzymes that destroy hydrogen peroxide and superoxide anions. These enzymes can neutralize the internally generated hydrogen peroxide and superoxide anions, but not the externally generated compounds that are already in culture medium. Microaerophilic conditions where the oxygen concentration is reduced from 20 to 5–6% lowers the oxygen available for the external generation of the toxic compounds. Certain chemicals can be added to the culture medium to destroy or prevent the formation of these compounds to toxic levels.

Catalase and superoxide dismutase can be added to culture medium to destroy these toxic compounds. A mixture of ferrous sulfate, sodium metabisulfite, and sodium pyruvate can be added to the culture medium to destroy and prevent the accumulation of these compounds to toxic levels. Ferrous ions and sodium bisulfite form a complex that destroys or quenches superoxide anions, while pyruvate destroys hydrogen peroxide. When these compounds are present in the culture medium, *C. fetus* grows at much higher oxygen concentrations than 5–6%, and also grows faster than in unsupplemented medium.

Campylobacter fetus infection in humans is a newly recognized one that has stimulated the interest of medical microbiologists. This interesting organism has specific cultural conditions (5% oxygen atmosphere) that must be met in order to isolate the organism from clinical specimens. The true incidence of campylobacter infection in humans will be known only after much work has been done in clinical laboratories and only when most clinical laboratories train their staff in the necessary isolation techniques.   [ROBERT SMIBERT]

Bibliography: J. P. Butzler et al., *J. Pediat.*, 82: 493–495, 1973; J. P. Butzler and M. B. Skirrow, *Clin. Gastroenterol.*, 8:737–765, 1979; H. A. George et al., *J. Clin. Microbiol.*, 8:36–41, 1978; P. S. Hoffman et al., *Can. J. Microbiol.*, 25:8–16, 1979; P. S. Hoffman, N. R. Krieg, and R. M. Smibert, *Can. J. Microbiol.*, 25:1–7, 1979; M. B. Skirrow, *Brit. Med. J.*, 2:9–11, 1977; R. M. Smibert, *Annu. Rev. Microbiol.*, 32:673–709, 1978.

Cell physiology

Calmodulin is a multifunctional, calcium-dependent modulatory protein found in all eukaryotic cells. It was discovered in the late 1960s as an activator of cyclic 3′,5′-nucleotide phosphodiesterase, an enzyme that specifically degrades cyclic nucleotides into 5′-nucleosides.

Calcium ion (Ca^{2+}) exerts a profound influence on many biological processes such as cell motility, muscle contraction, axonal flow, cytoplasmic streaming, chromosome movement, neurotransmitter release, endocytosis, and exocytosis. Yet, because of the paucity of information about the Ca^{2+} receptors, the mechanism of Ca^{2+} action in many of these processes has remained obscure. Evidence acquired over the past several years suggests that many of the actions of Ca^{2+} are mediated through a homologous class of calcium-binding proteins. Calmodulin, the most widely distributed and versatile of these, appears to be a primary receptor of this important divalent cation.

Biochemical and biological properties. Calmodulin is a heat-stable, globular protein consisting of 148 amino acids with a molecular weight of 16,700. About one-third of the amino acid residues are aspartic and glutamic, accounting for an isoelectric point of about 4. The protein contains no cysteine, hydroxyproline, or tryptophan and only two tyrosines. The absence of these amino

acids provides a tertiary structure highly flexible to interact with the various calmodulin-regulated proteins. The low content of tyrosine (two residues) compared to phenylalanine (eight residues) gives an unusual ultraviolet absorption spectrum with peaks at 253, 259, 265, and 269 nm due to phenylalanine, and at 277 nm due to tyrosine. Calmodulin does not contain phosphate or carbohydrate but does have a trimethylated lysine at position 115. Its amino terminus is an acetylated alanine.

Figure 1 shows the complete amino acid sequence of bovine brain calmodulin. Calmodulin is very similar to another calcium-binding protein, troponin c, and they have approximately 70% conservative and 50% direct sequence homology. While calmodulin effectively substitutes for troponin c in some instances, the latter usually does not substitute for calmodulin. Calmodulin harbors four calcium-binding sites with dissociation constants in the micromolar range. These sites consist of acidic amino acids, and they furnish carboxylate groups which bind Ca^{2+} reversibly and selectively in the presence of millimolar Mg^{2+}. Binding of Ca^{2+} to calmodulin results in an increase in the α-helical content, giving the active species.

Ubiquitous throughout the eukaryotes, calmodulin lacks both tissue and species specificity. Intracellularly, it is present in the cytosol and membranes. The proteins isolated from all sources stimulate brain phosphodiesterase; moreover, calmodulin proteins from divergent species crossreact with one another. In addition, the amino acid sequences of calmodulin from many species are essentially identical. This suggests the the protein is ancient, with its primary structure highly conserved, an attribute not unexpected of a fundamental regulatory protein.

Calmodulin has no intrinsic enzymic activity, but it regulates numerous enzymes and reactions, some of which coordinate fundamental cellular activities. Calmodulin controls the metabolism of cyclic nucleotides, NAD (nicotinamide adenine dinucleotide), glycogen, prostaglandins, the contraction of smooth muscle and nonmuscle cells, phosphorylation of membranes, release of neurotransmitters, the disassembly of microtubules, and even the intracellular level of Ca^{2+}. In addition, there are numerous calmodulin-binding proteins, and these may be additional calmodulin-dependent enzymes whose functions remain to be identified. In view of the extensive involvement of Ca^{2+} in cell function, it would not be surprising if future studies extend the role of calmodulin to other cellular processes.

Mode of Ca^{2+} action. The mode of calcium action is outlined in Fig. 2. The experimental basis for this scheme comes mainly from available information on calmodulin and troponin c, the Ca^{2+} receptor subunit of the troponin system. In mammalian cells, the steady-state concentration of Ca^{2+} in the cytosol ranges from 10^{-8} to 10^{-7} M, and is the limiting factor. Stimulation of the cell causes a transient increase of Ca^{2+} to 10^{-6} M or higher, a level sufficient to allow the formation of an active Ca^{2+}-binding protein complex. The complex in turn combines with the target apoenzyme or effector protein to trigger a biochemical reaction, culminating in a physiological response.

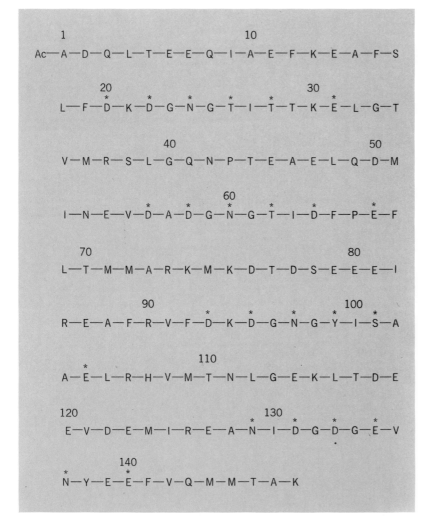

Fig. 1. Amino acid sequence of bovine brain calmodulin. The asterisks indicate putative Ca^{2+}-binding groups clustered into four domains. A indicates alanine; D, aspartate; E, glutamate; F, phenylalanine; G, glycine; H, histidine; I, isoleucine, K, lysine, L, leucine; M, methionine; N, asparagine; P, proline; Q, glycine; R, arginine; S, serine; T, threonine; V, valine; Y, tyrosine. The lysine at position 115 is trimethylated; the alanine at position 1 is acetylated.

The mechanism by which calmodulin acts has been examined in some detail with phosphodiesterase, adenylate cyclase, Ca^{2+}-ATPase, NAD kinase, and myosin light-chain kinase. As shown by the following scheme (the asterisk indicates a

$$Ca^{2+} + \text{calmodulin} \longleftrightarrow Ca^{2+} \cdot \text{calmodulin*}$$
$$\text{(inactive)} \qquad\qquad \text{(active)}$$

$$Ca^{2+} \cdot \text{calmodulin} + \text{enzyme} \longleftrightarrow$$
$$\text{(inactive)}$$
$$Ca^{2+} \cdot \text{calmodulin*} \cdot \text{enzyme*}$$
$$\text{(activated)}$$

new conformation), the binding of Ca^{2+} to calmodulin brings about a conformational change which allows the $Ca^{2+} \cdot$ calmodulin complex to interact with an apoenzyme to form a ternary complex, the active species. The reaction is reversible and is governed by the availability of cellular Ca^{2+}. The scheme should not be regarded as the only mode by which calmodulin acts. Phosphorylase kinase contains calmodulin as an integral subunit and cannot be removed from the enzyme by ethylene glycol bis(β-aminoethyl ether)N,N,N',N',-tetraacetic acid (EGTA), a specific Ca^{2+} chelator. Further, calmodulin inhibits adenylate cyclase of glioma cells and rat brain under certain conditions. In addition, calmodulin inhibits a 15-hydroxy prostaglandin dehydrogenase, and the mechanism of inhibition remains to be elucidated.

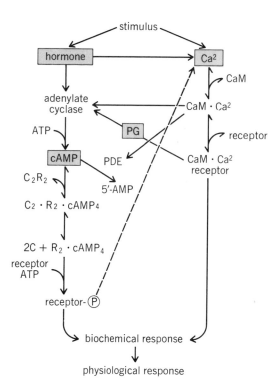

Fig. 3. Integration of cellular regulators by calmodulin. Stimulation of a cell leads to the release of hormones or Ca^{2+} or both. Catecholamines, certain peptide hormones, and prostaglandins activate adenylate cyclase and cause the increase of intracellular cAMP. Upon binding cAMP, the regulatory subunit (R) of protein kinase (C_2R_2) is dissociated from the catalytic subunit (C), which becomes active, and catalyzes the phosphorylation of a protein.

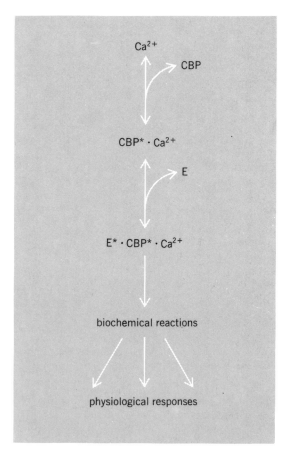

Fig. 2. Simplified scheme for the mechanism of Ca^{2+} action. CBP depicts a Ca^{2+}-binding protein; E, an apoenzyme or a receptor; and the asterisk, a new conformation.

Criteria for calmodulin-regulated functions.
The following set of criteria has been formulated as a guideline to evaluate whether a given reaction is governed by calmodulin: (1) the tissue or cell possesses sufficient calmodulin in the appropriate locale; (2) depletion of endogenous calmodulin from the experimental system by appropriate means alters the activity of a given reaction, and the system becomes responsive to an exogenous calmodulin; (3) since calmodulin requires Ca^{2+} for activity, sequestering Ca^{2+} in the reaction system by an appropriate chelator, such as EGTA, returns the calmodulin-induced activity to the basal level; (4) calmodulin avidly binds trifluoperazine (a phenylthiazine derivative with antipsychotic activity) in the presence of Ca^{2+} and then becomes biologically inactive; the addition of trifluoperazine returns the calmodulin-dependent activity to the steady-state level; and (5) the effect of calmodulin is reversed, at least in theory, by its antibody, provided that the affinity of the antigen-antibody complex is comparable to or greater than that of calmodulin for its receptor.

Integration of cellular regulators.
The effects and metabolism of peptide hormones, catecholamines, prostaglandins, cyclic AMP (adenosine 3', 5'-monophosphate), and Ca^{2+} are often intertwined (Fig. 3). Stimulation of the cell leads to the release of hormones or Ca^{2+}, or both. Many of the Ca^{2+} effects are mediated through calmodulin, which

regulates not only the metabolism of cyclic AMP and cyclic GMP (guanosine 3′,5′-monoposphate) but also that of prostaglandins, prostacyclin, and Ca^{2+}. Cyclic AMP in turn may control the uptake or release of Ca^{2+} by certain organelles. One cellular regulator may function independently or may modulate the effect of one of the others. Calmodulin integrates these regulatory pathways on a molecular basis.

Between the Ca^{2+}- and cyclic AMP–regulated pathways, the response of the Ca^{2+} signal appears inherently faster. The hormonal signal is transduced through adenylate cyclase, which catalyzes the synthesis of cyclic AMP. Cyclic AMP activates protein kinase, and the latter catalyzes the phosphorylation of a receptor protein, probably yet another enzyme. Calcium ion, on the other hand, only needs to be made available upon stimuli. It interacts with calmodulin, which may directly regulate its receptor, as in the case of myosin light-chain kinase. The smaller number of intermediate steps in Ca^{2+}-regulated pathways affords a faster biological response.

In summary, the extensive role of Ca^{2+} in cellular regulation has long been recognized; its molecular mode of action has been partially unraveled in the past several years. Ca^{2+} in itself is not active; its diverse effects are mediated through a homologous class of calcium-binding proteins, and calmodulin, the most ubiquitous and versatile of these, appears to be the principal intracellular receptor. Moreover, calmodulin integrates the major cellular regulators on a molecular basis.

For background information *see* CALCIUM METABOLISM in the McGraw-Hill Encyclopedia of Science and Technology.

[WAI YIU CHEUNG]

Bibliography: W. Y. Cheung, *Science*, 207:19–27, 1980; W. Y. Cheung (ed.), *Calcium and Cell Function*, vol. 1, *Calmodulin*, 1980; C. B. Klee, T. H. Crouch, and P. Richman, *Annu. Rev. Biochem.*, 49:489–515, 1980; J. H. Wang and D. M. Waisman, *Curr. Top. Cell Regulat.*, 15:47–107, 1979.

Chloroplast

Recent studies on chloroplast development have been concerned with the structure of chloroplast deoxyribonucleic acid (DNA) and the components coded for by the chloroplast genome; and also the transport of proteins into chloroplasts.

Chloroplast DNA. The presence of DNA in chloroplasts was first suggested by light-microscopic studies which showed that specific areas in the chloroplast became colored with the DNA-specific Feuglen stain. The electron-microscopic observation of these DNA-containing regions of the chloroplast showed the presence of filaments resembling bacterial DNA which, like DNA, could be stained by the binding of uranyl ions. Subsequently the DNA species in the chloroplast was shown to differ from the DNA of the nucleus in the ratios of DNA components and to constitute about 4% of the total DNA in the cell. Purification of chloroplasts and removal of any contaminating nuclear DNA by specific enzymic digestion showed that large circular DNA molecules constitute the bulk of the total chloroplast DNA. This chloroplast DNA is in the form of supercoiled, double-stranded, closed circular molecules with no breaks in either strand. All the available experimental evidence points to all the circular chloroplast DNA molecules in a given plant having the same nucleotide sequence and therefore coding for the same proteins. Since a chloroplast contains more DNA than is represented by one chloroplast DNA molecule, the chloroplast could contain as many as 20–60 copies of chloroplast DNA per chloroplast. Each chloroplast DNA molecule is large enough to encode for about 100–150 average-size proteins, but so far only a few of these proteins have been identified.

Although chloroplasts contain all the components necessary for biological systems to be autonomous—DNA, DNA polymerase, RNA polymerase, and protein-synthesizing machinery—these components neither synthesize nor code for all of the chloroplast proteins, but there is not enough DNA to code for all of the proteins of the chloroplast. Many chloroplast proteins are synthesized on cytoplasmic ribosomes, and many of the genes directly involved in chloroplast structure and function are located in the nucleus. One approach to the question of which chloroplast proteins are encoded by chloroplast DNA is to determine the proteins synthesized by isolated chloroplasts. By using light-derived ATP energy to drive protein synthesis, the possibility that some proteins are being synthesized by a contaminating cytoplasmic protein-synthesizing system is eliminated, because intact chloroplasts can generate the photosynthetically derived energy essential for protein synthesis. A major polypeptide shown by this method to be synthesized by isolated chloroplasts is the large subunit of ribulose biphosphate carboxylase. This enzyme catalyses the carbon dioxide–fixing reaction in photosynthesis, and is probably the most abundant protein in nature, accounting for up to 65% of the total soluble protein in leaf extracts. The enzyme is made up of eight copies each of the large (mol wt 55,000) and small (mol wt 12,000) subunits, and whereas the large subunit is synthesized in the chloroplast, the small subunit is coded for by the nuclear genome and synthesized on cytoplasmic ribosomes. Another protein with dissimilar subunits is chloroplast coupling factor, so called because it couples light-induced electron flow to adenosinetriphosphate formation, and in this case three of the five constituent subunits of the protein are synthesized by isolated chloroplasts. Besides the large subunit of ribulose biphosphate carboxylase and some of the subunits for chloroplast coupling factor, isolated chloroplasts synthesize several of the constituent proteins of the chloroplast membranes.

The investigation of the products of protein synthesis in intact organelles assumes that the messenger RNA present in the chloroplast has been transcribed from chloroplast DNA and not from nuclear DNA. Recent advances in technique have made it possible to employ a more direct approach: the extraction of chloroplast RNA and its translation in a cell-free protein-synthesizing system. Essentially, a cell-free protein-synthesizing system contains all the components necessary for protein synthesis in culture except the message, so that when messenger RNA is added it is translated. A preparation from the bacterium *Escherichia coli*

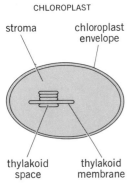

CHLOROPLAST

stroma

chloroplast envelope

thylakoid space

thylakoid membrane

Fig. 1. Structural regions of the chloroplast.

has been shown to translate chloroplast RNA, the major product being the large subunit of ribulose biphosphate carboxylase. Recently the system has been refined still further so that when chloroplast DNA is added to an extract of *E. coli*, the DNA is transcribed and translated; thus, by using this system, it should be possible ultimately to identify all the proteins coded for by the chloroplast DNA.

Transport of proteins into chloroplasts. Electron micrographs of chloroplasts in higher plants and green algae show three major structural regions (Fig. 1): highly organized internal saclike compressed vesicles, termed thylakoids; an amorphous background rich in soluble proteins, the stroma; and a pair of outer membranes constituting the chloroplast envelope. The lack of continuity between the inner envelope membrane and the thylakoid membrane means that in the chloroplast there are three separate compartments: the stroma, the thylakoid space, and the intermembrane space. The many chloroplast proteins synthesized in the cytoplasm must therefore traverse membrane barriers before reaching their ultimate location in the chloroplast. For example, a constituent protein of the stroma synthesized on cytoplasmic ribosomes would need to be transported through the chloroplast envelope membranes before reaching its final destination. Progress in understanding the mechanism of protein transport across membranes in other biological systems has facilitated investigation of the uptake of proteins into chloroplasts. The signal hypothesis to explain the transport of secretory proteins across the endoplasmic reticulum proposes that all secretory proteins are synthesized as precursors, which contain at their N termini a short-chain extension, the signal peptide, which specifically binds to receptors on the endoplasmic reticulum membrane and facilitates transport of the protein into the cisternae of the endoplasmic reticulum. The signal peptide is removed by an enzyme bound to the endoplasmic reticulum to give the mature secretory protein. When the messenger RNA for a secretory protein is translated in a cell-free system, in the absence of the enzyme that cleaves the signal, the product is the precursor molecule larger than the corresponding mature secretory protein. Similarly, translation in a cell-free system of the message for the small subunit of ribulose biphosphate carboxylase produces a precursor that is larger than the small subunit of the assembled enzyme molecule. However, whereas secretory proteins are translated on polysomes bound to the endoplasmic reticulum and are transported into the endoplasmic reticulum cisternae only during translation, the small subunit of ribulose biphosphate carboxylase is translated on free ribosomes so that entry into the chloroplast occurs only after translation is completed. When the radioactively labeled precursor of the small subunit is incubated with isolated, highly purified intact chloroplasts, it is recovered as the small subunit of the assembled ribulose biphosphate carboxylase molecule. The precursor form of the small subunit is undetected in chloroplasts, suggesting that processing occurs either during transport or soon afterward. The enzyme that processes the precursor to the mature small subunit is not membrane-associated, but is soluble and located in the stroma of the chloroplast. The

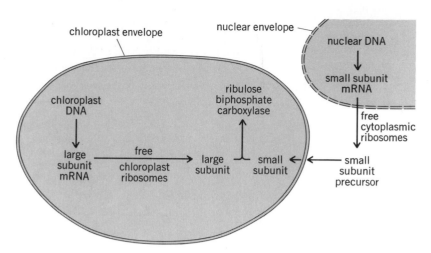

Fig. 2. Sequence of events in the synthesis of ribulose biphosphate carboxylase.

sequence of events in the synthesis and assembly of ribulose biphosphate carboxylase is shown in Fig. 2.

The extensions found on precursors of organelle proteins synthesized by free cytoplasmic ribosomes have been termed transit peptides to distinguish them from the signal peptide of secretory protein precursors. It appears likely that other chloroplast proteins synthesized in the cytoplasm, including thylakoid membrane proteins, will be transported into chloroplasts by the same type of posttranslational mechanism as that operating for the small subunit of ribulose biphosphate carboxylase. The chlorophyll-protein complex 11 has an important role in harvesting light energy, and the protein moiety, which accounts for a large proportion of the total thylakoid membrane protein, is synthesized as a larger precursor on cytoplasmic ribosomes. Similarly, when the message for ferredoxin, a peripheral thylakoid membrane protein, is translated in a cell-free system, a precursor larger than the mature protein results.

Summary. Several lines of experimental evidence support the concept that the entire chloroplast genome is represented by the sequence of a circular chloroplast DNA molecule. Many of the chloroplast proteins are coded for by nuclear DNA and synthesized on free ribosomes in the cytoplasm. These proteins are synthesized as precursors which contain a short-chain extension termed a transit peptide. During or soon after transport of the precursor protein into the chloroplast, this transit peptide is cleaved by a chloroplast enzyme to give the mature protein.

For background information *see* CELL PLASTIDS in the McGraw-Hill Encyclopedia of Science and Technology.

[MICHAEL J. MERRETT]

Bibliography: P. N. Campbell and G. Blobel, *FEBS Lett.*, 72:215–226, 1976; N. H. Chua and G. W. Schmidt, *J. Cell. Biol.*, 81:461–483, 1979; R. J. Ellis, *Trends Biochem. Sci.*, 4:241–244, 1979.

Cladding

The inlay clad process is a recent refinement of the age-old cladding process. Cladding is the joining of two or more metals metallurgically, providing a combination of properties not obtainable by a single metal or alloy. Clad metals historically have been limited to applications in the jewelry industry, but in the last 50 years have found uses in the electronics (thermostatic bimetals), automotive (automobile trim stock), and consumer-product (copper–stainless steel cookware) markets.

The use of inlay clad material for electrical contacts has emerged in the past 10 years, and growth of this practice has closely paralleled the increase in the cost of precious metals. Higher costs of gold and silver have forced designers to seek ways to conserve contact materials. Previous designs utilizing overall gold plating or thick, solid gold contact buttons which were acceptable at $40/oz gold markets became prohibitive at $600–800/oz markets. Compounding the higher-cost problem is the dramatic growth in the electronics marketplace for electrical contacts in connectors, dry circuit switches, relays, and other components. The inlay clad product has enabled the designers to conserve precious metal without sacrificing performance.

The most versatile inlay process is known as the skive inlay (Fig. 1). Starting with heavy-gage (typically 0.100-in. or 2.5-mm) base metals, grooves are skived or shaved in the base metal into which precious-metal strips are placed, and the resulting sandwich is fed into a bonding mill, where the materials are reduced in thickness a minimum of 50% to assure a good mechanical bond. Subsequent processing includes strand annealing and rolling similar to standard mill products procedures. The precious metal is usually backed

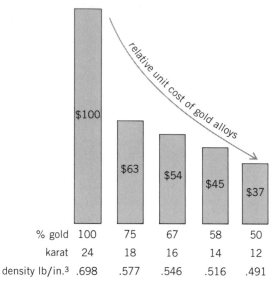

Fig. 2. Relative effect of a lower percentage of gold in a gold-silver alloy upon gold cost. Market value of gold is constant.

with a high-purity, ductile nickel which serves as a diffusion barrier between the base metal and precious metal. During the annealing cycle the metallurgical bond is achieved through diffusion of the atoms at the interface of the dissimilar metals.

The final operations of slitting and inspection result in large, continuous coils of material suitable for stamping into components for electrical contacts.

Economic advantages. There are numerous economic advantages that result from utilizing inlay cladding. Precious-metal strips can be precisely located in the required area for the contact. Up to six strips of precious metals can be located on one or both surfaces of the base metal. The precious metal can be as thin as 0.000010 in. or 0.25 μm and as narrow as 0.060 in., located to ± .015 in. Inlays, as opposed to electroplating, are not limited to pure precious metals. In fact, inlays are usually precious-metal alloys such as 18-kt gold (75% Au, 25% Ag), palladium-silver (60% Pd, 40% Ag), or coin silver (90% Ag, 10% Cu). These alloy combinations not only offer considerable economic savings by virtue of alloying lower-cost noble metals with high-priced precious metals, but also provide technical advantages impossible with pure noble metals (Fig. 2). Inlays enable the fabricator to produce contact parts from a single material and eliminate the need for secondary operations such as plating, riveting, or welding of precious-metal contacts to the base metal.

Technical advantages. In addition to the economic advantages, there are a number of technical advantages involving formability, porosity, hardness and wear, and design flexibility. Clad inlay precious-metal strips, being wrought metals, exhibit a superior ability to withstand forming without fracture as compared to electrodeposits. The clad inlay technique offers a wide selection of gold alloys ranging in hardness from Knopp 60 to 280. Clad inlays can be supplied in a wide variety of base metal–precious metal combinations. Non-

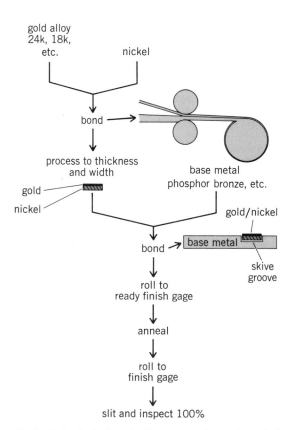

Fig. 1. In the typical production process to produce clad inlays, bonding is in two parts: first gold to nickel, then gold-nickel to base metal.

precious-metal inlays are also possible to provide welding and solderability characteristics.

For background information *see* ALLOY; CLADDING; METALLURGY in the McGraw-Hill Encyclopedia of Science and Technology.

[CRAIG B. HARLAN]

Bibliography: American Electroplaters' Society, *5th Plating in the Electronics Industry Symposium*, March 1975; R. J. Russell, Properties of inlay clad wrought gold alloys, *7th Annual Connector Symposium*, October 1974.

Coal

Coal is a solid hydrocarbon fuel which, although readily available in large quantities, has shown limited growth for use as a boiler fuel and as a coal-oil slurry, and limited potential for use as the base for a magnetohydrodynamic (MHD) power generation industry. This restrictive outlook is due to its relatively high sulfur and mineral content. When raw or physically cleaned coal is burned as boiler fuel or as coal-oil slurry fuel or for MHD power generation, the sulfur and mineral components are released as sulfur oxides, ash, and trace-element pollutants, which require that the effluent gases must be scrubbed and filtered prior to release to the atmosphere. These pollutants also cause corrosion problems in the operation of furnaces, boilers, MHD channels, and so forth.

The sulfur and mineral content of coal may be classed as either associated or inherent. Associated sulfur and minerals (composed of pyrite, silica, and clays) are mineral matter which can be released by crushing and then separated from the coal by physical cleaning in coal washeries. This has been a commercial practice for a number of years. Inherent sulfur, the form either bonded into the organic coal matrix (known as organic sulfur) or present in finely dispersed pyrite form (both usually totaling 1–2% of the weight of coal), and inherent mineral matter, which is the finely disseminated mineral matter (3–5% by weight of coal), can be removed only by chemical coal cleaning.

A photomicrograph of inherent sulfur-containing mineral matter in a coal pore is shown in Fig. 1. It can be seen that even size reduction of coal to the pulverized boiler-fuel particle-size range of 50–100 μm would not free these inherent components. However, chemical extraction or leaching processes can penetrate the coal pores and remove the inherent materials by chemical reaction. The removal of both associated and inherent sulfur and minerals from coal would: allow production of compliance fuel for boilers; decrease corrosion in boilers; all but eliminate the technical problems of retrofitting of oil-fired boilers to coal or coal-oil slurries; and provide a fuel for open-cycle MHD development which would eliminate equipment corrosion and erosion, as well as the need for regeneration of MHD seed prior to recycle.

No commercially demonstrated technology exists for the chemical extraction of inherent sulfur or minerals from coal. The Meyers process, which was demonstrated at test-plant scale in 1977, while efficiently removing the pyrite sulfur mineral component, does not remove organic sulfur. However, there are a number of potential methods for removing inherent sulfur and minerals, and several of these are under active development.

The overall goal of chemical coal cleaning is to develop chemical technology for the selective removal of nonhydrocarbon elements from coal, so that the coal fuel takes on the properties of a solid hydrocarbon raw material. A summary of the status of the major processes is shown in the table.

The U.S. Department of Energy has established a Multi-Use Fuel and Energy Processes Test Plant (abbreviated MEP) to serve as a general text unit for chemical desulfurization processes showing exceptional promise in laboratory or bench-scale testing. The MEP (Fig. 2) was originally built for testing of the Meyers process; however, modifications implemented by the Department of Energy have greatly expanded its potential use. The plant

1 μm

Fig. 1. Scanning electron photomicrographs of inherent minerals (pyrite) in coal pore opening.

now is capable of testing solid-liquid-gas processes at temperatures up to 250°C and 1000 psig (6.9 MPa). There are several process trains ranging in capacity from a few pounds per hour to one-third of a metric ton (333 kg) per hour.

In 1980 the department of energy sponsored testing in the MEP of three chemical desulfurization technologies which are grouped under the term oxydesulfurization. These technologies, known as the PETC, Ames, and Kennecott processes, utilize air or oxygen at temperatures of 130–200°C to remove all of the pyrite and some of the organic sulfur and mineral matter from coal. The three oxydesulfurization processes were tested in the MEP at a throughput of 30–40 lb/h (14–18 kg/h) during 1980 to obtain material balance data. Full-up pilot plant testing of these processes is planned in the MEP utilizing the 333 kg/h circuit. This will allow production of tonnage

Coal technologies summary

Technology	Current status	Removal of inherent		Removal of total sulfur*	Near-term plan/problems
		sulfur	minerals		
Oxydesulfurization: PETC, Ames, and Kennecott processes	MEP small-scale continuous test, 1980	20–40%	Some	60–70%	Plan MEP pilot plant level test, 1981–1982
TRW Gravimelt process	Laboratory tests, 1980	75%	90%	85–95%	Plan MEP small-scale continuous test, 1981–1982
JPL chlorinolysis	Small-scale continuous test, 1980	30–50%	–	70–80%	Solve chlorine and reproducibility problems
GE microwave	Small-scale continuous test, 1980	50–75%	Some	70–80%	Cost of microwave energy
KVB process	Laboratory tests, 1980	30–50%	Some	70–80%	Elimination of nitration problem

*Cumulative sulfur removal or the sum of associated mineral sulfur removed by physical cleaning and inherent sulfur removed by chemical leaching.

quantities of desulfurized coal for user evaluation.

The TRW Gravimelt process utilizes the effects of both gravity and chemical action in melted caustic media. Laboratory tests have already demonstrated 75% removal of inherent organic sulfur and essentially all of the inherent pyritic sulfur (which amounts to 90% removal of the combined associated and inherent sulfur in coal as mined, when an initial physical cleaning is performed). Removal of up to 90% of the inherent mineral matter has also been achieved. The Department of Energy is sponsoring a laboratory study of the chemistry of the process. The next step in the development of this technology will be continuous testing in the MEP. The product from this process should be especially useful for meeting the latest Clean Air Act Standards, providing a superior blending coal for coal-oil slurries for minimum retrofit

Fig. 2. Multi-Use Fuel and Energy Processes Plant (MUP) at TRW Inc. test site in San Juan Capistrano, CA.

problems, and providing a superior open-cycle MHD feed. The Gravimelt process uses caustic alkali metal salts, such as potassium hydroxide, to desulfurize the coal, while MHD seed-crystal material is also potassium-based. Thus it may be possible to meld the two technologies so that the Gravimelt process provides a desulfurized and de-ashed coal containing enough residual potassium to act as the feed to an MHD channel.

The JPL chlorinolysis process utilizes chlorine gas dissolved in halogenated hydrocarbon liquids, followed by a low-temperature carbonization step to remove pyritic and organic sulfur from coal. Continuous bench-scale testing has been initiated; previous laboratory removal results have been encouraging for some coals, but not as encouraging for others, while residual chlorine reactant on the coal is a variable problem according to coal type and processing. Eventual further testing of this process depends on solving these two problems.

The GE microwave process operates by microwave pyrolysis of coal containing an impregnated caustic chemical. This process has shown a high level of sulfur removal with very short reaction times. A major question in this process involves the cost of the microwave apparatus and the required microwave energy. Engineering and economic evaluation results of the small-scale testing will determine whether this activity is continued.

The KVB (Research Cottrell) process utilizes an initial oxidative nitration step followed by caustic treatment. This process has shown promise for removal of inherent sulfur from coal, but present test results show a high level of exothermic nitration coal. Potential solutions to this problem have been suggested, and will be tested in the laboratory.

The TRW Meyers process is not shown in the table, and further investigation is not planned at this time. The chemical extraction portion of the Meyers process has already been tested at the pilot-plant level in the MEP and shown to successfully operate in three-shift operation for conversion of nearly all of the pyritic sulfur in coal to elemental sulfur and soluble sulfates. A second step for removal and collection of the process-generated elemental sulfur was not included in the test plant project. Current applications for air-pollution control, coal-oil slurry, and MHD fuel require nearly total desulfurization, which can be achieved only through removal of the organic sulfur. Thus the Meyers process experience serves the purpose of demonstrating that chemical coal cleaning can operate in commercial equipment, reliably and to meet specifications, but no further testing is needed at this time.

For background information see COAL in the McGraw Hill Encyclopedia of Science and Technology.

[ROBERT A. MEYERS]

Bibliography: R. T. Greer, in R. A. Meyers (ed.), The Handbook of Coal Technology, 1981; R. A. Meyers, Coal Desulfurization, 1977; Reactor Test Project for Chemical Removal of Pyritic Sulfur From Coal, EPA-600/7-79-013a and b, 1979; T. D. Wheelock (ed.), Coal Desulfurization Chemical and Physical Methods, American Chemical Society, 1977.

Coelenterata

Hydra is a small polyp inhabiting ponds and streams. Because it is among the simplest of multicellular animals, it is being intensively studied to learn about the basic properties of cells and tissues. A recent branch of hydra investigation involves constructing chimeric hydra. A chimera is an organism constructed by combining cells or tissues from two genetically different strains of the same or related species. This approach allows the researcher to deduce which cell or tissue types are involved in various biological processes, since a chimera will resemble one or the other of the parental types depending on which cells or tissues determine those processes.

Anatomy. In order to understand how chimeric hydra can be constructed, and what their significance is, it is necessary to grasp the basic structure and developmental biology of this animal. Hydra is tubular in form, with a whorl of tentacles arising at one end of the column (Fig. 1). The entire structure is a bilayer of two tissues: a sheet of ectoderm on the outside and a sheet of endoderm on the inside. In both tissue layers, between the large epithelial cells which form the brickwork of the animal, are strewn numerous tiny cells with specialized functions: nerve cells, gland cells, nematocytes (stinging cells used to capture prey and to hold onto substrates), and gametes or sex cells. Also, a large population of undifferentiated cells, termed interstitial cells, continuously produce more of the specialized cells by the transformation processes called differentiation.

Steady-state growth. In contrast to other animals that grow to a mature size and then stop, hydra enters into an ageless condition where it grows continuously by cell proliferation and, at exactly the same rate, loses tissue by casting off small individuals called buds. Extensive and continual cell migrations occur to balance these two processes. As a result, the hydra does not change size even though growth and differentiation are occurring. Under conditions of constant laboratory culture, the hydra remains in a balanced steady state in which, over periods of months or years, it remains unchanged, despite the fact that cell turnover and

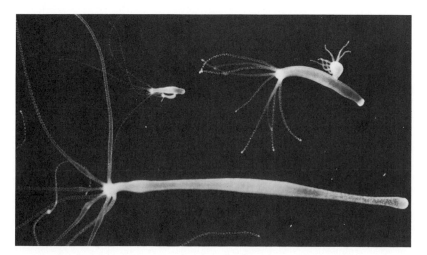

Fig. 1. Mutant hydra strains designated maxi (bottom) and mini (upper left), developed by T. Sugiyama, are compared with a normal hydra (upper right).

replenishment is occurring every few days.

There are three implications of the steady-state condition for these studies. First, analysis of developmental mechanisms can be done at any time or continuously over long periods of time. This is a great advantage over embryos, where structures are changing rapidly and therefore analyses must be precisely timed and can only be carried out once on a single individual. Second, since hydra reproduces asexually by budding, large numbers of genetically identical polyps can be easily produced for types of analyses that require killing the polyps or require massive numbers of them. Finally, since the hydra cells and tissues are normally in a state of great flux, the animal is relatively undisturbed by experimental manipulations, and many types of damage or perturbation are quickly evened out by the normal growth processes.

Cell-lineage recombination. During growth, cell mitosis (division) is restricted to three cell types: ectodermal and endodermal epithelial cells and interstitial cells. These three cell types transform into all of the other cell types. Ectodermal epithelial cells differentiate into the various specialized structural cells, such as the sticky foot cells. Endodermal epithelial cells transform into enormous, swollen elements that fill the tentacles and perhaps hold them rigid. Interstitial cells differentiate into nerves and into the other nonepithelial cells. These three lineages of proliferating cells are independent in that they never transform into one another.

Stable chimeras can be produced if the entire cell lineages are recombined from different parental strains. The most extensive experiments of this type have involved removing the interstitial cell lineage from one hydra and introducing this lineage from another hydra. The interstitial cell lineage can be removed by treating the hydra with colchicine or with a variety of cytotoxic agents such as x-irradiation or nitrogen mustard. The resulting hydra, consisting solely of epithelial cells (hence interesting as the only animal that can live and reproduce in a completely nerve-free condition), can be repopulated with interstitial cells from another strain of hydra. After a short time the cellular composition becomes stabilized, with epithelial cells of one genetic strain and nerve and interstitial cells from another genetic strain. The extent to which the chimeras resemble the epithelial-cell or the nerve-cell parent, respectively, will indicate the relative roles of the two cell classes in regulating morphogenesis. These studies have been enormously benefited by the genetic program of T. Sugiyama, in which numerous mutant strains of hydra have been produced that differ in various aspects of development and cell differentiation (Fig. 1).

These chimeras show rather consistent patterns: in morphological attributes, they strongly resemble the epithelial-cell parents; in details of the differentiation of nerve and interstitial cells, they strongly resemble the interstitial-cell parent. An example showing that size is controlled predominantly by epithelial cells involves chimeras constructed of cells from Sugiyama's mini and maxi strains of hydra (Fig. 1). Both of these strains are normal in all respects except size; maxi hydra are about 10 times as large as mini hydra. Chimeras in which epithelial cells are mini and nerve (and interstitial) cells are maxi have the typical appearance of mini hydra (Fig. 2). Chimeras constructed of maxi epithelial cells and mini nerve and interstitial cells have the appearance of maxi hydra. Clearly the epithelial cell types control the size of the body column. Other morphological traits for which this holds true include the number of tentacles, the position at which buds are formed, the growth and budding rates, the arrangement of tentacles on buds, and the smoothness of the column. In contrast, traits involving the details of differentiation of particular cell types are determined by the cell types involved. Hence nerve-cell density, the coiling pattern of the nematocyte thread, the relative abundances of the four classes of nematocytes, and the temperature sensitivity of interstitial cells all reflect the interstitial-cell parent in chimeras. The color of the epithelial cells in chimeras is determined by the epithelial cell parent.

Tissue-layer recombination. Another method for constructing chimeric hydra is to recombine the two tissue layers, ectoderm and endoderm, from different hydra. When complementary halves of two hydra are joined by grafting, the graft junction of the ectoderm cells sometimes becomes displaced relative to that of the endoderm cells. The intervening region thus is composed of ectoderm derived from one parent and endoderm derived from the other parent. If this narrow region is removed and allowed to grow and regenerate, a chimeric hydra is produced. By using this method with a variety of hydra strains, it has been shown that the developmental interactions between the two layers are very great indeed; that is, they have overlapping domains of effect. Both tissue layers control budding rates, column shape, size, column proportions, and probable growth rate. In different strains the two layers may play unequal roles in each function. For example, the shape of the conical mound of tissue at the top of the hydra, called

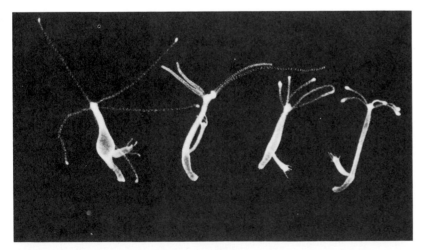

Fig. 2. Chimeras showing that epithelial cells control size. At left is a mini hydra. The other three polyps are chimeras consisting of mini epithelial cells and, left to right, respectively, mini, normal, and maxi interstitial cells. All are of mini size. Similarly, chimeras containing maxi epithelial cells are large, regardless of the interstitial cells they contain.

the hypostome and corresponding to the lips of other animals, seems to be under strict control of the endoderm in some chimeras, while in other chimeras the ectoderm controls hypostome shape.

Sexual chimeras. Sex determination represents another current area in which hydra chimeras are being used to study basic developmental phenomena. In humans the sex of an individual is determined by the chromosomes of the cells, but the complex events leading to sexual expression involve intricate interactions between the newly immigrated germ cells and the resident cells of the incipient gonad. In hydra the sex seems to be determined by factors other than chromosomes, and individual polyps can be hermaphroditic (forming both eggs and sperm simultaneously) or can alter sex from time to time. Current experimentation involving the production of sexual chimeras should reveal which cell type determines sex: the interstitial cells, which would correspond to the germ-cell line in vertebrates, or the epithelial cells, which could correspond to the gonad of vertebrates. In hydra there are no sex chromosomes, so that the interactions between germ and somatic cells probably control sex. Sexual chimeras are now being used to elucidate these interactions.

Unstable chimeras. Other methods produce unstable chimeras. Although these hydra are more difficult to analyze, since one cannot take advantage of a steady-state (unchanging) growth character but rather must analyse the hydra during periods of change, some interesting information has been derived from them. A strain of giant polyps was isolated which lack the ability to produce buds. Chimeras produced by simply grafting complementary parts of normal and nonbudding strains together are temporarily normal, but after a few weeks they invariably acquire the giant phenotype. The simplest interpretation of these chimeras (called heterocytes) is that they are composed of a mixture of normal and mutant cells, and that the mutant cells somehow have a competitive advantage and therefore always eventually replace the normal cells. This reinforces the belief that these hydra, which cannot differentiate normally to produce buds, represent a primitive form of cancer which can be studied with great simplicity.

Model system. The study of hydra chimeras has numerous avenues of relevance to other areas of biology, including health-related disciplines. Because hydra cells undergo continuous migrations and undergo frequent differentiations, they provide an obvious model for vertebrate cancers. The regeneration of arms and legs by vertebrates is another field in which hydra chimeric analysis offers some possible understanding, because limb regeneration is dependent upon intricate interactions between nerve cells and nonnervous cells of the limb. Hydra chimeric analysis provides a system in which such interactions can be easily studied. Finally, many congenital defects owe their inception to faulty inductive interactions between tissues during early embryogenesis when the organs are just forming. The mammalian embryo is so inaccessible for study and difficult to time, however, that little is known about these interactions. Chimeric hydra provide a model system where tissue interactions can be easily studied.

For background information *see* COELENTERATA; MOSAICISM AND CHIMERISM in the McGraw-Hill Encyclopedia of Science and Technology.

[RICHARD D. CAMPBELL]

Bibliography: P. Brien, *Bull. Acad. Roy. Belge Classe Sci.*, 48:825–847, 1962; R. D. Campbell, *Symposium of the Society for Developmental Biology*, vol. 37, pp. 267–293, 1979; B. A. Marcum and R. D. Campbell, *J. Cell Sci.*, 32:233–247, 1978; T. Sugiyama and T. Fujisawa, *J. Cell Sci.*, 32: 215–232, 1978.

Comet

The presence of interplanetary dust in the solar system manifests itself visually in the form of zodiacal light, a faint diffuse glow in the ecliptic plane that can be seen from extremely dark locations on Earth shortly after twilight at sunset or before twilight at sunrise. From such observations of sunlight reflected by the dust particles, the spatial distribution of the dust, concentrated in the ecliptic plane and decreasing in density with increasing distance from the Sun, is inferred. There is also an enhancement of zodiacal light in the antisolar direction, named the gegenschein, resulting from the strong backscattering efficiency of the dust grains. In recent years, knowledge of the distribution of the interplanetary grains has been enhanced by optical observations made from Earth-orbiting satellites and interplanetary probes and by direct measurements of dust particle impacts on interplanetary spacecraft over a range of distances from the Sun, extending from the orbit of Mercury to that of Jupiter. Interplanetary grains entering the Earth's atmosphere have been collected by sounding rockets, balloons, and high-altitude aircraft. The study of these particles provides information about their composition and size distribution.

Interplanetary dust cloud. Like the planets, the individual grains that make up the interplanetary dust cloud travel in prograde orbits around the Sun. However, because of their small mass and size (typically of the order of micrometers), these particles are affected by the Sun's radiation pressure as well as its gravity and so have a finite lifetime which is considerably less than the age of the solar system. Depending on shape and size, the small grains can either be blown out of the solar system or be forced to spiral inward toward the Sun (the Poynting-Robertson effect), where they are destroyed by vaporization or sputtering. The typical lifetime for micrometer-size grains is of the order of 10^4 years, and the estimated rate of mass loss from the entire dust cloud is 10 tons/second. If there is a steady state to the interplanetary dust (the evidence for this exists only over a period of a few centuries), the mass lost must be constantly replenished.

Dust sources. Comets have been suggested as the logical source of new dust in the solar system as they are emitters of dust grains and are often new visitors to the inner solar system. The currently accepted model for the cometary nucleus is that of a dirty snowball consisting of water ice with molecular and dust impurities embedded in it. As the nucleus approaches the Sun the solar radiation vaporizes the ice, with the molecules released or produced by photodissociation forming the coma, while the dust carried along with the gas produces the broad diffuse tail known as the dust or type II tail. The type I or ion tail results from the interaction of the solar wind with the gaseous coma. Both types of tails are seen to extend in the direction opposite the Sun as a result of solar radiation pressure acting on both the ions and the dust grains, and an analysis of the brightness contours of the dust tail yields a measure of both the size distribution of the emitted dust grains and the rate at which these particles are being emitted by the cometary nucleus. In addition, some comets, most recently Comet Kohoutek (1973 XII), display an antitail, a component of dust extending toward the Sun, which is believed to contain more massive dust particles that are less subject to radiation pressure and are thus less able to follow the rapid change in direction of the tail as the comet flies by its point of closest approach to the Sun (perihelion).

Comets are believed to be condensed from the volatiles and dust present in the primordial solar nebula near the orbit of Jupiter. They then are ejected into orbits at very large distance from the Sun, where they exist in a cloud (the Oort Cloud) of some 10^{11} comets that are essentially unchanged from the time of their formation. Under gravitational perturbations by nearby stars, individual comet orbits are altered to bring them much closer to the Sun, and the so-called new comets following these orbits are often the spectacular objects that appear at the rate of a few every century. Perturbations by the major planets can further alter the orbit, and the comet can remain near the inner solar system as a periodic comet. As a comet can lose up to 1% of its mass on each perihelion passage, periodic comets, especially short-period comets which have been observed at several apparitions since their initial discovery, tend to be smaller, with less gas and dust evolution, and consequently have a less impressive appearance than new comets. The dust emitted during these many traversals of the inner solar system is found to be spread out along the same elliptical orbit path as the comet, and the intersections of such dust streams with the Earth's orbit plane give rise to a number of annual meteor showers, each of which can be correlated with a given periodic comet. Dust in these streams, through collision with interplanetary grains in nearly circular orbits about the Sun, is removed from the stream and added to the interplanetary cloud.

Attempts to quantify the above arguments have not been completely successful. Estimates of the dust production rate of the known short-period comets are a factor of 10 smaller than the calculated rate of dust loss. The new comets appear to be insufficient in terms of the total amount of dust evolved during their brief traversal of the inner solar system. It has also been argued that the new comets have orbits isotropically distributed in space while the interplanetary dust particles and most of the periodic comets are found to have orbits close to the ecliptic plane. One possible solution that has been advanced is that the presently observed dust grains are the result of the apparition of a giant comet prior to recorded history, the

magnitude of which has not been seen since. It has also been suggested that Comet Encke, the periodic comet with the shortest known period, 3.3 years, may be the remains of that giant comet. It is also possible that the frequency of new comets was appreciably greater in the not too distant past. These ideas are rather speculative and do not provide a definitive answer to the question of whether comets are the source of the interplanetary dust. The discovery by the Voyager missions to Jupiter in 1979 of intense volcanic activity on the Jovian satellite Io raises the possibility of an additional source of new dust in the solar system which must be investigated.

For background information see COMET; SOLAR WIND in the McGraw-Hill Encyclopedia of Science and Technology.

[PAUL D. FELDMAN]

Bibliography: H. Elsässer and H. Fechtig (eds.), *Interplanetary Dust and Zodiacal Light*, Lecture Notes in Physics, vol. 48, 1976; C. Leinert, *Space Sci. Rev.*, 18:281–339, 1975.

Community

Ecologists use the term community to refer to a collection of animal and plant species living and interacting with each other in the same area. The term is usually not rigorously defined, and is often merely used to describe whatever is being studied. Thus one ecologist might refer to all the bird species censused on an island as a "bird community," even though many bird species do not interact with each other (for example, hummingbirds and seagulls), and many interactions occur between birds and nonbirds. Another ecologist might refer to a "predator-prey community" consisting of a bird species and the insects it eats, this time ignoring all the other bird species in the vicinity.

Despite the looseness with which the term is used, most community ecologists are concerned with determining the number and kinds of species that exist in a given area. The theories that have been advanced to explain observed patterns of biological communities are diverse, ranging from purely random processes to intricate cooperative relationships that knit species into one large community "superorganism."

Species interactions. Perhaps the most fundamental question to answer about communities is the extent to which the patterns are determined by interactions between species, as opposed to purely random processes of colonization and extinction. For example, it has long been known that large islands have more species than small islands, and that islands close to a mainland have more species than those more distant. In 1966 R. H. MacArthur and E. O. Wilson offered an amazingly simple explanation for these patterns. The island could be thought of as a target for species dispersing from the mainland. Large, close islands make a better target than small, distant islands, and therefore receive more species. At the same time, species become extinct at a certain rate, which depends on the size of the island and the number of species already present. The result is an equilibrium number of species that varies with island size and distance from the mainland, precisely as observed.

Many community ecologists feel uncomfortable

with MacArthur and Wilson's theory, because aside from an unspecified relationship between species number and extinction rate, the actual interactions between species appear to be irrelevant. In fact, one of the basic predictions of the theory is that species number is in a dynamic equilibrium — in other words, if one censuses an island at two different times, the number of species should stay approximately the same, but the species themselves should differ, as some become extinct and are replaced by others. In an experiment that involved fumigating mangrove islands and recording the recolonization by arthropods, D. S. Simberloff and Wilson demonstrated that this was indeed the case. In fact, long after species number equilibrated on the islands, individual species were colonizing and becoming extinct at the rate of 0.5 per day. By using this kind of data, it has been difficult to demonstrate that community patterns depend on what the species actually do. The patterns are equally well explained by assuming the species to be so many different colored marbles tossed about at random.

Single community. On the other hand, studies that examine a group of species in great detail often reveal them to be highly organized. The species are not a transient random sample drawn from a larger pool. Rather, they are carefully selected and modified by evolution to interact with each other in a special way. One of the best demonstrations of this process concerns hummingbirds, flowers, and mites. Various plant species produce flowers that are pollinated by hummingbirds. If all the plant species were visited by a single hummingbird species, their pollen would become mixed. Evolution has therefore given rise to a community of hummingbird species that accomplishes pollen transfer more efficiently. It turns out that there are two main "dimensions" along which hummingbirds can differ in ways that restrict their access to flowers. First, their bills may be short or long, and straight or curved. A plant with a curved flower will be visited only by hummingbirds with curved bills, and therefore its pollen will not be wasted on plant species with straight flowers. Second, the spatial distribution of the plants may be patchy or even. If the flowers occur in dense clumps, an individual hummingbird will attempt to monopolize it, chasing others away. However, the aerial fights involved in defending a resource requires a certain specialized body type, with short wings for rapid maneuvers. The body type that defends a resource best is actually poorly adapted for the long flights required for visiting widely spaced flowers that are too sparse to be defended. The pollen of a densely clumped plant will therefore not be wasted on the flowers of evenly spaced plants.

If one looks at the community of plants and hummingbirds at a given locality, it is not a random sample of species with different bill and flight characteristics (for hummingbirds) or flower types and spatial distributions (for plants). Rather, such a community contains representatives from each category, as one would expect from a community designed for efficient pollen transfer. Furthermore, because the best morphologies for defending a resource and visiting evenly spaced flowers both

depend upon the density of the air, what is optimal at one altitude is suboptimal at another. Hummingbird species are therefore restricted to altitudinal zones, and a species that visits evenly spaced flowers at one altitude may defend clumps at another.

As if this were not enough, a community of mites live and reproduce within the flowers, and disperse among flowers on the bills of the hummingbirds. The environment of the mites can be characterized by the different kinds of flowers, their seasonal occurrence, and the probabilities of transfer on hummingbird bills. Once again, the mite community is not at all randomly organized. It is highly predictable from a knowledge of natural selection in the context of that particular environment.

Other detailed studies of communities demonstrate an equal amount of organization. There are therefore two methods of investigating communities — one that compares species numbers and composition on islands, and the other that investigates the nature of single communities in detail. The two methods differ radically in their conclusions about how communities are organized. The detailed investigation of single communities seems to be the most reliable approach. Possibly the simple comparison of species numbers and composition on islands lumps so many interactions that the overall pattern cannot be distinguished from randomness.

Superorganism concept. Another major question asked by community ecologists concerns the degree to which species cooperate with each other in modifying their environment and obtaining necessary resources. Human communities are sophisticated economic associations in which most individuals perform functions that are beneficial to the group. Possibly biological communities are similarly organized, so that each species has a "profession" that contributes to the welfare of other species. In this way soil organisms could be construed as specially adapted to retain nutrients in the system and deliver it to the roots of plants. In return the plants might feed the soil organism with their own shed leaves. Other species might have special roles for defense, regulators of population size, and so on. In this way the whole community takes on the characteristics of a superorganism, with individual species becoming the equivalents of organs and appendages.

The idea of biological communities as superorganisms is very controversial. The problem is that evolution produces "selfish" animals and plants that behave in ways that maximize their reproductive success. Evolution is not directly sensitive to the welfare of whole communities, and if an organism can have more offspring by actually exploiting the community, such adaptations will evolve. Most modern ecologists therefore believe that the superorganism concept is false, and that adaptations in species that benefit their community are coincidental by-products of more narrowly selfish behavior.

However, this way of thinking neglects the fact that all species in a community are simultaneously evolving, and if one species acquires adaptations that exploit the community, other species will evolve adaptations to censure the exploiter. After all, many human individuals are selfish, yet this does not prevent human communities from turning into superorganisms in which individuals are constrained to benefit others, even when they are not spontaneously motivated to do so. The possibility remains that biological communities also evolve into superorganisms, when the evolution of all species is taken into account.

For background information *see* COMMUNITY; ECOLOGICAL INTERACTIONS; ISLAND FLORA AND FAUNA in the McGraw-Hill Encyclopedia of Science and Technology.   [DAVID SLOAN WILSON]

Bibliography: P. Feinsinger et al., *Amer. Natur.*, 113:481–493, 1979; R. H. MacArthur and E. O. Wilson, *The Theory of Island Biogeography*, 1967; D. S. Simberloff and E. O. Wilson, *Ecology*, 50: 278–296, 1969; D. S. Wilson, *The Natural Selection of Populations and Communities*, 1980.

Computer

It has recently been realized that cosmic-ray particles can cause noise bursts in microelectronic circuits that result in spontaneous changes in the information stored in computer memories. These changes are called soft fails. This sensitivity to cosmic rays is one of the unanticipated results of the ever-decreasing size of the components of integrated microelectronic circuits, and it presents new considerations in the development of very-large-scale integrated circuits. The problem is not necessarily catastrophic, since modern computers are usually made to continue to work properly, despite errors, by internal correction of electronic mistakes. However, the creation of soft fails adds an additional load to any internal correction scheme.

Microelectronics. The transistor, invented in 1948, was initially utilized as a separate component with which to build electronic circuits by using the same methods as had been developed for vacuum tube circuits. There was one obvious difference between vacuum tubes and transistors, however: transistors are inherently much smaller. This led to the realization that one can build integrated circuits, that is, electronic devices, with transistors, resistors, capacitors, and so forth, all made at the same time on one chip of silicon by using photolithographic techniques.

Since the first integrated circuit in 1958 (Fig. 1), the size of components in integrated circuits has decreased at a remarkable rate—almost a factor of 2 for every year. Now integrated circuits are available which have hundreds of thousands of components built on a single chip only a few millimeters across (Fig. 2). A typical state-of-the-art example is a 64 K (1K = 1024 bits) dynamic random-access computer memory (d-RAM) where binary information may be stored by charging (or not charging) a small capacitor. This capacitor stores approximately 1,500,000 electrons when charged, and is made of a thin film of conducting metal with an area of about 100 μm² on a thin insulator on a single-crystal silicon substrate. It is the very small amount of stored charge, along with the design of the capacitor, which makes such devices susceptible to the loss of bits of information by ionizing particles penetrating the underlying silicon.

Ionization-induced soft fails. The spontaneous flipping of a bit stored in a computer memory is referred to as a soft fail, as distinguished from a hard fail in which a circuit component is permanently damaged and must be replaced. Integrated-circuit storage memories are normally extraordinarily reliable. The usual industry reliability unit is in failures per million hours per chip with nominal reliability being one fail per million hours. That means that if a chip stores 64,000 bits of binary information, the mean time to fail for each bit is 7,500,000 years!

In 1978 a new and unexpected source of soft fails became apparent to the microelectronics industry. Memory circuits which had been designed based on seemingly reasonable extrapolations of previous devices had measured soft-fail rates far above expectation. The source of these unexpected upsets was traced to α-particles (helium nuclei) being emitted by naturally occurring radionuclides which had been inadvertently introduced as part of a ceramic support for the memory. The dramatic effect of α-particles on electronic memories is shown by the experimental array in Fig. 3. The charge-coupled device (CCD) memory matrix was filled with 1's, and then an α-particle from a radioactive source hit the memory at an angle of 20°. An array of hits spontaneously flipped to zeros in the pattern shown. This α-particle problem rapidly led to a study of the sources of the α-emitting contaminations and to the introduction of low background techniques in the semiconductor industry.

The discovery that α-particles from naturally occurring radionuclides could cause soft fails of computer memories led to the consideration of possible effects of another well-known source of naturally occurring radiation, cosmic rays. The average fluxes of various types of cosmic-ray particles at sea level are shown in Fig. 4. Investigations have demonstrated that sea-level cosmic rays have important effects which limit the reliability of memory devices.

The mechanism which causes these soft fails is familiar to all nuclear scientists because it is the basis for one of the most common types of radiation detectors, the solid-state ionization detector. While the detailed geometry, voltages, and so forth, vary considerably with memory circuit design, the underlying principle used to store information in integrated-circuit computer memories is always the same. Binary information is stored by the presence of charge on some element of the integrated circuit. In the case of d-RAM memories, the charge is on a capacitor.

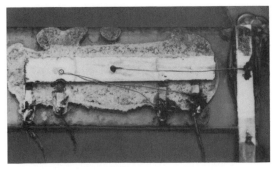

Fig. 1. One of the first integrated circuits, made in 1958. (*Texas Instruments*)

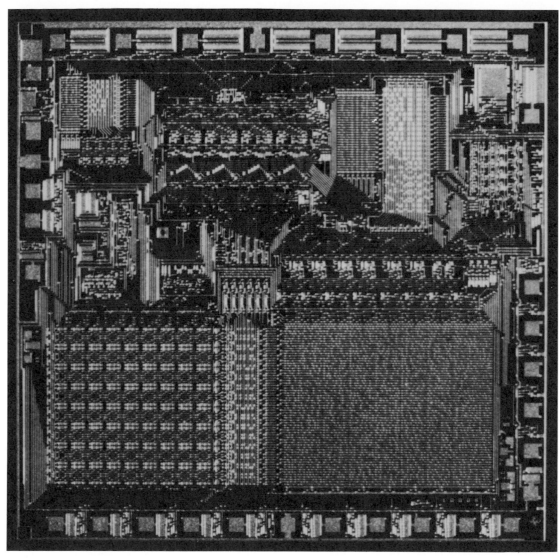

Fig. 2. Complete microcomputer built on one chip a few millimeters across. (*Texas Instruments*)

In the case of charged-coupled devices, the charge is stored in a potential well at the interface between crystal silicon and a covering insulator such as silicon dioxide (SiO_2). This charge then migrates in a "racetrack" and stores information by the sequence of charges stored in this racetrack. In all such devices, the stored charges have to be periodically refreshed because of the slow but continual leakage of charge through the semiconducting silicon substrate.

The presence of charge results in an electric field in the silicon so that any electrons or holes created in this biased region are rapidly collected at the "anodes" and "cathodes," respectively. These are just the conditions needed to create a solid-state ionization chamber used as a particle detector. In the detector case, the charge pulse of electron-hole pairs created by ionizing radiation passing through the biased region of the silicon is used as a signal to measure the amount of ionization created by the particle. In the integrated-circuit memory case, the charge pulse of electron-hole pairs created by ionizing radiation decreases the stored charge (for example, on the capacitor in

```
1 1 1 1 1 1 1 1 1 1 1 1 1
1 1 1 1 1 1 1 1 1 1 1 1 1
1 1 1 1 1 1 1 1 1 1 1 1 1
1 1 1 1 1 1 1 0 0 1 1 1 1
1 1 1 1 0 0 0 0 0 0 1 1 1
1 1 1 1 1 1 1 1 1 1 1 1 1
1 1 1 1 1 1 1 1 1 1 1 1 1
1 1 1 1 1 1 1 1 1 1 1 1 1
1 1 1 1 1 1 1 1 1 1 1 1 1
1 1 1 1 1 1 1 1 1 1 1 1 1
1 1 1 1 1 1 1 1 1 1 1 1 1
```

Fig. 3. Effect of α-particle ionization on a CCD memory matrix. (*Institute of Electrical and Electronics Engineers*)

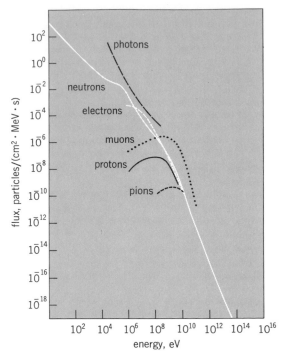

Fig. 4. Flux of cosmic-ray particles at sea level and geomagnetic latitude 45°N. These curves are average values, and large fluctuations exist, attributed to magnetic latitude, time of day, season, solar cycle, angle of incidence, and so forth. (*From J. Ziegler and W. A. Lanford. Effects of cosmic rays on computer memories. Science, 206:776–788, copyright 1979 by the American Association for the Advancement of Science*)

a d-RAM) and may result in the loss of the bit of information represented by this stored charge.

Ionization density thresholds. As the size of microelectronic circuits has decreased, the amount of charge used to store a bit of information has also decreased. For example, a 64K d-RAM may have a stored charge of order 1,500,000 electrons, and a 64K CCD has a stored charge an order of magnitude smaller. Table 1 lists some typical dimensions and charges for such devices. If this charge is decreased by an amount called $Q_{critical}$, the bit will be misread and a soft fail occurs. $Q_{critical}$ is of order 0.2 times the stored charge for most devices. While $Q_{critical}$ decreases with the size of the device, it also becomes progressively more difficult to deposit charge in the smaller active volume as dimensions decrease. For example, suppose $Q_{critical}$ is 500,000 electrons, and the active volume of the device has a mean diameter of order 10 μm. An ionizing particle must deposit charge within the sensitive volume at a rate of at least 500,000 electron-hole pairs per 10 μm = 50,000

Table 1. Device parameters for model computer memories

Parameter	64K d-RAM	64K CCD
Active area, μm²	100	200
Stored charge, e^-	1,500,000	180,000
$Q_{critical}$, e^-	300,000	36,000
Mean collection diameter, μm	12	16
Bits per chip	65,536	65,536

electron-hole pairs per micrometer. Such a device would be insensitive to the ionization wakes of protons, muons, or electrons, because the maximum ionization density of any charge-one particle is 37,000 electron-hole pairs per micrometer. On the other hand, α-particles (which have charge number $Z = 2$) have a maximum ionization density of 100,000 electron-hole pairs per micrometer and, hence, could cause soft fails in such a device. The ionization wake density of electron-hole pairs in silicon following the passage of various charged particles is shown in Fig. 5.

While there are many more detailed considerations, the concept of thresholds in ionization density is central to understanding when soft fails may occur. When, as a consequence of minimization, devices were made with $Q_{critical}$ below the threshold of maximum α-particle ionization density, the α-particle soft-fail problem became important. Most early devices had $Q_{critical}$ so large that even α-particles could not deposit enough charge in the sensitive column to cause a soft fail. Such devices may, however, be susceptible to heavy-ion cosmic rays, which, while not present at sea level, are present above the Earth's atmosphere. Indeed, such heavy-ion-induced upsets were discovered in satellite computers.

Soft fail rates for typical devices. Calculations for some model devices have been carried out by assuming the characterizations given in Table 1. The results for the estimated fail rates for the mechanisms considered are shown in Table 2. For the 64K d-RAM, the single most important cause of soft fails is the production of α-particles by the interaction of cosmic-ray neutrons with silicon nuclei. The 64K CCD has such a low $Q_{critical}$ that it is sensitive to the primary ionization wake of cosmic-ray muons.

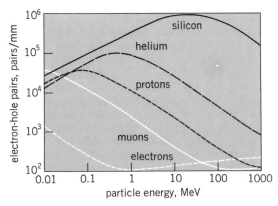

Fig. 5. Ionization wake density of electron-hole pairs in silicon following the passage of various charged particles. (*From J. F. Ziegler and W. A. Lanford, Effects of cosmic rays on computer memories, Science, 206:776–788, copyright 1979 by the American Association for the Advancement of Science*)

As can be seen in Table 2, both these model devices have fail rates larger than the traditional reliability standard of 1 fail per million hours with the 64K CCD memory orders of magnitude above this rate. These results are representative in the sense that, as $Q_{critical}$ falls below about 100,000 electrons

Table 2. Cosmic-ray-induced soft fails of model computer memories chip fails per million hours

Mechanism	64K d-RAM	64K CCD
Electron ionization wake	0	0
Proton ionization wake	0	140
Muon ionization wake	0	330
Si (silicon) recoils from electron scattering	0	<1
Si recoils from proton scattering	<1	<1
Si recoils from neutron scattering	1	100
Si recoils from muon scattering	<1	3
Proton + Si → α-particles	<1	<1
Neutron + Si → α-particles	6	22
Muon capture → nuclear disintegration	<1	7
TOTAL	~7	~600

for a device with dimension about 10 μm, it becomes sensitive to the ionization wake of $Z = 1$ particles and, consequently, has a rather large soft-fail rate. Because muons are so penetrating, it is impractical to avoid this problem by shielding, and if high reliability is needed, either the device design has to be changed or error-correcting codes must be used to detect and correct soft fails as they occur.

There is another area where integrated-circuit technology is employed and where cosmic rays are causing serious concern. As indicated above, the conditions in a single memory cell are very similar to those in solid-state ionization detectors. This fact has led to the development of large-area imaging detectors consisting of arrays of CCD memory cells. Such CCD cameras have the potential of becoming important low-level light detectors for use in astronomy. However, the importance of cosmic-ray-induced "background" events in CCD cameras has been demonstrated.

In summary, it has become clear that cosmic-ray-induced soft fails present a new unanticipated problem to the future minimization of microelectronic circuits. However, the microelectronics industry has faced several seemingly more difficult problems in the past which it has successfully solved, and it is anticipated that suitable solutions will be found to the problem of cosmic-ray soft fails.

For background information *see* CHARGED PARTICLE BEAMS; COSMIC RAYS; INTEGRATED CIRCUITS; JUNCTION DETECTOR in the McGraw-Hill Encyclopedia of Science and Technology.

[WILLIAM LANFORD; JAMES ZIEGLER]

Bibliography: J. F. Ziegler and W. A. Lanford, *IEEE Elec. Dev. Trans.*, 1980; J. F. Ziegler and W. A. Lanford, *IEEE 1980 International Solid State Circuits Conference Proc.*, pp. 70–80, 1980; J. F. Ziegler and W. A. Lanford, *Science*, 206:776–788, 1979.

Computerized tomography

Computerized tomography has been hailed as being as momentous for medical diagnosis as W. C. Roentgen's discovery of the x-ray. It is the process of producing a picture of human body organs in cross section by first electronically detecting the variation in x-ray transmission through the body section at different angles, and then using this information in a digital computer to reconstruct the x-ray absorption of the tissues at an array of points representing the cross section. The array of such absorption coefficients forms an image of the cross section when displayed on a television tube as varying gray-level values proportional to the values of the coefficients. The great importance of computerized tomography is that it presents the physician with images of soft-tissue structures in all organs of the body, including the brain, lungs, pancreas, liver, and kidneys, thereby enabling the direct diagnosis of tumors, cysts, and other lesions.

Medical tomographic x-ray equipment that produced cross-sectional images of the body was developed in the early 1920s. Pictures were fuzzy, and images of the brain could not be made through the skull. The concept of computerized tomography was initially formulated by A. M. Cormack. In 1963 he demonstrated the improved capabilities that could be obtained by the application of mathematical computation. The first commercial instrument capable of obtaining medically usable cross-sectional images of the brain through the skull was designed by G. N. Hounsfield and first marketed in 1973. However, this scanner used an extremely slow, iterative computer algorithm, had low resolution (80 × 80 points), and was limited to scanning from the ears to the top of the head because it required a water bag around the part of the head being scanned. The first whole-body computerized tomograph was developed by R. S. Ledley and put into clinical operation at the Georgetown University Hospital in early 1974. This scanner was also the first to have high resolution, to use the convolution algorithm for reconstructing the cross-sectional points, and to have a television display of the picture. Present-day scanners all have these features.

Scanning modes. A number of methods of scanning the body, using various configurations and motions of the x-ray beam and of the detectors, are employed.

Translation-rotation mode. The early computerized tomographic scanners used both a translational and a rotational movement. First, the translation arm of the mechanical scanner moved the highly collimated x-ray beam across the body section being scanned. Then the frame carrying the translation arm was rotated 1°. This translating and rotating motion continues through 180°, so that each point in the cross section is scanned from 180 different directions (Fig. 1).

The x-ray beam passes through the body section being scanned, which partially absorbs it; then the remaining unabsorbed photons are collected by a scintillating crystal coupled to a photomultiplier tube that converts the attenuated x-rays into light photons and then into the electronic output signals (Fig. 2). The crystal used is generally calcium fluoride, which has a high detection efficiency. Another possible scintillator is bismuth germinate, $Bi_4Ge_3O_{12}$, which has an excellent quantum efficiency but gives only one-fifth to one-tenth the light output of calcium fluoride.

Fan-beam mode with modified translation motion. Another mode of scanning utilizes a fan beam

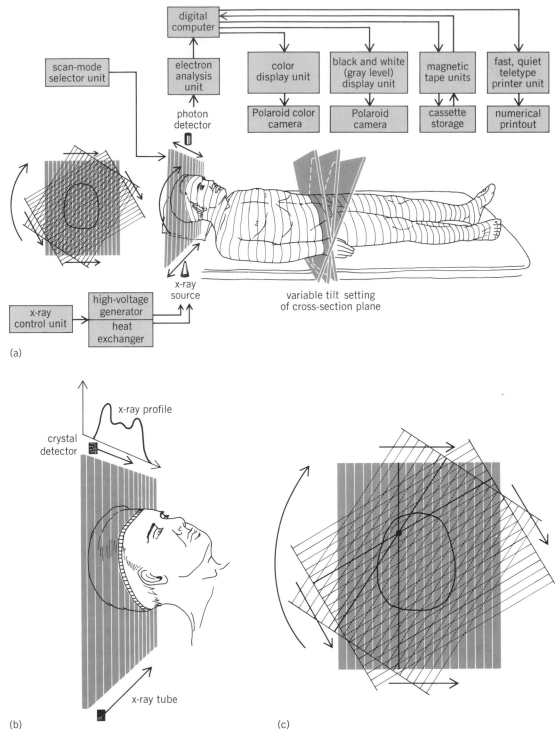

Fig. 1. Automated computerized transverse axial (ACTA) scanner system. (*a*) Block diagram. (*b*) Scan pass producing an x-ray profile. (*c*) Pattern produced by scans following successive 1° rotations of the carriage.

as the source and a line of detectors that are translated together. If the line of detectors intersect an angle θ of the fan beam, the rotation before the next translation can be up to θ degrees. The advantage of this combined fan-beam−translation mode is that the parallel algorithm can still be used (that is, the algorithm for translations with a 1° rotation pattern), but the number of tomslations is minimized because of the rotation through the angle θ between translations. If, for example, the fan angle is 30°, only six translations need be made for the complete 180° rotation. In practice, it turns out that it is better to rotate more than 180° and to take somewhat more translations than the minimum, because redundancy in the data assists in eliminating noise.

The scanning time for this type of system can be as small as 20 s. This short scanning time has the

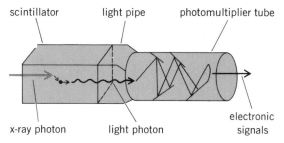

Fig. 2. Detector system coupling a scintillator crystal and a photomultiplier tube.

advantage that the patient can hold his or her breath, and hence reduce motion artifacts to a minimum. The detector system used in this type of scanning mode is similar to that of the translation-rotation mode, namely, calcium fluoride crystals coupled with photomultiplier tubes.

Rotation mode using a large fan beam. An alternative mode of scanning consists in using only a fan beam with no translation (Fig. 3). Here the frame holding the x-ray tube need only rotate, but the rotation must be a full 360°. The advantages of the rotation-only system are that it is mechanically simpler and the scan can be completed rapidly, in less than 5 s. However, the reconstruction algorithm is more complex, requiring additional computing.

Because there is no linear translation, finer spatial resolution is required in order to collect the necessary information for picture reconstruction. There are two possible configurations in constructing a rotating-only scanning frame. In the first type (Fig. 4), the x-ray tube and the detector system line up at opposite sides of the subject, and both rotate around him or her during the scanning. Some manufacturers use many pressurized xenon ionization chambers as detectors. Since such ionization chambers do not require photomultiplier tubes,

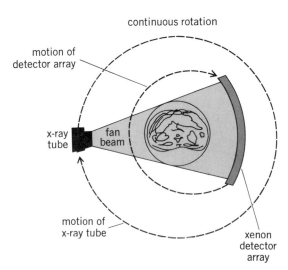

Fig. 4. Rotation scanning mode with rotating detector array. The detector array, usually pressurized xenon ionization chambers for compactness, rotates with the x-ray tube as a unit.

they may have the advantage of lighter weight and bulk, but the extremely high pressures used (20–30 atm or 2–3 MPa) can be hazardous. Also, there is a possibility that the xenon ionization chamber may introduce inaccuracies due to the time response of the chamber (particularly if the slower ionic charge is collected) and also due to the saturation characteristics of the chamber at high flux.

In the second configuration (Fig. 5), only the x-ray tube is in motion; the detector array, which forms the outside perimeter of the scanning frame, remains stationary. Since the detector array does not have to rotate around the subject, compactness does not become a problem and scintillation detectors are used. As many as 1000 detectors can be utilized in the array. The disadvantage of this configuration is that, as the x-ray source revolves, it continuously changes its distance and direction from each detector, and hence the signals from the detectors must be continuously corrected.

Algorithms. The method, or algorithm, used in the computer programs for the translating computerized tomographic scanner is called a parallel algorithm because it uses a profile formed by successive parallel x-ray beams passing through the subject during the translation. The parallel algorithm used by Cormack depended on evaluating a series of integrals. Cormack's process, however, was tedious and not well suited for digital computer processing. Another type of parallel algorithm, the iterative algorithm, is extremely time-consuming and does not converge well for large variations in density.

The type of parallel algorithm most widely used now is the convolution method, and at the present time the term parallel algorithm usually refers to the convolution method. The first use of the convolution method was in Ledley's automatic computerized transverse axial (ACTA) scanner. The convolution method enables the machine to scan every place on the body and to compute in real time during the scanning, so that the picture appears immediately upon completion of the mechanical scan movement. The computation here consists in first evaluating the convolution for every profile, and then back-projecting each convolution independently onto the picture matrix. Thus, as each scan pass is made, the convolution of the profile can be constructed and back-projected onto the picture matrix, in a cumulative fashion. In the original ACTA scanner, the time required for convolving and back-projecting each scan pass was slightly less than the mechanical time required for each pass, so that the picture was essentially completed when the scan was completed. For the very fast, 5-s computerized tomographic scanners, the computations have to be carried out after completion of the mechanical scan. Even though there are now extremely fast special-purpose convolving and back-projecting computers, the mechanical improvements have outpaced the electronic improvements.

In the computer a convolution is performed by taking each point on a profile and subtracting from it fractional parts of all other points on the profile. For each point, the fraction (called the weighting coefficient) of any other point that is subtracted

COMPUTERIZED TOMOGRAPHY

x-ray profile

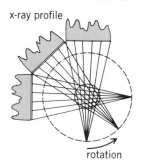

Fig. 3. Fan-beam scanning.

depends on the distance between the two points, and usually diminishes as this distance increases.

It is obvious from the convolution method that the quality of the reconstructed picture depends a great deal on the selection of the weighting coefficients which define the convolution. Various reconstruction artifacts due to different coefficient sets selected appear in the reconstructed image. Many studies have been done to simulate the artifacts by inputting different coefficient sets, and as a result, reconstructed images have been improved tremendously.

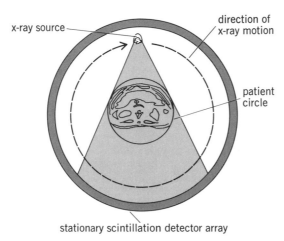

Fig. 5. Rotation scanning mode with stationary scintillation detector array.

The convolution algorithm for a fan-beam system is more complicated, both in computing the convolution and in back-projecting. First, each point of the fan profile is reduced by the cosine of the angle from the centerline to the individual ray which produced the point, before the weighting coefficients are applied. Then the convolution is formed, but with the weighting coefficients adjusted for the angular configuration. Finally, the back-projection is performed along the lines of the generating x-rays (that is, converging at the virtual x-ray source); but now the projected convolution at any point must be weighted with the inverse square of the relative distance of the point from the virtual x-ray source.

Medical application. The whole-body computerized tomographic scanner, such as the ACTA scanner, has virtually unlimited potential in the evaluation of any part of the human body. This instrument has enabled the diagnosis of tumors in the tissues surrounding the eyes, blood clots in the nasal sinuses, and tumors, cysts, hemorrhages, strokes, and other abnormalities of the brain. Syringomyelia (a cyst of the spinal cord) has been demonstrated, and tumors of the larynx, pharynx, thyroid, and parathyroid can be evaluated. It is possible to observe malignancies of the lungs and breasts, pancreatic or ovarian lesions and cysts, and tumors and abnormalities of the kidneys and liver. In the heart there is further potential for observing cardiac chamber size, hypertrophy of ventricular or arterial walls, and ventricular or aortic aneurysms, and possibly for recognizing damaged

myocardial tissue immediately or some time after an infarction.

Computerized tomographic scanners can simplify diagnostic procedures, thus eliminating the need for a number of tests that require hospitalization and are costly and carry a risk to the patient. The basis for treatment of many kinds of tumors and other lesions or abnormalities can be readily reevaluated on a periodic basis, which aids in assessing the effectiveness of treatment. Scans can be performed on an outpatient basis, conveniently, efficiently, and without patient discomfort.

Figure 6 shows examples of ACTA scans of normal and pathological living patients. Figure 6a is a brain scan showing a metastasis with accompanying edema. Figure 6b is a scan through the orbits and the base of the skull, showing with great clarity the nasal bone and eyeballs (top), the details of bony and soft tissue structures, and the foramen magnum (bottom center). Figure 6c is a scan through the chest area showing the lungs and heart, as well as surrounding body structures. Through the use of two simultaneous windows, low-density details of the lungs, the hilus and the bronchi, can be clearly distinguished from the higher-density structure of the large vessels, vertebrae, muscles, and ribs. The denser area in the posterior portion of the lungs is caused by the slight increase in the fluid content of this region attendant upon the subject lying on his or her back. Figure 6d is also a scan through the chest area, showing a right pulmonary mass whose shading matches that of the muscle in the scan, indicating that the mass has approximately the same density as the muscle. Figure 6e is a scan through the renal-pelvis area, showing the kidneys and the descending aorta as small circular masses in the center above the vertebrae. The renal vessels appear as white areas because of contrast material in the kidneys. Figure 6f is a scan through the knees showing details of the bony joints, including patellas and cancellous bone structure of the femoral condyles.

Image processing. If a serial sequence of closely spaced computerized tomographic scans is made, images in other than cross-sectional planes can be constructed. Saggital, coronal, and oblique sections can be made that can often demonstrate the medical aspects of the imaging more clearly than just axial cross sections. The computer of the computerized tomographic scanner can be used for such image processing, and the computer programs are now commercially available. More advanced techniques are available to construct curvilinear (or curviplanar) sectons, which can, for example, follow the contour of the optic nerve to observe pathological enlargements. From successive computerized tomographic scans, the volumes of tumors can be observed and compared with previous volume measurements, for the assessment of treatment progress. Cartesian coordinates can be superimposed on computerized tomographic scans by the computer, and dimensional measurements can be made of body tissues in preparation for surgery or for other purposes.

Another important image-processing method is the development from successive computerized tomographic scans of the molded-surface image of

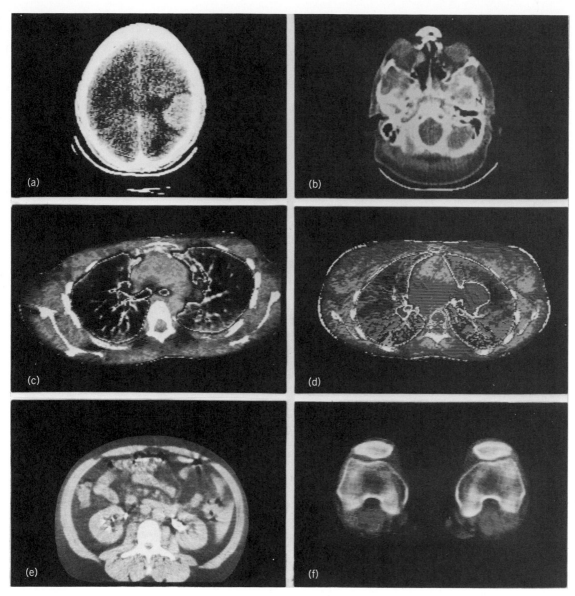

Fig. 6. Computerized tomographic scans. (a) Brain scan showing metastasis with associated edema. (b) Scan through orbits and base of skull. (c) Scan through chest area showing lungs, large blood vessels, and surrounding body structures. (d) Scan through chest area, showing a right pulmonary mass. (e) Scan through renal-pelvis area. (f) Scan through knees.

a body organ or bone. This surface view appears on the screen as if the organ had been excised for examination; the image can be revolved and tilted to obtain the optimum perspective of the shape. In this manner, for example, the size of the spinal canal or the intervertebral foramena can be assessed. All of this computer processing is accomplished with the computer of the computerized tomographic scanner itself.

Radiation therapy planning. An important nondiagnostic application of computerized tomographic scanning is as an aid to radiation therapy planning. Here the body section to be irradiated is overlaid on the television display with isodose contour lines for a particular pattern of radiation exposure. The radiation exposure pattern is adjusted so that the dose to the tumor is maximized, while that to normal tissues is minimized. Prior to the advent of computerized tomography, such planning and manipulations were accomplished by first drawing the subject's cross section by manual means, usually guessing at the exact location and size of the tumor, and then computing the dose distributions manually or with the aid of a calculator or computer. However, the use of computerized tomography not only makes the process much more precise, but usually enables many more radiation patterns to be viewed before the final choice for therapy is made.

Biopsy needle control. Recently computerized tomographic scanners have been used to aid in precisely positioning the needle used for a needle biopsy. The needle is inserted into the body, and a computerized tomographic scan is made to locate its position exactly; the needle is then adjusted as indicated and its location checked with another computerized tomographic scan, and so forth. Such procedures can avoid major exploratory surgery while still enabling the sampling of tissue abnormalities for microscopic examination with a minimum of trauma to the subject.

Economics. The advent of the computerized tomographic scanner just a half-dozen years ago has made a significant impact on improving radiological diagnoses, and hence medical care. Unfortunately, the lay press has wrongly emphasized the increasingly high cost of computerized tomographic scanners (from about $300,000 in 1974 to over $700,000 in 1980). But if one takes into account inflation (of almost 50% over that time interval) and the greatly improved technological capabilities (for example, from a scan time of 4.5 min in 1974 to a present-day scan time of under 5 s), the increase in cost is very modest indeed. Furthermore, many controlled studies have all corroborated the fact that in terms of the medical procedures and hospital days eliminated by the use of computerized tomographic scanning (not even evaluating lives saved and extreme discomfort avoided), the computerized tomographic scanner has contributed significantly to decreasing medical costs.

For background information see INTEGRAL TRANSFORM; IONIZATION CHAMBER; RADIOGRAPHY; SCINTILLATION COUNTER in the McGraw-Hill Encyclopedia of Science and Technology.

[ROBERT S. LEDLEY]

Bibliography: A. M. Cormack, *J. Appl. Phys.*, 34:2722–2727, September 1963; H. K. Huang and R. S. Ledley, *Comput. Biol. Med.*, 5(3):165–170, 1975; W. E. Kiker, T. W. Hinz, and R. S. Ledley, *Med. Phys.*, 3(1):42–44, January–February 1976; R. S. Ledley, *Comput. Biol. Med.*, 6(4):246–329, October 1976; R. S. Ledley et al., *Comput. Biol. Med.*, 4(2):145–155, December 1974; R. S. Ledley et al., *Science*, 186:207–212, Oct. 18, 1974; R. S. Ledley, C. M. Park, and R. D. Ray, *Comput. Tomog.*, 3(1):57–69, 1979; A. C. Scheer and R. S. Ledley, *Comput. Tomog.*, 1(4):275–282, 1977.

Coordination chemistry

Under the Earth's oxidizing atmosphere, iron exists primarily in the ferric (3+) state in aqueous solution. At physiological pH, Fe^{3+} is quantitatively insoluble as the hydroxide. Therefore, even though iron is one of the most abundant metals in the Earth's crust, its availability to most life-forms is highly restricted. Since iron enzymes are essential to virtually every organism, microbes have

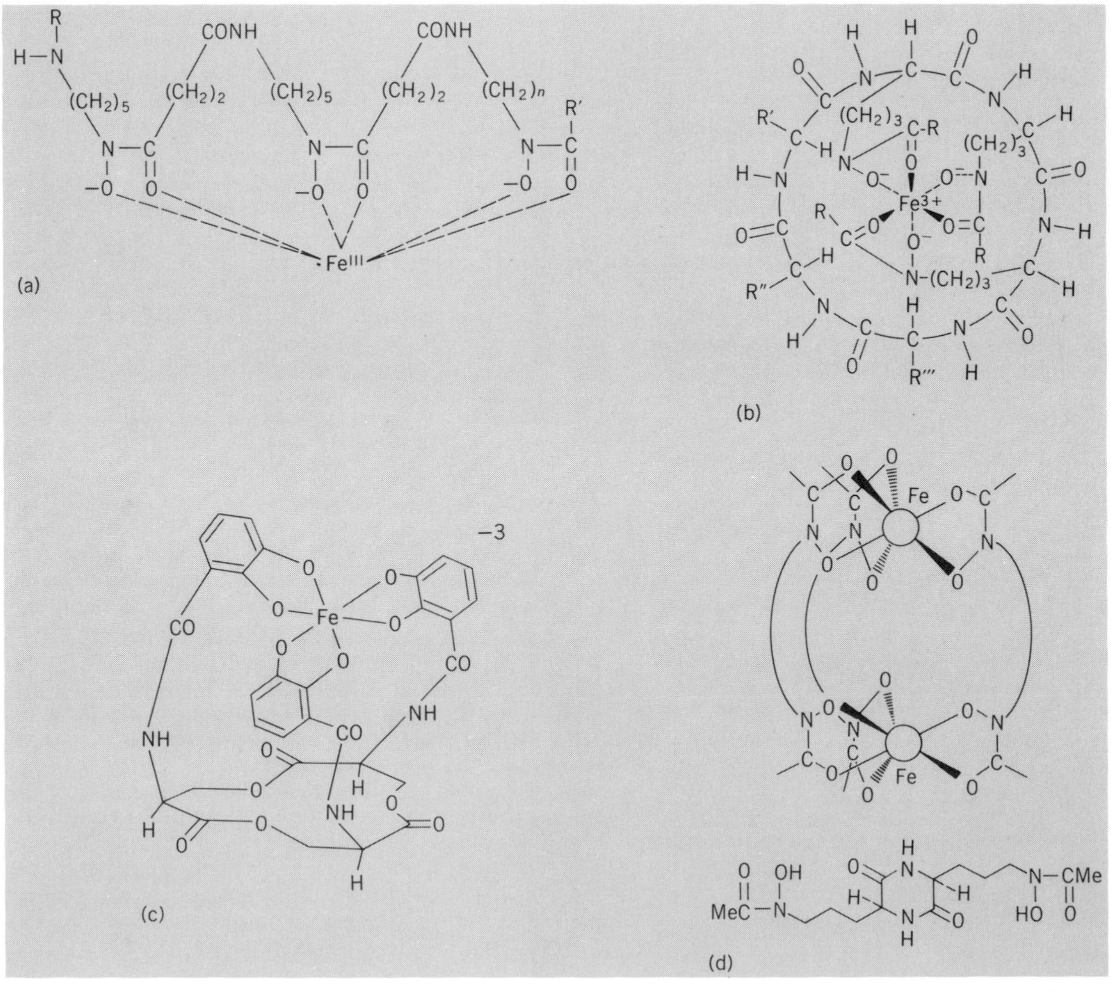

Fig. 1. Ferric siderophores. (*a*) Ferrioxamine B. (*b*) Ferrichrome. (*c*) Ferric enterobactin. (*d*) Ferric rhodotorulic acid (dimer).

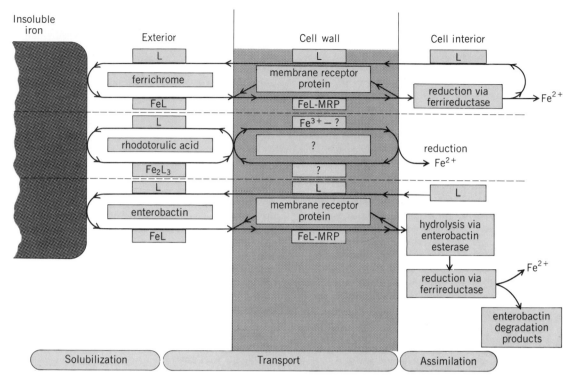

Fig. 2.   The three limiting mechanisms of iron acquisition by microbes. L = ligand.

responded to this requirement for iron through the production of low-molecular-weight organic compounds, called siderophores, which sequester ferric ion and facilitate its transport into the cell.

In contrast to microorganisms, humans have effective mechanisms enabling them to minimize the loss of iron once it has been acquired. The presence of a complex iron-storage system and the very effective recycling of iron in the body leads to the danger that, in excess, ferric ion can be extremely toxic. Chronic iron overload, leading to death by cardiac and hepatic failure, is associated with the treatment of blood disorders such as sickle-cell and Cooley's anemias. By investigating the coordination chemistry of metal siderophore complexes, chemists and biologists can understand better the mechanisms of iron acquisition and metabolism in biological systems. Furthermore, a biomimetic approach, in which metal-siderophore complexes are used as models for synthetic molecules, has been suggested for the therapy of iron-overload syndromes.

Ferric siderophores. A sampling of the siderophores produced by various yeasts, fungi, molds, and true bacteria is illustrated in Fig. 1. There is a similarity of coordination environment of the metal ion in each case: siderophores coordinate ferric ion in a high-spin (five unpaired d electrons), octahedral configuration by use of six oxygens either from three hydroxamate (N-hydroxyamide) or catechol (1,2-dihydroxybenzene) moieties. With the exception of rhodotorulic acid (Fig. 1d), the siderophores in Fig. 1 can satisfy the six-coordinate nature of the ferric ion by forming a 1:1 complex. One measure of the strength of complexation is the stability constant, where a large number indicates strong binding. Ferrioxamine B (Fig. 1a) has a sta-

bility constant with ferric ion of 4×10^{30}, and that for enterobactin (also called enterochelin, Fig. 1c) is estimated to be 10^{52}. Another consequence of the conserved nature of the iron coordination environment is that all siderophores are specific for ferric ion, as opposed to other biologically abundant metals. The relative stability of the ferric compared to the ferrous (2+) oxidation state provides a mechanism for iron release.

While the coordination site has been highly conserved, the general architecture of the siderophores is quite varied; linear (Fig. 1a), ring (Fig. 1b and c), and cyclic and dimeric forms (Fig. 1d) are all known. There are even examples of lipophilic molecules (mycobactins) which bind iron while embedded in a lipid membrane. Although the overall structure of the siderophore is not critical to metal complexation, the mechanism of iron transport is affected. Figure 2 illustrates the three primary modes of iron transport and assimilation in microorganisms.

The transport of iron into the cell can be accomplished by uptake of the intact ferric-siderophore complex (for example, ferrichrome, enterobactin). Alternatively, an extracellular siderophore (for example, rhodotorulic acid) may shuttle iron to a membrane-bound siderophore or protein, which then transports the metal to the cell interior. Once the iron has entered the cell, it must be removed from the complex. Microbes take advantage of the vast stability difference between ferrous and ferric sequestering agents; hydroxamate siderophores exhibit redox potentials at the limit of biological reductants, and apparently release iron by reduction. In contrast, the catecholate siderophore enterobactin has a potential which is far too negative for biological reduction. An enzyme has been reported that cleaves the backbone of this ligand, which then raises the redox potential into the physiological range. As a consequence, enterobactin may be used to sequester iron only once, whereas hydroxamate siderophores, which are excreted intact, are continually reused. This gives the following current picture: low-affinity transport uses hydroxamate siderophores and occurs under moderate iron stress; enteric bacteria in addition have a high-affinity mechanism (enterobactin), which is a last resort in extreme iron deprivation.

Clinical sequestering agents. Ferrioxamine B (Desferal) is among the first and most successful compounds for iron chelation therapy. This drug is almost always given to the patient via subcutaneous infusion, with the coadministration of vitamin C. Only after long-term treatment under these drastic conditions can the amount of iron being excreted exceed the amount which is absorbed (negative iron balance). Among the drawbacks of Desferal therapy is a lack of oral efficacy and an inability to compete with biological ligands such as

transferrin for iron. For these reasons the design, synthesis, and evaluation of effective new iron-sequestering agents for clinical use has become a major goal of many research groups. A biomimetic approach which retains the coordination environment of the siderophore, while tailoring the general architecture of the ligand to increase oral efficacy, is one promising path.

In a ranking of ligand strength, stability constants may be misleading when all equilibria are considered. A calculation of the free-metal concentration, $[Fe^{3+}]$, based on specified conditions of pH and total metal and ligand concentrations yields a number which ranks compounds by coordination strength: the smaller the value for the free-metal concentration, the more effective is the sequestering agent. For example, enterobactin and ferrioxamine B have $[Fe^{3+}]$ of $10^{-35.5}$ and $10^{-26.6}$, respectively, at pH 7.4, in systems containing micromolar total Fe^{3+} concentration and 10 micromolar in ligand.

The human iron-transport protein transferrin

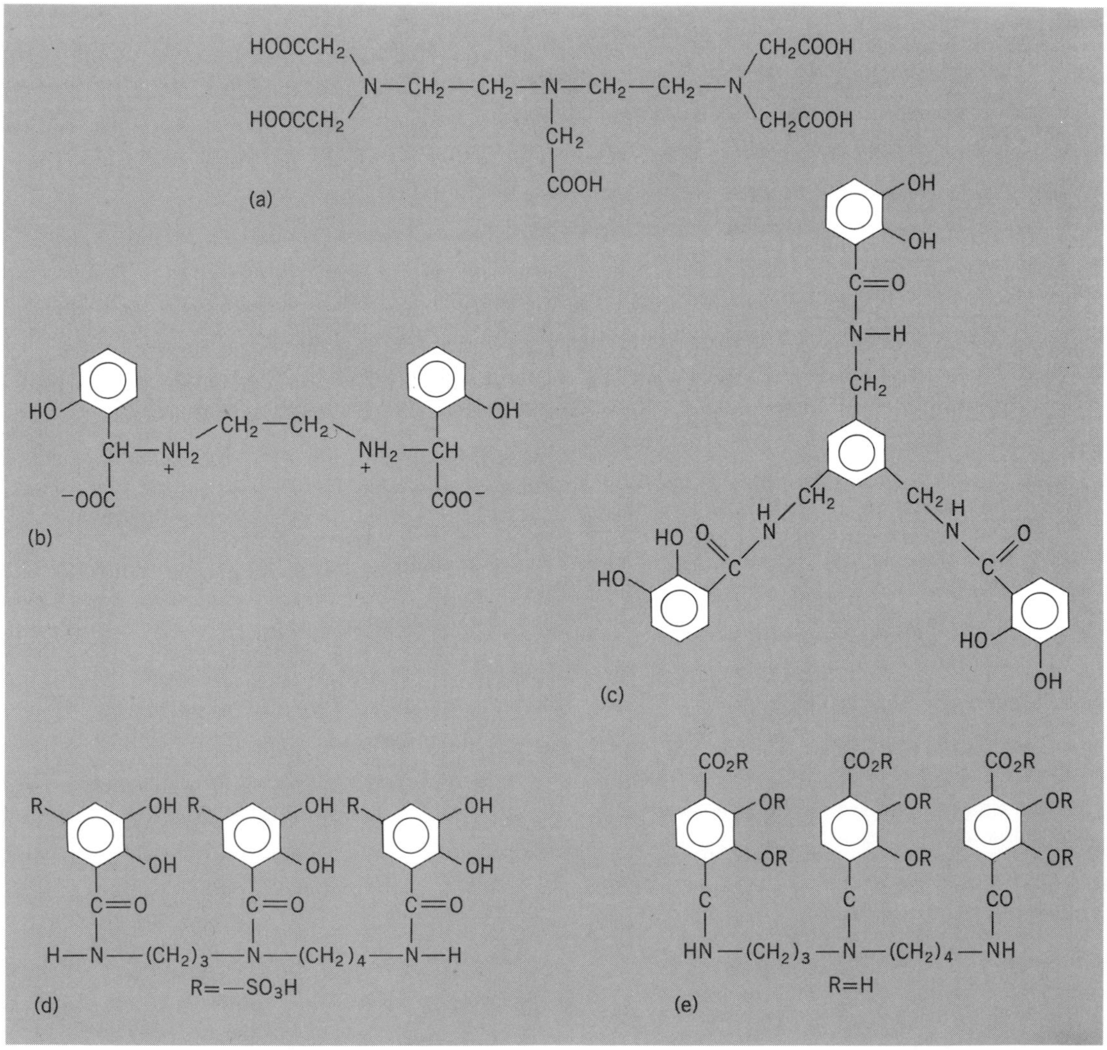

Fig. 3. Synthetic ligands which have been examined for clinical efficacy in the treatment of iron-overload syndromes. (a) Diethylenetriaminepentaacetic acid (DTPA). (b) Ethylene bis(o-hydroxyphenylglycine) (EHPG). (c) N, N′,N″-tris(2,3-dihydroxybenzoyl)-1,3,5-triaminomethyl- benzene (MECAM). (d) N,N′,N″-tris(2,3-dihydroxy-5-sulfobenzoyl)-1,3,9-triazadecane (3,4-LICAMS). (e) N, N′,N″-tris(2,3-dihydroxy-4-carboxybenzoyl)-1,3,9-triaza-decane (3,4,-LICAMC).

has [Fe^{3+}] equal to $10^{-23.6}$. If a ligand is to effectively compete with transferrin for iron, the drug should have [Fe^{3+}] less than 10^{-24}. The [Fe^{3+}] values for some ligands in Fig. 3 are: DTPA, $10^{-24.7}$; EHPG, $10^{-26.4}$; MECAM, $10^{-29.4}$; and 3,4-LICAMS, $10^{-28.5}$. The [Fe^{3+}] values show that all of these ligands can remove iron from transferrin; however, this process has been shown to be extremely slow.

A major structural difference observed for DTPA and EHPG is the inclusion of nitrogen into the coordination sphere of the metal ion. Although this does not drastically reduce the ability of these components to sequester ferric ion, the general specificity of binding is lost. Thus DTPA can also bind Ca^{2+} very well, and mice die after a short period of ingestion due to depletion of divalent cations. In contrast, no such side effects have been observed in Desferal therapy.

A few years ago a program was begun to design iron-chelating agents by using enterobactin as a prototype. Enterobactin itself is unsuitable for clinical use, since it readily hydrolyzes and is not orally effective. There is also a potentially dangerous side effect, since enterobactin acts as a growth factor for enteric bacteria. Thus catechol oxygens were used to bind ferric ion, while substituents were added which would increase the aqueous solubility, lengthen the biological half-life, and impart oral efficacy to the ligand.

The first ligand synthesized was MECAM (Fig. 3c), which uses three catechol moieties as does enterobactin, but has the added advantage that the benzene ring does not hydrolyze. As discussed earlier, this compound is quite an effective iron-sequestering agent, suggesting that the enterobactin coordination site had indeed been mimicked. However, since the central benzenoid platform is less hydrophilic than a triester ring, MECAM is unsuitable for iron chelation therapy; MECAM dissolves in basic solution only when it is deprotonated and, under these conditions, it is quite air-sensitive. Thus the next step was to increase the aqueous solubility and decrease the potential for air oxidation.

Both these goals were accomplished by sulfonation of the catechol ring. The linear catechol analog, 3,4-LICAMS (Fig. 3d), is extremely soluble in water and is air-stable. Furthermore, it was found that this and similar catechol ligands have the exciting property of rapidly removing iron from human transferrin. In the presence of vitamin C they also remove ferric ion from the iron-storage protein ferritin more rapidly than Desferal. However, the sulfonated catecholates are not orally effective.

The only ligand which has shown any degree of efficacy by oral administration is the catecholate 2,3-dihydroxybenzoic acid. Recently the carboxylate derivative 3,4-LICAMC (Fig. 3e) has been synthesized. Qualitative thermodynamic results indicate that this ligand forms strong iron complexes and that the kinetics of iron mobilization processes parallel the sulfonated derivatives. However, unlike previous compounds tested, animal experiments indicate that this ligand may be orally effective.

In summary, a rational approach to the design of agents to be used in iron chelation therapy begins by studying related compounds of biological origin.

Siderophores exhibit a high and very specific affinity for ferric ion, and have in common a ferric ion coordination site of six basic oxygen atoms that remains essentially constant. However, modification of the general molecular architecture allows diverse modes of membrane transport. The chemist can mimic this organization by retaining the general coordination environment about the metal ion, with specific alteration of the ligand's structure to suit requirements. A biomimetic approach, based on analogy to enterobactin, has generated ligands with a high affinity for ferric ion. Moreover, these catecholate sequestering agents can remove iron from transferrin at a reasonable rate. Further experiments with carboxylate derivatives may pave the way for orally active compounds. Future studies of the efficacy and toxicity of these iron-sequestering agents in the body will determine their value in the treatment of human iron-overload syndromes.

For background information *see* CHELATION; COMPLEX COMPOUNDS; COORDINATION CHEMISTRY in the McGraw-Hill Encyclopedia of Science and Technology.

[K. N. RAYMOND; V. L. PECORARO]

Bibliography: W. F. Andersen and M. C. Hiller (eds.), *Proceedings of a Symposium on Development of Iron Chelators for Clinical Use*, U.S. Department of Health, Education and Welfare, National Institute of Health, Bethesda, 1975; C. J. Carrano and K. N. Raymond, *J. Amer. Chem. Soc.*, 101:5401–5404, 1979; J. B. Neilands (ed.), *Microbial Iron Transport: A Comprehensive Treatise*, 1974; K. N. Raymond and C. J. Carrano, *Acc. Chem. Res.*, 12:183–190, 1979.

Cosmic thermometers

The volatile-element inventory of planetary material provides information on the origin of bodies in the solar system and the conditions that existed in the nebula during accretion. In principle, the problem appears tractable; with decreasing accretion temperatures, bodies become more richly endowed with volatile elements. But in practice, there are obstacles. At one extreme, techniques must be devised to separate, analyze, and characterize extraordinarily tiny pieces of matter (10^{-4} to 10^{-6} g) extracted from meteorites to search for elusive carriers rich in volatile elements. At the same time, sensitive instruments must be designed to function as they drop through the hot, murky Venusian atmosphere and continually measure its chemical composition.

Condensation. Besides precise analytical data on volatile-element content and distribution, it is also necessary to understand the condensation behavior of the elements in the primitive solar nebula. In most previous studies the nebular gas was assumed to have the composition of the Sun and to be thoroughly homogenized. The discovery of isotopic anomalies in some meteorites, coupled with unusual chemical and mineralogical features in others, has prompted numerous suggestions that inhomogeneities may have existed. Minor variations in the abundances of most elements have little effect on the overall condensation pattern. There are two exceptions, oxygen and carbon, the third and fourth most abundant elements. It has been pointed out that if the carbon-to-oxygen (C/O) ratio were only slightly higher than the solar ratio of 0.6, virtually the entire condensation sequence would change. Among the more pertinent changes would be that the stable condensates would take on a more reduced character and that nominally volatile elements such as nitrogen and sulfur would condense at very high temperatures as titanium nitride (TiN) and calcium sulfide (CaS). Graphite is also predicted to condense as the most stable form of carbon, but metastable carbynes (triple-bonded carbon compounds with physical properties and stability fields that fall between graphite and diamond) may form instead. Interestingly, one group of primitive meteorites with a highly reduced mineralogy contains CaS, TiN, and graphite, suggesting formation in a region of the nebula where the C/O ratio was greater than the solar value.

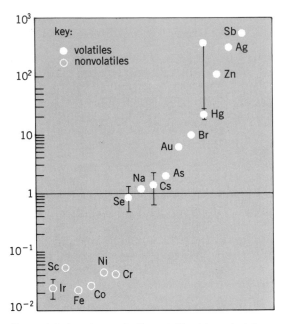

Elemental abundances in the volatile-rich material extracted from the Abee meteorite. The material is strongly enriched in volatile elements relative to abundances in the Sun or Cl carbonaceous chondrites (1 on the vertical scale). (*From R. Ganapathy and J. W. Larimer, A meteoritic component rich in volatile elements: Its characterization and implications, Science, 207:57–59, 1980*)

Meteorites. The chondrites, the most abundant type of meteorite which falls to Earth, contain the nonvolatile elements in roughly solar proportions, which strongly implies that they are little-altered samples of the original condensate. For this reason they continue to be carefully examined for information bearing on the question of how planetary material acquired its volatile-element inventory. The acquisition of noble gases, which occur in only minute amounts in all samples of planetary material, has proved to be a difficult puzzle whose pieces are very slowly falling into place. The bulk of the noble gases in meteorites resides in an insoluble residue amounting to less than 0.5 wt % of the total meteorite. The small sample size has so far frus-

trated attempts to characterize the material as to composition and mineralogy. However, most of each sample has now been shown to consist of carbynes, which must have existed metastably since the origin of the solar system.

Extensive studies of other volatile elements in meteorites have led to a broadly consistent pattern in which the oxidation state, mineralogy, and volatile-element contents all point to the same range of accretion conditions. But during the course of these studies it was also discovered that a few meteorites contain large excesses of volatile elements. The carrier material proved especially difficult to isolate and was given the name mysterite. While researchers were attempting to extract some of the unusual minerals mentioned above (CaS, TiN, and so on), some small (approximately 5×10^{-6} g) strands of material were discovered in the enstatite chondrite, the Abee meteorite, that are more enriched in volatile elements than any known sample of planetary matter. Such extreme enrichments have never been observed before in primitive solar system material. Some volatile elements (such as Sb, Ag, Zn, and Hg) are enriched by factors of 10^3 and to 10^4 relative to the nonvolatile elements (Fe, Ni, Co, and so on). The enrichment pattern tends to parallel that observed in other meteorite samples suspected to be enriched in mysterite, though the samples in this earlier work must have contained less than 10% of the pure material. Besides the volatile elements, the material extracted from Abee is largely made up of carbon, approximately 90 wt %. Conceivably this material is also a carbyne, although the definitive tests have yet to be made (see illustration).

Planets. It has long been suspected that the volatile-element inventory of planetary matter displayed a systematic increase in abundance with increasing radial distance from the Sun, culminating in the huge inventory manifest in Jupiter and the other giant planets. However, as more information accumulates, this simple pattern no longer appears to apply. First the Moon turned out to be highly depleted in volatiles relative to the Earth, and then Mars was discovered to more nearly resemble the Moon in volatile-element content rather than being more enriched than the Earth, as it was expected to be. Now it looks as if Venus, which among these planets is the closest to the Sun, may contain the highest proportion of volatiles.

An important piece of data used for comparative purposes is the ratio of two isotopes of argon, ^{36}Ar and ^{40}Ar. This ratio can be measured in the atmosphere of a planetary body with mass spectrometers on board a spacecraft dropping through the atmosphere on its way to the surface. The two isotopes of argon have different origins: most of the ^{36}Ar must have been acquired when the body accreted, while virtually all of the ^{40}Ar was derived from the decay of the potassium isotope ^{40}K long after the bodies accreted. The ^{36}Ar abundance is therefore thought to reflect a body's initial volatile-element inventory, while the abundance of ^{40}Ar is linked to the abundance of potassium in the body. The ^{36}Ar/^{40}Ar ratios in the atmospheres of Venus, Earth, and Mars are approximately 2, 300, 3000, respectively. If all three bodies have roughly the same potassium contents, their atmospheric composition of argon indicates that Earth is more enriched in volatiles than Mars by a factor of 10 and Venus is more enriched than Earth by a factor of 100. This suggests that relative to potassium, the volatile-enrichment pattern decreases, rather than increases, with increasing distance from the Sun. Several explanations remain viable. Conceivably the temperature throughout the inner solar system was nearly constant, and pressure effects (decreasing with increasing radial distance from the Sun) controlled the volatile contents. Alternatively, the potassium contents may vary widely, or the extent to which ^{36}Ar and ^{40}Ar are outgassed from planetary bodies may differ in some unknown way.

For background information *see* COSMOGONY; METEORITE; PLANET in the McGraw-Hill Encyclopedia of Science and Technology.

[JOHN W. LARIMER]

Bibliography: R. Ganapathy and J. W. Larimer, *Science*, 207:57–59, 1980; J. W. Larimer and M. Bartholomay, *Geochim. Cosmochim. Acta*, 43: 1455–1466, 1979; Pioneer Venus results, *Science*, 205:1–130; A. G. Whittaker et al., *Science*, publication pending.

Cosmology

Developments in cosmology and elementary particle physics have reached the point where advances in one field place important constraints on the other field. This is particularly true with regard to the very early history of the universe, where the entire universe was a sea of elementary particles. Thus the basic elementary particle properties determined the nature of the universe at that time. Developments in cosmology for setting limits on the number of fundamental particle types are especially significant. Another important development has been an explanation of the origin of matter in the early universe utilizing the grand unification of the fundamental forces. Finally there is the startling possibility that a particular type of elementary particle, the neutrino, may be the dominant component of the universe.

Big bang. The major development that has enabled the close interplay of cosmology and elementary particle physics has been the establishment of the "big bang" model of the universe. Astrophysicists are now confident that the universe was at one time extraordinarily hot and dense and that it expanded from such a state. This knowledge is based on the observation of the universal background radiation at a temperature of 3 K, discovered by A. Penzias and R. Wilson. Their discovery showed that the universe was at one time hot and dense enough to thermalize electromagnetic radiation throughout all space. G. Gamow and his collaborators in the 1940s predicted this radiation as a standard feature of the big bang, and it cannot be naturally explained in any other cosmological model. In making the prediction, Gamow pointed out that the big bang nuclear reactions would have occurred throughout space and would have required temperatures in excess of 10^9 K.

These basic arguments were refined and developed in the 1960s and 1970s by J. Peebles, R. Wagoner, W. Fowler, F. Hoyle, and others, who showed that the standard big bang model predicts that about one-quarter of the mass of the universe should be in the form ^{4}He. Observations now show that this is correct, and thus another major point in support of the big bang is the abundance of helium. Hydrogen, deuterium, lithium-7, and helium-3 are also produced in interesting quantities in the big bang.

If one continues to extrapolate toward progressively earlier times, the classical big bang model predicts infinite temperatures at time zero; however, the ability to extrapolate breaks down when the universe reaches the epoch where quantum gravitational effects should be important. This breakdown would be at a time only 10^{-43} s after the classic infinite temperature singularity and would occur at a temperature of 10^{32} K. It is impossible at the present time to scientifically know anything that happened earlier than this time or at temperatures higher than this temperature. However, since this temperature is so enormous and the standard model is basically sound until this point in time, it is clear that the universe has experienced an enormous range of temperatures and equivalent energies. Since it is known that elementary particle effects become dominant at high energies, it is clear that the very early universe was dominated by these effects. In particular, when the temperature of the universe was higher than about 10^{12} K (about 1 μs after the initial event), matter became much denser than the inside of an atomic nucleus, and as such, the neutrons and protons were squeezed together so tightly that they lost their identity. It is now known that neutrons and protons are made of quarks. Thus the very early universe was a "quark soup."

Fundamental particles. High-energy physicists working on accelerators have determined that matter is made up of certain fundamental particles. In particular, standard matter is made of two types of particles, quarks and leptons. The quarks taken in groups of three make up neutrons and protons which in turn form nuclei of atoms, which form the bulk of normal matter. The leptons in normal matter consist of electrons. These are the particles which orbit the nuclei and give chemical structure and chemical properties to atoms. Electrons carry one unit of negative electric charge, whereas protons carry one positive unit, and neutrons are electrically neutral. Quarks have electric charges which are $-1/3$ or $+2/3$ of a unit. Electrons, which are approximately 1/2000 of the mass of a neutron or proton, make up very little of the mass of normal matter even though their orbits take up most of the space of matter. Another kind of lepton is the neutrino. Neutrinos have no electric charge and either are massless or at most have a mass that is less than 1/10,000 of the mass of an electron.

From experiments it is known that quarks come in a variety of types called flavors. These flavors are: up (u) and down (d) which make normal neutrons and protons; strange (s) and charm (c); and top (t) and bottom (b). All but top have now been discovered experimentally. Leptons come in three families, electron (e), muon (μ), and tauon (τ), each lepton family having an associated neutrino (ν) as well as a charged particle (see the table). Each of the quark flavors comes in three so-called colors, which serve the nuclear strong interaction between quarks in the same way that electric charge serves for electromagnetic interactions. A proton is made of two ups and one down, and a neutron has one up and two downs. Each of the three quarks in a neutron or proton has a different color. From all present experiments it appears that these particles are truly fundamental and that they seem to have no physical dimension; that is, they interact as though they have zero radius and are merely points of matter. See QUARKS.

Four basic forces. It has been found that the fundamental particles described above interact with each other by four basic forces: the electromagnetic force, the nuclear weak force, the nuclear strong force, and the gravitational force. Actions that are observed in the universe are manifestations of these four basic forces.

One of the most important recent developments in physics has been the unification of the electromagnetic force and the nuclear weak force. S. Weinberg, A. Salam, and S. Glashow showed that these two forces were really the same basic interaction, the only difference being that the weak force has a massive exchange particle, and the electromagnetic force has a massless exchange particle. At very high energies, when the mass of the weak interaction exchange particles is negligible, the two forces have their symmetry restored and appear to be identical, whereas at the normal energies encountered in everyday life, the symmetry is spontaneously broken.

With the great success of this unification, physicists have been trying to unify the strong interaction with the weak and electromagnetic. However, in such a unification, one would require the leptons and the quarks to behave similarly, whereas ordinarily the quarks are the only particles which interact by the strong interaction, and the leptons can only interact by the weak and electromagnetic. To unify these three basic interactions would require an exchange particle that could switch a quark into a lepton or vice versa. The prediction that such a particle exists and that such an exchange can occur has enabled physicists and astrophysicists to begin to understand the question of how the matter in the universe originated. The unification of the fourth interaction with the other three seems to be the hardest task of all, since the understanding of gravity at the single-particle quantum level has not yet been accomplished, and these successes in unification seem all to be occurring at the quantum level.

Number of fundamental particles. In 1977 J. Gunn, G. Steigman, and D. Schramm showed that one could determine a limit to the number of quarks and leptons based on the amount of helium produced in the big bang. The illustration shows a graph of helium production in the big bang as a function of the present density of quark matter in the universe. (Big bang nucleosynthesis is sensitive only to the mass density of quark matter, not to the mass density of lepton matter.) The different curves are for different numbers of neutrino types. As the number of neutrino types increase, the amount of helium increases for a given density. It is known from observations of helium in various objects that the mass fraction of helium made in the big bang must have been less than 25%. In fact, current estimates place this number at about 23%. It is also known from the dynamics of galaxies that the mass density in the universe is greater than about 2×10^{-31} g/cm^3. However, it is not clear whether that density is in normal quark matter or whether it is in some other form very different from that ordinarily encountered, such as lepton matter. If the density is assumed to be normal quark-dominated matter, then the number of neutrino types must be less than four, or the universe would have produced too much helium. Also since each neutrino type corresponds to two quark flavors (see the table), there can be at most eight quark flavors. In fact, the best answer to the problem is that there are only three neutrino types and thus only six quark flavors. Thus, all of the neutrino types may have already been found and there may be only one undiscovered quark flavor remaining.

However, it is also conceivable that this lower limit on the mass density is not a lower limit on the quark matter density; a somewhat lower value for the density of quark matter would mean more neutrino types. An estimate of the limit on the mass density that is clearly quark matter, approximately 5×10^{-32} gm/cm^3, corresponds to a limit of eight neutrino types. This limit would correspond to 16 quark flavors. In either case, there is still a restric-

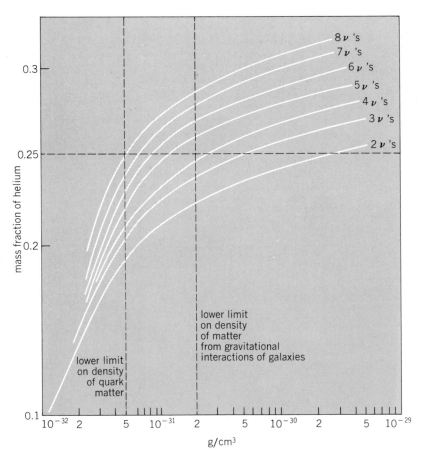

Helium production in the big bang as a function of the present density of quark matter in the universe for various numbers of neutrino (ν) types.

Fundamental particles

Quarks*	Leptons
$\begin{pmatrix} u \\ d \end{pmatrix}$	$\begin{pmatrix} e \\ \nu_e \end{pmatrix}$
$\begin{pmatrix} s \\ c \end{pmatrix}$	$\begin{pmatrix} \mu \\ \nu_\mu \end{pmatrix}$
$\begin{pmatrix} t \\ b \end{pmatrix}$	$\begin{pmatrix} \tau \\ \nu_\tau \end{pmatrix}$

*Each flavor listed comes in three colors.

tion on the number of fundamental particles from events in the early universe.

Origin of matter. As mentioned above, new attempts at unifying the strong, weak, and electromagnetic interactions require that there be a particle that can change quarks into leptons, and vice versa. Thus, in the very early universe, when the temperature of the universe was so high that the symmetry was restored between these three interactions and they were all identical, these exchange particles would have been ubiquitous. Quarks and leptons would have been constantly changing into each other, as opposed to the present era, when the quarks are not able to change into leptons and thus neutrons and protons are quite distinct from electrons and neutrinos. In addition, it is also known that there is not an exact symmetry between matter and antimatter with regard to decay rates into particular decay modes. This asymmetry has been observed in the experiments of V. Fitch and J. Cronin with K mesons. If such an asymmetry took place in the early universe when quarks and leptons were identical, it is possible that an excess of quark matter over antiquark matter could have been generated, and this could explain why the universe seems to be all one kind of matter instead of an equal amount of matter and antimatter. This production might also explain why the ratio of matter to radiation has its present value of approximately 10^9 photons per quark. This idea has been developed in detail by several scientists over the last 2 years. However, the first person to notice that there may be some connection between the grand unification ideas and the origin of matter was the Soviet scientist A. Sakharov.

Mass of neutrinos. One of the most important ideas at the boundary of particle physics and astrophysics has been the possibility that the neutrinos, previously thought to be massless, may have a mass. Since the number of neutrinos produced in the big bang is approximately equal to the number of photons in the universe, the ratio of neutrinos to quarks, and thus to neutrons and protons, would be approximately 10^9. If neutrinos have a mass that is greater than 10^{-9} that of a neutron or proton, then they would be contributing more to the mass of the universe than normal quark matter. Recent experiments by F. Reines and his group and by a group in the Soviet Union suggest that the neutrino will have a mass greater than several times 10^{-9} neutron or proton masses. Therefore the bulk of the mass of the universe could be in the form of netrinos rather than neutrons and protons.

Another important related problem in astrophysics is the fact that the amount of matter inferred from the gravitational interaction of galaxies seems to be much larger than the amount of matter that is actually seen to be radiating light. Since normal quark matter could radiate light, this allows for the possibility of a large amount of matter that is not quark matter. In fact, the upper limit on the amount of quark matter inferred from the synthesis of helium and deuterium in the big bang shows that the amount of quark matter seems to be less than the amount of matter gravitationally associated with galaxies.

Schramm and Steigman have used this point to argue that perhaps the so-called missing mass associated with galaxies is not quark matter at all but is instead massive neutrinos. It was shown by Gunn and S. Tremaine following the early work of R. Cowsik and his collaborators, that massive neutrinos could cluster with galaxies and form halos to the galaxies that would not be concentrated in the central visible regions but would instead add to the total mass when the galaxy interacted with another galaxy, thus being a perfect candidate for this missing mass. It is even possible that, despite the fact that there are not enough quarks to stop the present expansion of the universe, as was shown by Gunn, R. Gott, B. Tinsley, and Schramm, there may be enough neutrinos to do so. Thus, knowledge of the future of the universe—whether it will expand forever or eventually collapse—is now more uncertain than it was a few years ago, because of the new possibility that neutrinos may have mass.

For background information *see* COSMOLOGY; ELEMENTARY PARTICLE; INTERACTIONS, FUNDAMENTAL; QUARKS in the McGraw-Hill Encyclopedia of Science and Technology.

[DAVID N. SCHRAMM]

Bibliography: J. R. Gott el al., *Sci. Amer.*, 234(3): 62–79, March 1976; D. N. Schramm and G. Steigman, *Gravity Award Essay*, Gravity Research Foundation, 1980; M. S. Turner and D. N. Schramm, *Phys. Today*, 32(9):42–28, September 1979; S. Weinberg, *The First Three Minutes*, 1977.

Earthquake

The New Madrid earthquake zone is so named because New Madrid, MO, was the closest settlement to the epicenters of the series of earthquakes that occurred in the winter of 1811/1812. At that time St. Louis, Louisville, Cincinnati, and New Orleans had a population of a few thousand each, and Memphis, Chicago, and Detroit were not yet founded. At least three of the earthquakes of the series were felt throughout the entire United States, as it existed in those days, and the ground waves produced by the earthquakes were felt even by inhabitants of Quebec at a distance of 2000 km. In a 3-month interval from December 16, 1811, through March 15, 1812, an amateur scientist at Louisville, 400 km from the epicenters, observed 1874 distinct earthquakes of varying degrees of intensity. In Washington, DC, 1300 km from the epicenters, at least 18 of these earthquakes were felt. Insofar as the number, severity, felt ground waves, and damage areas of the earthquakes are concerned, this series is unique with respect to the United States and probably the world. Fortunately the earthquakes occurred when the area was sparsely inhabited and practically undeveloped, so that loss of life and property damage were not great. Questions naturally arise as to the potential for a recurrence of such earthquakes and of their socioeconomic impact on the United States. Thus the current research, which is devoted to understanding the causes of the earthquakes, their recurrence rates, and the character of the ground shaking produced by them, is of much practical as well as scientific interest.

In the late 1970s there was a concentrated effort by a number of universities, state geological surveys, the U.S. Geological Survey, the U.S. Nuclear Regulatory Commission, and the National Science Foundation to obtain a better understanding of the causes and effects of earthquakes in the New Madrid seismic zone. Although much remains to be learned, the research accomplishments are impressive. This article describes some of the more significant results.

Fault trace. Figure 1 shows the locations of earthquakes which have been detected and located from January 1976 through December 1978 by the Saint Louis University seismograph network.

Many of these earthquakes are so small as to be unnoticeable. These microearthquakes, in the magnitude range 1–2.5, are useful because they delineate the areal extent of the New Madrid Fault. This is an important matter, because the fault trace, unlike that of faults in California, is covered by alluvial sediments and cannot be seen at the Earth's surface. The microearthquake studies indicate that the New Madrid Fault consists of three segments: a southwest branch, a shorter central branch trending slightly west of north, and a northeast branch. The southern terminus, as deduced from present-day earthquake activity, is about 50 km northwest of Memphis. The northern terminus is near the junction of the Ohio and Mississippi rivers. Overall the fault length is about 200–250 km.

Motions and geological processes. Studies of the earthquake wave radiation from selected events indicate that the present-day motion on the southern and northern branches is right-lateral strike slip; that is, the northern side is sliding horizontally to the northeast with respect to the southern side. The central part of the fault zone is characterized by thrust faulting, in which one side of the fault rises up over the other. For the New Madrid Fault the observed present-day earthquake motions can in general be explained by a compressive stress field which is directed approximately east-west.

Aeromagnetic, gravity, and seismic reflection surveys carried out by the U.S. Geological Survey, help to explain the relation of the New Madrid earthquakes to geological processes. Aeromagnetic and gravity studies define a rift zone in the Precambrian rocks which is approximately 50 km wide and 200 km long (Fig. 2). The southern branch of the New Madrid Fault, as delineated by earthquakes, lies along the central axis of the rift zone. The northern portion of the New Madrid Fault, however, lies along the western edge of the rift zone. The aeromagnetic and gravity studies also indicate that a number of deep-seated intrusive geologic bodies, called plutons, lie on the borders or within the rift zone. Although there is no 1-to-1 correspondence between the existence of plutons and the location of present-day earthquakes, plutons are believed to be capable of locally modifying the regional stress field and thus perhaps producing conditions favorable for the occurrence of earthquakes.

A number of seismic reflection profiles conducted for the U.S. Geological Survey, on both the southern and central portions of the fault zone (Fig. 2), conclusively demonstrate the existence of subsurface faults. The major fault zone which they define is of pre–Late Cretaceous age, and there is evidence of recurrent tectonics in the earthquake zones.

Recurrence period. In general there are two distinct ways for estimating the average recurrence time of major earthquakes in a fault zone. One makes use of the displacement of geologically recent soils and sediments, and the other involves an extrapolation of the historical earthquake record. The former method, when employed in western Tennessee, showed evidence of three major and distinct earthquakes within the past 2000 years. The latter gives a return period of major

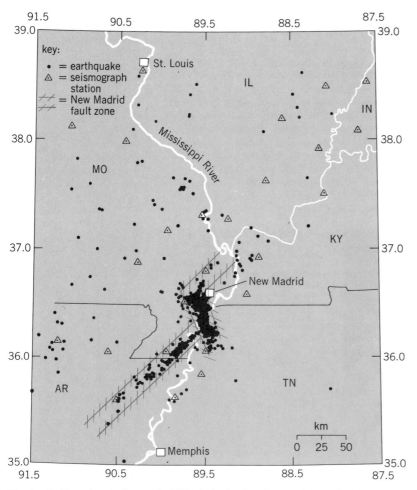

Fig. 1. Earthquakes in the central Mississippi valley for the time period January 1, 1976, through December 31, 1978.

earthquakes of 400–700 years, based upon the earthquake activity in the past 200 years. Thus these two independent methods for obtaining the recurrence time of major earthquakes give values which are quite similar, approximately 600 years. For comparison, the recurrence time of major earthquakes on the San Andreas Fault of California is estimated to be approximately 100 years.

The relatively long average recurrence period of major earthquakes along the New Madrid Fault is no cause for complacency. There is about a 6% probability that a severe earthquake will occur in any given 50-year time interval. Furthermore, somewhat smaller earthquakes, of moderate to major size, can be expected to occur more frequently. For them the probability of occurrence is about 50% in any given 50-year time interval. Earthquakes of this size occurred in 1843 and 1895, the former near the southern end of the New Madrid fault zone and the latter at the northern end.

Damage areas. Another reason for not being complacent about the hazard is that earthquakes east of the Rocky Mountains have much larger damage areas than earthquakes of similar magnitude to the west, because of smaller anelastic attenuation of the destructive high-frequency ground waves in the eastern region. Thus cities and towns at distances of up to several hundred kilometers from a moderate-to-major New Madrid earthquake have suffered damage in the past, and can be expected to do so in the future. The large increase in population in the central Mississippi valley in the last 50 years makes the problem more acute. The 1975 population residing within the area which suffered major damage from the New Madrid earthquakes of 1811/1812 is 12,600,000, compared to 4,000,000 residing in the major damage area of the 1906 San Francisco earthquake. Thus the relatively large damage areas of the New Madrid earthquakes tend to compensate for their large recurrence time. That is, there will be fewer damaging earthquakes along the New Madrid Fault than in areas such as California, but the amount of destruction averaged over a long time interval will be similar for the two.

Comparison with Tangshan earthquakes. The Tangshan, China, earthquakes of July 28, 1976, provide some interesting points of comparison with the New Madrid earthquakes of 1811/1812. The principal Chinese earthquake, of surface-wave magnitude 7.8, together with its large aftershocks, caused a great loss of life. First reports placed it at 650,000 people, but a later official report stated it was 250,000. In addition, more than 750,000 people were injured. The city of Tangshan bore the brunt of the damage, but a significant number of buildings in Tientsin, 100 km away, collapsed or were seriously damaged. By comparison, Memphis is 50 km away from the southern end of the New Madrid fault zone. Tangshan is situated near a coastal plain which has a thick surficial alluvial deposit of sandy gravels, sands, and clayey soils. The New Madrid fault zone is located at the northern end of the Mississippi Embayment, and also contains thick alluvial layers of similar composition. In both the 1811/1812 and 1976 earthquakes there was widespread evidence of soil failure, in the form of sandblows, landslides, liquefaction,

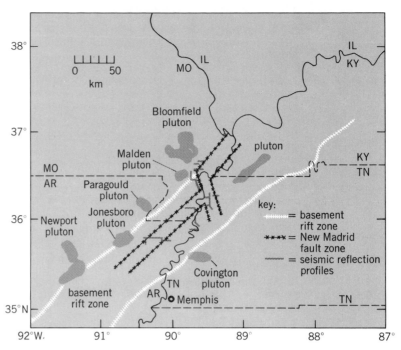

Fig. 2. Location of basement rift zone, plutons, New Madrid fault zone, and seismic reflection profiles conducted for the U.S. Geological Survey.

and large relative uplift and subsidence of the ground. The area of severe damage (intensity IX and greater) for the 1976 earthquake was 1800 km², whereas for the December 16, 1811, earthquake it is estimated to be about 50,000 km². Corresponding figures for areas of major damage (intensity VII and greater) are 33,000 and 500,000 km², respectively. The predominant type of buildings in Tangshan, as in the cities of the central Mississippi valley, is unreinforced brick. In Tangshan there was significant damage to bridges, highways, and pipelines over a large area. Damage to pipelines is of particular concern for the New Madrid region, since the gas pipelines which supply most of the north-central and northeastern United States pass through the Mississippi Embayment, close to the New Madrid Fault.

Earthquake prediction has not yet developed to the point where one can state with confidence when the next large earthquake will occur along the New Madrid Fault. Evidence both from surficial soil displacements and from the ongoing earthquake activity assure that large earthquakes will occur again there. Prudence dictates that efforts should be made to minimize the risk associated with such phenomena.

For background information *see* EARTHQUAKE; FAULT AND FAULT STRUCTURES in the McGraw-Hill Encyclopedia of Science and Technology.

[OTTO W. NUTTLI]

Bibliography: F. A. McKeown, Investigations of the New Madrid earthquake region: Overview and discussion, in F. A. McKeown and L. C. Pakiger (eds.), *Investigations of the New Madrid Earthquake Region*, Prof. Pap. U.S.G.S., 1981; O. W. Nuttli, *Bull. Seismol. Soc. Amer.*, 63:227–248, 1973; W. Stauder et al., *Bull. Seismol. Soc. Amer.*, 66:1953–1964, 1976; M. D. Zoback et al., Recurrent intraplate tectonism in the New Madrid seismic zone, *Science*, 209:971–976, 1980.

Ecological interactions

Chemoreception is a sense that is fundamental to virtually all living organisms. This is true for bacteria and unicellular algae, as well as for invertebrates and vertebrates. Chemoreception can be important for such essential functions as predator avoidance and the location of food, mates, and shelter. A fascinating area of research is the use of chemoreception by symbiotic invertebrates in the location of their hosts.

Symbiosis and chemoreception. Some species have evolved a special relationship in which they are primarily or solely found in association with each other; this relationship is called symbiosis. The host provides the symbiont with a habitat which can supply any of a number of necessary resources, including food and shelter. Symbiosis can be divided into three categories: mutualism, commensalism, and parasitism. In mutualism, both organisms receive benefit from the association. Mutualistic associations include some of the

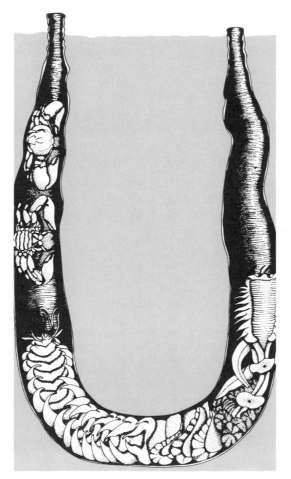

Symbiotic association between marine worm and crabs. This section of the U-shaped parchment tube of the polychaete worm *Chaetopterus* is shown as it naturally occurs in the sand. The worm is in the lower part of the tube, and two symbiotic crabs (*Polyonyx*) are in the left arms of the tube. (*From A. S. Pearse, On the habits of the crustaceans found in Chaetopterus tubes at Woods Hole. Massachusetts, Biol. Bull., 24:104–114, 1912, Fig. 7, Plate I. Publisher: Marine Biological Laboratory, Woods Hole, MA*)

most amazing examples of symbiosis, such as the association between certain hermit crabs and the sea anemones that live on their shells. The hermit crab goes through an elaborate ritual of stroking the anemone in an attempt to induce the anemone to attach onto the crab's shell. The anemone can provide the crab with both camouflage and protection from predators, while the messy eating habits of the crab can provide food for the anemone. In commensal associations, one member derives benefit while the other member is neither harmed nor helped. In parasitism, the benefit derived by one member of the association is detrimental to the other member. It is often difficult to determine the precise functional relationship between the two associated organisms; therefore, unless this relationship is adequately analyzed, it is best to reserve for it the more general term of symbiosis.

Many symbionts are so dependent on their hosts that they die if separated from their hosts. An effective mechanism for finding a host is therefore essential for the survival of the symbiont. Chemoreception is one such mechanism. There are many intriguing examples of the use of chemoreception in specialized symbiotic associations involving insects and plants. One example is the intricately interrelated life of certain bees and orchids. The flower mimics the body shape and color as well as the odor of female bees. When a male bee of the particular species attempts to copulate with the orchid flower, he becomes covered with pollen. When the male bee realizes his mistake and tries his luck with another flower, he pollinates the second flower.

Marine invertebrates are also involved in many curious symbiotic assications; however it was not until the work of Demorest Davenport, beginning in the 1950s, that the importance of chemoreception in these associations was realized. Davenport demonstrated that the search of many symbiotic polychaete worms for hosts is not random, but that chemicals emanating from these hosts could be sensed by the worms and used as cues in host location. Chemoreception is now known to be important in host location by marine invertebrate symbionts of many different taxonomic groups, such as sea anemones, flatworms, bivalves, snails, polychaete worms, copepods, shrimp, and crabs.

Pea crab—bivalve symbiosis. Pea crabs (*Pinnotheres maculatus*) are small marine crabs that live in association with a variety of hosts, including in such bivalves as mussels and scallops, and in the tubes of polychaete worms (see illustration). Their ability to inhabit several different host species makes them host generalists, in contrast to host specialists that can inhabit only one host species. Pea crabs are sexually dimorphic, with the female being much larger and more adapted for a symbiotic life-style than the male. It is the round shape of the body of the female that gives this species its common name.

Host location is important at several stages in the life cycle of pea crabs. After the eggs hatch, the young crabs, called larvae, live freely in the open water. When the larvae develop into the "invasive stage," they enter a host. They live in their hosts and go through several more developmental stages until they reach the "hard stage," where-upon both males and females leave their hosts, engage in open-water copulatory swarming, and then resettle into hosts again. From this description of their life cycle, it is obvious that host location is essential for pea crabs at the invasive and hard stages, as well as at any stage when the host might die.

It is interesting to consider how the crabs find their hosts and how they can differentiate between host and nonhost species. Vision may be a partial answer. The larvae of many crabs are known to change their response to light as they prepare to settle out of the open water. Whereas the early larval stages are attracted to light (a behavior called positive phototaxis), the later larval stages avoid light. This negative phototaxis may direct the invasive stage to the ocean floor, where the hosts are to be found. The crabs still have the task of locating bivalves and distinguishing between host and nonhost species. To examine the possible importance of chemoreception in host location by these crabs, C. Derby and J. Atema observed how subadult and adult crabs collected from the blue mussel, *Mytilus edulis*, responded to odors from several species of live bivalves. They found that the crabs were attracted to an odor from blue mussels but not to odors from other bivalves such as bay scallops, ocean scallops, or ribbed mussels. This is interesting because pea crabs are known to be able to inhabit most of these bivalve species.

One hypothesis to explain these results is that once a crab initially invades a host, it develops a strong attraction to the odor from individuals of that host species. This phenomenon has been called host induction or conditioning. This hypothesis was tested by allowing crabs that had been removed from blue mussels to associate with bay scallops for 2 to 5 weeks. During the course of this induction process, the crabs developed an increased attraction to bay scallop odor and even acquired a preference for bay scallop odor over blue mussel odor. This experiment therefore provides evidence for the involvement of chemically mediated host induction in the symbiotic association between pea crabs and bivalves. The adaptive significance of this phenomenon is that the crabs would be more likely to reinvade their host or other members of the host species, which is important in the hard stage following copulatory swarming. The crabs may also reinvade hosts when they are dislodged or when the hosts are damaged or die. Unfortunately, due to difficulty in raising pea crabs from eggs, little is known about the use of chemoreception by the larval and invasive stages. The larvae of some marine species, such as polychaete worms, snails, and barnacles, are known to use specific chemical cues in determining where to begin their adult lives. It is likely that the larval and invasive stages of pea crabs use such a mechanism, especially considering the chemoreceptive acuity of the older crabs. However, such a conclusion awaits further investigation.

Nature of chemical messages. A great deal is known concerning the structure of chemicals used by insects in intra- and interspecific relationships. These chemicals include pheromones, which are used in sexual recognition of male and female insects of the same species, and secondary plant compounds, which can be important in determining species-specific host or food preferences of insects. However, there are very few cases in which anything is known about the chemical nature of substances involved in host location by marine invertebrates. Perhaps the best-studied system in this regard is the association of the red abalone, *Haliotis rufescens*, and certain species of red algae: abalone larvae settle and begin metamorphosis after contact with the red algae due to the presence of γ-aminobutyric acid. In a few other cases, only the molecular weights of the attractants are known. Mixtures of chemicals rather than just single compounds are likely to be important in chemically mediated host location by marine invertebrates, as is now known from several insect pheromone systems. This would be especially interesting to know in such cases as the pea crab–bivalve association in which responses of the symbiont to host odors can be modified. The plasticity of the responses to host odors indicates that there may not be just one chemical compound that triggers the attraction, but rather that different chemical constituents of the odor can be used by the symbiont at different times.

For background information *see* CHEMORECEPTION; ECOLOGICAL INTERACTIONS in the McGraw-Hill Encyclopedia of Science and Technology.

[CHARLES D. DERBY; JELLE ATEMA]

Bibliography: P. Castro, *J. Exp. Mar. Biol. Ecol.*, 34:259–270, 1978; C. D. Derby and J. Atema, *Biol. Bull.*, 158:26–33; P. T. Grant and A. M. Mackie (eds.), *Chemoreception in Marine Organisms*, 1974; D. E. Morse et al., *Science*, 204:407–410, 1979; W. B. Vernberg (ed.), *Symbiosis in the Sea*, 1974.

Electric power generation

With the increasing cost and finite supplies of fossil fuels becoming an issue of widespread concern, the use of renewable, nondepleting energy sources such as solar for electric power generation is being carefully examined. A wide variety of solar electric energy systems are being developed, with applications ranging from large, centralized installations to small, distributed installations suitable for dispersed application and remote locations. In addition to offering the potential opportunity for a broad range of applications and a reduction in fossil fuel consumption, solar electric power systems may offer the benefit of reduced environmental impacts as compared to conventional power sources.

Solar energy originates from the Sun and travels to the earth in the form of electromagnetic radiation. At the Earth's surface, this energy is readily perceived as light and heat; additional manifestations of solar energy are wind currents, temperature gradients in water bodies, and plant growth. A key characteristic of solar energy is that it is both variable in quantity and unpredictable in nature; the intensity of solar energy varies with time of day, day of year, and location, and it is also subject to interruptions due to weather effects. Because this variability is not compatible with societal requirements for continual supplies of electricity, the need for energy storage becomes an integral consideration of solar electric power. Another key characteristic of solar energy is diffuseness (bright sunlight intensity is about 1 kW/m²; a typical average total insolation value on a horizontal surface for the United States is 183 W/m² for a 24-h day), meaning that large areas, and hence large collection devices, are required in order to acquire significant amounts of power.

Solar thermal power. The process of solar ther-

mal conversion has been the subject of sustained development since the initial, systematic technical and economic evaluation in 1974. This generic conversion process, which has many design-specific forms, involves the collection and concentration of solar energy incident upon a surface, its conversion to thermal energy, and the use of a heat engine to drive an electric generator. The concentration of solar energy is required to achieve elevated temperatures compatible with efficient heat engine operation. The collection/concentration

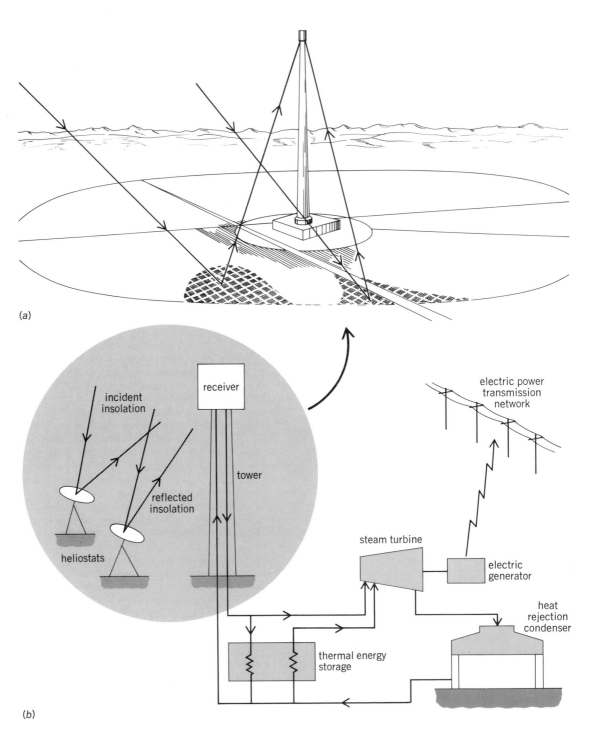

(a)

(b)

Fig. 1. Solar thermal central receiver system. (a) Configuration. (b) Block diagram of components.

phase of the process either can be centralized, in which sunlight from a large area is concentrated and converted to high-temperature heat (1000–2000°F or 540–1090°C) at a single location, or distributed, wherein the sunlight from smaller areas is collected and converted to moderate-temperature heat (500–1000°F or 260–540°C) at a series of locations. The centralized concept is suitable for large, utility-scale applications of solar electric power, while smaller, special-application situations favor distributed energy collection.

Central receiver. A central receiver system typical of those offering promise for electric utility applications for plants up to approximately 100 MW of electric power consists of several major components (Fig. 1). Large numbers of heliostats (Sun-tracking mirrors) are dispersed on the ground, each redirecting the Sun's image to a solar receiver located atop a tower. Multiple Sun images are overlaid upon one another, resulting in a high-energy density (500–1000 kW/m²) on the receiver. This intense radiant energy is absorbed by the receiver and converted to thermal energy (for the case illustrated, the thermal energy is in the form of high-temperature, high-pressure steam) which in turn drives a turbine. Other system components include an electric generator, heat rejection equipment, and controls; of these principal components, it is the heliostats and receiver that are novel and are the subject of current technology and hardware development programs. As a result of various U.S. Department of Energy and Electric Power Research Institute programs, equipment representative of both first- and second-generation designs has been fabricated and is undergoing performance testing. Other systems that utilize molten salts and liquid sodium as heat transfer fluids are also under development. The largest single project involves the construction of a pilot plant in Barstow, CA; this plant, scheduled for 1982 operation, will use 10.5-MPa 515°C (1520-psia 960°F) steam to generate 10 MW of electric power. Significant commercial applications of solar thermal central receiver systems in the early 1990s are possible as the technologies mature and as mass production of components reduces system costs.

Distributed systems. For small power systems (equal to or less than 1 MW of electric power), distributed solar thermal systems offer advantages related to less complex solar receiver designs, less severe operating temperatures, and modular collector field design. The principal component of a typical distributed solar electric power system is the parabolic trough collector (Fig. 2), which is used in lieu of the heliostats and central receiver described above. The trough rotates to track the Sun's apparent motion, redirecting and concentrating the Sun's image to a line focus on an absorber tube located along the trough longitudinal axis; a heat transfer fluid is forced through the absorber tube to carry off the resultant heat for use in a heat engine. The large surface area of the trough compared to the absorber tube results in intensities of approximately 60 kW/m² in current system designs. Because the energy intensities are much less than those achievable in the central receiver system, working-fluid temperatures are

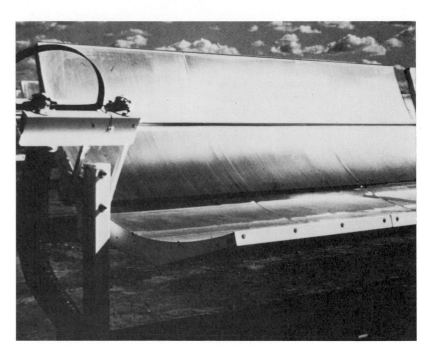

Fig. 2. Parabolic trough solar collector.

limited to around 315°C (600°F). This moderate temperature requires the use of special turbomachinery (such as organic fluid Rankine cycle turbines) and restricts the thermodynamic efficiency of the power generation process. When substantial amounts of power generation are required, this efficiency limitation overwhelms the apparent advantages of the distributed system, making the central receiver system more cost-effective.

Parabolic dish collectors represent another design approach to distributed system technologies. Dish collector systems concentrate redirected sunlight to a single-point focus, thus making high temperatures (about 538°C or 1000°F) and improved thermodynamic cycle efficiencies attainable. Designs under development include a receiver and heat engine colocated at the focal point of each dish collector, with both Brayton and Stirling cycle engines being evaluated at the power converter/electric generator.

Photovoltaics. Photovoltaic cells or "solar cells" (Fig. 3) convert sunlight to direct-current electricity through the photovoltaic effect. This effect occurs when the energy from light produces electron-hole pairs in semiconductor devices in which the separation of the positive (hole) and negative (electron) electric charges occurs, allowing electric current to flow; the semiconductors are made principally of materials such as silicon, cadmium sulfide, or gallium arsenide. Because the peak output of individual cells is typically 0.5 V at 2–3 A, large cell banks with various series and parallel electrical connections are required to achieve even modest electric power ratings.

Photovoltaic systems can be configured in two basic ways: concentrating and nonconcentrating. The nonconcentrating systems, the less complex and more developed of the two options, employ light incident directly from the Sun at peak levels

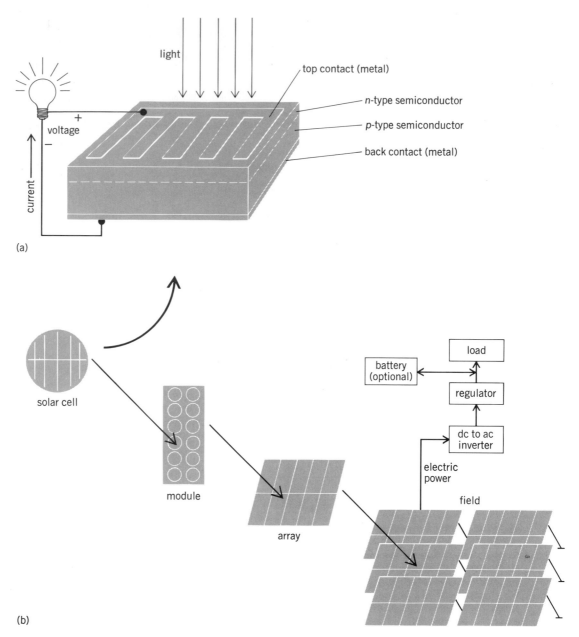

Fig. 3. Photovoltaic cell and system. (a) Structure of cell. Top contact is fingered or grid-shaped to allow light to reach the semiconductor surface. (b) Organization of components into system.

of approximately 1 kW/m². Alternatively, concentrating systems use a variety of devices such as mirrors or a Fresnel lens to achieve more intense radiation on the solar cells. The concentrating systems, while more complex, due to higher operating temperatures and requirements to cool the cells, are being pursued because of the potential for reductions in system cost via efficiency improvements and smaller numbers of cells. Photovoltaic systems, currently about twice as costly as solar thermal electric systems, are utilized primarily for special remote, low-power requirement applications; significant technology breakthroughs will be required to achieve the cost effectiveness and the attendant commercial applications forecast for the 1990s.

Wind power. Wind currents develop as a result of the differential heating and cooling responses of various earth masses (for example, land versus water) to solar energy. The kinetic energy in wind can be extracted by wind turbines that convert it to mechanical energy and then to electric energy. The power output of such a system would directly reflect the variable nature of wind currents, which are generally strongest during the daylight hours and during the spring.

Wind turbine technology has made substantial gains in recent years, with a variety of prototype machine designs currently undergoing field testing. Of the many design concepts, major emphasis to date has been placed on horizontal axis machines. The largest horizontal axis machines in current production have a rated capacity of 2.5 MW of electric power, with a 91-m-diameter (300-ft) rotor located atop a 61-m (200-ft) steel tower; three of these machines are scheduled for installation and testing near Goldendale, WA, in 1980 – 1981. Because of size limitations on the rotor blades imposed by stress/strain and fabricability considerations, the maximum expected capacity of individual wind turbines is 3–5 MW of electric power, though multiple turbines could be grouped in wind energy farms to create larger-capacity sources.

Ocean thermal. Ocean thermal energy conversion (OTEC) is a process that utilizes the solar energy collected by oceans in the form of warm surface waters. Thermal energy is extracted from the warm surface waters and is used to vaporize a working fluid such as ammonia, which drives a turbine to generate electricity. After passing through the turbine, the working fluid is condensed for recycling and the cycle waste heat rejected to cold subsurface water. Since the vertical temperature gradient in the oceans is essentially constant, this solar energy option offers baseload generation capability without the need for storage; it is expected that plants would have ratings of 100–400 MW of electric power.

This system would be housed on an oceangoing platform; a variety of concepts including barges, ships, and spheres have been considered for this floating platform. Studies have indicated that a temperature gradient of about 20°C (36°F) is required for plant operation. To employ this temperature difference as means of driving the electric generation process, cold water from approximately 1000 m (3300 ft) deep must be pumped to the surface in large quantities (approximately 3800 liters/s/MWe or 1000 gal/s/MWe). The engineering development of the large cold-water pipe, the very large heat exchangers, the special low-temperature turbines, oceangoing survival demands, and large-capacity submarine power transmission cables present substantial development challenges for this technology. Significant commercial application of these systems is limited by the number of suitable sites at which the required temperature gradient is both accessible and in proximity to an electrical load center.

Biomass. Biomass consists of materials that have been created by sunlight through photosynthesis. Although fossil fuels are metamorphosed biomass, the extreme length of time associated with their creation classifies them as nonrenewable. In contrast, biomass is renewable and makes use of materials such as wood, grain, crop residue, and manure. Because the solar energy is stored in a chemical form within these materials, biomass does not present energy storage problems as in the case of solar thermal or photovoltaic systems.

Biomass can be used by way of direct combustion (as in burning wood chips) or by way of gasification to low-, medium-, or high-Btu gases. These fuels can generally be used in conventional power plant equipment (steam boilers, gas turbines, and so forth) with relatively minor modifications. Although the technologies for converting biomass materials to gaseous fuels are well understood, experience with large-scale, long-term operations is lacking. If large-scale operations and substantial biomass feedstock supplies prove to be technically and economically feasible, gasified biomass could be an important element of commercial electric

power generation systems in the 1990s.

Storage. Electric power demands, while varying with time of day, are continual. If significant portions of that electrical demand are to be fulfilled by solar energy, storage is required to provide an energy source during solar outages (such as nighttime, cloud passage, or absence of wind). While energy may be stored as heat, chemical energy, or electric energy, the technologies for large-scale, long-term, cost-effective storage are not well developed at this time and present a major obstacle to the widespread commercial implementation of solar thermal, wind, and photovoltaic systems. Because the generation capability of these systems is subject to interruption, backup generation capacity is required to ensure that load demands can be met; this imposes large additional expenditures on electric utilities and thus reduces the overall economic viability of solar electric power systems. As a result of these considerations, the probable role of most solar technologies is to serve fractions of the peak and perhaps intermediate loads, while relying upon the geographic dispersion of generating stations to minimize the impacts of weather effects and on other power generating technologies to satisfy the majority of electric power demand.

Solar electric power generation can be in commercial use by 2000. The mix of solar technologies is likely to include solar thermal, wind, biomass, ocean thermal, and photovoltaic systems, in varying amounts that will be determined by economics and performance characteristics. However, the inherent nature of solar energy and its resultant technologies indicates that its proper role will be to supplement, rather than to replace, baseload power systems such as nuclear and coal-fired power plants.

For background information see SOLAR BATTERY; SOLAR ENERGY; WIND POWER in the McGraw-Hill Encyclopedia of Science and Technology.   [JOHN KINTIGH]

Bibliography: J. C. Powell and J. C. Grosskreutz, *Dynamic Conversion of Solar Generated Heat to Electricity*, NASA CR-134724, August 1974.

Electrical utility industry

Peak-load growth of electrical utilities in 1980 rebounded to its highest level since 1977. Driven by record heat throughout the south and central regions of the United States, summer peak demand rose 6.6%. This demand occurred in spite of a deep recession that slashed industrial activity. Energy sales, in fact, rose only 0.7%, dragged down by a negative industrial growth of −2%.

The combination of high peak and low sales tended to confirm what some forecasters had feared for some time: price-induced conservation would suppress electrical energy consumption but, under extreme conditions such as the heat that occurred in summer 1980, the consumer would abandon conservation for the duration of the extreme conditions. This produces a "needle" peak—that is, a very sharp spike of demand of relatively short duration.

This has strongly negative implications for utilities. Companies must construct facilities to supply peak demand, but low usage throughout the rest of the year would make it financially difficult to justify the investment required. There is evident, therefore, a growing inclination for utilities to plan their construction budgets not on what they need to meet peak demand, but on what they can reasonably finance. Buttressing their thinking is the fact that in 1980, utility equity issues brought to market sold at an average of 78% of book value.

The ongoing situation at the Three Mile Island nuclear plant in Pennsylvania reached another critical turning point in 1980. Workers entered the containment vessel for the first time since the accident that caused partial core meltdown and extensive release of radiation within the building in March 1979. They found that overall damage and the condition of the equipment were better than had been anticipated. Complete cleanup, however, will not be finished until about 1985.

Another nuclear milestone was reached in late September when a referendum in Maine was rejected. The referendum called for the shutdown of the nuclear plant at Wiscassett, which supplies almost a third of the electric energy used annually in the state.

Also in 1980, utilities set out to implement the new regulations of the Energy Conservation Service program, required under the National Energy Conservation Policy Act. These rules require that utilities actively encourage a wide range of consumer activities, including providing energy audits. Consumer response to the program to date has been unenthusiastic.

Ownership. Ownership of electrical utility facilities in the United States is pluralistic, being shared by private investors, customer-owned cooperatives, and public bodies on city, district, state, and Federal levels. Investor-owned companies constitute by far the major portion of the industry. They serve 69,378,000 customers, representing a 77.5% share of all electricity customers, and own 78.3% of the installed generating capacity. Cooperatives serve 10.2% of the total, but only own about 2% of the generating capacity. Public bodies at all levels serve 12.3% of the total electricity customers, and own 19.7% of the installed generating capacity. And, of this amount, Federal agencies hold 9.5%, and all others, 10.2%.

The small amount of generating capacity owned by the cooperatives reflects the fact that most such organizations are distribution companies which buy their power either from investor-owned utilities at wholesale rates, from special generation and transmission cooperatives, or from publicly owned utilities.

There is a growing tendency for cooperatives and, to some extent, municipal utilities to purchase shares in large generating units built by investor-owned utilities. This arrangement permits small utilities to share in the economies of scale of very large units, and in the lower costs of nuclear units. For the investor-owned builder, the arrangement eases the financial drain, since cooperatives and public entities have access to lower-cost financing and answer antitrust requirements. Typical is the joint ownership of Black Fox nuclear plant now under construction in Oklahoma. Public Service Company of Oklahoma owns a 700-MW share, Associated Electric Cooperative owns 250 MW, and Western Farmers Electric Cooperative the remaining 200 MW.

There is also a trend toward construction by cooperatives of large-base-load generating units specifically planned to supply energy to privately owned utilities. Cajun Electric Power Cooperative

United States electric power industry statistics for 1980*

Parameter	Amount	Increase or decrease compared with 1979, %
Generating capability, ×10³ kW		
Conventional hydro	65,585	4.0
Pumped-storage hydro	12,274	2.6
Fossil-fueled steam	529,590	4.2
Nuclear steam	63,116	17.4
Combustion turbine and		
internal combustion	56,795	1.3
TOTAL	627,630	4.9
Energy production, ×10⁶ kWh	2,296,000	0.6
Energy sales, ×10⁶ kWh		
Residential	704,400	1.5
Commercial	504,400	2.0
Industrial	797,700	−1.9
Miscellaneous	77,000	1.4
TOTAL	2,083,500	0.27
Revenues, total; ×10⁶ dollars†	86,600	2.3
Capital expenditures, total;		
×10⁶ dollars†	39,205	11.4
Customers, ×10³		
Residential	81,600	2.8
TOTAL	91,950	2.5
Residential usage, kWh (average)	8,751	−0.8
Residential bill, ¢/kWh† (average)	4.82	2.1

*From 31st annual electrical industry forecast. *Elec. World*, 194(6):55–70, Sept. 15, 1980; 1980 statistical report, *Elec. World*, 193(6):49–80, Mar. 15, 1980; and extrapolations from Edison Electric Institute monthly data.
†In 1980 dollars.

in Louisiana, for example, with a total system demand of only 945 MW, is now constructing three 500-MW coal-fired units, with another planned. The Seminole Electric Cooperative, with only 11 wholesale customers, has contracted for two 500-MW units, the output of which will be sold to the other Florida utilities.

Capacity additions. Utilities had a total generating capability of 597,523 MW at the end of 1979, having added 17,075 MW during that year. By the end of 1980, industry capability will have increased to 627,630 MW (see table). The annual summer peak demand for the entire country in 1980 was 436,100 MW which, with the substantial capacity additions in 1980, gave a new reserve margin of 32.6%. This is a slight decrease from the 36.3% recorded in 1979. Normal target is taken by most utilities to be 25%, though the norm can vary regionally from 13 to 28%.

The 30,107 MW of capacity added during 1980 consisted of 17,191 MW of fossil-fuel capacity, 9372 MW of nuclear units, 2496 MW of conventional hydroelectric, 307 MW of pumped-storage hydroelectric, 716 MW of combustion turbines, and only 25 MW of diesels.

Added capacity by type of ownership was 22,373 MW by investor-owned utilities, 2675 MW by cooperatives, 3315 MW by Federal agencies, and 1744 MW by public bodies.

The composition of total plant as of the end of 1980 was 429,590 MW of fossil-fueled (of this fossil-fired capacity, 61.0% was coal, 25.7% was oil-fired, and 13.3% was gas-fired), 9% nuclear, 10.8% conventional hydroelectric, 2.1% pumped-storage hydroelectric, 8.5% combustion turbines, and 0.9% of internal combustion engines such as diesels (see illustration).

Fossil-fueled capacity. Fossil-fuel units constitute 57.1% of the total new capacity added in 1980. Thirty-five individual units went into service, of which 33 were coal-fired, 2 were oil-fired, and none was fueled by gas. Four diesels, each rated at 6000 kW, came on line in municipal installations, and two geothermal plants with a total of 245 MW were started up.

The $11,900,000,000 expended in 1980 for fossil-fired construction was up from the $11,100,000,000 expended in 1979. This was actually a decrease of 1.6% in real terms.

Nuclear power. Utilities added 8 more nuclear units in 1980 to bring the total of reactors now operating in the United States to 80. Total capacity of the plants brought into service during the year was 9372 MW, raising the total now operating to 63,116 MW. Of the units added, 1 was a boiling water reactor (BWR) and 7 were pressurized water reactors (PWR). There are now 48 PWRs and 27 BWRs operating in the United States: the remaining 5 units use other technologies. During 1980, 80 units were postponed or delayed for periods of from 1 year to indefinitely. Because of lower growth financial pressures and licensing concerns, it is now probable that no new orders will be placed until 1986.

The 113 nuclear units now planned for the future or in construction have a total capacity of 129,284 MW which, if current plans hold, will bring nuclear capacity to about 15% of all installed

capacity by 1995. During 1979, nuclear units generated a total of 255.4×10^9 kWh, accounting for 13% of that year's total.

Combustion turbines. Combustion turbines, because of the quick-start capabilities, that is, the ability to go from cold to full load in 2–3 min, and their low initial capital cost of about $250/kW, have served admirably as peak-load units. Utilities keep an average of about 9% of their peak demand in gas turbine capacity, using them for several hundred hours a year to meet annual peak loads, or to go on line quickly to supply load in emergency situations. However, because of the uncertainty of the future supply of the distillate oil or gas that these machines burn, and the uncertainty of national policies concerning the permissible use of petroleum fuels and permissible levels of the nitrogen oxides that these machines produce in their

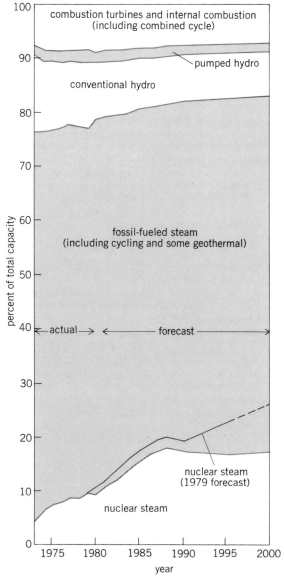

Probable mix of net generating capacity. (*From 31st annual electrical industry forecast, Elec. World, 194(6): 55–70, Sept. 15, 1980*)

exhaust gases, this percentage will undoubtedly decline substantially.

Utilities brought 716 MW of combustion turbines into service in 1980. This was made up of 11 individual units. The total installed combustion turbine capacity for the entire industry is now 51,278 MW, or 8.2% of total capacity. Of the total, however, 1315 MW represents the combustion turbine portion of combined-cycle installations, in which the 900–1000°F (482–538°C) exhaust gas of the turbines is used to generate steam for a conventional steam turbine. Utilities spent $465,000,000 on combustion turbine construction in 1980.

Hydroelectric installations. Installation of conventional hydroelectric capacity continued, with 2496 MW coming on line in 1980, and a total of about 12,400 MW additional capacity planned for the future. Installed capacity, amounting to 62,507 MW, provided 10.0% of total industry capacity in 1980. However, as sites become increasingly more difficult to find and develop, this percentage is expected to decline to 9% by 1985. During 1980 the industry spent $531,500,000 on hydroelectric projects, of which $298,500,000 was by Federal agencies and $172,000,000 by investor-owned utilities. The future of hydroelectric power will depend on economical development of low-head, or run-of-river, turbines, since environmentally acceptable high dam sites are almost nonexistent. The U.S. Department of Energy has singled out these small hydroelectric plants for special handling under the Public Utilities Regulatory Policy Act (PURPA) of 1978. The upper limit of size for such classification is 15 MW, and the Army Corps of Engineers has identified 50,000 small dams in the United States that could be developed for this capability.

Pumped storage. Pumped storage represents one of the few methods by which electric energy can be stored. In the mechanical analog, water is pumped to an elevated reservoir during off-peak periods and is released through hydraulic turbines during the subsequent peak period, recovering 65% of the original fuel energy. Although the 13,102 MW installed at the end of 1980 was only 2% of the industry's total capacity, another 12,000 MW is planned for the future, and utilities continue to spend about $400,000,000 annually on such projects. The world's largest pumped-storage installation is Ludington Station on Lake Michigan, rated at 970 MW.

Demand. Peak demand rose 6.6% in 1980, driven by an extraordinarily hot summer throughout the South and Midwest. Total noncoincident peak reached 436,100 MW in July. Growth due to the abnormally high summer temperatures was at least in part offset by the depressed economy. After adjusting for these two factors, real growth was approximately 2.5%.

This low value of real growth reinforced the case of those who see substantially lower growth in demand than that experienced previously. Pre-oil-embargo growth had averaged 8% and the 1974–1979 period experienced annualized growth of 3.7%. The consensus at this time, however, is that long-term growth will average about 4–4.5% annually.

There was wide regional disparity in growth of demand. The Texas and south-central regions, which were insulated from the economic recession because of a combination of population growth and energy industries, and which experienced more than a month of continuous temperatures of higher than 100°F (37.8°C), grew most strongly. Utilities in Texas and the mid-south-central regions averaged about 14%. The north-central areas, however, though also experiencing severe heat, were badly afflicted by the recession. These areas grew about 4.0% which, temperature-corrected, was effectively a zero or slightly negative real growth. The West Coast, normally an area of high growth, in 1980 exhibited the lowest growth of all regions. This was due in part to relatively normal summer temperatures, and to severe conservation measures imposed on utilities and business in general by policy in the southern regions, and by tight hydroelectric supply in the north. The region grew only 2.5% over 1979 as a result. Other regional peak demand growth rates were: Mid-Atlantic region, 8.3%; New England, 5.8%; Southeast, 8.2%.

Usage. Sales of electric energy lagged considerably behind peak growth, rising only 0.3% over 1979, to 2.0834×10^{12} kWh. Output rose from 2.2828×10^{12} kWh in 1979 to 2.2959×10^{12} kWh in 1980. The primary reasons for the low growth in sales were twofold: a warm winter offset the hot summer, and the economic recession markedly suppressed industrial consumption.

Residential consumption rose from 693.9×10^9 kWh in 1979 to 704.4×10^9 kWh in 1980, a 1.5% increase. Heating sales, however, dropped from 140.5×10^9 kWh in 1979 to only 135.2×10^9 kWh in 1980 because of the warm winter. Normally, sales for heating rise about 6 to 10×10^9 kWh annually. Roughly half of all new housing units being built each year are electrically heated. Average use per customer dropped, also, from 8828 to 8751 kWh per year. Despite the drop, the average annual bill per residential customer rose from $417 to $422, because of rate increases.

Commercial consumption rose 2%, from 494.5×10^9 kWh in 1979 to 504.4×10^9 kWh in 1980.

Industrial consumption sagged, dropping 2% from 813.6×10^9 kWh in 1979 to only 797.7×10^9 kWh in 1980. Generation by industrial plants other than those of central station utilities also dropped. Such plants generated 59.6×10^9 kWh in 1980 as against 65.3×10^9 kWh in 1979. Such generation has experienced a long-term decline in both absolute and percentage terms. This trend should continue, despite Federal encouragement of cogeneration by such plants. The high capital costs of such plants divert investment from productive equipment, and increasingly stringent pollution control requirements offer a strong disincentive to the industrialist.

Sales other than the above, comprising such uses as street lighting, railways, public authority use, and interdepartmental sales, rose from 75.9×10^9 kWh in 1979 to 77.0×10^9 kWh in 1980.

Fuels. The use of coal rose from 481.6×10^6 tons (437.0×10^6 metric tons) in 1978 to 529.1×10^6 tons (480.1×10^6 metric tons) in 1979, a rise of 9.9%.

Oil, however, dropped from 635.8×10^6 bbl (82.7×10^6 metric tons) in 1978 to 523.5×10^6 bbl (68.1×10^6 metric tons) in 1979, a sharp decrease of 17.7%. Gas use, however, rose from 3.1884×10^{12} ft³ (90.3×10^9 m³) in 1978 to 3.4903×10^{12} ft³

$(98.8 \times 10^9$ m³) in 1979, a 9.5% increase. As price and governmental pressures force utilities to shift away from oil as fuel, they are substituting gas in existing installations.

Energy generated by each of these major fuels and their percentage share of total generation were as follows for 1979: coal, 1.0755×10^{12} kWh (54.8%); oil, 303.0×10^9 kWh (15.4%); gas, 329.5×10^9 kWh (16.8%); and nuclear, 255.4×10^9 kWh (13.0%).

Transmission. Utilities spent $4,252,700,000 on transmission construction in 1980. This included $1,013,000,000 for overhead lines below 345 kV, $978,300,000 for overhead lines at 345 kV and above, $59,200,000 for underground construction, and $1,256,600 for substations. During 1980, utilities energized for the first time 5783 mi (9300 km) of overhead lines at 345 kV and above, and 8832 mi (14,200 km) at lower voltages. Only 100 mi (161 km) of underground cables were installed, primarily because of the 8:1 ratio of underground-to-overhead costs. Utilities also installed a total of 66.1 GVA of substation capacity during the year. Maintaining existing lines cost utilities $527,840,000 in 1980.

Distribution. Distribution facilities required the expenditure of $5,601,000,000 in 1980. Of this, $1,746,300,000 was spent to build 27,700 mi (44,520 km) of three-phase equivalent overhead primary lines ranging from 5 to 69 kV, with the majority at 15 kV. Of all overhead lines constructed in 1980, 3% was rated at 4 kV, 80% at 15 kV, 10% at 25 kV, and 7% at 34.5 kV. Expenditures for underground primary distribution lines amounted to $920,951,000 in 1980. In underground construction, of the 9700 mi (15,570 km) of three-phase equivalent lines built, 4.6% was rated at 5 kV, 76.3% at 15 kV, 12.2% at 25 kV, and 6.9% at 34.5 kV. Utilities energized 27,300 MVA of substation distribution capacity in 1980 at a total cost of $780,980,000. Maintenance costs for distribution were $1,825,740,000 in 1980.

Capital expenditures. Utilities increased their capital expenditures in 1980 to $39,205,000,000, up from $35,181,500,000 in 1979. Of this total, $27,084,000,000 was for generation, $4,252,700,000 for transmission, $5,601,109,000 for distribution, and $2,267,500,000 for miscellaneous uses, such as headquarters buildings, services, and vehicles, which cannot be directly posted to the other categories. Total assets for the investor-owned segment of the industry rose from $210,072,000,000 in 1978 to $232,750,000,000 in 1979, the last year for which figures are available. Total assets for the cooperative segment of the utility industry rose from $19,930,000,000 in 1978 to $23,728,730,000 in 1979.

For background information *see* ELECTRIC POWER GENERATION; ELECTRIC POWER SYSTEMS; ENERGY SOURCES; TRANSMISSION LINES in the McGraw-Hill Encyclopedia of Science and Technology.   [WILLIAM C. HAYES]

Bibliography: Edison Electric Institute, *Statistical Yearbook of the Electric Utility Industry*, 1980; 1980 statistical report, *Elec. World*, 193(6):49–80, Mar. 15, 1980; 31st annual electrical industry forecast, *Elec. World*, 194(6):55–70, Sept. 15, 1980; 21st steam station cost survey, *Elec. World*, 192(10):55–70, Nov. 15, 1979.

Electronics

A variety of new consumer electronic products will become available within the next few years. The products will simulate human characteristics of listening, talking, amusing, and educating. Among these microprocessor-controlled devices will be hand-held language translators, programmable personal computers, educational talking toys, intelligent television receivers, and portable remote terminals that use the telephone to gain access to various computer data bases.

These electronic devices are made possible through electronic microminiaturization called large-scale integration (LSI). Large-scale integration allows more features to be implemented in compact, relatively inexpensive products that were bulky and expensive only a few years ago. Circuits that had to be fabricated with large components and much wiring are now formed on smooth silicon wafers that are split up into "chips" or integrated circuits. These are then interconnected on a printed circuit board that has prefabricated silver connections for the circuits. These integrated circuits are programmable, have the capacity to store data, and perform needed calculations in a fraction of a second. Hence it is left to the imagination of the designer and the economic good sense of the marketeer to develop salable consumer electronic products. Some of the major products under development will be discussed.

Language translators. Portable language translators register keyboard entry of phrases in one language and speak the translated phrase in another. Each language for translation is preserved in

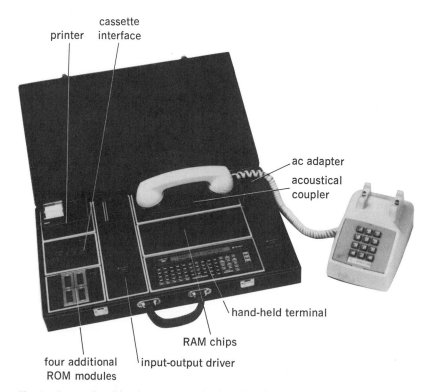

Fig. 1. Quasar hand-held computer. The hand-held terminal, which contains read-only-memory (ROM) modules for translation between languages, fits into a briefcase containing various optional components, including an acoustical telephone coupler to facilitate access to remote data bases by dialing the appropriate code.

memory modules called read-only-memory (ROM), which are basically tables storing corresponding words in both languages. One type of translator uses a speech synthesizer chip that was originally developed as an educational toy. Plug-in ROM modules are available for English, Spanish, French, German, and Russian.

Computers. Hand-held computers have been developed that fit into a briefcase whch has receptacles for acoustically coupling the computer with any telephone (Fig. 1). A variety of information can thereby be accessed while the user is out of reach of data bases. The interface protocol is compatible with most major main-frame computers. Some companies have combined the translator with a computer in the same portable terminal. One terminal has a capacity to hold up to eight language modules simultaneously. A random-access memory (RAM) is used for storing up to 500 characters. In addition, the terminal also acts as a calculator and a clock.

Electronic games. Two types of electronic games exist: those that measure users' dexterity and those that challenge their intelligence. Dexterity games are primarily aimed at the younger generation; they include popular games in electronic form, such as anagram, word matching, and sound matching, and sports games such as baseball and football. Intelligence games such as chess and backgammon build on the users' intelligence and ingenuity and hence appeal to a wider age group. Both types of games can be played alone against the computer or between two players.

Television receivers. Television receivers are now constructed almost entirely of integrated circuits with only the picture tube remaining as a reminder of electron tube technology. All receiver manufacturers have models that incorporate electronic tuners that lock precisely to the channel

frequency. Colors have become more vivid, and displays show less color distortion. Delay line/ comb filter integrated circuits that improve luminance resolution and reduce luminance-chroma crosstalk have been incorporated into the receiver. One type of delay line is based on charge-coupled-device technology.

In the National Television System Committee (NTSC) color television system, luminance and chrominance information is interlaced on a common channel. The two components have to be separated for enhanced picture quality and sharpness. The comb filter separates luminance signals by adding two composite video signals, one of which is delayed by one horizontal scan line relative to the other. Up to now this procedure involved adding a delayed and undelayed signal external to the charge-coupled-device integrated circuit. This causes delay variations that can affect the filter characteristics significantly. In the new comb filter integrated circuit, the delay and the filter functions are combined, thus ensuring that the delay between the two signals is 63.55 microseconds, exactly one horizontal scan line.

For separating chrominance signals, the composite video signals must be subtracted rather than added. Subtraction is accomplished by inverting the composite signal prior to the chrominance circuit in the receiver. By separating both the luminance and chrominance, picture quality and sharpness are considerably enhanced. The outputs of the processing circuits contain components that are free from "dot-crawl" and cross-color contaminations.

In a television receiver system (Fig. 2), the comb filter is interposed between the intermediate frequency (i-f) section and the luminance and chrominance processors. In general in a color television the received signal is fed to the tuner, where it is

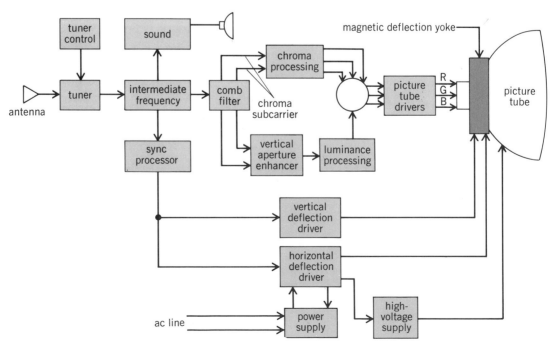

Fig. 2.   Color television receiver.

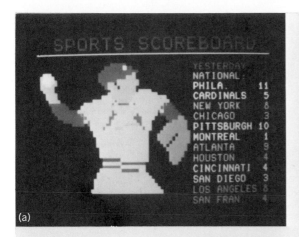

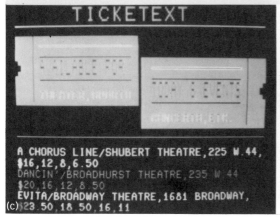

Fig. 3. Examples of teletext displays. (a) Real-time sports scores. (b) Constantly updated market indicators. (c) Theater ticket list. (d) Captioning of regular television programs for the hearing-impaired.

amplified and converted to the intermediate frequency (45.75 MHz). A wide variety of tuners are still used ranging from a microprocessor-controlled synthesizer to mechanical switching. The intermediate-frequency section recovers the baseband composite video signal and also produces the sound signal. The inputs to the comb filter are composite video from the intermediate-frequency section and the regenerated chroma subcarrier from the chroma oscillator. The comb filter outputs are the combed luminance, vertical detail, and combed chrominance signals. The luminance processing circuits ensure that the display is turned off during those portions when no picture information is present. A color subcarrier and chrominance signals drive three synchronous detectors which produce three color-difference signals: R-Y, G-Y, and B-Y. The luminance signal (Y) is subtracted from these, leaving the three red (R), green (G), and blue (B) signals which are applied to the cathodes of a picture tube. After being modulated in intensity each beam passes through a shadow mask that ensures it will strike only the phosphors on the faceplate which can glow with the appropriate color. The vertical and horizontal deflection drivers scan the beams simultaneously from left to right beginning at the top center; 525 horizontal lines are scanned in two fields during one frame, and each frame takes one-thirtieth of a second to scan.

Another improvement in the television receiver that is a direct result of LSI chips is a quasistereophonic system that simulates true stereo sound by generating two spectrally distant sound channels from the monophonic audio source of the receiver. Each sound channel drives one of two separate full-range loudspeakers located symmetrically about the television screen.

Home data retrieval. Decoders are being developed that will enable text and graphics to be retrieved on the home television screen. At first the decoders will be top-mounted on the set, and as LSI chips incorporate more functions, the decoders will be mounted inside the sets. Prototype decoders are already being displayed. As soon as the Federal Communications Commission establishes standards, United States television manufacturers will begin producing sets with built-in decoders.

The image, whether text or graphics, is synthetic video that is created locally. The display will show up to 24 rows, with 40 characters in each row, of constantly updated data on such diverse subjects as news, weather, sports, cooking instructions, child-adoption eligibility, and theater listings (Fig. 3).

The information is received by the television set as part of the regular broadcast but cannot be seen by the viewer unless a decoder is used. The decoder transforms information taken from the unseen

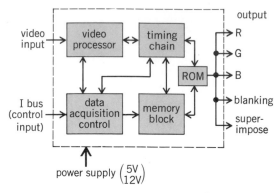

Fig. 4. Block diagram of one type of teletext decoder. *(From T. Suzuki et al., Television receiver design aspects for employing teletext LSI, IEEE Trans. Consumer Electr., CE-25(3):400–405, 1979)*

(blanking) portion of the screen to the visible portion. Between each frame on the screen are about 10 blanking lines, not used by the picture, but needed to account for the time during which the vertical retrace of the electronic beam travels from the bottom right side of the display back to the top left side for the next frame. A couple of the lines have been used in the past to carry captions for the hearing-impaired. Special different decoders are needed so that these people can see the captions.

Set-top versus built-in decoders. Set-top decoders have the disadvantage that their outputs are fed into the receiver together with the video signal from the antenna. As a result, distortions are observed on the display because the decoding signal must pass through all other circuits in the television set, causing interference patterns. The built-in decoders, however, are connected directly to the three color drivers controlling the cathodes of the picture tube. Thus the signal bypasses most of the causes for distortion, and a clear representation of text and graphics can be observed on the screen.

Viewdata. A more sophisticated system, called viewdata, is an outgrowth of the business community. Here the television again serves as the receiver of data and graphics and displays them when the decoder is activated; but the information reaches the set through the telephone. By modulating the voice line on the telephone, digital bit streams are sent through these circuits to the decoder. As a result, different data bases can be accessed by dialing their respective codes. This practice has long been utilized by businesses with on-line time-sharing systems. Soon, with decoders and modems available at affordable prices to the consumer, these same data bases can be accessed from the home. The services provided by viewdata systems are similar to teletext services with one major difference. In teletext a remote keypad is used to access the pages as they are broadcast so that, for example, if the viewer knows that the theater guide is on page 40 the viewer presses that number on the keypad, and after the strobe signal picks up that command it will then switch to that page on the next transmission. (Teletext data are transmitted constantly and repeated af-

ter each transmission.) This switching process takes a few seconds and limits the viewer's participation in the service.

In viewdata, on the other hand, the keypad is used as a two-way transmitter. Hence the same entry on page 40 in a viewdata system can be accessed instantly, provided the page for the category is known. Moreover, the keypad can be used to engage in a two-way conversation with the system. For instance, where the subject of child-adoption eligibility is being examined, consecutive frames might ask the viewer questions to be answered through the keypad. In the end it might even appear as if viewdata decides whether the viewer is eligible to adopt. In the future, viewdata systems dispersed around the country might even be used to play interactive video games such as national chess tournaments.

Implementation. Teletext decoders are already available for receivers in Great Britain, where this technology was established several years ago (Fig. 4). The decoder is a 160×120 mm circuit board containing four LSI chips and two 4000-bit RAMs for storing one page of data locally. When the decoder receives the remote control signal (Ibus), it extracts the teletext data signal from the video detector output and obtains red, blue, and green signals. These are synchronized with the television picture. The decoder also generates the necessary blanking signal so that the picture is invisible while teletext information is being displayed.

A number of semiconductor manufacturers are ready to supply chips for teletext and viewdata decoders as soon as international standards are established. The success of these information retrieval systems for home use will then depend only on consumers' appreciation of this new form of information dissemination and general acceptance of this valuable service.

For background information *see* COLOR TELEVISION; COMPUTER; COMPUTER, MULTI-ACCESS; INTEGRATED CIRCUITS; TELEVISION RECEIVER in the McGraw-Hill Encyclopedia of Science and Technology.   [NICOLAS MOKHOFF]

Bibliography: Color TV, special issue of *RCA Eng.* 25(6):1–88, 1980; Consumer Text Display Systems, special issue of *IEEE Trans. Consumer Electr.*, CE-25(3):233–423, 1979; Videotext and Teletext Systems, special issue of *IEEE Trans. Consumer Electr.*, CE-26(3), 1980.

Electrophoresis

The unique environment of space promises to make possible manufacturing processes and products unthinkable 10 or 20 years ago. By using the long-duration, weightless environment of an orbiting spacecraft—away from such gravity-induced phenomena as natural convection and sedimentation—new manufacturing processes may be able to produce products vastly superior to those manufactured on the ground, as well as entirely new classes of products. The McDonnell Douglas Corporation has for several years been working with such a process: continuous-flow electrophoresis, a separation process based on the relative motion of charged particles in an electrical field. In the microgravity environment of space, it is believed, electrophoresis can produce certain pharmaceuticals in much larger quantities and with much higher purities than is possible on Earth. The work to date has focused primarily on processing human plasma, which contains such medically important serum proteins as clotting factors, alpha antitrypsin, and albumin, and on isolating specific human cell types, which would make possible large-scale production of many important natural body materials.

Although many types of electrophoresis are widely used in diagnosis and analysis because of their ability to separate mixtures of proteins or cells, one form — continuous-flow electrophoresis — is best suited for use as a production system because it can process larger quantities of material.

Continuous-flow method. In continuous-flow electrophoresis, particles (including soluble materials such as proteins) are separated by an electrical field across a continuous flow of a carrier buffer (Fig. 1). The field produces a lateral force — proportional to the electrical charge of the particles and the strength of the electrical field — that causes the particles to migrate laterally as the carrier flow moves them vertically. Different kinds of particles are pulled different distances across the field, and they separate as they move through the device. They exit through different outlet ports at the top, where they are collected.

Gravity effects. The problem is that output quantities are limited by the effects of gravity. Specifically, gravity limits the allowable concentration of protein or cells (the sample) that can be processed in the carrier flow. In Fig. 2a the sample

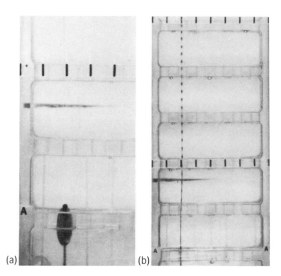

Fig. 2. Effects of density difference: (a) the heavier sample stream collapses; (b) the lighter one breaks into beads.

has a high concentration of protein, making it heavier than the surrounding buffer. In this case, where the chamber flow is upward, gravity causes the sample flow to collapse around the sample inlet port. Figure 2b shows what happens to a sample that is lighter than the surrounding buffer. The buoyant force breaks the sample column into beads, which float upward through the surrounding buffer. If the direction of the chamber flow is reversed — down rather than up — the phenomena are simply reversed. The heavy sample breaks up into drops like the stream from a dripping faucet, and the lighter sample floats up around the sample inlet port. These gravity-induced buoyancy effects have limited any attempts to increase output by increasing the sample concentration. Actual experimental results have shown that the concentration of protein can be only about 0.1% of the buffer by weight.

Space processing. In space, however, without gravity effects, these problems would be eliminated — the concentration of protein could be much higher, and the output much greater. Experimental data have shown that serum proteins remain soluble at concentrations in excess of 10%, indicating that space-processing output could be at least 100 times greater than is possible on the ground. To confirm this increase, McDonnell Douglas used a three-dimensional mathematical model of electrophoresis. With its gravity-dependent terms set to zero and the sample concentration increased to the limit of solubility, the model verified the predicted improvements.

In addition to increased output, electrophoresis in space also promises improved product purity by decreasing or eliminating overlap between the output streams. On Earth this overlap is caused by gravity. Because the sample is heavier than the carrier buffer, it will not rise through the device as quickly as the carrier flow. However, viscous shear forces at the boundary between the buffer and the sample periphery cause the sample at the periphery to move at a velocity closer to that of the buffer. This means that the center of the sample, moving more slowly than the periphery, spends more time under the influence of the electrical field. The difference in time exposed to the lateral force of the field, in turn, creates a difference in lateral motion that distorts and widens the sample stream. As the streams widen, they overlap and mix together, decreasing output purity.

In downward flow the situation is reversed — the center of a sample, heavier than the surrounding buffer, has a higher velocity and spends less time under the influence of the electrical field than the periphery. Again, the difference in time means a difference in lateral motion, causing the sample streams to overlap and decreasing the purity of separated fractions.

Processing in space could solve this problem as well. Again with the gravity-dependent terms in the equation set to zero and the sample input concentrated to the limit of solubility, the mathematical model showed that the velocity differences between the center and edges of the sample due to gravity are virtually eliminated, improving purity by as much as a factor of 5.

McDonnell Douglas's work on electrophoresis has included building and testing laboratory prototypes and conducting computerized analyses of sophisticated mathematical models of the process. Although these analyses predict improved processing in the absence of gravity, the only way to prove the predictions is by testing in space. This testing will now be possible because of a new form

ELECTROPHORESIS

output fractions

buffer flow

sample input

Fig. 1. Electrophoretic device. The sample, typically cells or proteins, is introduced into the upward flow of carrier buffer through an inlet port at the bottom.

of working agreement, called a Joint Endeavor, between McDonnell Douglas and the National Aeronautics and Space Administration (NASA). Through this agreement, which was consummated in January 1980, McDonnell Douglas will develop electrophoresis equipment that will operate in space to prove the advantages of weightlessness, and NASA will provide the shuttle flights for these tests. The tests are expected to start in late 1982 or early 1983. If they are successful, McDonnell Douglas is expected to commercialize the process.

For background information *see* ELECTROPHORESIS; MANNED SPACE FLIGHT in the McGraw-Hill Encyclopedia of Science and Technology.

[JAMES T. ROSE; RONALD A. WEISS; DAVID W. RICHMAN; J. WAYNE LANHAM]

Endocrine system (invertebrate)

The tobacco hornworm, *Manduca sexta*, is fast becoming a major research animal in laboratories throughout the world. Ease of rearing, large size, and short life cycle are only three of a number of reasons that this insect is increasing in popularity. Recent studies using time-lapse cinematography have analyzed and recorded the onset and duration of specific events throughout the life cycle of *M. sexta*. Such studies, when coupled with closely controlled environmental and dietary conditions, give precise time studies of behavior and morphological change. It is these changes that are triggered by endocrine events, and it is the predictable control of the onset of behavioral and morphological events that allows the critical investigations of the endocrine system which controls growth, development, and reproduction. Insects usually have comparatively rapid growth rates during one or more periods of the life cycle, and physiological and morphological changes can take place very quickly. The onset and duration of various stages are mostly dependent on a number of external factors, prinicipally nutritional and thermal.

Behavior. Time-lapse techniques were used in studies of *M. sexta* under controlled nutritional and environmental conditions (such as a high wheat-germ diet, 25°C, 40% relative humidity, 12 hours light: 12 hours dark). The table combines onset times of major events (observed with time-lapse techniques) with the weight of *M. sexta* during weight-stable stages from egg through pupation.

This insect hatches after 4 days and immediately consumes its egg case before taking other diet. Eating continues through both light and dark cycles. Of the five larval instars, the first has the highest growth rate, and the last instar the slowest. All but the last larval instar have similar behavior (as shown by time lapse), but the duration of the event series increases during each consecutive instar.

At a particular time in each of the first four instars, the larvae stop eating and become effectively motionless for 14–24 h, depending on the instar, as they complete the next pharate larval stage. Upon eclosion, the larvae shed the old skin within 5 min and usually consume it before feeding again on the diet within an hour or two.

Last larval instar. The fifth instar is more complex. Approximately 12 events were observed between the fourth to fifth and larval-pupal ecdyses. The first noticeable event after the onset of feeding is the appearance of fecal pellets coated with white deposits of uric acid as the larvae begin to purge

Time of onset of events in days for most physiologically advanced specimens* of laboratory-reared Manduca sexta and their weight during weight-stable stages at 25°C

Event	Time at onset of event, days†	N	Mean weight of insect during event, grams†	N
Oviposition	0.00 ± 0.01	11	0.0014 ± 0.0010	20
First-instar hatch (L1)	4.03 ± 0.02	4		
Molt sleep	5.78 ± 0.02	4	0.0076 ± 0.0013	20
L1–L2 ecdysis	6.37 ± 0.02	11		
Molt sleep	7.82 ± 0.06	10	0.030 ± 0.003	8
L2–L3 ecdysis	8.41 ± 0.06	10		
Molt sleep	9.97 ± 0.10	8	0.20 ± 0.02	21
L3–L4 ecdysis	10.65 ± 0.10	8		
Molt sleep	12.68 ± 0.09	7	1.08 ± 0.09	20
L4–L5 ecdysis	13.71 ± 0.11	11		
Coated fecal pellets	15.76 ± 0.04	4		
Cessation of feeding	17.61 ± 0.15	11	9.23 ± 0.53	10
Heart exposure	17.72 ± 0.12	5		
Body wetting	17.81 ± 0.08	11		
Wandering	17.92 ± 0.08	11		
Dorsal pigmentation	17.98 ± 0.20	3		
Burrowing	18.27 ± 0.19	11		
Fluid excretion	18.61 ± 0.22	7		
Reduced movement	19.23 ± 0.20	11		
Stationary stage	19.75 ± 0.10	11		
Metathoracic bars	22.11 ± 0.20	11	5.02 ± 0.63	10
Larval-pupal ecdysis	22.77 ± 0.20	11	4.71 ± 0.62	30

*About 10% of the colony were advanced specimens under rearing conditions of 40% RH and 12-h scotophase, starting at 6 P.M. Standard colony developed about 9% slower.

†Data represent the mean and standard deviation for numbers of insects (N) as indicated.

their bodies of nitrogenous waste in preparation for the pupal stage. This is followed by an abrupt cessation of feeding 2 days later. Within 8 h after feeding stops, the larvae begin the curious phenomenon of wetting their bodies with an oral secretion. The insects initiate body wetting at the thorax and proceed to cover the surface of the body to the end of the abdomen on one side and then repeat the process on the other side, taking 5–15 min to complete each side. When the insects are physiologically capable of this behavior a few hours before the onset of the dark phase, body wetting is delayed until just after the onset of the scotophase. This was the only instance noted when the photoperiod, in this case the onset of darkness, had an immediate effect on the behavior of the larvae.

Body wetting activity appears about the same time as the dorsal aorta becomes visible through the body wall. Heart exposure is apparently triggered by the first small release of molting hormone, ecdysone, during the fifth instar as the insect prepares for the pupal stage.

Immediately after body wetting, the larvae begin a very active stage known as wandering, an activity in which the insects search for a pupation site. Continued confinement to a 120-ml cup containing leftover food and frass induces burrowing activity in which a pupal chamber is formed from this material. During this time, the larvae show a pinkish pigmentation on the dorsal surface, which lasts a little more than 1 day. Shortly after burrowing activity begins, fluid is excreted from the anus six or seven times over a 20–35-h period, resulting in a loss of 40% or more of the maximum weight and 25% of their 8-cm length.

Pupation. By the time the insects have completed fluid excretion, their gut lumens have become purged of ingesta, and they have lost almost all locomotor ability. About 16 h prior to pupation, a pair of metathoracic bars located on the dorsal surface perpendicular to each side of the heart begins to melanize. Pupation occurs 23 days after eggs are oviposited.

Adult moth. Nondiapause pupae (reared under 15 h light:9 h dark) eclose about 20 days after pupation, while duration of the diapausing pupal stage varies widely (97 ± 30 days). Eclosion (adult emergence) and the subsequent expansion and folding of the wings to the tentlike moth position are composed of a number of endocrine-triggered activities that are controlled in part by environmental conditions. The adult can cast its pupal skin in less than 30 s and immediately discharge approximately 1 ml of accumulated nitrogenous waste (meconium). If not confined, the moths will find a wing-spreading site in about 11 min, and 33 min later will have fully expanded wings. By 1½ h after eclosion, the wings are abruptly folded from the butterfly position to the tent position and are first tried about 3 or 4 h after eclosion. Ovipositional activity usually begins on the third night after eclosion.

Growth rate control. This brief summary of major events in the life cycle of *M. sexta* is based on time-lapse cinematographic observations of insects reared on a stable diet and controlled environment. Temperature plays a vital role in the predictability of the growth rate of cold-blooded,

short-lived animals. If all other factors remain constant, temperature can be used to control the growth rate precisely. By using the time of event onset *(E)* at 25°C in the table and any temperature between 18 and 33°C *(T)*, the change in time of occurrence for the same event at the new temperature can be estimated closely by the equation below. Solving for *Y* gives the change of time to be

$$Y = E[0.0061 + 0.0980(25 - T) + 0.0085(25 - T)^2]$$

added to or subtracted from *E*. The illustration shows the observed and calculated times of occurrence of major activities at different temperatures.

Greater control of *M. sexta* for research purposes must take into consideration the variation of the specimens. Some insects will obviously lack proper vigor and growth. Still others will fall a day

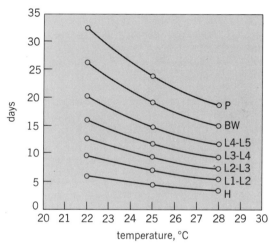

Temperature effect on growth rate of *Manduca sexta* as depicted by direct observations (circles) at 22°, 25°, and 28°C and the use of the equation (solid lines) in the text. H = hatch; L1 – L2 = second ecdysis, and so on; BW = body wetting; P = pupation.

behind, and yet give all other appearances of being healthy. These late insects apparently are not competent to release a hormone from the brain that controls the release of the molting hormone needed for continued development. The release of this brain hormone, called prothoracicotropic hormone (PTTH), is regulated by a photosensitive circadian clock. The clock interacts with the photoperiod and restricts the release of PTTH to a specific portion of the light-dark cycle. These specific times of light regimen are referred to as temporal gates, and insects not competent to release hormone before the temporal gate closes must await the opening of the gate on the following day. Larvae that are competent to release PTTH either before or after a gate closes are called gate I and gate II larvae, respectively.

Summary. Superior insect-rearing techniques are increasing rapidly under current pressure to rear specific insects for biological control and for more critical research programs. The needed control of laboratory-reared insects relies not only on diet and environment but also on an understanding of the behavioral activities. Research studies of the life cycle of *M. sexta* using time-lapse cinematography techniques and controlled environmental conditions have yielded information which makes this insect one of the most useful research specimens available to science today.

For background information *see* ENDOCRINE SYSTEM (INVERTEBRATE); INSECT PHYSIOLOGY in the McGraw-Hill Encyclopedia of Science and Technology.

[JOHN P. REINECKE]

Bibliography: A. H. Baumhover et al., *An Improved Method for Mass Rearing the Tobacco Hornworm*, USDA Bull. S-157, 1977; A. H. Madden and F. S. Chamberlin, *Biology of the Tobacco Hornworm in Southern Cigar-Tobacco District*, USDA Tech. Bull. 896, 1945; J. P. Reinecke, J. S. Buckner, and S. R. Grugel, *Biol. Bull.*, 158: 129–140, 1980; J. W. Truman and L. M. Riddiford, in P. L. Munson, et al. (eds.), *Vitamins and Hormones*, vol. 35, pp. 283–315, 1977.

Erosion

In recent years there has been much research concerning the problem of restoring fertility to soils whose productivity has been decreased by soil erosion. The effect of soil erosion on soil productivity depends on the nature of the soil, the level of technology available, and the crop to be grown.

Productivity loss. On soils that are shallow to rock, loss of the surface soil by erosion may result in the land being unfit for any vegetation. Deep deposits of medium-textured soils, such as loess (windblown silt), can lose several feet of surface and still be productive if necessary fertilizers, usually nitrogen and phosphorus, are added. If no fertilizers are available, all eroded soils produce poor yields of grain crops. Some forage crops and trees, however, flourish on severely eroded soils even when no fertilizers are used. The surface soil contains the soil organic matter. Loss of surface soil through erosion removes the soil organic matter and with it, most of the plant-available nitrogen and phosphorus. The presence of organic matter also makes the soil more mellow and easy to till. Soils low in organic matter tend to be cloddy, have low water-intake rates, and, in general, tend to be much more difficult to till than the high-organic-matter soils.

Fertility restoration. Fertility can, however, be restored, and some soils are reasonably easy to till even when low in organic matter. During the past 20 years numerous studies have been made on this subject in the United States. These studies have been mostly on calcareous or neutral soils of the Midwest and West, and have shown that high yields of corn and wheat can be obtained on soils devoid of surface soil. The findings of one study illustrate the type of results frequently found. On a calcareous loess in western Iowa, on which some 7 ft (2.1 m) of surface soil had been removed, fertilizers plus manure restored corn yields the first year after soil removal to a level as high as on the undisturbed soil. First-year corn yield, under irrigation, was 163 bu/acre (14.2 m³/ha).

In the more humid regions, soils are commonly

acid and subsoils are poorly suited for crop growth. Such soils are frequently shallow to clay, to a hard pan, or to a zone of soluble aluminum concentrations that are toxic to plant growth. Crop yields are usually severely reduced by erosion on such soils. If the cause of yield reduction resulting from erosion can be identified and corrected, be it increased droughtiness, lower fertility, or soil toxicity, the yielding capacity of the soils can be restored. However, aside from restoring fertility elements, it is seldom economically feasible to amend the soil sufficiently to compensate for the damage caused by erosion. On such soils, the usual practice is to shift to a lower value and more tolerant crop. Growth of southern pine on old cultivated fields in the southern United States is a widespread example of such a practice.

On soils where rooting depth is limited to the surface soil, any loss in surface is likely to result in reduced yields. On soils with high aluminum concentration in the subsoil, any loss of surface soil reduces the water-holding capacity of the soil and lower yields are liable to result. On such soils, even when they have as much as 20 in. (50 cm) of surface soil, water-holding capacity may be less than a third as much as is held in a good corn soil in the Midwest. One study in North Carolina showed, for example, that an uneroded soil with a 20-in. (50-cm) surface should hold about 3 in. (8 cm) of water in a plant-available form as compared to 10 in. (25 cm) in a good central Illinois soil. Three in. (8 cm) of water is enough to supply needs of a corn crop for about a week in July or early August. In the North Carolina study, a reduction in surface soil depth from 20 to 12 in. reduced corn yields about 35%.

Organic matter in the surface soil is very important in maintaining soils in a state of good tilth. When the surface erodes, tilth deteriorates, water infiltration is slowed, and thus less water enters an eroded soil as compared to an uneroded soil. Loss of water by increased runoff can be serious on any soil, but it is especially serious on a soil of low water-holding capacity.

The increased droughtiness resulting from erosion could be corrected by irrigation, or the plant rooting depth and thus soil water-holding capacity could be increased by extensive and very expensive subsoil plowing, liming, and fertilizing. Erosion on these soils results either in drastically reduced crop yields, in greatly increased costs, or in shifting to the production of a more tolerant and usually lower-value crop. The probable effect of erosion on production can be predicted with fair accuracy providing enough is known about the nature of the soil in question. A recent study in west Tennessee involved three widely differing soil conditions. One soil, Memphis silt loam, was deep, had no restrictive zones in the plant root zone, and had a high water-holding capacity. The second soil, Grenada silt loam, had a compacted zone in the subsoil that limited root penetration to the upper 30 in. (75 cm) of the soil. The third soil, Brandon silt loam, was shallow to gravel.

When eroded, the Memphis soil had a higher runoff and was more difficult to till, but when fertilized, yields of corn were reduced about 15% and grass about 10%. The Grenada soil had a 27% reduction in corn yields and a 10% reduction in grass yields as a result of erosion. In this same study, the soil that was shallow to gravel lost almost half of its ability to produce corn and a fourth of its worth as a grassland soil when eroded.

These results could have been anticipated from the nature of the soils. Even when eroded, the Memphis soil held plenty of moisture. The main effect of erosion was loss of fertility, mostly nitrogen and phosphorus. When these were added, good yields were obtained.

The other soils were critically lacking in water-holding capacity even when uneroded, so any loss in capacity or decrease in infiltration of water into the soil could be expected to result in lower yields, regardless of the amount of fertilizers supplied.

A thorough knowledge of the characteristics of different types of soils makes valid predictions of erosion effects possible. A knowledge of growth habits of different plant species is an aid in selecting the most appropriate crop for an eroded area.

Economic effects. In most instances, the economic value of an erosion-tolerant crop is lower than a crop that is severely affected by soil erosion. In most cases, erosion increases costs for all crops on all soils. Widespread use of fertilizers in the United States has greatly reduced the deleterious effect of erosion on crop yield on some soils. However, even with adequate fertilization there is a moderate to severe reduction in crop yields of the principal food crops as a result of erosion on most soils in the United States and throughout the world.

For background information see SOIL; SOIL CONSERVATION in the McGraw-Hill Encyclopedia of Science and Technology.   [W. D. SHRADER]

Bibliography: G. T. Buntley and F. F. Bell, *Yield Estimates for the Major Crops Grown on Soils of West Tennessee*, Tenn. Agr. Exper. Sta. Bull. 501, 1976; W. C. Moldenhauer and C. A. Onstad, *J. Soil Water Conserv.*, 30(4):166–168, 1975; D. J. Thomas and D. K. Cassel, *J. Soil Water Conserv.*, 31(1): 20–24, 1979.

Fat digestion

Fat, "nature's gasoline," requires enzymatic hydrolysis before it can be efficiently absorbed by the epithelial cells of animal digestive systems. The best-studied fat-splitting enzyme is the one found in the mammalian pancreas and called pancreatic lipase. Enzymes with similar substrate specificity and pH optima have been found in many simpler forms of life, including bacteria and fungi. Until recently, understanding of fat digestion has been derived almost entirely from measurements of chemical quantities and activities. Now it has been shown that fat digestion can be observed directly by light microscopy. The products of lipase hydrolysis, fatty acids and monoglycerides, form distinct and visible product phases which emanate from the digesting fat droplet. By watching fat digestion directly through the light microscope, it has been possible to study enzyme activity on single fat droplets and to measure the flow of fat-soluble chemicals from nondigested fat into the hydrolyzed products. At high concentrations of lipase, such as those found in animal digestive tracts, the digestion of fat droplets between a cover slip and a

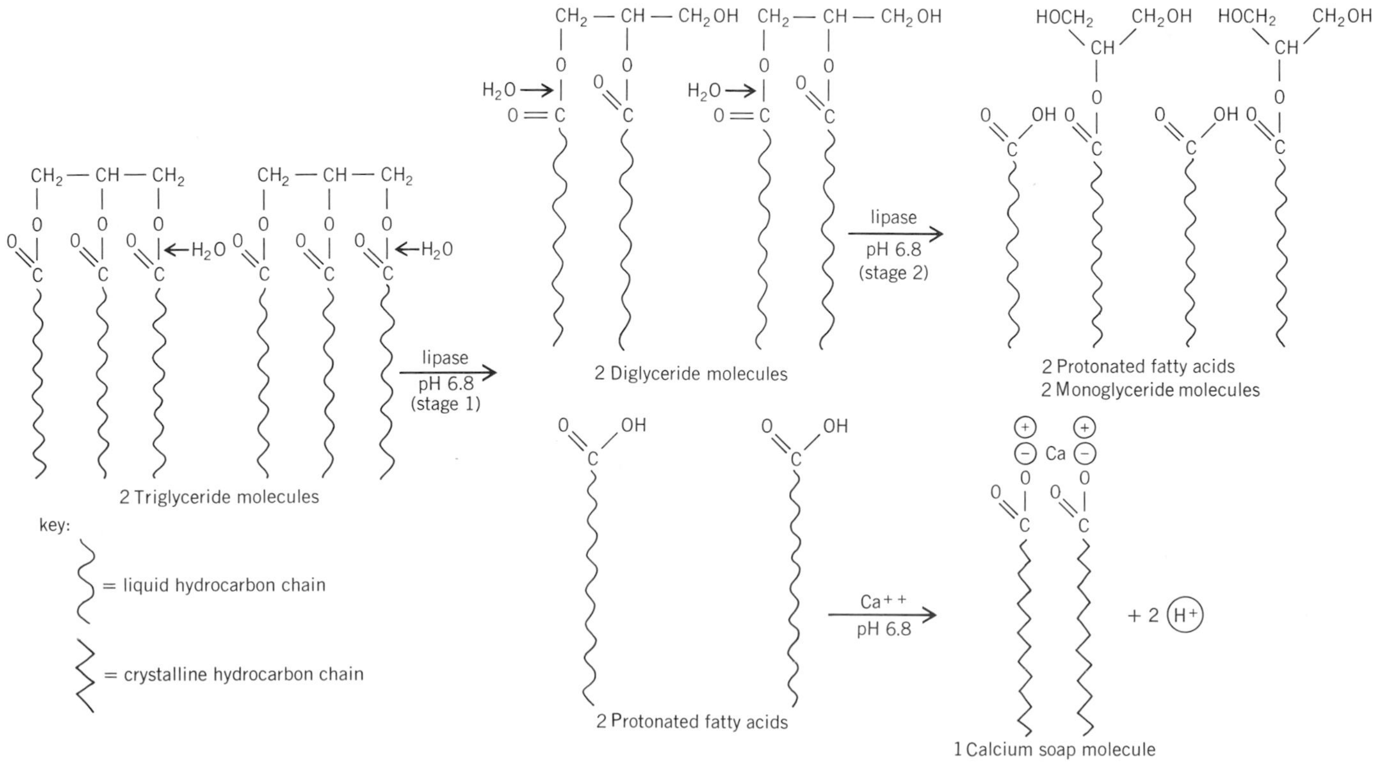

Fig. 1. Schematic representation of long chain triglyceride (fat) digestion by pancreatic lipase. Lipase sequentially hydrolyzes the outside ester bonds of the triglyceride molecule.

microscope slide proceeds rapidly, within minutes to completion.

Quantitatively, fat digestion is the digestion of energy-rich triglyceride. Qualitatively, fat digestion is a process which enables a vast array of fat-soluble chemicals to be efficiently absorbed by animals. Among the chemicals that may "ride in on the coattails" of triglyceride digestion are some of the essential vitamins and nutrients, along with many pollutants, drugs, food additives, and carcinogens. Most fat digestion in vertebrates occurs rapidly in the upper small intestine through the integrated action of pancreatic lipase, its protein cofactor colipase, and bile. It is thought that colipase binds to the surface of fat droplets and provides an attachment site for lipase. Lipase sequentially attacks the two outside ester bonds of the triglyceride molecules, producing first a molecule of diglyceride and one of fatty acid, and then two molecules of fatty acid and one of monoglyceride (Fig. 1). During this reaction a molecule of water is split and added to the separate halves of the cleaved ester bond. Stage 1 lipase reactions dominate at first, and produce protonated fatty acids and diglycerides. Diglycerides remain with triglyceride in the oil phase. The protonated fatty acids may become ionized and complexed with calcium to form a soap phase. Stage 2 lipase reactions increase in frequency with time, and produce monoglycerides as well as protonated fatty acids. Monoglycerides prevent calcium soap formation by fatty acids. The lipase reactions are reversed by reducing the amount of water in the system. Calcium soaps are hydrolyzed by lowering the pH. Bile salts, which do not emulsify fats, are considered to incorporate these insoluble products of lipase hydrolysis into mixed micelles from which fat absorption presumably occurs.

Four-phase system. Prior to the microscope studies, fat digestion (lipolysis) was viewed as a two-phase system consisting of oil droplets and mixed micelles of bile salts and lipolytic products. Mixed micelles cannot be seen by light microscopy. When the known components of fat digestion were combined on a microscope slide under simulated physiological conditions, it was quickly evident that the process was at least a four-phase system, with formation of other additional phases possible, depending on reaction conditions. The first visible product phase that forms is calcium soap. This brittle, dense phase, containing two ionized fatty acids for every calcium molecule (Fig. 2), is produced early in the reaction when stage 1 lipase reactions dominate (Fig. 1). Its crystalline structure has been determined by x-ray diffraction. With time, phase 2 lipase reactions produce enough monoglyceride to block further soap formation, and the second product phase, the viscous isotropic phase, begins to appear (Fig. 2). This phase is the dominant phase produced under physiological conditions. The structure of the viscous isotropic phase is not known for certain; however, freeze-etch micrographs indicate that it contains variable amounts of water between bilayer stacks of lipolytic products. The formation of the viscous isotropic phase continues until the entire fat droplet is digested. A small remnant particle is left behind when the reaction is over.

Fate of enzyme molecules. The question of what happens to the enzyme molecules during the digestion of a fat droplet has been answered indirectly. As the droplet shrinks in size and the surface area diminishes, enzyme molecules appear to be left behind, distributed within the product phases. Evidence that this occurs comes from two observations. First, if the enzyme activity on an individual droplet is determined by measuring the change in dimensions of the droplet with time, it is clear that total activity drops precipitously as the droplet is digested. This suggests that enzyme either is being displaced from the triglyceride surface or is becoming inactive. The second observation that indicates enzyme is distributed within the product phases is based on the fact that the lipase reaction is completely reversible if the water content of the system is reduced. This can be accomplished with the microscope preparation by simply allowing the water to evaporate at the edges of the cover slip. If lipase molecules were concentrated at the remnant, reappearance of the oil phase would be expected to originate at this site upon desiccation of the system. In fact, reappearance of the oil phase during desiccation occurs throughout the viscous isotropic phase as numerous small droplets. Thus both the activity calculations and the resynthesis observations suggest that, during fat digestion, lipase molecules become distributed within the product phases they produce.

Fate of water-insoluble chemicals. The fate of hydrophobic or nonpolar chemicals dissolved in triglyceride droplets during hydrolysis of the tri-

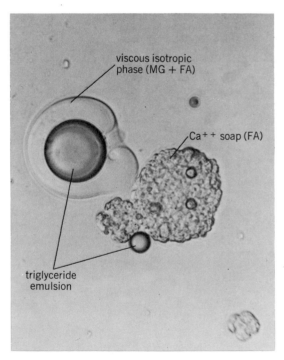

Fig. 2. Photomicrograph taken during the digestion of olive oil triglyceride by human pancratic lipase, showing two product phases, the calcium soap phase and viscous isotropic phase. MG = monoglyceride; FA = fatty acid. The diameter of the large triglyceride sphere is about 50 μm. (*From J. S. Patton and M. C. Carey. Watching fat digestion, Science, 204:145–149, copyright 1979 by the American Association for the Advancement of Science*)

Micellar Phase   Viscous Isotropic   Oil Phase
                                        Phase

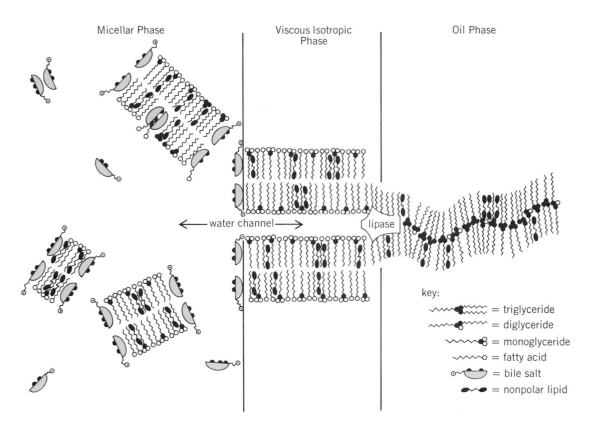

key:

〰〰●〰〰 = triglyceride

〰〰○〰 = diglyceride

〰〰○ = monoglyceride

〰〰○ = fatty acid

○〰⌣ = bile salt

●〰● = nonpolar lipid

Fig. 3. Schematic "zipper" model of fat digestion show-ing the preservation of the hydrocarbon domain during lipase hydrolysis of oil (triglyceride) and the subsequent solubilization of the viscous isotropic phase by bile salts.

glyceride could also be studied with the light microscope. There has never been a satisfactory explanation of how such water-insoluble chemicals (aqueous solubilities of 10^{-10} to 10^{-14} M) could be dispersed during digestion, and because of this, it has generally been assumed that they are poorly absorbed. Watching digestion with the microscope revealed how high concentrations of nonpolar molecules are dispersed during fat digestion. Fluorescent nonpolar molecules were dissolved in triglyceride (fat droplets), and the enzymatic digestion of the droplets was followed by fluorescence microscopy. The fluorescent fat droplets were converted directly into fluorescent product phases. These observations indicate that there is a continuous hydrocarbon domain in triglyceride which remains intact during the enzymatic conversion of the triglyceride to fatty acid and monoglyceride. The preservation of this hydrocarbon domain during fat digestion allows nonpolar molecules that are dissolved within it to flow from nondispersible triglyceride to the products of fat digestion which are dispersed by bile salts and absorbed. The concept of an uninterrupted hydrocarbon domain during fat digestion that can carry many different molecules has profound significance for human health and disease; it explains how a large family of biologically important molecules are dispersed in the intestine and thereby absorbed; it offers a mechanism to explain food-chain concentration of hydrophobic pollutants and toxins; it is the logical pathway of intestinal entry of nonpolar carcinogens; and it reveals a rapid and simple method of drug delivery. Figure 3 shows a schematic "zipper" model of the reversible fat hydrolysis reaction and the hydrocarbon domain.

The light microscope has proved to be a valuable tool for studying an enzyme reaction where the products form visible crystalline or liquid crystalline phases. Since most food constituents are visible by light microscopy (starch granules, muscle fibers, and so forth), it should also be possible to follow their digestion by light microscopy.

For background information see ENZYME; LIPID METABOLISM; MICROSCOPE, OPTICAL in the McGraw-Hill Encyclopedia of Science and Technology.   [JOHN S. PATTON]

Bibliography: B. Borgström, in D. H. Smyth (ed.), *Biomembranes*, vol. 4B: *Fat Digestion and Absorption*, 1974; J. S. Patton, in L. R. Johnson et al. (eds.), *Physiology of the Digestive Tract: Gastrointestinal Lipid Digestion*, 1980; J. S. Patton and M. C. Carey, *Science*, 204:145–149, 1979; M. Semeriva and P. Desnuelle, *Advan. Enzymol.*, 48: 319–370, 1979.

Fertilizer

Slow-release fertilizers are commonly used on horticultural crops and turfgrasses, and the recent expansion in commercial production of sulfur-coated urea (SCU) has given growers an alternative to other commonly used slow-release nitrogen sources. Because of a lower cost and a relatively higher efficiency than some of the other slow-release nitrogen sources, SCU has found a spot in the marketplace, and the increasing use of SCU is expected to continue.

Slow- and fast-release nitrogen fertilizers. Slow-release nitrogen fertilizers can be classified according to the process by which nitrogen is released: Microbial decomposition releases nitrogen from natural organic fertilizers and ureaform, a synthetic organic made by reacting urea and formaldehyde. Low solubility and a very slow rate of dissolution control the release of nitrogen from isobutylidene diurea (IBDU) and magnesium ammonium phosphate. Coatings (such as plastic, wax, asphalt, or sulfur) on or mixed with soluble nitrogen sources act as physical barriers that delay or slow the dissolution of nitrogen; examples are plastic-coated fertilizers and sulfur-coated urea.

Examples of fast-release (soluble) fertilizers are urea, ammonium sulfate, and ammonium nitrate. They give a rapid response after application, and split applications are normally recommended to minimize overstimulation of growth and injury due to fertilizer salts (fertilizer burn). Soluble sources

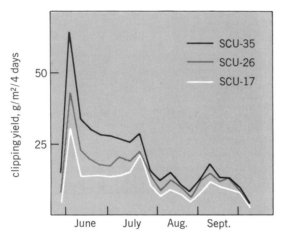

Fig. 1. Clipping yields of Kentucky bluegrass reflect differences in nitrogen release from single spring applications (195 kg N/ha) of SCU formulations having dissolution rates of 17, 26, and 35%. (*From D. V. Waddington and T. R. Turner, Evaluation of sufur-coated urea fertilizers on Merion Kentucky bluegrass, Soil Sci. Soc. Amer. J., 44: 413–417, 1980*)

are readily leached, particularly in areas of high rainfall or irrigation and on sandy soils. Compared to the soluble sources, slow-release nitrogen fertilizers offer the advantages of longer duration of nitrogen release, fewer applications, lower leaching losses, and greater safety from the standpoint of fertilizer burn. Disadvantages of the slow-release materials are the higher cost of nitrogen (two to four times as great) and the low efficiency (relating to the percentage of applied nitrogen utilized by the plants) that often occurs in the first year or two of use.

Production of sulfur-coated urea. To manufacture SCU, urea granules are preheated and then sprayed with molten sulfur in a rotating drum. A sealant, such as wax or a mixture of polyethylene and heavy-weight oil, is sprayed over the sulfur coating to seal imperfections in the sulfur coat. A diatomaceous earth conditioner is used to counteract the stickiness caused by the sealant and gives the product good handling characteristics. The total weight of sulfur applied is usually in the range of 13–22% of the final weight. Both the sealant

weight and conditioner weight total about 2%. Thus, the final nitrogen content of SCU is in the range of 32–38%, as compared to 45% in the urea. When urea prills are coated, a greater weight of coating is needed to obtain a given coating thickness and release rate because the prills are smaller and have more surface area per unit weight than granules.

Release of nitrogen. The coating on SCU delays the dissolution of urea (a water-soluble fertilizer), and different release rates can be obtained by varying the coating weight (thickness) and coating technique. Release rates are characterized by a laboratory 7-day dissolution rate, which indicates the percent of urea that goes into solution when a sample of the product is immersed in water at 100°F (38°C) for 7 days. Values for commercial products fall within the range of 25–35%. In Fig. 1, turfgrass clipping yields indicate the availability of nitrogen from three SCU products having different 7-day dissolution rates. Nitrogen was released more rapidly as the dissolution rate increased (coating weight decreased).

Nitrogen is released by degradation of the coating or diffusion of soluble nitrogen through pores (pinholes) in the coating. Microbial activity is not considered to be a major factor affecting release; however, on some well-coated particles sulfur-oxidizing bacteria no doubt aid in degradation of the coating. Coatings are not identical on each particle. Imperfectly coated or cracked particles release nitrogen very rapidly when placed in or on the soil. Particles with thin coatings or with imperfections in the sulfur coating sealed by the sealant have an intermediate release rate. Thickly coated particles without imperfections may be very slow to release nitrogen, and some have been observed to be intact even after several years. It is the variability in particles that provides the controlled-release feature, and the release characteristics of an SCU product reflect the combined effect of properties of individual particles. The release from an individual particle has been shown to be very rapid once it has been initiated.

Release of nitrogen from SCU increases as temperature increases, and it appears that materials with low 7-day dissolution rates (thick coatings) are better suited to warmer climates. Under waterlogged conditions, release from SCU has been observed to be decreased due to a sealing of the pellets by an iron sulfide precipitate that forms under anaerobic conditions.

Use of sulfur-coated urea. Compared to the use of urea, benefits of SCU include: longer-lasting effectiveness; fewer applications; less fertilizer burn; lower leaching losses; less loss of nitrogen by volatilization as urea is hydrolyzed; and increased efficiency in the use of nitrogen by a crop. The major disadvantage of SCU is the higher cost per unit of nitrogen. The sulfur in the coating becomes available for plant use when it is converted to sulfate by sulfur-oxidizing bacteria.

Compared to other slow-release sources, SCU has advantages of lower cost of nitrogen and, in most cases, better efficiency of applied nitrogen. As with other slow-release sources, SCU can be used alone or in combination with soluble sources in mixed fertilizers. Response following application of SCU is usually quicker and more intense than with other slow-release sources. At high rates of application, this characteristic could cause excessive growth.

Yield and color can be used to indicate availability of fertilizer nitrogen on turfgrass areas. The yield and color responses of Kentucky bluegrass in Figs. 2 and 3 indicate release differences between ammonium sulfate (a soluble fertilizer), ureaform (a slow-release fertilizer dependent on microbial decomposition for nitrogen release), and SCU. Following fertilization, greater and more rapid response was obtained with ammonium sulfate. This effect was not long-lasting, and response dropped below that of the slow-release fertilizers in the summer. Manufacturers of turfgrass fertilizers often include soluble and slow-release nitrogen sources in order to obtain both a quick response and a residual effect. The results in these figures also show a greater efficiency of applied nitrogen with SCU than with ureaform. The results are from the second year of use on the area, and differences between the efficiencies of these two slow-release fertilizers would be expected to decrease in following years of use because the efficiency of ureaform increases as soil levels increase due to continued use.

On some turf grass areas, a spottiness in response has been observed. This effect has been attributed to the relatively large particle size of SCU and to movement and concentration of particles by rain or irrigation. On close-cut turf, such as on golf greens, problems with SCU are breakage and pick-up by mowing equipment. Use of the finer SCU products minimizes this problem.

Use of SCU has been suggested for crops such

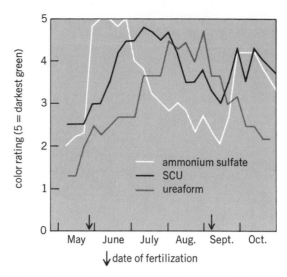

Fig. 3. Color ratings of Kentucky bluegrass reflect differences in nitrogen release characteristics of ammonium sulfate, SCU, and ureaform applied in the spring and fall at a rate on 98 kg N/ha (2 lb N/1000 ft²). (*From N. W. Hummel, Evaluation of Slow-Release Nitrogen Sources for Turfgrass Fertilization, M.S. thesis, Pennsylvania State University, 1980*)

as rice, sugarcane, pineapple, horticultural crops, turfgrass, forage grasses, and other crops that would normally require multiple applications of a soluble nitrogen fertilizer. Minimal benefits from SCU have been obtained on crops such as corn, sorghum, cotton, wheat, and barley that take up most of their nitrogen requirement over a relatively short period during the growing season. The greatest commercial use of SCU has been on turfgrass and horticultural crop areas where other slow-release fertilizers have been used successfully in the past. On these areas the concept of slow-release fertilizers is not new, and past experience has shown that the benefits can justify the additional cost for slow-release nitrogen.

For background information *see* FERTILIZER; UREA in the McGraw-Hill Encyclopedia of Science and Technology. [DONALD V. WADDINGTON]

Bibliography: S. E. Allen and D. A. Mays, Coated and other slow-release fertilizers for forages, in D. A. Mays (ed.) *Forage Fertilization*, American Society of Agronomy, Madison, WI, 1974; L. H. Davies, *Slow-Release Fertilizers, Particularly Sulphur-Coated Urea*, Proc. Fertiliser Soc., London, no. 153, 1976; E. O. Huffman et al., *Experience of TVA with Sulfur-Coated Urea and Other Controlled-Release Fertilizers*, TVA Circ. Z-59, 1975; D. N. Maynard and O. A. Lorenz, *Controlled-Release Fertilizer and Nitrification Inhibitors for Horticultural Crops (An Indexed Reference List)*, Vegetable Crops Series 196, Department of Vegetable Crops, University of California, Davis, July 1978; R. M. Scheib and G. H. McClellan, *Sulphur Inst. J.*, 12(1):2–5, 1976.

Fluorocarbon

Carbon is believed to be capable of forming an infinite number of compounds, since carbon atoms can bond with each other, forming chains and rings with single, double, or triple bonds. Hydro-

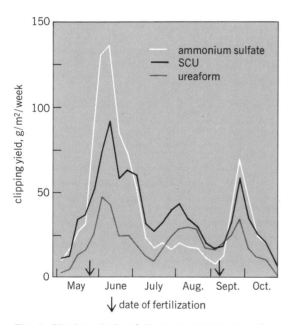

Fig. 2. Clipping yields of Kentucky bluegrass reflect differences in nitrogen release characteristics of ammonium sulfate, SCU, and ureaform applied in the spring and fall at a rate of 98 kg N/ha (2 lb N/1000 ft²). (*From N. W. Hummel, Evaluation of Slow-Release Nitrogen Sources for Turfgrass Fertilization, M.S. thesis, Pennsylvania State University, 1980*)

gen frequently enters into combination with carbon, and if it were the only element forming the other bonds, an infinite number of hydrogen compounds would be possible as well. Fluorine, unlike chlorine, oxygen, or nitrogen, can replace hydrogen partially or completely. Therefore an infinite number of fluorine compounds is also possible.

Originally, it was found that some combinations of carbon, hydrogen, and fluorine resisted synthesis, even simple molecules such as trifluoromethanol (CF_3OH) and trifluoromethylamine (CF_3NH_2). Yet parent species methanol (CH_3OH) and methylamine (CH_3NH_2) are well known, and even CF_3OF and CF_3NF_2 were prepared in the 1950s. A suggested explanation for the lack of success in the synthesis of CF_3OH and CF_3NH_2 was that their structure contained hydrogen and fluorine in close proximity, with the hydrogen being kinetically activated as an acid hydrogen, favoring HF elimination. Success in isolating both species was achieved by developing a reaction sequence where this elimination could be avoided.

In 1976 it was found that hydrolysis of $ClOSF_5$ at low temperatures yields the very unstable $HOSF_5$, pentafluorosulfuric acid VI [reaction (1)].

$$ClOSF_5 + H_2O \longrightarrow HOCl + HOSF_5 \qquad (1)$$

But such reactions with water are very difficult to control at low temperatures. A more convenient route was found in the reaction with HCl, since HCl is liquid at $-100°C$ [reaction (2)]. The reaction

$$HCl + ClOSF_5 \longrightarrow Cl_2 + HOSF_5 \qquad (2)$$

principle is the recombination of partially negative chlorine in HCl and partially positive chlorine in $ClOSF_5$. Synthesis of CF_3OH was achieved by attempting a similar reaction with $ClOCF_3$ [reaction (3)]. The formation of CF_3OH is highly exo-

$$HCl + ClOCF_3 \xrightarrow{-140°} Cl_2 + CF_3OH \qquad (3)$$

thermic, and yet has to be carried out at very low temperature. Therefore, only small amounts can be prepared in one run. Trifluoromethanol has been identified by spectroscopic methods, for example, ^{19}F, ^{1}H, ^{13}C nuclear magnetic resonance (NMR) spectroscopy. Although CF_3OH decomposes into CF_2O and HF above $-30°C$, gas-phase infrared spectra can be obtained at room temperature. The loss of HF as a β-elimination does not take place in a unimolecular mechanism. In contrast to methanol, CF_3OH is very volatile and its boiling point has been estimated as roughly $-20°C$, indicating much weaker hydrogen bonding than in CH_3OH. The name trifluoromethanol does not imply that CF_3OH is a strong acid, forming salts with trimethylamine and such. It could be called trifluoro-*ortho*-carbonic acid as well. CF_3OH is not the only perfluorinated alcohol known. $(CF_3)_3C—OH$ is also a commercially available, stable material. However, HF elimination from this species would be a less favored γ-elimination, resulting in an unstable species,

$$(CF_3)_2C—O$$
$$\diagdown \diagup$$
$$CF_2$$

Another stable perfluorinated alcohol is known:

$$F_2C \begin{array}{c} CF_2 \\ \diagup \quad \diagdown \\ \quad \quad CFOH \\ \diagdown \quad \diagup \\ CF_2 \end{array}$$

It can be prepared from

$$F_2C \begin{array}{c} CF_2 \\ \diagup \quad \diagdown \\ \quad \quad C=O \\ \diagdown \quad \diagup \\ CF_2 \end{array}$$

by adding HF. Its stability is enhanced by the fact that the small angles in a four-ring compound make a carbonyl function less favored. Normally carbonyl groups have bond angles close to 120° in contrast to alcohols, which tend to have angles at 108° at the carbon atom.

Many of these stability arguments hold for perfluorinated amines as well, especially primary amines. Tertiary amines such as $(CF_3)_3N$ are very stable. Even secondary amines do not lose HF so readily. The reaction $CF_3—N{=}CF_2 + HF \rightleftharpoons (CF_3)_2NH$ may be considered a true equilibrium.

However, primary perfluorinated amines were still lacking, but the same reaction procedure that worked successfully for CF_3OH can be used to prepare CF_3NH_2 [reaction (4)]. Another route to CF_3NH_2 is given in reaction (5).

$$CF_3—NCl_2 + 3HCl \xrightarrow{-78°C}$$

$$2Cl_2 + CF_3—NH_2 \cdot HCl \xrightarrow{base} CF_3NH_2 \qquad (4)$$

$$CF_3—NH—COO—t—Bu + 2HCl \xrightarrow{-25°C}$$

$$t—C_4H_9Cl + CO_2 + CF_3—NH_2 \cdot HCl \qquad (5)$$

The reaction between HCl and CF_3NCl_2 is much less exothermic than that between HCl and CF_3OH. It is interesting to note that CF_3NH_2 is still a base in spite of the electron withdrawal capacity of the CF_3 group, but a weaker base than normal organic nitrogen bases such as $(CH_3)_3N$ or quinoline, since the full amine is liberated easily from its hydrochloride with these bases.

Why CF_3NH_2 is less volatile than CH_3NH_2 is still not clearly understood, in spite of the opposite behavior of CF_3OH and CH_3OH. Detailed investigations of this fact are difficult because of the instability of the substances.

CF_3OH and CF_3NH_2 have analogs in inorganic chemistry: SF_5OH and SF_5NH_2. While SF_5OH, the only hexacoordinated sulfur acid, is prepared like CF_3OH [see reaction (3)], the amine SF_5NH_2 is obtained by reaction (6). The similarity between

$$NSF_3 + 2HF \rightarrow SF_5NH_2 \qquad (6)$$

these two classes of components becomes even more evident when the decomposition of CF_3NH_2 in the gas phase is studied [reaction (7)].

$$CF_3—NH_2 \rightarrow F—C{\equiv}N + 2HF \qquad (7)$$

The procedure of allowing —OCl or —NCl_2 groups to react with HCl at low temperatures probably can be applied in general to the preparation of unknown —OH or —NH_2 compounds, provided the corresponding chloro derivatives are available. The successful preparation of CF_3OH and CF_3NH_2 shows how simply some very delicate problems of preparative chemistry may be solved.

For background information *see* FLUOROCARBON; ORGANIC CHEMICAL SYNTHESIS; STERIC EFFECT (CHEMISTRY) in the McGraw-Hill Encyclopedia of Science and Technology.

[KONRAD SEPPELT]

Bibliography: S. Andreades and D. C. England, *J. Amer. Chem. Soc.*, 83:4670, 1961; G. Kloter et al., *Angew. Chem. Internat. Edit.*, 16:707, 1977; G. Kloter and K. Seppelt, *J. Amer. Chem. Soc.*, 101:347, 1979; R. S. Porter and G. H. Cady, *J. Amer. Chem. Soc.*, 79:5628, 1957; K. Seppelt, *Angew. Chem. Internat. Edit.*, 15:44, 1976; K. Seppelt, *Z. Anorg. Allg. Chem.*, 428:35, 1977.

Fluoroxysulfates

The fluoroxysulfate ion, SO_4F^-, is the only known example of a hypofluorite ion, that is, an ion containing an oxygen-fluorine bond. This ion was probably first prepared unwittingly in 1926 by F. Fichter, who bubbled F_2 through aqueous solutions of alkali sulfates and observed the formation of a powerful but short-lived oxidizing species, which he termed *vergängliches Oxidationsmittel*, or "ephemeral oxidant." However, Fichter was unable to isolate or identify this oxidant, and his observations have generally been ignored or discounted. It was not until 1979 that E. H. Appelman and coworkers showed that when F_2 was passed through cold, concentrated solutions of rubidium or cesium sulfates, stable precipitates could be obtained with compositions corresponding to the formulas $RbSO_4F$ and $CsSO_4F$, respectively. An x-ray diffraction study of the rubidium salt confirmed that the anion was indeed a hypofluorite (Fig. 1), containing three short sulfur-oxygen double bonds and one longer sulfur-oxygen single bond to the oxygen that was also bonded to fluorine.

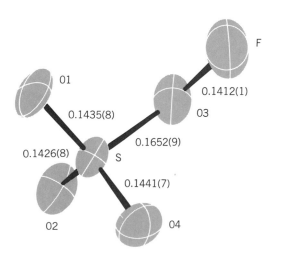

Fig. 1. Structure of the fluoroxysulfate ion, SO_4F^-, from x-ray crystallography. Bond lengths are in nanometers, with uncertainties indicated in parentheses. (*From E. Gebert, E. H. Appelman, and A. H. Reis, Jr., Crystal structure of rubidium fluoroxysulfate: Characterization of the fluoroxysulfate anion, Inorg. Chem., 18:2465–2468, 1979*)

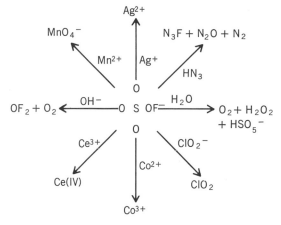

Fig. 2. Preparation of fluoroxysulfate and its reactions in aqueous solution.

The infrared and Raman spectra of the fluoroxysulfate ion are similar to those of the isoelectronic molecule ClO_4F, fluorine perchlorate. Both of these species have rather low symmetry, their only symmetry element being a single plane, and this results in complex molecular spectra involving 12 fundamental vibration frequencies.

The fluorine-19 nuclear magnetic resonance (NMR) spectrum of the fluoroxysulfate ion in acetonitrile shows a single line at -132.3 ppm relative to $CFCl_3$. This may be compared with a line at -37.5 ppm for the aqueous fluorosulfate ion, SO_3F^-, in which the fluorine is bonded to sulfur. The difference indicates reduced shielding of the fluorine nucleus in the fluoroxysulfate, which is to be expected for a fluorine bonded to oxygen.

The rubidium and cesium fluoroxysulfates are fairly stable at room temperature, although gradual decomposition at the rate of a few percent per month has been observed. Explosion of a cesium fluoroxysulfate sample at room temperature has been reported, but this appears to have been an isolated occurrence, and the sample involved may have been contaminated in some way. A mild detonation does take place when either of the salts is heated to near 100°C. This thermal decomposition proceeds primarily in accordance with reaction (1),

$$MSO_4F \rightarrow MSO_3F + \tfrac{1}{2}O_2 \qquad (1)$$

but small amounts of SO_2, SOF_2, and SO_2F_2 are also formed.

The heat of formation of $CsSO_4F$ has been determined from calorimetric measurements to be 1005 kJ/mole, and the standard electrode potential of aqueous fluoroxysulfate has been estimated at 2.46 V. This makes the fluoroxysulfate ion one of the most potent oxidants known, and is entirely in accord with its vigorous oxidizing behavior in aqueous solution, as shown in Fig. 2.

Aqueous fluoroxysulfate solutions are not stable, but decompose gradually to produce oxygen, hydrogen peroxide, and peroxymonosulfate, HSO_5^-.

Reaction of SO_4F^- with aromatic compounds in acetonitrile

| | | Product yields, % of $(SO_4F^-)_0$ | | | |
| Substrate | Substrate consumed, % of $(SO_4F^-)_0$ | Fluoroaromatics | | | |
		Ortho	Meta	Para	Other
Phenol	71	55	<0.2	4	Quinone (1), difluorophenols (5)
Anisole	67	39	<0.2	12	Quinone (5)
Toluene	50	6	<0.3	1	Benzyl fluoride (40), difluorotoluene (2)
Biphenyl	35	10	0.3	5	Tetraphenyl
Benzene	27	Monofluorobenzene (12)			Ring opening
Fluorobenzene	12	2	<0.3	4	Polar
Methyl benzoate	?	2	2	0.7	?
Benzonitrile	?	0.4	0.4	0.4	?
Nitrobenzene	?	0.6	0.9	0.2	?
Naphthalene	65	19 (α)	0.3 (β)		Polar

At 15°C the half-life of SO_4F^- in acidic solution is about 35 min. Isotopic tracer studies have shown that the H_2O_2 and O_2 formed each contain one oxygen atom from the solvent and one from the fluoroxysulfate. In alkaline solution, decomposition takes place extremely rapidly to produce oxygen and some OF_2.

Rates of oxidation by aqueous fluoroxysulfate vary considerably, ranging from the sluggish oxidation of Ce^{3+} to the very rapid reaction with Ag^+ and ClO_2^-. The reaction with ClO_2^- provides a convenient and specific means of analyzing for fluoroxysulfate, since the chlorine dioxide product can be readily determined spectrophotometrically.

Because of the ease with which it is oxidized, Ag^+ can be used as a catalyst for the oxidation of other substances by SO_4F^-. For example, in the absence of a catalyst, Cr^{3+} is not oxidized at all by fluoroxysulfate, but oxidation to dichromate proceeds smoothly in the presence of small quantities of Ag^+.

Oxidations by aqueous fluoroxysulfate are thought to proceed by F atom transfer, with the intermediate formation of the SO_4^- radical ion. Evidence for such a mechanism comes from the oxidation of Cr^{2+}, which produces an approximately equimolar mixture of fluorochromium(III) and hexaaquochromium(III), and therefore most likely involves reactions (2) and (3).

$$Cr(H_2O)_6^{2+} + SO_4F^- \rightarrow (H_2O)_5CrF^{2+} + SO_4^- + H_2O \quad (2)$$

$$Cr(H_2O)_6^{2+} + SO_4^- \rightarrow Cr(H_2O)_6^{3+} + SO_4^{2-} \quad (3)$$

The rubidium and cesium fluoroxysulfates are somewhat soluble in acetonitrile. These solutions are relatively stable and are convenient media in which to study the reactions of fluoroxysulfate with organic compounds. The reactions of a variety of aromatic compounds with SO_4F^- are described in the table. Substantial fluorine substitution takes place, especially with the highly reactive aromatics such as phenol and anisole, but other reactions also occur, including oxidation and degradation of the aromatic ring. It is interesting that, in the case of toluene, fluorination of the side chain to yield benzyl fluoride is the principal reaction. This tends to indicate a free-radical process.

Very few reagents can substitute fluorine for hydrogen on an aromatic ring, and the results shown in the table have considerable fundamental significance. Consideration of the product distribution and the relative reactivities of the various aromatic species leads to the conclusion that, in its initial reaction with aromatic compounds, the fluoroxysulfate ion is acting as an electrophilic fluorinating agent, that is, as a donor of positive fluorine. However, subsequent steps in the reaction can lead to the formation of free radicals and can thereby generate the variety of observed products. Fluoroxysulfate is rather unusual in that it is an electrophilic anion.

Fluoroxysulfate reacts rapidly with olefins. The initial products have yet to be characterized, but they are most likely addition products of the type indicated by reaction (4).

$$\underset{R}{\overset{R}{C}}=\underset{R}{\overset{R}{C}} + SO_4F^- \rightarrow R-\underset{F}{\overset{R}{C}}-\underset{\underset{SO_3^-}{\overset{|}{O}}}{\overset{R}{C}}-R \quad (4)$$

The fluoroxysulfates differ from other known hypofluorites in that they are relatively stable salts that can be easily prepared and conveniently stored prior to use. Therefore they may find significant application as reagents for both inorganic and organic chemistry.

For background information *see* FLUORINE in the McGraw-Hill Encyclopedia of Science and Technology. [EVAN H. APPELMAN]

Bibliography: E. H. Appelman, L. J. Basile, and R. C. Thompson, *J. Amer. Chem. Soc.*, 101: 3384–3385, 1979; F. Fichter and K. Humpert, *Helv. Chim. Acta*, 9:602–611, 1926; E. Gebert, E. H. Appelman, and A. H. Reis, Jr., *Inorg. Chem.*, 18:2465–2468, 1979; R. C. Thompson and E. H. Appelman, *Inorg. Chem.*, 19:3248–3253, 1980.

Forest management

Nutrients that are stored in the soil, humus, litter, and living plants of forest ecosystems concern forest managers. If nutrients are lacking, additions are needed to promote increased or more rapid forest growth. Nutrients are lost from a site through physical removal by harvesting and through basic changes in the nutrient cycle after harvesting. Concern has increased in recent years because of the widespread adoption of clear-cutting as a harvest method, more complete utilization of biomass from harvest sites, and improved ability to measure nutrient gains and losses in forest ecosystems.

Measuring nutrient losses. For the past 20 years small watersheds have been important for studying the nutrient status and cycles of forest ecosystems. Two or more similar watersheds ranging in size from 5 to 100 hectares are selected to represent a particular forest type, land use, geology, and soil. Elements that are deposited on the watersheds by precipitation and dry fallout are measured, as are losses of elements in stream water and windborne and eroded matter. These measurements are the basis of input-output budgets for nutrient-element cycling. Individual aspects of the nutrient cycle may also be examined, such as: rates of nutrient uptake by plants; rates at which organic matter decays and renews the supply of nutrients available for plant growth; and pathways and mechanisms for leaching of nutrients to streams and groundwater.

The watersheds are usually studied in an undisturbed state for several years to obtain baseline data and establish relationships among them. Then one watershed is selected as a control, and one or more of the remaining watersheds are subjected to a treatment or change in vegetative cover such as a harvesting. Data from the treated watershed are compared with those from the control watershed both before and after treatment periods to determine nutrient losses from the ecosystem and how the treatment affected various aspects of the nutrient cycle.

Changes in stream water nutrients. Harvesting causes several important changes in nutrient cycles of forest ecosystems. One of the most obvious and easily detected changes is an increase in the concentration of nutrient ions in streams draining harvested watersheds. Such increases accelerate loss of nutrients from the harvested site. The best example is nitrogen, or more specifically the nitrate ion (NO_3^-), a major form in which nitrogen passes into and out of forest ecosystems. Probably because of the climate and soil types, the maximum response of nitrate concentrations to forest cutting has occurred in northern hardwood stands in northeastern United States. Watershed studies on the White Mountain National Forest in New Hampshire illustrate the magnitude of the increases. Clear-cutting of all merchantable trees raised stream concentrations of nitrate from the usual baseline level of 2 mg/liter (synonymous with 2 parts nitrate to 1,000,000 parts water) to a maximum of 28 mg/liter and an annual average of 18 mg/liter by the second year after harvest. The elevated level of nitrate receded quickly with regeneration of a new stand and returned to pretreatment level within 4–5 years after cutting. Disturbances to forest and rangeland ecosystems at other locations in the United States have also caused measured responses in the nitrate ion in streams. The responses are not usually as great as those in the New Hampshire studies, but they may last longer.

Why nitrate increases in streams draining cutover watersheds is not completely understood, although the increases appear to result from basic changes in the nitrogen cycle. In mature forests free of recent disturbance, the major source of nitrogen is organic material accumulated in the forest soil. Organic matter undergoes decomposition by assorted microorganisms freeing nitrogen in several forms, including ammonium (NH_4^+). Ammonium is used by trees and is also held tightly by negatively charged soil and organic particles. A small amount of ammonium is converted to nitrate (NO_3^+) by soil microorganisms in a process called nitrification. Nitrate is only weakly attracted by the charge on soil particles and readily leaches from the soil into streams. Research suggests that mature forests have some mechanism for inhibiting nitrification. The low concentrations of nitrate in streams draining mature forests substantiates this idea. One possibility is that tree roots secrete exudates that have allelopathic or inhibiting effects on nitrifying bacteria, although this idea is unproved.

Several factors are thought to interact to change the above sequence and increase the nitrate concentration of soil water and stream water when the forest is cut: (1) The forest floor is exposed to greater than normal amounts of heat, light, and moisture. This accelerates decomposition and

Fig. 1. Experimental watersheds at the Hubbard Brook Experimental Forest in New Hampshire. The watershed on the left is shown during one of three separate strip cuttings that led to complete clear-cutting.

mineralization of organic matter and release of available nitrogen in the form of ammonium. (2) The uptake of nitrogen is temporarily blocked. (3) The nitrification process is accelerated in the forest soil, thus increasing the availability of nitrate. Studies in New Hampshire have linked the increased nitrification with increases in the population of nitrifying bacteria after cutting. (4) More moisture moves through the soil because transpiration is reduced after cutting. Watershed studies from many locations have shown that reduced transpiration after clear-cutting increases annual streamflow by as much as 20–40%. This final factor is especially important because soil water is the transport mechanism for nitrate and other nutrients leaching from the soil into streams. All four of these factors operate at increased rates within weeks after forest cutting and persist for variable lengths of time, depending upon climate, soil factors, and speed of revegetation.

Effects of increased nitrification. The increased production and leaching of nitrate has both beneficial and adverse implications for forest and stream ecosystems. The increase of available nitrogen in soils soon after harvest is beneficial; it provides a nutrient-enriched solution for plant development at a critical period of stand regeneration. Also, the higher levels of nitrate in streams might be beneficial in stimulating productivity at various levels of the aquatic food chain. Streams from forests free of recent disturbance are usually low in nutrients and have correspondingly low productivity. However, the change in stream productivity may be adverse and increase the numbers and kinds of undesirable species. Also, there is a remote possibility that concentrations of nutrient ions could increase to levels that are toxic to aquatic biota or exceed safe drinking-water standards.

Probably the most important concern is that the increased nitrate concentrations in streams represent a loss of nitrogen from the forest ecosystem. Watershed studies conducted as part of the Hubbard Brook ecosystem studies in New Hampshire indicate that an additional 100–250 kg of nitrogen are lost to streams from each hectare of cutover forest in the 5-year period immediately after clear-cutting. This is less than 5% of the overall nitrogen capital of the sites. However, it represents nitrogen in a form that is most readily available for plant use as opposed to the bulk of nitrogen which is bound in organic matter and not readily available to plants. Thus, the nitrogen lost to streams would be better conserved on site.

Watershed studies in progress at Hubbard Brook show that forest managers can take steps to moderate nitrogen losses after clear-cutting. One such step is to vary the cutting configuration. In an experiment several years ago, a watershed was completely clear-cut in the usual fashion of one large block. At the same time, a nearby watershed was harvested by cutting a series of 25-m-wide

Fig. 2. Whole-tree harvester at work in a spruce-fir stand in Maine. Trees are sheared at the base, and then the entire tree is lifted onto the rear bed for transport to roadside.

strips alternating with 50-m-wide uncut strips (Fig. 1). The remaining uncut strip was harvested in 25-m widths at 2-year intervals to complete clear-cutting of the watershed in 4 years. Nitrate concentrations in streams draining the watershed cut in strips reached a maximum of only 9 mg/liter as opposed to a maximum of 27 mg/liter from the watershed cut in one block. Nitrogen losses to streamflow from the strip-cutting were less than one-half those from the block clear-cutting.

Nutrient removal in harvested products. In past years there was not much concern about nutrients that were removed from a forest site in the harvested products. Normally only the merchantable bole wood was removed, and tops of trees including branch wood and leaf material were left on site to decay and provide nutrients for future trees. Also, the time between harvests was long and allowed nutrient cycles to recover from cutting disturbances.

In the past few years, however, harvesting intensity and projected time between harvests have undergone significant changes. Rising demands for all forms of forest products, improved methods for utilizing more tree species and smaller-size trees, and the rapid development and deployment of mechanized harvesters (Fig. 2) have resulted in much greater forest utilization and have shortened the time between harvests. Whole-tree harvesting, in which the entire aboveground portion of the tree is utilized, has become commonplace in many areas. Research reported at a recent symposium on the impact of intensive harvest on nutrient cycling indicated that whole-tree harvesting can more than double the nutrient removal of bole-only harvests (Fig. 3). A whole-tree harvest in a stand of northern hardwoods in New Hampshire could remove an estimated 390 kg of nitrogen per hectare as opposed to 165 kg/ha for a bole-only harvest. Corresponding values for a loblolly pine plantation in North Carolina are 321 kg/ha of nitrogen removed by whole-tree harvesting versus 104 kg/ha for bole-only harvests.

Nutrient losses to streams and harvests are not limited to nitrogen but also include calcium, phosphorus, and potassium. Calcium is a special concern because sizable amounts are removed in harvested products, and replacement from organic matter and soil sources can be slow. There are also other forms of nutrient losses from forest ecosystems, and a major one is denitrification. This process occurs when soils are anaerobic or nearly saturated with water, and soil bacteria use nitrate as an oxygen source and emit gaseous forms of nitrogen into the atmosphere as waste products. Detection of denitrification in forests is difficult, and losses have rarely been quantified. However, such losses could be significant in areas where the water table is at or near the soil surface.

Concern about effects of nutrient losses and removals on future productivity of forest stands is resulting in more intensive research of forest ecosystems and their nutrient cycles. Rates at which nutrients are replenished by means such as mineralization and precipitation are being studied to determine the time needed between harvests to maintain optimum productivity. Research also continues on methods of replacing nutrients by

Fig. 3. Harvesting methods. (a) An area that was a bole-only harvest; branchwood and tree tops are left on site. (b) Considerably more biomass and associated nutrients are removed from an area that was a whole-tree harvest.

using commercial fertilizers or by recycling waste products through forest ecosystems.

For background information *see* FOREST ECOLOGY; FOREST SOIL in the McGraw-Hill Encyclopedia of Science and Technology.

[JAMES W. HORNBECK]

Bibliography: G. E. Likens et al., *Biogeochemistry of a Forested Ecosystem*, 1977; C. W. Martin and R. S. Pierce, *J. Forest.*, 78:268–272, 1980; *Proc. Impact of Intensive Harvesting on Forest Nutrient Cycling*, State University of New York, 1979; P. M. Vitousek et al., *Science*, 204:469–474, 1979.

Galactic antiprotons

The role of antimatter in the physical processes of the universe has been a topic of discussion since the discovery of the positron (a particle with the mass of an electron but carrying a positive charge) in 1932, and the antiproton (a particle with the mass of a proton but carrying a negative charge) in 1952. These particles, together with the antineutron, could be used to form the antimatter equivalents to the elements known on Earth. Entire stars could be constructed from antihydrogen (an atom comprising an antiproton and a positron) and antihelium. The light from such a star would be indistinguishable from the light of an ordinary star. This is because the masses of the atomic constituents would be identical for both stars. The polarity of the charges of the subatomic particles would not affect the light emitted by the star. There is reason to believe that there may be large quantities of antimatter somewhere in the universe. Until the last 10 years it was speculated that perhaps some neighboring stars were composed of antimatter.

Laboratory experiments have shown that particles of matter (like the proton) and antimatter (like the antiproton) can combine, causing both to disappear and to produce energy in their place. Similarly, it is possible to convert energy into matter and antimatter. Equal numbers of matter and antimatter particles are produced in this case. In the most widely accepted scientific theory of the creation of the universe, the big bang theory, the universe began as a fireball of radiant energy. As the fireball expanded and cooled, matter was formed, and from it the universe as it is now known evolved. If the present understanding of the relationships between matter, antimatter, and energy pertained during the big bang, equal amounts of matter and antimatter were produced. Similar conclusions are reached for the alternative theories of the evolution of the universe. The question then naturally arises as to the location of the antimatter. Is every other star made of antimatter? As mentioned above, the question cannot be answered by visual observation of the stars.

Search for antimatter in cosmic rays. One of the very best ways to find out if one of the neighboring stars is antimatter, is to gather a sample. Since it would take 8 years to get to the nearest star and back at the speed of light, a mission to sample a nearby star would be hopelessly long at ordinary rocket speeds (10,000 times slower than light). Fortunately, a natural sample is available in the form of cosmic rays. Stars much larger than the Sun end their lives in a cataclysmic explosion called a supernova. By a process not yet clearly understood, the supernova causes atomic particles in the vicinity to be accelerated to velocities very near the speed of light. These particles, called cosmic rays, travel throughout the Galaxy. They are composed mostly of protons, helium nuclei, and more complex atomic nuclei such as those of carbon, oxygen, and iron. In the early 1960s it became apparent to a number of scientists, including Luis Alvarez, that instruments to see whether or not the cosmic rays included both matter and antimatter could be built and flown aboard satellites or high-altitude research balloons. Two research programs grew from Alvarez's ideas. In 1969–1979 a number of balloon flights were made to search for antihelium and heavier antielements in the cosmic rays. To date, no such heavy antielements have been observed, even though more than a quarter of a million cosmic rays have been analyzed. These results imply that there are probably no antistars in the Galaxy.

Production of antiprotons in the Galaxy. In the early 1970s one of the antimatter research teams (then located at NASA's Johnson Space Center and presently at New Mexico State University) realized that, according to the results of terrestrial nuclear physics experiments, there should be a small number of antiprotons continuously manufactured in the Galaxy. The process producing the antiprotons would be nuclear collisions between cosmic rays and the gas molecules between the stars. Observation of this small number of antiprotons would confirm that the laws of physics as they are known in the laboratory do hold true throughout the Galaxy, and that the antiproton is indeed a stable subatomic particle like the proton. With this motivation, the team at Johnson Space Center, with the sponsorship of NASA, began to develop the specialized instrumentation to observe the antiproton.

Experimental apparatus. A superconducting magnet forms the heart of the apparatus for the antiproton experiment (Fig. 1). The magnet coil is about 0.7 m in diameter. It has 13,000 turns of copperclad niobium-titanium superconductor. When charged to its normal 120-A operating current, the magnet produces a magnetic field of from 10 to 40 kG (1–4 T) in the area where the cosmic-ray detectors are located. Cosmic rays which traverse the magnetic field are studied to determine whether or not they are antiprotons. As each particle traverses the magnetic field, devices called

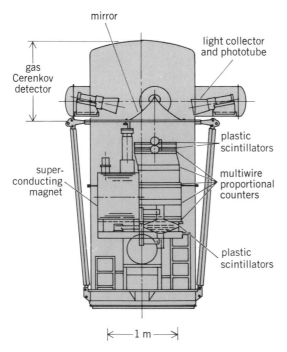

Fig. 1. Experimental apparatus for observation of antiprotons in cosmic rays.

multiwire proportional counters trace the trajectories. In general, charged particles (such as protons or antiprotons) are deflected as they traverse a magnetic field. Positive particles (like protons) would be deflected in the opposite direction of negative particles (like antiprotons). Other radiation detectors (plastic scintillators and a gas Cerenkov detector) are used to distinguish the protons and antiprotons from other components of the cosmic radiation such as helium nuclei.

In order for the instrument to make its observations, it had to be lifted above most of the atmosphere. This is because collisions of the cosmic rays in the atmosphere can also produce antiprotons in sufficient number to obscure the small number being produced in the Galaxy. A helium-filled high-altitude balloon was used to accomplish this objective. The 28,000,000-ft³ (7.9 × 10⁵ m³) balloon was launched by the National Scientific Ballooning Facility at Palestine, TX, on June 21, 1979. The 5800-lb (2600-kg) instrument flew at altitudes near 120,000 ft (37 km) for 8 h as it drifted slowly westward in the stratospheric winds. During the flight, measurements of the 20 cosmic rays/s traversing the experiment were sent by telemetry to a computer system operated by the research team. Adjustment instructions were sent back to the experiment by telemetry to keep it operating at maximum sensitivity. Recordings of the telemetered observations were made at the launch site and at a remote recording station in Pecos, TX. Almost a million cosmic rays were observed. At the end of the flight, a tracking plane commanded the balloon (by radio) to gradually release its helium. As the experiment descended through 90,000 ft (27 km), a command was sent to the experiment which caused the apparatus to separate from the balloon and float to Earth on a parachute.

Results. Analysis of the data recordings took nearly 3 months; the results are shown in Fig. 2. The protons in the cosmic rays are observed as the large bump of particles on the right of the centerline. To the far left of the figure are particles produced by collisions of cosmic rays with the atmosphere. The small bump to the left of center was interpreted as the antiprotons from the Galaxy. Theoretical estimates predicted from 1 to 4 antiprotons would be discovered for every 10,000 protons observed. The investigators placed the observed flux of antiprotons to be 5.2 ± 1.5 antiprotons for every 10,000 protons.

Implications. The implications of the observation are that the antiproton is a stable particle like the proton. In addition, the laboratory experiments which lead to the predicted flux of antiprotons adequately describe the properties of antimatter in the Galaxy. These answers lead to some questions. If the antiproton is stable and the physics of the production of antiprotons is understood, antimatter should have been produced during the big bang and should still exist. Either the laws of physics were different during the big bang, or the antimatter is somewhere outside the Galaxy. Some scientists have speculated that the physics governing production of antimatter may have been different because the energies involved in the reactions producing antimatter during the big bang were much higher than those which have been attained

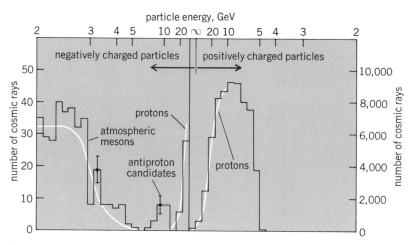

Fig. 2. Number of cosmic rays observed at various energies in an experiment to detect antiprotons.

in the laboratory. Other scientists believe that there are equal amounts of matter and antimatter, and somehow they became separated on a very large scale, causing the universe to be divided into a relatively small number of zones which are either pure matter or pure antimatter.

As with any new observation, there is a degree of controversy regarding the results. However, questions raised regarding possible experimental biases have been answered to the satisfaction of most scientists in the field. Efforts are now under way to confirm and extend these observations. A research group at the California Institute of Technology has recently performed a balloon flight to search for antiprotons, and data analysis is under way. Further balloon flights and space shuttle experiments have been proposed and are being considered by NASA.

For background information see ANTIMATTER; ANTIPROTON; COSMIC RAYS; COSMOLOGY; PARTICLE DETECTOR in the McGraw-Hill Encyclopedia of Science and Technology. [ROBERT L. GOLDEN]

Bibliography: G. D. Badhwar et al., Astrophys. Space Sci., 37(2):283–300, February 1976; T. K. Gaisser and R. H. Maurer, Phys. Rev. Lett., 30(25):1264–1267, June 18, 1973; R. L. Golden et al., Phys. Rev. Lett., 43(16):1196–1199, Oct. 15, 1979; G. A. Stiegman, Annu. Rev. Astron. Astrophys., 14(2):339–372, February 1976.

Galaxy

Recent developments in radio and infrared astronomical instrumentation have permitted the first detailed observations of the center of the Milky Way Galaxy. The observations show that the center has a distinct and bright core, or nucleus, several light-years (1 ly = 9.5 × 10¹² km) in diameter, containing (1) the highest-density stellar cluster in the Galaxy, (2) a very small yet bright radio-emitting object of unknown nature, and (3) a collection of hot, compact, high-velocity clouds of unusual characteristics. There is evidence that in the past violent events have occurred in or near the nucleus. However, at present the nucleus, though turbulent, is relatively quiescent and is far less luminous than similarly sized, active nuclei of some external galaxies.

Observations. The center of the Galaxy is roughly 30,000 light-years from the Sun in the direction of the constellation Sagittarius. This distance, although large compared with distances to familiar objects within the Galaxy, is very small compared with distances to other galaxies. For example, the nearest external galaxy containing a well-defined center, the Andromeda Galaxy (M31), is already some 70 times farther away. Thus in the Sun's vicinity the opportunity exists for relatively detailed study of the center of a normal spiral galaxy. Because both the Sun and the galactic center are located in the plane of the Galaxy, with several dusty spiral arms intervening, the nucleus cannot be de-

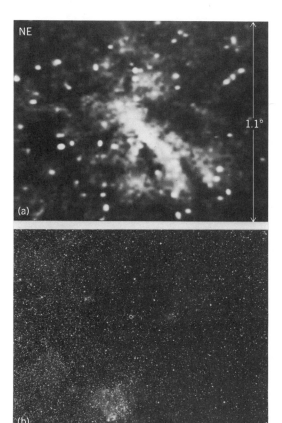

Fig. 1. Central 1° (500 light-years) of the Galaxy. (a) Near-infrared contour map presented as a shaded black and white photograph. Pointlike objects are stars in foreground. (b) Visual photograph. (From E. E. Becklin and G. Neugebauer, 2.2 micron map of the central 1° of the galactic center. Publ. Astron. Soc. Pac., 90:657–660. 1978)

tected at visual wavelengths. Infrared and radio radiation, however, penetrate the interstellar medium, so that at those wavelengths the nucleus can be studied.

Most of the radiation which reaches the Earth from the galactic center can be identified as originating in stars, interstellar dust, or interstellar gas. In general, each of these three components radiates in different wavelength bands, and hence each can be observed separately. The shortest-wavelength infrared radiation which is detectable

from the center is predominantly starlight. Its distribution, observed by E. Becklin and G. Neugebauer (Fig. 1), indicates that the density of stars peaks strongly at the nucleus. It is estimated that several million stars lie within 2 light-years of the center. (For comparison, the distance to the nearest star from the Sun is 4 light-years.)

The mass of interstellar dust at the center is negligible compared to the stellar mass there. The dust, however, is highly efficient at absorbing visual and ultraviolet starlight and reradiating it at middle- and long-infrared wavelengths. In fact, dust proves to be the dominant observable source of radiation in the nucleus. Infrared observations at wavelengths between 3.5 and 20 μm by G. H. Rieke and F. J. Low and by Becklin and Neugebauer have demonstrated that the warm dust in the nucleus is not distributed like the stars, but instead is concentrated in clumps of size 0.3–1.5 light-years (Fig. 2).

Interstellar gas is the source of most of the radio emission observed from the nucleus. The emitting gas is ionized and at a temperature of about 10,000 K. Accurate radio measurements of its spatial distribution are under way at the recently completed Very Large Array near Socorro, NM. However, recent studies of an infrared spectral line of ionized neon, a constituent of the ionized gas, already have provided a large amount of basic information not only about the distribution of the gas, but also about its motion. These studies, by J. H. Lacy, C. H. Townes, and their collaborators, show that most of the ionized gas belongs to the same group of compact clouds as the dust. The 14 clouds they identified within 5 light-years of the center have masses comparable to that of the Sun and velocities of typically a few hundred kilometers per second (Fig. 2). Based on additional infrared line and radio continuum observations by several groups of astronomers, the clouds all appear to have similar low-ionization states, all requiring stars (or other luminous sources) having temperatures no higher than 35,000 K to heat them. The cloud sizes and internal velocity dispersions suggest that each cloud likely dissipates or is dispersed by a cloud-cloud collision in roughly 10^4 years. If the present number of clouds and their apparent destruction rate represent a steady state maintained in the nucleus, then new clouds must form at the average rate of one every 10^3 years.

Possible black hole. A position for the center of the Galaxy can be determined from both the peak of the stellar distribution and from the spatial distribution and velocities of the compact clouds of dust and ionized gas. The two independent estimates agree to better than 0.5 light-year (Fig. 2). Analysis of the distribution of gas velocities about this nominal center suggests that in addition to several million solar masses of distributed stars, the nucleus may contain a central condensation of comparable mass. If it exists, such a compact object presumably would be a black hole. A massive black hole could be partially or totally responsible for the heating of the surrounding interstellar gas by ultraviolet radiation from its accretion disk as material in the disk spirals inward. If there is no black hole, the gas in the nucleus probably is ionized by a small number of hot and massive stars which, if they exist, must have formed in the nucleus less than 10,000,000 years ago.

The nucleus also contains a bright and extremely compact source of radio emission, first observed in 1974 by R. Brown and B. Balick. This source is located very close to the center (Fig. 2). Radio interferometric measurements by K. Kellerman and coworkers imply that the core of the compact object is roughly 0.0001 light-year across. Its luminosity is roughly 10^{-6} of that found in the nuclei of nearby bright radio galaxies. The nature of the object is not well understood: some of its properties resemble those of radio-emitting binary stars in which the companion is losing mass; however, J. Oort has suggested that it may contain a massive black hole. Study of the compact radio source is continuing.

Nature of compact clouds. In their sizes, masses, and luminosities, the clouds of dust and ionized gas in the nucleus resemble the compact clouds found in galactic spiral arms. Such clouds surround, and are heated by, newly formed massive stars. It is highly unlikely, however, given the apparently rapid cloud disruption rate in the galactic nucleus, that the clouds seen there are created by star formation or are gravitationally bound to their ionizing source or sources. Lacy, Townes, and D. Hollenbach have suggested several explanations for their existence. If the clouds are condensations of material, they may be either the remnants of planetary nebulae or the remains of disruptive stellar collisions within the dense nuclear cluster. The rates at which each of these phenomena proceed in the nucleus are uncertain, but it appears that either or both might account for the apparent cloud generation rate. It is also possible that the clouds actually are not condensations at all, but stand out simply because each hot, luminous star which is embedded in the larger nuclear gas cloud heats up the gas in its vicinity.

Self-regulating nucleus. Planetary nebulae and stellar collisions, as well as other phenomena, add interstellar material to the nucleus. What becomes of this material is not known, but it cannot build up indefinitely. Possible mechanisms for its removal are ejection from the nucleus (for example, by radiation pressure), star formation, and accretion by a massive compact object within the nucleus. One can speculate that in the long term a rough equilibrium is maintained. For example, a buildup in interstellar material might lead to an increased nuclear luminosity due to either the formation of new stars or to a higher accretion rate of the proposed black hole. The higher luminosity, in turn, would tend to drive away some of the remaining interstellar material and thus reduce the accretion or star formation rate to the earlier levels. Eventually astronomers hope to learn to what extent such a self-regulatory mechanism actually operates in the galactic nucleus, and whether it operates smoothly or occasionally in violent events. In fact, for many years astronomers have known that there are several massive clouds, presently located far from the galactic center, which are moving outward from the center at high velocities. The existence of these clouds indicates an earlier stage of consider-

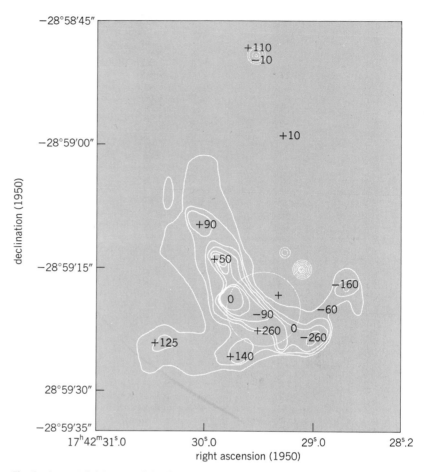

Fig. 2.   Inner 5 light-years of the Galaxy. Contours refer to intensity of infrared emission by warm dust. Numbers are radial velocities in kilometers per second, located at positions of ionized gas clouds. Large circle denotes probable location of center from neon measurements. Cross indicates position of compact radio source and approximate position of highest stellar density. (*From J. H. Lacy et al., Observations of the motion and distribution of the ionized gas in the central parsec of the Galaxy, II, Astrophys. J., 241:132–146, 1980*)

able upheaval in the center and hints that the nucleus of the Galaxy is not always in the relatively calm state observed today.

For background information *see* BLACK HOLE; GALAXY; GALAXY, EXTERNAL; INTERSTELLAR MATTER in the McGraw-Hill Encyclopedia of Science and Technology.   [THOMAS R. GEBALLE]

Bibliography: T. R. Geballe, *Sci. Amer.*, 241(1): 60–70, July 1979; J. H. Lacy et al., *Astrophys. J.*, 241:132–146, 1980; J. H. Oort. *Annu. Rev. Astron. Astrophys.*, 15:295–362, 1977; G. Wynn-Williams, *Mercury*, 8:97–100, 1979.

Girder

A concrete box girder is a rigid structural element. It is inherently strong in resisting bending, twisting, and asymmetrical forces. For these reasons, it is commonly used for bridges with long and short spans, and in buildings to span larger distances. The introduction of prestressing and higher-strength concrete, together with new construction methods, has made it possible to build more slender, larger, and longer concrete box girders, overcoming many environmental constraints. Because

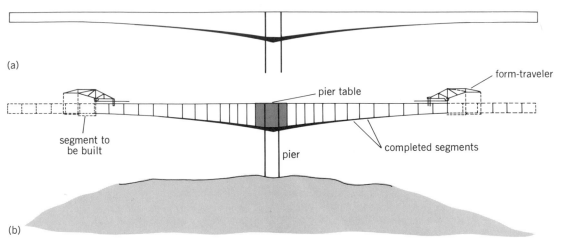

(a)

(b)

Fig. 1. Cantilever construction method. (a) Longitudinal section. (b) Construction scheme.

the concrete box girder can now be made virtually watertight, it can also be used for water flumes or submerged highway tunnels.

Prestressed concrete. Concrete can resist high compressive stresses but lacks tensile capacity. Plain concrete cracks easily under tension. Reinforcing steel embedded in concrete can carry the tensile stresses that exist in a concrete element, but this does not prevent the concrete from cracking, because it cannot elongate as much as the steel.

The principle of prestressing is to use very-high-strength steel in the form of strands, wires, or bars (called tendons) to precompress the concrete. The precompression provides the concrete with a capacity to withstand both compressive and tensile stresses. The tendons are placed mostly inside the concrete in thin metal tubings. They are tensioned by means of hydraulic rams to obtain the predetermined prestressing force for the concrete. Cement grout is then injected to fill the void inside the tubings.

Construction methods. The formwork for a box girder is most commonly supported on falsework. This support system can be steel, concrete, or timber, depending on what is most economically available to the builder. Also, such a main structural member as a box girder is more efficiently constructed when it is separated into smaller segments. The interface between these elements is called a construction joint. Joints are generally vertical, although the girder can also be sliced into horizontal elements. The typical concrete placement sequence is: first, pour the bottom slab up to the lower edge of the web forms; next, cast the webs to their full height; and finally, close the cell by pouring the top slab. Dividing the structure into smaller units is highly economical. The labor to make the forms, install steel, and place concrete becomes efficient because of the shorter learning period necessary for this more repetitive type of work. Then, too, the amount of form material is small and amortized over a large area of the member being constructed.

Prevailing site conditions will often prohibit the use of falsework. Bridging navigational streams, railways, highways, and deep water and construc-

tion in mountainous terrain, areas with poor soil conditions, and urban areas are a few examples. These circumstances suggest the selection of cantilever construction. Ulrich Finsterwalder developed what is considered the first functional segmental, free cantilever structure—the Lahn Bridge in Germany. All of the building for this type of structure, with the exception of the foundation and piers, takes place from the top of the bridge, eliminating falsework and consequently causing minimum disruption to the immediate environment. The superstructure is balanced in increments off the pier, creating a giant tee. The tees are extended between piers until they touch at mid-span, joining the bridge between supports. The formtraveler is the construction mechanism that erects the span, segment by segment (Fig. 1). Cast-in-place segments are usually 10–16.5 ft (3–5 m) in length.

A variation on this sequence is to precast the

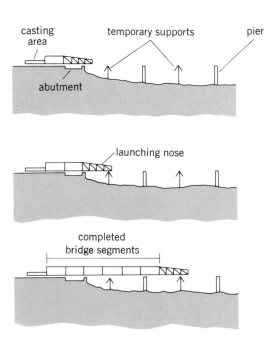

Fig. 2. Incremental launching method.

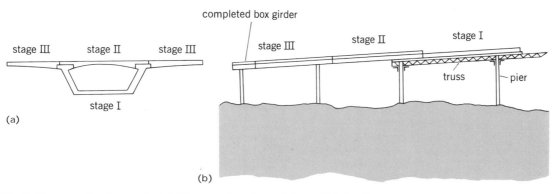

Fig. 3. Stage construction method. (a) Cross section of box girder and (b) elevation.

segments at a ground site and transport them to the point of erection by truck. Because of transportation and hoisting, segments tend to be smaller than for cast-in-place construction. Nevertheless, certain situations permit barge delivery, and therefore large precast pieces are possible. These elements are also locked and stressed together, permitting the structure to evolve sequentially. To assure that the construction joints are tight, a system called match casting is utilized. One unit is cast against its preceding match piece in order to provide a full contact joint. To further assist in this regard, the surface at the joints is coated with high-strength epoxy at the time of erection. The epoxy, aside from sealing the joint, acts as a lubricant to facilitate fine adjustment of the mated pieces. Because of the handling requirements associated with precast pieces, this type of bridge is generally suitable for intermediate-range spans, whereas for long spans it is usually preferable to use the cast-in-place method.

The introduction of special equipment has increased the efficiency of cantilever construction. Overhead gantry trusses are used to transfer form-travelers and materials from pier to pier, eliminating the necessity of dismantling, relocating, and reassembling this equipment. The gantry also contributes to improved productivity by transporting segments across the top of the bridge.

Another method which has similar advantages but which is mainly for short-span bridges is the incremental launching method. This method was developed by F. Leonhardt in West Germany. The bridge girder is divided into 40–80-ft (12–24-m) segments which are all cast at the abutment. After each segment is cast, the girder is pushed out from the abutment so that the next segment can be cast against the previous one. As shown in Fig. 2, a temporary launching nose is attached to the front end segment. A simple launching method is to first lift the bridge slightly off the abutment by a hydraulic jack and then use another hydraulic jack to push the bridge forward. Pulling devices have also been employed on some recent bridges. Teflon pads and steel plates are used at the piers to reduce friction. There are certain restrictions for this type of construction: the foundation at the abutment must be very sturdy, the superstructure must be either completely straight or have a constant curvature, temporary supports are required

for spans over about 160 ft (49 m), and the bridge girder should be slightly deeper than a comparable bridge built by other methods.

All the systems discussed to this point fragment the box girder by the use of vertical construction joints. M.-C. Tang developed a construction system called stage casting, which was first applied to the Denny Creek Bridge in the state of Washington. This system deviates from all the others in that the girder is separated into three span-by-span, longitudinal stages. This is a medium-span bridge technique which uses a light truss system to establish the first segmental longitudinal element. Because this element represents only about one-third the total weight of the girder, the truss is correspondingly lighter. The first element, which is self-supporting, has a bottom slab and two webs which effect a U-shaped cross section. After the U has been posttensioned, the truss is advanced to the next span (Fig. 3). The top slab of the U will be placed on a movable form which follows inside the U one span behind the truss. The wing slabs are then cast by another form-carrier waiting two spans behind the truss. This concept permits three stages of the cross section to be manufactured simultaneously at contiguous spans.

Other applications. Although mainly box girders have been discussed, it is important to recognize that for bridge-size spans this same type of member has other highly successful applications. In Oberstdorf, West Germany, the ski jump for the 1973 World Championship was a box girder. This inclined box beam, cantilevering from a rock base, is 328 ft (100 m) long and 9.8 ft (3 m) wide. The box girder has also been employed to provide bent-frame roofs for stadiums, auditoriums, sports arenas, airplane hangars, tunnels, pipelines, water flumes, and underwater vehicular tunnels.

For background information *see* BRIDGE; CONCRETE; GIRDER, PLATE in the McGraw-Hill Encyclopedia of Science and Technology.

[MAN-CHUNG TANG]

Granite

Although H. H. Read's dictum, enunciated more than 30 years ago, that "there are granites and granites" still holds, it has a very different meaning today than it did in 1948 when granitization was widely accepted as an alternative to magmatism for the origin of granitic rocks. Experiments with silicate melts of different compositions (including varying water content) over wide ranges of temperature and pressure have yielded results that correspond closely to the patterns observed in natural granitic rocks. Lacking comparable support, few geologists now consider granitization to be a tenable alternative mechanism for the generation of granitic rocks and batholiths. Today the great problems involve magma: how and where it forms, how and why it rises, how it makes room for itself, how the great diversity of rock types that make up batholiths arises, and above all how batholiths relate to plate tectonics and especially to convergent plate boundaries.

Types of granitic rocks. One of the more important advances made during the 1970s was the recognition by B. W. Chappell and A. J. R. White of I-type and S-type granitic rocks. Although both types are presumed to have magmatic origins, I-type magmas were formed from igneous source materials, and S-type magmas from sedimentary source materials that have been through a weathering cycle and have lost sodium to sea water and calcium to carbonates, and consequently have been residually enriched in aluminum.

Although nowhere explicitly stated, this twofold division is implicit in some recent European literature. For example, Jean Didier distinguishes granitic rocks that contain hornblende-bearing mafic inclusions from those that contain sedimentary inclusions. The criteria for distinguishing between I- and S-type granitic rocks are largely chemical but include field and modal criteria (see table). This division of the granitic rocks has received wide attention, and modifications and additions are already appearing. White has proposed designating the low-potassium granitic rocks of continental margins, which chemically and isotopically resemble the volcanic rocks of island arcs, as M types (mantle-deprived), and D. R. Wones the alkaline rocks as A types.

I-type granitic rocks predominate among the huge Mesozoic batholiths of the west coast of North and South America and the Caledonian granitic rocks of Europe, whereas S-type granitic rocks predominate in Southeast Asia and among the Hercynian granitic rocks of Europe. Both types

occur in eastern Australia and Japan. Tin and greisen-type tin-tungsten deposits are typically associated with S types, and porphyry copper deposits with I types. According to Wones, I-type magmas are characterized by high oxygen fugacities, high CaO activities, and moderate H_2O fugacities, whereas S-type magmas have low oxygen fugacities, high Al_2O_3 activities, and generally high H_2O fugacities. The presence of higher-temperature mineral assemblages among I-type than among S-type granitic rocks suggests that I types were generated at greater depths.

Shunso Ishihara has proposed a somewhat similar plan according to which magnetite-series granitic rocks containing 0.1 to 2.0 vol % of magnetite are distinguished from ilmenite-series granitic rocks, which lack magnetite and contain less than 0.1 vol % of ilmenite. All the magnetite-series granitic rocks correspond to I types, and most of the ilmenite-series rocks to S types, although a few ilmenite-series rocks are I types. Ishihara attributes the absence of magnetite in ilmenite-series granitic rocks to the reducing effect of carbon present in the associated sedimentary rocks.

Tectonic setting. Most phanerozoic batholiths are now conceived to have been generated at or near plate boundaries: I types where oceanic crust is subducted beneath continental crust, and S types where continental plates collide. Wallace Pitcher characterizes these orogenic environments as Alpinotype and Hercynotype, respectively. The region immediately adjacent to a subduction zone commonly is characterized by blueschist metamorphism, the occurrence of ultramafic rocks, and an absence of batholiths. I-type batholiths occur some distance in the down-dip direction from the surface trace of the subduction zone, where they may be localized along a major fault or suture. The best example of such structural control is the coastal batholith of Peru, which is localized along a fault system within the Precambrian shield, well inland from the trench that marks the surface position of the subduction zone. The Sierra Nevada batholith is situated across the edge of the Precambrian continental crust within a belt of complex faulting that may involve far-traveled miniplates.

Criteria for I- and S-type granitic rocks

I types	S types
Relatively high Na_2O content, normally >3.2 wt % in felsic varieties, decreasing to >2.2 wt % in mafic varieties	Relatively low Na_2O content, normally <3.2 wt % in rocks with ~5 wt % K_2O, decreasing to <2.2 wt % in rocks with ~2 wt % K_2O
Mol $Al_2O_3/(Na_2O + K_2O + CaO) < 1.1$	Mol $Al_2O_3/(Na_2O + K_2O + CaO) > 1.1$
Broad spectrum of compositions, ranging from quartz diorite or tonalite to granite. SiO_2 ranges from ~58 to 78 wt %	Relatively restricted compositions in the range of granite and biotite granodiorite; SiO_2 content high, generally >65 wt %
Regular interelement variations within plutons; linear or near-linear variation diagrams	Variation diagrams somewhat irregular
CIPW normative diopside or <1 wt % normative corundum	>1 wt % normative corundum
Hornblende common in granodiorite, tonalite, and quartz diorite	Hornblende absent, aluminum silicates (andalusite, silmanite, kyanite), garnet, and cordierite may be present; muscovite may be abundant in felsic varieties
Mafic hornblende-bearing inclusions common in hornblende-bearing rocks	Hornblende-bearing mafic inclusions absent; sedimentary inclusions may be present and abundant
Initial $^{87}Sr/^{86}Sr$ ratio generally but not invariably low (<0.706)	Initial $^{87}Sr/^{86}Sr$ ratio high (>0.706)

Source regions of granitic magmas. The close association of batholiths with continental crust has persuaded most geologists that crustal materials must form a large component of most granitic magmas. If S-type magmas are generated from sedimentary sequences, an origin in the crust seems certain. The source region of I-type magmas, however, is less certain. Presently, the most favored model is that mantle-derived mafic material rises (generally from a subduction zone) into the lower crust where it partially melts and mixes with crustal materials. A common initial strontium-87 to strontium-86 ratio ($^{87}Sr/^{86}Sr$) of less than 0.706 in tonalite and accompanying trondhjemite suggests that the parent magmas of these rocks may have been generated entirely within the mantle, although the magmas may also have been derived largely from crustal materials that were extracted from the mantle only shortly before the parent magmas were generated. Granodioritic and granitic magmas may contain either larger amounts of crustal material or older crustal material in which the $^{87}Sr/^{86}Sr$ ratio has increased.

A less popular view is that granitic magmas are generated by fractional melting along subduction zones, from either subducted crustal material or adjacent zones in the mantle, and that the potassium content of the magma increases with increasing depth of magma generation. The principal argument in favor of this model is that the potassium content increases eastward across the Sierra Nevada and other Mesozoic batholiths of western North America in the down-dip direction of a presumed former subduction zone. Increase of potassium across island arcs has been correlated with the vertical distance to underlying seismically active zones where some workers presume the magmas were generated. Several relations, however, favor lateral change in the composition of the source materials over depth to a subduction zone as the cause of the eastward increase of potassium. The eastward increase in potassium involves granitic rocks ranging in age from Triassic to Cretaceous, and a similar, though smaller, increase occurs in the country rocks. R. W. Kistler and Zell Peterman reported that initial $^{87}Sr/^{86}Sr$ increases and $^{206}Pb/^{204}Pb$ decreases eastward, and that a sharp transition occurs along a line corresponding to the former edge of the Precambrian shield.

Rise of magma. Magma is thought to rise buoyantly because of its lesser density and increased volume in comparison with the rocks from which it was formed; an analogy with salt domes is frequently made. The viscosity contrast between the magma and its wall rocks governs the behavior of the magma. According to Pitcher, at deeper levels, where the viscosity contrast is smaller, magmas rise by squeezing their walls aside; at higher levels in the zone of brittle fracture, however, they mechanically displace the wall rocks and roof rocks by stopping and uplift. Although a circular intrusive form requires the least energy, structural inhomogeneities in the intruded rocks commonly cause elongate forms. Several experiments by Hans Ramberg show analogs of magmas mushrooming as they approach the surface; other workers postulate teardrop-shaped bodies of magma

becoming detached from their source region as they move upward to some critical level where they form a composite layer floored by country rocks. Although this model is attractive, the bottoms of granite plutons have not been observed, and inward-dipping contacts are scarce.

Differentiation of magmas. Geologists have long recognized that granitic rocks in a given area may constitute consanguineous suites or sequences, even though they include rocks differing widely in composition. Contamination at the exposed levels generally fails to explain these compositional variations, and the common model involves crystal fractionation, by which higher-temperature minerals are subtracted in one way or another and successively more felsic residual magmas are left.

Paul Bateman and Bruce Chappell have proposed that the diversity of rock types in the concentrically zoned Tuolumne Intrusive Suite is due to crystal fractionation during solidification. According to this model, fractionation occurred because residual material (restite) carried upward in the magma from its place of origin, together with minerals precipitating from the melt phase, was progressively cleared from the central, more fluid parts of the magma by settling, by preferential accretion of crystalline material to the solidifying margins of the magma chamber, or by both processes operating together. The kind, composition, and proportion of crystals precipitating, and the composition of the melt phase, changed with falling temperature. Bateman and Chappell envisage that the simplest kind of comagmatic plutonic sequence is a concentrically zoned pluton in which relatively mafic rocks composed of high-temperature mineral assemblages in the margins pass inward without discontinuities into more felsic rocks composed of lower-temperature assemblages. More complex sequences are caused by surges of the less highly crystallized core magma breaking through the solidifying carapace, generally repeatedly.

Although this model is attractive and appears to explain the general distribution of mineral phases and major elements in granitic rocks, studies of minor elements and isotopic ratios suggest that all or part of the diversity of the rocks may result from heterogeneity at depth in the composition of parent magma. Increase of the initial $^{87}Sr/^{86}Sr$ ratio inward in the concentrically zoned Griffel pluton of Scotland led W. E. Stephens and A. N. Halliday to postulate the mixing of two source magmas. Others have postulated the progressive contamination of andesitic or basaltic magma from the mantle by lower-temperature feldspathic rocks of the lower crust. The relative importance of crystal fractionation and of changing composition of the parent magma at depth remains unsettled.

For background information *see* BATHOLITH; GRANITE; PLATE TECTONICS in the McGraw-Hill Encyclopedia of Science and Technology.

[PAUL C. BATEMAN]

Bibliography: P. C. Bateman and B. W. Chappell, *Geol. Soc. Amer. Bull.*, 90:465–482, 1979; B. W. Chappell and A. J. R. White, *Pacific Geol.*, 8:173–174, 1974; R. W. Kistler and Z. E. Peterman, *Geol. Soc. Amer. Bull.*, 84:3489–3512, 1973; W. S. Pitcher, *J. Geol. Soc. London*, 136:627–662, 1979.

Graph theory

One of the most widely known problems in mathematics is the four-color problem: that in any map drawn on a sphere (or in the plane), each country can be colored with one of four colors so that no two neighboring countries with a boundary line in common are colored the same, although they may have one or even a finite number of points in common. The map-coloring problem remained unsolved from 1852, when it was announced by Francies Guthrie, until 1976, when it was solved by Kenneth Appel, Wolfgang Haken, and John Koch with the assistance of a computer. That four colors are necessary can be easily seen from Fig. 1.

Because numerous mathematicians have attempted the four-color problem, there are a large number of alternative ways to look at it. For various surfaces other than the sphere, the problem of exactly how many colors are needed to color an arbitrarily drawn map was settled some time ago. Such surfaces may be topologically more complex than the sphere, yet the problem may be simpler to solve. For example, seven colors have been shown to be necessary and sufficient for coloring maps on the torus. However, the four-color problem for the sphere itself remained unsolved until attacked with the most advanced computers. The proof was long (more than 100 pages), covered a huge number of cases (nearly 1400), and required several hundred hours (originally, 1200) by computer. An alternative proof of the problem would be much desired, and the search for it continues.

Contiguity of regions. It can easily be shown that five points in the plane or on a sphere cannot be connected by arcs in pairs in such a way that two arcs meet only at one of the five points. There must always be at least one pair whose arcs intersect at a point other than the five points. For example, two of the 10 arcs connecting pairs of the five points A, B, C, D, E, in Fig. 2 intersect at X. This fact about five points can be easily related to five countries, as will be shown below. It is impossible to draw a map on a sphere that involves five regions, each of which has a boundary in common with every one of the other four. However, this does not furnish a proof of the four-color problem. The question of the contiguity of regions is not the same as the question of colorability. For example, the map of Fig. 3 requires four colors, yet there are

no groups of four countries, every one of which touches all others.

Kempe's attempted proof. In 1879 A. B. Kempe published what he thought was "proof" of the problem, and it was generally believed that he had settled the matter. But in 1890 P. J. Heawood found a flaw in it. Fortunately he was able to salvage Kempe's approach to prove that five colors would be sufficient to color any map on the sphere. Kempe's ideas will now be discussed, and used to illustrate how the computer was employed to solve the problem.

Dual graphs. A point (called a vertex) may be associated with each country (usually called a region) of a map. The point may represent the capital of that region. Then each pair of vertices may be joined by an arc if their countries have a common boundary line. The result is called the dual graph of the map (Fig. 4).

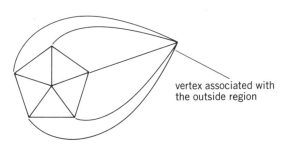

vertex associated with the outside region

Fig. 4. Dual graph of the map in Fig. 3.

A planar graph is a graph which can be drawn on the sphere or in the plane so that the only points where arcs meet are vertices and hence there are no crossings of arcs. The dual graph to a planar map is itself planar.

Coloring the countries of the original map is equivalent to coloring the vertices of its dual graph so that no two vertices that are joined by an arc have the same color.

Plane triangulation. Consider any graph, particularly the dual graph of a planar map with no multiple connections between the same pair of vertices, and no bridges, that is, arcs that do not bound a region. (Both multiple arcs and bridges can be easily eliminated.) By adding arcs to connect pairs of vertices, it is possible to produce another graph, each of whose regions is bounded by three sides. The result is called a plane triangulation. Figure 5*b* is a triangulation of 5*a*.

A planar map is maximal planar if no new arcs can be added without forcing crossings and hence violating planarity. It can be proved that a graph is maximal planar if and only if every region has three sides. It follows that the triangulation in Fig. 5*b* is maximal planar. Triangulation is a convenient standardization of graphs.

If the vertices of a plane triangulation of a graph can be four-colored, the vertices of the original graph can also be four-colored by removing the new arcs, which of course would have imposed additional constraints on coloring and adjacency that are not in the original graph. Thus if the vertices of an arbitrary maximal planar graph can be

GRAPH THEORY

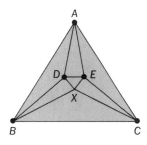

Fig. 1. Map demonstrating that four colors are necessary to distinguish neighboring countries.

GRAPH THEORY

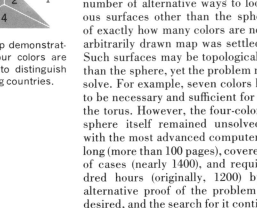

Fig. 2. Graph with five vertices and arcs connecting each pair of vertices. Two of the arcs intersect at X.

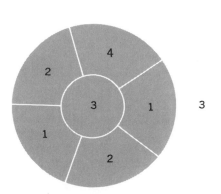

Fig. 3. Map demonstrating difference between contiguity of regions and colorability.

four-colored, so can the vertices of any planar graph.

Early in the history of the problem, people searched for counter-examples that are maximal planar but which required more than four colors. Of course that effort had to fail, but the solution revolves around this kind of graph.

Existence of vertex of degree 5 or less. Leonhard Euler proved that the following relationship exists among the number of vertices n, arcs m, and regions r of any planar map: $n - m + r = 2$.

The degree of a vertex is the number of arcs which have that vertex as an end point. In a planar triangulation there is always one vertex whose degree is at most 5; this can be proved by contradiction. If every vertex is of degree at least 6, then $6n < 2m$. Since, in a planar triangulation, $3r = 2m$, if these two relations are substituted in Euler's formula, the inequality

$$\frac{2m}{6} - m + \frac{2m}{3} \geq 2 \quad \text{or} \quad 0 \geq 2$$

obtained leads to a contradiction. Thus some vertex has degree at most 5. In other words, every planar triangulation contains at least one of the configurations in Fig. 6, in which a vertex may be of degree 2, 3, 4, or 5.

Elimination of vertices of degree 2 or 3. Kempe (and also Appel and Haken) argued by contradiction as follows: If there is a dual graph whose vertices are not four-colorable, then there is one with a minimum number of vertices. If it is not a triangulation, extra arcs can be added to turn it into a triangulation T without increasing the number of its vertices. Every graph which has fewer vertices than T is four-colorable. Now T cannot contain the first two configurations of the previous figure: If it does, v can be removed. Since the resulting graph has fewer vertices than T, it is four-colorable, and by restoring v, the resulting graph can still be four-colored, since v has common arcs with at most three other vertices. This contradicts the fact that it cannot be four-colored.

Sufficiency of five colors. The theorem that five colors are sufficient for coloring the regions of a planar map can be proved by induction on the number of vertices of the dual. Assume the theorem is true for $n - 1$ vertices; it will be proved for n vertices.

A graph with n vertices has at least one vertex v whose degree is equal to or less than 5. If v is removed from this graph, the resulting graph can be colored with five colors by the induction hypothesis. Consider the worst possibility, that is, that five vertices $v_1, \ldots, v_5$ are adjacent to v and arranged in clockwise direction. Suppose that the coloring of the graph without v (when v is removed, the arcs incident with it are also removed) assigns a different color to each of these vertices (otherwise one of the remaining colors could be assigned to v and the theorem is proved). Let the respective colors be $c_1, \ldots, c_5$. It will now be shown that it is possible to reallocate the colors so that v will receive a color different from the vertices with which it is incident; that is, at least two of these vertices will be assigned the same color. Consider the subgraph (called a Kempe chain) of vertices colored with c_1 and c_3 (the same colors as

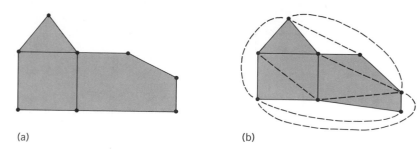

Fig. 5. Operation of plane triangulation, illustrated by (a) a planar graph and (b) its plane triangulation.

those of v_1 and v_3). If v_1 and v_3 are not connected (that is, there is no chain between them) in this subgraph, then the vertices colored with c_1 of the component which contains v_1 can be colored with c_3 and those colored with c_3 in that component are now colored with c_1. In this manner both v_1 and v_3 receive the color c_3 and v can be colored with c_1. If, on the other hand, v_1 and v_3 are connected in the subgraph of vertices colored with c_1 and c_3, then, for example, v_2 and v_4 cannot be connected in the subgraph of vertices colored with c_2 and c_4. Otherwise, the chain connecting them must meet the chain connecting v_1 and v_3, and the vertex where the two chains meet will have assigned to it one color from one subgraph and a second color from the second subgraph. This subgraph which includes v_2 and v_4 can be recolored in the same manner as the disconnected case for v_1 and v_3 above, and hence v_2 and v_4 receive the same color. The other color is then assigned to v.

Sufficiency of four colors. This sort of proof can be used to show that if T contains the configuration of Fig. 6c, then again removing v and restoring it would lead to the four-colorability of T. The problem would be solved; that is, there cannot exist a minimal T that is not four-colorable, if it were not for the possibility that T may contain the configuration of Fig. 6d, for which this approach breaks down. But it may be possible to carry out the proof through a related approach that involves listing a complete set of configurations (called an unavoidable set) of which every graph has samples, and showing that all these configurations are reducible, which means that the four-colorability of any planar graph containing such a configuration can be deduced from the four-colorability of planar graphs with fewer vertices. Thus, reducible configurations cannot be contained in a minimum counterexample to the four-color problem, or any counterexample containing any of them cannot be minimal and can be reduced to a smaller counterexample.

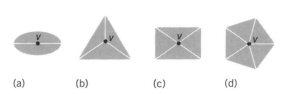

Fig. 6. Configurations with vertices of degree (a) 2, (b) 3, (c) 4, and (d) 5.

The configuration of Fig. 6d has one vertex inside and is bordered by a ring of five vertices. Each of Appel and Haken's approximately 1400 configurations has several vertices inside but all of them are bordered by rings of no more than 14 vertices.

An increase of ring size by one vertex results in about a threefold increase in the number of possible four-colorings of the ring, requiring much greater computer time. By interacting with the computer to determine the set of unavoidable configurations, Appel and Haken gained insight into the likelihood of their reducibility. They were also able to select or modify configurations to one requiring reasonable computer time for proving reducibility. They benefited from more complicated reducibility algorithms on the subject involving the discharging of vertices by H. Heesch. Thus they proved the four-colorability of all planar maps.

Surfaces other than the sphere. For (orientable) surfaces other than the sphere whose genus (indicated by the number of holes) is p, Euler's formula is $n - m + r = 2 - 2p$. Thus for the torus $p = 1$ and $n - m + r = 0$. If a is the average number of edges per vertex in a triangulation of a map on such a surface, then there must be some vertices whose degree is less than or equal to a. It is to such a vertex and its neighbors that reduction is applied.

The equation $an = 2m = 3r$, substituted in Euler's formula, gives $a = 6 + 12\,(p-1)/n$. Reduction will work if there is one extra color to assign to the vertex, should its neighbors be assigned all a colors. Thus, if 1 is added to both sides of the equation, and the number of colors is denoted by X, then the inequality X < n gives X < $7 + 12\,(p-1)/X$. One can solve this quadratic inequality in X for $p \geq 1$, use the positive root, and take the nearest integer less than the quantity in brackets below to obtain Heawood's formula for the number of colors (which is known to be) sufficient for coloring maps on an orientable surface of genus p:

$$\left[\frac{7 + \sqrt{1 + 48p}}{2} \right] \qquad p \geq 1$$

The result for nonorientable surfaces is obtained by replacing $2p$ by q.

Now that the answer to the four-color problem is known, it is hoped that a more manageable proof will eventually be found.

For background information *see* GRAPH THEORY in the McGraw-Hill Encyclopedia of Science and Technology. [THOMAS L. SAATY]

Bibliography: T. L. Saaty and P. Kainen, *The Four Color Problem: Assaults and Conquest*, 1977.

Halogenated hydrocarbon

Concern about pollution of the environment is no recent phenomenon. In the last century diseases transmitted by contaminated water supplies, such as cholera and typhoid, were eliminated from major industrial countries by public health measures. Similarly pollution of the air by combustion processes has been curbed by clean-air legislation. But now there is a new problem: one caused by the widespread use of chemicals. Pollution of the environment through poor industrial practice, the dumping of toxic waste chemicals, and the indis-

criminate use of chemical products is a major concern of regulatory authorities. Three chemicals — dioxins, polychlorinated biphenyls (PCBs), and dichlorodiphenyltrichloroethane (DDT) — illustrate the problem and the measures taken to contain it.

Dioxins. Dioxins (polychlorinated dibenzo-*p*-dioxins) are exceedingly toxic contaminants produced in the manufacture of chlorinated phenols. Two of the better-known phenols are 2,4,5-trichlorophenol and pentachlorophenol. It is in the course of the manufacture of the former that the most toxic of the contaminants, 2,3,7,8-tetrachlorodibenzo-*p*-dioxin (TCDD) (Figs. 1 and 2), is produced.

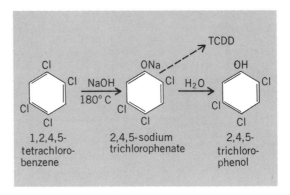

Fig. 1. Synthesis of 2,4,5-trichlorophenol.

TCDD is very stable, requiring a temperature in excess of 800° C for thermal decomposition. Bacteria can degrade TCDD; its half-life in soil varies between 220 days and 3 years, depending on the type of bacteria, their concentration, and the quantity of the chemical present. Higher chlorinated dioxins with six to eight chlorine atoms are made in the synthesis of pentachlorophenol (Fig. 3).

Reaction conditions are such that TCDD is an unavoidable contaminant of 2,4,5-trichlorophenol production. In consequence, trace amounts of dioxins are found in the products made from the phenols. 2,4,5-trichlorophenol is used to make the herbicide 2,4,5-trichlorophenoxyacetic acid (2,4,5-

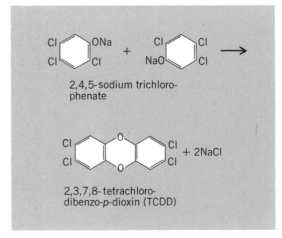

Fig. 2. Formation of 2,3,7,8-tetrachlorodibenzo-*p*-dioxin (TCDD).

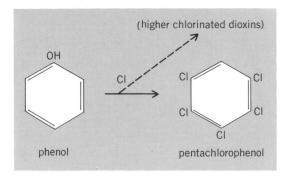

Fig. 3. Synthesis of pentachlorophenol.

T) and the bactericide hexachlorophene (Fig. 4a and b). Pentachlorophenol preparations are used as wood preservatives, as fungicides, and in the curing of leathers.

Two events brought dioxins into the public arena. The first was the acknowledgment that preparations of 2,4,5-T used during the Vietnam War were teratogenic when tested in animal species; the contaminant present in the herbicide and responsible for the malformed fetuses was TCDD. The second incident was the venting directly into the atmosphere of a TCDD-contaminated mixture from a reactor manufacturing 2,4,5-trichlorophenol near the Italian town of Séveso on July 10, 1976.

Agent Orange. An estimated 100 kg of TCDD was deposited on Vietnam in the course of the military herbicide spraying missions from 1962 to 1970. The TCDD was present in a formulation of the herbicide code-named Agent Orange. A 1:1 mixture of 2,4,5-T and the related herbicide 2,4-dichlorophenoxyacetic acid (2,4-D), Agent Orange contain-

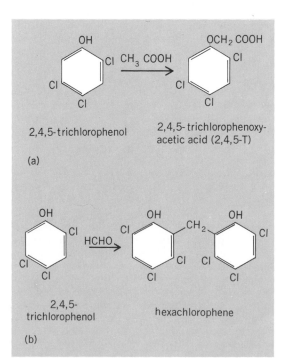

Fig. 4. Syntheses from 2,4,5-trichlorophenol. (a) Formation of the herbicide 2,4,5-T. (b) Formation of the bactericide hexachlorophene.

ed 0.05–47 ppm of dioxin. Current formulations of 2,4,5-T sold in Western Europe must, by law, have TCDD concentrations of 0.1 ppm or less.

According to Vietnamese scientists, the widespread use of herbicides during the war was responsible for an increase both in the incidence of spontaneous abortions and in the number of children born malformed in areas of the country which were heavily sprayed. Claims have also been made by other Vietnamese scientists that TCDD in the herbicide could be responsible for an increase they have noted in the incidence of liver cancer. Although TCDD is a known teratogen and carcinogen in animals, evidence for it causing these problems in humans is inconclusive.

Good hospital records are an essential prerequisite for establishing increases in the rate of malformation. The dislocations of the populations in the course of the war has meant that records are sparse and inadequate for any kind of assessment. For this reason it is impossible to substantiate the charge that TCDD has led to an increase in the number of malformed children in Vietnam.

On the question of whether TCDD has caused an increase in the incidence of liver cancer in Vietnam, many scientists remain skeptical. The latency period between exposure to a carcinogen and the development of a tumor is usually 20–30 years. TCDD, if it behaves like known chemical carcinogens, has really not been present in Vietnam long enough to be the cause of liver cancer. The link between the two is at best circumstantial.

Séveso incident. Clearly observable effects of exposure to TCDD have been observed in the residents of Séveso following the discharge of some 250 g of the chemical from the reactor over a populated area. At least 100 children developed mild to moderately severe forms of the skin disease chloracne. A positive sign of dioxin poisoning, chloracne in its milder form resembles adolescent acne with blackheads and straw-colored cysts around the eyes and on the ears. More severe cases develop pustules which, on resolution of the skin condition, will form scars. Not only is chloracne unsightly, it is also very persistent. Workers exposed to TCDD more than 30 years ago still have the disease.

Skin complaints were the most visible symptoms of dioxin exposure on the residents of Séveso. Other less quantifiable symptoms, such as liver abnormalities and neurological disorders, have also been reported in some of the residents. But the real tragedy of Séveso was borne by the women who were pregnant at the time that the accident occurred. With TCDD known to be a teratogen, many had to make the agonizing decision of whether or not to terminate their pregnancies. Thirty-four Séveso women decided this was the best course and had abortions legally in Italy. Twice this number are thought to have had an abortion in another country, or illegally in Italy.

For those who chose to see their pregnancies through to term, the outcome was a considerable relief. The incidence of malformed children born to Séveso mothers did not rise. Both the incidence of malformation and the type of abnormalities seen were similar to the picture for the rest of Italy and for Western Europe as a whole.

TCDD's effect on the residents of Séveso in the short term was far less serious than had been anticipated. Few will predict its effect in the long term. With the chemical established as a carcinogen in animals, many fear that it will cause cancer in humans. Epidemiological studies to test this hypothesis are currently in progress. The investigations include mortality studies of workers exposed to TCDD as a result of accidents in the chemical industry 25–30 years ago.

The evidence for an increased risk of TCDD-exposed men developing cancer is at best equivocal. One study suggests that they may have an increased risk from gastrointestinal cancer, while a second suggests that the total incidence of cancer in TCDD-exposed men is no greater than would be expected in the population at large. However, these studies have one severe limitation: they are based on only a small number of cases. More information from other dioxin incidents is required before the long-term risks of exposure to this chemical can be assessed.

Polychlorinated biphenyls. Equally serious perhaps is the long-term effect of exposure to PCBs. These chemicals, because of their heat-resistant, nonflammable, and electrically nonconducting properties, enjoyed widespread use in industry from 1929 until the early 1970s. Although their most common application was as dielectrics in transformers and capacitors, PCBs were also used in lubricating oils, in heat-exchanger fluids, and as plasticizers in paints. Adverse reports in the late 1960s of their toxic properties changed this, and stringent controls on the use and disposal of PCBs were imposed.

The uncontrolled discharge of PCBs directly into rivers and seepage of these chemicals from insubstantial landfills has led to widespread dispersal of these chemicals. Many animal species, including humans, have been surveyed and found to contain trace amounts of PCBs in adipose tissue. Commercial mixtures of these chemicals cause cancer in rodents.

Evidence of cancer in humans from exposure to PCBs is restricted to skin cancer in workers exposed in the course of their employment. The biological effects of exposure to these compounds are thought to be due to induction and inhibition of enzymes involved in a variety of biochemical processes. These probably account for the deleterious effects of PCBs on reproduction, growth, development, and immune defense systems in animals. Adverse health effects observed in individuals occupationally or accidentally exposed to them could mean that the widespread pollution caused by these chemicals does present a risk to humans.

DDT. Organochlorine insecticides, and particularly DDT, serve as examples of the indiscriminate use of chemicals. Legislative controls have severely curtailed this use. In the United States, for example, the Environmental Protection Agency stopped the use of DDT in agriculture in 1972. Even though DDT, a suspected carcinogen, and its metabolites have been detected in human tissue, the risk to humans is still debatable. What is clear, however, is that this chemical has been of immense value in controlling and even eradicating insect-borne diseases. Some tropical countries, such as Vietnam, still use the insecticide as part of a public health program to control malaria-transmitting mosquitoes.

Clearly this poses an acute problem for those dealing with chemicals. Few would deny that the pollution they cause is a serious problem. Similarly, few would fail to recognize that many of these chemicals are of proved worth. Strict controls on their production, use, and disposal are necessary but, in the final analysis, those restrictions must be seen to be fair.

For background information *see* HALOGENATED HYDROCARBON in the McGraw-Hill Encyclopedia of Science and Technology.   [ALASTAIR HAY]

Bibliography: *Co-ordination of Epidemiological Studies on the Long Term Hazards of Chlorinated Dibenzo-Dioxins/Chlorinated Dibenzofurans*, Intern. Tech. Rep. no. 78/001, International Agency for Research on Cancer, March 1978; M. Wassermann et al., *Ann. N.Y. Acad. Sci.*, 320:69–124, 1979; J. A. Zack and R. R. Suskind, *J. Occupational Med.*, 22:11–14, 1980.

Helicopter

During the last decade, civil applications of helicopters have grown significantly in the areas of resource exploration and development, construction, forestry management, agriculture, public police and fire services, and air transportation. This growth has been stimulated largely from technological improvements in the helicopter derived from research and development investments for military utilization. Most of these technological improvements were modest evolutions from proved designs. To accelerate development of the improvements necessary for further exploration of the unique capabilities of helicopters, the National Aeronautics and Space Administration (NASA), jointly with the U.S. Army, has recently developed two research aircraft—the rotor systems research aircraft and the tilt rotor research aircraft. These research rotorcraft will provide the capability to explore improvements in those areas that presently inhibit growth in utilization and acceptance of helicopters, namely reducing vibration, reducing both external and internal noise, increasing cruise speed, and increasing lift capability, all with associated improvements in safety, reliability, and efficiency.

Rotor systems research aircraft. Two versions of the rotor systems research aircraft (RSRA) have been developed and delivered to NASA's Ames Research Center to provide a versatile research facility capable of precise in-flight measurement and control of rotor forces and of aircraft maneuvering flight parameters over a broad range of operating conditions. One RSRA is configured as a conventional helicopter (Fig. 1) and the other as a compound helicopter (Fig. 2) with wings and auxiliary propulsion engines. Both the conventional helicopter and the compound configuration have the capability of accepting a variety of experimental rotors or rotor components such as advanced airfoils, advanced blade tip shapes, blades or hubs with aeroelastically or aerodynamically tailored physical properties, advanced controls mechanisms, or other advanced rotor concepts. Both configurations also have an instrumented balance system mounted between the main rotor and the airframe.

Balance systems. In the past, the net effects of the rotor in combination with the airframe have been determined by conventional flight testing methods, as with conventional aircraft. Tests of rotors on whirl stands or in wind tunnels have provided the forces and moments produced by the rotor isolated from its airframe. However, the interactions between the rotor and the airframe (for example, the vertical drag on the fuselage from the rotor and the ground cushion effect of the fuselage on the rotor) in the complex and highly interactive aerodynamic environment of the rotor have not been previously investigated in sufficient depth to adequately understand them. The balance systems that are installed in the RSRA are similar to balance systems used in wind tunnels. They supply the capability to investigate aerodynamic interactions by providing accurate measurement of the six components of the forces and moments produced by the rotor in the interactive environment of flight. The balance system presently installed in the RSRA compound configuration consists of a set of relatively stiff load cells mounted to an adapter plate between the main transmission and the airframe. The balance system presently installed in the RSRA helicopter configuration provides the additional feature of isolating rotor vibrations from the airframe by means of an active isolation balance system.

Active isolation balance system (AIBS). Instead of employing stiff load cells, the AIBS consists of a set of hydropneumatic units that, in effect, place air-oil accumulators between the transmission and the airframe. These units act as soft springs that isolate the airframe from the relatively high-frequency (10 Hz and above) vibratory forces induced by the rotor. A hydraulic displacement feedback servo provides the "active" feature of maintaining a constant length of the isolator units, which prevents relative displacement between the transmission and the airframe during low-frequency maneuvers. Pressure transducers in the hydraulic units provide load measurements similar to those provided by load cells. Both the load cell and the AIBS measurement systems provide the capability of accurate measurement of rotor loads over a wide range of lift and drag variations.

Variable incidence wings. In a conventional helicopter, the ability to vary lift on the rotor is essentially limited to that which can be achieved by varying payload within the limits of minimum operating weight and maximum gross weight. The wing in the compound configuration, together with wing flaps, provides the capability of sharing the lift between the wing and the rotor, which on one extreme makes it possible to provide zero or even negative lift on the rotor by progressively distributing more load to the wing. Conversely, a very high lift rotor can be tested to its maximum lift limit by unloading or even downloading the wing—a capability provided by wing incidence variability from +15 to −9°. The wing is mounted to the airframe through a set of load cells that provide precise measurement of the lift, drag, and pitching moment contributions of the wing.

Auxiliary propulsion engines. Conventional ro-

Fig. 1. Helicopter version of rotor system research aircraft (RSRA). (*NASA Ames Research Center*)

torcraft are similarly limited in the range over which rotor forces along the propulsive force-drag axis can be varied. The auxiliary propulsion engines, together with drag brakes, provide the capability of varying rotor propulsive forces over the complete range between rotor upper stall limit and autorotation.

Electronic flight control system. Flight testing of conventional rotorcraft often requires that the test pilot make prespecified control inputs and that the instrumentation system record the resulting effects on aircraft motions. Other test conditions may require the pilot to hold constant a specified rotor variable (such as thrust or pitching moment) while performing other maneuvers (such as spiraling turns). Although test pilots have become quite skillful at these control tasks, the human factor often results in a lack of precision or repeatability of these control inputs. The electronic flight control system (EFCS) in the RSRA provides the necessary precise control of test conditions through a digital computer that drives control actuators to superimpose computer-command control motions on pilot-commanded motions. Other features of the EFCS provide stability augmentation, pro-

Fig. 2. Compound helicopter version of rotor system research aircraft (RSRA). (*NASA Ames Research Center*)

grammable artificial force feed, autopilot functions, and a variety of control tasks available by software programming of the flight computer.

RSRA applications. The RSRA is a research aircraft dedicated exclusively to conducting flight investigations of advanced rotor concepts and improvements in design predictive analyses. It is not intended as a proof-of-concept demonstrator or prototype for future mission applications. The flight investigations on the RSRA are expected to produce a more detailed understanding of the complex aerodynamics of rotorcraft, from which improved design prediction methodologies and advanced rotor concepts can be developed.

Tilt rotor research aircraft. The helicopter has the unique advantage over the airplane of being able to take off and land vertically and hover over a fixed ground reference. The airplane has an advantage of higher speeds where the conventional helicopter rotor begins to encounter stall on the retreating blade, generally resulting in an increase in vibration levels and structural loads, a degradation in handling qualities, and low lift-to-drag ratios that are not fuel-efficient. The tilt rotor is a concept for applying the advantages and avoiding the disadvantages in both flight regimes. With the rotors deployed horizontally (lifting vertically), the aircraft performs as a helicopter to achieve vertical takeoff and landing and the ability to hover. When the plane of the rotor is tilted forward to the verti-

cal, the rotors act as propellers to provide forward thrust as an airplane. The tilt rotor research aircraft (TRRA) project, designated XV-15 (Fig. 3), was formulated as a proof-of-concept demonstrator to investigate the technology and operational characteristics of the tilt rotor aircraft. This concept uses large-diameter, highly twisted prop-rotors mounted on turboprop engines at the wing tips. The rotors and engines, which are coupled by cross-shafting in case of engine failure, are rotated forward as an assembly in the airplane mode of operation. The anticipated benefits of this aircraft type include efficient hover comparable to the helicopter and high-speed capability comparable to turboprop aircraft. Very low noise levels in the airplane mode and high energy efficiency in all flight modes are inherent in the concept. Because of the promise of this V/STOL (vertical/short takeoff and landing) aircraft concept for future civil and military applications, the project is funded and managed under a NASA/Army/Navy joint agreement.

Status of development. The TRRA program achieved a major milestone in 1979. Full conversions from helicopter to the airplane mode were accomplished successfully, and an initial safe flight envelope of pylon angle versus airspeed was established. The aircraft has flown at airspeeds from hover to air taxi to 135 m/s (302 mph) and 3048 m (10,000 ft) altitude. Various turn and bank angles and ascents and descents to 15 m/s (3000 ft/min)

Fig. 3.   Tilt rotor research aircraft showing in-flight conversion from airplane to helicopter modes. (*Bell Helicopter Textron*)

were demonstrated. Investigations of the behavior of the aircraft in flight included a survey to investigate correction of empennage buffet problems that were revealed during wind tunnel tests, emergency procedures such as simulated engine failures, and dynamic stability tests designed to identify structural resonances. None of these tests showed significant limitations on the performance or handling qualities of the aircraft. Several NASA evaluation flights have been flown, and the aircraft has accumulated about 50 flight hours. The aircraft is extremely quiet in the airplane mode. An engine failure during a recent flight demonstrated the value of the engine-out capability of the aircraft. Power was transmitted smoothly through the aircraft cross-shafting and a normal run-on landing was accomplished. The next phase of testing will be an evaluation program that will investigate potential civil and military applications, using both civil and military operational pilots in mission-oriented flight profiles.

TRRA applications. The potential applications of the tilt rotor concept that will be examined in the evaluation program will include operations from aircraft carriers and offshore oil rigs as well as from more conventional heliports. Some of the potential missions and applications being studied range from the Army or Marine Corps assault transport and the Army advanced attack helicopter to a 30-passenger commercial transport.

For background information *see* CONVERTIPLANE; FLIGHT CONTROLS; HELICOPTER; VERTICAL TAKEOFF AND LANDING (VTOL) in the McGraw-Hill Encyclopedia of Science and Technology.

[SAMUEL WHITE]

Bibliography: E. J. Bulban, *Aviat. Week Space Technol.*, 112(16):89–92, Apr. 21, 1980; R. J. Huston, J. L. Jenkins, Jr., and J. L. Shipley, *The Rotor Systems Research Aircraft: A New Step in the Technology and Rotor System Verification Cycle*, Pap. 18 presented at the AGARD Flight Mechanics Panel Symposium on Rotorcraft Design, Ames Research Center, Moffett Field, CA, May 16–19, 1977, AGARD Conf. Proc. no. 233, January 1978; R. Letchworth and G. W. Condon, *Rotor Systems Research Aircraft (RSRA)*, Pap. 12 presented at the AGARD Flight Mechanics Panel Symposium, Valloire, France, June 9–12, 1975, AGARD Conf. Proc. no. 187, April 1976; S. White, Jr., and G. W. Condon, *Flight Research Capabilities of the NASA/Army Rotor Systems Research Aircraft*, NASA TM 78522, Pap. 27 presented at the 4th European Rotorcraft and Powered Lift Aircraft Forum, Stresa, Italy, Sept. 13–15, 1978.

Hemoglobin

Hemoglobin has been extensively studied at the molecular level. The structure of many animal and human hemoglobins is known and the organization of the normal hemoglobin genes determined by genetic and biochemical data. In addition, abnormal hemoglobins and quantitative disorders of hemoglobin production (the thalassemias) provide mutations of nature which allow correlations between hemoglobin strucure and function. Very recently a new area of research, best described as molecular genetics, has emerged and has permitted study of the hemoglobin genes directly. The major components in this new research include the ability to: synthesize specific radioactive nucleic acid probes which distinguish one hemoglobin gene from another; cleave DNA at specific nucleotide sequences by using enzymes called restriction endonucleases; and clone or completely isolate and purify individual genes. With these techniques the organization of the human globin genes has been extensively analyzed, and the specific nucleotide defects defined in certain anemias of humans, including some of the thalassemias (Cooley's anemia) and sickle-cell anemia.

Human hemoglobins. Most animal hemoglobins are tetramers containing two α- and two non-α-polypeptide (or globin) chains. The α-globin genes in humans are on chromosome 16, while the non-α genes are linked on chromosome 11. An α-like gene, the ζ gene, also exists and is expressed in fetal life. In most species there are four or more non-α-globin genes. In humans there are at least five such genes on a single 40-kilobase (kb) stretch of DNA. Reading from the 5' (leftward) to 3' (rightward) direction, these include: (1) ϵ gene (a component of two embryonic hemoglobins, hemoglobins Gower I, $\epsilon_2\zeta_2$, and Gower II, $\epsilon_2\alpha_2$; (2) two γ genes, a $^G\gamma$ and $^A\gamma$, differing in an alanine (A) or glycine (G) at position 136 (these γ-globins combine with α-globins to form the predominant hemoglobin in fetal life, fetal hemoglobin or hemoglobin F); (3) a δ gene contained in a minor hemoglobin, hemoglobin A$_2$; and (4) the β gene, a part of the major hemoglobin, hemoglobin A, $\alpha_2\beta_2$.

Probes for globin genes. Hemoglobin-producing cells contain large amounts of globin messenger RNAs (mRNAs), since the major protein synthesized in these cells is hemoglobin. This globin mRNA can be transcribed into a DNA copy of itself called complementary DNA (cDNA) by an enzyme called reverse transcriptase. This cDNA under specific conditions will bind or hybridize only to its specific nucleotide complement. Although there are over a million nucleotide sequences comparable in size to the globin genes in human DNA, cDNA will hybridize only with the globin genes and no others. In addition, the specificity of these cDNA probes is so complete that α, β, and γ cDNAs will not cross-react with the genes for each other, permitting specific identification of these genes in cellular DNA. The cDNA probes are synthesized in the presence of highly radioactive nucleotides, and the radioactivity is used to localize and measure the globin genes in cellular DNA. The radioactivity can be detected either by subsequent exposure of the bound probes to x-ray film, a process known as radioautography, or by measurement of the amount of radioactivity in a radioactive counter. More recently each of the globin cDNAs has been further purified by joining it to a bacterial plasmid; this recombinant DNA is then used to infect bacteria, and bacteria containing the specific cDNA can be isolated as purified material in large amounts.

Human globin gene organization. By their ability to cleave cellular DNA at specific sites, the restriction enzymes generate a discrete set of DNA fragments for analysis. These fragments can then be separated by agarose gel electrophoresis. Subsequently the fragments are transferred to nitrocellulose filter paper by a technique called blotting. This transfer of the cellular DNA to nitrocellulose paper facilitates the subsequent hybridization to radio-labeled cDNA probes. After hybridization, the nitrocellulose filters are washed and exposed to x-ray film, and specific DNA bands are identified corresponding to the size of the DNA fragments containing the globin genes. By comparing the size and number of DNA fragments generated by different restriction endonucleases, the organization of the human globin genes has been established. Two γ genes, 3.5 kb apart, are located 5′ to the δ- and β-globin genes. There are 15 kb of DNA between the γ and δ genes, and 6 kb between the δ and β genes. Restriction-enzyme analysis also has identified sequences within the globin genes which are not represented in globin mRNA. These are so-called intervening sequences (IVS) which had previously been found in rabbit and mouse globin genes. These IVS of DNA are transcribed into RNA, and are subsequently cleaved or spliced out of the globin RNA precursor as it is processed to mature globin mRNA.

Changes in human diseases. The technique of restriction endonuclease analysis of human DNA has been applied to the diagnosis of certain abnormalities in human diseases of hemoglobin. Prenatal diagnosis can be performed by using amniocentesis fluid cells, since these are fetal cells and enough amniocentesis fluid can be obtained between the sixteenth and twentieth weeks of gestation to permit DNA analysis by the use of restriction enzymes as described above. In the α-thalassemias there is decreased or absent α-globin, often associated with deletion of α-globin DNA, and specific deletions of α-globin sequences have been used to diagnose the α-thalassemias prenatally. Deletions in the δ- and β-globin gene complex have also been detected by using this technique. In a rare type of thalassemia called δβ-thalassemia, there is extensive deletion of most of the δ- and β-globin genes. In a related disorder called hereditary persistence of fetal hemoglobin (HPFH), no δ- or β-globin gene material is detectable. It is of interest that in these disorders there is a remarkable compensation for the lack of δ- and β-globin gene material by increased γ-globin and fetal hemoglobin production. The more common types of β-thalassemia are β+-thalassemia, in which a small amount of β-globin is produced, and β0-thalassemia, in which β-globin is absent. In most cases of β+- and β0-thalassemia, no deletion of β-globin gene fragments are detectable. These disorders are associated with severe anemia, whereas δβ-thalassemia and HPFH are not. It has been postulated on the basis of these analyses that deletion of material in the δ and β gene region somehow increases the expression of γ-globin genes.

Prenatal diagnosis of sickle-cell anemia is now possible by using restriction-enzyme analysis of DNA. The enzyme Hpa I generates a larger DNA fragment than normal in about half the cases, when the sickle or β^s gene is present. This large fragment can then be used to identify the sickle-cell gene. The change in DNA associated with this unique restriction site is not at the site of the change in DNA resulting in β^s-globin (in the sixth position of the β-globin gene), but rather far outside (3′ to) the β-globin structural gene. This type of change or polymorphism in a nucleotide sequence, either within or surrounding genes, may be extremely useful in providing other markers for other disorders in the near future.

Cloning of cellular globin genes. The most extraordinary advance over the past several years in DNA analysis has been the ability to clone specific portions of animal and human genes or entire genes in bacteria. Recombinant DNAs have been formed by ligating cellular DNA to specific bacterial phages or plasmids. After infection of bacteria, plasmids or phages are analyzed for their content of specific genes by the use of specific probes. In the case of the human globin genes, α, β, and γ cDNAs have been used to isolate clones of phage which contain the cellular α, β, γ, δ, and ε genes and their flanking sequences. These clones have then been more completely analyzed by restriction endonuclease digestion and the direct nucleotide sequencing of the isolated purified DNA. The results of these studies have confirmed and extended the details of the organization of the human globin genes obtained by restriction. Analysis of clones has shown that the δ- and β-globin genes have two IVSs, one a small IVS between codons 30 and 31 of approximately 130 base pairs and the other a large IVS between codons 104 and 105 of approximately 900–950 base pairs. It has further been shown that the large IVS of the δ- and β-globin genes are largely nonhomologous. The γ-globin genes both contain small and large IVSs in comparable positions to the β-globin genes, although there is no homology between the γ and δ or β IVSs. Additionally, the two γ IVSs are almost precisely identical. The α-globin genes also have two IVSs; however, the second IVS is much smaller than that of the non-α genes. Extensive analysis of the rabbit and mouse globin genes by the same methods have indicated that both of these species have several non-α globin genes, and all of these genes are interrupted by IVSs at precisely the same positions as each other and as the human globin genes. The meaning and function of the IVSs within globin genes, or any other genes for that matter, are as yet speculative.

The availability of purified isolated genes for analysis should provide significant clues to the sequences in DNA responsible for defects in function in human disorders of hemoglobin in the near future. These cloned genes also provide material for the direct study of the function of isolated globin genes in cells by transformation experiments in which DNA is added to cells and its function analyzed within cell cultures or intact animals.

For background information *see* DEOXYRIBONUCLEIC ACID (DNA); GENE; HEMOGLOBIN in the McGraw-Hill Encyclopedia of Science and Technology.

[ARTHUR BANK]

Bibliography: A. Bank, J. G. Mears, and F. Ramirez, *Science*, 207:486–493, 1980; H. F. Bunn, B. G. Forget, and H. M. Ranney, *Hemoglobinopathies*, vol. 12, 1977; Y. W. Kan and A. M. Dozy, *Proc. Nat. Acad. Sci.*, 75:5631–5635, 1978; T. Maniatis et al., *Cell*, 15:687–701, 1978.

Hemoproteins

Hemoproteins are a class of biomolecules in which the protein contains as its active site some derivative of iron porphyrin (Fig. 1).

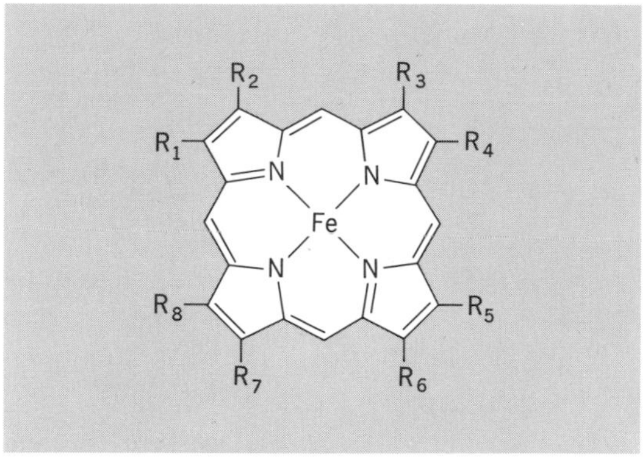

Fig. 1.   Iron porphyrin (Fe^{II} or Fe^{III}).

Porphyrin reactivities. The reactivities of these Fe^{II} or Fe^{III} porphyrin derivatives are differentiated by the addition of fifth and sixth iron ligands, furnished by the protein, to afford hemoproteins which carry out a wide variety of important biochemical processes.

For example, using the same iron porphyrin, protoheme (R$_1$, R$_8$ = CH$_2$CH$_2$COOH; R$_2$, R$_4$, R$_6$, R$_7$ = CH$_3$; R$_3$, R$_5$ = CH$_2$=CH—), the proteins cytochrome c and cytochrome b_5 act as single electron oxidation-reduction catalysts, furnishing electrons for other redox biochemistry [reaction (1)]. They

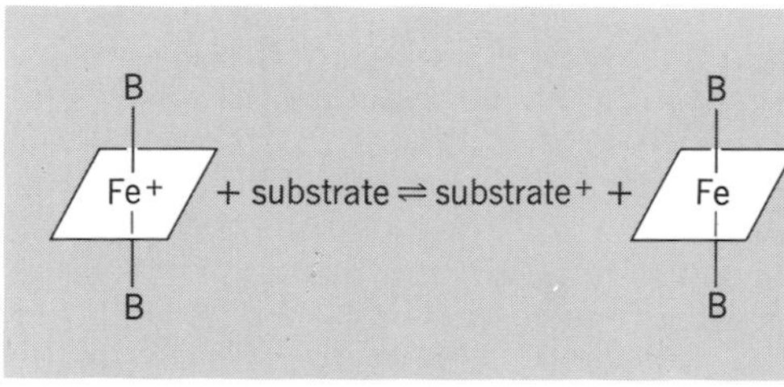

are different from other heme proteins in that both the fifth and sixth positions on iron have ligands permanently attached (B=imidazole, imidazole in cytochrome b_5 or imidazole, thioether in cytochrome c).

A second class of heme proteins, of which hemoglobin and myoglobin are examples, has the ability to reversibly bind dioxygen using this same protoheme in its Fe^{II} state [reaction (2)], where B=imidazole. This remarkable process is accomplished by maintaining the heme in a five-coordinated state, only one imidazole group being supplied by the protein. These proteins also prevent oxidation to Fe^{III} by methods which are only partly clear.

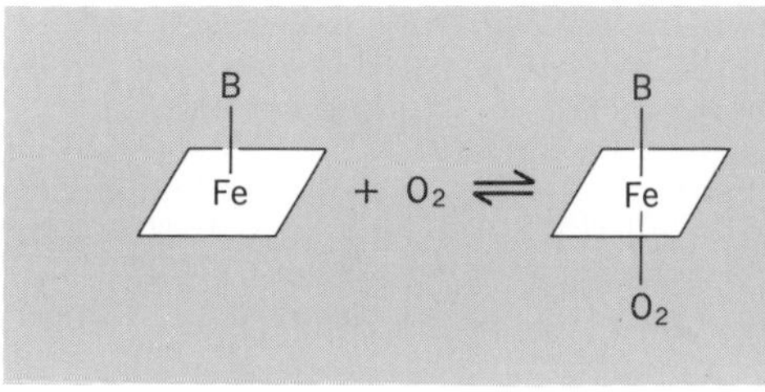

Hemoglobin is composed of four protein subunits, each of which contains one heme group. Interactions between these subgroups result in cooperative binding of dioxygen by which the first of these four heme groups binds dioxygen with low affinity and the last with high affinity. In this way hemoglobin can quantitatively absorb dioxygen in the lungs at normal pressures (1/5 atm or 20 kPa) and release most of it to myoglobin, a heme protein with a dioxygen affinity similar to that in the last step and much greater than that in the first step in hemoglobin-dioxygen binding.

A third class of heme proteins, which includes peroxidases and oxidases such as horseradish peroxidase, cytochrome c peroxidase, cytochrome oxidase, catalase, and the important hydroxylase, cytochrome P-450, also employs a five-coordinated iron. However, these proteins exist in the Fe^{III} form and use the empty sixth position to catalytically carry out reactions which break the O-O bond in O$_2$ or H$_2$O$_2$. This produces species which either oxidize a substrate or reduce O$_2$ to H$_2$O. For example, horseradish peroxidase (HRP) catalyzes the otherwise very slow reaction of hydrogen peroxide with phenols [reaction (3)]. Cytochrome P-

450 catalyzes the extraordinary reaction in which one of the atoms in dioxygen appears as an alcohol by hydrocarbon hydroxylation [reaction (4)]. It is

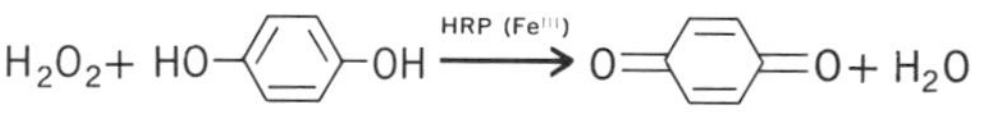

this last reaction which allows drugs to be detoxified in the liver and steroids to be synthesized.

Model systems. Recently chemists have become fascinated with the wide spectrum of reactivities imparted to essentially the same iron porphyrin by these relatively simple manipulations of the fifth and sixth ligands on the iron and by other minor changes in the environment of this heme group. Because one of the principal differences in these three classes of hemoproteins is the maintenance of five- or six-coordination at iron, one of the first objectives in modeling these protein reactions using simple protein-free heme compounds was to achieve pure five- or six-coordination as desired.

Several methods have been attempted with varying degrees of success. Thus a protected heme in which one face is covered with some inert structure would prevent the sixth position from being occupied by a base [reaction (5)]. This method frequently affects the O$_2$ or CO binding and thwarts the objective. Similar problems arise with the use of a sterically hindered base, in which case a bulky imidazole binds to the fifth but not the sixth position [reaction (6)]. This bulky base interferes with the attainment of planarity of the heme and thus reduces the O$_2$ or CO affinity.

The most successful method of maintaining the five-coordination in model compounds involves the

$$\text{protected heme} \quad \underset{\text{excess } B}{\rightleftharpoons} \quad \cancel{\to} \qquad (5)$$

$$-Fe- \; + \; \text{(imidazole)} \; \rightleftharpoons \; -Fe- \; \cancel{\to} \; -Fe- \qquad (6)$$

covalent attachment of the base to a side chain of the heme itself. By choosing the proper chain length to ensure strain-free ligand binding, a five-coordinated heme complex is obtained. Because this method holds the fifth (or fifth and sixth) base in a position for further iron chelation, such compounds have been called chelated hemes. Two examples, imidazole chelated protoheme and mercaptide chelated protoheme, are shown in Fig. 2.

These two model compounds display visible, ultraviolet, and nuclear magnetic resonance (NMR) spectra of their various forms (Fe^{II}-CO complex, and so on) which are remarkably close to those of the two proteins which they were designed to model, hemoglobin and cytochrome P-450, respectively. Indeed, the ^{13}CO complex of mercaptide chelated protoheme, when compared to that of cytochrome P-450 itself, provides strong evidence for this unusual Fe-$\bar{S}$-R structure in a hemoprotein.

The purpose of preparing such chelated hemes is to provide simple compounds with which hemoprotein functions can be duplicated and structure-function relationships studied. A comparison of the kinetics, equilibria, and the associated thermodynamic parameters for dioxygen and carbon monoxide binding to imidazole chelated protoheme (Ch. Hm) and to the high-affinity (last step, R-state) hemoglobin (Hb) is shown in reactions (7) and (8).

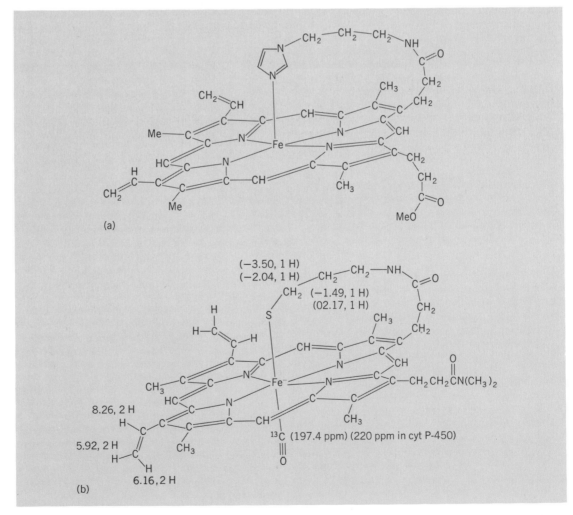

Fig. 2.   Chelated hemes. (a) Imidazole chelated protoheme. (b) Mercaptide chelated protoheme. NMR shifts of protons or ^{13}C relative to tetramethylsilane are shown adjacent to the atom.

$$\text{Heme} + O_2 \xrightleftharpoons[\substack{\text{Hb, } 10-40 \text{ s}^{-1} \\ \text{Ch. Hm, } 45 \text{ s}^{-1}}]{\substack{\text{Hb, } 6\times10^7 \, M^{-1}\text{s}^{-1} \\ \text{Ch. Hm, } 3\times10^7 \, M^{-1}\text{s}^{-1}}} \text{heme}-O_2 \quad (7)$$

$$\text{Heme} + CO \xrightleftharpoons[\substack{\text{Hb, } 0.009 \text{ s}^{-1} \\ \text{Ch. Hm, } 0.007 \text{ s}^{-1}}]{\substack{\text{Hb, } 6\times10^6 \, M^{-1}\text{s}^{-1} \\ \text{Ch. Hm, } 4\times10^6 \, M^{-1}\text{s}^{-1}}} \text{heme}-CO \quad (8)$$

$\triangle G^{CO} = -17.5 \text{ kcal}/M$ or $-73.2 \text{ kJ}/M$ (Hb),
   $-17.5 \text{ kcal}/M$
      or $-73.2 \text{ kJ}/M$ (chelated protoheme)
$\triangle G^{O_2} = -14 \text{ kcal}/M$ or $-58.6 \text{ kJ}/M$ (Hb),
   $-13.5 \text{ kcal}/M$
      or $-56.4 \text{ kJ}/M$ (chelated protoheme)

It is clear from this close correspondence that the high-affinity hemoglobin reactivity is duplicated by a compound which comprises essentially the hemoglobin site removed from the protein and placed in an aqueous suspension. This suggests that, in this protein, the function of the protein is simply to prevent Fe^{II} oxidation and to suspend the heme in water. However, other hemoproteins, having different affinities, are not duplicated by this model. The reasons for these differences are probed by changing this model system.

Structure-function relationships. The kinetics and equilibria of O_2 and CO binding to imidazole chelated protoheme in various solvents reveal little effect of solvent polarity on either the rates or equilibria of CO binding. In contrast, changing to more polar solvents decreases the rate of dissociation of dioxygen resulting in increased O_2 affinity as the solvent polarity increases. This suggests more charge separation in the $\overset{\delta^+}{Fe}\text{-}\overset{\delta^-}{O}\text{-}O$ bond than in the Fe-CO bond, as might be anticipated from the greater electronegativity of dioxygen.

Electronic effects in chelated hemes upon O_2 and CO binding have been studied by replacing the $CH_2{=}CH-$ groups in imidazole chelated protoheme with ethyl or $CH_3\overset{\overset{\textstyle O}{\|}}{C}-$ groups, or alternatively by replacing the covalently bound imidazole with a covalently bound pyridine. Such changes, $CH_3\overset{\overset{\textstyle O}{\|}}{C}-$ to $CH_2{=}CH$ to C_2H_5- or pyridine to imidazole, would tend to increase electron density at the iron [reaction (9), where $L = O_2$ or CO and $K^L = CO$ or O_2 binding constant].

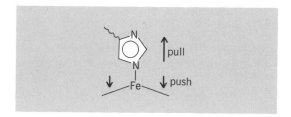

Increasing the electron density in these ways has relatively little effect on the kinetics or equilibria of CO binding. (It should be noted, however, that if B is negatively charged, for example,

$$^-N{\diagup}{\diagdown}N{-}Fe \text{ or } R\overset{-}{S}{-}Fe$$

the CO affinity is greatly reduced, up to 100 times compared to the uncharged form.) The dioxygen dissociation rate is decreased by the increase in electron density introduced with the $CH_3\overset{\overset{\textstyle O}{\|}}{C}-$ to C_2H_5- or the pyridine to imidazole changes, with a consequent increase in O_2 affinity. This is consistent with the polar character of the $\overset{\delta^+}{Fe}\text{-}\overset{\delta^-}{O}\text{-}O$ bond.

One of the most interesting suggestions in hemoprotein reactivity is the Hoard-Perutz theory that the low-affinity form of hemoglobin achieves its low affinity by preventing the heme from becoming planar as a result of protein pull on the proximal imidazole (Fig. 3). This has been shown to be a

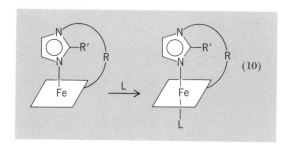

Fig. 3. Protein pull on the proximal imidazole.

possible mechanism in chelated hemes by observing the effect of shortening the length of the side chain holding the chelation arm or by introducing bulky groups into this arm [reaction (10)].

If R' or R contain bulky groups or R is a shortened chain, the O_2 and CO affinities are reduced. In the case of O_2, this reduced affinity is a result of a faster O_2 dissociation rate, a phenomenon also noticed in the low-affinity compared to high-affinity forms of hemoglobin.

But other kinds of steric effects have also been suggested in hemoprotein reactions. Thus, myoglobin and peroxidases have lower CO affinities than does high-affinity hemoglobin. Although this reduction might be explained by the proximal pull effect just described, the isonitrile binding to hemoproteins, studied 30 years ago by Linus Pauling, cannot. All hemoproteins studied display the property of binding isonitriles with much reduced affinities as compared to chelated protoheme. More importantly, this reduced affinity shows an inverse relationship with the size of the isonitrile. For example, the equilibrium constants for binding n-butyl-NC and t-butyl-NC to imidazole chelated protoheme are $4\times10^8 \, M^{-1}$ and $2\times10^8 \, M^{-1}$, respectively, both very similar to the constant for CO binding (4×10^8 M^{-1}). However, with myoglobin the corresponding constants for Me-NC, n-BuNC and t-BuNC are 10^6 M^{-1}, 6×10^4 M^{-1}, and 800 M^{-1}, respectively. This large difference between n-BuNC and t-BuNC in myoglobin, hemoglobin, and many other hemoproteins, which does not appear in chelated hemes, is surely due to a steric effect which prevents the ligand (isonitrile, and so on) from approaching the iron.

This steric effect has been examined in model compounds by determining the affinities of the hindered anthracene cyclophane heme (Fig. 4). The striking result is that the binding constants for CO, n-BuNC, and t-BuNC are 6×10^5 M^{-1}, 6×10^4 M^{-1}, and 150 M^{-1}, respectively, similar to those found in myoglobin and much reduced as compared to chelated protoheme. But a striking new kinetic effect appeared in this study. The reduced affinities in both the models and the hemoproteins, engendered by this distal side steric effect, reside entirely in the association rates. This contrasts the high- to low-affinity change in hemoglobin (and the proximal pull in models) in which the effect appears in both association and dissociation rates. It has therefore become possible to use kinetic studies of model systems to establish the mechanisms by which hemoproteins alter the reactivities of their active sites.

Chelated hemins as peroxidase models. Chelated heme compounds display very effective peroxidase activity. For example, imidazole chelated mesoheme [Et (ethyl) instead of CH_2=CH of chelated protoheme] catalyzes the oxidation of tri-t-butyl phenol to the stable phenoxy radical in a reaction which closely resembles catalysis by horseradish peroxidase [reaction (11)].

Studies of the third (oxidase) class of hemoprotein model compounds is only beginning, and much future research will probably be directed toward studies of heme-catalyzed oxidations. In addition to these studies, chelated heme compounds containing two hemes have been prepared. These show promise as cooperative heme systems,

$$\underset{(10^{-4}\,M)}{\text{Cl–C}_6\text{H}_4\text{–C(=O')–OOH}} + 2\,\underset{(10^{-3}\,M)}{\text{(t-Bu)}_2\text{C}_6\text{H}_2\text{–OH}} \xrightarrow[\text{(10}^4\,M)]{\text{chelated heme}}$$

$$2\,\underset{(2 \times 10^{-4}\,M)}{\text{(t-Bu)}_2\text{C}_6\text{H}_2\text{–O·}} + \underset{(10^{-4}\,M)}{\text{Cl–C}_6\text{H}_4\text{–C(=O)–OH}} \qquad (11)$$

and future work will surely involve studies of cooperative binding to model systems.

For background information *see* CHELATION; COORDINATION CHEMISTRY; COORDINATION NUMBER in the McGraw-Hill Encyclopedia of Science and Technology.   [T. G. TRAYLOR]

Bibliography: E. Antonini and M. Brunori, *Hemoglobin and Myoglobin in Their Reactions with Ligands*, 1971; J. Geibel et al., *J. Amer. Chem. Soc.*, 100:3575–3585, 1978; T. Mashiko et al., *J. Amer. Chem. Soc.*, 101:3653–3655, 1979; T. G. Traylor, Hemoprotein oxygen transport: Models and mechanism, in E. E. van Tamelen (ed.), *Bioorganic Chemistry*, vol. 4, 1978; T. G. Traylor et al., *The Chemical Basis of Variations in Hemoglobin Reactivity*, Advances in Chemistry Series, 1980; T. G. Traylor et al., Cyclophane hemes, 3. Magnitudes of distal side steric effects in hemes and hemoproteins, *J. Amer. Chem. Soc.*, 1980; T. G. Traylor et al., *J. Amer. Chem. Soc.*, 101:6716–6731, 1979; T. G. Traylor and A. P. Berzinis, The binding of O_2 and CO to hemes and hemoproteins, *Proc. Nat. Acad. Sci. U.S.A.*, 1980.

High-pressure phenomena

Materials in the Earth's interior are subjected to a large range of pressures, reaching a high of approximately 3,600,000 atm (360 GPa) at the Earth's center. Under such extreme conditions, the electronic orbitals on some atoms are greatly modified, sometimes leading to radical changes in the macroscopic properties of the materials they make up. In turn, these changes may have important consequences on the distribution of heat sources and trace elements within the Earth.

Electronic transitions. In 1948 P. W. Bridgman found that when cesium is compressed to approximately 42,000 atm (4.2 GPa) its volume collapses discontinuously without any apparent crystallographic changes. E. Fermi suggested that the cesium atoms collapsed by promoting an electron from a $6s$ state to a $5d$ state. (Electrons in $5d$ states are, on the average, closer to the nucleus than electrons in $6s$ states.) Theoretical calculations have demonstrated that Fermi's suggestion is essentially correct, although the details of the electronic transition are somewhat more complicated.

Just after Bridgman's pioneering work, an electronic transition was proposed to explain a discontinuous drop in the electrical resistivity of cerium; here again it was found that the $5d$ states played an important role. Since then, numerous other elec-

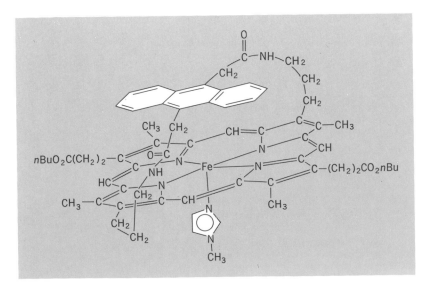

Fig. 4. Anthracene cyclophaneheme.

tronic transitions have been found: some in metals, such as potassium, rubidium, calcium, strontium, and barium, and some in compounds of atoms with available *d* states, for example, europium oxide and samarium sulfide. These compounds are insulators at low pressures and become metallic as a consequence of the electronic transitions.

Electrons in atoms can have only certain discrete energies. Electrons with such energies are said to occupy stationary states. When atoms are brought together to form a crystal, the discrete energy states are converted to continuous bands of energies, separated from bands arising from other states by energy gaps. It is not possible for an electron to have an energy that falls in one of the gaps. Also, because of Pauli's exclusion principle, no more than two electrons can occupy the same state.

Each energy band contains enough states to accommodate 2*N* electrons, where *N* is the number of atoms in the crystal. Consider the building of the crystal by first locating the nuclei at their fixed sites, and then adding the electrons one by one. The first two electrons are put into the lowest energy state of the lowest energy band; the second pair goes into the second-lowest state, and so on until the first band is filled. Then the process is continued by placing electrons in the second band, and so on. The process stops when all the electrons are exhausted; the crystal is now in its lowest energy configuration. An insulator is obtained if all the bands are either completely filled or completely empty, and a metal, or semimetal, if some of the bands are only partly filled. The latter case may arise if each atom has an odd number of electrons, or if some of the energy bands overlap.

When a crystal is compressed, the electrons are brought closer to each other and to the ions. The electrostatic interactions get stronger, the kinetic energy of the electrons rises, and consequently the various energy bands change positions relative to each other. Some bands move apart, some come closer, and some may be forced to overlap. It is this relative motion of bands that is responsible for the electronic transitions.

The *d* states play a prominent role in electronic transitions. In crystals, *d* states are replaced by *d* bands. In many materials, compression causes the energy of the *d* bands to be lowered relative to that of other bands. For example, in the heavy-alkali metals potassium, rubidium, and cesium, the highest occupied band is only half filled at normal conditions. The lowest unfilled band is a *d* band. Upon compression, the *d* band increases its overlap with the occupied band. At some pressure, electrons are transferred into the *d* band. The observed decrease in volume is related to the fact that electrons in *d* states, on the average, are closer to the nucleus, thus effectively decreasing the size of the atoms.

The same mechanism is responsible for the electronic transitions in insulators like europium oxide; the only difference is that the highest occupied band starts out being completely filled. When a *d* conduction band is lowered and overlaps with the filled band, two partially filled bands result and the insulator turns metallic.

Distribution of heat sources. Heat is produced in the Earth's interior by the decay of radioactive isotopes of uranium, thorium, and potassium. Since most of the Earth's interior is inaccessible, there is no direct evidence about the absolute, or even relative, abundance of these elements. This lack can be overcome by postulating a model Earth that has the same chemical composition as chondritic meteorites. This class of meteorites is thought to be representative of the chemical composition of the solar nebula from which the Earth accreted. If a measurement is made of the total heat that leaves the Earth's surface each second, it is found to be very close to the heat produced by the radioactive elements in the model chondritic Earth. This near equality has been called the chondritic coincidence.

Measurements of the abundance of radioactive isotopes in a broad class of rocks have established that, relative to refractory elements like uranium, the Earth's crust and uppermost mantle are depleted in potassium. If, as is commonly assumed, most of the potassium is in the outermost layer of the Earth, it follows that the Earth is depleted in potassium relative to the chondrites, and the chondritic coincidence is just that. However, the assumption that potassium has a strong affinity for crust-forming minerals may not be valid under all conditions.

Potassium is thought to have a strong affinity for the crust because it has a large ionic radius. It is difficult to accommodate a large ion in a close-packed structure of smaller ions, such as those found in the mantle. Thus, when a magma forms by partial melting of mantle rocks, ions with large radii tend to escape from the solid to the fluid magma, which eventually takes them to the surface. Consequently, if the whole mantle (the layer of the Earth between the crust and the core) participates in convective motion, much of its potassium may have been partitioned into the crust. However, it is possible that a good portion of the Earth's potassium was withdrawn into the core soon after the Earth's formation.

Most Earth scientists believe that the Earth started out as a more or less homogeneous aggregate of all minerals. Heating by radioactive decay led to the melting of iron with some impurities; the mixture, being denser than the surrounding material, sank to the Earth's center.

There is some evidence which suggests that the major impurity in the core is sulfur. It is well known that the transition metals (metals with partially filled *d* states) have a strong affinity to sulfur. Since the electronic transition discussed earlier converts potassium into a monovalent transition metal, it may be argued that potassium was also included in the iron-sulfur melt at depths below approximately 800 km. The net result would be a transfer of potassium from the lower mantle into the core. A significant consequence of this process is an ample supply of energy needed to generate the magnetic field of the Earth. As little as 0.1% of potassium in the core (less than half the total content of a chondritic Earth) would be more than sufficient.

As with most problems in geophysics, the ideas described above are rather speculative. Since there is no direct access to the Earth's interior, geophysicists must frequently rely on theory and laboratory measurements of properties of plausible candidate Earth-forming materials. Low-pressure measurements of the distribution of potassium between coexisting iron melts and solid silicates have produced conflicting data. A rigorous assessment of the theory must await the completion of similar experiments under high-pressure conditions. Because the electronic transition in potassium occurs somewhere between 200,000 and 300,000 atm (20 and 30 GPa), it may be necessary to start with measurements on analog materials like cesium, which undergoes a transformation at a more easily accessible pressure.

Trace-element distribution. Potassium is not the only element whose chemical affinities may be altered by pressure. The sizes and compressibilities of ions vary greatly across the periodic chart, and many atoms and crystals have energy levels whose order may be modified by high pressures. For example, when europium oxide and samarium sulfide become metallic, the valence of europium and samarium changes from 2 to 3. Similar behavior is predicted for the oxides of calcium, strontium, and barium, and has been observed in samarium selenide and telluride. Thus valency changes, and therefore probable changes in chemical affinity, may be rather common among elements with d states. Since many of the trace elements in the Earth fall in that category, measurements of their relative abundances in magmas may not provide a reliable estimate of the abundances in the mantle.

Perhaps the only firm conclusion that can be drawn is that the rather limited understanding of the Earth's deep interior will continue to evolve for many years to come. In the meantime, techniques of high-pressure petrology must be developed to supply the necessary data.

For background information *see* EARTH, INTERIOR OF; HIGH-PRESSURE PHENOMENA in the McGraw-Hill Encyclopedia of Science and Technology.   [MARK S. T. BUKOWINSKI]

Bibliography: M. S. T. Bukowinski, *Geophys. Res. Lett.*, 3(8):491–494, 1976; M. S. T. Bukowinski and R. Jeanloz, *Geophys. Res. Lett.*, 7(4):227–230, 1980; H. G. Drickamer, *Solid State Phys.*, 17:1–133, 1965; A. Jayaraman, *Phys. Rev. Lett.*, 29:1674–1676, 1972.

Hubble constant

Some recent measurements of distances to remote galaxies have cast doubt on the ability of the simplest of the general relativistic "big bang" cosmologies to describe the history of the universe.

Early determinations of the constant. A half century ago Edwin Hubble discovered that the universe expands. Remote galaxies have line-of-sight velocities, revealed by their Doppler-shifted spectra, which are in proportion to their distances. Astronomers have worked since then to determine distances to galaxies and thereby establish the value of the constant of proportionality, the Hubble constant H (expressed in units of kilometers per second per megaparsec; 1 megaparsec = 3.26×10^6 light-years = 3.09×10^{19} km). In standard big bang cosmologies based on Einstein's general theory of relativity, the age of the universe is deduced by extrapolating the expansion back in time, to be $10^{12}/H$ years, or less if deceleration of the expansion has not been negligible.

Hubble estimated the distances to the nearest galaxies by comparing the observed brightnesses of variable stars contained in them to the absolute luminosities of similar stars in the Milky Way Galaxy. The nearest galaxies are not suitable for determining the value of H because their random velocities are not negligible compared to their velocities caused by the general expansion. Hubble extended his distance scale by determining the absolute luminosities of the brightest stars in the galaxies of known distance and comparing these with the brightest observed stars in more remote galaxies whose variable stars were too faint to be seen from Earth. The distances and velocities of the remote galaxies implied a value of H of approximately 500 km s^{-1} mpc^{-1}, and an age of the universe of 2×10^9 years. This was in conflict with the known age of the Earth.

During the following 4 decades several revisions to Hubble's distance scale led astronomers to lower their estimates of the value of H and remove the age discrepancy. By 1975 A. R. Sandage and G. A. Tammann had concluded from extensive studies of cepheid varable stars, gaseous nebulae, and the bulk properties of the galaxies themselves that $H = 50$ km s^{-1} mpc^{-1} and that the universe is 2×10^{10} years old.

Recent work. Several teams of astronomers, working independently, have recently derived new distances to galaxies and larger values of H. Their methods rely on comparisons of stars in the Galaxy and nearby galaxies to establish distances to the latter, and comparisons of bulk properties of the nearby and remote galaxies to measure larger distances. G. de Vaucouleurs has carried out a calibration of distances based on statistics of variable stars, novae, and diameters and luminosities of galaxies. In 1979 he and G. Bollinger concluded that $H = 100$ km s^{-1} mpc^{-1}. Another approach has been based on the discovery by J. R. Fisher and R. B. Tully in 1976 that the luminosities of spiral galaxies are closely correlated with their rotation rates, as determined by observations with radio telescopes of Doppler-shifted photons of 21-cm wavelength which are emitted by the rotating interstellar gas. Distances to the nearest galaxies are determined in the usual way, and distances to more remote galaxies are then found by comparing their rotation rates and observed brightnesses to those of the nearby galaxies. Fisher and Tully encountered the difficulty that while rotation rates can be measured unambiguously only for galaxies whose disks are oriented edge-on with respect to the line of sight from the Earth, the brightnesses of edge-on galaxies are affected by absorption of light by interstellar dust particles contained in the disks. M. Aaronson, J. Huchra, J. Mould, and their collaborators have avoided this difficulty by measuring instead the infrared brightnesses of galaxies, for which the effects of dust are almost negligible. In 1980 they announced that $H = 95$ km s^{-1} mpc^{-1}, in close agreement with Fisher and Tully's original conclusions.

These new results imply that the age of the universe is only 10^{10} years or less. A new age discrepancy arises because the oldest stars in the Galaxy are estimated on the basis of stellar evolution theory to be at least 1.5×10^{10} years old. If the universal expansion age is less than the age of any star, commonly accepted theories of the evolution of the universe (or the stars) will need to be revised.

Independent method. When a massive star completes its evolution, it explodes as a supernova and temporarily attains the luminosity of 10^{10} suns. Early in the 1970s astronomers learned how to interpret the optical spectra of supernovae and realized that a new way to determine extragalactic distances had become possible. The radius of a supernova at a given time is the product of the time elapsed since the explosion and the velocity of ejected material. The velocity, typically 10,000 km s^{-1}, is determined from Doppler-shifted spectral lines. The absolute luminosity of the supernova depends on its radius and temperature, which is determined from the observed colors. Comparison of the calculated luminosity and the observed apparent brightness gives the distance. This method is preferable in principle to the ones discussed above because it is independent of all intermediate distance calibrations, and it does not rely on the assumption that remote stars and galaxies can be correctly matched to identical ones nearby. D. R. Branch, R. P. Kirshner, and their collaborators have found that this method implies distance consistent with $H = 50$ km s^{-1} mpc^{-1}.

Prospects. All of the distance determinations discussed above have expected uncertainties due to random errors of only 20% or less, yet the results differ by a factor of 2. At least some of these methods must be affected by systematic errors which are as yet unidentified. Nevertheless, now that several independent methods are available, astronomers hope that further careful measurements of stars, supernovae, and galaxies from ground-based and orbiting observatories will soon produce a consensus regarding the distances to the galaxies and a reliable determination of the value of Hubble's constant. When this has been achieved, the question of whether the universe can have been evolving in the simplest possible way will be resolved.

For background information *see* COSMOLOGY; GALAXY, EXTERNAL; HUBBLE CONSTANT in the McGraw-Hill Encyclopedia of Science and Technology.   [DAVID BRANCH]

Bibliography: J. R. Fisher and R. B. Tully, *Comment. Astrophys.*, 7:85–93, 1977; B. K. Hartline, *Science*, 207:167–169, 1980; A. Sandage, *Quart. J. Roy. Astron. Soc.*, 13:282–296, 1972; S. van den Bergh, *Nature*, 225:503–505, 1970.

Hydrogen

According to theory, hydrogen gas should solidify and eventually possess the structure of an alkali metal at ultrahigh pressure. Metallic hydrogen potentially would be a very useful material, because its properties are predicted to be those of an elevated-temperature superconductor. In addition, metallic hydrogen could be used as an efficient fuel for a nuclear fusion generator, as a conventional rocket fuel with extremely high energy content, or as a powerful explosive. Until 1975, the pressures required for making metallic hydrogen were far beyond the reach of static experimental techniques, and knowledge of metallic hydrogen was limited to theoretical calculations.

Recent advances in high-pressure technology have been made which extend the range of routine experiments to megabar (megabar = 100 GPa) pressures. Properties of molecular hydrogen have been measured quantitatively to much higher pressures than ever before, in order to furnish crucial data for theoretical studies of the metallic transition. Even though synthesis of true alkali-metal-type metallic hydrogen has yet to be demonstrated, major progress in understanding the behavior of solid hydrogen at high pressures has been made, and thus in the near future metallic hydrogen may be a reality.

Metallization. Hydrogen has a unique position in the periodic table, since it can be assigned either to the alkali-metal group or to the halide group. Under atmospheric pressure, hydrogen is a molecular gas at room temperature, becomes a fluid at 20.4 K, and solidifies to an insulating molecular solid at 14 K. The diatomic hydrogen molecule, covalently bonded by two electrons, has physical properties consistent with the halide group.

At sufficiently high pressures, the density of solid hydrogen increases manyfold, and the intermolecular distance becomes comparable to the intramolecular atomic distance. The hydrogen molecules should then dissociate into hydrogen atoms. The monatomic hydrogen, with one valence electron per atom forming a half-filled band, should be a metal with physical properties consistent with metals of the alkali group.

Recent theoretical predictions indicate that there may be another metallization transition occurring below the dissociation pressure. The diatomic molecular hydrogen may itself become a metal while still retaining its molecular structure if compression results in a closing of the gap that exists between the valence and conduction bands. Molecular metallic hydrogen would be similar to the metallized molecular halides such as I_2, which is different from an alkali metal.

Significance of metallic hydrogen. The hydrogen atom is the simplest possible atom. The properties of monatomic metallic hydrogen can be calculated in great detail from first principles of physics. Synthesis of metallic hydrogen experimentally and cross-checking of its properties with theoretical calculations would be valuable for determining the accuracy of a fundamental block of physics.

Hydrogen is a major component of the Jovian planets. Jupiter and Saturn, having bulk hydrogen compositions and high internal pressures, are mainly made of metallic hydrogen. Uranus and Neptune, on the other hand, have hydrogen concentrated mainly in the outer envelopes, where the pressures may not be high enough to form metallic hydrogen. Knowledge of the properties of molecular and metallic hydrogen under high pressure is crucial for the modeling of the interior of Jovian planets.

Monatomic metallic hydrogen potentially also has practical usefulness. The possibility exists that it may be metastable for a significant length of time after the release of pressure. As a result of its high Debye temperature ($\sim 2000-3000$ K), it may be an elevated-temperature (possibly as high as room temperature) superconductor. Its high density characterizes metallic hydrogen as a nuclear fusion fuel with high efficiency. If it can be preserved metastably and can be introduced as small grains, metallic hydrogen fusion fuel may eliminate the debris problem in laser fusion technology. To be used as a conventional fuel, on the other hand, metallic hydrogen would have a specific impulse (1400 s) four times as high as rocket fuel, and an energy content (50 kcal/g or 209 kJ/g) 300 times greater than the best available aircraft fuel. To be used as a conventional explosive, its energy content of 50 kcal/g (209 kJ/g) is 35 times more powerful than TNT.

In the molecular form, metallic hydrogen is not expected to be metastable or to be superconductive, and thus does not have the potential applications that are predicted for the monatomic form. The usefulness of monatomic metallic hydrogen depends on whether it can be synthesized in the laboratory, whether it is metastable after releasing pressure, and whether it possesses extraordinary properties such as high-temperature superconductivity. Theoretical and experimental research is focused on solving these problems.

Metallic transition pressure. The transition pressure between the molecular and monatomic metallic hydrogen is the pressure at which the Gibbs free energies of the two phases are equal. Since the free energy as a function of pressure can be calculated directly from the equation of state, the problem of calculating the transition pressure becomes the problem of finding equations of state of the two phases.

The equation of state of molecular hydrogen is calculated on the basis of an assumed pair potential describing the interaction between two H_2 molecules. Once the function is specified, quantum crystal theory and liquid hard-sphere perturbation theory can be used to calculate isothermal and Hugoniot (shock-wave) equations of state for comparison with experiment. At present, the available experimental data are an accurate solid isotherm up to 25 kbar (2.5 GPa) and a few less accurate Hugoniot points up to 1 Mbar (100 GPa). The reliability of the calculated theoretical equation of state is limited by the reliability and extent of available experimental data.

The equation of state of the monatomic metallic hydrogen can be calculated by using any one of the four commonly used band structure methods for metals. They are the augmented phase wave (APW), linear combination of atomic orbitals (LCAO), free electron perturbation theory (PERT), and the Wigner-Seitz (WS) methods. The results of the four quite different methods are relatively close, indicating the computation for metallic hydrogen is insensitive to the particular method. Without experimental data, however, the absolute accuracy is difficult to evaluate.

The most important source of error for the theoretical calculation of metallic hydrogen probably is the correlation energy. Unfortunately, the theory of electron correlation is a poorly understood quantum-mechanical effect, and the numerical results may have an error of a factor of 2 or 3, which could cause an error in transition pressure of a few megabars. In summary, calculation of the transition pressure from the molecular to monatomic metallic hydrogen is generally in the range of 2 to 7 Mbar (200 to 700 GPa) with extreme values from 0.7 to 20 Mbar (70 to 2000 GPa).

The LCAO calculations and the mixed basis calculations for the molecular hydrogen both indicate that, at pressures below the dissociation of the H_2 molecules (0.3–2 Mbar or 30–200 GPa), the electrons in the fully occupied first Brillouin zone will begin to occupy states in the second zone, and hydrogen will become a molecular metallic crystal. Such band overlap metallization results in a continuous change from an insulator to a metal. The transition at the point where the band gap becomes zero is second-order and is spontaneously reversible.

Monatomic metallic hydrogen. Because of the small mass and the large zero-point motion of hydrogen atoms, the possibility has been raised that monatomic hydrogen might be a quantum liquid even at zero temperature. The recent Monte Carlo variational calculations of K. K. Mon, G. V. Chester, and N. W. Ashcroft indicate that at high densities (r_s, the radius of the electron sphere, lower than 1.488 bohr), the crystalline phase is clearly more stable than liquid metallic hydrogen. At very low densities near zero pressure ($r_s = 1.6$ bohr), however, the energy differences between solid and liquid are smaller than the error, and thus it is not feasible to determine the relative stability of the solid and liquid.

Application of the standard weak-coupling Bardeen-Cooper-Schrieffer (BCS) theory to monatomic metallic hydrogen suggests that it will be a high-temperature superconductor. Depending on various approaches for calculating Debye temperature, electron-phonon coupling, and electron-electron coupling, the calculated superconducting temperature for solid monatomic hydrogen is generally between 100 K and room temperature. Even if metallic hydrogen is a liquid, J. Jaffe and Ashcroft indicate that since the electron-phonon coupling is strong, hydrogen may be a novel ordered state of matter, namely, a liquid superconductor.

Liquid metallic hydrogen will not be metastable. The crystalline metallic hydrogen is estimated to be metastable for a fraction of a second. The results of recent investigations indicate that maintaining metallic hydrogen for long periods of time may involve keeping it tightly enclosed in a pressurized vessel at a constant volume, in order to prevent evaporation and recombination.

Experimental research. There are two approaches in experimental research of the metallic hydrogen: the qualitative study of generating extremely high pressures in order to make metallic hydrogen, and the quantitative study of measuring various properties of hydrogen at high pressures in order to furnish the basic data for theoretical calculations and predictions.

L. F. Vereshchagin in the Soviet Union, using a carbonado diamond indentor, and N. Kawai in Japan, using a split-sphere high-pressure apparatus, both reported a 6- to 8-order decrease in electrical resistance of hydrogen at pressures of the order of 1 Mbar (100 GPa). Unfortunately their pressures were poorly calibrated, and their resistance measurements were subject to possible electrical shorting, especially because they reported similar resistance decreases in many wide-gap insulators such as NaCl, MgO, and Al_2O_3, and the decreases could not be reproduced in other, better-calibrated nonshorting experiments. Their results were met with considerable skepticism from other scientists.

Recent dynamic studies using isentropic compression in a magnetic-flux compression device found that hydrogen became a conductor at a density of about 1.06 g/cm^3 and a calculated pressure of about 2 Mbar (200 GPa). It is uncertain whether the conductive phase is monatomic or molecular metallic hydrogen.

With the development of the megabar diamond cell (MBC), a static pressure of 1.7 Mbar (170 GPa) has been generated and calibrated. In this cell, properties of hydrogen could be measured directly at ultrahigh pressures. Hydrogen has been solidified at room temperature and 57 kbar (5.7 GPa). The density of hydrogen has been determined to be 0.6–0.7 g/cm^3 at 360 kbar (36 GPa). Hydrogen solid has been compressed up to 0.9 Mbar (90 GPa) without metallization. Brillouin scattering of hydrogen has been measured to determine the equation of state. Raman frequency shift of hydrogen molecules has been studied up to 0.7 Mbar (70 GPa), and it has been found to reach a maximum at 0.32 Mbar (32 GPa). The Raman shift decreases in frequency at higher pressures, indicating softening of the molecular bonding.

All measurements are progressing rapidly, and the results will lead to better understanding of the metallic transition.

For background information see BAND THEORY OF SOLIDS; FREE ENERGY; HIGH-PRESSURE PHYSICS; PLANETARY PHYSICS; RAMAN EFFECT; SCATTERING OF ELECTROMAGNETIC RADIATION; SUPERCONDUCTIVITY in the McGraw-Hill Encyclopedia of Science and Technology.

[AGNES L. MAO; H. K. MAO]

Bibliography: P. S. Hawke et al., Phys. Rev. Lett., 41:994–997, 1978; K. K. Mon, G. V. Chester, and N. W. Ashcroft, Phys. Rev., B21:2641–2646, 1980; M. Ross and C. Shishkevish, Molecular and Metallic Hydrogen, ARPA Rep. no. R-2056, Rand Corp., Santa Monica, 1977; S. K. Sharma, H. K. Mao, and P. M. Bell, Phys. Rev. Lett., 44:886–888, 1980.

Integrated circuits

Gallium arsenide, GaAs, is a compound semiconductor whose unusual electronic properties make it attractive for applications to microwave semiconductor devices (Gunn, IMPATT, and Schottky diodes and field-effect transistors) and optoelectronic devices (light-emitting diodes and photodiodes). Over the past decade, significant advances in material technology and fabrication methods such as ion implantation, projection alignment, plasma etching, and ion milling techniques have made it possible to begin to fabricate gallium arsenide digital and analog integrated circuits. These circuits have provided very high switching speeds. Propagation delay per logic gate as low as 65 picoseconds has been observed on circuits fabricated by optical lithography, while electron-beam lithographic techniques have yielded circuits with delay as low as 34 ps per gate. Speed can also be exchanged for increased power dissipation. Power-delay products under 10 femtojoules (10^{-14} J)/gate have been reported for minimum-area ring oscillators. Current research is directed toward higher speed, lower power, and large-scale integration (LSI) of gallium arsenide digital circuits and analog circuit integration.

Clearly there would be applications possible at all levels of integration for circuits exhibiting the performance described above. At the small-scale integration (SSI) level (under 20 gates), high-speed frequency dividers, sample-hold amplifiers, latches, and comparators have been demonstrated in several laboratories. Frequency dividers which operate at input frequencies above 4 GHz and sample-hold circuits with acquisition times less than 300 ps have been reported. High-speed medium-scale integration (MSI) circuits (20–100 gates) such as 1 gigabit/s multiplexers, demultiplexers, and shift registers, multipliers, and 1.6-GHz variable-modulus dividers have all been demonstrated in the laboratory, and have operated at data rates consistent with the 100–200 ps/gate propagation delays expected for MSI/LSI gallium arsenide integrated circuits and the chosen circuit architecture. At the large-scale integration complexity level (approximately 1000 gates), a much wider range of potential applications becomes available, covering computational, signal processing, frequency synthesis, data conversion, and other areas. Low power-delay products are essential to the realization of LSI high-speed gallium arsenide integrated circuits because total power dissipation per chip must be less than 1–2 W if a practical, reliable circuit is to be obtained.

In addition, in order to effectively utilize high-speed, low-power integrated circuits, it is essential to build at the highest possible level of integration so that power-consuming interconnects and interconnect delays will be minimized. The optimum level of integration is that in which all input and output data are provided at a relatively low rate while the chip itself operates on those data at the maximum possible speed.

There are three major gallium arsenide digital integrated circuit research areas being actively addressed. First, the above-mentioned LSI design and fabrication area is being pursued using the Schottky diode field-effect-transistor logic (SDFL) planar approach described below (Fig. 1). Second, ultrahigh-speed circuits are being developed with the use of electron-beam lithography to provide submicrometer device dimensions. Here all emphasis has been on speed, with little effort being directed toward minimizing power dissipation. Finally, gallium arsenide integrated circuits with propagation delays less than 500 ps are being developed with emphasis on ultralow-power dissipation by use of normally-off field-effect transistor (FET) devices.

There is one further area of research being pursued in the gallium arsenide integrated circuit area, which is an outgrowth of extensive gallium

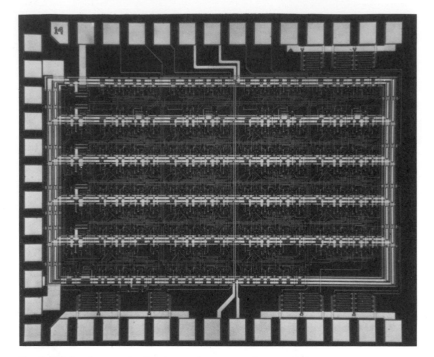

Fig. 1. Photomicrograph of 250-gate (large-scale integration) gallium arsenide integrated circuit fabricated with the planar Schottky diode field-effect-transistor logic (SDFL) technology.

arsenide microwave field-effect transistor development and application over the past decade. This is the development of high-speed analog circuits and monolithic microwave integrated circuits (MMIC). These circuits will ultimately find use in high-speed data acquisition and microwave receiver types of applications. Sample-hold amplifiers, comparators, and operational amplifiers are being developed for future integration into gigabit analog-to-digital converters. Also, integrated broadband microwave low-noise amplifiers, mixers, local oscillators, and intermediate-frequency amplifiers, which will ultimately lead to complete microwave receivers on a single chip, have been demonstrated. This will lead to significant cost savings and may prove to be essential for the economics of consumer satellite television receiver (cheap downlink receivers to convert 12-GHz satellite television broadcasts to the UHF range).

Gallium arsenide versus silicon. The same properties which make gallium arsenide attractive as a microwave device material also make it desirable for high-speed logic circuits. First, the electron mobility of gallium arsenide (4000 cm²/V-s at 1×10^{17} donors/cm³) is 8–10 times greater than that of silicon. Second, the peak electron velocity (2×10^7 cm/s) is reached at an electric field of only 3.5 kV/cm in gallium arsenide. The electron velocity of silicon continues to increase as higher electric fields are applied, and slowly approaches a saturated drift velocity of 6.5×10^6 cm/s at very high fields, while the GaAs electron velocity decreases slowly at high electric fields and ultimately approaches 1 to 1.4×10^7 cm/s.

These very divergent properties between gallium arsenide and silicon are quite significant for very high-speed, low-power digital circuits. In order to minimize power dissipation, a low logic volt-

age swing (ΔV) is desired since dynamic power increases as ΔV^2. For high speeds, high electron velocities are needed since the gain-bandwidth product of the field-effect transistor is inversely proportional to the transit time under the gate electrode. Gallium arsenide, because of its electron-transport properties, yields its maximum peak velocity at relatively low bias voltages, thus providing the capability for achieving high speed and low power in the same device. Silicon, on the other hand, requires large electric fields to achieve high electron velocity. Therefore high-speed silicon devices dissipate much greater power per logic gate than gallium arsenide.

Finally, gallium arsenide possesses a high-resistivity (greater than 10^8 Ω cm) semiinsulating substrate which provides isolation between devices without the need for pn junction or heteroepitaxial (silicon-on-sapphire; SOS) isolation. Therefore circuit density can be greater than with comparable silicon integrated circuit approaches. Also, the insulating substrate reduces interconnect line parasitic capacitance, thereby increasing speed and decreasing dynamic power dissipation.

MESFET. The Schottky-barrier gate metal-semiconductor field-effect transistor (MESFET) is the main active device used in gallium arsenide integrated circuits to provide current gain and inversion. Figure 2 shows a modern planar, ion-implanted gallium arsenide MESFET fabricated by localized implantations of dopants into a semiinsulating gallium arsenide substrate. The source and drain regions receive a deep, low-sheet-resistance n-type implant to reduce parasitic channel and contact resistance. The channel region is implanted with a shallow (100- to 200-nm) n-type implant. Drain-source current scales directly with channel width. The metal gate electrode, which is typically 0.5 to 1μm long, forms a Schottky barrier on the 3-μm-long transistor channel, in which it is centered. This gate can be used to pinch off the

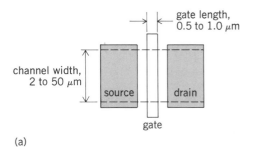

(a)

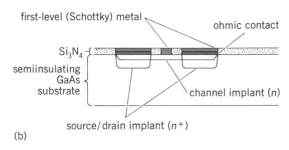

(b)

Fig. 2. Planar ion-implanted gallium arsenide MESFET. (a) Top view. (b) Cross section.

channel (depletion mode) when reverse-biased. In the case where the channel receives a lighter implant dose and is pinched off at gate-source voltage $V_{gs} = 0$, the gate enables current flow to occur as it is forward-biased (enhancement mode).

The gallium arsenide MESFET is well suited for high-speed integrated circuit applications because of its high current gain–bandwidth product (15–20 GHz for a 1-μm-gate-length transistor) and the relative ease of fabricating MESFET with micrometer or submicrometer gate lengths. Gate alignment tolerances for gallium arsenide MESFET are relatively less demanding than silicon metal oxide semiconductor field-effect transistors (MOSFET) of equivalent gate length, and the requirement for the extremely thin (25-nm) gate oxide is eliminated. Threshold control is achieved in the MESFET by precise control of the channel implant.

GaAs integrated circuit approaches. Three main logic gate designs, the buffered FET logic (BFL), Schottky diode FET logic (SDFL), and direct-coupled FET logic (DCFL) approaches have been employed in gallium arsenide digital integrated circuits. NOR gates using these three designs are illustrated in Fig. 3. The BFL and SDFL approaches are applicable for depletion-mode MESFET and therefore require two power supplies, V_{DD} and $-V_{SS}$. Level-shift diodes are necessary so that input and output voltage levels are compatible. In the BFL approach this level shift is achieved at the output by Schottky-barrier diodes as shown in Fig. 3a. Either NOR or NAND functions are achieved at the input by parallel or series combinations of MESFET. For practical purposes, a NOR fan-in of 4 or NAND fan-in of 2 is the maximum usable with BFL. BFL circuits have achieved very low propagation delays (34 ps/gate for 0.5-μm-gate-length MESFET), but their power dissipation (40 mW/gate) is too large for even most MSI applications. Lower-power (5.5 mW/gate) BFL gates have been designed, but with some sacrifice in speed (82 ps/gate). Two-level logic gates (NAND/NOR) can also be realized in BFL. MSI frequency dividers fabricated with two-level BFL gates have operated as high as 4.5 GHz. Other MSI circuits, such as multiplexers, have also been demonstrated.

The SDFL circuit (Fig. 3b) uses very small-area Schottky diodes at the gate input to provide the logical OR function (diodes D_A–D_D) and level shifting (diode D_S). Schottky diodes make very-high-speed logic elements in gallium arsenide because of their extremely high cutoff frequencies. MESFET Q1 provides current gain and inversion; 1-μm SDFL gates (62 ps/gate) have also demonstrated low propagaion delays, while maintaining very-low-power dissipations as well (less than 1 mW/gate). Power-delay products as low as 16 fJ have been obtained. This approach has the greatest potential for LSI; 500–1000 gate circuits were expected in 1980. Multiple-level gates have also been achieved in SDFL (OR/NAND, OR/NAND/AND) and can be effectively used to further increase circuit speed and reduce power dissipation.

The DCFL circuit (Fig. 3c) utilizes enhancement-mode field-effect transistors to achieve the NOR functions. Because the gate-source voltage V_{gs} is greater than 0 for these devices, no level shifting

is required, and only one power supply (V_{DD}) is needed. Thus the logic gate requires fewer components than those required in the depletion-mode logic approaches, and will, in most cases, require less wafer area than a BFL gate. Very-small-area ring oscillators have demonstrated propagation delays below 200 ps with power-delay products of 4

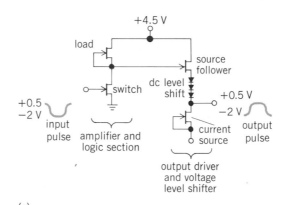

(a)

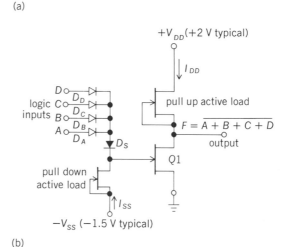

(b)

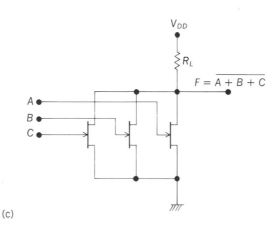

(c)

Fig. 3. Three types of NOR gate commonly utilized in gallium arsenide integrated circuits. (a) Buffered FET logic (depletion mode) (*from R. L. Van Tuyl et al., GaAs MESFET logic with 4-GHz clock rate, IEEE J. Solid-State Circuits, SC-12:485–496, 1977*). (b) Schottky diode FET logic (depletion mode); (c) direct-coupled FET logic (enhancement mode) (*from R. C. Eden et al., The prospects for ultrahigh-speed VLSI GaAs digital logic, IEEE J. Solid-State Circuits, SC-14:221–239, 1979*).

fJ. Frequency dividers have also been fabricated with DCFL NOR gates, but with somewhat higher propagation delay (400–500 ps) due to fan-in and fan-out loading.

Monolithic microwave integrated circuits have all focused on the use of the MESFET as the primary active circuit element. Schottky diodes have also been used used for switching and mixers. Impedance-matching networks are often included in the chip so that inputs and outputs can directly drive a transmission line. Circuit designs have generally emphasized lumped element rather than distributed matching networks because of the smaller chip area requirements and broader bandwidth of the lumped element approach. These passive circuit elements include capacitors (interdigitated, metal-insulator-metal, or Schottky diode), spiral and simple loop inductors, and implanted or mesa resistors. Broadband, low-noise, single-stage amplifiers which provide 6.0 ± 0.5 dB gain from 8 to 18 GHz using these techniques have been reported. Feedback amplifiers providing essentially flat gain from direct current to 4 GHz, and monolitic signal generator chips have been made. Operation of a 1-W monolithic X-band amplifier chip has also been reported. Further efforts are expected to achieve complete receiver front-end integration, including low-noise amplifier, mixer, local oscillator, and intermediate-frequency amplifier in the near future.

For background information *see* INTEGRATED CIRCUITS; MICROWAVE SOLID-STATE DEVICES; TRANSISTOR in the McGraw-Hill Encyclopedia of Science and Technology.   [STEPHEN I. LONG]

Bibliography: D. A. Abbott et al., *Microwave Syst. News*, 9(8):72–96, August 1979; P. Greiling et al., *Microwave Syst. News*, 10(1):48–60, January 1980; C. A. Liechti, *Microwaves*, 17(10):44–49, October 1978; S. I. Long et al., *IEEE Trans. Microwave Theory Techniques*, MTT-28(5):466–471, May 1980.

Laser

Recent advances in laser technology have included the further development and commercial application of semiconductor diode lasers, and the first demonstration of a free-electron laser.

SEMICONDUCTOR DIODE LASERS

Semiconductor diode lasers based on the group IV–VI lead salt compounds, which presently cover the 2.5- to 32-μm-wavelength range, have been under development for over a decade. Their narrow spectral width and wavelength tunability have made them unique tools in infrared molecular spectroscopy and the monitoring of atmospheric pollutants. Applications range from isotope analysis to proposed studies of stratospheric chemistry. Following the initial development and application of these lasers at the MIT Lincoln Laboratory, a number of efforts have been undertaken elsewhere to further develop the laser technology and to incorporate the devices into commercial instruments. Recent developments include continuous-wave (cw) operation at temperatures up to 120 K and longitudinal mode control by distributed feedback, as well as improvements in efficiency and output power.

Laser wavelength ranges for alloys based on lead salt compounds

Alloy system	Composition range	Laser wavelength range at 4.2 K, μm
$Pb_{1-x}Ge_xTe$	$0 \leq x \leq 0.05$	6.5–4.4
$Pb_{1-x}Cd_xS$	$0 \leq x \leq 0.058$	4.1–2.5
$Pb_{1-x}Sn_xTe$	$0 \leq x \leq 0.32$	6.5–32
$Pb_{1-x}Sn_xSe$	$0 \leq x \leq 0.10$	8.4–32
$PbS_{1-x}Se_x$	$0 \leq x \leq 1.0$	4.1–8.4

Materials. The IV–VI lead salt compounds have direct semiconductor energy gaps, which make possible highly efficient radiative electron-hole recombination and laser emission. Of particular interest are ternary alloys based on these materials, since their energy gaps and hence laser emission wavelengths can be altered by changing the alloy composition. The table gives the approximate wavelength ranges that can presently be covered by changing the composition of each of the five alloy systems that have been investigated.

Bulk single crystals of these materials can be grown with relative ease by several techniques; epitaxial techniques are used for growing thin layers on single-crystal substrates.

Fabrication. These diode laser structures are similar to those of gallium arsenide (GaAs) lasers, which operate at shorter wavelengths (0.85 μm). Figure 1 shows a stripe-geometry laser which has

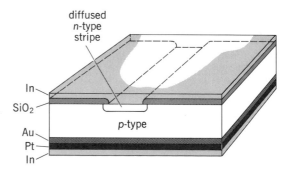

Fig. 1. Stripe-geometry lead salt diode laser structure.

been formed by selectively diffusing a stripe *pn* junction into a lead salt substrate. A Fabry-Perot cavity is formed by the cleaved end faces of the structure. Metallic contacts are made to both the stripe and the substrate, and the diode is mounted in a heat-sink package. These lasers require low temperatures for their operation, generally below 77 K, the temperature of liquid nitrogen. Recent research has been aimed at increasing the operating temperature by developing more sophisticated laser structures, such as double heterostructures, where a thin active region is sandwiched between two layers of a different composition alloy with a wider energy gap and lower refractive index. Such structures, produced by liquid-phase, vapor-phase, or molecular-beam epitaxy, have operated in the continuous-wave mode at temperatures up to 120 K. To ensure operation in a single longitudinal mode, distributed feedback lasers have been developed in which optical feedback is provided by a grating fabricated along the active region rather than by the Fabry-Perot cavity. Elimination of the Fabry-Perot cavity will also facilitate the future incorporation of such lasers into integrated optical circuits, such as integrated infrared heterodyne receivers.

Characteristics. These lasers are generally low-power devices with single-mode continous-wave output usually in the 0.01–1 mW range. In some lead sulfide (PbS) lasers, single-mode powers up to 50 mW with 24% external quantum efficiency have been achieved. For many spectroscopy applications, narrow-line powers of 0.01–0.1 mW are in fact quite adequate, while for others (saturation spectroscopy) 10–50-mW powers are needed. The radiation from stripe-geometry lasers is emitted in a beam that typically measures 5° in the junction plane and 8° perpendicular to the junction.

For spectroscopy, one of the most important parameters is the spectral width of a single mode. Accurate line-width measurements have been made by observing the spectrum of the beat frequency between a diode laser and a stable carbon dioxide (CO_2) laser in a heterodyne experiment. For a $Pb_{1-x}Sn_xTe$ laser emitting at $10\mu m$, a line width of 54 kHz was obtained for a mode with a power of 0.24 mW. The line width was inversely proportional to the power in the mode, indicating that the laser line width is limited by the fundamental quantum phase noise, as predicted by theory.

The wavelength of IV–VI diode lasers can be tuned by changing the temperature or by applying a magnetic field, hydrostatic pressure, or uniaxial stress. Temperature tuning, the simplest and most widely used technique, is based on the temperature dependence of both the energy gap and the index of refraction of the alloys. Thus a $Pb_{1-x}Sn_xTe$ laser has been temperature-tuned from 15.9 to 8.5 μm (an energy gap change from 0.078 to 0.146 eV) by increasing the temperature of a closed-cycle cooler from 10 to 114 K. Fine tuning can be accomplished by adjusting the diode current. This produces a small variation in temperature, the effect of which is illustrated by Fig. 2. Here the continuous tuning of each mode of the Fabry-Perot cavity results from a change in the index of refraction, while the discontinuous jumps from one mode to another are caused by the wavelength shift of the laser gain, associated with a change in energy gap. A distributed feedback laser can generally operate only in one mode, which is determined by the spacing of the grating. However, the continuous tuning range of this mode is substantially larger, 7 cm⁻¹ (210 GHz), compared to 1 cm⁻¹ for a typical Fabry-Perot laser.

In a magnetic field, recombination occurs between field-dependent magnetic energy levels in the valence and conduction bands. Thus a field of 10 kilogauss (1 tesla) has been used to change the wavelength of a $Pb_{1-x}Sn_xTe$ laser from 15.4 to 14.5 μm. A very broad tuning range can be achieved by applying hydrostatic pressure, for example, a lead selenide (PbSe) laser operating at 77 K has been tuned from 7.5 to 22 μm by a hydrostatic pressure change of 14 kbar (1.4×10^9 gigapascals).

Applications. The resolution achievable with a laser spectrometer is limited by the laser line width, and hence can be better than 10^5 Hz. In comparison, the resolution of a high-quality laboratory grating spectrometer is about 10^9 Hz, which is insufficient to resolve the approximately 10^8-Hz-wide Doppler-broadened absorption lines for gases at room temperature.

The experimental techniques include simple transmission spectroscopy, where a laser beam is detected after it has passed through a gas sample; heterodyne spectroscopy, where radiation from a thermal source is mixed in a wide-band infrared

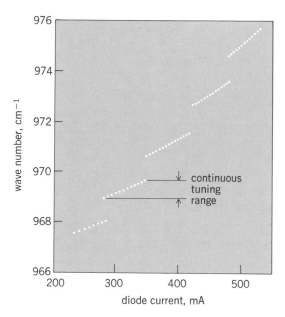

Fig. 2. Fine tuning of a $Pb_{1-x}Sn_xTe$ diode laser by varying the diode current. The tuning is continuous within each of the five modes of the Fabry-Perot cavity. *(From E. D. Hinkley, K. W. Nill, and F. A. Blum, Infrared spectroscopy with tunable laser, in H. Walther, ed., Topics in Applied Physics, vol. 2, pp. 127–190, Springer Verlag, 1976)*

detector with radiation from a tunable laser; and saturation spectroscopy, which uses a high-power saturating beam and a low-power probe beam.

The first observation of nuclear hyperfine splitting in the fully resolved infrared spectrum of a nitrogen oxide (NO) vibration-rotation absorption line was made with a $Pb_{1-x}Sn_xTe$ laser (Fig. 3). The splitting into symmetric pairs of lines is the so-called lambda-type doubling, and the further splitting of each line is caused by coupling of electrons to nuclear spin.

Laser spectroscopic techniques also have very significant advantages in high-precision real-time detection of atmospheric pollutant gases. The techniques range from the monitoring of pollution sources by optoacoustic detectors to remote detection of gases in the upper atmosphere. The utility of long-path monitoring techniques was first demonstrated by measuring the variation in ethylene (C_2H_4) concentration over a parking lot with changes in traffic volume. Heterodyne spectroscopy, although less developed than transmission spectroscopy, holds promise for numerous applications, such as high-resolution studies of atmos-

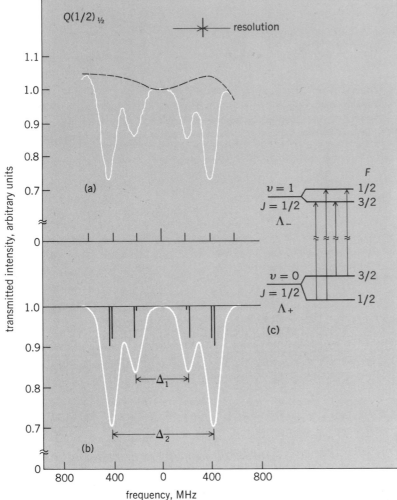

Fig. 3. Spectrum of the $Q(1/2)_{1/2}$ line of nitrogen oxide (NO). (a) Absorption spectrum, measured with a $Pb_{1-x}Sn_xTe$ diode laser, showing nuclear hyperfine splitting. (b) Theoretical spectrum. (c) Energy levels and associated transitions for a lambda-doublet pair. (From F. A. Blum et al., Observation of nuclear hyperfine splitting in the infrared vibration-rotation spectrum of the molecule, Chem. Phys. Lett., 15:144–146, 1972)

pheric chemistry at various altitudes from space platforms and of emission from astronomical sources.

Analytical instruments for various applications based on lead salt lasers have been commercially available for some time. Significant improvements in system size and simplicity can be anticipated as the laser operating temperature is further increased, permitting the use of simpler closed-cycle coolers.   [IVARS MELNGAILIS]

FREE-ELECTRON LASERS

In a free-electron laser, a beam of relativistic electrons passes through a static periodic magnetic field to amplify a superimposed coherent optical wave (Fig 4).

Here the lasing process has been reduced to its most fundamental form, and is manifestly classical in nature. This point is at the root of many of the free-electron laser's potential advantages over conventional atomic lasers; several properties of atomic lasers, such as efficiency, are limited by

quantum mechanics. The new laser is free from the bonds constraining atomic lasers to a particular wavelength, and therefore is continuously tunable. The optical cavity contains only light, radiating electrons, and the magnetic field, so that intense optical fields may propagate without the degrading nonlinear effects (self-focusing and so forth) of denser media. The advanced technology of high-energy electron accelerators and storage rings promises efficient recirculation of beam energy.

Development. The earliest coherent radiation sources, radar and microwave electron tubes, used classical nonrelativistic electron beams to amplify long-wavelength radiation (10–0.1 cm). These devices satisfied a wide range of applications with hundreds of varied designs, but it was not possible to generate shorter wavelengths until the early 1960s, when atomic and molecular lasers were developed. A necessary technical advance at that time was the replacement of "closed" microwave cavities with "open" optical resonators. J. M. J. Madey's conception of the free-electron laser in 1971 showed how relativistic electrons and "open" resonators could extend the advantages of electron tubes to the optical regime.

Madey and collaborators at Stanford University demonstrated free-electron laser amplification in 1976 and laser oscillation in 1977. In the oscillator experiments (Fig. 4), a nearly monoenergetic 43-MeV electron beam from a superconducting linear accelerator was passed through a 5.2-m-long helical magnet that produced a transverse periodic magnetic field with a field strength of $B_0 = 2.4$ kG (0.24 T) wavelength $\lambda_0 = 3.2$ cm. Short 4-picosecond electron pulses of approximately 1 A peak current produced 2000 kW of peak optical power at $\lambda = 3.4$ μm wavelength with circular polarization.

Operating principles. The fundamental physics of free-electron lasers is now well understood; several theoretical viewpoints adequately describe its behavior. Semiclassical quantum theory, or quantum electrodynamics, explains the laser action as stimulated Compton backscattering of the virtual photons in the periodic magnet, or equivalently, as stimulated magnetic bremsstrahlung. In this view the finite-length magnet and the resulting electron kinematics allow stimulated emission to exceed absorption. Viewed classically, the electron beam is a cold relativistic plasma; dispersion relations from the Boltzmann equation can properly characterize the evolution of the electron distribution in the optical wave. The most fruitful and widely used theory calculates the dynamics of individual electrons as they are affected by the fields in the laser cavity; the total transverse current then drives Maxwell's nonlinear wave equation.

To simplify the discussion, only the essential physics of the Stanford oscillator experiment are explained. Mirrors are placed at each end of the interaction region to store radiation. A partially transmitting mirror allows useful coherent radiation to escape. The electrons are removed after each pass, and fresh electrons are either supplied continuously or injected to overlap the rebounding optical pulse. As electrons enter the laser cavity, they are acted on by the static magnetic field, and the oscillating electric and magnetic components of the nearly free optical plane wave; interparticle

Coulomb forces are small for the high-energy, low-density beam of the Stanford experiment. The magnet guides an electron through N periodic oscillations as it travels the length of the magnet $L = N\lambda_0$ with z-velocity (velocity component parallel to the magnet axis) $\beta_z c$ ($\beta_z \approx 1$). The small transverse accelerations produce a small amount of spontaneous radiation carrying the polarization of the magnet geometry: circular polarization for a helical magnet, and linear polarization for alternating poles. The emission is confined to within an angle of approximately $1/2\gamma$ (γmc^2 is the electron energy) about the forward motion, and within a narrow (approximately $1/2N$) spectral line width about the fundamental wavelength $\lambda = \lambda_0 (1 - \beta_z) \approx \lambda_0(1 + K^2)/2\gamma^2$ for $\gamma \gg 1$, where $K = eB_0\lambda_0/2\pi mc^2$ in cgs units, and $K = eB_0\lambda_0/2\pi mc$ in mksa or SI units, e and m are the electron charge and mass, and c is the speed of light. If $K \approx 1$, as is usually the case, there will be a small amount of emission into a few well-separated higher harmonics. The Stanford experiment gives typical values for these parameters, and has demonstrated the tunable characteristic of the laser frequency by varying the accelerator energy. In future machines the tunable wavelength range is estimated to be about a decade; this is primarily determined by the dynamic range of the electron source.

The radiation from multiple passes of the electron beam is stored in the resonant cavity. The magnet alone does no work on electrons (neglecting spontaneous emission), but does give a small transverse velocity, $\vec{\beta}_\perp$. The radiation fields alone have no significant effect on either the electron trajectory or energy, since the forces due to the optical electric and magnetic fields nearly cancel. In combination, the magnetic field guides electrons through a transverse path so that the radiation electric field $\vec{E}$ can do work on an electron

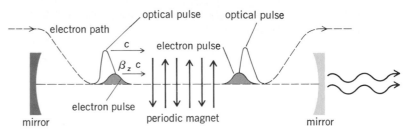

Fig. 4.   Free-electron laser oscillator configuration.

according to the equation $\dot{\gamma} = (e/mc)\vec{\beta}_\perp \cdot \vec{E}$. The fundamental emission frequency is such that one wavelength of light passes over an electron as it passes through one magnet wavelength; therefore the transverse velocity $\vec{\beta}_\perp$ retains its orientation relative to $\vec{E}$ over many magnet periods, and the energy exchange persists in the same direction. The direction of energy flow ($\dot{\gamma}$ being positive or negative) is determined by the electron's phase $\zeta = 2\pi[(\lambda^{-1} + \lambda_0^{-1}) z(t) - \lambda^{-1} ct]$ within sections of the electron beam, each an optical wavelength long. Evolution of electrons in the ζ-coordinate space (the "resonant frame") is slow and simple; for low gain, ζ is approximately governed by the pendulum equation $\ddot{\zeta} = \Omega^2 \cos(\zeta + \phi)$, where $\Omega^2 = 2e^2B_0E/(\gamma mc)^2$ or $\Omega^2 = 2e^2B_0cE/(\gamma mc)^2$ in mksa or SI units, and ϕ is the optical phase.

Since any practical electron beam is many optical wavelengths long, the potential $V(\zeta) = -\Omega^2 \sin(\zeta + \phi)$ is uniformly populated with electrons along the ζ axis. If electrons enter at the resonant velocity $c\lambda_0/(\lambda_0 + \lambda)$, they are initially stationary on the $V(\zeta)$-surface [$\dot{\zeta}(0) = 0$]; an equal number of particles "roll" ahead and back, exchanging equal amounts of energy with the optical wave. There is no gain in this case. If electrons enter at a slightly

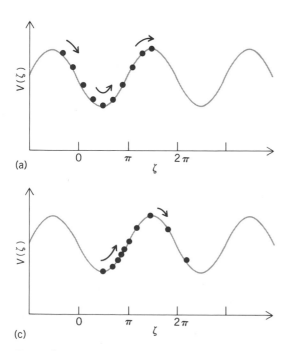

(a)

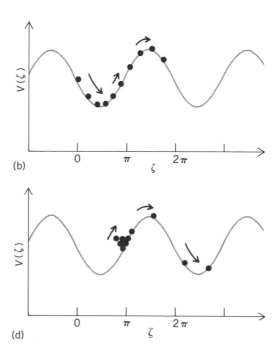

(b)

(c)

(d)

Fig. 5.   Evolution of sample electrons (spanning one optical wavelength) in the potential $V(\zeta)$ in the resonant beam frame. (a) Time $t = 0$; (b) $t = \frac{1}{3}L/c$; (c) $t = \frac{2}{3}L/c$; (d) $t = L/c$.

higher velocity, then all electrons are initially "rolling" along the ζ axis of the corrugated V-surface. For optimum gain conditions, the "rolling" is slow, so that none of the electrons will "roll" past more than one crest during the interaction time. The peaks in $V(\zeta)$ then cause spatial bunching and decelerate the particles, so that they give up their energy to the optical beam (Fig. 5). This is the gain mechanism. [The potential $V(\zeta)$ shown in Fig. 5 actually grows in amplitude and shifts in phase self-consistently, and almost imperceptibly, according to Maxwell's wave equation; this shows the simple pendulum to be an accurate analogy.] During amplification the initially monoenergetic, uniform beam becomes bunched at the optical wavelength and spread in energy; the fractional energy spread is $\delta\gamma/\gamma \approx \Omega L/\pi Nc$ for weak fields and approximately $1/2N$ at saturation. Maximum gain for weak fields is $G_{max} = 0.27 e^4 B_0^2 \rho \lambda_0 \cdot (L/\gamma mc^2)^3$ in cgs units, or $G_{max} = 0.27 e^4 B_0^2 c^2 \rho \lambda_0 \cdot (L/\gamma mc^2)^3/4\pi\epsilon_0$ in mksa or SI units, when the "rolling" velocity is $\dot\zeta(0) = 2.6c/L$, and the beam density is ρ. Absorption is predicted and observed for $\dot\zeta(0)L/c \leq 0$. For typical parameters, each ampere of beam current within the optical mode cross section gives a few percent gain; 1–100 A peak current can be provided by accelerators or storage rings. The useful energy range is roughly 2 to several hundred MeV; this spans a range of wavelengths from submillimeter to x-rays. Higher energies (with the best feasible magnets) result in very low gain.

The character of the gain process is analogous to the energy exchanged between two weakly coupled pendula (the optical wave and electron beam). For very short times, little energy can be transferred, and for very long times, the exchange averages to zero. But for the appropriate finite time, defined by $\dot\zeta(0)L/c$ in this case, energy flows in one direction only, giving a net transfer to the optical wave. The energy density in a relativistic electron beam can be quite large; any reasonable fraction that can be transferred to an optical wave produces a sizable laser field.

After many passes of the electron beam, the intracavity optical amplitude becomes large, $V(\zeta)$ is large, and saturation occurs. When $\Omega L/c \gg 2.6$, there is no value of $\dot\zeta(0)$ which can prevent the nearly symmetric falling of particles into the potential troughs, the electrons become "trapped," and gain decreases. As the gain decreases to equal the cavity losses, the laser runs in steady state. In the Stanford experiment, only a small fraction (approximately $1/2N$ or 0.5%) of the beam energy is extracted prior to saturation.

Modified magnet geometries. In future experiments, the deep troughs may be put to an advantage, increasing the energy extracted from the electron beam and extending the laser performance. At large field strengths, electrons are trapped in the beginning stages of the magnet; the magnet (called a tapered wiggler) is designed with a slowly decreasing wavelength, so that $V(\zeta)$ moves to the left in the resonant frame. Computer simulations show that about half the electrons remain trapped in the deep decelerating "buckets," with approximately 10% (to possibly 50%) energy extraction. This is the same mechanism used in linear accelerators; in fact, a periodic magnet with a slowly increasing wavelength and a powerful laser pulse may be used as a particle accelerator. The possibility of modified magnet geometries is an important flexibility in free-electron laser design; in an atomic laser, this would correspond to altering the atomic structure, seen by an excited electron, during the emission process.

Collective gain process. For high-density, low-energy electron beams (where ρ/γ^3 is greater than about 10^2 times Stanford's parameters), interparticle Coulomb forces can influence a particle's motion in competition with $V(\zeta)$. The gain process can then be collective; many electrons oscillate together due to spatial beam instabilities, and amplification is nonlinear in the current. Still, relativistic electrons are necessary to reach short wavelengths, and electron bunching is the key to gain; the emitted wavelength is generally related to the system parameters through dispersion relations containing the electron density. Free-electron maser action has been demonstrated in the collective regime with moderate energy beams ($\gamma \approx 2$).

System configurations. At present, free-electron laser development is in its infancy; only the Stanford laser has operated in the short-wavelength regime. Several experiments are now under way in the United States and Europe, and many new designs are being considered. The necessity of high-current, high-energy electron beams appears to dictate that, for the near future at least, free-electron lasers will be large machines. But these facilities will be unique in that they are continuously tunable with high average power and high efficiency. Some basic configurations currently under investigation are diagrammed in Fig. 6. A specific single-pass arrangement (Fig. 6a) uses an induction linac (50 MeV and 2 kA peak current) as the electron source for a tapered wiggler mag-

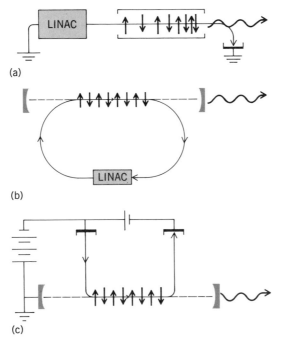

Fig. 6. Some basic free-electron laser system configurations. (a) Single-pass. (b) Storage ring. (c) Electrostatic.

net; collective effects are important for this beam. With moderate extraction, an impressive optical pulse is developed; the degraded electron pulse would then be discarded.

With less energy extraction, the electron-beam quality can be maintained and recirculated in a storage ring (Fig. 6b); the electron energy lost per pass is replaced with a radio-frequency linac in the ring. However, even small beam degradation per pass can build up over many cycles until synchrotron radiation damping eventually allows steady-state operation; analysis predicts that the available laser power will then be only a small fraction (about $1/2N$) of the synchrotron power. Several possible "cures" for this "ailment" have been devised, and now it appears that the laser output can greatly exceed the necessary synchrotron damping. One method alters the periodic magnet structure to diminish the beam degradation; in another the magnet structure is specifically designed to operate with an energy-broadened electron beam; and another method recycles the beam energy without recycling the electrons. Electrostatic accelerators can also be used efficiently with lower-energy beams ($\gamma \leq 10$). The electron current (about 10 A) and energy can be recovered by electrostatic deceleration of the beam after a pass through the laser (Fig. 6c).

Several conceptual designs indicate that 20% (to possibly 50%) "wall-plug" efficiency is possible, the greatest losses coming from bending magnets and power supplies. It is the classical and relativistic nature of the electron beam which, in principle, allows efficient flow of energy into the electrons and then the optical wave. Efficient free-electron lasers with an average optical output from kilowatts to megawatts and optical wavelengths from submillimeter to submicrometer have been proposed.

Prospects. Numerous free-electron laser schemes have been proposed. The above ideas just give the flavor of the research. Scientific, industrial, and military applications look promising. Solid-state, atomic, and chemical spectroscopy can reach tunable wavelengths not previously accessible. The military needs powerful far-reaching beams for communications, radar, and weapons; in particular, space applications require high efficiency. Industrial photochemical processing looks promising in that free-electron laser light appears relatively inexpensive. Judging from the more mature technologies, electron masers and atomic lasers, several free-electron laser configurations will be required to satisfy differing needs.

For background information see LASER; PARTICLE ACCELERATOR; SEMICONDUCTOR in the McGraw-Hill Encyclopedia of Science and Technology.   [WILLIAM B. COLSON]

Bibliography: D. A. G. Deacon et al., Phys. Rev. Lett., 38:892–894, 1977; T. C. Harman and I. Melngailis, in R. Wolfe (ed.), Applied Solid State Science, vol. 4, pp. 1–94, 1974; E. D. Hinkely, K. W. Nill, and F. A. Blum, in H. Walther (ed.), Topics in Applied Physics, vol. 2, pp. 127–190, 1976; S. F. Jacobs et al. (eds.), Quantum Electronics, vol. 5, 1978, and vol. 7, 1980; J. M. J. Madey, J. Appl. Phys., 42:1906–1913, 1971; H. Preier, Appl. Phys., 20:189–206, 1979.

Leukocytes

One of the major host defenses against microbial diseases is the unidirectional movement of blood leukocytes from the vasculature toward the site of inflammation. This aspect of white cell locomotion is called chemotaxis. The migration of leukocytes is triggered and sustained by chemical compounds which are present at the inflammatory area. These factors are known as chemoattractants or cytotaxins, and appear to be derived primarily from the complement system, or from antigen-stimulated lymphoid cells. They are extremely potent, since only very small amounts (nanomoles or picomoles) are required for activity. Compounds which inhibit chemotaxis are also known to exist.

There is considerable evidence that the chemoattractants bind to specific receptors on the plasma membrane of granulocytes (neutrophils) and monocytes, and that this particular interaction initiates a series of cellular events which lead to the locomotion of the cells toward increasing concentration of the attractant. Since granulocytes are highly specialized cells whose primary function is to ingest and kill microorganisms, their rapid recruitment to the site of infection in large numbers is important for the initiation of a massive attack on invading bacteria. Thus the chemotactic phenomenon plays a central role in the defense against microbial diseases. In individuals with suppressed chemotactic migration, which can be due to intrinsic neutrophil defects, chemoattractant deficiency, presence of chemotactic inhibitors, and other causes, recurrent and prolonged infections are the rule. In general, the clinical application of techniques for assessment of leukocyte chemotaxis has disclosed that abnormalities of this function are associated with numerous human diseases, including cancer.

Chemotactic factors. A large number and variety of substances have been reported to be chemotactic for leukocytes, especially neutrophils. The heterogeneity, complexity, and unknown structure of these compounds made it impossible to systematically investigate their interaction with the neutrophil surface at the molecular level. Some of these are found in bacterial cultures and activated serum. Simple, synthetic N-formyl methionine peptides are reported to be chemotactic for neutrophils and macrophages. These peptides may be related to bacterial products known to be potent chemoattractants. They are active at very low concentrations (10^{-11} M) in laboratory cultures and induce lysosomal enzyme release in neutrophils, an event contributing to inflammation. The presence of the N-formyl group and a nonpolar residue in the C-terminal position seem to dramatically enhance activity. Experimental evidence suggests that the peptides bind to a stereospecific receptor at the cell surface. Furthermore, it has been postulated that the neutrophil receptor may have two functional sites, one for binding the attractant and the other for allowing the receptor to be freed of ligand by hydrolysis. This mechanism may be involved in permitting the cell to continuously detect the attractant gradient concentration. Recently evidence was presented against a strict steric specificity for binding.

Another source of chemotactic compounds

which has been extensively studied is the plasma complement (C) system. Chemotactic activity is exhibited by the following C activation products: $\overline{C567}$ macromolecular complex, C3a and C5a peptides, and $\overline{C3B}$ from the properdin pathway. Compelling evidence indicates that C5a is the major chemotactic factor for neutrophils present in human serum; the sequence of its 74 amino acids has been established. An oligosaccharide is attached to residue 64. The C-terminal arginyl residue appears to augment chemotactic activity. A C5a-derived peptide (6000 daltons molecular weight) has been shown to exhibit chemotactic activity for tumor cells. C5a has also been shown to display specific structural features for binding to human polymorphonuclear (PMN) leukocytes. The number of C5a binding sites per cell was estimated at $1-3 \times 10^5$.

Other compounds implicated in chemotaxis are cytophilic antibodies (antigen-dependent chemotaxis), lymphocyte-produced lymphokines, lipids such as hydroxy fatty acids (HETE), ascorbic acid, proteins from fibroblast cultures, Thromboxane B2, and factors of more complex composition.

Measurement of chemotaxis. Methods for the measurement of the unidirectional movement of leukocytes in the presence of chemotactic factors fall into two general categories: those where single cells are observed visually in motion, and those where the migration of the cells in bulk is determined indirectly after a fixed time interval, usually a few hours. Phase-contrast microscopy and cinematography are utilized in the visual assay to record the migration tracts of individual cells moving toward a chemoattractant source, such as blastospores of *Candida albicans*, chemoattractant-loaded Sephadex beads, or a well containing chemoattractants. By analysis of the paths taken by cells, a detailed description of movement direction, cell velocity, turn behavior, and so forth can be obtained.

Another group of assays involves cells migrating as a population toward a gradient of chemoattractant. The most frequently used techniques are those utilizing: two compartment chambers separated by a micropore filter known as the Boyden chamber; and migration of cells under agarose gels in tissue-culture plates. In the Boyden technique, the lower chamber contains the chemoattractant and the upper contains cells which try to pass through the pores of the filter in response to chemotactic migration. After a certain period of incubation at 37°C, the filter is fixed and the cells present on its lower side are enumerated microscopically after staining, or by gamma counting if radioactively labeled. In the agarose method, the cells are fixed and stained after appropriate incubation, and the linear distance of migration from the margin of the cell well toward the chemotactic factor is determined by microprojection.

Most biological and clinical studies have been performed with some variation of the Boyden assay, although increasing activity in controlling the variables of the agarose method has been reported very recently. Undoubtedly the single-cell cinematographic techniques are more informative, but require sophisticated instrumentation and data-processing procedures.

Cellular events. The stimultion of leukocytes by chemoattractants is a multistep process which is just beginning to receive focused attention. Of interest are the molecular events that follow the surface binding of these compounds, and the mechanism of the morphological alterations which lead to directional cell locomotion. Ultrastructural studies of chemotactic neutrophils have revealed that cell elongation and polarization take place, with pseudopods appearing at the leading end of the cells and nuclei toward the back. The centrioles and their radial array of microtubules orient between the nuclei and pseudopods, while the microfilaments are localized in the leading pseudopods. Cytochalasin B and colchicine have been used to assess the role of microfilaments and microtubules in cell orientation. They demonstrated that cell orientation and direction finding are stabilized by microtubule function, whereas the microfilaments are essential for locomotion. Electrophysiological studies of ionic flux events have also revealed attractant-induced membrane potential changes with increased permeability of Na$^+$, K$^+$, and Ca^{++}. These events may play a role in the condensation of actin filaments in the leading edge of the cell where locomotion is initiated. Transient neutrophil aggregation, swelling, and adhesion have also been studied in the presence of chemoattractants and as influenced by a number of variables.

Modulators and disorders. Local chemotactic leukocyte responses in the living body may be modulated by a number of factors so as to produce enhancement or inhibition. Immunological reactions and their effector pathways may produce various manifestations of control, depending on the type of the reaction and stimulus specificity. Other modulators may include various pharmacologic agents, cyclic nucleotides, histamine, and contents of neutrophil granules. The last-mentioned can activate the complement system via the alternate pathway, generate a chemotactic fragment from C5, and destroy the chemotactic activity of preformed C5a. Serum inhibitors of chemotaxis can also act as modulators by inactivating the chemotactic factors or by exhibiting a direct inhibitory effect on the responding cell. A number of humoral factors have been described which are capable of suppressing chemotaxis by one of the above mechanisms. Of importance is the finding that tumor cells produce a chemotactic factor inactivator (CFI), and the findings involving abnormal monocyte chemotaxis in patients with cancer.

Chemotactic disorders can be fundamentally divided into chemotactic factor defects and cellular defects. The former disorders have been well characterized at the molecular level, whereas the latter remain poorly defined because of a lack of understanding of the cellular pathophysiology of the leukocyte. Basic research in this area, coupled with clinical studies, may provide a rational foundation for the treatment of chemotactic disorders.

For background information *see* IMMUNOLOGY, CELLULAR in the McGraw-Hill Encyclopedia of Science and Technology.

[NICHOLAS CATSIMPOOLAS]

Bibliography: J. A. Bellanti and D. H. Dayton (eds.), *The Phagocytic Cell in Host Resistance*, 1975; J. I. Gallin and P. G. Quie (eds.), *Leukocyte*

Chemotaxis: Methods, Physiology and Clinical Implications, 1978; E. Sorkin (ed.), *Antibiotics and Chemotherapy*, vol. 19: *Chemotaxis: Its Biology and Biochemistry*, 1974; P. C. Wilkinson, *Chemotaxis and Inflammation*, 1974.

Liquid crystal

The anisotropic nature of sound propagation in certain media reflects their microscopic symmetry. For example, sound has been widely used to study changes of crystal structure in solids. The relation between certain sound propagation characteristics of liquid crystals and their microscopic symmetry has been a subject of intense study in the past 10 years. In particular, the following observations were made, which signify a dramatic difference in the symmetry of liquid crystals from ordinary fluids: sound attenuation anisotropy in the liquid crystals called nematics, anisotropy in both velocity and attenuation in smectics, and a diffraction effect in cholesterics. In order to understand these effects, the symmetry of liquid crystals must first be discussed.

Liquid crystalline symmetry. Liquid crystals are intermediary between liquids and crystalline solids. They have a consistency ranging from water to molasses, yet on a microscopic level there is a long-range molecular order. They are organic compounds with a molecular weight of order 300 or more. Generally the molecules have an elongated cigarlike shape, and there is a tendency for these molecules to line up parallel. According to the symmetry of molecular ordering (that is, the way molecules are spatially arranged), liquid crystals are grouped into three classes (mesophases): nematic, cholesteric, and smectic (Fig. 1).

In nematics, the molecules are nearly parallel to each other (with an average direction denoted by $\vec{n}$ in Fig. 1a) but free to move anywhere in the sample, just as in an ordinary fluid (uniaxial symmetry). When the overall molecular orientation $\vec{n}$ rotates spatially along the direction perpendicular to $\vec{n}$ (denoted by p in Fig. 1b), one has cholesterics (translational symmetry with chirality). The cholesterics are thus "twisted nematics." In smectics, the molecules are confined in layers (Fig. 1c), but the layers glide over each other freely with a relatively small viscosity. Three subclasses of the smectics are known. In smectic-A, molecules are free to move within layers and are parallel to the layer normal (uniaxial and translational symmetry). When the molecules are parallel to the layer normal but arranged in a close-packed lattice, one has smectic-B (uniaxial and translational symmetry). In smectic-C, the molecules are free to move but their axes are tilted away from the layer normal (a "tilted smectic-A" with biaxial and translational symmetry).

Nematics. Imagine a cube of nematic liquid crystal. One can apply a stress to this box from the side perpendicular to $\vec{n}$ or from the side parallel to it, keeping all other sides fixed in position. The former situation corresponds to sound propagation along $\vec{n}$, and the latter perpendicular to it. Because the molecules are not spatially restricted, they redistribute themselves to conform to the new shape of the box. The final state has, therefore, the same density ρ and pressure p regardless of

the direction in which the stress is applied. As is the case in the ordinary fluid, the restoring force constant is then given by the adiabatic compressibility $\rho(\partial p/\partial \rho)_s$ (where the subscript s indicates that the derivative is taken at constant entropy), and the sound velocity V is given by Eq. (1), which

$$V = \sqrt{(\partial p/\partial \rho)_s} \qquad (1)$$

is independent of the propagation direction. The molecules, however, undergo different paths to reach the final states, and thus the energy lost in each process is different. The main contribution to the energy loss is due to viscosity. In general, three independent viscosity coefficients are involved when the sound propagates in the direction

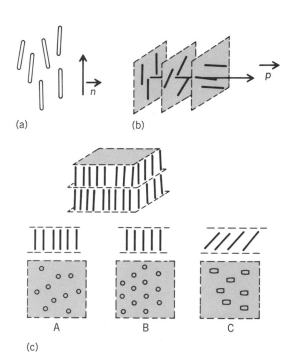

Fig. 1. Molecular arrangement in liquid crystal mesophases. Bars depict molecules. (a) Nematic. (b) Cholesteric. (c) Smectic, showing subclasses A, B, and C.

making an angle θ from $\vec{n}$. The attenuation α is anisotropic and has the form of Eq. (2), where ω is

$$\alpha = \frac{\omega^2}{2\rho V^3}[\eta_a \sin^2\theta + \eta_b \cos^2\theta$$
$$+ \eta_c \sin^2\theta \cos^2\theta] \qquad (2)$$

the angular frequency of the sound and η_a, η_b, and η_c are three independent viscosity coefficients. Equation (2) is the most general form of attenuation for a uniaxial system and, therefore, it also describes sound attenuation in smectic-A and -B.

Smectic-A. Consider a cartesian coordinate system (Fig. 2a) where the z axis is along the layer normal (bars depict molecules). The displacement of layers along the z direction is written as u_z, and thus the strain of the layer structure is $\partial u_z/\partial z \equiv u_{zz}$. Because the molecules are restricted in planes, the density ρ and the layer deformation u_{zz} are independent variables. For example, the stress along

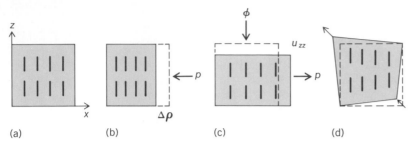

Fig. 2. Deformations of smectic-A structure. (a) Undeformed structure. (b) Deformation along x axis, parallel to layer. (c) Deformation along z axis, perpendicular to layer. (d) Diagonal deformation.

the x axis (Fig. 2b) changes the density, while the layer spacing is constant. The force constant for this deformation is thus $\rho(\partial p/\partial \rho)_{s,u_{zz}}$. On the other hand, in Fig. 2c layers are deformed but ρ is held constant. The associated stress is denoted as ϕ, and the force constant is then $(\partial \phi/\partial u_{zz})_{s,\rho}$. When the sound propagates in the z direction, u_{zz} and ρ vary simultaneously. Since u_{zz} affects the pressure (Fig. 2c), there is one more force constant in this process, $(\partial p/\partial u_{zz})_{s,\rho}$. In general, when the sound propagates in the direction at an angle θ from the z axis, the sound velocity is given by Eq. (3). The

$$V = \left[\left(\frac{\partial p}{\partial \rho}\right)_{s,u_{zz}} - \frac{2}{\rho}\left(\frac{\partial p}{\partial u_{zz}}\right)_{s,\rho} \cos^2 \theta \right.$$
$$\left. + \frac{1}{\rho}\left(\frac{\partial \phi}{\partial u_{zz}}\right)_{s,\rho} \cos^4 \theta \right]^{1/2} \quad (3)$$

sound attenuation anisotropy has the same form as Eq. (2).

So far only a longitudinal (compressional) sound wave has been considered. The fact that the material responds to the shape change leads to another type of sound propagation. When the sample is deformed diagonally (Fig. 2d), the density is constant but u_{zz} is positive on one end and negative on the other. This leads to a shear wave (transverse sound) propagating perpendicular to the deformation direction. The shear wave velocity anisotropy is given by Eq. (4). That the wave does not propa-

$$V = \left[\frac{1}{\rho}\left(\frac{\partial \phi}{\partial u_{zz}}\right)_{s,\rho} \sin^2 \theta \cos^2 \theta \right]^{1/2} \quad (4)$$

gate along the z axis or perpendicular to it is easily understood because the transverse deformation

along these directions does not involve u_{zz} and hence no restoring force results. A propagating shear wave is the consequence of the additional variable u_{zz}. This situation is mathematically similar to the appearance of the second sound in superfluid liquid ^{4}He. Therefore, this shear mode is sometimes referred to as the second sound.

Smectic-B. A smectic-B is a stack of loosely coupled hexagonal crystalline layers. Therefore, a shear wave polarized in the smectic plane can propagate in any direction within the plane just as in a solid. If there is a fixed phase difference between shear waves in adjacent layers, the phase front of the entire wave propagates across the layers. Thus this mode can propagate in all directions except along the z axis. Other than this additional mode, the longitudinal wave and the transverse wave polarized in the plane of the z axis and the propagation direction are of the same forms as in smectic-A.

Smectic-C. Consider a coordinate system where the z axis is along the layer normal and the molecules are tilted in the x-z plane. Even though the molecules are tilted, the restoring force for a strain in the x-y plane is isotropic for the same reason as in the nematic case. Therefore, sound velocities of both longitudinal and transverse modes have the same forms as in smectic-A. However, again using the same argument as was used in the nematic case, the sound attenuation is anisotropic in the x-y plane. The sound velocity profiles as a function of the propagation direction for nematics, and smectic-A,-B, and -C are illustrated in Fig. 3. Typical longitudinal sound velocity at $\theta = 0$ is about 1.5×10^3 m/s.

Cholesterics. The periodicity of the twist of the molecular axis in cholesterics is usually between 0.1 and 1 μm. In some materials, however, this pitch length is highly temperature-dependent and can be as long as 1 mm. By sweeping the temperature, one can match the wavelength of sound to the pitch length. When a wave enters into a structure with periodic rigidity, the diffraction effect prevents a wave of matched wavelength from propagating. In cholesterics, this diffraction effect occurs in attenuation. Because cholesterics are twisted nematics, the viscosity has the same periodic structure as the molecular axis rotation. Thus, it acts as a viscous grating. Therefore, when the wavelength matches the periodicity, certain attenuation values are forbidden and the attenuation changes discontinuously as the pitch length is varied. Mathematically this problem is identical to that of electrons in crystals; electrons of certain wavelengths are forbidden in crystalline solids.

For background information *see* BAND THEORY OF SOLIDS; SECOND SOUND; SOUND; VISCOSITY OF LIQUIDS in the McGraw-Hill Encyclopedia of Science and Technology.   [K. MIYANO]

Bibliography: S. Candau and S. V. Letcher, *Adv. Liq. Cryst.*, 3:168–235, 1978; S. Candau and P. Martinoty, in *Proceedings of the International School of Physics, "Enrico Fermi" Course LXIII: New Directions in Physical Acoustics*, 1976; K. Miyano and J. B. Ketterson, *Phys. Acoust.*, 14:93–178, 1979; G. G. Natale, *J. Acoust. Soc. Amer.*, 63:1265–1278, 1677–1693, 1978.

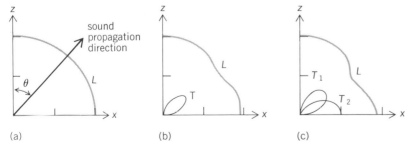

Fig. 3. Sound velocity anisotropy in (a) nematics, (b) smectics-A and -C, and (c) smectic-B. L stands for longitudinal wave and T for transverse wave. T_1 and T_2 are the two transverse modes in smectic-B.

Liquid helium

Recent developments in the study of liquid helium include the discovery of a new mode of sound propagation, known as fifth sound, in superfluid helium-4, and the observation of new unique aspects of the hydrodynamics of superfluid helium-3.

DISCOVERY OF FIFTH SOUND

A new type of sound which propagates in superfluid helium was discovered in 1979. The wave mode, known as fifth sound, joins four previously known sound modes that can propagate in liquid helium. A study of its properties has yielded new insights into the hydrodynamics of quantum fluids.

Two-fluid model. The basis for understanding the acoustic properties of superfluid helium is the two-fluid model, in which the liquid helium is considered to be composed of two parts: a superfluid component with density ρ_s, and a normal fluid component with density ρ_n. These densities add to give the total density of the liquid, $\rho = \rho_s + \rho_n$. Both ρ_s and ρ_n vary with temperature as the liquid ^{4}He is cooled below the transition temperature $T_\lambda = 2.17$ K. The superfluid fraction ρ_s/ρ increases from zero at T_λ to nearly unity at temperatures below 1 K, and conversely the normal fluid fraction ρ_n/ρ varies from unity at T_λ to nearly zero below 1 K.

The "super" properties of the superfluid are that its viscosity and entropy are identically zero. The zero viscosity means that the superfluid component can flow through the tiniest of capillary tubes without friction, whereas the viscous normal fluid is completely immobilized in such a situation. The zero entropy of the superfluid gives rise to the unusual property that the superfluid flows in response to temperature differences in the liquid, as well as to pressure differences. The superfluid component is accelerated toward hotter regions of the liquid and away from colder areas. It is this property which leads to the existence of waves in which the temperature of the liquid oscillates.

Types of sound. Based on predictions of the two-fluid model, four types of sound were known to propagate in superfluid helium. The velocities of some of these modes as a function of temperature are shown in Fig. 1. First sound is similar to the usual sound waves found in ordinary fluids. It is a wave in which the pressure and density oscillate. The velocity of the first sound, C_1, is about 240 m/s and is determined by the compressibility of the liquid. Second sound is unique to the superfluid, being a wave in which the temperature of the liquid oscillates. The superfluid and normal fluid flow in opposite directions in the wave, and hence there are no density oscillations. The velocity C_2 is zero at T_λ and then increases to a value of about 20 m/s at 1.5 K. Third sound is a thickness wave in very thin films of liquid helium. The velocity is determined by the van der Waals force attracting the atoms of the helium film to the underlying substrate. Fourth sound is a pressure wave which propagates when the helium is contained in a porous material such as a tightly packed powder. In such a geometry, only the superfluid component can move, because of its zero viscosity, while the

viscous normal fluid is immobilized. (The packed powder is often referred to as a superleak since it offers no resistance to the superfluid.) The ability of the superfluid to move gives rise to the pressure oscillations of the fourth sound wave, and the velocity C_4 is related to the first sound velocity C_1 by the relation $C_4 = (\rho_s/\rho)^{1/2}\, C_1$. At low temperatures where $\rho_s/\rho \simeq 1$, the fourth sound velocity equals that of first sound. However, as the temperature is increased, ρ_s/ρ decreases and hence C_4 decreases. The velocity falls to zero at T_λ since $\rho_s/\rho = 0$ at that temperature.

Fifth sound. It was realized that in addition to the above modes there could exist a further independent sound mode, fifth sound. Fifth sound is a

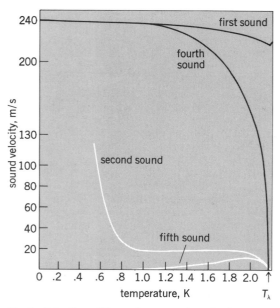

Fig. 1. Velocities of the various sound modes in superfluid helium.

temperature wave which propagates in the superfluid helium when it is contained in a superleak. The fifth sound is related to second sound (both involve temperature oscillations), but is different because the normal fluid component in the superleak is held fixed and cannot participate in the wave motion. This modifies the velocity predicted by the two-fluid model to the value in Eq. (1),

$$C_5 = (\rho_n/\rho)^{1/2}\, C_2 \qquad (1)$$

where C_2 is the second sound velocity (Fig. 1). C_5 is zero at T_λ since $C_2 = 0$ there, rises to a maximum value of 12 m/s at 1.9 K, and then decreases toward zero at $T = 0$ where ρ_n/ρ becomes zero. A condition for the fifth sound to be observable is that there be no pressure oscillations of the liquid in the superleak because then fourth sound would also propagate, and it would dominate because its velocity is so much larger than that of fifth sound. This pressure-release condition can be met by only partially filling the superleak with liquid, leaving the free surface within the powder. Instead of a pressure oscillation building up, the free surface can mound and relieve the pressure. This eliminates the fourth sound and allows only the temperature oscillations of the fifth sound to propagate.

Experiments. The experimental apparatus used in fifth sound measurements is shown in Fig. 2.

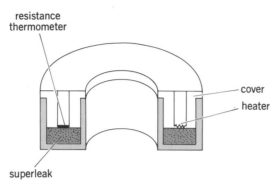

Fig. 2. Cross-sectional view of annular resonator used in fifth-sound measurements. (*From R. Rosenbaum et al., Surface tension sound in He II films adsorbed on alumina powder, J. Low Temp. Phys., 37: 663–678, 1979*)

This is an annular acoustic resonator packed with superleak powder particles (1-μm-diameter aluminum oxide powder). Pulses of current in a heater wire on one side of the annulus generate the fifth sound waves, and they are detected on the opposite side by a carbon resistance thermometer. The sound velocity can be determined by observing the resonant frequencies of the annulus.

In this apparatus the pressure-release condition is not exact because the curvature of the liquid helium coating the powder particles leads to surface-tension pressures in the wave. Fortunately these are small compared to the fifth sound part of the wave, and it is possible to subtract the surface-tension contribution. The corrected experimental values of C_5 for pure ^{4}He are shown along the lower curve of Fig. 3. The curve is the theoretical predic-

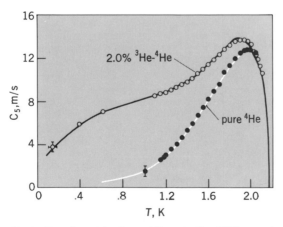

Fig. 3. Experimental values of the velocity of fifth sound in pure ^{4}He and in a 2% ^{3}He-^{4}He mixture. The curves are the theoretical prediction. (*From G. A. Williams, R. Rosenbaum, and I. Rudnick, Fifth sound in superfluid ^{4}He and ^{3}He-^{4}He mixtures, Phys. Rev. Lett., 42:1282–1285, 1979*)

tion of the above equation, and it can be seen that the agreement between theory and experiment is excellent. As a further check, the measurements were also carried out on a sample of helium containing 2% of the isotope ^{3}He. The ^{3}He increases both the normal fluid fraction ρ_n/ρ and the second sound velocity C_2, and by the equation above this increases C_5. The results are shown along the upper curve in Fig. 3, and again the agreement with theory is excellent. These results confirm the existence of the fifth sound mode and verify in detail the predictions of the two-fluid hydrodynamics.

The fifth sound was also independently observed in a different geometry consisting of a thin helium film on a flat substrate. This eliminates the surface-tension correction, but because the films are very thin (approximately 10 nm), the van der Waals force is large, and the wave is a combination of third sound and fifth sound. By making careful measurements of the sound velocity as a function of film thickness, researchers were able to subtract the third sound component and obtain the fifth sound velocity. The results again were in excellent agreement with theory.

Future research. The existence of the independent fifth sound mode is experimentally well established. The fifth sound can now be used as a tool for further investigations into the hydrodynamic and thermodynamic properties of superfluid helium. Currently, studies of the superfluid to nonsuperfluid phase transition in partially filled superleaks are in progress utilizing the new sound mode. The fifth sound should also be useful for studying the properties of persistent currents in the superfluid. In these experiments the flow velocities in the helium can be measured by observing the Doppler shift of the fifth sound, and this allows high-resolution studies of the dynamics of the currents.

It may also be possible to observe the fifth sound in the superfluid phases which occur in liquid ^{3}He at ultralow temperatures (below 0.003 K). Second sound has not been observed in the ^{3}He because of attenuation of the sound by the very high viscosity of the normal fluid. For fifth sound, however, this is not a problem because the viscous normal fluid is immobilized. The fifth sound may be the only way that a propagating temperature wave can be observed in the exotic phases of the superfluid ^{3}He.

[GARY A. WILLIAMS]

SUPERFLUID HELIUM-3

In 1971 D. Osheroff, R. Richardson, and D. Lee discovered two new phases of liquid helium-3 which appeared at temperatures below 2.6 mK. The A and B phases, as they were called, are now known to be two forms of a superfluid state closely related to the superconducting state of electrons in a metal. Like those electrons, the atoms of the isotope ^{3}He are fermions. In the superfluid state, the ^{3}He atoms bind weakly together to form large moleculelike Cooper pairs with relative angular momentum one (p-wave-paired) and total spin one. This section reviews some new aspects of the hydrodynamics of these unusual fluids and discusses several of the most important probes of their behavior.

Understanding superfluid ^{3}He is important for

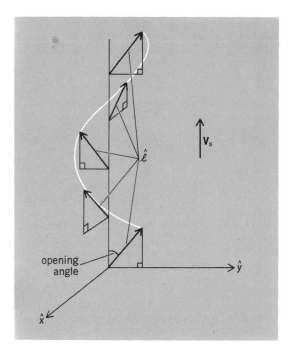

Fig. 4. A helical texture in the A phase of superfluid helium-3 in presence of a superflow V_s. One wavelength of the helix is illustrated. The orientation of l is the same for all values of the coordinates x and y.

many reasons. The neutrons in a neutron star, for example, are expected to form a p-wave-paired superfluid whose properties should be related to those of ^{3}He-A and B. In fact, the nucleons in conventional nuclei are expected to show significant p-wave pairing, and it would not be at all surprising if characteristic behavior in superfluid ^{3}He had analogies in nuclear physics. More abstractly, the topological shape of the A phase (or more precisely of its order parameter, the energy gap) is equivalent to that of the ordered state of a liquid crystal. The low-energy excitations of both systems should show important similarities. Practical applications are still years off, but the complex ordered state of superfluid helium has tremendous potential.

Superfluid hydrodynamics. The A phase has received special attention during the last few years because the angular momentum of each of the Cooper pairs in that case lines up along a single direction $\hat{l}$. Most measurable properties of the fluid depend on the orientation of this inherently quantum-mechanical variable. Its direction can vary over distances large compared with a Cooper pair radius (approximately 10 nm) under the competing influence of boundaries, flow, and magnetic field. The resulting configuration or texture of the $\hat{l}$ vector can be quite complicated and even time-dependent.

As in superfluid ^{4}He, the hydrodynamic behavior of superfluid ^{3}He can be described in terms of two interpenetrating fluids, one a superfluid component flowing without viscosity and one a normal component. The unique aspect of ^{3}He-A is that the dynamics of the two fluids are coupled to the orientation and motion of the $\hat{l}$ vector. The resulting behavior can be quite complex.

Bulk samples of ^{3}He-A, in the absence of flow

and magnetic field, take on a uniform texture with $\hat{l}$ pointing in the same direction everywhere. Theory predicts that the application of a small superflow (approximately 1 mm/s) will cause this uniform texture to undergo a second-order phase transition to a spatially periodic helical structure in which the $\hat{l}$ vector winds around the direction of superflow with some fixed opening angle and a well-defined spatial period (Fig. 4). The appearance of helical textures will cause nuclear magnetic resonance (NMR) frequencies to split, and two groups are preparing to search for these states.

Further stressing of the superfluid with increased flow or magnetic field is expected to cause the helical texture itself to become unstable. The ensuing behavior is thought to involve time-dependent, periodic motion of $\hat{l}$. This prediction is borne out by numerical simulations of the hydrodynamics by J. Hook and H. Hall. Experimental observations of flowing ^{3}He-A by E. Flint, R. Mueller, and D. Adams have shown the splitting of NMR resonances, as well as complicated time-dependent NMR response. Much work remains before even this simple configuration is understood.

Rotating samples of superfluid ^{4}He are known to contain a triangular lattice of quantized vortex lines running along the rotation axis. These vortex lines have singular cores in which the superfluid properties vanish. In ^{3}He-A, the extra freedom of motion of $\hat{l}$ allows nonsingular vortices to form, but there are still conflicting predictions as to whether rotating a sample will produce a lattice of small vortices or perhaps just a single large vortex. Two groups are presently constructing rotating cryostats to investigate these possibilities. Such experiments should also resolve whether persistent currents are possible in ^{3}He-A. It may be that the A phase can "unwind" circulation through distortions of the $\hat{l}$ vector.

Recent experiments by C. Gould show how soliton textures can be generated in the A phase that are analogous to magnetic domains in a crystal. The orientation of the Cooper pair spins in ^{3}He-A is described by a unit vector $\hat{d}$ along which the pairs always have zero spin projection. The experiments begin with a sample of the superfluid in which the $\hat{l}$ and $\hat{d}$ vectors are constant throughout, perpendicular to an applied magnetic field. A transverse rf magnetic field pulse flips the spin polarization within an NMR coil. This causes the $\hat{d}$ vector to begin rotating within the coil which generates the solitons shown in Fig. 5. These localized textures are detected by a characteristic NMR sig-

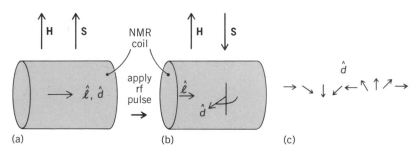

Fig. 5. Generation of solitons in the A phase of superfluid helium-3. (a) Initial state. (b) Generating solitons. (c) A soliton.

nal and are seen to move out of the region of creation at a well-defined velocity, later to be detected in a nearby coil.

M. Liu and M. Cross have studied another unique aspect of the hydrodynamics of ³He-A. They noted that a mechanical rotation of the fluid causes the phases of the superfluid to wind up, which induces a superflow along the axis of rotation. This "gauge wheel" effect could be detected by an oscillator composed of two disks attached to a torsion rod (Fig. 6). The usual torsional oscillation of the disks couples to a sloshing mode of the superfluid, and the resonant frequency will be shifted in a characteristic way.

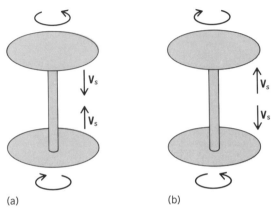

(a)   (b)

Fig. 6. The "gauge wheel" oscillator. V_s is the induced superflow. (a) Time $t = 0$; (b) $t = T/2$, where T is the period of the mode.

Probes of superfluid structure. The explanation of the observed shifts in NMR resonance frequencies in uniform samples of superfluid ³He was one of the first major successes of the Cooper pair model. That theory is now used to translate more complex NMR behavior into knowledge of nonuniform textures of the superfluid. In the same way, much early effort was spent understanding the drag force on ions moving through the superfluid phases. In particular, it was noted that the drag force depended on the angle between the ion's velocity and the $\hat{l}$ vector in the A phase. As a result, nonuniform textures focus sheets of moving ions into characteristic patterns that can be used to reveal the underlying structure. Studying the absorption of different sound modes continues to reveal new collective excitations of the fluid.

The anisotropy of the dielectric susceptibility ϵ in the A phase has been measured by D. Packard, who finds that it obeys Eq. (2), where $\mathbf{E}$ is the electric field and T_c is the transition temperature of

$$\epsilon(\hat{l} \perp \mathbf{E}) - \epsilon(\hat{l} \parallel \mathbf{E}) = 2.8 \pm 0.08 \times 10^{-10} (1 - T/T_c) \quad (2)$$

tric field and T_c is the transition temperature of 2.6 mK. This measurement is the first observation of any dielectric behavior in superfluid ³He.

Solid ³He. The solid state of ³He near the melting pressure has recently been found to be undergoing a first-order phase transition at 1 mK to an antiferromagnetic state in which alternating pairs of crystal planes have their spins aligned up, up and then down, down. This "uudd" phase was identified by its unique NMR signal.

For background information *see* HELIUM, LIQUID; SECOND SOUND; SUPERCONDUCTIVITY in the McGraw-Hill Encyclopedia of Science and Technology.   [DAVID BROMLEY]

Bibliography: A. L. Fetter, *Phys. Rev.*, B20:303, 1979; C. Gould, Solitons in ³He-A, *J. Low Temp. Phys.*, publication pending; G. J. Jelatis, J. A. Roth, and J. D. Maynard, *Phys. Rev. Lett.*, 42:1285–1288, 1979; A. J. Legget, *Rev. Med. Phys.*, 47:331, 1975; I. Rudnick et al., *Phys. Rev.*, B20:1934–1937, 1979; J. C. Wheatley, *Rev. Mod. Phys.*, 47:331, 1975; G. A. Williams, R. Rosenbaum, and I. Rudnick, *Phys. Rev. Lett.*, 42:1282–1285, 1979.

Lizard

The family Gekkonidae is distributed circumglobally between 50°N and 50°S, and its more than 650 species are currently arranged in four subfamilies. Three of these include members with digits remarkably modified for climbing. The digits have expanded pads bearing setae on the ventral surface. These minute, hairlike outgrowths (Figs. 1*k* and 4) are epidermal derivatives and thus part of the skin. They are composed of β-keratin (a tough protein substance also found in feathers) and produce the bond between organism and locomotor surface. Research has indicated that these climbing pads have been developed independently in the different groups of geckos, providing a good example of parallel evolution. Recent interest has centered upon complex internal modifications associated with the external structures and aspects of the adhesive process.

External form of digits. The pads are expansions beneath various parts of the digit, depending upon the genus (Fig. 1). The expansions are, in essence, modified scales but also possess some deeper components. The pads are composed of lamellae (Fig. 2*c*) numbering from a single terminal pair to from two to many divided or undivided overlapping plates (Fig. 1). The setae, which cover the pad surfaces, have spatulate terminal tips (Figs. 1*k* and 4) and may be branched. Typical dimensions of these setae are 0.5–5.0 μm in diameter at the widest point of the shaft, 0.2–0.75 μm in width across the spatula, and 12–100 μm in length, depending on the species and the degree of branching (Fig. 1*k*). In the tokay gecko, a well-known species from southeastern Asia, it has been estimated that there are about a million setae on all 20 digits combined. With the profuse branching of the setal tips (Fig. 1*k*) this gives rise to many millions of spatulae potentially able to make contact with the locomotor substratum.

The setae are elaborations of the fine spinules that are present over the entire body surface of geckos (Fig. 4*e*). The spinules are visible with the scanning electron microscope, and they are generally less than a micrometer in height. The setae, being larger, are fewer in number per unit area than the spinules. Almost all other lizards have a nonspinulate epidermis. An exception is the anoline lizards (family Iguanidae), which have both a spinulate epidermis and climbing pads (Fig. 4*f*) that are structurally and mechanically very similar to those of geckos. In this respect geckos and anoles represent a major example of convergent evolution.

Other epidermal modifications, in the form of

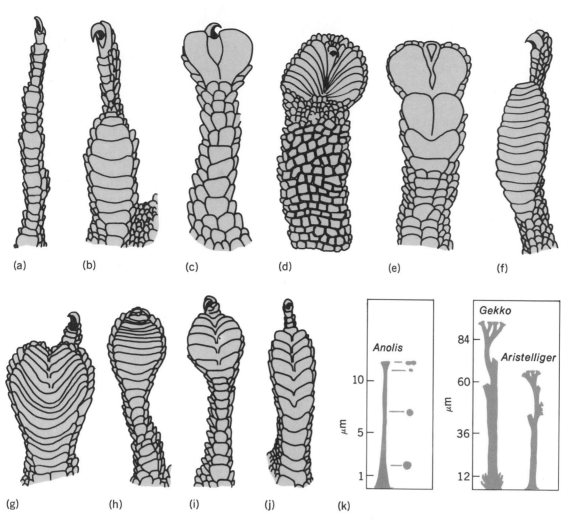

Fig. 1. Variation in the external form of the digits of geckos (subfamily Gekkoninae); digits are not drawn to scale, and all represent the fourth toe of the hindfoot (*a, b, d, e, f, h,* and *i* from the left side; *c, g,* and *j* from the right). (*a*) *Pristurus carteri;* (*b*) *Cyrtodactylus brevipalma-tus;* (*c*) *Phyllodactylus marmoratus;* (*d*) *Uroplatus ebenaui;* (*e*) *Afroedura transvaalica;* (*f*) *Aristelliger lar;* (*g*) *Perochirus articulatus;* (*h*) *Phelsuma barbouri;* (*i*) *Lygodactylus bonsi;* (*j*) *Hemidactylus garnotii.* (*k*) Relative size and complexity of setae from two geckos and an anole.

setalike mechanoreceptors (Fig. 4*d, e*), are also present on the scales subtending the lamellae, and the digits are unusually sensitive to pressure, vibration, and foreign material.

Internal modifications of limbs and feet. A number of modifications of the dermal, muscular, skeletal, and circulatory systems of the limbs and feet aid in the gecko's effective use of the setae. When functioning as adhesive mechanisms, the setae are under tensile stress and such stress must be transmitted to a resistant area. The dermis (deep layer of the skin) of the ventral surface of the lamellae is modified so that the epidermis sits directly on a tightly knit collagenous layer rather than on a loose network typical of the outer dermal portion of most reptilian scales. This collagenous layer links up with stout lateral tendons which follow the contours of the pad and link up with the skeleton at the joint capsule between the metacarpal/tarsal and the first phalanx of that digit. Thus, tensile stresses are passed from setae to collagen to tendon to skeleton via a joint capsule. The linkage system is governed by sets of muscles in the foot and lower limb and also

via the joint capsules. This control system permits varying load-bearing conditions to be met and ensures adequate contact with the locomotor surface. The flexor muscles, by way of this linkage system, pull the lamellae onto the surface when contact is made, and various check ligaments ensure that such pull is optimal. The lamellar system (Fig. 2*c*) permits individual control over relatively small fields of setae, and the inherent problems of application and removal of a continuous sheet of such structures are thus overcome.

The skeleton of the feet and digits is specialized in a number of ways. The arrangement of the carpal and tarsal elements has given rise to a secondarily symmetrical foot (Fig. 2*a*) with digits radiating around a semicircle. This contrasts with the markedly asymmetrical feet of most other lizards (Fig. 2*b*). The phalanges (finger bones) also have a much modified form, being broad and flattened dorsoventrally, rather than cylindrical. The joints between the phalanges are extremely mobile and play an important role in the application of the lamellae and their removal from the substratum. The digital muscles are also markedly developed in

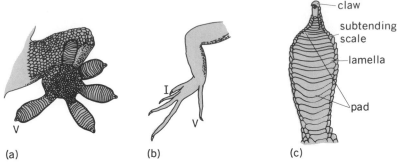

Fig. 2. Digit and foot modifications. (a) Ventral view of the left hindfoot of the tokay (Gekko gecko) showing the digital lamellae and the secondary symmetry of the foot. (b) Dorsal view of the left hindfoot of *Uta stansburiana*, an iguanid lizard, showing the more typical asymmetrical nature of the hindfoot in lizards (digits numbered I–V). (c) Ventral view of the fourth digit of the left hindfoot of the tokay (G. gecko); the pad consists of a series of lamellae.

many species, the extensor set traversing all of the phalanges as far as the claw instead of terminating at the base of the digits. Combined modifications of the phalanges and muscles permit the phenomenon of hyperextension of the digits.

Another significant feature of the digits is seen in the enormous proliferation of the blood system. Previous work often emphasized the importance of digital blood sinuses, but recent investigations indicate that these are only a small part of a much more extensive blood lacunar system which ramifies the core of each lamella. The blood system of the pads acts as a hydrostatic skeleton and enhances the ability of the lamellae to conform with the substratum and each other.

Behavioral modifications. Certain aspects of locomotor behavior have been mentioned above, especially the phenomenon of digital hyperextension. In lizards that climb without the aid of pilose pads, the claws and certain inflections of the digits are of great importance in maintenance of the grip.

First the claws contact the surface and create the grip, which is maintained until the moment of release. The digits are often considerably contorted as the force stroke of the limb cycle reaches its conclusion. When the claws are finally released, the limb is carried rapidly through its recovery cycle for placement of the claws again.

In geckos with pads, a dramatic difference is evident. The digits are carried through the recovery cycle in a hyperextended state (that is, rolled back on themselves) and are placed onto the substratum by unrolling from base to tip, with the claw being the last part of the digit to make contact. The force stroke of the limb proceeds without any notable distortion of the digits, which are finally released in a direct reversal of their placement pattern, being rolled up from tip to base. The claw is thus the first part of the digit to lose contact.

Such behavioral modifications, in association with the functioning of the setae, have led to some remarkable occurrences. Some genera of geckos have lost their claws totally, and thus rely for grip on the setae alone (Fig. 1h). These are the only instances in lizards of claws being forsaken. The efficiency of the system has also enabled locomotor gaits typical of terrestrial lizards to be employed on vertical surfaces (Fig. 3).

Theories of adhesion. One of the most active lines of research with respect to these modifications has been investigation into the mechanism of adhesion. A wide array of theories have been put forward intimating a number of physical phenomena. Early theories attempted to explain the process of adhesion as resulting from secretions, but there is no evidence of any glands on the subdigital surfaces. Later theories, advanced around the turn of the century, proposed that negative pressure was involved, either by air being driven out from between the lamellae or by the setae forming minute suction cups. The latter idea was briefly revived in the late 1960s following a study with the scanning electron microscope. Previous and subsequent

Fig. 3. Stages in the locomotion (from left to right) of *Gekko gecko* on a vertical glass surface. Only the feet in contact with the substratum are shown; those limbs in which the foot is not shown are undergoing their recovery stroke. This represents a complete locomotor cycle of all four limbs, which in life takes about 0.166 second. (*From A. P. Russell, A contribution to the functional analysis of the foot of the tokay, Gekko gecko (Reptilia: Gekkonidae), J. Zool. London, 176:437–476, 1976*)

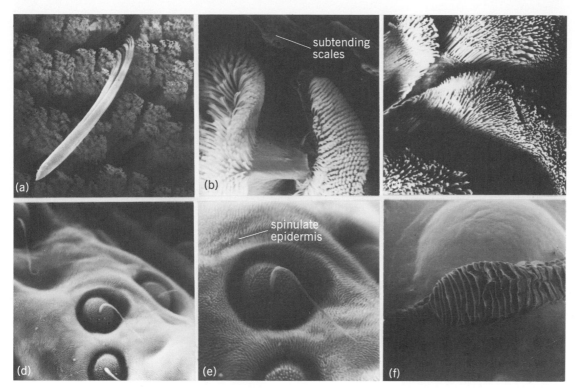

subtending
scales

spinulate
epidermis

(a) (b)
(d) (e) (f)

Fig. 4. Scanning electron micrographs of aspects of digital structure. (a) A detached seta of *Gekko gecko* above a field of setae still attached to lamellae *(from A. P. Russell, A contribution to the functional analysis of the foot of the tokay, Gekko gecko (Reptilia: Gekkonidae), J. Zool. London, 176:437–476. 1976).* (b) Two lamellae of *G. gecko* with associated setae. Subtending scales with mechanoreceptors can be seen at the top. (c) Fields of setae on the divided lamellae of *Hemidactylus brooki.* (d) Hairlike mechanoreceptors of G. *gecko* on a subtending scale. (e) A hairlike mechanoreceptor and surrounding spinulate epidermis typical of the general body surface. (f) Subdigital lamellae of the iguanid *Anolis carolinensis* showing the structural similarity to the foot pads of geckos.

work indicating that gecko feet can adhere quite effectively in a vacuum, however, lend little credence to these ideas. Other theories in the early part of this century proposed electrostatic and frictional forces.

Current understanding of the mechanism of adhesion is the result of a synthesis of several approaches, and it is becoming apparent that a combination of factors may be involved. From 1934 to the present, a growing body of information has been accumulating based on anatomical and experimental approaches. Recent work has concentrated upon the types of surfaces used in nature rather than on the artificial substrata employed in the laboratory. In the late 1960s considerable experimental work indicated that high free-surface energy (surface tension) is conducive to adhesion. Experiments leading to this conclusion were based on testing the grip of animals on polyethylene sheeting subjected to different levels of coronal discharge—the higher the charge, the better the adhesive ability. Correlating these data with field data, however, leads to some problems. It appears that in nature setae may largely be employed on leaf surfaces. The waxy cuticles covering these have very low free-surface energy, yet adhesion is effective. Such conflicts have led to reassessment and compromise, and it is now being suggested that both free-surface energy and the catching of setae on minute irregularities (frictional forces) may play a part at different times. Thus setae may

have a dual mode of functioning. This area is still one of open debate, and further research into the nature of surfaces and β-keratin, as well as further behavioral observations, is required.

Summary. The foot pads of gekkonid lizards (and the similar structures of anoline lizards) are aids to climbing on a variety of natural and artificial surfaces. The ultimate agents in the adhesive process are microscopic setae borne on the underside of subdigital lamellae. The setae are composed of β-keratin, and mechanical connection to the skeleton is accomplished via the rest of the epidermis, dermal collagen, tendons, and joint capsules. Many modifications of the skeleton, blood system, and musculature of the feet, and locomotor behavior enable the setae to be employed effectively. The actual mechanism of adhesion remains somewhat enigmatic, but recent evidence suggests that both free-surface energy and microscopic interdigitation may be involved at different times, or simultaneously at different contact points on a heterogeneous substratum.

For background information *see* GECKO; SQUAMATA in the McGraw-Hill Encyclopedia of Science and Technology.

[ANTHONY P. RUSSELL]

Bibliography: U. Hiller, *Morphol. Tiere*, 62: 307–362, 1968; P. F. A. Maderson, *Forma et Functio*, 3:179–204, 1970; A. P. Russell, *Copeia*, 1979: 1–21, 1979; A. P. Russell, *J. Zool. London*, 176: 437–476, 1975.

Lymphokines

Lymphokines are nonspecific mediators (that is, chemicals that are secreted by one group of cells in response to various stimuli and act on other cells) which have powerful effects on cells of the immune system. Two recent developments have injected new excitement into the study of these substances. First, immunologists have begun using them to generate defined populations of leukocytes in cell cultures. There are now standard protocols for cloning and propagating immunocompetent T lymphocytes by using one of the lymphokines. This opens the way to cellular and molecular studies and also provides a rational approach to specific antitumor immunotherapy. In one case, T lymphocytes cytotoxic to a syngeneic tumor in mice have been grown in large numbers in cell culture, using a lymphokine. These cells have then been used to destroy the same tumor in mice. The second development is the progress in producing and characterizing at least partially purified lymphokines, an essential step in understanding and utilizing these mediators. This work has been facilitated enormously by the availability of clonal cell lines which produce lymphokines.

Definition. The lymphoid system of higher animals exists in a state of balanced tension. It responds vigorously to recognizably foreign agents by a rapid expansion of specific clones of reactive cells. However, it must not react to self antigens. Moreover, any response must be of precisely gaged magnitude and duration, to minimize damage to host tissue and to avoid wasting immunological resources. Obviously, the immune system is also an information network for control signals. What are the mechanisms of intercellular communication?

An early and recurring observation has been that various types of leukocytes, and subsets of each type, influence one another's behavior by soluble effectors. Leukocyte products (other than the already well-defined and understood immunoglobulins) acting on other leukocytes to modify their activities are called lymphokines.

Lymphocyte products affecting macrophages. Some of the earliest and most thoroughly studied effects involve alterations in macrophage functions induced by products of activated lymphocytes. These effects include: inhibition of macrophage movement, or migration; activation of macrophages to increased phagocytosis, synthesis of characteristic enzymes, digestion of ingested bacteria, and increased tumoricidal activity; induction of proliferation; movement of macrophages toward the source of an effector (chemotaxis); agglutination, or clumping; and production of macrophage products which affect lymphocytes. The last effect illustrates the recurring chicken-and-egg theme in lymphocyte-macrophage interactions.

Each of these effects might have been predicted by a teleologist, because it makes biological sense. In immune responses, it is advantageous to recruit large numbers of mature macrophages to the site of invasion, and to activate them. The subsequent return of an activation signal to the neighboring lymphocytes also serves a useful purpose. Interesting points to be considered are: whether the observed effects are biologically significant, that is, do they make a contribution to immune regulation in the living animal; whether each observed effect is caused by a different molecular agent (lymphokine); and what are the molecular and cellular mechanisms that generate the lymphokine and explain its effects.

The migration-inhibiting effect has been ascribed to a migration-inhibiting factor (MIF), a protein of molecular weight in the range 30,000–60,000. This effect was first observed in leukocytes from individuals showing delayed-type hypersensitivity to sepcific antigens; these leukocytes were then exposed to the same antigens in cell culture. MIF is also generated upon exposure to nonspecific activators such as the mitogenic plant lectins, concanavalin A and phytohemagglutinin. These agents mimic many effects of specific antigens, but do so polyclonally; that is, they activate most, if not all, leukocytes in a population. By contrast, the cells responsive to a specific antigen may represent less than 1% of the total number.

MIF is also found in the circulation of animals given a strong challenge with an antigen to which they exhibit delayed-type hypersensitivity. This suggests that MIF indeed plays a role in certain immune responses. Moreover, alterations in macrophage distribution in the living organism are also consistent with the actions of MIF in laboratory cultures.

The question of the possible multiplicity of actions of MIF is not yet answerable. Neither MIF nor its mechanisms of production or action are clearly defined. There is evidence, however, that a macrophage-aggregating factor and one inducing mitogenesis are each different from MIF. One suspects, then, that several lymphokines in this category will eventually be purified and defined.

A notable example of a well-defined product acting on monocytes and macrophages to regulate their growth and differentiation is colony-stimulating factor (CSF). CSF has been purified to homogeneity and molecularly defined. Surprisingly, perhaps, it is produced by numerous types of cells, including leukocytes and fibroblasts. A number of molecular variants of CSF exist. An interesting mechanism of feedback control on CSF is its inhibition by prostaglandins secreted by mature macrophages.

CSF also induces macrophages to produce another factor, which functions principally as a lymphocytes activator. In the past this factor was generally known as lymphocyte-activating factor, and now, in an attempt to systematize and simplify the nomenclature, it is called interleukin 1 (IL 1).

Macrophage product affecting lymphocytes. The stimulation of lymphocytes to undergo proliferation, lymphokine secretion, or antigen-specific reactions is dependent on macrophages. Macrophages can sometimes be replaced by their soluble products such as IL 1. The IL 1 secreted by normal or cell line macrophages is of molecular weight near 16,000 and is a protein containing one polypeptide chain. In fulfilling the need for macrophages in mitogenic or antigenic T cell responses within the organism, IL 1 obligatorily induces the production of another lymphokine, interleukin 2 (IL 2). The requirements for a quantitative biochemical and cellular study of IL 1 are already met — linear dose-response assays can be performed, using clonal cell lines. It seems likely that IL 1 will soon be purified to homogeneity.

The secretion of IL 1 involves an activation cascade which begins with the stimulation by antigen of specific receptor-bearing lymphocytes. These lymphocytes produce MIF and macrophage-activating, -chemotactic, and -agglutinating factors which are largely responsible for the accumulation of activated macrophages at the site of an immune reaction. The differentiation of mature macrophages induced by CSF enhances local macrophage activity as well. The activated macrophages then generate IL 1.

Lymphocyte products activating lymphocytes. Different sets of lymphocytes affect each other's activities. Some of these effects can be obtained with factors produced by the relevant cells. (Some soluble factors carry receptors either for the foreign antigen to which a response is being generated or for self antigens. As such, they are not generally considered to be lymphokines, but possibly they carry a lymphokine as an effector component.) Two molecular products of lymphocytes which affect lymphocytes are immune-type interferon and IL 2. When T lymphocytes of the helper category are stimulated with the specific antigen, plus IL 1, they secrete IL 2. Murine IL 2 contains two polypeptides of molecular weight about 16,000 each, whereas the analogous human and rat factors are isolated as single polypeptides of this molecular weight. Mouse, rat, and human IL 2 have been partially purified and characterized. They are potent molecules, being active at less than $10^{-10}M$. Since cell lines producing IL 2 are now available, purification to homogeneity can be expected. IL 2 has two effects on T lymphocytes of the cytotoxic lineage: it provides a necessary signal for induction when antigen first appears, and it stimulates the continued expansion of the properly induced clones. As such, it has become a useful tool for the quantitative cellular and chemical analysis of functional T lymphocytes, since these can now be cloned and cultured in large numbers.

The early effect of IL 2 is complicated by the simultaneous generation of leukocyte interferon. The activation of cytotoxic T lymphocytes may involve the two mediators, IL 2 and interferon, acting sequentially. The continued growth of T lymphocyte lines apparently requires only IL 2; indeed, this is its definitive biological activity.

Other cellular targets. Some lymphokines affect nonlymphoid cells. An activity which may be the same as the one acting on macrophages affects the movement of eosinophils and neutrophils. A factor called lymphotoxin which is secreted by activated T lymphocytes kills various target cells, including nonlymphoid ones. It is, however, not the mechanism whereby cytotoxic T lymphocytes mediate target destruction.

Problems and possibilities. It is not known where nonspecific factors such as lymphokines fit into specific immune responses. One possible model is outlined in the illustration. The key feature is the associations among antigen-specific T lymphocytes, antigen, and macrophages. The specificity is imparted by receptors on the lympho-

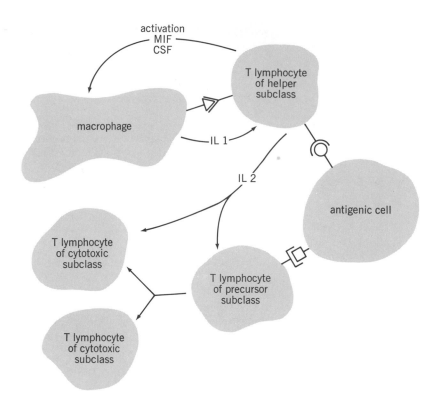

Model depicting regulation of a cytotoxic T cell response by lymphokines.

cytes; the intercellular communication includes nonspecific mediators. It is known that receptors for IL 2 are expressed only on lymphocytes which have been triggered by antigen. Thus, IL 2 stimulated only cells activated by the relevant antigen, and is removed from the lymphoid organ by them. In this way, the magnitude and duration of the response are limited by the generation of IL 2, which stops when the antigenic perturbation is successfully removed.

For background information *see* IMMUNOLOGY, CELLULAR in the McGraw-Hill Encyclopedia of Science and Technology.

[VERNER PAETKAU]

Bibliography: L. A. Aarden et al., *J. Immunol.*, 123:2928–2929, 1979; G. B. Mills, G. Carlson, and V. Paetkau, *J. Immunol.*, November 1980; S. B. Mizel, *J. Immunol.*, 122:2167–2172, 1979; G. Möller (ed.), *Immunological Reviews*, vol. 51., 1980.

Magnetic sense

The idea that animals use the Earth's magnetic field as a cue in their orientation during migrations has been suggested frequently. Yet this has always seemed rather doubtful because neither mammals nor birds have any organ known to be sensitive to magnetic fields. Very recently the magnetic mineral magnetite was found in substantial amounts in a variety of organisms, including bacteria, chitons, honeybees, and even pigeons. Chitons probably use it to scrape algae off rocks; salt-marsh bacteria use it to orient in the Earth's magnetic field; but its function in bees and pigeons is unknown. However, it seems possible that magnetite is involved in them too with the detection of the Earth's mag-

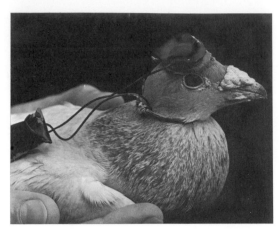

Fig. 1. Pigeon equipped with a pair of coils; one is glued to the top of its head, and the other fits around its neck.

netic field. If so, this would support evidence from recent studies which strongly indicates that homing pigeons use the Earth's magnetic field in returning to their lofts. How they know the direction toward home is not understood. This is the central problem of bird orientation which has been variously called the map, navigation, or position fixing. Once a bird has determined that home lies in a certain direction, it could use a compass system to fly in that direction. *See* BACTERIA.

Compass systems. There are, in fact, at least

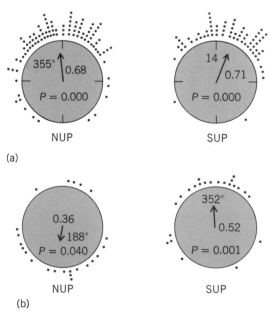

(a)

(b)

Fig. 2. Effect of the magnetic field from paired coils: (a) under sunny conditions (from C. Walcott, Magnetic fields and the orientation of homing pigeons under sun. J. Exp. Biol., 70:105–123. 1977); (b) under overcast skies. Each dot is the direction in which a pigeon vanished from the release point; home is at the top. The arrow in the center points in the average direction that the pigeons chose; its length is proportional to the degree of clumping of the vanishing bearings. P is the probability that the observed distribution of vanishing bearings arose by chance (from C. Walcott, The homing of pigeons. Amer. Sci., 62:543–552. 1974).

two compass systems known in pigeons. Under sunny conditions they use the Sun; but when the Sun is not visible, pigeons seem to use a magnetic compass. This conclusion comes from a series of experiments in which pigeons with small bar magnets glued to their backs were disoriented under overcast skies, but were only slightly disturbed when the Sun was visible. C. Walcott and colleagues used a pair of coils around the pigeon's head to apply a weak magnetic field of approximately the same strength as that of the Earth (Fig. 1). With current flowing in one direction through the coils, a magnetic field with its south pole up (SUP) was generated. Pigeons released under overcast skies with such a SUP field oriented toward home. Pigeons with the same coils and current flowing in the opposite direction (NUP) flew 180° away from home (Fig. 2). Under sunny conditions there is only a small difference in the angle of the departure bearings. The interpretation of these experiments is that the magnets did not interfere with the position fixing or with the Sun compass, but disturbed only a magnetic compass that the pigeons used under overcast conditions. This finding agrees with the conclusions of F. W. Merkel and W. R. Wiltschko, who showed that European robins also have a magnetic compass. In experiments in orientation cages, the robins use magnetic fields to orient in the seasonally appropriate direction. When night migratory birds see the sky, apparently they can also use the stars as a compass.

Sensory cues. However, a bird needs more than one or two compasses to reach home. It needs the information from a position-fixing or map system. The Earth's magnetic field is often mentioned as the possible sensory cue for this position fixing. Yet until recently the overwhelming experimental evidence that magnetism had little effect on the orientation of pigeons under sunny conditions and the lack of any known sense organ for magnetic fields seemed to rule this out.

It has long been known that for pigeons flown on days when there are large numbers of sunspots or when the variability of the Earth's magnetic field (known as the K index) is high, homing is very slow and many birds are lost. A correlation has been demonstrated between the average direction that pigeons fly at a release point and the cumulative K index over the preceding 12 homes. This is surprising because the variability of the Earth's magnetic field as measured by these K indices is so very small—at most 1000 γ (1 microtesla) out of a total Earth magnetic field strength of about 50,000γ (50 μT).

Anomalies of Earth's field. If pigeons' orientation was upset by these tiny changes in the Earth's magnetic field, what would happen to birds released in areas where natural deposits of magnetic minerals greatly disturb the Earth's field? Experiments have shown that the more irregular the magnetic field, the less accurate the orientation toward home. Again it was observed that a weak magnetic disturbance had an effect. The strongest anomaly reported was 3000 γ (about 6% of the normal Earth field). A pigeon using a magnetic compass should not be bothered by such a weak fluctuation. Even a simple magnetic hand compass at the strongest of the anomalies showed an error of at most 30°, not enough to mislead a pigeon significantly. An unusual aspect of the results is that the pigeons were disoriented on sunny days (Fig. 3). This seemed to imply either that the anomalous magnetic field was upsetting the Sun compass or that the pigeon's map or position fixing depends on the Earth's magnetic field. It is difficult to under-

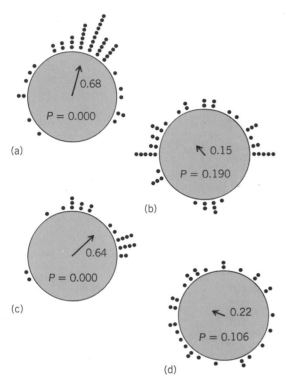

(a)

(b)

(c)

(d)

Fig. 3. Homing of pigeons released under sunny skies (a. c) at magnetically normal sites and (b. d) at magnetic anomalies. Birds at normal sites were well oriented; at magnetic anomalies the pigeons flew off in random directions. Symbols are the same as those in Fig. 2. *(Parts a and b from C. Walcott. Anomalies in the Earth's magnetic field increase the scatter of pigeon vanishing bearings. in K. Schmidt-Koenig and W. T. Keeton. eds.. Animal Migration. Navigation and Homing. Springer-Verlag. 1978)*

stand how a magnetic field could upset a Sun compass system, and since the experiments with magnets under sun seemed to effectively dispose of the magnetic map idea, a problem remained. Yet there is an easy way to resolve this problem: the assumption behind the magnet experiments was that the pigeons could surely not detect the tiny 100- or even 1000-γ fields of magnetic storms or the magnetic field gradient of the Earth in the presence of the relatively huge (100,000-γ) field of the magnet. The magnetic anomalies provided an opportunity to test this idea.

It had already been determined that pigeons were disoriented at magnetic anomalies. Since the amount of disorientation was proportional to the magnetic irregularity at the release site, it was the disturbance of the magnetic field that caused the disorientation. In addition, it had been demonstrated that bar magnets on pigeons at magnetically

normal sites had little effect on orientation. The question now was whether pigeons carrying bar magnets at the anomaly would be disoriented. If they were, it would mean that pigeons can detect the small irregular field of the anomaly, even in the presence of the much stronger field of the magnet. This was confirmed experimentally. A likely inference is that pigeons may therefore have two magnetic detection systems: a compass, used under overcast, that is disturbed by magnets; and a position-fixing magnetic system, used under both sun and overcast, which is immune to the effects of magnets but is disturbed by magnetic storms and by anomalies in the Earth's field.

It is possible to use the magnetic storm and anomaly data to estimate how sensitive pigeons are to the field. Analysis of the data provides some evidence for an effect with a variation in field strength of about 10 γ (10 nT). In studying the variation of the Earth's total field strength, a gradient of about 10 γ/mi (6 nT/km) is found by moving roughly magnetic north or south. If it is assumed that pigeons use this gradient as part of their map, pigeon navigation might be accurate to about a mile. This conclusion agrees quite well with the data.

Sense organ. An unsolved problem remains concerning the sense organ with which pigeons detect the Earth's field. However, the discovery of magnetite in animals as diverse as bacteria, honeybees, chitons, and pigeons seemed to indicate that this mineral may be involved in the detection of the Earth's magnetic field. Too little is known about how this mineral is arranged in the tissues of the animal to speculate about how a magnetic sense organ might work; nor is there any direct proof that the magnetite is even involved in the animal's sensitivity to magnetic fields. Yet the discovery that some animals contain magnetite provides a place to start investigations, and it may lead to a better understanding of how animals can detect the Earth's magnetic field.

For background information *see* GEOMAGNETIC STORM; GEOMAGNETIC VARIATIONS, TRANSIENT; MAGNETIC FIELD; MIGRATORY BEHAVIOR in the McGraw-Hill Encyclopedia of Science and Technology. [CHARLES WALCOTT]

Bibliography: K. Schmidt-Koenig, *Avian Orientation and Navigation*, 1979; K. Schmidt-Koenig and W. T. Keeton (eds.), *Animal Migration, Navigation and Homing*, 1978; C. Walcott, Magnetic orientation in homing pigeons, *IEEE Trans. Magnetics*, September 1980.

Manufacturing engineering

Manufacturing engineering is a function of industry that is now becoming an academic discipline. Originating in tool engineering, it has evolved into a wide-ranging, highly sophisticated profession that deals with all aspects of the production process.

Function. The function of manufacturing engineering is to develop and optimize the production process. This can best be understood by considering the total process through which a designer's concept becomes a marketable product: (1) The product designer (or product design department) conceptualizes a product. Drawings and one or

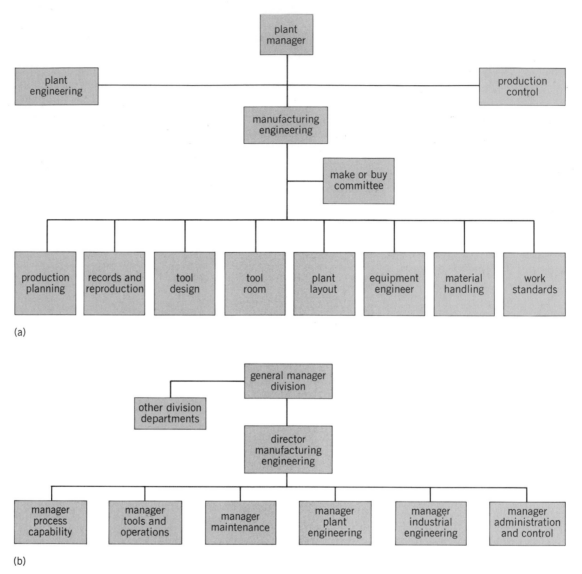

(a)

(b)

Manufacturing engineering organization varies with the needs of the company, as is seen in two markedly different organizational charts. (a) Ford Aeronutronics (from H. W. Wage, Manufacturing Engineering, McGraw-Hill, 1963). (b) Whirlpool Corporation (from R. E. Finley and H. R. Ziobro, eds., The Manufacturing Man and His Job, American Management Association, 1966).

more prototypes of this product are produced, usually by manual methods, but increasingly by computer-aided design and manufacturing. (2) The finalized prototype and its part drawings are released to the manufacturing engineering department, which starts designing and building an economically justifiable process by which the product will be produced. (3) When the manufacturing process developed by manufacturing engineering has been thoroughly tried and proved workable, it is turned over to the production department, which then assumes responsibility for product manufacture.

In brief, the manufacturing engineering department bridges the gap between product design and full production. It discharges responsibilities in a number of subordinate fields of engineering: process engineering, tool engineering, material handling, plant engineering, and standards and methods.

It must be noted that this list of component disciplines, while typical, is not universal. In virtually all manufacturing facilities the composition of subordinate engineering functions is adjusted to meet the needs of the product to be made. As an example, the manufacturing engineering structure in an aerospace plant would be very different from that of an automobile plant. A single company, such as Westinghouse or General Electric, will show marked differences in the composition of manufacturing engineering within its different branches (see illustration). In all cases, however, the objective of manufacturing engineering is the same: to design and develop the manufacturing process in the most economical manner possible.

Process engineering. The function of this component is to develop a logical sequence of manufacturing operations for each product part. A gear is a typical, extremely simple example. In one of the several methods by which a gear may be made, the gear blank must be sawed from round bar stock; its thickness, outside and inside diameters, and teeth must be machined; it must go through a heat-treatment process to obtain requisite hardness, after which all surfaces, including the teeth, must be ground. It is the function of process engineering to determine the correct sequence of operations, interposing the appropriate number and types of inspection operations between operations. Each operation must be assigned to an appropriate machine; preliminary estimates of the amount of time required for each operation must be made. All of this information is recorded on the process sheet, the most important document used in the entire manufacturing engineering process.

Tool engineering. The process sheets specify the need for tools by which the various operations are performed. In this context, the term tool refers to an array of industrial equipment such as dies, jigs, fixtures, gages, drills, reamers, counterbores, taps, milling cutters, and single-point turning and boring tools. Many of these tools—drills, for example—are standardized. As such, they can easily be procured through commercial outlets. Other tools, such as dies, jigs, and fixtures, are not standard; they must be made to suit the needs of the individual product component. Thus the first assignment of tool engineering is to design the nonstandard tools.

With tool design complete, tool engineering must determine which tools can be produced internally, that is, in the company's own toolroom, and which must be produced externally, in contract tool shops. The decision is usually based on economic considerations (the bid process in contract shops assures lower prices), on toolroom load (most plants do not have sufficient toolroom capacity to handle an entire tooling program), and on union contracts (many contracts place severe constraints on the release of work to outside vendors). Whether the tooling is produced internally or externally, tool engineering bears responsibility for its fabrication according to company standards and for its ability to perform its specified assignments in production.

Material handling. This third component of manufacturing engineering also starts work with the process sheet. The sheet specifies that certain operations will be performed at certain machines or at specified points in the assembly process. This implies the movement of parts in process over extended distances. Part movement from one plant to another is not uncommon, sometimes over widely separated geographical areas. Thus material handling is faced with the problem of moving parts in a time context. A failure in this regard means parts piling up in one area while production comes to a halt in another due to lack of materials. In brief, the flow of materials must be constant and of correct quantity. Material handling must establish the routing and the schedules.

Plant engineering. Closely allied to the material handling function is plant layout or plant engineering. This function involves the correct locating and positioning of machine tools and other pieces of production equipment. The importance of this function can best be appreciated by considering the vast changes made in the automobile plants of the United States during the downsizing opera-

tions of the late 1970s and early 1980s. The abrupt change in automobile size, plus significant changes in the materials used, necessitated drastic alteration in the physical layout of each plant. Plant engineering effected these changes.

Smaller changes in plant layout are part of an ongoing process. Many are made in response to changes in product design. But other changes in plant layout are made to optimize the production process. The objective is to reduce manufacturing costs to a minimum. This, above all else, is the corporate objective of manufacturing engineering.

Standards and methods. Every manufacturing concern of any size finds it necessary to establish a standards department. Closely allied to standards is methods engineering. Standards comprise work standards and tool and machine standards. A work standard is the scientifically established amount of time required to perform a given work assignment. It is one of the most socially sensitive aspects of modern engineering, and of critical importance. In mass production the entire system operates by the coordinated efforts of hundreds, perhaps thousands, of persons. Each worker must perform his or her task in an allotted period of time. Accordingly it is imperative that work standards (that is, time standards) be established.

Tool and machine standards pertain to the physical characteristics of plant hardware and material. These standards are especially prevalent in tooling. As an example, stamping dies are built in die sets, which must be adaptable to the company's presses. To assure their adaptability, the standards department develops appropriate dimensions, which the tool engineering department must adhere to in the design of tooling. Similarly experience may have proved that a certain type of steel exhibits superior performance in given tooling applications. The standards department then establishes that steel as a company standard for use in similar applications. This process of standardization reaches into every aspect of production, the objective being to simplify the work and thus reduce production costs.

Through methods engineering, the individual jobs and the entire workplace are made more efficient. This department studies the methods by which work is done. (Does a worker have to reach too far to pick up an incoming workpiece? Could a worker perform more efficiently if the workplace were better lighted? Does the worker have efficient access to tools?) Thus methods engineering and standards engineering work hand in hand to simplify work and reduce its time requirement.

Other functions. Manufacturing engineering is involved in numerous other aspects of the production process, often by committee approach. A key committee is the make-or-buy. Every component must be evaluated as to whether it should be made in-house or purchased from vendors. The economics of each decision is based largely on capabilities and costs. In its make-or-buy role, manufacturing engineering performs a function that cannot be handled via routine purchasing department practice.

Valuable feedback to product design is provided by manufacturing engineering's value analysis committee. It is not uncommon for product design to increase production costs unnecessarily through certain design requirements. After performing a value analysis of each component, manufacturing engineering can often obtain design concessions that lead to production-cost reductions while in no way affecting part performance.

Academic roots. Manufacturing engineering has evolved in response to a need—that of bridging the gap between product design and full production. It is now developing academic roots that extend deeply into the conventional engineering subjects of mathematics, chemistry, physics, and the like. These are followed with intensive study of the newer and more exotic technologies, namely, the laser, robotics, numerical control, and all aspects of electrical machining. Additionally, students are given intensive indoctrination in the use of the computer as the primary tool of manufacturing. These curricula, many of which are still in the formative state, are currently being given impetus through the educational activities of the Society of Manufacturing Engineers (SME) and through the financial support of its Manufacturing Engineering Education Foundation.

Mechanical engineering. This well-established engineering discipline has evolved in response to a more generalized industrialized need. Mechanical engineers are especially proficient in all aspects of machine design, of industrial research, and in the development of various forms of industrial equipment. Many mechanical engineers have gravitated into key posts in manufacturing engineering, where their academic training has enabled them to adapt to the demands of this related profession. Mechanical engineers are given the usual basic courses in mathematics and the sciences, followed by intensive indoctrination in thermodynamics, strength of materials, fluid dynamics, chemistry encompassing qualitative (and sometimes quantitative) analysis, plus advanced courses in electricity.

Industrial engineering. This engineering function was originally established by Frederick Winslow Taylor, a pioneer American engineer and inventor, and the conceptualizer of scientific management. As conceived by Taylor, industrial engineering was concerned with the development of work standards and methods. As with manufacturing engineering, the role of industrial engineering is largely determined by individual companies. In some concerns, manufacturing engineering and industrial engineering perform parallel functions; in others, industrial engineering retains its original function (standards and methods), but has been enlarged to include the plant engineering function. Increasingly, however, industrial engineering is—like tool engineering, process engineering, and the others—a subordinate branch of the more inclusive function and discipline of manufacturing engineering.

[DANIEL B. DALLAS]

Bibliography: D. B. Dallas, *Manufacturing Engineering Defined*, Spec. Rep. Soc. Manuf. Eng., 1978; R. E. Finley and H. R. Ziobro (eds.), *The Manufacturing Man and His Job*, American Management Association, 1966; *The Manufacturing Engineer—Past, Present and Future*, Battelle Memorial Institute, 1979; H. W. Wage, *Manufacturing Engineering*, 1963.

Marine containers

During the past 20 years, the ocean transportation of general cargo or freight has undergone a revolutionary technological change. The innovation was containerization—the development of standard marine cargo containers for consolidating packages into units of interchange between ships, docks, trucks, and railcars, and the development of special ships and handling systems to transport these containers at sea. This innovation brought economies of scale to marine cargo-handling operations, introduced capital-intensive processes to the labor-oriented stevedoring tasks, reduced cargo theft and damage, reduced the time a ship spent in port, and provided the means for efficient intermodal transport of cargoes.

Basic to the change to containerization was a new approach to loading and unloading cargoes. Instead of using nets or slings to lift individual bales, boxes, sacks, or pallets of cargo in and out of the ship's holds, the new systems employ standard cargo containers and special handling equipment to place the containers aboard. The containers are loaded or stuffed at an inland factory or terminal and usually are never opened until they reach their ultimate destination. The ship carrying these containers becomes an extension of a truckline or a railroad.

Intermodability, moving containerized cargo by more than one mode of transport without the need for intermediate reloading, is the essence of this technology. The containers can be moved to the ship via regular highway trailers or trailer chassis, or by flatbed railcar or rail piggyback. Once at the marine terminal, they can be lifted or rolled on the ships. Upon arrival at the discharge port, all land and water modes are again available to move them, with cargo undisturbed, to the consignee.

Development. Cargo containers are as old as commerce itself. Mixed shipments of small packages have been assembled in large boxes since the days of the Phoenicians. Various steamship companies have experimented with cargo boxes in various forms through the years. The familiar tractor-and-trailer rig of the trucking industry is a well-established form of containerization. Years ago railroads inaugurated a form of containerization in their piggyback or roll-on/roll-off service. Airlines too have adopted containerization in the form of specially designed cargo boxes that fit the shape of an airplane's fuselage. But containerization is now a term more widely used to refer to the intermodal movement of standard (demountable) containers by truck, rail, sea, and to a lesser extent air.

Marine containers, similar to those widely used in intermodal transport of cargoes today, were initially developed by Sea-Land Service, Inc., in 1956 for that company's east and Gulf coasts domestic trades. Two years later, Matson Navigation Company introduced containerization in its west coast to Hawaii trade. The impetus for these developments was the then steadily rising costs of conventional shipping which threatened the future of American-flag ship operators, particularly unsubsidized lines in the domestic trades.

These pioneering efforts were extraordinarily

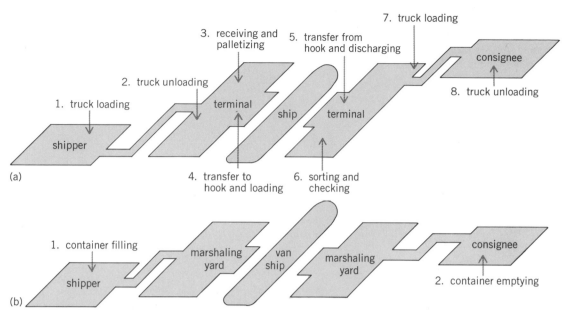

Fig. 1. Cargo-handling operations. (*a*) Conventional cargo-handling system. (*b*) Container system.

successful. The substantial savings in shipping costs and time which resulted from use of these container systems prompted the explosive worldwide growth of containerization during the 1960s and 1970s, as increasing numbers of domestic and international shipping companies introduced this new technology. By January 1980 there were an estimated 1560 container-carrying ships in service or on order worldwide, with a capacity of 1,240,000 twenty-foot equivalent units (TFEUs; equivalent to standard $20 \times 8 \times 8$ ft or $6.10 \times 2.44 \times 2.44$ m containers). It has also been estimated that in 1980 there were more than 3,000,000 containers in service worldwide, and that by 1985 this inventory of container equipment owned by shipping (and leasing) companies would increase to 4,500,000 units.

Benefits of containerization. An examination (Fig. 1) of the physical flow of marine cargo movement from shipper to consignee demonstrates the primary benefits of containerization. Figure 1*a* shows in a schematic way the various facilities through which general cargo traveled in a conventional shipping system. The arrows point out the various locations where separate handling of the cargo itself took place. All of these operations were predominately manual, with only a low degree of mechanization; for example, a forklift truck. In contrast, Fig. 1*b* illustrates a simplified container system. The important point is that the number of times the merchandise was physically handled (arrows) drops from eight to two. It is that drop which provides the primary benefits, in terms of reduced handling costs and reduced pilferage, of containerization.

The older, conventional (break-bulk) cargo-handling methods (Fig. 2) also created other problems. The slow stevedoring process resulted in inefficient utilization of ships which typically spent as much time in port as at sea. A longshore gang could usually handle no more than 15 tons (1 ton = 0.9 metric ton) per hour. With a complement of ten gangs, the cargo-loading rate could be 150 tons per

hour. In contrast, a modern container-handling system employing cranes similar to the one shown in Fig. 3 can load twenty 20-ton containers per

Fig. 2. Conventional break-bulk cargo-handling operation.

hour. Thus, with a normal operation employing two cranes, 800 tons per hour can be loaded. As a result, a container ship can achieve a 4:1 sea-to-port-time ratio instead of the 1:1 ratio of conventional break-bulk ships. The impact on ship economics is dramatic: if freight rates remain unchanged, potential vessel revenue for container ships can be four times that of a conventional ship with the same cargo deadweight (tonnage capacity).

Container equipment. The successful introduction of containerization has required substantial capital investment by shipping companies, terminal operators, and ports in new containers, container ships, container-handling equipment, and marine container terminals. Large inventories of containers specially designed to transport a variety of commodities and withstand the rigors of the intended service had to be developed. Ships had to be designed to stow these containers efficiently. Equipment had to be developed to lift the containers on and off trucks and railcars, as well as load and discharge the ships. Special port and terminal facilities had to be developed to accommodate the containers, ships, and handling equipment.

Container design. The pioneering companies (Sea-Land and Matson) developed container sizes suited to their particular trades. Lacking industry standards, each company initially selected different sizes after analyzing the expected cargo mix, ship stowage constraints, and highway weight limitations applicable to their operations. Sea-Land

developed a 35 × 8 × 8 ft (10.67 × 2.44 × 2.44 m) container, while Matson chose a 24 × 8 × 8.5 ft (7.32 × 2.44 × 2.59 m) unit for their systems. Different sizes were selected by other companies. As the introduction of containerization became more widespread in other trades, the need for interchangeability among carriers prompted the industry to adopt 20 × 8 × 8 ft (6.10 × 2.44 × 2.44 m) and 40 × 8 × 8 ft (12.19 × 2.44 × 2.44 m) sizes as an international standard. Today these are the most prevalent sizes.

Construction materials include steel, aluminum, and to a lesser extent fiber glass. Containers generally consist of a watertight or weatherproof steel, aluminum, or fiber-glass outer skin fastened to a strong steel inner frame, necessary for lifting and stacking them up to six high in ships' holds. Attached to each container's eight corners are special castings designed to allow lifting, lashing, and interconnecting the units from these points.

A typical 40-ft (12-m) aluminum dry-cargo container weighs 6000 lb (2720 kg) empty, and has a maximum load capacity of about 60,000 lb (27,200 kg) or a total of 33 short tons (29.9 metric tons). If this forty-footer must travel United States' highways at either end of its journey, its load must be limited to about 48,000 lb (21,800 kg). Its internal capacity (bale) is about 2300 ft³ (65 m³). These load and bale limitations are roughly similar to 40-ft highway semitrailers.

Of all the containers, those most often used (90% of them) are dry (nonrefrigerated) cargo vans with double doors at one end. Special types include refrigerated vans, insulated vans, open-top models for overhead loading of heavy gear, tank containers for liquid cargo, carriers for live animals, and frames for carrying automobiles.

Terminals and handling equipment. Containerization also required the development of radically new marine terminals and cargo-handling equipment. Old piers and wharfs, as well as traditional shipboard cargo-handling gear, were inadequate. Large pierside container yards were required to marshal the hundreds and frequently thousands of containers prior to loading them or after discharging them from a ship. Huge gantry cranes (Fig. 3) had to be constructed to lift the containers on or off the ships in an efficient manner.

Various marine container terminal designs have been developed over the years; however, there are basically two types: decked and wheeled systems. In a decked system all the containers are lifted from their highway chassis after receipt and set in rows or stacked one to four high on the yard pavement. When they are to be loaded aboard ship, they are transferred from the storage area to the gantry crane at the pier. In this case, the transfer is normally done with special mobile cranes such as the straddle carrier, also shown in Fig. 3. In a wheeled or chassis system all containers stored in the yard remain mounted on their highway chassis and are transferred to the ship loading crane by yard tractors.

A typical marine container terminal may consist of a pier with two berths served by two or three large gantry cranes and up to 50 acres (20 hectares) of paved yard area adjacent the pier. Such a

Fig. 3. Container crane and straddle carrier.

Fig. 4. Modern container ship.

terminal could load and discharge 150–200 container ships per year, and in the process handle 60,000–80,000 containers. As containerization has grown in international shipping, ports and terminals worldwide have faced increasing congestion problems. The limited availability of waterfront land suitable for terminals' expansion and decreasing productivity caused by this congestion are leading to the development of completely new types of container-handling systems designed to increase space utilization and productivity.

Container ships. When Sea-Land Service, Inc., and Matson Navigation Company began their container services, the containers were carried on the deck of conventional cargo ships. Sea-Land began such a service in 1956, carrying 58 containers on deck of the *Ideal-X*. Matson's first shipment was made in 1958, when a deck load of 22 containers was carried to Honolulu from San Francisco on the *Hawaiian Merchant*. These companies soon converted conventional cargo vessels to full container ships. The first full container ship, Sea-Land's *Gateway City*, was outfitted to carry 226 35-ft (10.67-m) containers in 1957. Matson's first full container ship, the *Hawaiian Citizen*, could carry up to 488 24-ft (7.32-m) containers, and made its maiden voyage as a container ship in 1960. Both of these vessels were relatively small and slow (16–18 knots or 8–9 m/s) by today's standards.

During the 20 years since these pioneering steps were taken, ship design technology has advanced to meet the demand for larger, faster, and more efficient container ships. The first vessels designed from their inception as container ships appeared in the mid 1960s. By the late 1970s, ships with capacities of 1200–1600 TFEUs and design speeds of 20–25 knots (10–13 m/s) were common. Sea-Land's *Sealand Patriot* (Fig. 4) is representative of modern container-ship designs.

For background information *see* SHIP, MERCHANT in the McGraw-Hill Encyclopedia of Science and Technology.   [JOHN C. COUCH]

Bibliography: J. R. Barker and R. Brandwein, *The United States Merchant Marine in National Perspective*, 1970; R. F. Gibney (ed.), *Containerisation International Yearbook*, 1979; L. A. Harlander, *Engineering Development of a Container System for the West Coast Hawaii Trade*, Transactions of the Society of Naval Architects and Marine Engineers, 68:1052–1088, 1960.

Marine ecosystem

Ecologists have long recognized the importance of detritus in the nutrition and production of marine food chains, but only recently have investigated the dynamics of detritus-based food chains. Detritus designates all forms of nonliving organic carbon derived from dead plants and animals or from egestion, excretion, and secretions of organisms. Danish workers believed detritus, especially that derived from the decay of sea grasses, was important to coastal bottom organisms. These organisms in turn are the food for bottom-feeding fish.

These ideas were not followed through, in part because of the weakness of the concept of trophic levels when dealing with detritus feeders other than as decomposers. R. Lindeman's concept of trophic (feeding) levels emphasized a food chain of primary production by plants that are grazed upon by herbivores, which are subsequently subject to predation by carnivores. Decomposers or saprophages (for example, bacteria and fungi) then mineralize the organic matter back to inorganic nutrients. There was no exposition of production of macroconsumers feeding on detritus. Early microbiologists suggested that bacteria might be a food for marine animals, but this was not widely accepted.

Another reason for the rudimentary knowledge of detritus-based systems was the emphasis of a simplistic view of "grazing" food chains in the oceans. With the development of techniques for measuring primary production, such as the carbon-14 tracer method, there was a great burst of studies on marine pelagic food chains of phytoplankton-zooplankton-fish. This growing food chain concept was popularized in standard ecology tests, and the importance of detritus-based production was overlooked.

Early view of role of microbes.
Historically, when work on detritus feeders began in the mid-1960s, aging of detritus was linked to a buildup of microbial biomass on detrital particles; this supposedly was the actual food for deposit feeders. Deposit feeders reingested fecal pellets (either their own or those of other animals) on which microbes had grown, a process called coprophagy. The microbes were stripped from the fecal pellets in the gut of the deposit feeder, but the fecal pellet was egested essentially unchanged. These pellets would then be recolonized by microbes, and the process repeated. Because fecal pellets (as well as many other sources of detritus) are usually low in nitrogen, aging is viewed essentially as protein enrichment by microbial biomass. Thus aging improved the nutritional quality by lowering the carbon-nitrogen ratio.

Nutritional value of detritus.
This emphasis on protein enrichment ignores the possible nutritive value of the detritus per se, and seems an oversimplification of the role of aging and microbial decomposition of detritus. Recent studies are delineating differences in biochemical composition due to detrital source and state of decomposition that regulate availability and nutritional value to macroconsumers.

The major inputs of detritus to coastal marine systems are from terrestrial runoff, fecal pellets, and decaying macrophytes (both vascular plants, such as sea and marsh grasses, and seaweeds). These detrital types differ in nutritional composition of protein and available caloric content. Detritus derived from vascular plants is usually low in organic nitrogen and available calories. In contrast, detritus derived from seaweeds is usually higher in organic nitrogen and available calories. The availability of the calories must be stressed because detritus is composed in part of highly complexed structural materials that most detritus feeders cannot assimilate. For example, many detritivores lack digestive enzymes that hydrolize structural carbohydrates and intractable lignins. An analogy would be the difference between a person eating sugar or sawdust—both high in calories! Thus the fraction of calories available to the macroconsumer assumes importance in controlling the availability and nutritional value of detritus.

Aging of detritus.
There are three processes affecting detrital utilization: mechanical breakdown of particles, leaching and sorption, and microbial decomposition. The breakdown of large recognizable particles of plant material of fecal pellets into amorphous material is a function of mechanical activity of physical processes such as wave action, sediment abrasion, and bioturbation (sediment reworking) by animals. Sediments are transported vertically from depth to surface by the feeding activities of sediment-ingesting deposit feeders. Burrowing activity also reworks sediment, and pumping for respiration increases water circulation. These activities facilitate the exchange of sediment pore water with the overlying water and the flushing of metabolites out of the sediment. This enhances microbial activity on the surfaces of the detritus particles. Particle size is an important factor affecting microbial activity and availability of detritus to ingestion by detritivores. During the last decade much research has centered on the mechanics and behavior of particle size selection by deposit feeders.

The decomposition and changes in nutritional value of aging detritus depends largely on the chemical composition of the source material. Leachable and hydrolyzable components, intractable structural materials such as lignins, nitrogen content, and phenolic residues affect decomposition rate. Initially new detritus rapidly loses weight. Readily soluble organic and inorganic components are leached quickly. Autolysis further increases the soluble and hydrolyzable substances. Thus as much as 60% of the initial weight of seaweeds and 40% of sea grasses may be lost in the early days of decomposition of detritus.

Detritus types differ with respect to rates of microbial decomposition. Detritus derived from phytoplankton and seaweeds decays quickly. In contrast, detritus derived from vascular plants decomposes slowly. This slow rate of decay is caused by the difficulty of decomposing the complex structural carbohydrates and lignins characteristic of detritus derived from vascular plants.

In addition, detritus derived from marine plants contains phenolic compounds and others that can be antibiotic and thus slow decomposition. These substances may also make otherwise nutritionally valuable detritus unpalatable to macroconsumers.

Because various components of a detritus pool age and are utilized by macroconsumers at different rates of time, food availability is spread over a longer period of time in a detritus-based system.

Trophic activity in detritus-based systems.
Biotic interactions can affect the efficiency of macroconsumer production in detrital systems. The transfer of energy from primary producer (plant) to herbivore in grazing food chains involves metabolic partitioning of food assimilated by the herbivore into that used for metabolic activity versus that going into production of new biomass. When dealing with detritus-based food chains, besides the energy partitioning of the detritivore, there is also energy expended and biomass produced by microbes, ciliates, and meiofauna (small benthic organisms) that exploit varying proportions of the detritus resource. This exploitation can

either be competitive (that is, the detritus is mineralized without the macroconsumer being able to use it) or beneficial (the macroconsumer eats the smaller organisms). The smaller organisms (ciliates and meiofauna) associated with detritus seem, based on recent experiments, to increase the availability of detritus to macroconsumers by enhancing microbial production and by serving as a food source themselves for macroconsumers.

For background information *see* FOOD CHAIN; MARINE ECOSYSTEM in the McGraw-Hill Encyclopedia of Science and Technology.

[KENNETH R. TENORE]

Bibliography: T. M. Fenchel and B. Barket Jorgensen, Detritus food chains of aquatic ecosystems: The role of bacteria, in M. Alexander (ed.), *Advances in Microbial Ecology*, vol. 1, 1977; K. R. Tenore and D. L. Rice, A review of trophic factors affecting secondary production of deposit feeders, in K. R. Tenore and B. Coull (eds.), *Marine Benthic Dynamics*, 1980.

Marine navigation

The first marine computer-based radar anticollision systems were developed around 1970 to warn watch officers of merchant vessels of potential danger of collision. Capabilities of these systems include: automatic processing of time coordinates of radar echo signals into space coordinates in digital form; automatic determining of consecutive coordinates and motion parameters of targets by means of a tracking-while-scanning method; predicting closest point of approach (CPA) and time to the closest point of approach (TCPA), based on present positions and motion parameters of tracked targets; presenting calculated motion parameters (course and speed), CPA, and TCPA in graphic or alphanumeric form on the radar display screen or on a special display; and automatic switching on of audible and visual alarms if the CPA or TCPA is less than values that have been preset by the operator.

Since these functions amount to automatization of radar plotting, the anticollision systems are called automatic radar plotting aids (ARPAs). Two methods of displaying information have been developed: by means of motion vectors only and by means of motion vectors with predicted areas of danger. The second method will be discussed in this article.

Objectives of radar plotting. Marine radar is used for two main purposes: coastal navigation and recognition of collision hazards. Radar plotting indicates whether a danger of collision exists, and determines CPA and TCPA values. It also provides approximate determination of the course and speed values of other vessels from previous observations.

All this information is required for planning and taking any necessary remedial action. To enable the navigator to plan and take avoiding action, the radar observer provides this information in a report that follows a standard format. The report consists of two parts: (1) Last bearing, passing ahead or astern respectively; last range, decreasing or increasing; nearest approach (CPA) as forecast; time interval to the nearest approach from the last observation (TCPA). (2) True course and aspect of target; speed of target.

The report can be calculated by plotting on a special sheet of paper or directly onto the reflection plotter, a special device that covers the screen of the plan position indicator (PPI) display. Two types of plot are used: the relative motion plot and the true motion plot.

Relative motion plot. The relative motion plot represents the motion of the echo of the target relative to a fixed reference point that corresponds to the position of the navigator's ship. Consecutive positions of the echo, observed at constant time intervals on a north-up relative-motion radar display, are plotted. Based on these consecutive positions of the echo, it is possible to determine the CPA, and by considering the time interval between consecutive positions to be a unit, it is also possible to calculate the TCPA. Knowing the values of two sides of a triangle, representing the distance covered by the target and the distance covered by the navigator's ship in the time interval between consecutive observations, makes it possible to determine the third side of the triangle, which represents the true motion vector of target.

True motion plot. The true motion plot is the true motion of the navigator's ship plotted against the true motion of the target. Consecutive positions of the target and of the navigator's ship are plotted on the reflection plotter and represent motion vectors. These two vectors determine the direction and speed of the relative motion of the target, which in turn determine the CPA and TCPA.

Computer-based systems. The laboriousness of manual plotting and the time necessary to do it motivated the design of different plotting devices, from quite simple ones to very sophisticated computer-based systems. Computer-based ARPAs that have been developed and improved during the last 10 years show the predicted relative or true motion by means of vectors directly on the radar screen. The vector's length is proportional to a time controlled by the operator (Fig. 1).

The computer-based Sperry collision avoidance systems (CAS) display the true vectors drawn from present positions of target echoes to the point of possible collision (PPC) with an elliptical or hexagonal potential area of danger (PAD; Fig. 2). Readout of data in digital form, automatic danger alarms, and trial maneuver facilities are also available in most computer-based systems. In computer-based ARPA systems all processes, from detection of targets to measurement of their coordinates, target tracking, and calculation of anticipated CPA and TCPA, are automatically realized, partially with electronic logic circuits and mainly by computer. Second-generation ARPAs are based on hardware associated with microprocessor-based computing systems. Echo selection to track in order of degree of risk is initiated automatically or has to be initiated separately for each echo by the operator. The number of targets that can be simultaneously tracked varies from one system to another. The results of data computation, displayed as relative or true motion vectors or by means of true motion lines with PADs, contain the

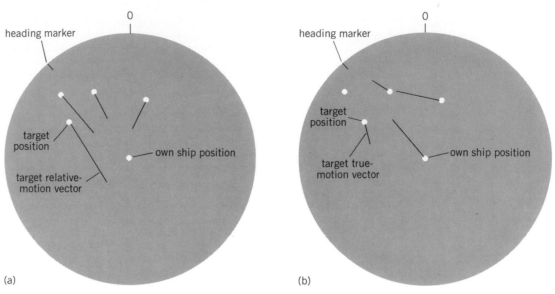

Fig. 1.   Example of vectorial display format: (a) relative vectors; (b) true vectors.

essential data of the radar observer's report.

Sperry CAS and Sperry CAS II are unique systems in which the PPCs and PADs show directly the dangerous sectors of the ship's projected course, making it possible to gain a clearer picture of the situation. In the opinion of users of the system, a watch officer trained in one method of ARPA information display may have difficulties or make mistakes using ARPAs with other methods of information display. Aiming at systems unification, the Intergovernmental Maritime Consultative Organization (IMCO) has issued draft performance standards for ARPAs. A final determination of the most serviceable method of display has not been made.

Results of operational tests. The Sperry CAS and Sperry CAS II compare favorably with the

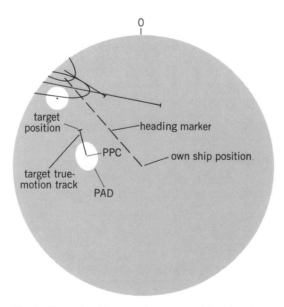

Fig. 2.   Example of true motion tracks with point of possible collision (PPCs) and potential areas of danger (PADs).

other systems. The Sperry system does not use relative vectors of targets, which are needed only to estimate the risk of collision. These relative vectors have been replaced by true vectors whose terminals indicate PPCs. PADs, which give readily interpreted information on the dangerous sectors of the ship's course, are an additional aid. The ease with which a new course can be estimated, without the need of simulating maneuvers, is an additional advantage.

In the period from 1975 to 1979 the simplest of the models offered by Sperry CAS was operated on board the merchant ship Kapitan Ledóchowski. The graphic presentation of a situation by means of true vectors that are terminated at the PPCs with elliptical PADs is very easy to interpret and clearly shows target motion in most of the conditions that are met in practice. However, there are special cases of great traffic density or concentrations of fishing vessels that move different directions and at different—chiefly low—speeds. The difficulties in interpretation which have arisen in these cases can be overcome only with experience and a good knowledge of the system.

Observations of the system's operation in different conditions have shown that certain features that were initially estimated as faults arose from physical limitations. These limitations must be well known to users and must be taken into consideration for proper appreciation of the situation. There were cases in which the heading line intersected the elliptical PAD even though the distance to the target at the CPA was sufficient to avoid any danger of collision, and then the instant digital readout of CPA and TCP values was needed.

Automatic acquisition was often lacking in the defined area. This situation arose most often in areas of high traffic density where the echo of a relatively speedy target would be masked by graphic symbols. Delayed visual detection of such an echo resulted in delayed target acquisition by the operator.

In areas with relatively low traffic density, where the probability of meeting other vessels was minimal, a guard zone with controlled location and dimensions for automatic acquisition was very useful.

In areas with great traffic density (many tests were made passing through the Straits of Dover), simultaneous manual plotting of a great number of targets would be impossible. Under these conditions the information delivered by CAS would be attainable in no other manner. In such areas CAS-facilitated ship steering was used, even with excellent visibility.

In situations of great traffic density, limiting the number of simultaneously tracked targets to 20 proved inconvenient. Difficulty arose when it became necessary to cancel less dangerous targets in order to acquire targets that seemed more dangerous. The time required for cancellation and acquisition delayed the availability of information.

Targets appearing at similar bearings required increased time intervals for calculations of their motion parameters. In these cases the delay in the appearance of a vector accompanied by a PAD often reached several minutes. The reasons for this phenomenon were understood, but the delays interfered with estimation of the situation.

Masking of echoes by contiguous PADs could interfere with visualization of targets on the radar display screen. In these cases a device for temporarily switching off PADs on request, while retaining vectors, was needed.

The manner of displaying information in the Sperry CAS has proved very convenient for choosing collision avoidance maneuvers in normal conditions of navigation. Modifications that have been introduced in the second-generation Sperry CAS II eliminate the inconveniences discussed above.

The trained and experienced navigator should understand the basic principles of the ARPA operation, both its capabilities and its limitations, and have a good knowledge of how the system responds in various situations, including any errors that might occur.

For background information *see* MARINE NAVIGATION; NAVIGATION SYSTEMS, ELECTRONIC; RADAR in the McGraw-Hill Encyclopedia of Science and Technology.   [JERZY FEDOROWSKI]

Bibliography: W. Burger, *Radar Observer's Handbook for Merchant Navy Officers*, 1975; R. F. Riggs and J. P. O. Sullivan, *J. Navig.*, 33(2):259–283, 1980; F. J. Wylie, *J. Navig.*, 27(3):298–304, 1974.

Marine sediments

Siliceous and windblown marine sediments can provide fundamental information on the history of the Earth with respect to climatic change, plate movements, organic productivity, and circulation of ocean bottom waters through time.

SILICEOUS SEDIMENTS

Deep-sea siliceous sediments consist primarily of the skeletal remains of siliceous algae (diatoms) and protozoans (radiolarians). These skeletal tests are constructed of hydrated amorphous silica, opal. The primary source of this silica in the oceans is dissolved silica in continental runoff. Recent studies of solutions emerging from hydrothermal vents at mid-ocean ridges suggest that these solutions may also contribute a significant amount of dissolved silica to sea water. The organisms that secrete siliceous tests are capable of extracting virtually all of the silica available to them in the surface water, so they are abundant only in areas where silica is plentiful in surface water—such as nearshore areas and zones of upwelling or divergence where nutrients regenerated beneath the thermocline are supplied to the surface. The siliceous skeletal material is most concentrated in sediments at abyssal depths beneath biologically productive surface waters such as those around Antarctica and in the North Pacific. Although nearshore environments and areas of nearshore upwelling such as the Peru and Benguela currents probably support large populations of silica-secreting organisms, siliceous remains are not quantitatively important in the sediments beneath these areas as a result of dilution by terrigenous detritus. Other important areas of siliceous sediment accumulation are the equatorial Pacific, South Atlantic, and the equatorial Indian oceans, although the siliceous sediment in some of these areas is diluted by large concentrations of calcareous microfossil skeletal material.

Siliceous productivity. The present-day accumulation of siliceous sediments represents the balance between supply of opaline skeletal material from surface waters and dissolution of this skeletal material by ocean bottom waters which are highly undersaturated with respect to opal. The effect of this dissolution is seen clearly in the enrichment of bottom waters in silica as they circulate over highly siliceous sediments. The balance between supply and dissolution is responsible for the gross distribution of siliceous sediments on the sea floor. The Atlantic Ocean contains fewer zones where siliceous plankton dominate surface biological productivity, and has silica-poor bottom water. As a result the Atlantic contains less siliceous sediment than the Pacific Ocean, which contains important zones of siliceous plankton productivity and more silica-rich bottom water. The Indian Ocean is intermediate in siliceous productivity and bottom water silica content and hence is intermediate in siliceous sediment content.

Diagenesis. Once incorporated into the sediment, the amorphous biogenic silica in microfossil tests may undergo diagenetic reactions transforming it to a disordered crystalline silica phase, opal-CT. During these reactions the sediments generally become lithified, forming rocks called porcellanites. The silica may be transformed further to cryptocrystalline quartz, forming rocks called cherts. The fate of siliceous skeletal material once it has been buried has been shown by M. Kastner to be highly dependent on the nature of the nonsiliceous sediment. The diagenetic reactions transforming siliceous oozes to porcellanites and cherts are favored in calcium carbonate–rich sediment. Thus siliceous oozes are more likely to become lithified if they are deposited in sediments which also contain a high proportion of calcareous microfossil material.

Variation through time. Changes in climate and in continental positions with time are known to have caused substantial changes in the circulation of the oceans. Because the location and importance of siliceous sediments are highly dependent on surface water processes, the distribution and accumulation rate of biologically precipitated silica can be expected to change in response to climate change and in response to the circulation changes caused by rearrangement of the continents. In addition, for the ocean to be at steady state, the amount of silica buried in the sediments must balance the amount contributed by continental weathering and that emerging in hydrothermal solutions at sea floor spreading centers. Because the residence time of silica in sea water is only 18,000 years (based on the rate of supply from rivers and the rate of incorporation of silica into sediment), any alteration in the rate of supply of silica to the oceans or rearrangement of zones of deposition will be reflected in siliceous sedimentation quickly, in a geological sense.

Short-term fluctuations. Extensive studies of deep-sea sedimentation during the last glacial maximum (18,000 years ago) and the last full interglacial (120,000 years ago) by the CLIMAP project have established a scenario for changes in the global distribution and accumulation of opal effected by climate change. The CLIMAP studies suggest that in response to an increase in the Equator-to-pole gradient of ocean surface water temperature during glacial stages, the belts of siliceous biological productivity associated with oceanic fronts and upwelling zones shifted toward the Equator. The width of equatorial upwelling zones became narrower, and the intensity of surface water mixing upon which the biological productivity depends also appears to have increased. The rate of accumulation of opal in almost all locations studied by the CLIMAP project increased during glacial periods, suggesting that the supply of silica to the oceans may have been greater during glacial periods. The geologically short fluctuations in sedimentation characteristic of the Pleistocene Epoch (the last 1.8 million years) appear to extend as far back as the late Miocene Epoch (7–10 million years before present [MY BP]), although they are reduced in magnitude.

Long-term trends. Superimposed on the short-time-scale fluctuations are long-term trends in siliceous sediment deposition. The sediment in cores recovered by the Deep Sea Drilling Project provides a reasonable representation of deep ocean sedimentation for the last 50 MY of the Cenozoic Period (0–65 MY BP). Since much of the older sea floor has been destroyed by processes which thrust it under the continents and ocean at deep-sea trenches, the earlier sedimentary record is not represented by many cores and is very sketchy. The data for deposits older than 50 MY that are available show that few unlithified siliceous oozes are preserved in sediments deposited before 50 MY BP. Porcellanites and cherts, the diagenetic products of siliceous sediments are common, however. Unfortunately, because no satisfactory analogy between rates of silica accumulation in siliceous ooze and those represented by the lithified porcellanites and cherts has been drawn, it is difficult to draw comparisons between the rate of silica supply to those ancient siliceous deposits and to those of the present.

Cenozoic. The Cenozoic and late Mesozoic (65–90 MY BP) sediment record does show several dramatic changes in the sites of siliceous sedimentation through time. During the early Cenozoic more of the deep-sea sediments of both the Atlantic and Pacific oceans were dominated by siliceous sediment than today. Both oceans showed wide belts of sediment rich in radiolarian tests along the Equator. Early Cenozoic sediments deposited in low-latitude nearshore basin environments which are now uplifted and incorporated into continental rocks in the western United States and in Europe also contain thick sequences of radiolarian and diatom cherts and porcellanites deposited during the middle to late Eocene (40–50 MY BP). Few cores are available which contain material deposited in temperate and polar regions of the abyssal ocean during the early Cenozoic, but those that are available show little evidence of siliceous sedimentation like that present today at high latitudes. During the early Cenozoic the accumulation of siliceous sediment in the equatorial Atlantic and Pacific oceans decreased, and the belts themselves became progressively narrower.

During the last half of the Cenozoic the rate of biogenic silica accumulation in the narrow equatorial belts increased to a maximum about 5–10 MY BP and has since decreased. During the same time, high-latitude siliceous sediment accumulation became important, first around the margins of the North Pacific and then in the Antarctic. At present these areas are quantitatively the most important sinks of siliceous sediment in the world ocean: almost 70% of the siliceous biological material being deposited today is deposited around Antarctica. The change from equatorial to polar silica deposition cannot be accounted for by a simple exchange in the locus of deep-sea siliceous sediment deposition. Available data suggest that the total silica accumulation at abyssal depths in the equatorial Atlantic and Pacific during the early Cenozoic is not as large as the present rate of accumulation in the abyssal North Pacific and Antarctic. This difference implies either that other important early Cenozoic sinks of siliceous sediment occur which are unknown or which have not been included in deep-sea estimates (for example, the continental shelf and nearshore basins), or that a change in the supply of silica to the ocean has occurred between the early Cenozoic and the late Cenozoic. Similar decreases in the supply of calcium carbonate to the deep ocean relative to continental shelves, resulting in a decrease in the preservation of calcareous sediment in the deep ocean, have been documented during the late Eocene and the middle Miocene.

Theories of location changes. Such shifting back and forth between shallow and deep depositional sites could account for accumulation rates changes in lieu of changes in the rate of supply of silica to sea water. They cannot account for the change from low- to high-latitude dominance in siliceous sedimentation. Several theories have been suggested relating this change to oceanic circulation, including the gradual decay of low-latitude circumglobal circulation through the Mesozoic and early Cenozoic caused by the northward movement of Australia, South America, and Africa, and the gradual evolution of high-latitude circumglobal

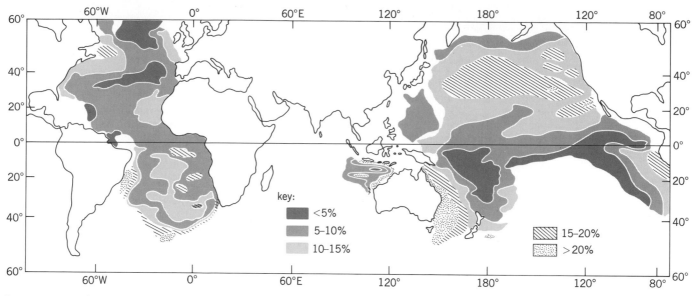

Quartz content of deep-sea sediment. Contours show weight percent quartz in nonbiogenic sediment (to eliminate effects of dilution).

circulation around Antarctica caused by the same continental movements. Other theories suggest that the change in location of siliceous deposition occurred in response to the ecological evolution of high-latitude siliceous plankton. Still others argue, in analogy to Pleistocene siliceous sediment variation, that the long-term constriction of the equatorial belts of siliceous productivity point to climatic controls on silica distribution. An evaluation of the changes in location and rate of accumulation of siliceous sediments must await more quantitative information on siliceous sedimentation in continental shelf and nearshore basin environments as well as more detailed information from the abyssal sediment record.

WINDBLOWN SEDIMENTS

One of the most important transport paths through which nonbiogenic material enters the ocean is the wind. Both terrigenous dust and sporadic volcanic contributions are carried by the wind. The dust cloud over the central Atlantic emanating from the Sahara Desert has been known to scientists for over a century, and C. Darwin correctly speculated that the dust which falls out of this cloud could be detected in the sediments of the Atlantic Ocean. Windblown or eolian dust can be carried over the ocean for thousands of kilometers before it falls out or is scrubbed out of the atmosphere by rainfall. For example, Saharan dust carried west by the trade winds has been detected in the atmosphere as far west of the African coast as Barbados.

Composition. Analyses of dust collected from the atmosphere indicate that it is largely composed of varying proportions of clay minerals and quartz with smaller contributions of other common rock-forming minerals. The quartz content of worldwide marine sediments is shown in the illustration. Two minor components of eolian dust, but ones which clearly pinpoint its continental origin, are the microscopic skeletons of fresh-water siliceous algae

(diatoms) and phytoliths, microscopic siliceous elements from the leaves of grass, sedge, and rushes. Volcanic ash is a significant portion of the atmospheric aerosol near sites of explosive volcanic activity, but the deep-sea sediment record also demonstrates that large eruptions can result in ash layers that are several centimeters thick at distances of even a thousand kilometers from the site of the eruption.

Particle size. Windblown dust is very fine-grained. For example, near Asia the particles of dust in the atmosphere average about 8 μm in diameter. The grain size decreases with increasing distance from the source. Eolian particles carried by the westerlies into the central North Pacific from Asia average less than 3 μm in diameter. The fine-grained terrigenous material carried by winds to the ocean accumulates very slowly (about 1 m/MY). It is therefore easily masked by other detrital sediment except where the sea floor is protected from the more rapidly accumulating terrigenous sediments transported by continental runoff or by turbidity currents. The eolian dust can be diluted also by rapidly accumulating biogenic debris. Where the deep-ocean floor is protected from these other sources of sediment, however, such as in the North Pacific Ocean, almost all of the fine-grained crystalline component of the sediment is eolian.

Sources. Although most continental regions contribute some crystalline material to the atmospheric dust and hence to oceanic sediment, the largest sources of dust are arid regions where surface material is not held in place by vegetation and where particles are not quickly removed from the atmosphere by rainfall. Therefore, eolian sediments are also regionally important adjacent to arid regions such as the Sahara and the deserts of Australia, southwestern North America, and Arabia, where the quartz component of the windblown dust alone can account for up to 20% of the sediment by weight.

Because of its resistance to chemical and mechanical weathering, quartz is enriched in soils and is therefore an important component of terrigenous eolian dust. Quartz is stable in sea water and is resistant to both chemical and mechanical destruction on the sea floor. The concentration of quartz in deep-sea sediments can also be determined quantitatively by x-ray diffraction analysis. For these reasons, most inferences about the pattern and intensity of atmospheric circulation have drawn heavily on quartz as an indicator of wind activity. Studies of the topmost sediments on the ocean floor in the Pacific and the Indian oceans show that the pattern of quartz distribution in these sediments is very closely related to atmospheric circulation. Although the pattern of atmospheric circulation recorded on a day-to-day basis on weather maps is very complicated, studies of eolian sediments show that they record only the more general, and more stable, average zonal wind distributions. For example, in the North Pacific Ocean the concentration of quartz in surface sediments is greatest in a latitudinal band between 25° and 40° N and in a lobe extending southwest from Baja California. The band between 25° and 40° N reflects deposition of eolian material carried by high-altitude (jet stream) westerlies from source regions in Asia. The lobe extending southwest from Baja California reflects deposition of material carried by the northeast trade winds from the Sonora Desert and other arid regions in southwestern North America. Sharp gradients in the quartz concentration of Pacific and Indian ocean sediments occur at the margins of the wind belts. The pattern of quartz distribution in the Atlantic Ocean does show the influence of the northeast trade winds, but because of the abundance of turbidity-current-transported terrigenous material and the active reworking of sediment by bottom currents in the Atlantic, the pattern of quartz distribution does not reflect the other zonal wind belts as clearly as it does in the Indian and Pacific Oceans.

Deposition. The sharply defined patterns of eolian quartz distribution which coincide with wind belts imply that eolian sediment must settle very quickly through ocean water to the bottom. If eolian sediment settles as single particles 2–8 μm in diameter, it would take up to 2 weeks to settle from the surface to the bottom of the ocean. During this time it would be moved large distances by currents and would show a more homogeneous distribution sediment. It is therefore assumed that most eolian dust is incorporated in and sinks with the fecal pellets of microscopic marine organisms.

Paleoclimatic inferences. The clarity with which the eolian sediment, particularly quartz, reveals the present atmospheric circulation pattern indicates that eolian sediment can be used to evaluate older atmospheric circulation patterns and to infer paleoclimatic information like the aridity of the source region and strength of the prevailing winds. Fluctuations in the position of wind systems between the present and the past are clearly revealed by comparing modern sediments with their ancient counterparts. Eolian quartz deposited in the North Pacific during glacial maxima shows a southward shift of the westerlies toward the Equator. Similar studies of quartz in glacial sediment from the South Pacific and Atlantic also show shifts of the major zonal wind belts toward the Equator. Studies of the size distribution of atmospherically transported particles in cores off the west coast of Africa have demonstrated that the vigor of atmospheric circulation was also greater during the last glacial maximum than it is today. Finally, the quantity of eolian material reaching deep-sea sites during glacial times was much greater than that being deposited today. The increase in eolian deposition during glacial times has been attributed to the combined effects of stronger winds, the removal of vegetative cover by glacial erosion, and a drier climate that reduced vegetation and allowed particles to remain in the atmosphere longer before being removed by rainfall.

Very few studies have examined eolian sedimentation in the ocean before 1 or 2 MY BP. Studies of sediments dating back 70 MY in the central North Pacific indicate that the amount of eolian sediment carried to the North Pacific increased by two orders of magnitude after the onset of Northern Hemisphere glaciation 3.5 MY BP. Although almost all of the detrital sediment deposited in the North Pacific before that time was eolian, as it is today, the total quantity was vastly smaller. In addition, the North Pacific sediments indicate that eolian deposition dropped to exceedingly low levels 40–50 MY BP. This time interval is one during which semitropical vegetation extended as far north as southern Alaska and as far south as New Zealand. This time interval was probably one of the warmest and most humid during the last 70 MY, and although there are not enough available data to map the pattern of zonal wind circulation at the time, it is clear that the amount and size of atmospherically transported particles was minimal.

For background information *see* MARINE SEDIMENTS; PALEOCLIMATOLOGY; SEDIMENTATION (GEOLOGY) in the McGraw-Hill Encyclopedia of Science and Technology.

[MARGARET LEINEN]

Bibliography: N. A. Brewster, *Geol. Soc. Amer. Bull.*, vol. 91, pt. 1, no. 6, pp. 337–347, 1980; R. M. Cline and J. D. Hays (eds.), *Investigation of Late Quaternary Paleoceanography and Paleoclimatology*, Geol. Soc. Amer. Mem. 145, 1976; G. R. Heath, Dissolved silica and deep-sea sediments, in W. W. Hay (ed.), *Studies in Paleo-Oceanography*, Soc. Econ. Paleontol. Mineral. Spec. Publ. no. 20, pp. 77–93, 1974; M. Kastner, J. B. Keene, and J. M. Gieskes, *Geochim. Cosmochim. Acta*, 41(8): 1041–1060, 1977; V. Kolla, P. E. Biscaye, and A. F. Hanley, *Quaternary Res.* 11:261–277, 1979; M. Leinen, *Geol. Soc. Amer. Bull.* vol. 90, pt. 2, p. 1301, 1979; M. Leinen and G. R. Heath, Sedimentary indicators of atmospheric activity in the Northern Hemisphere during the Cenozoic, *Paleoecology, Paleogeography, Paleoclimatology*, 1981; D. W. Parkin, Trade winds during the glacial cycles, *Proceedings of the Royal Society of London*, ser. A, vol. 337, pp. 73–100, 1974; J. Thiede, *Geology*, 7:259–262, 1979; H. L. Windom, *J. Sediment. Petrol.*, 45:520–529, 1975.

Meadowfoam

The common name meadowfoam, used for all members of the plant genus *Limnanthes* (family Limnanthaceae), describes the showy floral display of these plants, which are found in large stands along vernal streams, pools, and meadows in California and southern Oregon. All *Limnanthes* spp. are herbaceous winter-spring annuals with dissected leaves, a 5-merous (except 4-merous in *L. vinculans*) perianth, 6–10 free stamens, and a superior gynoecium of 3–5 distinct ovaries, each maturing with a one-seeded nutlet. *Limnanthes douglasii* is widely known as an ornamental in Europe and North America. Nearly 18 different taxa are known, but several biosystematic studies are likely to provide new information.

Agricultural potential. An extensive survey of plants as sources of industrially useful oils suggested that *Limnanthes* might possibly be a good new crop plant. Its seed oil is unique, with at least 95% of the fatty acids being long chains (C_{20} and C_{22}) and the C_{22} component having unique positions of unsaturation. The only comparable and currently much-researched source of long-chain compounds is the liquid wax from jojoba in which, instead of triglycerides of *Limnanthes*, fatty acids occur with fatty alcohols. Organic chemists suggested that meadowfoam oil could be readily converted into a product substantially identical to liquid jojoba wax and thus could become useful for producing high-quality lubricants, detergents, and plasticizers; in addition, the double bond might allow synthesis of many new chemical compounds. Interest in jojoba and meadowfoam mainly derives from their ability to provide substitutes for the sperm whale oil, a matter of special interest to the conservationists. Meadowfoam also offers the advantages of easy cultivation as an annual crop.

Domestication. In order to develop meadowfoam into a highly productive crop species, selection of cultivars is aimed at upright plant growth with synchronous and high seed set per flower, seed retention at maturity, and overall high seed and oil yield per hectare (current yields range up to 1600 kg, with oil percentage in the range of 22–26%). Different taxa and their diverse populations represent a very wide range of variability for agronomically desirable traits; however, on the basis of seed retention and yield, adaptability to cultivation, and growth characteristics, primary attention is being given to the breeding work in *Limnanthes alba*, *L. douglasii*, and their hybrids. A large collection of population samples and their synthesized gene pools represent useful genetic resources for breeding work, and programs for their conservation have been initiated. Researchers have explored optimum planting and harvesting procedures; some of the critical problems arise in relation to early establishment, good weed control, and mechanical harvesting. Marketing and industrial processing of oil have not yet begun. In addition to oil as an economic product, the seed meal has 21–34% protein with lysine and methionine in amounts comparable to the legumes, but its use in animal feed will require research to remove certain toxic thioglucoside-derived isothiocynates.

Biosystematics and evolution. The *Limnanthes* plant also has excellent features as a research organism. Different populations occur in islandlike vernal pools or meadows, so that island models of evolution and biogeography are readily testable; all taxa are diploid ($2n = 10$), easy to grow in large numbers experimentally with short generation time, and have flowers easy to manipulate for controlled selfing and cross pollination. Assays of allozyme variation and variation in breeding system and in life history patterns at hand provide useful tools for research in biosystematics and ecological aspects of evolution. Significantly enough, evolutionists have discovered several new taxa within the last decade. A study of nutlet morphology using a scanning electron microscope provided another useful scheme for grouping species and their habitats in relation to the role of nutlet characteristics in germination.

Conservation and future prospects. Since several taxa of *Limnanthes* are considered endangered or highly threatened, conservation programs have stimulated population biological research on vernal pool flora. Genetic resources in *Limnanthes* might also provide a paradigm for testing strategies of gene conservation. While there are worldwide efforts under way for exploring domestication, albeit on a small scale, crop development will most likely occur with some dramatic change in the industrial supply-and-demand or a sociopolitical crisis, as one often finds in the history of domesticates that became important within the last 100 years (such as rapeseed, soybean, rubber, and guayule).   [SUBODH K. JAIN]

Bibliography: C. R. Brown, H. Hauptli, and S. K. Jain, *Econ. Bot.*, 33:267–274, 1979; H. S. Gentry and R. W. Miller, *Econ. Bot.*, 19:25–32, 1965; J. J. Higgins et al., *Econ. Bot.*, 25:44–54, 1971; S. K. Jain, in D. Wood (ed.), *Modern Concepts of Breeding*, American Society of Agronomy, 1980; S. K. Jain, R. O. Pierce, and H. Hauptli, *Calif. Agr.*, 31:18–20, 1977; C. T. Mason, *Univ. Calif. Publ. Bot.*, 25:455–512, 1952.

Memory

When a crime or accident occurs, the police typically come to the scene and try to obtain eyewitness accounts. If a trial occurs as a result of the incident, these eyewitnesses may be asked to testify in court. The accuracy of their memories is critical. Jurors in the trial will listen to their testimony, and as they do, they will construct in their minds an image of the incident that was not witnessed by any one of them. When all of the evidence has been presented, they will begin their deliberations and eventually—in most cases—will reach a verdict. The accuracy of jurors' memories thus is also critical.

Memory of witnesses. In the event of a crime or an accident, precise memory suddenly becomes crucial. Small details assume enormous importance. Did the assailant have a mustache, or was he clean-shaven? Was the light red, or was it green? A case often rests on such fine details, and these details can be hard to obtain. To be mistaken about details is not the result of a bad memory, but of the normal functioning of human memory.

Memory failures occur for a number of different reasons. Perhaps the information was not perceived in the first place—failure at the acquisition stage. Or the information may have been accurately perceived, but then lost or interfered with—failure at the retention stage, the time between the acquisition and the recollection of the information. Finally, information may have been accurately perceived and retained, but cannot be recalled during questioning—failure at the retrieval stage.

Acquisition stage. People are characteristically upset when they see a crime or an accident, and this is the time when the information that makes up memory is acquired. Many people think that their memory becomes more precise at these times. "I was so frightened that I'll never forget that face" is a common remark of victims of serious crimes. In fact, people who witness arousing events remember the details of them less accurately than they recall ordinary happenings, since stress or fear disrupts perception, and therefore memory.

Stress can also affect a person's ability to recall something observed or learned during a period of relative tranquility. For example, in one experiment servicemen flying in an airplane were told that the plane was going to crash, and were instructed to prepare for an emergency landing. Then they were put under even more stress by being asked to make two lists: in the first, they disposed of their personal belongings; in the second, they described the routine emergency procedures they were to follow. The servicemen made many more errors than a second group of unstressed servicemen who were asked for the same information.

Retention stage. After information is perceived, it does not just lie passively, waiting to be recalled. Many things happen to the witness during this phase: time passes, the memory fades, and more crucially, the witness may be exposed to new information that adds to or alters the memory. The influence of new information has been shown clearly in a study in which college students watched films of automobile accidents and immediately afterward answered a series of questions, some of which were deliberately misleading. For example, some students were asked how fast the white sports car was going when it passed the barn while traveling along the country road. No barn had been shown in the film. Others were asked simply how fast the white sports car was going while traveling along the country road. Later all the students were asked if they had seen a barn in the film. More than 17% of the students who had been asked the misleading question reported having seen the nonexistent barn; less than 3% of the other group said that they had seen it. The misleading question contained information—in this case, false information—that supplemented and became integrated into some students' remembrance of the event.

In another experiment a nonexistent object was introduced into memory without even mentioning the object. Students viewed films of accidents and then answered questions about what they saw. The question "About how fast were the cars going when they smashed into each other?" was put to some of the students and elicited a higher estimate of speed than the same question put to others with the verbs *collided, bumped, contacted,* or *hit* in place of *smashed. Smashed* had other implications as well. A week later, the same students were asked whether they had seen any broken glass. Those who had been asked the question with the verb *smash* were more likely to answer in the affirmative than the others.

Why do people come to recollect things that never existed? A person sees a car accident and forms some representation in memory. Later, new information supplements that recollection. For example, when an investigator asks, "About how fast were the cars going when they smashed into each other?", the witness's recollection and the extra information in the investigator's question are integrated, and the witness recalls the accident as being more severe than it was. Because there is usually broken glass when two cars smash into each other, the witness is likely to think he or she saw broken glass.

Retrieval stage. At an accident or crime scene, after the witness tells the police all he or she can, the witness may be asked to come to the police station to look through a set of photographs or to observe a lineup. A different aspect of human memory is involved here—in essence, the witness is now performing a recognition test. In a recognition test either a single item (in this case a photograph) or a set of items is shown, and the witness indicates whether he or she has seen any of them before.

Obviously the composition of the set of photos or people in the lineup is crucial: the number of people in it, what they look like, what they are wearing. A lineup must be as free as possible from suggestive influences or it loses its value. If the suspect is a large, bearded man, the lineup should contain other large, bearded men. Unless people resembling the suspect are included in the lineup, the suspect may be picked by default—not recognition.

Memory of jurors. Jurors must ultimately consider the testimony, and from it arrive at a verdict. Great faith is placed in jurors' ability to remember numerous facts and details as they listen to evidence, as they deliberate, and as they move toward a decision. But since the memories of eyewitnesses can so easily be altered or amended, might not the same problem hold for the memories of jurors? Research on jurors indicates that occasionally they do indeed misremember facts that were presented to them. In one study, subject-jurors read a description of a hypothetical case about a man accused of burglary and then deliberated until they reached a verdict. Many distortions of fact were made by jurors in the course of the deliberations. Slightly over 40% of the distortions of fact were corrected by other group members, leaving 59% uncorrected. This study suggests that errors of memory are made by jurors, and that a substantial portion of errors may go uncorrected.

These errors occur for several reasons. The juror is faced with the unenviable task of first taking in large amounts of sometimes quite complex information. The information must be retained in memory until a verdict is reached. Furthermore, the testimony presented during a trial can itself lead to distortions in jurors' memories. Lawyers have long been aware of the fact that jurors' impressions of an event can be manipulated through the skillful use of language.

Emotionally charged words probably affect the construction of the incident in the minds of jurors and thereby affect their verdict. In an experiment subjects were recruited to act as jurors who would read information about a case and then reach a verdict based on the evidence presented. The subjects received either an unbiased, neutral version of the case or a version containing violent words intended to evoke an emotional response. For example, jurors read that a witness was asked "How much of the incident did you see?" versus "How much of the struggle did you see?" The results showed that those subject-jurors who read the violent version were almost twice as likely to return a guilty verdict as those reading the neutral version. Furthermore, when those subjects who read the violent account were later asked what they remembered about the case, they related a much more violent account than did the neutral group—even though the only difference between the two groups involved the use of various types of words. This outcome supports the fact that the language used in the courtroom can affect the construction of an incident in the mind of a juror and thereby affect the verdict in a trial. Results such as these raise important issues about jury functioning, such as to what extent a verdict is based on fact and to what extent on distortions of memory

about these facts. Researchers are far from a complete understanding of the processes by which jurors arrive at verdicts. The research thus far can only fuel further enquiry into this essential element of the judicial system.

For background information *see* MEMORY in the McGraw-Hill Encyclopedia of Science and Technology. [ELIZABETH F. LOFTUS; GLEN R. SCOTT]

Merchant ship

The spread of industrialization and an increase in the international transfer of new technologies have developed the need for the movement of large, high-value cargoes. Freight costs for a single unit of cargo may be more than $1,000,000, yet this will probably be only a small amount of the total cost of a project. Shipments have included power-generating equipment for both nuclear and steam installations, factory components, heavy machinery for construction, offshore oil equipment, locomotives, and barges. To meet this need, ships capable of loading and discharging individual cargo units of up to 1000 metric tons in both existing and remote ports have been developed. Modern heavy-lift ships provide some of the most sophisticated methods for the movement of large, high-value cargo units over great distances.

Types of heavy-lift ships. The world fleet of heavy-lift ships can be divided into three types: conventional cargo ships with a heavy-lift capability, barge-carrying cargo ships, and specialized heavy-lift ships. Conventional cargo ships with a

Fig. 1.   Heavy-lift derricks discharging a 620-metric-ton nuclear reactor. *(From H. W. Janecke and W. F. Muir, Modern heavy lift ships (state of the art), Soc. Nav. Architects Mar. Eng. Trans., 86:347–374, 1978)*

heavy-lift capability generally operate on a specific trade route and, by virtue of their size, are limited to developed ports with existing pier facilities. These ships are normally engaged in nonheavy-lift trade, and carry heavy-lift cargoes only when the market is favorable.

Barge carriers transport preloaded barges, which are towed to or from the deep-sea liner at its regular port of call nearest the origin or destination of the cargo. These ships are versatile; a variety of cargoes can be carried at one time. The majority of barge carriers in service today are either the Lykes Seabee type, which can lift two 1000-metric-ton barge units simultaneously on a submersible elevator to a self-propelled transporter that carries each barge to its stowage position, or the LASH (lighter aboard ship) type, which uses a gantry crane to lift 400-metric-ton barge units over the stern of the ship to stowage positions. The only major drawback of this system is its reliance on shore-side cranes at the point of origin and final destination of the cargo. Conventional cargo ships and barge carriers have capacities which range from 8000 to 20,000 DWT (deadweight metric tons). Main propulsion is provided by steam turbines or diesel engines. The service speeds of these ships range from 13 to 19 knots (7 to 10 m/s).

Specialized heavy-lift ships are generally smaller than the previously mentioned ships. Their capacity ranges from 1000 to 7000 DWT. Main propulsion is provided by automated diesel engines which drive either single or twin propellers. The length/beam ratio of these ships is lower than that of conventional cargo ships. This is due to the combined requirements of shallow draft and large deck area on a ship of this size. Service speeds range from 11 to 14 knots (6 to 7 m/s). Large-capacity ballast pumps are provided for counterballasting as cargos are transferred. Usually specialized heavy-lift ships are used exclusively for heavy-lift cargoes, and either lie idle or travel in ballast (empty) between cargoes.

Movement of heavy-lift cargoes. The movement of heavy-lift cargoes often requires months of preplanning. The availability of a particular heavy-lift ships may dictate the maximum size of the unit which can be shipped at one time. Special care must be taken to ensure that the cargo can be brought to and from the ship without damage to bridges, pipelines, and so forth. In remote areas it may be necessary to establish new road beds or beachheads capable of receiving heavy-lift cargo. The most critical time in the movement of heavy-lift cargo by ship occurs during the transfer of cargo from shore to ship and back again. On the smaller specialized ships this may involve cargo whose weight approaches and in some cases exceeds 30% of the ship's deadweight. Modern heavy-lift ships use the following methods to transfer cargoes: lift on/lift off; roll on/roll off; float on/float off; slide on/slide off; or combinations thereof.

Lift on/lift off. The majority of heavy-lift ships transfer cargo via the lift on/lift off method (Fig. 1). Four types of lifting gear are used: nonrotating derricks, rotating derricks, gantry cranes, and submersible elevators. Conventional cargo ships with heavy-lift capabilities use rotating or nonrotating derricks mounted on the centerline of the ship. The majority of these vessels have heavy-lift capabilities serving only part of the vessel. As noted above, barge carriers use either submersible elevators or gantry cranes in order to lift cargo aboard.

Gantry cranes travel fore and aft on rails on the main deck and straddle a ship's cargo holds. Cargo is usually lifted over the stern of the vessel, since outreach over the vessel's side dramatically reduces a gantry crane's capacity. The majority of specialized heavy-lift ships use matched sets of derricks mounted at either end of a single cargo hold to handle cargo lifts. Smaller derricks are provided for lifting hatch covers and assisting in various cargo operations.

Often heavy-lift derricks are mounted on one side of a vessel to provide the maximum clear deck area. However, when this is done, additional derrick boom length must be provided to achieve the same outreach as that of a derrick mounted on centerline.

Roll on/roll off. Heavy-lift ships which use the roll on/roll off method to transfer cargo are usually fitted with bow or stern ramps (or both) to load cargo. Cargo is driven on board on special auxiliary heavy-lift equipment such as self-propelled wheeled vehicles or tracked crawlers. Bow visors or bow doors are often provided to improve access. Deckhouses are located to one side of the vessel or designed in a "bridge type" manner, again to improve access. Most specialized heavy-lift ships are of shallow draft, and are specially strengthened (to prevent bottom damage) to allow the transfer of cargoes directly to shore via ramps.

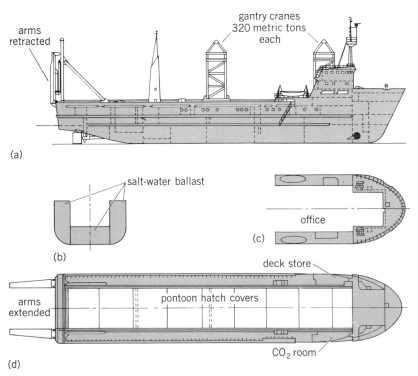

Fig. 2. MV *Dock Lift 1*, which combines float on/float off, lift on/lift off, and roll on/roll off features. (a) Profile. (b) Typical section. (c) Forecastle deck. (d) Upper deck. (*From H. W. Janecke and W. F. Muir, Heavy lift ships (state of the art), Soc. Nav. Architects Mar. Eng. Trans., 86:347–374, 1978*)

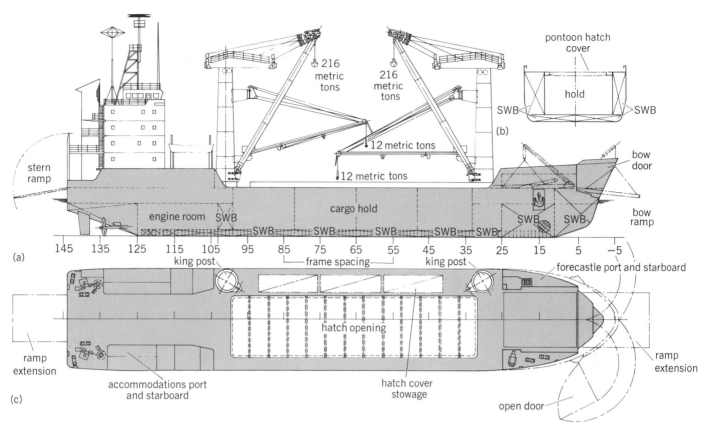

Fig. 3. MS *John Henry*. (a) Profile. (b) Typical section. (c) Main deck. SWB = salt water ballast. (*From H. W. Janecke* and *W. F. Muir, Modern heavy lift ships (state of the art), Soc. Nav. Architects Mar. Eng. Trans., 86:347–374, 1978*)

Float on/float off. The transfer of cargo by the float on/float off method is accomplished by two types of heavy-lift ships. Both of these types ballast (take on water in tanks) to submerge their cargo decks. One type can float cargo aboard only through a stern door into a well between the ship's sides. The stern door is then closed and the cargo well is pumped dry. The other type provides a large, open, barge-type area to accept cargo floated on over the vessel's stern or side. Deckhouses on both types are located well forward to maximize available cargo areas. The main difficulty with this type of system is that relatively deep water is needed for loading and discharging cargo.

Slide on/slide off. Ships using the slide on/slide off method transfer cargo by the use of tracks on retractable girders with specially prepared low-friction surfaces. Cargo is slid from a dock over the vessel's side, where it is stowed and secured with the assistance of additional ship gear.

Versatile heavy-lift ships. The versatility of many heavy-lift ships is increased by having two independent means of loading cargo which can be combined as necessary. The most common combination in specialized heavy-lift ships is the roll on/roll off, lift on/lift off ship. A heavy-lift cargo can be rolled aboard one of these vessels via a bow ramp, and be lifted off its transporter, which then exits via the stern ramp. The cargo is then lowered and secured for the voyage. On another type of heavy-lift ship, a loaded barge can be floated into a cargo well where a gantry crane lifts the cargo off

the barge, which is then removed. The cargo well is pumped dry, and the cargo lowered and secured for the voyage (Fig. 2).

MS John Henry and MS Paul Bunyan. Recently two specialized heavy-lift ships, the first vessels of this type of United States flag registry, entered service. They are typical examples of modern heavy-lift combination (lift on/lift off and roll on/roll off) vessels. These sister ships, the MS *John Henry* (Fig. 3) and MS *Paul Bunyan*, are modified versions of the MS *Stahlek*, which has been in operation for Hansa Line since 1977. These ships have a single cargo hold which measures 133.8 ft (40.8 m) long and 38.7 ft (11.8 m) wide, with a height of 21.3 ft (6.5 m). This hold has 13 watertight hatch covers which are designed to withstand the same load as the main deck. A pair of cargo derricks mounted on the port side of the ship (to maximize clear deck area) serve the cargo hold with a single-unit lifting ability of 432 metric tons. Two 12-metric-ton derricks are provided for lifting hatch covers and light cargoes.

When the bow doors are swung clear, a bow ramp capable of accepting cargo units of up to 1000 metric tons can be put in place. A tunnel through the deckhouse and a stern ramp allow cargo transport vehicles to leave the loaded ship. Main propulsion is provided by twin diesel engines which are located aft and turn twin propellers. This arrangement, in combination with twin rudders and a bow thruster, provides excellent maneuvering characteristics. These ships are of relatively shallow draft, and can discharge cargo directly to shore via the bow ramp. Four anchors, two forward and two aft, are provided to assist in such operations. Weight saving are achieved by the use of aluminum in the upper deck house and high-strength steel in the cargo derricks. As cargo is brought aboard, two large centrifugal pumps transfer ballast quickly to keep the vessels trim (fore-and-aft angle with the horizontal) and heel (side-to-side angle with the horizontal) within acceptable limits. During all cargo movements, the ship's master, ballast control operator, and lifting gear operators are in constant communication via radio.

For background information *see* SHIP, MERCHANT in the McGraw-Hill Encyclopedia of Science and Technology.

[CHARLES ZEIEN]

Microwave landing system

Microwave landing systems (MLS) intended to improve upon the instrument landing system (ILS) for aircraft approach and landing guidance have been under development for over 30 years. There have been dozens of concepts proposed, developed, and in some cases implemented on a limited basis. One of these systems, pulse-coded scanning-beam (PCSB), used by the U.S. Navy, has now been adapted to the special needs of the space shuttle. However, that system was not judged suitable for international standardization for civil aircraft. In April 1978 the International Civil Aviation Organization (ICAO) selected the time-reference scanning-beam (TRSB) concept as its future standard for MLS. The selection climaxed 10 years of effort to narrow the field to a single choice.

Current standards. The ILS, which operates in the very-high-frequency/ultra-high-frequency (vhf/uhf) band, has been the ICAO standard since 1947. Despite its venerable vintage, the ILS is still a growing system. Hundreds of new installations are being added to the runways of the world each year. The motivations for its replacement are the anticipation of requirements for wider sectors of coverage, and the desire for a system that will not be as vulnerable to the effects of radio reflections.

The ILS, despite its shortcomings, has been serving the needs of the conventional civilian aircraft at most airports, and will continue to do so at least through the year 1995. This protection of the ILS as an international standard is also assured by ICAO resolutions. It is the nonconventional segments of the aviation community that have found the ILS less than adequate. For instance, little use is made of the ILS by helicopters. These machines make their approaches at steeper glide angles (6° rather than 3°) than do conventional aircraft, and often land at small heliports rather than on runways. The ILS provides only a single glide path angle and requires an extensive smooth reflecting surface in order to operate properly.

The military and naval services of the world have also found the ILS of limited utility because of its size and the time required for proper installation and alignment. Consequently the aid used most commonly by military aviation for guidance to approach and landing is the ground-controlled-approach (GCA) radar. This apparatus is used by ground operators who "talk" the aircraft to the runway on the basis of the returns they track on their radar screens. It was the search for a means to provide the guidance information directly to the pilot, without intermediate human intervention, that led to the original investigation of the scanning-beam concept of microwave landing systems about 25 years ago.

Pulse-coded scanning-beam. The first MLS put into widespread use was developed for the U.S. Navy for aircraft carriers. The system uses pulse-coded scanning beams to provide coverage in a vertical and a horizontal sector. These beams radiate from two separate antennas and have a thin broad form which sweeps past the aircraft position. One scanning antenna covers an azimuth sector around the centerline of approach, and is called the localizer. The second scanning antenna covers a vertical angle sector, and is called the glide path (Fig. 1). A code of transmitted pulses conveys the pointing angle of the antenna at the instant of reception in the aircraft. This coding is accomplished through a system of pairs of pulses in which the time between pulses in a pair conveys the function being transmitted. The time between pairs provides the beam-pointing information. When decoded, the transmissions provide measures of the angular displacements from desired centerline and glide path angle (Fig. 2).

The information is presented to the pilot on a standard cockpit display called a course deviation indicator (CDI). It has a vertical needle to show the aircraft alignment with the desired glide path angle. Similar information can be used in an autopilot which will automatically guide the aircraft along the desired path to the landing point.

The U.S. Navy system, which was also adopted by the Royal Swedish and the Finnish air forces, operates on several channels between 15.4 and 15.7 GHz. It has been a fully operational system since 1971. Almost 200 ground or ship stations have been produced. About 3000 aircraft are equipped to use this sytem, but less than 25 of them are in civilian hands.

A singular application of this PCSB system will be its use to guide the space shuttle through its approach and landing maneuvers. The shuttle requires glide path coverage up to 30°. Each end of

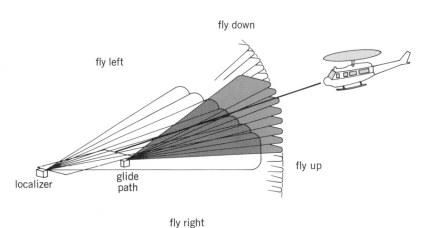

Fig. 1.   Scanning antenna patterns in scanning-beam microwave landing system.

the runways planned for shuttle use in Florida and California have been equipped with a dual system consisting of primary and backup equipment. Monitors which detect malfunction switch equipment automatically if the guidance does not meet the required standards.

The shuttles each have three navigation sets to receive the signals. Each set independently feeds a navigation computer, which in turn provides the inputs to the redundant automatic flight control systems. The performance of the shuttle microwave landing system must be assured, because the vehicle is unpowered when it returns to land. It therefore is committed to the chosen touchdown point from the time it leaves its orbit. *See* SPACE FLIGHT.

Time-reference scanning-beam. The civilian aviation authorities in the United States and other parts of the world embarked on a course of development for a new system in 1969, conceived as an international undertaking in which all nations could participate. Progress has been slow. The considered alternatives have been numerous and often strongly conflicting. The final selection narrowed when the U.S. Federal Aviation Agency (FAA) abandoned its own proposal and adopted a concept conceived in Australia and called Interscan. It evolved into the TRSB format, which competed with a concept that employed the Doppler principle developed in England. After considerable technical debate, the TRSB concept received the endorsement in the ICAO councils.

The set of specifications that shaped the development of the new ICAO standard system are based on the requirements of international air commerce. The system will have two hundred 300-kHz-wide channels in the frequency band from 5.030 to 5.091 GHz. Accommodation is made in the signal format for azimuth (localizer), elevation (glide path) courses, missed-approach azimuth (MAZ) to guide an aircraft during a missed-approach maneuver, flare elevation (FL) guidance, and a 360° azimuth signal to aid in terminal maneuvering.

The angular guidance is conveyed by a scanning beam with an angular velocity of 20,000° per second. It sweeps through the coverage sector in a "to" and then "fro" direction. The measured time interval between the "to" and "fro" beam passages is a measure of the angle at the reception point. To achieve the high angular velocity of the beam, the antennas employed for these functions are electronically scanned arrays.

A significant feature of the TRSB-type MLS is its message capability. This capability provides a means of transmitting to a receiving aircraft a complete range of information relating to the special characteristics of the particular ground system. The geometric relationships of the various elements of the system are also transmitted. These messages will allow sophisticated receiving and processing equipment in the aircraft to adapt the approach and landing maneuvers to a wide variety of conditions.

The TRSB signal format is made up of a series of transmissions from the various antennas of the ground subsystems which time-share the selected radio-frequency channel (Fig. 3). The transmis-

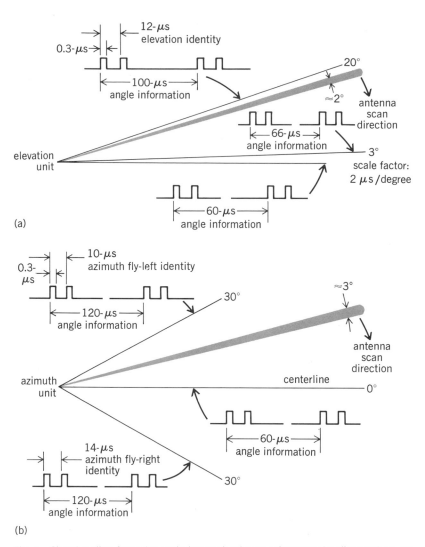

(a)

(b)

Fig. 2. Signal coding for pulse-coded scanning-beam microwave landing system. (*a*) Elevation guidance. (*b*) Azimuth guidance.

sions are grouped into words strung together in sequences. The sequences can vary, depending upon local requirements. The individual words can be either guidance words (angle words) or data words, but all begin with a 24-bit preamble transmitted from a fixed antenna. Digital modulation is differential phase-shift keying (DPSK). DPSK is characterized by a 180° phase shift of the carrier frequency to transmit a binary 1. Absence of a phase shift represents a binary 0. The preamble provides the necessary time references required to synchronize the air set with the ground transmissions, identifies the type of information to follow, and if it is a guidance word, provides the time base for the angular measurements. The transmission is switched to the electronically scanned antenna in a guidance word so that the "to" and "fro" beam sweeps can be generated. The transmission is continuous, but the phase control of the antenna elements creates a moving beam. If it is a data word, the DPSK modulation is continued from the fixed field antenna, and an additional 20 bits of information follow the preamble. It is also possible to transmit a 76-bit auxiliary word with additional data formats.

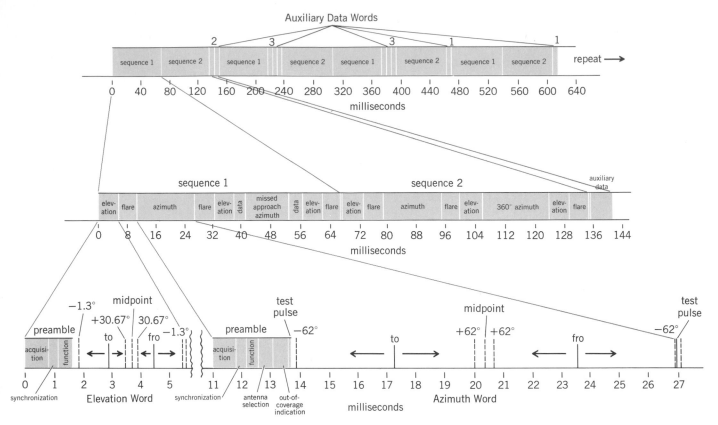

Fig. 3. Signal format of time-reference scanning-beam microwave landing system.

When received in the aircraft, the data parts of the messages are decoded. The "to" and "fro" beams appear as pulses about 25μs long. The elapsed time between the two pulses is the measure of the angle.

The sequences of the various words are arranged so that the guidance words occur with sufficient frequency to provide the data rates necessary to accommodate a high-performance aircraft under automatic pilot control. This frequency can be as high as 40 updates per second for the glide path functions.

The system is considered to be flexible enough to provide a wide variety of capabilities. These range from the simplest, least expensive versions of both ground and air equipment that could provide service about equivalent to the most basic ILS, to a full-scale configuration that would be applicable for automatic landings by advanced aircraft at major jetports under conditions of zero visibility. The more complex equipment is designed to provide guidance along paths that might be dog-legged or curved. This feature is deemed important for future crowded terminal areas and for noise-abatement procedures.

Future development. After final definition of the TRSB signal specification by an ICAO technical working group, it must be submitted for final ratification. Formal publication as a Standard and Recommended Practice (SARP) is done under Annex 10 of the Convention of International Civil Aviation, a special agency of the United Nations.

Implementation of the new system is at the discretion of the ICAO member states. It is not expected that any nation will rush to develop finished equipment. There are additional tests and refinements required, and the costs are substantial. However, nations or companies will be free to exploit the potential of the new system.

The benefits are interdependent — to the governments which install the ground stations, to the aircraft owners who must buy their own airborne sets, and to the manufacturers who must invest in the development of the equipment. All three must advance together for the system to come into common use. While this seems inevitable, now that the ICAO has acted, it will still be years before MLS has eliminated ILS from aviation.

For background information *see* GROUND-CONTROLLED-APPROACH (GCA) SYSTEM; INSTRUMENT LANDING SYSTEM (ILS).   [FREDERICK B. POGUST]

Bibliography: Department of Transportation, *Federal Aviation Administration Engineering Requirement*: *Microwave Landing System (MLS) Signal Format and System Level Functional Requirements*, FAA-ER-700-08 C, revised May 16, 1980; R. J. Kelly, H. M. Redlien, and J. L. Shagena, *IEEE Spect.*, 15(9):52–57, September 1978; F. Pogust, *IEEE Spect.*, 15(3):30–36, March 1978.

Military aircraft

The new United States Navy/Marine Corps F/A-18 aircraft, now undergoing full-scale flight development testing, was derived from the Northrop YF-17 prototype lightweight air combat fighter. After a competitive fly-off with the YF-16 in 1974 at Ed-

Fig. 1. Navy/Marine Corps F/A-18A Hornet aircraft in flight. (*McDonnell Aircraft Co.*)

wards Air Force Base, the YF-17 design baseline was modified to suit Navy requirements. These requirements were addressed to a dual fighter and attack aircraft replacement for the current Navy F-4 and A-7. The basic F/A-18 (Fig. 1) incorporates many changes from the YF-17 to satisfy Navy missions. Carrier suitability emphasized high lift for takeoff and landing, and strengthened wing, fuselage, and landing-gear structures for catapult launch and high-load arresting gear landings. Mission demands led to more internal fuel, upgraded engine thrust, and larger lifting surface areas than

in the YF-17. In addition, numerous onboard systems were incorporated to assure reliable and maintainable service once in the fleet.

F/A-18 configuration design. The F/A-18A is a single-place, twin-tail, twin-engine fighter/attack carrier-based aircraft. Extensive graphite-epoxide composite materials (13%) are incorporated into the wing skin, flaps, ailerons, stabilator, vertical tail, and many access doors. Each aircraft contains its own accessory drive and auxiliary power systems to start the engines and drive the fuel and hydraulic pumps and generators. These features enhance the self-sufficient capability of the aircraft, and essentially eliminate the need for ground-support carts. The two afterburning F404-GE-400 engines were derived from the YJ101 used in the YF-17, and the inlet design and interface are nearly the same as in the YF-17. The F404 develops 16,000 lb (71 kilonewtons) of afterburning thrust at sea-level static conditions, and weighs about 2000 lb (900 kg). Its low bypass (fan to core flow) ratio of 0.34 characterizes the engine as more of a turbojet than a turbofan. The engine has been designed with an emphasis on reliability, maintainability, and durability, and contains only 14,300 parts as compared to the J-79 turbojet with 22,000 parts and the F-100 turbofan in the F-15 and F-16 with 31,100 parts. Its maintainability is further enhanced by its accessibility in that an engine change can be made in the aircraft's shadow by disconnecting at only 11 points.

Digital avionics and control systems. To further enhance aircraft reliability and to ease the

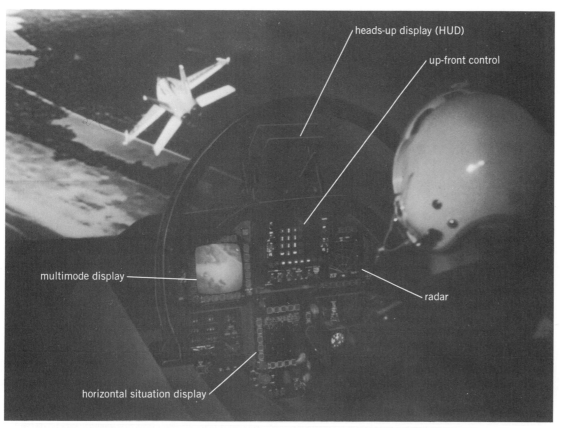

Fig. 2. F/A-18 air combat simulator. Note single pilot avionics, displays, and controls. (*McDonnell Aircraft Co.*)

single-pilot workload, a commitment was made to fully computer-integrate displays and controls. A digital flight-control system, special data management, systems monitoring, and control features were incorporated into the F/A-18 by use of digital computers and multiplexing techniques.

As shown in Fig. 2, very advanced cockpit displays and controls are available. The most prominent displays are the cathode-ray-tube (CRT) multimode radar, and horizontal situation displays at the pilot's left, right, and forward-of-stick positions, respectively. Directly above the horizontal situation display is a complex of displays and controls called the up-front control. All communications and identification avionics can be selected and controlled on this central panel by either of the pilot's hands. Above the up-front control and in the center of the cockpit field of view is the heads-up display (HUD), which presents low-distraction combat mission information to the pilot. These displays and controls are used by the pilot to monitor and control most of the functions on the aircraft. Figure 3 summarizes some of the functions which the F/A-18 computer and multiplex system controls and makes available to the pilot. The fully integrated system connects the component systems via a standardized multiplex bus and controls the complex computers. The system gives the single pilot of an F/A-18 superior mission flexibility, controllability, and survivability over most combat aircraft by easing the pilot's information-handling tasks. The pilot, by controlling the information presentations available, can access the communications, navigation, or combat systems indicated and take appropriate actions. In addition, electronic sensing devices and countermeasure conditions are also presented on the pilot's displays.

Digital flight-control system. A design philosophy similar to that used in the extensive applications of digital computers in the displays and

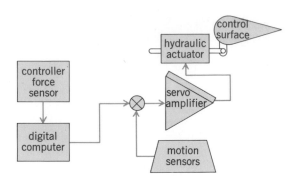

Fig. 4. Digital flight-control system.

avionics systems has been further extended in the F/A-18 to the aircraft flight control. Where previous combat aircraft have relied on hydraulics and mechanical devices for aircraft primary control, the F/A-18 used a digital computer to direct motions of the aircraft control surfaces. Figure 4 illustrates a simplified version of such a digital flight-control system. The pilot's inputs via an apparently conventional stick are transmitted via a force sensor to the digital computer, which directs the hydraulic actuator to make the appropriate control surface motion depending on the condition of the aircraft. The aircraft condition is sensed by a set of motion sensors on the aircraft, which transmit their information to the computers and the avionics system. The F/A-18 flight-control system is based on the principle illustrated in Fig. 4, but is, of course, considerably more sophisticated. Its system is characterized by quad redundancy, which means that there are essentially four levels of sensing and actuation for each control surface. The F/A-18 is the first combat aircraft to commit to digital fly-by-wire as the primary flight control. The digital system is backed up by an analog direct electrical linkage (DEL) for aileron and rudder control and a mechanical linkage for stabilizer back-up control. Reversion to the back-up control is automatic if each of the digital back-up combination of sensors should fail. An automatic control-system-status monitor tests the flight control by operating in the background of the normal in-flight modes and determining status for display to the pilot. This system is called the built-in test (BIT) system, and further assures pilot awareness of the flight-control system operability.

Full-scale flight test program. The full-scale development of the F/A-18 Hornet aircraft is being carried out at the Patuxent River Naval Air Test Center (NATC) in Maryland, where nine single-seat aircraft and two trainer (two-seat) aircraft are undergoing flight tests to prove the capabilities of the aircraft. Each aircraft is assigned particular systems and capabilities tasks which must be demonstrated. The 11 aircraft will accumulate nearly 5000 flight hours of testing during the approximate 3½ years of tests, which began in late 1978.

For background information see AIRCRAFT MILITARY; AIRCRAFT INSTRUMENTATION; FLIGHT CONTROLS in the McGraw-Hill Encyclopedia of Science and Technology.   [RONALD H. SMITH]

Bibliography: L. E. Fairbanks, R. P. Kurlak, and K. W. Ramby, in *3d Digital Avionics System Conference*, Fort Worth, TX, Nov. 7, 1979; L. J.

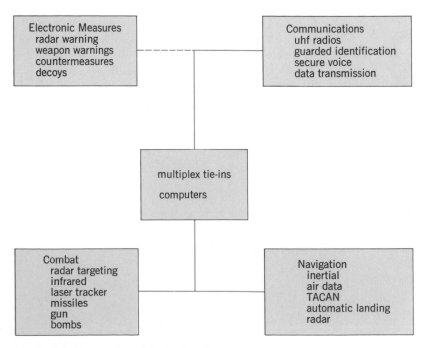

Fig. 3. F/A-18 controls and display functions.

Franchi, in *Military Electronics/Countermeasures*, December 1979; C. G. Gilson, *Flight Int.*, Aug. 21, 1975; R. L. Kisslinger, W. J. Momeno, and J. M. Urnes, in *3d Digital Avionics System Conference*, Fort Worth, TX, Nov. 7, 1979.

Morphogenesis

Morphogenesis is the progressive attainment of bodily form and structure during embryonic development. The process is the result of complex interactions and relationships between cells of the embryo. Recent investigations have developed computer simulations to unravel the physical interactions among embryonic cells. By expressing cell behavior in mathematical terms, a formal mathe-matical description of the flow of tissues in a developing embryo can be determined. This analysis has been called morphodynamics, and utilizes three areas of scientific investigation: experimental observation, computer simulation, and formal mathematics.

In this article, morphodynamics (computer simulation and mathematical analysis) and experiments on embryos are presented to illustrate the cellular interactions involved in morphogenetic processes of gastrulation of an embryo.

Gastrulation. Gastrulation is the first and most crucial step in the transformation of the cleaving egg into some semblance of the embryo-to-be. As a morphogenetic process, gastrulation is a highly

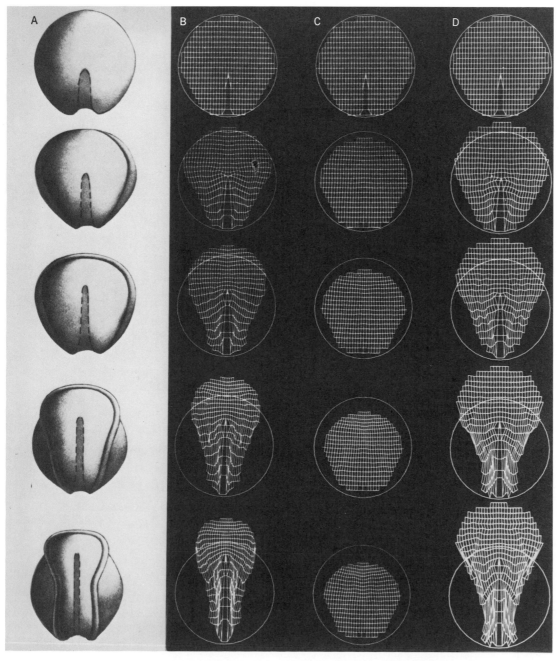

Development of the neural plate of an amphibian embryo. Normal development (A) is compared to computer simulations (B–D). *(From R. Gordon and A. G. Jacobson, Sci. Amer., 238:106–113, 1978)*

integrated activity that involves the embryo as a whole, and takes place in markedly different ways in different types of developing eggs. During gastrulation the embryo lays down the basic body structure in three primary germ layers: the ectoderm, mesoderm, and endoderm. Each layer has different potentialities, and each will give rise to different types of tissues and organs. During gastrulation, cells move as individuals, as small clusters, or in loose streams in which individual cells may be constantly shifting their contact relationships with each other. Some cells form confluent sheets which are firmly adherent to one another, with few or no apparent gaps between them.

Recent experiments have shown that some mesodermal cells follow paths similar to those of the overlying ectoderm. In older embryos the mesoderm forms a rodlike structure called the notochord which runs almost the entire length of the embryo under the spinal cord and brain. The notochord adheres tightly to the overlying ectoderm, and has been shown to induce the overlying ectoderm to form neural tissue. The ectoderm that overlies the notochord has been termed the supranotochordal region.

The central nervous system of the embryo is the first major morphogenetic change seen as a result of gastrulation. In the amphibian embryo the central nervous system begins as a hemispheric sheet one cell thick. It undergoes morphogenetic changes to form a keyhole-shaped, thickened neural plate, while remaining a monolayer of cells in the process.

The sequence of the development of the neural plate is shown in the illustration. Time-lapse motion pictures were analyzed with a stop-motion projector equipped with a frame counter, so that an accurate description and staging of embryonic development could be made. As can be seen in illustration column A, there is an increase in the height of the neural plate cells and a decrease in the surface area of the plate during gastrulation.

Computer simulation. In illustration columns B, C, and D, computer simulations are shown to test the adequacy of different modes of cell behavior and to explain the observed changes in the shape of the neural plate. The shaping of the tissue is represented as the distortion of a geometric grid placed over the embryo. Two computer simulations are represented which test the following modes of cell behavior: uniform shrinkage of neural plate cells, and elongation of the supranotochordal region. In illustration column C the only force is provided by cell shrinkage, and the resulting grid does not attain the keyhole shape of the neural plate. In illustration column D the only driving force is provided by supranotochordal elongation. Although the grid does develop the keyhole shape, the top is overlarge. Illustration column B shows a computer simulation that incorporates both forces: uniform shrinkage of the neural plate cells, and elongation of the supranotochordal region. The resulting shape of the transformed grid is virtually identical to that seen in the embryo. Thus it appears that elongation and uniform shrinkage can account for the cellular basis seen in the nobel grid transformation, and that these morphogenetic forces are operative in the formation of the neural plate of the embryo. This experiment could not have been done with living embryos, and shows that the key to understanding cell movements during neurulation is to be found in facts governing cell shape changes.

Developmental boundaries in embryo. Since the surface area of the neural plate is shrinking during gastrulation, cells of the surface ectoderm are changing cell neighbors. As a result of this movement, shear lines are formed at the interface between a sheet of cells whose surface area is shrinking and an adjacent sheet of cells whose surface area remains the same or actually increases. There are at least two regions in the neural plate where lines of shear are formed: in the supranotochordal region, and at the boundary between the neural plate and the epidermis (the rest of the outer layer of the embryo). Why abrupt boundaries arise between cell domains in the neural plate is unclear, but evidence suggests that shear may serve to alter the function of gap junctions between cells and lead to their functional separation.

The formation of the supranotochordal region of the neural plate and the notochord remains unexplained. It is not understood how cells in this part of the sheet can rearrange themselves in a coordinated manner to give rise to elongated structures. Recent evidence suggests that the answer may lie in surface-tension interactions between domains of cells in the neural plate and the epidermis.

An understanding of the development of an embryo offers one of the most intriguing challenges to scientists today. Morphodynamics has yielded quantitative information of how the behavior of cells in a single tissue leads to its change in shape. Information about other morphogenetic processes, however, is still lacking.

For background information *see* EMBRYOLOGY, EXPERIMENTAL in the McGraw-Hill Encyclopedia of Science and Technology.   [ROBERT W. KEANE]

Bibliography: R. Gordon and A. G. Jacobson, *Sci. Amer.*, 238:106–113, 1978; A. G. Jacobson and R. Gordon, *J. Exp. Zool.*, 197:191–246, 1976; J. P. Trinkhaus, in G. Poste and G. L. Nicolson (eds.), *Cell Surface Reviews*, vol. 1, pp. 225–329, 1976.

Nobel prizes

The Swedish Royal Academy announced 11 recipients of the Nobel prizes for 1980.

Medicine and physiology. This award was given to two Americans and a Frenchman for their independent work on a group of genes associated with the body's immune response known as the histocompatibility complex. Sharing the prize were George Snell of the Jackson Laboratory, Bar Harbor, ME, Baruj Benacerraf of Harvard University, and Jean Dausset of the Saint-Louis Hospital at the University of Paris. Their discoveries have greatly aided donor-recipient matching for transplant operations.

Chemistry. Half of this prize was awarded to American Paul Berg of Stanford University for his research into the biochemistry of DNA and for designing a successful technique for gene splicing. The other half was shared by Walter Gilbert of Harvard University and Frederick Sanger of Cambridge University, England, for developing separate methods of determining nucleotide sequence. The work of all three researchers has advanced the technology of DNA recombination, which may eventually lead to gene replacement for genetic disease.

Physics. Sharing the award were two Americans, Val L. Fritch of Princeton University and James W. Cronin of the University of Chicago. Their work has disproved the inviolability of a supposed fundamental law of nature: the time-reversal principle that a reaction in one direction can run with equal ease in the other direction. This discovery indicates that the universe is not as symmetrical as was once believed, and may answer some of the questions raised by the big bang theory of the origin of the universe.

Economics. American Lawrence R. Klein of the University of Pennsylvania received the prize for developing mathematical models to predict trends in national economies from raw statistics. Klein's work has led to the widespread use of econometric models in public administration, political organizations, and business enterprises.

Literature. Poet and novelist Czeslaw Milosz was honored for his writing, "many-voiced and dramatic, insistent and provocative." Milosz was active in the Polish resistance during World War II and served 4 years as a diplomat under the postwar Communist government. He fled Poland in 1951 and presently teaches at the University of California, Berkeley.

Peace. The peace prize was awarded to Adolfo Pérez Esquivel, Secretary General of Servico Paz y Justicia en América Latina, for his struggle against political oppression in his native Argentina. A human-rights activist and champion of the poor and the destitute, Esquivel accepted the award in the name of the people of Latin America.

Nondestructive testing

Lasers are increasingly becoming commonplace as inspection tools in the manufacture of precision parts. The development of small lasers linked to microprocessors that manipulate the workpiece as well as collect the data from the laser has permitted significant advances in the area of mechanical metallurgy.

A need was identified several years ago for a technology that could provide statistical quantities of in-reactor creep data for the development of constitutive equations to be used in design and performance analyses of liquid-metal fast breeder reactor (LMFBR) core components. Conventional testing technology, at that time, consisted of a few sequential tests per year; long-term tests precluded the performance of short-term tests. An expanded testing technology was needed, not only for the characterization of key reference materials, but also for the development and screening of new and untested alloys.

To meet these needs, a test program was developed based on highly accurate and rapid measurement of creep strain in small pressurized-tube specimens. This method enabled the exploration of a broad range of conditions which encompassed the operating temperatures, pressures, stresses, and neutron fluxes and fluences in LMFBR cores such as the Fast Test Reactor, the Clinch River Breeder Reactor, and conceptual design studies. This article describes the techniques used for measuring the creep strain in these pressurized-tube specimens.

LVDT model. The first model used in initiating the pressurized-tube creep program consisted of an LVDT (linear variable differential transformer) measurement station. The device (Fig. 1) is capable of 2.5 μm accuracy. However, its accuracy is very sensitive to adjustments in alignment of the specimen-positioning devices, wear of the contacting probes, and deviation of the specimens from idealized cylindrical geometry. Ten min is required to obtain 20 measurements at five axial locations and four positions of rotation on each specimen. These data are read from a digital display by the operator and recorded by pen in a logbook.

Fig. 1. Test station using LVDT probes to measure diameters of pressurized-tube creep specimens.

Laser contacting model. The second model used laser interferometry to enhance the accuracy and speed of testing. A heterodyning system measures relative displacement of two retroreflectors by dividing the beam into f_1 and f_2 components which are directed to a reference and a measurement retroreflector. The resultant beams are sent to the photodetector in the laser receiver. Relative motion between the retroreflectors causes a difference in the Doppler shifts in the return-beam frequencies, thus creating a difference between the beat frequency seen by the measurement photodetector and the reference photodetector. This difference is monitored by a subtractor, and accumulated in a fringe count register located in the instrumentation console. A digital calculator samples the accumulated value every 5 μs and performs a two-stage multiplication, one for refractive index correction and the other for conversion to inches or millimeters. The resulting value updates the display in the console.

Upon leaving the laser head, the beam is split and sent to the diameter/radius station. The beam is again split at the diameter/radius station into two optical components of 50% power each. Each component is then sent to its respective interferometer and retroreflector. Beams are split at each station in order to achieve a two-axis system with axes independent of each other. One axis is the measurement axis, and the other is the reference

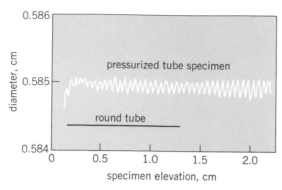

Fig. 2. Diameter profile showing eccentricity of a pressurized tube in contrast with a round tube.

axis. The two signals are sent to the subtractor and accumulated in a fringe count register located in the instrument console. This approach is needed to obtain a differential measurement of the specimen.

The signals received from each station can be summed mathematically to give a diameter reading or read individually at the diameter/radius station to give a radius measurement relative to the specimen centerline.

Two sapphire styli contact the rotating and translating specimen. The styli are suspended by a parallel reed arrangement to optimize alignment. The retroreflectors are also rigidly attached to the reeds, and thereby adjust the laser optical path in accordance with the diameter of the specimen. All operations of the measuring machine are con-

trolled by a minicomputer. The diameter measurements are digitally displayed while being recorded on magnetic tape.

Accuracy with this method was improved to better than 0.3 μm, while the number of measurements was increased to 1600, equally spaced in angle of rotation and length along the specimen without increasing the time required to complete the measurements. A complete profile of a specimen was attainable from the data taken from each specimen and recorded on magnetic tape (Fig. 2). The oscillatory nature of the profile represents the eccentricity of the specimen. The round standard specimen is routinely used to test the calibration of the system, and yields the uniform profile shown in Fig. 2.

Laser noncontacting model. The pressurized-tube creep test programs have matured to the point that the measurement of hundreds of specimens must be completed in a few days, thus allowing irradiation of the specimens to be resumed with minimal time lost. In order to accomplish this objective, a noncontacting laser measuring machine was developed which measures at a rate of 400 scans per second. The laser beam is scanned across the specimen diameter by a motor-driven mirror. The radially scanned beam is converted to a parallel scanning beam by means of a scanner lens, oriented so that its focal point coincides with the rotating mirror surface. The scanner lens is designed so that the parallel scanning beam has a constant scanning velocity as it sweeps across the specimen. The specimen interrupts the scanning beam (Fig. 3) for a period of time proportional to the diameter. The interrupted beam is collected by a receiver lens and focused onto a photodetector, which converts the light signal to a time-dependent signal. Averaging techniques are used to reduce system errors, such as vibration and instantaneous changes in the index of refraction in the air surrounding the specimen.

The specimen is held by conical chucks (Fig. 3), and is rotated and translated normal to the laser beam so that 2000 measurements are made on each specimen at five axial locations. The measurement of each specimen requires less than 1 min. All operations of specimen positioning and measurements are managed by a microcomputer. A copy of these data is printed for assembly in a data log. For measurement of radioactive pressurized specimens, only the measuring machine is placed within the shielded facility, leaving all other components available for hands-on maintenance.

Comparison of data. One of the requirements for changing from one set of measuring equipment to another is that the data be interchangeable. This requirement is dictated by the need to provide continuity of measurement on long-life specimens after retirement of a measuring device. Achievement of the requirement is demonstrated in the table. The agreement between the measuring devices is within 0.5 μm. If a specimen has an unusually irregular profile, greater disagreement is likely, since the measuring devices do not average over the identical surface areas. However, the number of specimens with irregular profiles is less than 10%, and thus the new equipment has been successfully incorporated in the program.

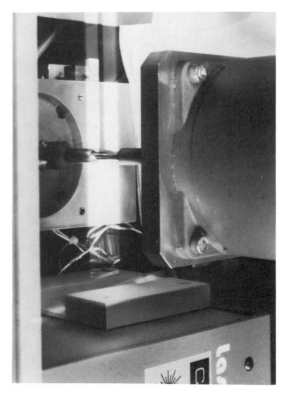

Fig. 3. Noncontacting laser with beam interrupted by the specimen.

Comparison of data from different diameter measuring devices

Specimen number	LVDT	Average diameter, cm	
		Laser contacting	Laser noncontacting
NU	.58533	.58527	—
EH	.58506	.58498	—
ND	.58533	.58527	—
ET	.58512	.58499	—
NC	.58538	.58533	—
ER	.58504	.58499	—
NT	.58542	.58536	—
FN11		.46084	.46086
FN19		.45808	.45806
FN18		.45783	.45781
FP32		.46269	.46270
FP37		.45751	.45750
HN37		.45697	.45696
HL03		.45748	.45746

The LVDT diameter results are consistently larger than the contacting laser by approximately 8 μm. The difference is attributable to the specimen alignment provided by the parallel reeds which support the sapphire styli. The cosine error due to misalignment in the LVDT system resulted in erroneously large diameter measurements.

For background information *see* LASER; NONDESTRUCTIVE TESTING in the McGraw-Hill Encyclopedia of Science and Technology.

[ROBERT GILBERT]

Nonlinear programming

The general area of linear and nonlinear programming is currently the subject of extremely widespread research. Of the many recent developments, three will be discussed here. The first development is the ellipsoidal algorithm (or Khachian's algorithm) for the solution of linear programming problems. The second effort concerns the convergence analysis of the so-called secant (or quasi-Newton) methods for unconstrained minimization, and the third activity is the extension of these secant methods to constrained optimization.

Khachian's algorithm. As is well known, although the simplex algorithm for the solution of linear programming problems is an effective algorithm, it is not a polynomial time algorithm. This means that the number of computations (equivalently the computation time) required to solve the problem cannot be described by a polynomial in the dimension of the problem. In fact, in its worst case the simplex algorithm requires exponential time. However, in practice this is essentially never the case. It has been an open question for many years whether or not an algorithm could be constructed which would solve the linear programming problem in polynomial time.

Very recently, L. G. Khachian demonstrated that a modification of N. Z. Shor's ellipsoidal algorithm could be implemented in polynomial time. P. Gacs and L. Lovasy showed how Khachian's algorithm could be applied to solve linear programming problems. Khachian's contribution was viewed by many to be of such significant proportions that it would render the simplex algorithm and many other algorithms obsolete. However, preliminary numerical investigations have demonstrated that the algorithm is quite inefficient on many problems and as yet is not competitive with existing algorithms. Further studies are needed before any final conclusions are drawn.

Essentially the basic ellipsoidal algorithm begins by constructing a spheroid with a radius so large that it must contain a solution to the problem if one exists. The algorithm then constructs a sequence of ellipsoids with decreasing volume. It can be shown that after a number of steps, which can be represented by a polynomial in the dimension of the problem, the algorithm must demonstrate that no solution exists or must construct an ellipse whose center is within a given tolerance of the solution.

Secant methods. The construction of secant methods for unconstrained nonlinear optimization was motivated by the desire to obtain an algorithm which behaved like Newton's method but did not require the calculation of second-order derivatives or the solution of a linear system at each iteration. The two most successful secant methods have been the DFP and the BFGS methods. While these methods were becoming increasingly popular and were replacing existing algorithms, their general convergence properties could not be established mathematically. At the time these methods were proposed, analysts were of the opinion that any successful Newton-like method must be consistent (that is, the approximate hessian matrices must converge to the true hessian at the solution). Moreover, essentially all of the existing convergence theory required consistency.

M. Powell constructed a numerical example which demonstrated that these secant methods were not necessarily consistent. This meant that a new theory was needed for the general convergence analysis of the secant methods. This new theory was developed by C. Broyden, J. Dennis, and J. Moré, and was used to rigorously establish the general convergence of the secant methods.

By the middle to late 1970s the theory and implementation of the secant methods for unconstrained nonlinear optimization were well understood and reasonably complete. However, the extension of these methods to constrained optimization was only beginning. For years researchers felt that constrained optimization should be handled by sequential unconstrained optimization, that is, by solving a sequence of unconstrained optimization problems which had the property that the solutions of the unconstrained problems would converge to the solution of the constrained problem. This philosophy or formulation had the advantage of requiring only algorithms for unconstrained optimization. Prior to 1970 the main formulation of this type was the penalty function method. From 1970 to 1975 the most popular sequential unconstrained optimization algorithm was the multiplier method. It was generally concluded by 1975 that solving a sequence of unconstrained optimization problems was inefficient and could certainly not be the optimal approach to constrained optimization. The well-known necessity conditions for constrained optimization were used to attack the constrained optimization problem by solving for a saddle point of the lagrangian functional. The DFP and the BFGS secant methods require that the hessian of the functional which is being optimized is not indefinite. Since the hessian of the lagrangian is always indefinite at a saddle point, it is clear that the DFP and BFGS secant methods cannot be applied directly to approximate this saddle point. Hence, they cannot be applied directly to constrained optimization. Perhaps it is exactly this observation which held up progress for so long.

Recently researchers have taken three seemingly distinct approaches to extending the DFP and the BFGS secant methods to constrained optimization. The first approach was to use the secant formulas to approximate only a piece of the hessian of the langrangian functional. This piece can be safely assumed to not be indefinite. The remaining pieces of the hessian of the lagrangian contain only first-order information and can be calculated exactly. This philosophy has been referred to as the extended problem approach and takes advantage of the structure of the problem.

The second approach consists of modifying the sequential unconstrained optimization philosophy referred to earlier as the multiplier method so that only one secant step is taken on each unconstrained optimization problem. After this one step the estimate of the Lagrange multiplier is updated and used in the new unconstrained optimization problem. This philosophy has been called the multiplier update approach.

The third philosophy consists of using the approximate hessian obtained from the secant formula to construct a quadratic approximation to the lagrangian functional. This quadratic functional is then optimized subject to a linearized form of the constraints of the original problem. The solution of the original constrained problem is then estimated by the solution of the quadratic problem. A new quadratic program is constructed using the secant update formula, and a new approximation to the solution of the original problem is obtained. This philosophy is referred to as the successive or recursive quadratic programming approach.

Using the Broyden-Dennis-More convergence theory, the proper convergence analysis has been obtained for the above three approaches. It is of interest that in the case of equality constraints, the extended problem approach, the multiplier update approach, and the successive quadratic programming approach are all equivalent.

For background information *see* LINEAR PROGRAMMING; NONLINEAR PROGRAMMING; OPTIMIZATION in the McGraw-Hill Encyclopedia of Science and Technology.

[RICHARD A. TAPIA]

Bibliography: C. G. Broyden, J. E. Dennis, and J. J. Moré, *J. Inst. Math. Appl.*, 12:223–246, 1973; J. E. Dennis and J. J. Moré, *SIAM Rev.*, 10:46–89, 1977; O. Mangasarian, R. Meyer, and S. Robinson, *Nonlinear Programming 3*, 1978; P. Wolfe, *A Bibliography for Ellipsoidal Algorithms*, IBM Research Center, Yorktown Heights, NY, 1980.

Nuclear reaction

Although nuclear reactions induced by heavy ions have been studied for over 3 decades, much still remains to be learned about the reaction mechanism and the effect that the structure of the involved nuclei has on the behavior of a many-body collision. In the last few years, new facilities and techniques have allowed much progress to be made in exploring new energy regions with higher precision. A much better overview has resulted of the systematic features present in the behavior of the various reaction processes as a function of bombarding energy and projectile-target combination. Such systematics in turn make it possible to observe deviations from the average behavior which are indicative of nuclear structure effects. Recent results from studies of light heavy-ion–induced reactions ($10 \leqslant A_{projectile} \leqslant 20$, where $A_{projectile}$ is the atomic mass number of the projectile) have given great insight into the influence of heavy-ion projectile properties on the fusion reaction process and on the distribution of total reaction strength. These results show that the fusion cross-section behavior is sensitive to the detailed structure of the nuclei involved and the properties of nuclei under extreme conditions of high excitation and large deformation.

Fusion and direct reactions. Based on early studies of nuclear reactions induced by light-particle projectiles, a simple picture of the reaction mechanism has evolved in which all reactions fall into one of two basic categories, depending on the time scale on which they occur. The reactions where the projectile and target nucleus "fuse" to form an intermediate or compound system which lives a long time (on the order of 10^{-16} s) relative to the time it takes for the projectiles and target to pass each other (on the order of 10^{-21} s) are called compound or fusion reactions. Those reactions which are characterized by the time it takes for the projectile and target to pass each other are designated as direct reactions.

As originally proposed by N. Bohr, the fusion reaction is envisioned in two distinct steps. In the first step, the intermediate compound nucleus is formed, and the kinetic energy of the projectile is shared among all the nucleons. All memory of the target and projectile is assumed to be lost. The compound nucleus is always in a highly excited unstable state, and will decay (as the second step) into a number of different exit channels. The decay may proceed by the statistical evaporation of nucleons or light particles; this continues until a bound state of some residual nucleus is formed and further particle evaporation is not energetically possible. The remaining energy is then removed by gamma-ray transitions, leaving the evaporation residue in its ground state. For heavier systems ($A \geqslant 100$), the decay of the compound nucleus may first proceed by fissioning into two fairly equal-mass fragments which then may decay by light-particle evaporation as discussed above.

In direct reactions, no such long-lived intermediate systems are envisioned as being formed. Rather it is assumed that the incident nucleon or particle does not interact with the target nucleus as a whole, but that it or some component of it interacts with the surface or the nucleons on the surface of the target nucleus. There are several direct processes that may occur; namely, elastic scattering where the projectile and target interact but remain intact; inelastic scattering where the two partners remain intact but where some of the kinetic energy has been converted into internal excitation of one or both of the partners; and transfer reactions where there is an exchange of nucleons between the target and projectile. In direct interactions, information about the target and projectile is not lost, and those reactions which excite nuclear states via inelastic scattering or transfer reactions possess information about the structures of these states.

Model for heavy-ion fusion reactions. While the mechanisms described above have not been completely successful in accounting for all the observed features in light-particle–induced reactions, they have provided a means for correlating and understanding many of the experimental facts. Since there are several properties of heavy ions as projectiles that distinguish them from light-particle projectiles, it is not clear that the same simple picture for the reaction mechanism will still be applicable without some revisions. In particular, the heavy ion brings into a reaction: (1) a large nuclear charge Z (which introduces a larger Coulomb interaction), (2) a large number of nucleons (which introduces more degrees of freedom and the possibility for the transfer of more nucleons), (3) a large energy E (which together with kinematic considerations allows for the transfer of larger excitation energy into a final nucleus), and (4) a large orbital angular momentum l (which introduces the possibility for population of higher spin states in the final nucleus).

Influence of entrance channel dynamics. The importance of properties of the heavy-ion projectile in determining the overall behavior of the fusion reaction can be seen by considering the effective real potential which describes the nucleus-nucleus interaction. Shown in Fig. 1 is the effective real potential, $V(r) = V_N + V_{Coul} + V_{cent}$, for the $^{16}O + ^{48}Ca$ system plotted as a function of the distance between the nucleus centers, where V_N is the nuclear interaction (taken to be of Wood-Saxon form), V_{Coul} is the contribution of the Coulomb interaction, and V_{cent} is the contribution of the centrifugal pseudopotential for the various orbital angular momenta l. For fusion to take place, it is assumed that the two nuclei must come into close contact. The critical distance of approach, R_{cri} (where dissipation of the relative kinetic energy into internal excitation sets in rapidly) is generally assumed to be roughly that of the sum of the two half-density radii. The detailed shape of the potential as a function of radius and angular momentum can be seen as a delicate balance between the attractive nuclear interaction and the repulsive Coulomb and centrifugal contributions. At lower bombarding energies, illustrated by E_1 ($< E_{cri}$) in Fig. 1, there is a well-developed pocket, and an interaction barrier located at a rather large radius must be surpassed in order to reach R_{cri}. As the bombarding energy is increased, the pocket begins to disappear. Moreover, at some energy, labeled as E_2 ($> E_{cri}$) in Fig. 1, the repulsive centrifugal contribution prevents the projectiles with large angular momentum from reaching R_{cri}, and surpassing the interaction barrier is no longer sufficient for fusion to occur. At this energy, fusion may still occur if sufficient kinetic energy and angular momentum are dissipated in the surface region so that the projectile can be trapped in

the pocket. This is indicated schematically in Fig. 1.

Based on the entrance-channel picture, one would predict that at lower bombarding energies ($< E_{cri}$), the fusion cross-section behavior is dominated by the interaction barrier and has an energy dependence that is roughly of the form given by Eq. (1), where E is the center of mass energy, R_B is

$$\sigma_{fus}(E) = \pi R_B^2 \left(1 - \frac{V_B}{E}\right) \qquad (1)$$

the interaction barrier radius, and V_B is associated with the value of the nuclear plus Coulomb potential at R_B. The interaction barrier depends on the shape of the nuclear potential in the surface region, and there is the question of whether the

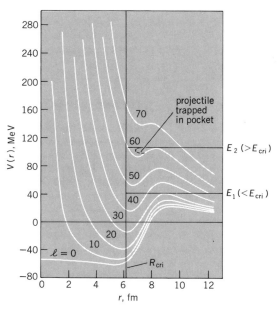

Fig. 1. Total real potential $V(r) = V_N + V_{Coul} + V_{cent}$ for the $^{16}O + ^{48}Ca$ system as function of distance r and orbital angular momentum l.

detailed structure of the interacting nuclei will affect the interaction potential sufficiently to influence the cross-section behavior.

At higher energies ($> E_{cri}$), there should be a change in the slope of the fusion cross-section behavior as a function of energy, reflecting the fact that overcoming the barrier is no longer a sufficient condition for fusion. In the simplest picture, where one assumes that no energy or angular momentum dissipation occurs before reaching R_{cri}, the fusion cross-section behavior at these higher energies will be similar in form to that of Eq. (1), and given by Eq. (2), where R_{cri} and V_{cri} correspond to the

$$\sigma_{fus}(E) = \pi R_{cri}^2 \left(1 - \frac{V_{cri}}{E}\right) \qquad (2)$$

critical radius and the value of the nuclear plus Coulomb potential at the critical radius, respectively. The value of the effective real potential in the radial region about R_{cri} depends sensitively on the balance between the repulsive and attractive contributions and is demonstrated rather dramatically for heavy systems (that is, $Z_1 Z_2 \geq 1000$), where the large Coulomb and centrifugal contribution to the potential prevents formation of a pocket and fusion does not occur even at low bombarding energies.

Influence of compound nucleus properties. Thus far the discussion has focused on the influence of the entrance channel reaction dynamics on the fusion process. It can also be argued that a limitation of the fusion cross sections at higher energies can be imposed by the properties of the compound nucleus. Specifically, if at a given excitation energy in the compound nucleus, the grazing angular momentum brought in is larger than the yrast spin at this excitation, then the compound nucleus cannot accommodate the angular momentum brought into it. The yrast band in any particular nucleus is defined by the sequence of nuclear states, each of which represents the lowest excitation energy for its specific spin J, and can be roughly described by the simple rotational formula, Eq. (3), where E_0 is

$$E_J = E_0 + \frac{\hbar^2}{2\mathscr{I}} J(J+1) \qquad (3)$$

the band head excitation energy, $\mathscr{I}$ is the moment of inertia of the nucleus, E_J is the excitation energy of the yrast state of spin J, and $\hbar$ is Planck's constant divided by 2π. If one assumes that the critical angular momentum is given by $l_{cri} = J$, the fusion cross section can then be written as Eq. (4),

$$\sigma_{fus}(E) = \pi \frac{\mathscr{I}}{\mu} \left[1 + \frac{(Q - E_0)}{E} \right] \qquad (4)$$

where Q is the ground-state Q-value for the fusion process, and μ is the reduced mass of the two nuclei. The expression is identical in form to that of the critical distance model. The difference is that in the former case the limitation on the fusion process was imposed by the entrance channel nucleus-nucleus potential, and in the present case by the properties of the compound nucleus. If the yrast line is indeed introducing the limitations on the fusion cross sections in this energy region, then these data will be providing information about the high excitation properties of the compound nucleus. Calculations based on the rotating liquid drop model of S. Cohen, F. Plasil, and W. Swiatecki indicate that a nucleus can accommodate a maximum spin J^{max} before it becomes unstable and fissions. The encountering of the liquid drop limit would be reflected in the fusion cross-section behavior by a sharp decrease in magnitude at the highest bombarding energies.

These simple considerations have provided the background against which the results of experimental studies have been viewed, and the simple semiclassical models have provided the framework in which the systematic features in the cross-section behaviors can be observed.

Experimental measurements. The most straightforward way of studying the various reaction processes occurring in a nucleus-nucleus collision is by detection of the charged-particle reaction products. Shown in Fig. 2 is a two-dimensional (ΔE versus E_{total}) spectrum obtained for the $^{16}O + ^{24}Mg$ reaction at a laboratory energy (for the ^{16}O projectile) of 138 MeV, using a silicon surface-barrier detector ΔE-E telescope. The ΔE-E telescope consists of two detectors placed one behind the

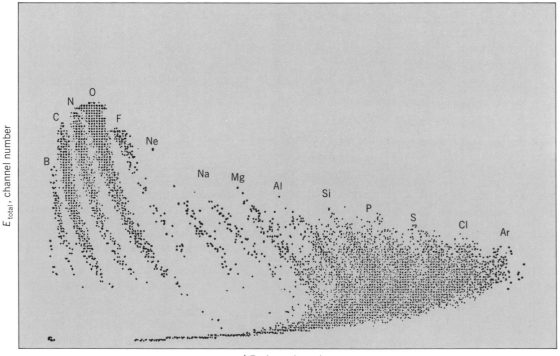

Fig. 2.   Two-dimensional ΔE versus E_{total} plot, obtained using a silicon surface-barrier detector ΔE-E telescope, for the reaction $^{16}O + ^{26}Mg$, at a laboratory energy for the ^{16}O projectile of 138 MeV and a laboratory scattering angle of 7°.

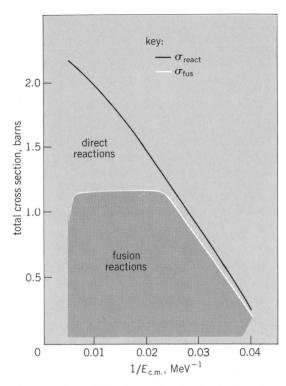

Fig. 3. Schematic plot illustrating the distribution of total reaction cross sections observed in heavy-ion–induced reactions in the mass range $10 \leq A_{\text{projectile}}$, $A_{\text{target}} \leq 40$. 1 barn $= 10^{-28}$ m^2.

other with the first thin detector (approximately 4 μm thick) used to measure the differential energy loss ΔE, and the second thicker detector (approximately 300 μm thick) used to stop the charged particle and measure the remaining energy E. (The total kinetic energy of the particle detected, E_{total}, is simply the sum of ΔE and E.) This spectrum presents a clear overview of the total distribution of reaction strength. The strength for the direct processes is concentrated in the carbon, nitrogen, and oxygen lines which appear at low values of ΔE near that of the projectile ^{16}O. While there is some strength in quasielastic processes (that is, those involving small losses of kinetic energy compared with elastic scattering), the great bulk of the strength is found in processes which are characterized as deeply inelastic because of the large loss of kinetic energy. The cross sections for transfer products with $Z > 8$ are weak. The fusion or evaporation residues (that is, the heavy fragments remaining after compound nucleus formation and subsequent particle emission) are contained in the intense region corresponding to low E_{total} and the highest values of ΔE near that of the compound nucleus ^{40}Ca. While the Z-resolution is not sufficient to unambiguously separate lines corresponding to different elements, these lines can be discerned in Fig. 2. The particles stopped in the front detector fall along the gently sloping straight line ($E_{\text{total}} = \Delta E$) near the bottom of the plot, with intercept at the origin, and represent a very small fraction of the total yield. To obtain total cross sections for the direct reaction products and the fusion reaction, angular distributions for the yields are measured and then integrated as a function of angle.

Distribution of total reaction strength. For the mass region $10 \leq A_{\text{projectile}}$, $A_{\text{target}} \leq 40$, the distribution of total reaction strength as a function of the inverse of the center-of-mass energy, $1/E_{\text{c.m.}}$, has been found to show the general behavior illustrated in Fig. 3. At low bombarding energies, the total fusion cross section, $\sigma_{\text{fus}}(E)$, grows linearly versus $1/E_{\text{c.m.}}$ and is found to account for approximately 90% of the total reaction cross section $\sigma_{\text{react}}(E)$. At a bombarding energy of roughly two to three times the Coulomb barrier energy, $\sigma_{\text{fus}}(E)$ is found to "saturate" and remains approximately constant (or decreases slowly for the lighter systems) as the energy increases, while $\sigma_{\text{react}}(E)$ continues to grow. At still higher energies, there is evidence that $\sigma_{\text{fus}}(E)$ begins to decrease rather sharply as the energy increases further. In the lower energy region, the direct reactions account for approximately 10% of $\sigma_{\text{react}}(E)$ and are found to be quasielastic in character. As one goes to higher energies, there is a sudden increase in the direct strength so that, as can be seen in Fig. 3, these processes account for 50% of $\sigma_{\text{react}}(E)$ at the higher energies. It is observed that in this energy region, the quasielastic processes have grown in strength by perhaps only a factor of 2 and that the great bulk of the direct strength resides in reactions characterized by large losses of kinetic energy. These deeply inelastic processes do not fit into the usual picture of a direct process, and the mechanism by which they arise is not clear. It does appear clear, however, that the mechanism is intimately related to the observed saturation of the fusion cross section.

Microscopic aspects of fusion reaction behavior. The cross-section behavior illustrated in Fig. 3 represents the average or "macroscopic" cross-section behavior, determined by studying a large

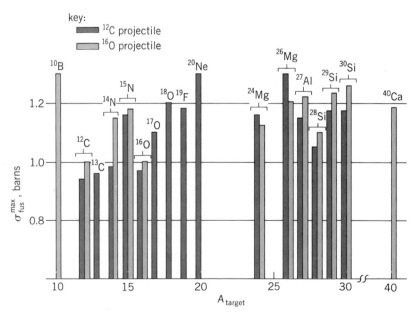

Fig. 4. Variations in the maximum total fusion cross sections $\sigma_{\text{fus}}^{\text{max}}$ observed in ^{12}C- and ^{16}O-induced reactions on a variety of targets. The average uncertainties in these cross sections are about ±5%. 1 barn $= 10^{-28}$ m^2.

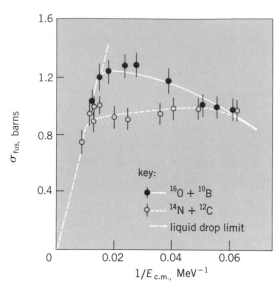

Fig. 5. Fusion cross sections σ_{fus} for the $^{16}O + ^{10}B$ and $^{14}N + ^{12}C$ systems plotted versus $1/E_{c.m.}$. The curves are meant to guide the eye. Straight line intersecting the origin gives the expected trend of the cross section for a constant maximum angular momentum of 27 $\hbar$, the liquid drop limit. 1 barn $= 10^{-28}$ m². *(From J. Gomez del Campo et al., Comparison of fusion cross sections for $^{10}B + ^{16}O \rightarrow ^{26}Al$ and $^{12}C + ^{14}N \rightarrow ^{26}Al$, Phys. Rev. Lett., 43: 26–29, 1979)*

number of projectile-target systems. The observed fusion cross-section behavior is similar to that predicted by the simple semiclassical models discussed above. When examined in detail, however, it is found that these models, which are based on the macroscopic properties of the interaction nuclei, cannot describe the observed behaviors. In particular, the studies reveal that deviations from the average behavior are present which show a dependence on the projectile-target combination. These differences manifest themselves in the shape and magnitude of the fusion cross-section behavior as a function of the bombarding energy. The two most convincing examples of this are the observation of significant differences in the magnitude of the maximum fusion cross sections measured for systems which differ by only a nucleon or two, and the observation of oscillatory or resonancelike structures in the fusion cross sections for some light systems.

Differences between similar systems. The maximum fusion cross sections, σ_{fus}^{max}, observed for reactions induced by ^{12}C and ^{16}O ions on light- to medium-weight nuclei (Fig. 4), are observed to change by as much as 200 millibarns (2×10^{-29} m²) for systems which differ by only a single nucleon. Perhaps even more interesting is the observation that for systems which form the same compound nucleus, large differences in σ_{fus}^{max} are found. This is illustrated in the study of J. Gomez del Campo and coworkers, who studied the $^{14}N + ^{12}C$ and $^{16}O + ^{10}B$ systems which form the compound nucleus ^{26}Al (Fig. 5). The fusion cross-section magnitudes differ by approximately 200 millibarns (2×10^{-29} m²). Macroscopic models which incorporate the measure half-density radii of the interacting nuclei predict fusion cross-section behaviors for the two systems which are very similar and which

closely follow the behavior observed for $^{14}N + ^{12}C$. The $^{16}O + ^{10}B$ results are anomalously larger and are not understood. These results and those of Fig. 4 are taken as evidence for the influence of individual nucleons on $\sigma_{fus}(E)$ above and beyond their static contribution to the mean-nucleon-density and interaction potential. The data in Fig. 5 show the sharp decrease in cross section at the highest energies measured, which has been interpreted as evidence of the liquid drop limit. The predicted cross-section behavior is indicated by the straight line intersecting the origin in Fig. 5.

Oscillatory structures. A second class of observations which indicate the influence of the detailed structure of the nuclei involved in the fusion process was first noted by P. Sperr and coworkers. In this study, it was shown that at energies well above the Coulomb barrier the total fusion cross-section excitation functions for the light systems $^{12}C + ^{12}C$ and $^{12}C + ^{16}O$ display oscillatory or resonancelike structures. Subsequent studies have shown that only one other system, $^{16}O + ^{16}O$, shows such prominent structure. In the $^{12}C + ^{12}C$ system, (Fig. 6) the structures are approximately 2–3 MeV wide, with spacings of roughly 6 MeV. These structures have been found to be correlated with similar structures in the excitation functions for elastic and inelastic scattering. The behavior is believed to be the unexpected continuity to higher energies of a phenomena observed by D. A. Bromley and coworkers in measurements at bombarding energies below and near the Coulomb barrier. The phenomena are believed to be quasimolecular excitations, in which the nucleus (^{24}Mg) appears to have the form of two smaller nuclei (^{12}C) orbiting about each other. The excitation functions, in this picture, are sensitive to and show the locations of these special configurations which lie at excitation energies of approximately 20–30 MeV in the ^{24}Mg nucleus.

Present status. When the experimental results obtained thus far are viewed in the context of the simple semiclassical fusion models that have been proposed, there are several conclusions that can be drawn. In the lower-energy region, the average cross-section behaviors observed appear to be rather well understood in terms of the simple entrance channel model, and there is little evidence

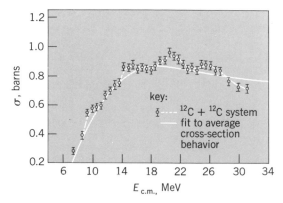

Fig. 6. Total fusion cross-section excitation function σ for the $^{12}C + ^{12}C$ system in which oscillatory structures are clearly seen. Curve through data points is meant to guide the eye. A fit to the average cross-section behavior is also shown.

for a strong dependence on the detailed structure of the interacting nuclei. The exceptions to this are the three systems which have been observed to show the oscillatory structure in their excitation functions. The presence of these structures in these systems but not others and the fact that they are correlated with similar structures in other reaction channels clearly indicate that the detailed properties of the interacting nuclei play an important role. At the higher energies, it is experimentally observed that the maximum fusion cross sections found in ^{12}C- and ^{16}O-induced reactions on light- and medium-weight targets show significant differences which are not consistent with the predicted behavior of the simple entrance channel model or with limitations imposed by the compound nuclei yrast line. More sophisticated entrance channel model calculations which include the effects of the dissipation of energy and angular momentum in the surface region have been able to reproduce rather well the basic trend experimentally observed for heavy-ion systems which span the periodic table; however, they have not been able to account for the variations of maximum fusion cross sections observed.

Experimental studies are under way to investigate in more detail the deeply inelastic direct processes which appear at the bombarding energy at which the fusion cross sections saturate. These studies will attempt to discover what role the detailed structures of the target and projectile have in dissipation of the kinetic energy and angular momentum in the surface region. Such dissipation effects will also influence the predicted limitations imposed by the yrast states in the compound nucleus. Whatever the explanation, the observed fusion cross-section behavior depends sensitively on the detailed structure of the nuclei involved, and there is the promise of extracting information about the properties of nuclei at high excitation. At the highest energies studied, there are observations that the total fusion cross section decreases sharply with increasing energy which has been interpreted as evidence of the liquid drop limit. While more experimental data are necessary to substantiate this behavior, such information about stability of nuclei at high excitation energy and large deformation would add greatly to knowledge of these systems.

The significance of the experimental results obtained thus far lies in the fact that the energy dependence of the total fusion cross sections (and the distributions of total reaction strength), while dominated for the most part by the macroscopic properties of the interacting nuclei, nevertheless shows the influence of the detailed structure of nuclei. This sensitivity makes these cross-section behaviors a probe of the properties of the nucleus which are inaccessible by any other means.

For background information see MOLECULE, NUCLEAR; NUCLEAR REACTION; NUCLEAR STRUCTURE in the McGraw-Hill Encyclopedia of Science and Technology.

[DENNIS G. KOVAR]

Bibliography: J. R. Birkelund et al., Phys. Rep., 56(3):107–166, 1979; D. G. Kovar et al., Phys. Rev., C20:1305–1331, 1979; B. A. Robson, (ed.), Lecture Notes in Physics 92: Nuclear Interactions, 1979.

Oil shale

Modified in situ (MIS) processing of oil shale to produce shale oil is receiving increasing recognition as a leading candidate for early commercialization among synthetic fuel technologies. There are two major field projects and a host of laboratory studies under way to improve, expand, and test the technology. The most significant recent experiment was Occidental Petroleum Corporation's Retort 6. This test showed that very large rubble piles can be created underground with acceptably small variations in permeability, but failure of a pillar during startup led to less than expected oil yield.

MIS process. To recover oil from shale, it is necessary to heat the shale to about 800°F (425°C). The organic component of the rock (kerogen) decomposes to produce a liquid fuel similar to petroleum and a hydrocarbon-containing gas. Residual carbon is also left on the mineral matrix. The basic problem is that oil shale is a low-grade resource, containing 80% or more unwanted rock. Mining this material and bringing it to the surface for processing, the conventional approach for producing oil from shale, requires moving huge tonnages of rock to produce significant quantities of oil. The key to the MIS process is to retort large volumes of rock with a minimum of materials handling.

Typical oil shale is hard and nonporous, and is generally unsuitable for in situ retorting without prior fracturing. The MIS concept involves creation of voids in the formation by conventional mining. These voids are then distributed through a large volume of rock by using explosives to fragment, or rubbleize, the formation around the mined voids. This procedure creates a dimensionally controlled underground chamber with permeability adequate to allow the flow of gases through the mass of broken shale. Generally the mined void fraction (the part of the retort that is empty space, air, or gas) and the resulting porosity within a retort are in the range of 15–25%. Because some mining is required to create the necessary permeability, the process is usually called modified in situ, as opposed to true in situ, where no conventional mining is needed.

The retorting is started by heating the rubble pile at the top, using an outside energy source. Once the top is hot, the use of fuel is discontinued and retorting is continued by passing air and diluent gas down through the rubble. The diluent gas is usually steam, which serves to control the temperature of the underground combustion.

An operating retort contains four major zones, as shown in the illustration. Within the first, or preheat, zone the inlet gas is heated by contact with hot spent shale. The hot gas then reaches a combustion zone in which the oxygen is consumed by burning residual carbon in the spent shale. Some of the product oil and gas may also be burned. Below the combustion zone is the retorting zone in which the hot combustion gases thermally decompose the organic component of the shale to produce oil and hydrocarbon gas. In the final zone the combustion and retorting gases are cooled as they flow downward, condensing some water and vaporized oil. The oil and water that flow from the bottom of the retort collect in a sump located be-

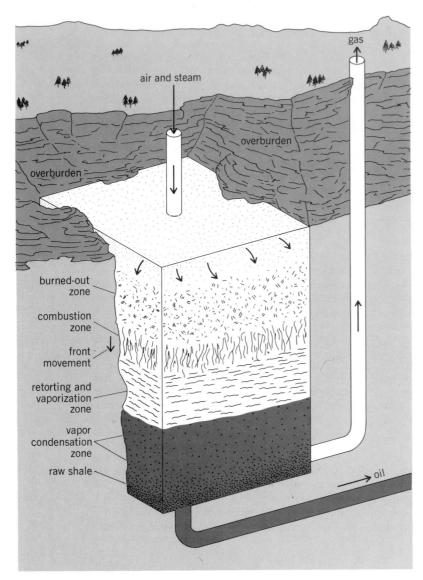

Modified in situ (MIS) oil shale retorting process, showing the four operating zones. (*Occidental Research Corp.*)

hind a gas-tight bulkhead at the base of the retort. After initial oil and water separation takes place, the products are pumped to the surface through separate systems. The operation is continued until all of the shale has been heated and the oil recovered.

Earlier field experiments. The most critical step in the development of the MIS process was to demonstrate in the field both rubblization and in situ retorting. No amount of theoretical modeling or pilot plant work could answer the basic question of field operability. The first field tests were performed by Occidental Petroleum beginning in the summer of 1972. In the first three tests, the retort cross sections were about 1000 ft² (93 m²), and the heights varied from 72 to 113 ft (22 to 34 m).

Although these earlier operations encountered startup problems, the results were very encouraging. The oil recoveries ranged from 75 to 90% of those predicted by mathematical process model-

ing, with the best results given in Retort 3, which was burned in 1975. Retorts 4, 5 and 6 were much larger. Retorts 4 and 5 were approximately 120 ft (37 m) square; Retort 6 was 162 ft (49 m) square. Retort height varied from 167 ft (51 m) for Retort 5 to over 250 ft (76 m) for Retorts 4 and 6. Retort 6 is close to the size planned for commercial operation.

These larger-scale experiments disclosed the need to deal with a number of scale-up problems. One problem has been nonuniform distribution of gas flow through the rubble. In Retort 4 some areas of the retort were not properly rubblized. Only about 60% of the maximum recoverable oil was produced. Retort 5 apparently had developed a major channel down the center of the retort. Efforts to control flow in spite of this channeling were not very effective, and recovery from this retort was about 30% of the expected yield.

Field Retort 6. Retort 6, ignited Aug. 28, 1978, was more successful. The operator recovered 48,000 bbl (7600 m³), about three-quarters of the expected oil production. Tracer tests showed the rubble pile created for Retort 6 to be significantly more uniform than in the previous large retorts. However, 16 days after ignition, the roof of the retort was breached, and the shale bed was irreparably damaged. At first, this event did not seem to seriously affect the overall operation. However, the higher oil production rates expected in February 1979 failed to materialize. It is likely that significant amounts of oil were lost because of regions of the bed being bypassed, or the product oil becoming coked as a result of the operational upset and subsequent loss of control of the inlet air flow distribution. The regions of the retort that operated normally apparently produced the expected amounts of oil.

Current field activity. Occidental started construction of its next two large, commercial-size, experimental retorts at Logan Wash in 1979. Like Retorts 5 and 6, this work is partially supported by Department of Energy (DOE) funds. Scheduled for ignition late in 1981, these retorts are similar in design to Retort 6, with some modification to the inlet and outlet in order to give higher sweep efficiencies (oil recovery) and avoid the retort roof problems that occurred in that experiment. They will be operated in parallel, the first time that multiple-retort operation has been attempted. Proposed commercial designs call for as many as 100 retorts on stream simultaneously.

Rio Blanco Oil Shale, a partnership of Gulf Oil and Standard Oil (Indiana), has purchased access to this technology for use in a development and testing program, and has planned field tests of the MIS process. The key difference in approach is Rio Blanco's plan to operate the retorts at faster processing rates. In order to achieve this, the void fraction will be greater, about 0.40. Air will be supplied from the surface directly to the top of the retorts, and thereby the need for a second mine level will be avoided. This is feasible because the overburden thickness on the Rio Blanco property is quite modest, only about 500 ft (150 m). If mining is confined to the bottom of the retorts, higher pressures might be possible in the upper parts of the rubble without endangering the mine workings by the leakage of toxic gases from the retort. The higher void fraction and operating pressure would give higher gas flow and faster retorting. Rio Blanco hopes that flow distribution across the retorts will also be more uniform.

Occidental Oil Shale, Inc., is a partner with Tenneco Shale Oil Company in commercial development of the Colorado C-b tract as the Cathedral Bluffs Oil Shale Project. Shaft sinking is well under way, and current plans call for initial light-off in 1985 and full production rates of over 50,000 bbl/day (8000 m³/day) by 1988.

Related laboratory work. While most of the key technological questions can be resolved only in the field, DOE is also funding a wide variety of support programs at major national laboratories and in universities. Sandia National Laboratories is developing and testing instrumentation for in situ processing; Los Alamos Scientific Laboratory is developing basic theory and data for rock fragmentation; and Lawrence Livermore Laboratory is studying retorting in the laboratory and pilot plant. Livermore has made contributions in reaction kinetics and process modeling in the past. Current activities there center on understanding the effects of bed nonuniformities, on determining retort control strategy from the compositions of the oil and gas being produced, and on improving the process model.

The effect of MIS processing on the environment is also being examined intensively, primarily with DOE funds. The most serious concern seems to be possible contamination of groundwater by contact with spent shale. The high operating temperatures and slow flame front movement in MIS retorts tend to minimize the soluble salts in spent shale. There are still unresolved questions about some trace elements such as fluorine, boron, and selenium, and about possible organic contaminants. It appears likely that the environmental impact can be controlled, but the cost will not be well defined until full-sized commercial facilities are in operation.

For background information *see* OIL SHALE in the McGraw-Hill Encyclopedia of Science and Technology.

[ROBERT E. LUMPKIN]

Bibliography: W. R. Chappell, in *12th Oil Shale Symposium Proceedings*, Colorado School of Mines, pp. 149–155, August 1979; M. C. T. Kuo et al., in *12th Oil Shale Symposium Proceedings*, Colorado School of Mines, pp. 81–93, August 1979; D. J. Murphy, *IGT Symposium: Synthetic Fuels from Oil Shale*, Atlanta, Dec. 3–6, 1979. A. J. Rothman, *LLL Oil Shale Project Quarterly Reports*, UCID 16986-79–1, 2, 3, and 4, 1979.

Optical communications

Optical fiber telecommunications systems have moved from concept to production in a period of 15 years, since 1966. Today they bring simplification to the telecommunications network in terms of fewer electronic amplifiers or repeaters, smaller cables, and greater freedom from cross talk or interference. In the future, they offer the prospect of new wide-band services to the home or business customer, during a period when energy problems will encourage reduction in travel, and associated

very-high-performance long-haul transmission systems for land or undersea application.

Early development. The first optical fiber communication systems are now entering the world's telecommunications networks. A number of key developments have led to this situation. In 1962 the first semiconductor laser was described, but it was 1970 before this had been developed to the point that it would run continuously at room temperature. In 1966 the first serious proposals to use glass fibers to carry optical signals in telecommunications were made, and estimates were given of the attenuation that it would be necessary to achieve in such fibers before the systems could be viable. These led to the general conclusion that an attenuation of about 20 dB/km was required. This figure was based upon the assumption that repeaters or amplifiers would have to be spaced at about 2-km intervals as with the then current systems.

Fiber structure. These early proposals assumed that the fiber would be of the monomode type. All glass fibers are characterized by a core of one type of glass surrounded by a cladding of a lower refractive index or dielectric constant glass. Light is guided predominantly in the core region. In the monomode fiber, the core is very small, typically 3–10 μm in diameter, and consequently these fibers are difficult ones with which to work because jointing and launching of energy into them requires very precise alignment. Early in the 1970s it was realized that a multimode fiber, having a very much larger core diameter of 50–60 μm, could give a perfectly adequate performance and greatly ease the problems of installation. To achieve this performance, however, the fiber had to be of the graded index type. In this fiber, the refractive index of the core glass is varied with respect to radius in a parabolic fashion (Fig. 1). The variation of refractive index with radius has to be very closely controlled, and this presents manufacturing problems, but today such fibers can be made with good yield and with exceptionally high performance.

Increased repeater spacing. By 1970 the first fiber having an attenuation of 20 dB/km had been made, but it was not until the mid-1970s that fibers

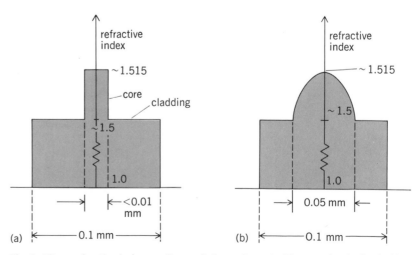

Fig. 1. Fiber refractive index profiles and dimensions. (a) Monomode. (b) Graded index.

with losses that were attractive for real-systems trials were readily available. By then, the fiber attenuation had fallen to between 4 and 5 dB/km, and this meant that repeaters could be spaced at 6–10-km intervals. From 1977 to 1979 a series of trial systems was demonstrated throughout Europe, North America, and Japan. The characteristics of these systems in terms of repeater spacing are shown in Fig. 2. The achievement of a fiber attenuation of around 4 dB/km allows the large repeater spacings shown, and the graded index fibers used had sufficiently low pulse spreading so that systems operating to as high as 140 megabits/s (1920 telephone conversations per fiber) were possible. For comparison purposes, Fig. 2 also shows

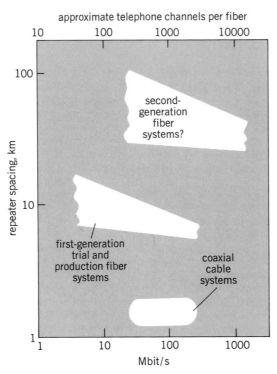

Fig. 2. Repeater spacings of fiber system compared to previous coaxial cable systems.

the repeater spacings achieved in the coaxial cable systems that these fiber systems are destined to replace, indicating that much greater repeater spacings are achieved by a factor of 4 to 5, and consequently the amount of electronics that has to be buried under the ground is dramatically reduced. Furthermore, for links between switching centers inside cities, the distances between these centers is such that optical fibers allow repeaterless connection with the result that here buried electronics will become a thing of the past.

Consequently, production systems are being installed. In the United Kingdom, systems operating at 8 Mbit/s, 34 Mbit/s, and 140 Mbit/s are in production. These digital bit rates correspond to 120, 480, and 1920 telephone channels per optical fiber, respectively. In North America and Japan, similar digital rates are being used to fit in with the standard digital hierarchies in those countries.

Principle of operation. In all these systems, the principle of operation is the same. A small semiconductor laser light source is switched off and on by the incoming digital signal, laser on for a 1 and off for a 0. The pulses of light so generated are carried along the optical fiber and are then detected, usually with a silicon avalanche detector, and converted back into electrical form in the next repeater, where they are amplified and squared off prior to retransmission by another light source. In addition to the very long distance that the light can travel before it needs to be amplified, the fact that stray light cannot enter an optical fiber cable in the way that electrical pick-up can enter a metallic cable means that these systems have an exceptional resistance to electrical interference. Consequently, they are particularly suitable for use in difficult electrical environments and as such are finding application for signaling alongside electrified railway lines or in numerous military or factory environments.

Long-wavelength systems. The fiber systems described above all operate with sources made from gallium arsenide. These emit light at about 850 nm wavelength, which is just beyond the red end of the visible spectrum. At this wavelength, glass has a minimum attenuation of about 2 dB/km set by the Rayleigh scattering. The fiber cables referred to above with losses of 4 dB/km thus have an additional 2 dB/km attenuation arising from impurity in the glass and from imperfections in the fiber and the lay of the fiber within the cable. However, the Rayleigh scattering decreases as the reciprocal of the fourth power of wavelength. Consequently, at a wavelength of 1550 nm, instead of 2 dB/km, the attenuation due to Rayleigh scattering is only 0.1–0.2 dB/km. Advance in fiber fabrication techniques over the last few years has allowed fiber having a total attenuation of 0.2–0.4 dB/km at this wavelength to become fairly commonly available. The implications of these very low losses are quite striking. A 10-km link of 4 dB/km of fiber corresponds to an attenuation of 40 dB. The same attenuation with 0.2 dB/km corresponds to a distance of 200 km. In practice, such low losses will probably not be achieved in real jointed fiber links, but even if the finished installed attenuation should be as high as 0.4 dB/km, one sees immediately that one could be considering distances of as much as 100 km of fiber without any intermediate electronics required.

Such immense distances are of great attraction to telecommunications systems designers for use in high-speed data links between cities and for use in submarine systems. The realistic prospect of repeater spacings of 50–100 km carrying fairly high data rates in the 100–1000 Mbit/s level implies a fiber having an exceptionally large bandwidth. It seems virtually certain that this will be achieved by using the monomode fiber already referred to. Consequently, there is today a very large research activity aimed at solving the problems of working with this fiber of very small core diameter in the practical environment that the system has to work in. Simultaneously with this, it has been necessary to develop both new sources and new detectors. Lasers based upon gallium arsenide can operate only at wavelengths to about 900 nm, and to achieve the longer-wavelength operation at about 1550 nm requires a laser fabricated from the quaternary materials system gallium indium arsenide phosphide. To complement the sources, detectors also have to be fabricated in a related material system, and a likely material is gallium indium arsenide in which very-high-performance PIN detectors have been demonstrated with adequate sensitivities for high performance systems. Another possibility is an avalanche photodiode fabricated in germanium as opposed to silicon used in the first systems. However, while the silicon avalanche photodiode is very nearly ideal for use at wavelengths shorter than 1 μm, the germanium avalanche photodiode is a very noisy device and consequently does not offer as good a sensitivity as the PIN detector fabricated from gallium indium arsenide. Furthermore, the avalanche photodiode requires a high voltage supply and a control circuit to control the gain, while the PIN detector, mounted in a hybrid package with a field-effect-transistor (FET) amplifier can run off a 15-V rail. It seems likely that the first trial systems of this advanced monomode type will be seen during the next 2 or 3 years.

Optical fibers in the home or business. Another area of intense activity concerns the use of optical fibers to the private home. Most activity in this area envisages the combination of familiar telecommunications services with the type of television service that is provided by the more advanced community antenna television (CATV) networks, further coupled with new telecommunications information services and possibly with a video phone. The fiber is seen as providing a wide-band cable into the private home, running from a central wide-band switch point. It thus leads to the development of a fully integrated wide-band distribution network which can provide to the home an almost unlimited range of services encompassing video, hi-fi audio, and data. It is fairly clear that the 1980s will be a period of great experimentation in the use of the services and in the impact that they have on a community. One of the greatest uncertainties in attempting to design such wide-band networks is the almost total lack of knowledge of how a community would in fact make use of such a service, since there is very little real experience to draw upon at present.

Implications. In the next 5 years, fiber systems will increasingly be used in place of metal cable systems, but to do much the same tasks. Toward the end of the 1980s, other effects may become apparent. The increasing penetration of fiber links in the telecommunications network, coupled with the very long repeater spacings they will allow, may lead to a telecommunications network that is much more versatile than today's, a trend also encouraged by an increasing use of digital transmission techniques. In the local network, leading to private homes and businesses, optical fibers will begin to make available low-cost wide-bandwidth connections, which will encourage the use of wide-band services such as high-speed facsimile, video teleconference, or wide-band information retrieval and display. This trend is likely to be encouraged by increasing energy and travel costs. Thus it seems probable that by the 1990s optical fibers

together with a wide range of digital switching, storage, display, and processing technology will make possible the much more widespread use of electronic communication in the place of direct human communication and the travel associated with it.

For background information *see* JUNCTION DIODE; LASER; OPTICAL COMMUNICATIONS; SCATTERING OF ELECTROMAGNETIC RADIATION in the McGraw-Hill Encyclopedia of Science and Technology.   [JOHN MIDWINTER]

Bibliography: J. E. Midwinter, *Optical Fibers for Transmission*, 1979; S. E. Miller and A. G. Chynoweth, *Optical Fiber Telecommunications*, 1979; G. H. B. Thompson, *Physics of Semiconductor Laser Diodes*, 1980.

Organic solids

Organic solids have been the subject of rapidly increasing research effort during the past decade because of their increasing prominence as model materials illustrating new and unexpected solid-state phenomena and their growing utilization as electronic and optical, as well as structural, materials. One special class of organic solids, namely, polymers, have enjoyed a renaissance in the interest shown in their electrical properties associated with new applications in the electrophotographic and electronics industries and with the quasi-one-dimensional nature of electron motion within the individual macromolecules which compose a polymer. An example of a polymer whose electrical properties make it technologically useful for trans-

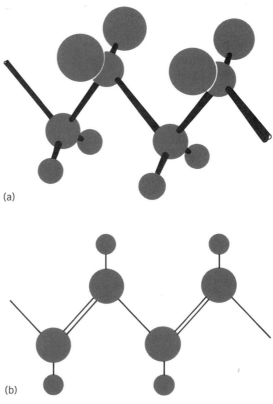

Fig. 1. Schematic indications of polymer structures: (*a*) poly(vinylidene fluoride), and (*b*) polyacetylene. Solid lines indicate chemical bonds.

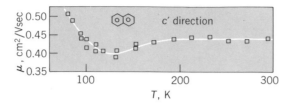

Fig. 2. Mobilities of electrons in the c' direction of naphthalene, $C_{10}H_8$, from 54 to 300 K. Below 100 K the mobility rises with decreasing temperature; above 100 K the mobility is almost temperature-independent. These data indicate that a band-to-hopping transition occurs near 100 K.

ducers is poly(vinylidene fluoride), or PVF_2 (Fig. 1a), whereas an example of one thought to exhibit properties uniquely associated with quasi-one-dimensional electron motion is polyacetylene, or $(CH)_x$ (Fig. 1b).

Electron localization. One important phenomenon characteristic of organic solids is the localization of electronic states due to the absence of periodic long-range order. The prominence of electron localization in these materials is intimately associated with their molecular nature. In the solid state, organic molecules retain their identity and interact only weakly, usually through van der Waals forces. The weakness of the intermolecular interactions in organic solids makes them prone to be disordered owing either to a noncrystalline array of molecules (as in amorphous thin films and polymers) or to the thermal motions of the molecules. Because the interactions between electrons on different molecules also are weak, even in molecular crystals the thermally induced disorder often suffices to localize injected charges on individual molecular constituents, thereby producing molecular anions and cations, rather than "electrons" and "holes" which are extended over many unit cells in the crystal and are characteristic of crystalline network semiconductors like silicon. In certain insulating van der Waals crystals like naphthalene, with increasing temperature T the thermal disorder induces a transition between extended-state motion of carriers in energy bands at low temperatures (T less than about 100 K) and thermally assisted hopping motion between localized states at high temperatures. An example of the behavior of the mobilities μ of electrons in the c' direction in naphthalene which mirror this transition is shown in Fig. 2. Carrier mobilities in most polymers are much smaller, $\mu < 10^{-4}$ cm²/Vs, and exhibit an activated behavior, $\mu = \mu_o \exp(-E_a/\kappa T)$, associated with an activation energy E_a caused by static disorder. The symbol κ designates Boltzmann's constant, and μ_o is a function of the distribution of localized electron sites in the material.

Collective behavior. Certain organic polymers and crystalline charge-transfer salts characterized by special segregated-stack molecular structures (Fig. 3) are thought to exhibit another new solid-state phenomenon: collective insulating behavior. This behavior is caused by the occurrence of a collective ground state called the Peierls insulator or charge-density-wave state. Such a state is associated with the one-dimensional character of electron motion along the macromolecules which com-

pose a polymer or along a line of closely spaced molecules in the segregated-stack charge-transfer crystal geometry. It is analogous in many ways to a superconducting ground state, but leads to insulating rather than superconducting electrical properties.

The expected features of the Peierls insulating state can be illustrated by the case of polyacetylene, $(CH)_x$, for which the macromolecular geometry is indicated in Fig. 1b. If all of the carbon-carbon bond lengths in $(CH)_x$ were equal, then in a one-electron model each individual macromolecule would exhibit metallic behavior because there is one additional electron per carbon atom, associated with p_z orbitals normal to the plane of the $(CH)_x$ chain, which form a half-filled band. This behavior does not occur in one dimension, however, because interactions between the electrons give rise to an insulator ground state associated either with a magnetic (spin-density-wave) or with a nonmagnetic (charge-density-wave) electron density, the periodicity of which is determined so that

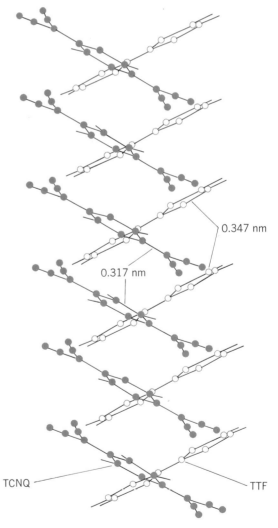

Fig. 3. Schematic diagram of the segregated-stack extended-linear-chain organic conductor (TTF)(TCNQ), consisting of equal compositions of tetrathiofulvalene (TTF) and 7,7,8,8-tetracyano-p-quinodimethane (TCNQ).

the one-electron excitation spectrum exhibits a gap at the Fermi energy. Of course, a model in which the carbon-carbon bond lengths are constrained to be equal is unrealistic for actual solids, because the interactions of the electrons with the normal-mode vibrations (phonons) of the macromolecule also must be considered. This extension of the model permits two new types of collective ground state to occur: the Peierls insulator state in which the development of a charge density wave is accompanied by a static lattice distortion so that the bond lengths alternate, and a superconducting state in which the bond lengths remain equal. By extending the model further to include electron interactions with other excitations and with impurities, additional types of collective ground state can be identified. For example, it has been speculated that excitonic superconductors could occur in organic solids, presumably stabilized by attractive electron-electron interactions caused by exciton (electron-hole-pair) exchange rather than phonon exchange between the interacting electrons. Of these many possibilities, the Peierls insulator ground state seems most appropriate for $(CH)_x$, because $(CH)_x$ exhibits insulating behavior and (probably) alternating bond lengths as indicated in Fig. 1b by alternating double (short) and single (long) bonds.

The existence of Peierls insulator ground states is difficult to confirm experimentally unless a metal-to-insulator transition is observed in which the energy of a normal mode of vibration of the metal lattice goes to zero as the transition is approached from the metallic region. Definitive evidence for such "soft" vibrational modes has proved difficult to acquire for organic solids, so that indications of collective insulating ground-state behavior have come from other sources, such as the use of x-ray scattering to determine the periodicity of the lattice relaxation accompanying the electronic charge density wave or the use of ac conductivity to observe manifestations of the movement of these charge density waves. The occurrence of collective ground states of either insulating or superconducting character in organic materials is still a matter of controversy and ongoing research. Nevertheless, the search for such states in those organic solids which are thought to represent quasi-one-dimensional systems has provided considerable stimulus for the recent resurgence of interest in the organic solid state.

Preparation and properties. Another reason for recent interest in organic solids is the increased availability of sample materials. During the past decade enormous strides have been made in the preparation of pure crystals of organic materials, especially charge-transfer salts which crystallize in the segregated-stack structure as illustrated in Fig. 3 for one of the most widely known compounds in this class, (TTF)(TCNQ). The synthesis and characterization of these crystals has been one of the major ingredients in recent efforts to study collective behavior in quasi-one-dimensional organic systems. Moreover, improved purification and handling techniques of such a traditional organic material as crystalline naphthalene were essential to the direct observation of the band-to-hopping transition illustrated in Fig. 2. Examples

of practical applications of improved crystal growth techniques include the preparation of phthalocyanine for solar cells and aggregate photoconductors, and of urea and 2-methyl-4-nitroaniline for use as nonlinear elements in optical circuits.

Important advances in polymer preparation during the past decade include new techniques to produce electrically active materials by doping and to fabricate polymers in thin-film form. For example, photoconducting polymer films have been synthesized by doping poly(N-vinyl carbazole) with 2,4,7-trinitro-9-fluorenone; p-type semiconductors by doping polyacetylene with arsenic pentafluoride; and p-type (hole) transport materials by doping polycarbonates with triphenylamine. Moreover, the fabrication of semiconducting films of doped polyacetylene was enabled by the development of methods to prepare polyacetylene in thin-film form. Similarly, piezoelectric films of poly(vinylidene fluoride) can now be prepared because of the invention of orientation and poling processes which generate electrically active thin-film sheets of this material. These techniques are appropriately viewed as recent steps in a long-standing trend of new process development for the preparation of thin-film polymers for electronics applications, including their well-established uses as dielectric spacers in capacitors, insulating layers for wires and cables, and photoresists in microelectronic circuit fabrication.

Applications. Traditionally, the various uses of polymers as structural materials, especially in the form of fibers and molded parts, have constituted the main applications of organic solids. Recently, however, the electrical properties of polymers have become increasingly significant for specialty applications in the electronics, electrophotographic, and communications industries. The use of polymers as dielectrics and photoresists in the electronics industry already has been noted; this is an arena of applications which is continually being expanded with the advent of electron-beam and x-ray resists, together with plasma processing for microelectronics fabrication. Historically utilized as the dry "ink" (that is, developer) materials in the electrophotographic industry, in recent years organic materials have become popular for photoreceptors as well. With the advent of optical communications and signal processing, organics are playing an expanded role in the communications industry as light guides and nonlinear circuit elements. Indeed, they also are increasingly employed in traditional optical systems such as lenses and filters, and specialty components such as cements and polarizers. The use of poly(vinylidene fluoride) in transducers and loudspeakers, as well as that of other organics in solar cells, was mentioned earlier. Given the modern trends toward less expensive computation via microelectronics, the use of renewable energy sources via solar cells, and the substitution of energy-efficient communication for energy-intensive travel, one can confidently anticipate a growing role of organic solids in microelectronics, photovoltaics, photochemistry, and optical communications, as well as in their traditional uses as lightweight, cost-effective structural materials.

For background information *see* BAND THEORY OF SOLIDS; EXCITON; POLYMER PROPERTIES; SUPERCONDUCTIVITY in the McGraw-Hill Encyclopedia of Science and Technology.

[CHARLES B. DUKE]

Bibliography: H. Bloch, *Advan. Polymer Sci.*, 33:93–174, 1979; M. J. Bowen, *Crit. Rev. Solid State Mater. Sci.*, 8:223–264, 1979; J. T. Devreese, R. P. Evrard, and V. E. van Doren (eds.), *Highly Conducting One-Dimensional Solids*, 1979; C. B. Duke and L. B. Schein, *Phys. Today*, 33(2): 42–48, 1980; J. S. Miller and A. J. Epstein (eds.), *Synthesis and Properties of Low-Dimensional Materials*, New York Academy of Sciences, 1978.

Osteichthyes

The paddlefish (*Polyodon spathula*), also known as the spoonbill or spoonbill catfish, is a survivor of a primitive, largely extinct family of fish, Polyodontidae.

Because of its mainly cartilaginous skeleton, scaleless skin, and heterocercal tail, the paddlefish was originally and incorrectly classified as a shark (Chondrichthyes, the cartilaginous fish). It is, however, a true bony fish (Osteichthyes), although it has few bones. Recent studies have demonstrated how little information is known about this important stem family of fish. Its cartilaginous skeleton and heterocercal tail probably help make the fish more buoyant. Buoyancy is apparently a problem since functional flotation structures such as swim bladders are lacking. Paddlefish rarely inhabit deep water, however, and therefore have little need of depth-regulating abilities.

The bony fish are divided into two subclasses based on the anatomy of their fins. One group, Sarcopterygii or lobe-finned fish, possess short, fleshy fins composed of bones surrounded by groups of muscles. This is a small group but of major evolutionary importance as they gave rise to terrestrial vertebrates; the fins evolved into limbs of terrestrial vertebrates, such as arms and legs. The other subclass, Actinopterygii or ray-finned fish, have fins with many thin, spiny rays connected by a thin membrane. The muscles used to move these fins are found inside the body of the fish, not in the fins themselves. Paddlefish (and sturgeons) belong to the infraclass Chondrostei, the oldest and most primitive group of actinopterygians. They gave rise to the infraclass Holostei, which in turn gave rise to the infraclass Teleostei.

The paddlefish make up the order Polyodontiformes. There are only two surviving species of this small order: *Polyodon*, the North American survivor, and *Psepherus gladiator*, the Chinese paddlefish. Both occupy large river systems: *Polyodon* is found in the Mississippi-Missouri river system, and *Psepherus* in the Yangtze River. *Polyodon* is currently distributed from Montana to Louisiana and as far east as Kentucky. Except for the Mississippi-Missouri river system, the only known habitats are the Mobile Bay drainage, some coastal rivers in Texas, and previously (and somewhat unsubstantiated) the Great Lakes and southern Canada. All of these North American localities may have been connected in ancient glacial periods. Exactly how this small order was distributed throughout the world is unclear. Either paddlefish can or could at one time tolerate salt water, or their ancestors were salt-water inhabiters and each surviving group independently occupied its respective river systems. Unfortunately, the few existing fossils of this order do not clarify this situation.

There are a number of similar characteristics between these two species. Besides the characteristic paddle (rostrum), both species possess numerous teeth. However, *Polyodon* loses its teeth at an early age. Both species have a heterocercal tail, a smooth, scaleless skin, a subterminal mouth, relatively large eyes, and two small barbels. *Polyodon* also possesses numerous small sensory organs on the anterior end of its body. The exact function of these organs is unknown but is thought to be electrosensory in nature. It is not known if *Psepherus* possesses them. Both species can reach considerable size. *Polyodon* can reach 2 m in length and 100 kg in weight. *Psepherus* is reported to reach 7 m in length and several hundreds of kilograms in weight. These sizes and weights are similar throughout the entire group of Chondrostei.

These two species differ in their feeding habits. *Psepherus* is a predacious fish-eating carnivore, while *Polyodon* is a filter feeder. *Polyodon* feeds by opening its large mouth and swimming through masses of zooplankton, especially *Daphnia*. These small organisms are caught on the numerous gill processes known as gill rakers and are then swallowed. The paddle or rostrum apparently functions as a plane during feeding and maintains the horizontal position of the fish (see illustration).

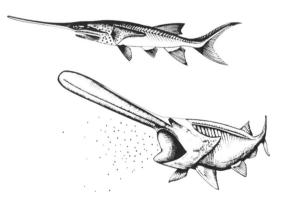

A paddlefish, side view and feeding on a group of *Daphnia*.

The future of the paddlefish is uncertain. The loss of habitat due to pollution is a major concern. River channelization and impoundment have removed natural breeding grounds. Several states (Missouri, South Dakota, and Montana) have developed artificial breeding programs in attempts to reduce the decreased propagation potentials. However, with newer environmental laws, public awareness, and concern, the paddlefish can continue its existence as one of the most primitive bony fish.

For background information *see* CHONDROSTEI; OSTEICHTHYES in the McGraw-Hill Encyclopedia of Science and Technology.   [TODD GEORGI]

Bibliography: T. A. Georgi and D. Beedle, *J. Fish Biol.*, 13:587–590, 1978; W. K. Gregory,

Trans. Amer. Phil. Soc. (N.S. no. 2), 23:7–481, 1933; J. Jorgenson, A. Flock, and J. Wersall, *Zellforsch. Mikrosk. Anat.*, 130:362–377, 1972; R. W. Pasch, P. A. Hackney, and J. A. Holbrook, II, *Trans. Amer. Soc. Fish.*, 109(2):157–167.

Paleomagnetics

The Earth's magnetic field owes its origin to motions throughout its fluid iron-nickel outer core. These motions help to maintain a small axial dipolar field (0.5 nanotesla) because the fluid core can act as an electromagnetic dynamo. In addition to the next axial dipolar component, which normally accounts for 80–90% of the geomagnetic field, the remaining 10–20% is nondipolar and can be modeled as being due to seven or eight weak radial dipoles caused by motions in the outermost part of the fluid core (Fig. 1). Secular variation is the time variation of the internal geomagnetic field as a whole, whereas paleosecular variation is the term given to the ancient history of secular variation as recorded in rocks, sediments, and archeological samples. It has recently become apparent that the paleosecular variation record from lake sediments can be used as a relative age-dating tool, especially for the proxy paleoclimatic fluctuations recorded in the same sediments; while some workers have suggested that certain paleoclimatic fluctuations and the coincidental paleosecular variations may even be causally linked by a mechanism which is as yet not clearly understood. This article describes the character of paleosecular variation and its paleomagnetic record, and then looks critically at the claimed correlations between paleomagnetic and paleoclimatic fluctuations.

Paleosecular variation. Time variation of the geomagnetic field vector can be demonstrated by plotting the instrumentally recorded values of geomagnetic inclination (the local dip below the horizon of the tip of a compass needle) versus geomagnetic declination (the local departure of the tip of a compass needle from true north). The resulting slow motion of the tip of the geomagnetic vector has an apparent periodicity of about 600 years, and is similar at nearby observatory pairs such as London-Paris or Boston-Baltimore, while constrasting variations are seen between the European and the North American observatory pairs. Such behavior seems to indicate that secular variation is caused by nondipole sources that are local in character with regard to intercontinental distances. E. C. Bullard and colleagues were among the first to explain recent secular variation (recorded globally between 1940 and 1950) as due to a constant dipole field and a slow (0.2° longitude per year) westward drift of the nondipole field.

In the last 30 years this simple picture of secular variation has undergone revisions. First, it has been clearly observed that the dipole component of the field also undergoes secular variation, both in intensity and in direction. Second, although some nondipole features exhibit classical westward drift, others can be stationary and some drift in an eastward direction. The present consensus is that secular variation contains a number of periodic components, and although the dipole field does take part in the time variation, the nondipole contribution to secular variation, expressed as east- or westward drift or stationary waxing and waning of regional magnetic highs or lows, is predominant.

One facet of paleosecular variation that has received much recent attention is geomagnetic excursion. It is well known that the geomagnetic field (which, it will be recalled, is 80–90% dipolar) undergoes 180° reversals roughly once every 1,000,000 years. Geomagnetic excursions appear to be aborted reversals when the geomagnetic pole departs sharply from its position (by as much as 135°) but returns to its previous orientation within 1000–10,000 years without actually attaining a fully (180°) reversed orientation. Such excursions have not been recorded historically, but are suspected to have occurred, since they are seen in the paleomagnetic record in rocks, sediments, and archeological samples.

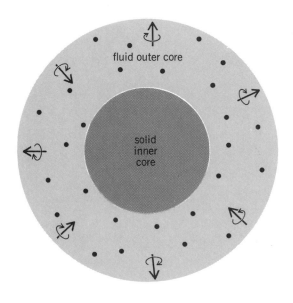

key:

• location of eddies which contribute to observed dipole field

⚓ radial dipoles and associated current loops which contribute to observed nondipole field

Fig. 1. Fluid motions in the Earth's outer core responsible for the Earth's total (dipole and nondipole components) magnetic field.

The record. The history of the geomagnetic field is recorded by three processes: thermoremanent magnetization, depositional or postdepositional remanent magnetization, and chemical remanent magnetization. The history of the Earth's magnetic field is considered in this article as recorded in lake sediments by the second process mentioned above: depositional or postdepositional alignment of a submicroscopic magnetic mineral (usually magnetite) close to the ancient geomagnetic field direction. By careful search, lake sediments with a continuous and high sedimentation rate and a good mix of organic matter (for radiocarbon dating) and magnetite content can be located. A sedimentation rate of about 2 mm/year is not uncommon. Thus a standard 2-cm-thick paleomagnetic sample of lake sediment can provide a record of the direction of the geomagnetic field averaged over only 10 years. The conventional radiocarbon method can provide ages accurate to ±100 years for the last 20,000 years, and before that with increasing errors to an age of about 40,000 years. Fission-track dating or the presence of known marker horizons (for example, volcanic ash) can allow age dating of older lake sediments.

Postglacial lakes contain sediments that are 15,000 to 16,000 years old at the base. Large inland lakes, such as the Biwa Lake in Japan, may contain a 200,000-year-old record in their sediments. Thus sediments cored from present-day lakes are very attractive for their potentially continuous paleomagnetic and proxy paleoclimatic (pollen, seeds, diatoms) records. There are also lake sediments which are presently dry and exposed. The southwestern and western regions of the United States have many such ancient lake deposits which can provide records as old as 1,000,000 years. The magnetic minerals in the wet sediments of present-day lakes may have lain undisturbed since their deposition, whereas the dry-lake sediments may have been exposed to nonuniform mechanical forces during drying and to a change in the chemical environment, both of which can add undesirable secondary components to their natural remanent magnetization (NRM). However, the dry-lake sediments can be sampled easily, and they can supply considerably older records. Under special circumstances that lead to relatively faster sedimentation rates, ocean sediments can compete with dry- and wet-lake sediments in providing continuous paleomagnetic and paleoclimatic records for the past few million years with a high time resolution.

Records of paleosecular variation for Europe have been gathered from postglacial lakes and from the Aegean Sea and the Black Sea. These show that, under favorable recording conditions,

declination and inclination values may show semiperiodic oscillations of 50 to 15° peak-to-peak values for the last 20,000 years. There are regional variations, however, and oscillations present in western Europe are absent in eastern and southeastern Europe. Some of the difference can be attributed to extremely local nondipolar sources, but inefficient or erratic recording must be responsible for some others.

Sharp, aperiodic variations such as geomagnetic excursions have also been seen, and potentially they can be very valuable in providing a framework for relative age dating of paleoclimatic features. K. Verosub and S. K. Banerjee have studied the list of proposed geomagnetic excursions and their paleomagnetic record and have concluded that while some of the proposed excursions may be based on faulty recording processes, there are others that are likely to prove helpful in magnetic age dating. Among the most probable real excursions are the following: at approximately 18,000 years B.P. (Before Present, that is, before 1950), 30,000 years B.P., 50,000 years B.P., and 100,000 years B.P.

Paleoclimatic and paleomagnetic fluctuations. Proxy records of past climatic fluctuations can be found in sediments in the form of tree and herb pollen, seeds, and microfossils such as foraminiferans. The foraminiferans can be analyzed for fluctuations in temperature-sensitive isotopes of oxygen ($\delta^{18}O$) for records of past temperature fluctuations. However, none of the proxy records are perfect, in that they always show minor fluctuations in response to factors other than temperature, so that only major fluctuations should be accepted at present.

A 200-m-long sediment core was obtained from Lake Biwa in Japan, and the upper 30 m, representing roughly the last 60,000 years, have been analyzed by N. Kawai and colleagues for paleomagnetic and paleoclimatic records. The NRM is proportional both to the ancient field intensity and to the concentration of magnetic minerals that contribute to the NRM. In order to remove the influence of the second factor, Kawai and colleagues applied a large, saturating magnetic field at room temperature to all the samples, and then measured the fluctuations in the saturation isothermal remanent magnetization (SIRM). They then divided the observed NRM at each sediment horizon with the corresponding SIRM to obtain a relative paleointensity record. In the absence of a comparative study of this and other available normalization approaches, the ratio NRM/SIRM is the best approximation to true paleointensity fluctuations in Lake Biwa sediments. It should be noted, however, that SIRM normalization has been shown to be an inferior approximation in some other lake sediment studies.

Pollen from herbs and trees representing warm and cool conditions were isolated in the sediments from Lake Biwa; Fig. 2 shows both the pollen and relative paleointensity (NRM/SIRM) fluctuations observed in the sediments. From a visual inspection, Kawai and colleagues concluded that there is a correlation between higher temperatures and higher values of NRM/SIRM. Banerjee and colleagues digitized 200 pairs of data points from the

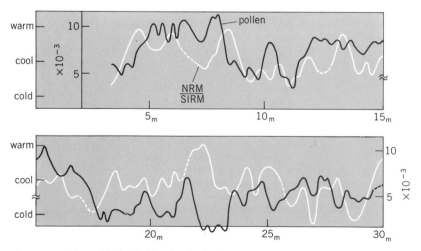

Fig. 2. Variation with depth (that is, time) of proxy records of temperature (from pollen types) and relative geomagnetic field intensity (NRM/ISRM). The time span is 400–60,000 years B.P. (*From N. Kawai et al., Voice of geomagnetism from Lake Biwa, in S. Horie, ed., Paleolimnology of Lake Biwa and the Japanese Pleistocene, vol. 3, Otsu Hydrobiological Station, Kyoto University, 1975*)

data shown and attempted a more quantitative cross correlation, which revealed a correlation coefficient of only +0.34. Thus their analysis confirms Kawai and others' statement that the correlation coefficient is positive in sign, but it is not possible to conclude that the correlation is significant. The correlation did not improve upon consideration of various values of phase lags between the two data sets. Banerjee and colleagues concluded therefore that if the data of Fig. 2 are accepted at face value, temperature and geomagnetic field intensity are directly correlated, albeit at a low statistical significance.

Ocean sediments with relatively high deposition rates have also been used in a search for correlation between paleomagnetic and paleoclimatic fluctuations. Even in the best circumstances, however, each 2-cm-thick ocean sediment sample provides an integrated record of 1000 years, as against 10 years for the corresponding figure for lake sediments. In ocean sediments, with one exception, efforts have been concentrated on the record of the present Brunhes magnetic polarity epoch, which covers the last 700,000 years. The paleoclimatic proxy records from these sediments can be obtained by a variety of methods: $\delta^{18}O$ ratios, abundance of a special type of foraminiferan, rate of accumulation of carbonates, and so forth. D. Kent and N. Opdyke performed power spectrum analysis of relative paleointensity data obtained from a 18-m-long sediment core from the Pacific Ocean. The relative paleointensity was obtained by normalizing NRM with anhysteretic remanent magnetization (ARM); the latter was shown to be a more appropriate laboratory magnetization than SIRM. A periodicity of 43,000 years was seen to be present in the relative paleointensity data. Instead of cross-correlating magnetic data with proxy paleoclimatic data from the same core, Kent and Opdyke pointed to the apparent similarity between the 43,000-year periodicity in relative paleointensity and the 41,000-year periodicity which is known to be superimposed upon the Earth's precessional motion (this latter motion has its own periodicity of 21,000 years). They concluded tentatively that a common mechanism may control both the 43,000-year periodicity in paleointensity and the 41,000-year periodicity which is superimposed on the Earth's precessional motion, which should have a perceptible climatic influence.

G. Wollin and colleagues have taken an alternative approach of comparing the paleoclimatic and paleomagnetic time series directly, as was done by Kawai and coworkers for the last 60,000 years. Figure 3 shows a correlation for the last 2,000,000 years presented by Wollin and colleagues. The fluctuations in the eccentricity of the Earth's orbit are known to have an even stronger effect on climate modulation than either the 41,000- or 21,000-year periodicity related to the precession of the Earth's rotation axis. Wollin claimed that the data of Fig. 3 show a negative correlation between geomagnetic field intensity and temperature for the last 2,000,000 years. This conclusion is exactly the opposite of Kawai. Recent studies in solar-terrestrial physics indicate, however, that on a shorter time scale (approximately days) solar-controlled field fluctuation and terrestrial weather may be

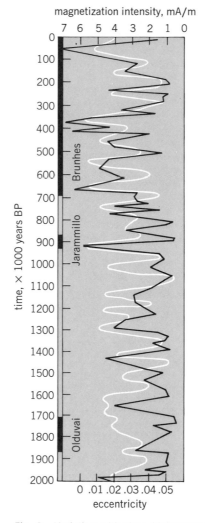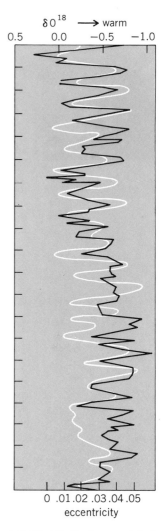

Fig. 3. Variation with time (determined from geomagnetic polarity time scale) of natural remanent magnetization intensity and $\delta^{18}O$ in a deep-sea sediment core from the North Atlantic Ocean. The broken line denotes the nearly 100,000-year periodicity in the eccentricity of the Earth's orbit. (From G. Wollin, W. B. F. Ryan, and D. B. Ericson, Climate changes . . ., Earth Planet. Sci. Lett., 41:395–397, 1978)

correlated negatively, that is, exactly as Wollin found it to be.

A. Chave and C. Denham have noted, perhaps quite justifiably, that the data of Fig. 3 are related to fluctuations in the raw NRM data without any normalization to take into account possible fluctuations in the magnetic mineral content. These later workers made a study of their own on a sediment core from the North Atlantic, and found that there is no correlation between temperature-sensitive microfossils and relative paleointensity obtained after normalizing with ARM. As the vitality of this research area indicates, the final word on correlation between paleomagnetic field and paleoclimate has not yet been said.

For background information see PALEOCLIMATOLOGY; PALEOMAGNETICS in the McGraw-Hill Encyclopedia of Science and Technology.

[SUBIR K. BANERJEE]

Bibliography: A. D. Chave and C. R. Denham, Earth Planet. Sci. Lett., 44:150–152, 1979; J. A. Jacobs, The Earth's Core, 1975; N. Kawai et al.,

Voice of geomagnetism from Lake Biwa, in S. Horie (ed.), *Paleolimnology of Lake Biwa and the Japanese Pleistocene*, vol. 3, Otsu Hydrobiological Station, Kyoto University, 1975; D. V. Kent and N. D. Opdyke, *Nature*, 266:156–159, 1977; K. Verosub and S. K. Banerjee, *Rev. Geophys. Space Phys.*, 15:145–155, 1977; G. Wollin, W. B. F. Ryan, and D. B. Ericson, Climate changes . . ., *Earth Planet. Sci. Lett.*, 41:395–397, 1978.

Paleontology

Evidence from the fossil record provided clues to those who pioneered in the development of the hypotheses of continental drift and plate tectonics in the early and middle 20th century. During the last few years, it has been demonstrated that the widening of the Atlantic Ocean had a direct and major effect on the degree of similarity among animals living on, or on the margins of, continents on opposite sides of the ocean and that this effect can be quantified.

Ocean basins as barriers. In the early part of the 20th century, those who advocated movement of continents cited the occurrence of identical fossils on landmasses now separated by ocean basins hundreds of kilometers wide. Most paleontologists recognized the similarities but ascribed them to the existence of land bridges between continents. These bridges, which were thought to have subsided later, would have provided migration routes for land-dwelling and fresh-water organisms and for those marine life-forms which lived in shallow water. The continental drift proponents often cited similarities among nonmarine organisms such as terrestrial or fresh-water reptiles and plants. The widespread distribution of these organisms was especially striking in the Southern Hemisphere. It was also recognized, however, that marine fossils, most of which lived in a shallow-water environment, also displayed transoceanic similarities. Although it is not as surprising to find marine fossils separated by wide ocean basins as it is to find remains of terrestrial organisms, deep ocean basins are also a barrier to most marine life which lives in the shallow environment, a habitat generally restricted to the continents and their margins.

Some of the organisms common in the fossil record could swim or float passively in ocean currents and thus achieve long-distance dispersal. Most marine invertebrates which are common fossils lived in shallow water on the sea floor. Some burrowed into the sediments. Many of them, such as corals, were firmly attached to the bottom. Bottom-dwelling forms which have some powers of movement are relatively slow, and are incapable of long-distance migration as adults. However, most marine invertebrates which live on the bottom as adults have floating larval stages. By floating in ocean currents these larvae can be dispersed, and some long-lived larval stages can be distributed over transoceanic distances, assuming that environmental factors such as temperature and salinity are favorable. The presence of islands and shallow areas of the sea floor such as the Mid-Atlantic Ridge would provide shallow-water migration routes for some organisms. In spite of the floating larval stages and the existence of shallow migration routes, the presence of distinctly different living faunal assemblages on opposite sides of oceans demonstrates the effectiveness of deep ocean basins as barriers.

Sea-floor spreading during the Mesozoic and Cenozoic eras has been demonstrated in the Atlantic Ocean using paleomagnetic, sedimentological, and other kinds of evidence. The chronology developed by these techniques indicates that the present Atlantic Ocean did not exist at the beginning of the Mesozoic Era, about 225,000,000 years ago. The North Atlantic began opening approximately 180,000,000 years ago and has been widening at the rate of several centimeters per year since. As the ocean grew wider, the barrier to migration of organisms evolving on each side of the North Atlantic should have had a greater and greater effect through time.

Even though sea-floor spreading may have had a stronger effect on the degree of transoceanic similarity among terrestrial and fresh-water organisms, the fact that marine organisms are much more abundant in the fossil record suggests that the latter might provide the better data base for quantifying the effect of continental separation on dispersal across ocean basins.

Fossil record. The *Treatise on Invertebrate Paleontology* provides information on thousands of invertebrate fossil genera, most of which are marine. Although *Treatise* volumes for all groups of fossil invertebrates have not yet appeared, cover-

Trans–North Atlantic fossil data and width of the ocean basin*

Time unit	Genera in North American area	Genera in European-African area	Genera common to both areas	Simpson coefficient	Average width of ocean basin, km†
Neogene	558	835	330	0.59	3009
Paleogene	616	1144	383	0.62	2602
Late Cretaceous	572	1386	394	0.69	1772
Early Cretaceous	338	977	263	0.78	832
Late Jurassic– Early Cretaceous	—	—	—	—	555
Late Jurassic	168	867	141	0.84	—
Middle Jurassic	142	833	118	0.83	101
Early Jurassic	112	710	95	0.85	0‡

*From W. C. Fallaw, *Geology*, 7:398–400, 1979.

†Sclater and others, from whose paleogeographic maps the width data were estimated, published a map for Late Jurassic–Early Cretaceous time but not a map for Late Jurassic time.

‡The Early Jurassic width value is based on the assumption that there was no deep ocean basin between the landmasses at that time.

age for most of the major categories common in the fossil record is available. The groups included in the study of transoceanic similarity are Foraminiferida, Porifera, Coelenterata, Brachiopoda, Mollusca (except Coleoidea and some gastropods), Arthropoda (except Hexapoda), Echinodermata (except Crinoidea), Graptolithina, and some miscellaneous groups. Information on the times during which they lived and the areas in which the fossils are found is included in the *Treatise*. Although the data are not as precise as might be desired, they are accurate enough to be useful in determining trends in the degree of fossil similarity through time. Of course, the incompleteness of the fossil record does not allow a precise determination of transoceanic similarities among organisms, and the various parts of the world have not been studied to the same degree by paleontologists. The biases in the data are probably not systematic enough to invalidate conclusions reached from the available information.

Data concerning the chronological and geographical distribution of genera were compiled from the *Treatise* for landmasses that are on the western side of the North Atlantic, including Greenland, the Caribbean, and those parts of North America east of the Rocky Mountains. Mexican occurrences were not tabulated. Occurrences were tabulated for areas east of the North Atlantic, including Europe, and northern and western Africa. Soviet occurrences were not tabulated. The data in the table indicate the number of genera listed in the *Treatise* as occurring in the western or North American area, the number occurring in the eastern or European-African area, and the number occurring on both sides of the North Atlantic.

Simpson coefficient. The degree of transoceanic fossil similarity can be quantified by various indices. The one used here is the Simpson coefficient, C/N_1, in which C is the number of genera common to the two areas being compared, and N_1 is the number of genera in the area having the smaller number of genera. The number of genera in the area having the larger number is not used because this would distort the degree of similarity if one area had very many more genera than the other. The fifth column in the table indicates the Simpson coefficient values for time units of the Mesozoic and Cenozoic during which the North Atlantic widened.

A series of paleogeographic maps published by J. G. Sclater and others in 1977 shows the width of the Atlantic Ocean at various times during the Mesozoic and Cenozoic. These maps were based on several types of geological evidence, including depth of the ocean floor, paleomagnetic evidence, and shapes of the continental blocks. The average width of the North Atlantic at various geological times can be estimated by measuring the distances across the ocean basin at several places along the continental margins on the paleogeographic maps. The average widths which correspond most closely to the times for which fossil data are tabulated from the *Treatise* are shown in the last column of the table.

The figure is a plot of the Simpson coefficient of fossil similarity values against the widths estimated from the paleogeographic maps. An almost straight line results from the plot. The linear cor-

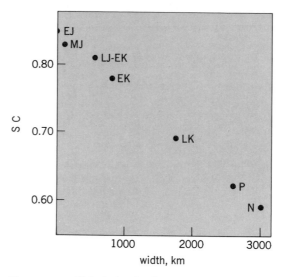

Simpson coefficient fossil similarity values plotted against the estimated average width of the North Atlantic ocean basin. The LJ-EK similarity value is an average of the values for those two epochs. The Early Jurassic width of zero is assumed. EJ= Early Jurassic, MJ= Middle Jurassic, LJ-EK= Late Jurassic-Early Cretaceous, LK=Late Cretaceous, P=Paleogene, N=Neogene. (*From W. C. Fallaw, Trans-North Atlantic similarity among Mesozoic and Cenozoic invertebrates correlated with widening of the ocean basin, Geol., 7:398–400, 1979*)

relation coefficient between the two variables is −0.998.

The relationship between average ocean basin width and fossil similarity is complicated by the fact that sea-floor spreading occurred between Greenland and Canada during the latter part of Mesozoic time and the early part of the Cenozoic, and spreading later developed between Greenland and Europe. Also, a relatively shallow-water environment now extends from eastern Canada, around southern Greenland, to Iceland and from there eastward to the British Isles and northern Europe. This geographical pattern, and the existence of the Mid-Atlantic Ridge, probably aided migration in the northern part of the Atlantic and therefore would have had an effect on transoceanic fossil similarity values. Ecological factors also affect the ability of organisms to attain a widespread distribution. In particular, if an organism is able to travel across an ocean, a suitable environment must be available for the organism to become established.

In spite of the presence of a shallow northern Atlantic migration route and other environmental components which affect dispersal, the high negative correlation between average ocean basin width and the Simpson coefficient fossil similarity values indicates that sea-floor spreading had a direct and strong effect on trans–North Atlantic dispersal of invertebrates during Mesozoic and Cenozoic time. [W. C. FALLAW]

Bibliography: W. C. Fallaw, *Geol.*, 7:398–400, 1979; A. Hallam, *Palaeogeogr. Palaeoclimatol. Palaeoecol.*, 3:201–241, 1967; R. C. Moore and C. Teichert (eds.), *Treatise on Invertebrate Paleontology*, 1953–1972; J. G. Sclater et al., *J. Geol.*, 85: 509–522, 1977.

Parity (quantum mechanics)

Experiments currently being carried out to measure parity nonconservation in hydrogen atoms are primarily motivated by a desire to further understanding of the weak interaction. These experiments are expected to provide unambiguous information concerning the validity at low energy of the unified gauge theory of weak and electromagnetic interactions proposed by S. Weinberg and A. Salam. If the results are consistent with the theory, they will be used to provide a more accurate value of its only free parameter, the Weinberg angle.

Weak interaction effects in atoms. Prior to 1973 all known examples of the weak interaction involved a change in the charges of the weakly interacting particles which, according to the theory, was due to the exchange of a charged boson, the $W^{\pm}$. An electron involved in such an interaction must turn into a neutrino, and therefore it seemed that the electrons of an atom could experience weak interactions only in second order. Since the weak coupling constant G_w is 2×10^{-14} atomic unit ($\hbar = m = e = 1$, where $\hbar$ is Planck's constant divided by 2π, and e and m are the charge and mass of the electron), this implied weak effects at the hopelessly minute level of 4×10^{-28} of the gross atomic features.

A major development occurred in 1967–1968 when Weinberg and Salam independently proposed a new theory which intertwined the weak and electromagnetic interactions and, as a natural consequence of the theory, predicted a new weak interaction, mediated by the neutral boson Z^0. The experimental observation in 1973–1974 of neutral weak events in high-energy neutrino scattering was the first of many successes for the Weinberg-Salam theory, which is now so widely accepted that some authors refer to the "electroweak" interaction.

In the light of the new neutral weak interaction, it becomes reasonable to expect first-order weak effects in atoms. Moreover, since the energies involved in an atom are so low compared with the masses of the particles, the application of the Weinberg-Salam model is quite straightforward, and atoms would seem to be an ideal testing ground for the theory.

Early experiments. The earliest weak interaction experiments on atoms (which are still being carried out) use the heavy elements bismuth, thallium, and cesium. These experiments take advantage of an effect first discussed by M. Bouchiat and C. Bouchiat, who pointed out that part of the weak interaction between the electron and the nucleus was enhanced by a factor of order Z^3 (where Z is the atomic number) compared with the dimensional value of 2×10^{-14} atomic unit. This enhancement alone is not expected to make the weak interaction detectable; the experiments rely on the additional fact that the weak neutral current should be parity-nonconserving, while the much stronger electromagnetic interaction conserves parity. Unfortunately, it is difficult to interpret these experiments very precisely because of the complexity of the many-body atomic theory of heavy atoms. An alternative to the heavy atoms is offered by the hydrogen atom and its isotopes,

which have the advantage that the atomic physics calculations can be done with precision.

Principles of hydrogen experiments. Groups at Yale and at the Universities of Washington and Michigan are currently working to measure effects in hydrogen due to the parity-nonconserving part of the weak electron-proton interaction. The key feature of hydrogen, which makes these experiments possible even though there is no Z^3 enhancement, is the near degeneracy of the $2S_{1/2}$ and $2P_{1/2}$ levels. The anticipated small parity-nonconserving weak interaction can produce a measurable $2P_{1/2}$ admixture into the wave function of the $2S_{1/2}$ state because the two states are separated in energy by only 1.5×10^{-7} atomic unit. This admixture of opposite-parity wave functions imparts to the $2S_{1/2}$ state of the atom a slight handedness which the experiments are designed to detect. The possibility of detecting parity-nonconserving effects in hydrogen was first discussed by F. C. Michel in 1964.

The technique used to detect the atomic handedness in the Yale experiment involves passing a beam of $2S_{1/2}$ (metastable) hydrogen atoms through a region of separate, coherent, oscillating magnetic and electric fields b and ϵ. Since b arises from a circulating current i, the fields taken together have a corkscrew sense which can be switched from left to right by reversing the sign of $\vec{b}.\vec{\epsilon}$ (Fig. 1). The

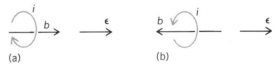

(a)   (b)

Fig. 1.   External fields used in the Yale experiment to detect an atomic handedness. (a) Right-handed combination in which pseudoscalar $\vec{b} \cdot \vec{\epsilon}$ is positive. (b) Left-handed case in which $\vec{b} \cdot \vec{\epsilon}$ is negative.

fields are used to drive a transition from the state of total angular momentum $F = 0$ to the state $F = 1$, and the handedness of the atom is detected as a difference between the transition rates when the sign of the pseudoscalar $\vec{b}.\vec{\epsilon}$ is reversed. When the transition probability is P, the parity-nonconserving fraction f of the transition rate is given by the equation below, in which θ_w is the Weinberg angle. It is anticipated that $\sin^2 \theta_w$ can be determined in this way with a precision of about 0.02.

$$f = 2 \times 10^{-8}(1 - 4\sin^2 \theta_w) P^{-1/2}$$

The experiments of W. Williams in Michigan and E. Adelberger in Washington are similar in that they also induce transitions in a beam of $2S_{1/2}$ hydrogen and measure a pseudoscalar contribution to the rate. The electromagnetic field configurations used are shown in Fig. 2, where E and B are static electric and magnetic fields and ϵ is a microwave electric field. The pseudoscalar of interest in the Michigan experiment (Fig. 2a) is $(\vec{\epsilon} \cdot \vec{E})$ $(\vec{\epsilon} \cdot \vec{B})$. In the Washington experiment (Fig. 2b) the pseudoscalar is $(\vec{\epsilon}_1 \cdot \vec{E}) (\vec{\epsilon}_2 \cdot \vec{B})$, where ϵ_1 and ϵ_2 are separated and coherent. The most striking difference between these experiments and the Yale work is the presence of the static magnetic field

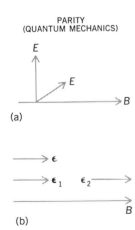

Fig. 2.   External fields used by (a) Michigan group and (b) Washington group.

B which is used to shift the $2S_{1/2}$ and $2P_{1/2}$ levels to a crossing. When B is varied close to the level crossing, the rapid change in energy difference, and hence in $2S_{1/2}-2P_{1/2}$ mixing, is manifested as a readily recognizable variation in the pseudoscalar contribution to the rate. Thus it is possible to discriminate against certain types of systematic error.

The three schemes will all, in principle, reach roughly the same level of sensitivity and, by virtue of their different transition regions, form a complementary set of experiments.

Apparatus. A single, general description of all the experiments can be given because the extreme experimental constraints have led all three groups to construct an apparatus (Fig. 3) with certain basic similarities.

Fig. 3. Block diagram of the beam apparatus used to measure parity nonconservation in hydrogen.

The equation above shows that f, the measured parity-nonconserving fraction of the transition rate, is enhanced by the use of small transition probability P. At the same time, statistical arguments show that f is more easily measured when a large number of atoms make the transition. In order to satisfy both requirements, it is obviously necessary to start with an intense beam of $2S_{1/2}$ hydrogen atoms. The experiments have all adopted a technology which was developed to produce intense beams of polarized ions for particle accelerators. Protons are extracted from a hydrogen plasma and passed through a pipe of cesium vapor in which the probability of electron capture into the $2S_{1/2}$ state of hydrogen is roughly 1:3. The method works because of an accidental near-degeneracy between the binding energies of $2S_{1/2}$ hydrogen and the cesium ground state, first pointed out by B. Donnally. In this way atomic beams can be made with as many as 5×10^{13} $2S_{1/2}$ hydrogen atoms per second in a useful solid angle.

The second and fourth sections of the apparatus are the state selector and state analyzer, which together make the atomic transition observable. Each of these sections contains a microwave cavity in which an oscillating electric field resonantly admixes the $2P_{1/2}$ state into chosen $2S_{1/2}$ hyperfine states, causing them to decay to the ground state. The state selector is tuned to transmit the initial state of the transition but not the final state, while the converse is true for the state analyzer. Thus the only $2S_{1/2}$ atoms emerging from the state analyzer are those which made the transition.

The last section of apparatus is the detector. For each $2S_{1/2}$ atom entering the detector, there are many ground-state atoms because the transition probability P is chosen to be small, typically 10^{-6}. Consequently the detector must be both sensitive to $2S_{1/2}$ atoms and extremely insensitive to the ground-state beam. The technique used is to pass the beam through an electric field which induces decay of the $2S_{1/2}$ atoms, and to detect the emitted 122-nm decay photons by using an ion chamber.

When the transition rate equals the detector background, it can be shown that it is no longer advantageous to reduce P, and thus the detector background is the reason for choosing P to be roughly 10^{-6}.

Theory. If it is assumed that the weak interaction lagrangian has a four-fermion vector (V) and axial vector (A) structure, there are two parity-nonconserving terms in the electron-proton (e-p) weak neutral interaction: $C_{1p} G_w V_p \cdot A_e$ and $C_{2p} G_w A_p \cdot V_e$. C_{1p} and C_{2p} are phenomenological coupling constants. The corresponding electron-neutron (e-n) coupling constants are known as C_{1n} and C_{2n}. The constant C_{1n} is particularly interesting because it is predicted in the Weinberg-Salam model to have the value $-\frac{1}{2}$, regardless of the value of the Weinberg angle. The other three constants are proportional to $(1-4\sin^2\theta_w)$ and are expected to be small.

The hydrogen experiments described here are primarily sensitive to C_{2p}, but the hope for the future is that experiments on hydrogen and deuterium, and perhaps tritium, will determine all four phenomenological coupling constants.

For background information *see* ATOMIC STRUCTURE AND SPECTRA; INTERACTIONS, FUNDAMENTAL; INTERACTIONS, WEAK NUCLEAR; ION SOURCES; PARITY (QUANTUM MECHANICS) in the McGraw-Hill Encyclopedia of Science and Technology.   [EDWARD HINDS]

Bibliography: R. W. Dunford, R. R. Lewis, and W. L. Williams, *Phys. Rev. A*, 18:2421–2436, 1978; E. A. Hinds, *Phys. Rev. Lett.*, 44:374–376, 1980; E. A. Hinds and V. W. Hughes, *Phys. Lett. B*, 67:487–488, 1977; W. L. Williams (ed.), *Proceedings of the International Workshop on Neutral Current Interactions in Atoms, September 1979*, National Science Foundation, 1979.

Petroleum

In recent years there have been significant discoveries of petroleum reserves in widely separated parts of the world. Major oil production is now taking place in Alaska and the North Sea. The Orinoco Oil Belt in Venezuela is considered to have tremendous potential for future production.

PRUDHOE BAY FIELD

In the first 2 centuries of Alaska's recorded history, petroleum production totaled just over 150,000 barrels (2.4×10^4 m³). The fabulous oil field at Prudhoe Bay on Alaska's far North Slope now produces this much in just over 2 hours. The largest oil field on the North American continent, it produces a maximum allowable rate of 1.5×10^6 bbl (2.4×10^5 m³) of oil per day, and should continue to do so for at least the next 5 years (Fig. 1). The price tag for such a mammoth operation already exceeds $14 billion (including $8 billion for the Trans Alaska Pipeline System, TAPS), and nearly $27 billion (including $2 billion for water injection) will be expended before it is over. In addition, the gas pipeline cost estimate has soared to a current estimate of some $30 billion, and is climbing.

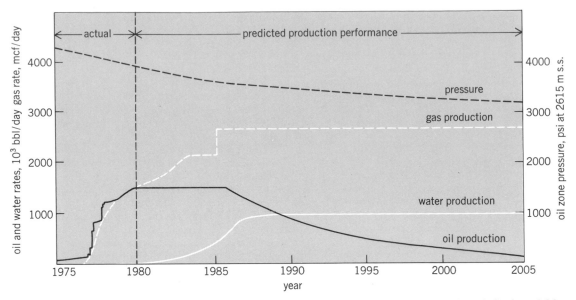

Fig. 1. Predicted production performance curve for Prudhoe Bay field. 10³ bbl/day = 159 m³/day; 10⁶ ft³/day = 2.83 × 10⁴ m³/day; 1 psi = 6.895 kPa.

Producing interval. The Prudhoe Bay Field actually contains three major producing horizons. The shallowest commercial zone is called the Kuparuk Sand of Cretaceous-Jurassic age and is found at about 5000 ft (1.5 km) in the western part of, and extending beyond, the unit area. The Sadlerochit Sand (the main interval of interest and the only one being produced at this time) is of Permian-Triassic age at an average depth of about 8500 ft (2.6 km) and is found throughout the unit area.

The Lisburne limestone/dolomite is of Pennsylvanian age and is found around 11,000 ft (3.4 km) in the eastern part of the unit area. The Sadlerochit interval is a series of sandstones, conglomerates, and interbedded shales making up an oil column over 440 ft (134 m) thick containing some 20.6×10^9 bbl (3.28×10^9 m³) of oil and 14.9×10^{12} ft³ (4.22×10^{11} m³) of gas in solution. This is overlain by a massive gas cap containing an additional 23.4×10^{12} ft³ (6.63×10^{11} m³) of gas. The estimated 9.5×10^9 bbl (1.5×10^9 m³) producible oil and condensate from the Sadlerochit alone represents one-third of the United States total oil reserves, which is indicative of its critical national importance.

Discovery. Oil exploration on the North Slope dates back to the turn of the century with sporadic attempts and dismal results. Following one such flurry of a dozen very expensive and very dry holes during the 1960s, Atlantic Richfield Company and Humble Oil and Refining (now Exxon USA) jointly spudded the Prudhoe Bay State #1. In the winter of 1968 oil and gas were found atop the large structure underlying the Prudhoe Bay area. The Sag River State #1, drilled 7 mi (11 km) to the southeast, confirmed the discovery of major consequences, even considering its remote location.

Climate. Well north of the Arctic Circle, Prudhoe is covered by ice and snow 9 months a year. With winter nights 3 months long and temperatures often exceeding 60°F (−51°C) below zero, only low-growing tundra-type vegetation exists. People and equipment fail more easily so that most things are automated and all things must be heated, even the oil entering the TAPS line. All flow lines are insulated with several inches (1 in = 2.5 cm) of polyurethane and jacketed with stainless steel to provide up to 2 weeks downtime (even in winter) without serious restart problems from cooling and gelling of the oil.

Permafrost. Even during the short summer, the ground is permanently frozen from a few inches down, to over 2500 ft (760 m) in some areas. This

Fig. 2. Trans Alaska Pipeline (TAPS) elevated aboveground on insulated supports. Special heat exchangers atop support posts prevent thawing of delicate permafrost. (*Alyeska Pipeline*)

soil, in which the moisture is ice, causes operational difficulties and has been a deterrent to proper interpretation of seismic surveys for years. But though it tenaciously guards the riches below, permafrost is fragile and great efforts are made to protect it. Buildings are constructed on pilings literally frozen in place with dirt and water, well casings are cemented with special arctic cement, and concentric pipes are insulated to prevent heat loss. Even much of the giant TAPS line is suspended on special scaffolding aboveground (Fig. 2).

TAPS. Over half of this $8 billion system is aboveground on its tortuous 800-mi path (Fig. 3) from Prudhoe, over the Brooks Range (hence the name North Slope), to its terminus at the all-weather port at Valdez (Fig. 4). This 4-ft- (1.2 m-) diameter pipeline is so large that even though Prudhoe produces more than any other field in the Western Hemisphere, it still requires some 6 days for the oil to traverse the line. Once in Valdez, oil is loaded onto tankers for shipment to refineries along the west coast and through the Panama Canal to the Gulf and east coast. On January 17, 1980, the one-billionth barrel of oil from Prudhoe had been shipped down this line, just 2½ years after it was commissioned in June 1977.

Wells. This field has already produced more oil than most of the fields in the United States have in place, and less than half of the estimated 540 ultimate producing wells have been completed. It will take nearly 4 more years to complete the remaining producing wells, together with the 175± proposed water injection wells, and the remainder of the 20± gas injection wells.

Producing wells are directionally drilled from native gravel pads, are cemented through the pay zone, selectively perforated, and flow through 5½- or 7-in. (14- or 18-cm) "tubing." Special cementing requirements protect the permafrost interval from thermal decay, and all wells are equipped with downhole safety valves, set below the permafrost, which close automatically if anything happens to the well head or flow line. All producing wells are currently flowing and should continue to do so for several more years, but eventually the producing wells will require artificial lift (probably in the form of gas-lift).

Gathering system and separation stations. Hundreds of miles of gathering system pipelines connect each of the wells to one of six strategically located separation stations. The well flow lines in the Western Operating Area are 6-in. (15-cm) individual lines to the gathering center, where they are manifolded for combined oil shipment to Pump Station #1 and gas to the Central Compressor Plant (Fig. 5). In the Eastern Operating Area flow lines are manifolded at the drill site, and fluids are moved to the flow station through 12- or 16-in. (30- or 41-cm) common lines. Gas lines connecting separation stations to the Central Compressor Plant range from 30 to 42 in. (76 to 107 cm), while trunk oil lines are 30 to 36 in. (76 to 91 cm).

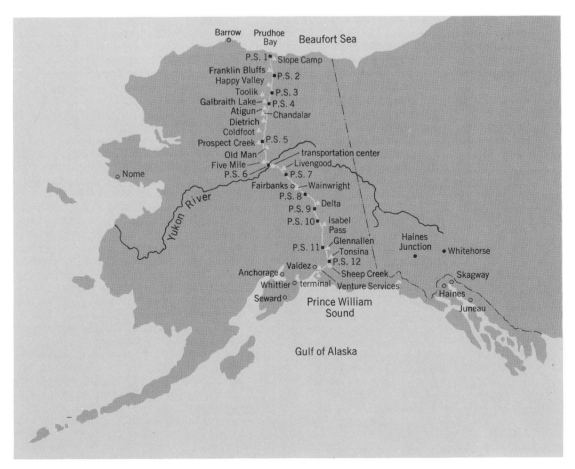

Fig. 3. Path of TAPS showing pump stations and various other reference points. (*Alyeska Pipeline*)

Fig. 4. Valdez pipeline terminal and marine loading facilities. (*Alyeska Pipeline*)

The oil is then remetered at Pump Station #1 and shipped to Valdez. Produced water is disposed of in shallow Cretaceous water-bearing sands (it will be reinjected in the Sadlerochit zone starting in 1981). Gas at the Central Compressor Plant is split into two streams, about 10% is consumed for fuel in field use and the first four TAPS pump stations, while the remainder is reinjected into the gas cap (up to 2×10^9 ft³ or 5.7×10^7 m³ per day will be sold in the future). The design capacity of the six metering stations is currently about 1.8×10^6 bbl (2.9×10^5 m³) of oil and about 3.4×10^9 ft³ (9.6×10^7 m³) of gas per day.

Compressor plant. The gas to be reinjected enters the Central Compressor Plant at some 600-psi (4.1 MPa) line pressure, and is boosted in two stages to approximately 4500 psi (31.0 MPa). A bank of 12 turbine compressors—8 first-stage compressors boost it to about 2075 psi (14.3 MPa), and the 4 second-stage compressors raise it to about 4500 psi—are six-step, barrel type with 25,000-hp (18.6 MW) single-shaft gas turbine prime movers. This plant has an injection capacity of just under 2 $\times 10^9$ ft³ (5.7×10^7 m³) per day.

Unit operations. The Prudhoe Bay Unit is so large (37 by 15 mi or 60 by 24 km, containing nearly a quarter-million acres or 10^5 hectares) that it has two unit operators. SOHIO operates the west half, and ARCO operates the east half. In addition, there are different participation factors for the "Oil Rim" and the "Gas Cap" because the original

ownership of the tracts constituting the bulk of each "area" is different.

The two operators have separate base camps with water and sewage disposal, crew quarters, theaters, maintenance shops, and so on. Several facilities, however, serve both sides of the field, and include a Crude Topping Unit (to produce diesel oil) the Airstrip, the Dock, a Communications Network, a 150-MW power plant, and the 300,000-hp (224 MW) Central Compressor Plant for gas reinjection.

Water injection. A mammoth source water injection project is planned for 1984 to supplement produced water injection. In addition to providing the necessary pressure maintenance to allow for future gas sales, reservoir analyses by both the operators and the Alaska Oil and Gas Conservation Commission (AOGCC) indicate that an additional 10^9 bbl of oil can be recovered from the proposed large-scale water injection project.

Surveillance. This field is the most closely watched, thoroughly engineered, economically scrutinized field in the history of the United States. Gas-oil contact and water-oil contact monitoring are conducted continuously and reviewed by the Conservation Commission. Numerous bottom hole pressure surveys and well tests for fluid ratio determinations are required year-round. Fluid profile surveys are also required for all wells. In addition, very large and sophisticated two- and three-dimensional reservoir simulators have been used by several of the owner oil companies and the AOGCC alike. These will continue to be used to monitor field operations to ensure that maximum ultimate recovery will be realized.

[JOSEPH GREEN]

NORTH SEA

Since the granting of the first North Sea licenses in 1964, there has been a continuing saga of exploration, discovery, and development of hydrocarbons. The period 1966–1969 witnessed the discovery and development of several major gas fields in the southern North Sea. In 1969 Phillips Petroleum made the first significant oil discovery, Ekofisk, in the Norwegian sector of the central North Sea, and this was followed in 1970 by the discoveries of Montrose and Forties in the United Kingdom sector of the central North Sea. In 1971 Shell/Esso discovered the Brent field in the Viking Graben of the northern North Sea. These three important hydrocarbon basins were the target for intensive exploration and development throughout the 1970s, and these activities will continue in the 1980s (Fig. 6). Smaller accumulations have also been found in the Dutch and Danish areas of the North Sea. The latest significant discovery is that large volumes of oil-in-place exist west of the Shet-

lands, although they still await volumetric and technical appraisal.

Present status. Table 1 shows the actual and estimated oil production by fields for the years 1979, 1980, and 1981. From this table, oil production in 1980 was expected to average about 1.82 and 0.65×10^6 bbl (2.89 and 1.03×10^5 m³) of oil per day from the United Kingdom and Norwegian sectors respectively. This means that in 1980 the United Kingdom was expected to meet the goal of oil self-sufficiency and to become, as Norway already is, a net exporter of oil. In addition to the oil, substantial volumes of gas were produced, 3.85 and 2.02×10^9 ft³ (10.9 and 5.7×10^7 m³) per day respectively from the United Kingdom and Norway in 1979.

Future development. There are reserves under development in areas of the United Kingdom and Norway.

United Kingdom. Reserves of the current 14 producing fields are of the order of 10 to 12×10^9 bbl, (1.6 to 1.9×10^9 m³), while fields under development or appraisal are expected to add another 6 to 8×10^9 bbl (1.0 to 1.3 m³). In addition, substantial volumes of oil-in-place are known to exist west of the Shetlands, and other large areas still remain to be explored. Given stable political and fiscal

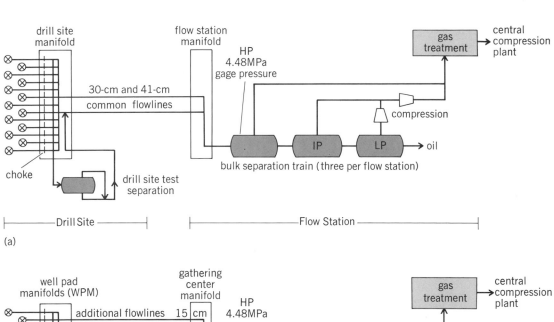

(a)

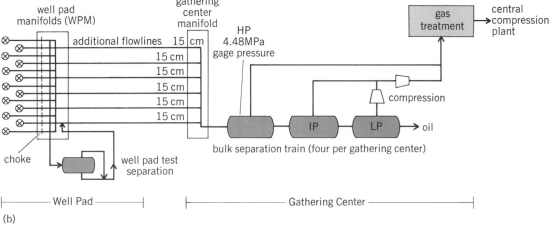

(b)

Fig. 5. Typical drill pad and separation station for the (a) Eastern Operating Area and (b) Western Operating Area.

conditions to encourage a continuing high level of exploration, it is possible to foresee continuing oil self-sufficiency for the United Kingdom through the end of the 20th century and into the next. Much will depend upon government policy with regard to taxation and control of the rate of depletion. What seems certain is that future development and production will be controlled to some degree, although the mechanism is not at all clear. It is likely that, with the present government at least, some level of agreed production will be negotiated between the government and the operating companies and that future development will be planned accordingly.

Norway. Oil reserves from the fields now under development, that is the Ekofisk complex, Statfjord, Murchison, and Valhall, amount to some 4.6×10^9 bbl (7.3×10^8 m³). There is little doubt that substantial further oil reserves exist and that Norway can look forward to a long future of independence as a consumer and of prosperity as an exporter of hydrocarbons. Significant gas discoveries have also been made in blocks 31/2 and 30/6.

It has already been clearly established by the

Table 1. North Sea oil production pattern

Field	Date of start-up	Average oil production (10³ bbl‡ per day) 1979	1980*	1981*
United Kingdom				
Argyll	June 1975	17	17	13
Auk	February 1976	18	12	9
Beatrice	Mid-1981*	—	—	40
Beryl	June 1976	94	95	90
Brent	November 1976	179	150	350
Buchan	Mid-1980*	—	35	45
Claymore	November 1977	80	110	110
Cormorant (S)	December 1979	1	40	60
Cormorant (N)	Mid-1983*	—	—	—
Dunlin	October 1978	116	95	120
Forties	September 1975	502	500	500
Fulmar	Mid-1982*	—	—	—
Heather	October 1978	17	40	50
N.W. Hutton	Mid-1983*	—	—	—
Magnus	Mid-1983*	—	—	—
Maureen	Mid-1982*	—	—	—
Montrose	July 1976	28	30	28
Murchison	Early 1981*	—	—	59
Ninian	December 1978	158	275	360
Piper	December 1976	274	240	180
Statfjord	November 1979	1	11	17
Tartan	Mid-1980*	—	40	77
Thistle	April 1978	80	150	180
Total United Kingdom		1565	1820	2288
Norway				
Ekofisk Area	July 1971	402	586	686
Murchison	Early 1981*	—	—	11
Stratfjord	November 1979	4	59	93
Valhall	Late 1981*	—	—	15
Total Norway		406	645	805
TOTAL North Sea		1971†	2465	3093

*Estimate   †Actual figures.
‡1 bbl = 0.159 m³.

Norwegian government that production and depletion will be closely controlled in the national interest, and that future licenses will be granted only in conjunction with Norwegian companies. An important step was taken in 1980 with the first permission to drill north of the 62d parallel.

Advances in technology. The growth of offshore technology continues into the 1980s with the decision by Conoco and its partners to develop the Hutton field by means of a tension leg platform (TLP). The concept could offer significant advantages for the exploitation of deeper-water fields, and this first development will be watched carefully by the industry. Shell and Esso announced that they plan to develop the Central Cormorant field completely with a subsea production system (SPS), and this again is a first step into deeper water for such systems. There has been a significant trend toward drilling wells through templates and completing them subsea. Templates have already been installed and subsea wells drilled in the following fields: Beatrice, Buchan, Fulmar, N.W. Hutton, Maureen, and S. Montrose.

It is clear that, for the North Sea at any rate, the heyday of large piled steel and concrete gravity platforms seems to be over, and that for development of smaller fields or fields in deeper-water conditions the industry is looking to the next technological generation of tension leg platforms, single well oil production systems (SWOPS), or subsea production systems. In Norway with the advance of the drilling frontier beyond the 62d parallel, new drilling rigs and techniques for deeper and more dangerous water conditions are being developed.

Economic conditions. Extremely large sums of money have been spent by operating companies in the search for hydrocarbons in the North Sea. Eight of the current projects, for example, have cost more than $1 billion each. Figure 7 is an estimate of total expenditure in the United Kingdom sector, and shows a current annual expenditure of about $6 billion. Figure 8 shows an estimate of the North Sea rates of return for different groups of fields as defined in Table 2. From this it is clear that the average rate of return which can be achieved in the United Kingdom North Sea has fallen with time and with size of reserves, and that within each group the outcome is likely to be lower than initially expected. However, the likelihood that in the long run the oil price will continue to increase in real terms will probably ensure that North Sea projects will continue to attract willing investors.   [J. BROWN]

VENEZUELA

Venezuela started to produce oil in 1914, and was the world's largest oil exporter for the period 1919–1970. The oil production during 1979 averaged 2.4×10^6 bbl (3.8×10^5 m³) per day, and proved reserves amount to 18.5×10^9 bbl (2.94×10^9 m³) of oil. The probable and possible reserves (excluding the Oil Belt) are estimated at 45×10^9 bbl (7.2×10^9 m³).

The Venezuelan oil explotation is very complex, comprising 9000 reservoirs and 30,000 wells. A great deal of experience has been obtained, particularly for the production of heavy oil (<22° API)

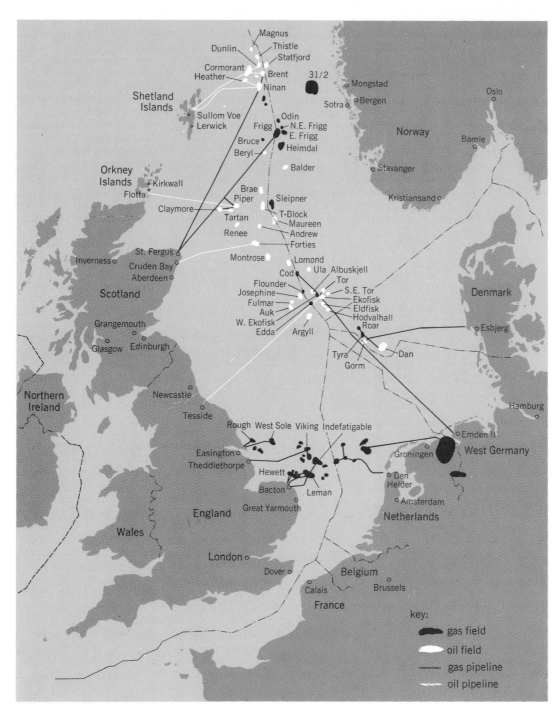

Fig. 6. North Sea developments and discoveries. (*Phillips Petroleum Company*)

with a total production of 725,000 bbl (115,000 m³) per day during 1979, from which more than 200,000 bbl (32,000 m³) per day were produced due to thermal projects, mainly by steam soak.

Current proved reserves will be depleted early next century, but they may be increased by adopting enhanced oil recovery; in addition, expectations have been raised by the exploration of new areas, both offshore and onshore. However, Venezuela has a virtually untapped and potentially gigantic energy resource in the Orinoco Oil Belt, which will allow the country to continue as a reli-

able oil exporter to the Western Hemisphere as it has been in the past, even during difficult situations such as World War II and the 1973–1974 embargo.

Orinoco Oil Belt. The Orinoco Oil Belt is located in the southern part of the eastern Venezuela basin and contains a large complex of hydrocarbon accumulations covering a total area of about 42,000 km² (slightly larger than Switzerland) (Fig. 9) It contains a wide range of reservoir characteristics and crude properties. Some crude has a gravity higher than 15° API, a fair amount ranges between 10 and

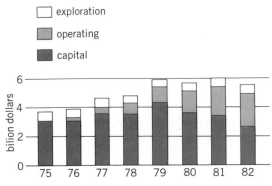

Fig. 7. Total expenditure in the United Kingdom sector.

15° API, and most of it is crude of less than 10° API.

The estimation of the oil-in-place is uncertain due to the paucity of information, the extension of the reservoirs, and the geological complexity of the area. The estimated volume of oil-in-place depends on the assumptions made by the geologists regarding the structural pattern, trapping characteristic, and sources and migrations of oil. With the information available about 1967 and assuming a discount factor of 50%, it was estimated that the oil-in-place would be about 700×10^9 bbl (1.1×10^{11} m³). In 1977 it was estimated that the volume of oil-in-place was 2×10^{12} bbl (5.7×10^7 m³).

Up to June 1980 a total of 240 wells had been drilled in the area, and the results confirmed the existence of giant oil accumulations with gravities ranging between 7 and 20° API.

Evaluation program. In order to make firm development plans for its exploitation and at the same time to have a thorough evaluation of these resources, Petróleos de Venezuela started during mid-1979 an intensive exploration campaign aimed at completing the regional evaluation of the Orinoco Belt by 1983. During this program, 16,000 km of seismic lines will be shot and 520 wells will be drilled. Taking into consideration the size of the area of interest and the development program envisaged for the medium-term future, it was necessary to establish priorities based on expected volume of oil per acre and oil quality.

Simultaneously, another drilling campaign was started during January 1980 with the objective of evaluating the most promising areas already delineated and to define more precisely the areas where the exploitation will take place. This program calls for the drilling of 380 additional wells in the period 1980–1984, which is the minimum

number of wells required to evaluate the two areas already selected to start the exploitation.

Production tests. Since the oil-bearing formations are nonconsolidated sandstones and the oil viscosity is high, production tests using portable pumping units and rods pumps are contemplated. Production tests made in 50 wells have shown very high productivity, with oil rates between 50 and 600 bbl (8 and 95 m³) per day per well per productive interval, without using any heating or diluent. Each well may have up to 10 different intervals. This fact confirms that even the heaviest crude in the area (7° API) is producible by primary, conventional means; therefore from the exploitation standpoint, this crude can be considered conventional. Furthermore, the oil productivity is very similar to the one in the Jobo, Morichal, Miga, Melones, Oveja, and Pilon fields located on the northern fringe of the Oil Belt, which are currently producing 140,000 bbl (22,000 m³) per day between 8 and 15° API.

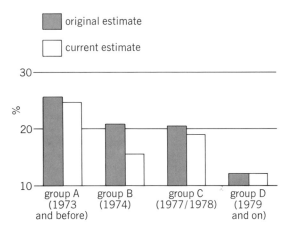

Fig. 8. North Sea rates of return.

Enhanced recovery pilot projects. Based on the production tests performed, exploitation can be initiated by primary means and steam stimulation. The oil recovery to be obtained using these production techniques will depend on the indigenous production mechanisms.

It has been estimated that if solution gas drive were the only production mechanism, the average oil recovery by primary and steam stimulation could be 8% STOIIP (stock tank oil initially in place), while the latter would increase up to 15% of the STOIIP if compaction happens to be an important driving mechanism. This means that the de-

Table 2. Trends in profitability in North Sea Fields

Group	First development	Average field size, 10⁹ bbl*	Total Capex, billion dollars		Total reserves, 10⁹ bbl*		Capex/barrel, dollars	
			Previous estimate	Present estimate	Previous estimate	Present estimate	Previous estimate	Present estimate
A	1973 and before	1.0	5.0	8.3	5.0	5.1	0.99	1.63
B	1974	0.4	5.6	8.3	3.6	3.2	1.57	2.56
C	1977 and 1978	0.3	3.2	3.4	1.4	1.3	2.25	2.62
D	1979 and on	0.3	0.3	7.9	1.8	1.8	4.43	4.43

*1 bbl = 0.159 m³.

velopment of the belt can be sustained for many years by using these well-known production techniques. This is confirmed by the excellent results obtained both with the primary production tests of the wells in the Orinoco Belt and with the steam soak pilot projects performed in the fields located on the northern fringe of the Oil Belt. In addition, steam drive will be applied after depletion of the reserves yielded by steam soak.

In the prospective areas which have already been selected to start the exploitation, the following processes will be tested: steam stimulation, continuous steam injection, and in-place combustion. In addition, other nonthermal processes may also be evaluated depending on laboratory test results.

Steam soak and steam drive pilot tests will be initiated during 1980.

Development of production. It was concluded during the 1st Unitar Conference that the exploitation of these resources does not face technological problems and therefore can make a significant contribution to the energy requirements in the medium-term future.

As far as the Orinoco Oil Belt is concerned, the program calls for the development of about 10^6 bbl per day by the end of this century. This program assumes that the conservation policy as defined by the government of Venezuela will aim at the development of a production capacity of only 2.8×10^6 bbl (4.5×10^5 m³) per day, and that a very limited degree of success is expected in the search for medium and light oil in other prospective areas in Venezuela. Therefore, the future development of any potential will be determined by both political and economic considerations defined by the Venezuelan government and, of course, by the lead time required to develop these resources. On a short-term basis, two objectives have been defined: to develop 125,000 bbl (20,000 m³) per day of upgraded crude in the southern portion of the state of Monagas by 1988, and to establish a production capacity of 50,000 bbl (8,000 m³) per day of non-upgraded crude by 1983 and 75,000 bbl (12,000 m³) by 1988 in the southern portion of the state of Anzoátegui.

The planning studies for the development of 125,000 bbl (20,000 m³) per day of upgraded crude were initiated in 1978, and the project is expected to go on stream by 1988. Construction of the infrastructure is expected in 1982 and the production/upgrading facilities in 1983. The investment has been estimated to be in the order of $7 billion, and it comprises more than 1000 wells located on an underdeveloped area, production facilities, a coking plant to upgrade the 8° API oil to about 30° API, a pipeline to the north and a terminal on the Caribbean Sea located at 300 km north of the producing fields, as well as the basic infrastructure and services to support the project.

The development of the 50,000 bbl (8,000 m³) per day of non-upgraded oil in the southern portion of Anzoátegui will be mainly less viscous oil (12–15° API) and will use the existing production facilities of the area.

Upgrading of the crudes. Since most of the crude contains a high concentration of metals (200–500 ppm of nickel plus vanadium) and sulfur, it requires upgrading to make the product accept-

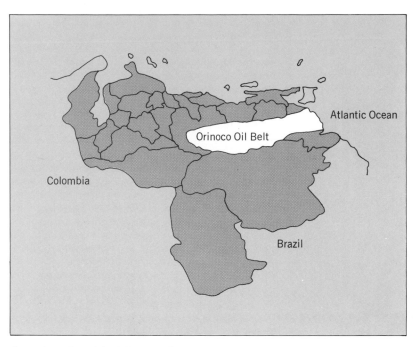

Fig. 9. Location of the Orinoco Oil Belt in eastern Venezuela.

able to conventional refineries. The first project being designed is a production-upgrading module of 125,000 bbl (20,000 m³) per day. However, since there is a definite trend of world production to become heavier and due to the importance of the giant accumulations of heavy oil, such as the Orinoco Oil Belt for the future supply of oil, a strong effort from the consumer countries to construct such upgrading facilities will be required. This fact is reinforced by the fact that the financial and human resources required to develop projects of the described magnitude makes improbable that the countries that own these large heavy-oil deposits can undertake the exploitation of very large volumes of upgrade crude. Therefore, there will be ample opportunities for the consuming countries to install refining facilities to permit the intake of the extra-heavy oil in its virgin state and convert it into the required products. In this fashion, larger quantities of these reserves could be incorporated faster into the world's hydrocarbon supply.

For background information *see* OIL AND GAS, OFFSHORE; OIL AND GAS FIELD EXPLOITATION; PIPELINE in the McGraw-Hill Encyclopedia of Science and Technology.

[CARLOS BORREGALES]

Bibliography: C. J. Borregales, Evaluation and development of the Orinoco Oil Belt, Venezuela, *Unitar Conference on Long Term Energy Resources*, Montreal, December 1979; C. J. Borregales, Production characteristics and oil recovery in the Orinoco Oil Belt, *Unitar Conference on the Future of Heavy Crude and Tar Sands*, Edmonton, June 1979; C. A. Edmondson, A history of Alaskan oil, *Alaska Petroleum and Industrial Directory*, p. 11, 1970–1971; F. Gutierrez et al., Technological needs for production-upgrading of the Orinoco extra heavy crude oils, *10th World Energy Conference*, Istanbul, September 1977; Unitar, *1st International Conference on the Future of Heavy Crude and Tar Sands*, June 1979.

Phloem

Since 1975 several instances of a fatal disease of palms associated with the trypanosomatid flagellate *Phytomonas* have been reported. The disease is essentially a type of wilt and has been reported from Surinam, Trinidad, Colombia, Equador, and Peru. The flagellate-associated disease has been called hartrot in Surinam, cedros wilt in Trinidad, and marchitez sorpresiva in Colombia, Ecuador, and Peru. The economically important coconut *(Cocos nucifera)* and the oil palm *(Elaeis guineensis)* have been mainly affected, but the disease also occasionally affects the maripa palm *(Maxmiliana maripa)* in Surinam. Because of its rapid spread, the disease seems to threaten the palm industry of Caribbean and South American countries. The palm disease, referred to here as hartrot (from the Dutch term for heartrot) was first noted in Surinam in 1908 and was subsequently confused with another palm disease, lethal yellowing, because of some similarities in the external symptoms. Lethal yellowing, however, is associated with a mycoplasmalike organism (mycoplasmas are prokaryotic organisms that lack a cell wall and belong to the class Mollicutes), whereas hartrot is associated with flagellates.

Biology of Phytomonas. The genus *Phytomonas* (formerly known as *Leptomonas* or *Herpetomonas*) is a protozoan belonging to the family Trypanoso-

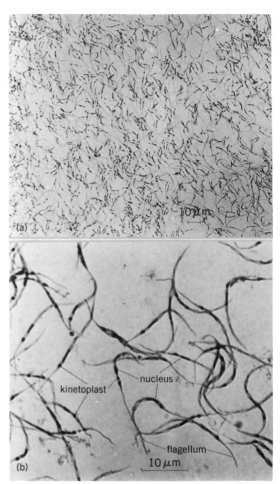

Fig. 2. *Phytomonas staheli* (*a*) in drop of phloem sap squeezed out from the inflorescence stalk of a coconut palm affected by hartrot; and (*b*) after Giemsa staining, showing flagellate bodies.

matidae in the order Kinetoplastida. Although the existence of the plant-infecting *Phytomonas* has been known for nearly 70 years, its biology is still not clearly understood. The flagellate has been found chiefly in latex-containing plants of the families Euphorbiaceae, Asclepiadaceae, and Moraceae, in which plants it is confined to the latex-bearing cells—the laticifers. The two best-known species of *Phytomonas* are *P. davidi* and *P. elmassiani*, which, respectively, commonly infect euphorbs *(Euphorbia gamella, E. hirta, E. hypericifolia, E. heterophylla)* and milkweeds (most often *Asclepias curassavica* and *A. syriaca*) respectively. *Phytomonas davidi* is about 20–24 μm long and about 1.5 μm wide, whereas *P. elmassiani* is about 13 μm long and about 1.75 μm wide. Protozoans can be seen with the light microscope in latex smears from infected plants stained with Giemsa stain (Fig. 1*a*). Giemsa stain helps to reveal the general morphology and stains the nucleus and the kinetoplast (a DNA-rich organelle associated with the mitochondrion). *Phytomonas elmassiani* is transmitted by the mealy bug *Oncopeltus fasciatus* (family Lygaeidae), and *P. davidi* appears to be transmitted by another insect of the same family, *Pachybrachus billobatus scutellatus*. The *Phytomonas* multiplies by binary or multiple fissions

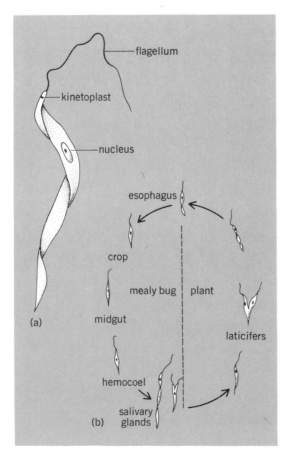

Fig. 1. *Phytomonas*. (*a*) Typical organism. (*b*) The life cycle of *P. elmassiani* that commonly infects the milkweed.

within its plant hosts. Figure 1*b* illustrates the life cycle of *P. elmassiani.*

Phytomonas davidi occurs almost worldwide wherever the host euphorb plants grow. *Phytomonas elmassiani* also has a widespread occurrence and has been reported as far north as 44° latitude. With the exception of two reports, all investigations in laticiferous plants infected with *Phytomonas* suggest that the flagellate is nonpathogenic to its hosts. This is not surprising because most laticifers apparently do not play a vital role in metabolism; they only accumulate metabolic by-products or substances that are regarded as excretory (such as terpenes and rubber). If the protozoan infects the food-conducting tissue (phloem), however, it is apparently fatal to its plant host.

Phytomonas associated with plant diseases. The first report of *Phytomonas* infection in a non-laticiferous plant was made nearly 50 years ago by G. Stahel, who reported that coffee plants *(Coffea liberica)* were affected by a phloem disease in Surinam. He demonstrated the presence of *Phytomonas* in the sieve tubes of affected coffee plants. He observed that phloem cells infected with flagellates were often necrotic, and proposed the name phloem necrosis for the disease. He also described the putative pathogen as a new species, *Phytomonas leptovasorum.* This species is considerably shorter (6−10 μm in length) than other species of the genus. Stahel noted an unusual "multiple division" that formed up to eight tangential walls in sieve tubes differentiating near the cambium. Subsequently, the sieve tubes near the cambium became necrotic. He concluded that the multiple division was the only anatomical criterion by which phloem necrosis could be distinguished from other similar diseases of coffee. Although there was strong circumstantial evidence to indicate that the flagellate was the causal agent of the coffee phloem necrosis, Stahel was not able to prove it scientifically. He was also unable to find the transmitting agent.

Forty-five years after Stahel's report, *Phytomonas* was discovered in the phloem of coconut and oil palms affected by hartrot disease in Surinam. The protozoan is confined to the sieve tubes in the infected phloem. The flagellates are 12−18 μm in length and 0.5−2 μm in diameter. Smears of sap squeezed from young flower stalks or from roots of infected palms usually contain hundreds of the flagellate in a small drop (Fig. 2*a*). Staining the sap smears with Giemsa stain clearly shows the twisted nature of the flagellate body, the nucleus, and the kinetoplast (Fig. 2*b*). The species of *Phytomonas* that infect coconut and oil palms in Surinam has been tentatively assigned the status of a new species, *Phytomonas staheli.* The flagellates are apparently narrow enough to pass from sieve element to sieve element through sieve plate pores that range from 0.5 μm to 1 μm in diameter. Although the organism has a flagellum, only a wiggling motion is noticeable and there is no obvious locomotion. It is likely that the organisms are passively translocated along with food materials from one part of the palm to another through the sieve tubes. The wiggling motion might help them pass through sieve plate pores. Infected phloem often has sieve tubes that are plugged with the flagellate (Fig. 3). Samples from

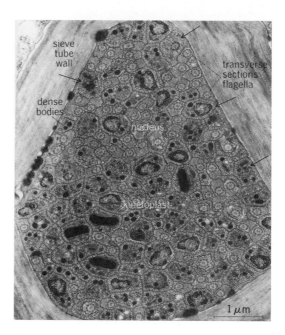

Fig. 3. Electromicrograph of a sieve tube in an infected inflorescence of coconut palm showing the flagellates plugging the lumen. Cross sections of the flagellate at various levels are indicated by arrows.

palms in advanced stages of the disease have flagellates in 10−100% of the sieve tubes. Coconut and oil palm trees can succumb to the disease 4 to 16 weeks after the earliest symptoms begin to appear. Younger palms are affected more rapidly than the taller and older palms. The absence of organisms other than *Phytomonas* during the earliest symptom of the disease and the correlated increase and spread of the flagellates in the sieve tubes as the disease progresses strongly suggest, but do not prove, that the flagellates are pathogenic to their hosts. No obvious difference in the structure or divisional patterns of the infected palm phloem has thus far been observed. Although the flagellate-associated palm disease has been reported only from five countries, it is likely that the flagellates have a relatively wide distribution in South America. Nothing is known about the transmission of *P. staheli.*

In preliminary experiments, the disease was not controlled by chemotherapy with antibiotics, aromatic amidines, quinoline pyrimidines, and phenanthiridinium compounds that are supposed to affect trypanosomatids, but these organisms are notorious for their resistance to chemotherapeutic agents. A practical solution to the problem of disease control might evolve from an understanding of the biology of the protozoan and from knowledge of the transmitting insect so that it could be controlled at a vulnerable period of its life cycle. A search for varieties of coconut and oil palms resistant to the flagellate-associated disease could also be useful. A team of scientists from Surinam and the United States is attempting to initiate a thorough study of the disease and its possible control.

For background information *see* PLANT DISEASE; TRYPANOSOMATIDAE in the McGraw-Hill Encyclopedia of Science and Technology.

[M. V. PARTHASARATHY]

Bibliography: R. B. McGhee and A. H. McGhee,

J. Protozool., 26:348–351, 1979; M. V. Parthasarathy, W. G. van Slobbe, and J. A. J. Hesen, *Principes*, 22:3–14, 1978; M. V. Parthasarathy, W. G. van Slobbe, and C. Soudant, *Science*, 192:1346–1348, 1976; G. Stahel, *Phytopathol. Z.*, 4:65–82, 1931.

Phonoreception

Infrasounds are atmospheric-pressure oscillations having a pitch so low that they are normally considered inaudible. The upper limit of pitch or frequency (which is expressed in units of hertz, or cycles per second) is defined by the limits of human sensitivity at moderate amplitudes. Humans are effectively "deaf" to sounds with frequencies below about 16 Hz. Infrasounds at this frequency and lower are common in the environment, and may be a potential source of information about the environment to animals able to detect them. Pigeons *(Columba livia)* are one species with infrasonic sensitivity well beyond that of humans.

Pigeon thresholds. Thresholds of detection of pigeons to infrasound have been measured by using a cardiac conditioning technique. A measured increase in the bird's heart rate gives an indication if the sound stimulus presented has been detected. Pigeons have thresholds of about 60 dB SPL (sound pressure level) at 5 Hz and 50 dB SPL at 10 Hz. (Zero dB SPL corresponds to the human threshold of detection at its most sensitive frequency range, 3000–4000 Hz, and provides the reference of SPL.) The sensitivity of pigeons in the infrasonic range can be compared to that of humans, which is about 115 dB SPL at 5 Hz and 104 dB SPL at 10 Hz (see illustration).

Sensory mechanism. Pigeons experiencing removal of both cochleas (the sensory hair cell structure of the inner ear which transduces sound vibration into nervous impulses) and middle-ear bones (called the columella in birds) could no longer detect infrasound even at very high amplitudes (above 100 dB SPL). Bilateral bisection and removal of a piece of the columella also resulted in a loss of sensitivity, raising the thresholds at 5 and

10 Hz to approximately 95 dB SPL. Such results suggest that infrasound is transduced by some structure involving the peripheral auditory apparatus rather than elsewhere on the pigeon's body.

Recent studies have concentrated on the role of the eardrum or tympanic membrane in the detection of infrasound by pigeons. Perforating the tympanic membranes of both ears with a small hole (1–2 mm²) results in a sensitivity deficit comparable to that seen in birds with bisected middle-ear bones. The losses in the pigeons with cut columellas are permanent, whereas the pigeons with the bilateral tympanic membrane perforations recover their ability to detect infrasound at normal levels after about 2 weeks. At that time examination of the tympanic membranes reveals that the small holes are healed over.

A pigeon with a hole in only one tympanic membrane also has the same sensory loss in infrasonic sensitivity as if both eardrums had been perforated. The intact ear does not function normally at infrasonic frequencies, although sensitivity to higher frequencies (500–2000 Hz) is not affected by the unilateral perforation. The profound effect that perforating one ear has on the other is at first somewhat surprising, until the anatomy of the avian middle ear is considered. The middle-ear cavities are the spaces inside the head behind the tympanic membranes. In birds, but not in mammals, these cavities are connected by a passageway. Not only does this connection allow air to flow, but even water dripped into one middle ear through a hole in the eardrum can flow through to the inside surface of the other ear's tympanic membrane. Infrasound detection is probably the result of vibration of the tympanic membranes. When one of these membranes is perforated, air can flow in through the connecting passage to the inside surface of the other, intact ear drum. This may act to equalize and cancel the pressure difference across the inner and outer surfaces of the tympanic membrane which would normally cause its vibration. A frequencies of thousands of hertz, there is probably not enough time for this equalization to occur, so that the membrane will vibrate and therefore transduce the sound normally. If the external ear canal of the perforated eardrum is sealed airtight, this pressure equalization cannot occur even at infrasonic frequencies. Pigeons with one tympanic membrane perforated, but having that external ear canal sealed, can detect infrasound at normal threshold levels. Removing the seal again results in a deficit of infrasonic sensitivity.

Infrasound detection in pigeons is more than just a general arousal to a stimulus; it represents a tuned system with frequency discrimination. D. B. Quine demonstrated the pigeons can detect frequency differences as low as 3% at 2, 5, and 10 Hz. This is about the same degree of discrimination shown by pigeons at higher frequencies in the conventional "audible" range.

Infrasound for information. Infrasonic pressure oscillations may be produced by a variety of sources, both natural and made by humans. These include thunderstorms, the shearing action of the jet stream, auroras, wind patterns over mountains, meteors, volcanoes, ocean waves, rockets, aircraft,

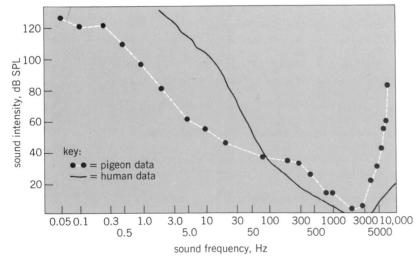

Pigeon and human audiograms showing thresholds of detection over a wide range of frequencies.

explosions, vibrating machinery, and ventilating systems. True infrasound travels at the characteristic speed of sound in air, about 344 m/s at 21°C. Since attenuation varies inversely with frequency, propagated infrasound can travel for thousands of kilometers without appreciable loss of energy. That means signals from very distant sources may be available to pigeons through infrasound. Such information may be valuable to birds as indicators of approaching weather systems. Monitoring areas of turbulence or updrafts might also be evaluated from infrasonic information. Particular patterns of infrasound emanating from a certain fixed location could potentially act as a beacon to migrating birds, aiding in their navigation.

Pseudosound. There remain several problems facing a bird using infrasound as a source of environmental information. One is the presence of pressure oscillations at infrasonic frequencies, called pseudosound, which are the result of air flow over surfaces. These turbulence-induced fluctuations in pressure are a local phenomenon and do not propagate over distances like true sound waves. It is not obvious how a pressure receptor like an avian ear could differentiate the usually very intense pseudosound noise from traveling infrasound waves.

Localization. Another difficulty lies in the ability of birds to localize the source of infrasound. This is critical if the infrasound is to be used as a navigational aid, or if a certain weather pattern or storm system is to be avoided. Sound localization is generally accomplished by comparing differences in phase, intensity, or time of arrival between the two ears (binaural differences). These cues are very much reduced as the wavelength of the sound increases relative to the distance between the ears. A 10-Hz sound has a wavelength of 34 m, which is very much larger than the size of a pigeon's head, and makes conventional binaural cues essentially useless.

One scheme proposed for localization of infrasound by a flying bird employs the Doppler shift phenomenon. To a bird flying toward the source of a sound, there will be an apparent increase in frequency; a bird flying away from the sound source will hear an apparent decrease in frequency or lowering of the pitch. Assuming the originating sound from the source is constant in frequency, and the frequency shifts caused by the bird's movement relative to the sound are within the bird's frequency discrimination capabilities, then it might be possible for a bird to determine the direction from which the sound is emanating. Quine's frequency discrimination experiments in the infrasonic range indicate that pigeons probably are able to make the necessary discrimination.

Still another possible method of localizing sounds of very long wavelengths may involve the connection between the middle ears in birds. This path is acoustically transparent so that sound may travel through the head and reach the inner surface of the tympanic membrane as well as the outer surface. This arrangement may act as a pressure-gradient receiver which is inherently directional.

Many questions about avian detection of infrasound remain unanswered. It is known that pigeons can detect infrasonic pressure oscillations under laboratory conditions, and that acoustic stimuli at these frequencies are present in the environment. Exactly how birds use this natural infrasound is unknown. Much remains to be explored about this ability of pigeons to detect sounds that humans cannot hear.

For background information *see* PHONORECEPTION in the McGraw-Hill Encyclopedia of Science and Technology.   [MARILYN L. YODLOWSKI]

Bibliography: M. L. Kreithen and D. B. Quine, *J. Comp. Physiol.*, 129:1–4, 1979; D. B. Quine, *Infrasound Detection and Frequency Discrimination by the Homing Pigeon*, Ph.D. dissertation, Cornell University, 1979; M. L. Yodlowski, *Infrasound Sensitivity in Pigeons*, Ph.D. dissertation, Rockefeller University, 1980; M. L. Yodlowski, M. L. Kreithen, and W. T. Keeton, *Nature*, 265: 725–726, 1977.

Photochemistry

Molecular architecture in monolayer assemblies is a specialty in itself. It has been developed to a refined elegance by Hans Kuhn and his coworkers in Germany. In a series of experiments dating back to the mid 1960s, they perfected the technique and investigated the photophysical as well as the photochemical phenomena in monolayer assemblies.

The concept of the monolayer assembly as a device to capture and convert light energy is surely an attempt to mimic photosynthesis. Photosynthesis occurs with a quantum yield near unity, and this exceptionally high quantum yield is viewed with envy by photochemists. Chlorophyll is present in the lamellae of green plants at high concentrations, perhaps as high as 0.1 M. A monomolecular film serves as a model for this obviously intricate living system. Thus it is hardly an accident that much of the early monolayer work of I. Langmuir, K. Blodgett, and V. J. Schaefer involved chlorophyll. A monolayer of chlorophyll is about 1.5 nm thick with each molecule occupying an area of about 1 nm^2. The result is a film which has a concentration near 1 M in chlorophyll. Allowing for lipid dilution in the plant chloroplast, a monomolecular film would seem to be a fairly reasonable first model.

However, well-defined experiments in monolayer assemblies began when Kuhn, D. Möbius, and coworkers investigated the fluorescence, fluorescence quenching, and energy transfer processes of dyes in monomolecular films. Later they extended these experiments to include the electron tunneling processes through monolayer films. These experiments are now culminating in the concept of the light-driven electron pump as formulated by Kuhn.

Light-driven electron pump. Fundamentally, the monolayer assembly must capture light energy via a dye molecule (D), then transfer this excitation energy by electron donation to an acceptor molecule (A), resulting in a temporary charge separation as the D$^+$ and A$^-$ pair. Ideally, the donor would be reduced back to the ground state quickly before back-transfer could occur, thereby permitting the electron to move forward to a collecting electrode. In bright sunlight a dye molecule might capture one photon per second. Each acceptor species might be considered as a holding tank for approximately 100 dye molecules; the acceptor could expect one electron to arrive every 10 ms; that is, its recovery time would have to be about 10 ms. Placing this process onto a time scale, the monolayer electron pump would absorb light on the picosecond scale, transfer an electron from the donor to the acceptor in less than a nanosecond, and hold the electron here temporarily on the millisecond scale before donating to the external circuit to form a steady-state current. This time sequence, too, is analogous to that seen in photosynthesis. On the basis of this concept the fundamental requirements for the realization of a light-harvesting monolayer assembly may be formulated.

First the back-tunneling time between the A^- and D^+ should be of the order of 10 ms. This criterion implies that the barrier should be quite wide, near 3 nm. Second, the barrier should not be very high (0.6–0.7 eV), so that electrons in the excited dye molecule, D^+, can surmount the barrier thermally, thereby yielding the temporary charge separation in D^+ and A^-. In addition, on the opposite side of the dye there should be a very narrow but high barrier. The resulting tunneling time to reduce the D^+ species then would be comparable to the lifetime of the excited state of the dye, and considerably shorter than the back-tunneling time from A^- to D^+. A barrier with a thickness of 0.6 nm has been suggested. Since the barrier is high, the excited dye would be unable to donate an electron in that direction. The dye donor is therefore in an asymmetric environment. This concept is shown in Fig. 1. By this sequence a vectorial charge separation is achieved. Precautions should be taken to ensure that the energy level of the acceptor species is only slightly lower than the donor excited state. A small energy drop will provide a kinetic drive while not wasting much of the energy. Indeed, a series of acceptors might be envisaged in which the donated electron is led down a staircase of acceptors, only a minute amount of energy being lost at each step.

Recent advances. The progress in electron transfer reaction has occurred largely with cyanine dyes and the chlorophylls. Work on synthetic porphyrins and highly colored ruthenium organometallic compounds is beginning and will doubtless become important in the future. For the limited purpose of this article, it is primarily the cyanine and chlorophyll work that will be cited.

This remarkable chemical and photochemical stability of the cyanine dyes has led Kuhn and collaborators to work mostly with these materials. The fact that these substances absorb in the ultraviolet and blue region of the spectrum means that photochemical changes can occur fairly easily, there being ample energy per photon. In typical methodical fashion this group first characterized the electrical properties of fatty acid mono- and multilayers prior to investigating electron transfer reactions. For example, they measured the conduction band level of the fatty acid layer to be 2 eV below vacuum. By working with a single fatty acid monolayer sandwiched between two metal electrodes, they showed that a tunneling current through the lipid layer depends exponentially on the length of the fatty acid chain. And so the model

grew, adding cyanine dye donors to the lipid matrix. They were then able to demonstrate electron donation to a biased electrode from the cyanine dye via thermal excitation over a lipid barrier. At low temperatures they demonstrated light-induced electron tunneling from the cyanine dye through a lipid barrier to a biased electrode.

The work of this group is at its most elegant in the recent efforts in which they incorporated both a cyanine donor as well as an azo acceptor into a monolayer assembly. Care was taken to retain a low barrier between donor and acceptor. The model is shown in Fig. 2. The presence of the

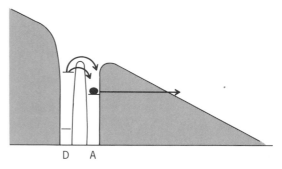

Fig. 2. Donor-acceptor system in which the cyanine dye (D) donates through and over a small barrier to an acceptor azo dye (A). The slope indicates the electrode bias.

azo acceptor caused a tenfold increase in the light-induced current from the cyanine dye to the biased electrode. The acceptor clearly behaves as a temporary stopover point for an electron in transit and permits the dye donor to be recycled. By these and other experiments Kuhn and coworkers have verified the conceptual validity of the light-induced electron pump.

Chlorophyll. While the stability of cyanine dyes is admirable, the fact that they absorb in the ultraviolet leaves them of doubtful value as models for light-harvesting antenna chlorophyll. In order to capture a significant amount of light of the solar spectrum the chromophore should absorb in the visible region of the spectrum. According to the detailed calculations of J. Bolton and R. Ross, the absorber should have its maximum absorption in the red region of the spectrum. In this sense chlorophyll is a more suitable agent, despite its reputation for stability problems. Nonetheless, F. Janzen and Bolton further developed the concept of Kuhn's molecular wire with chlorophyll. They found that chlorophyll monolayers sandwiched between an aluminum and a mercury electrode showed a vectorial charge separation, that is, a photovoltage and a photocurrent. Next they incorporated unsaturated polyisoprene compounds such as squalene, plastoquinone, or ubiquinone as conducting acceptor elements in their monolayer assembly. The result was a dramatic increase in photovoltage, a tenfold increase in current, and also a sharp reduction in internal resistance. The presence of double bonds in the isoprene compound produced localized wells in the lipid barriers and thereby increased the electron tunneling probability. This concept is summarized in Fig. 3. These unsaturated

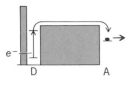

Fig. 1. Conceptual model of the light-driven electron pump.

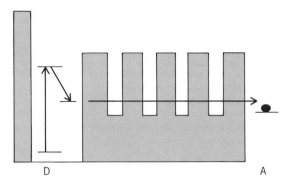

Fig. 3. Concept of the molecular wire. The unsaturated polyisoprene compounds promote electron tunneling through a lipid matrix from the donor (D), chlorophyll, to an acceptor (A). In the case of chlorophyll, the electron transfer probably occurs via the triplet state.

isoprene compounds are so widely distributed in nature that in living systems they may possibly also behave as molecular wires, promoting tunneling through lipids.

The work of T. Miyasaka and colleagues in Japan represents yet another dimension of the work. They were able to eliminate the metal/organic interface entirely. They placed the chlorophyll monolayers on the surface of SnO_2 semiconducting glass, and the assembly was then dipped into an aqueous solution containing sodium sulfate and hydroquinone. They observed light-induced currents and voltages with remarkably high quantum yields, about 12%, but fairly low photovoltages, about 18 mV. Both results are very likely due to the low conduction band in SnO_2. In this system the SnO_2/chlorophyll interface acts as the barrier between donor and acceptor, while the solution/chlorophyll interface acts as the narrow high barrier discussed earlier.

Future directions. Future developments will include the synthesis of very stable chromophores on the one hand and a further mimicking of the tricks of photosynthesis on the other. For example, R. Memming and F. Schröppel have carried out experiments with intensively studied ruthenium bipyridyls. Quantum yields near 15% were observed, with one electron ejected per adsorbed ruthenium complex molecule in less than 1 s. These compounds are interesting in spite of much lower absorption coefficients for light than chlorophyll.

Antenna chlorophyll in photosynthesis is used only for gathering light. Light energy conversion occurs only in the reaction center, where excitation energy in a special pair of chlorophyll molecules initiates charge separation. A mixed pigment system in which one pigment might be ideal for absorption and Förster energy transfer, while the other might be ideally suited to electron transfer, could mimic this situation. The synthesis work of P. Loach and of J. Bolton is particularly interesting in this regard. The molecule which Bolton's group has prepared incorporates a porphyrin donor, attached via a short chain to a quinone acceptor. With such a molecule, it may become possible to engineer the donor-acceptor distance by one carbon unit at a time. Such an approach might eliminate the requirement of constructing a barrier between donor and acceptor by the physical deposition of a barrier, and the entire donor-acceptor complex might then lie within one single monolayer.

For background information *see* CHLOROPHYLL; DYE; PHOTOCHEMISTRY; PHOTOSYNTHESIS in the McGraw-Hill Encyclopedia of Science and Technology. [A. FREDERICK JANZEN]

Bibliography: A. F. Janzen and J. R. Bolton, *J. Amer. Chem. Soc.*, 101:6342–6348, 1979; H. Kuhn, *J. Photochem.*, 10:111–132, 1979; H. Kuhn, D. Möbius, and H. Bücher, in A. Weissberger and B. W. Rossiter (eds.), *Physical Methods of Chemistry*, pt. 3B, pp. 577–702, 1972; T. Miyasaka et al., *J. Amer. Chem. Soc.* 100:6657–6665, 1978.

Pineal gland

Pineal organs are found in practically all vertebrates, and develop from an outpouching of the roof of the diencephalon. Before the mid-1950s many scientists believed that the pineal was a functionless vestige. Today, however, the pineal is recognized as an actively functioning organ which responds to photic stimuli, shows marked daily rhythms in a variety of biochemical events, acts as a "biological clock" in at least some animals, and influences the hypothalamo-pituitary-gonadal axis.

There is a considerable variety in the innervation and morphology of pineal organs in different kinds of vertebrates. In general, the pineal organs of lower vertebrates, such as fish, amphibians, and reptiles, can have a direct photosensory function since they contain identifiable photoreceptor cells which can give electrical responses to illumination. Lighting information can, therefore, be directly transduced into hormonal information by the pineal or it can be sent to the brain via nerves. The pineal organs of birds and mammals, in contrast, have a more glandular appearance and lack obvious photosensory cells. Lighting information reaches the mammalian pineal indirectly by a circuitous route which involves retinal photoreception.

Mammalian pineal. The mammalian pineal shows abundant evidence of secretory activity. The principal cell type of the mammalian (and avian) pineal is the pineal parenchymal cell or pinealocyte, which is believed to have evolved from the photosensory cells of the lower vertebrates. Pinealocytes show some structural resemblances to photosensory cells, particularly in neonatal mammals, but have modifications characteristic of metabolically active endocrine secretory cells.

Biochemistry and daily rhythms. The most intensively studied group of pineal compounds is the indoleamines; these biochemicals are derived from the amino acid tryptophan (Fig. 1). The first pineal indoleamine which was isolated and identified was 5-methoxy N-acetyltryptamine or melatonin. This compound was so named because it is a derivative of serotonin and has the ability to cause the dermal melanophores of larval amphibians to contract, so that the animals turn pale. Although the biochemical pathways for indoleamine synthesis have been elucidated most thoroughly in the mammal, indoleamine synthesis occurs in all vertebrate pineals and probably involves similar path-

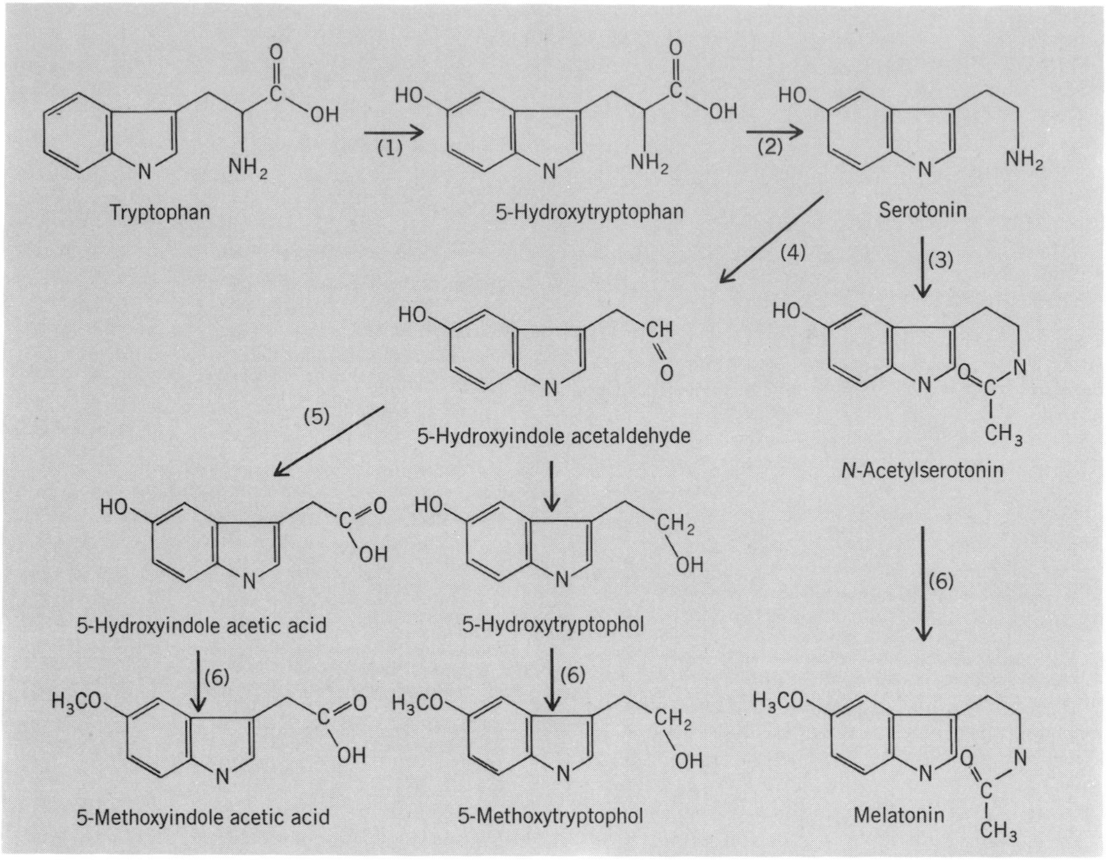

Fig. 1. Indoleamine metabolism within mammalian pineal organs. Enzymes: (1) tryptophan hydroxylase, (2) aromatic L-amino acid decarboxylase, (3) N-acetyl- transferase, (4) monoamine oxidase, (5) aldehyde dehydrogenase, (6) hydroxyindole-O-methyltransferase.

ways. A remarkable feature of pineal biochemistry is the large daily rhythms that occur in enzyme activities and the amounts of biochemicals within the pineal. For example, in the mammalian pineal a large daily rhythm occurs in the activity of the enzyme N-acetyltransferase (NAT), which is the rate-limiting enzyme in melatonin synthesis. Accordingly, a large daily rhythm in melatonin pineal content is also seen, with maximal levels occurring at night (Fig. 2). Melatonin is released into the blood from the pineal, and consequently blood melatonin levels show a daily rhythm which parallels the pineal melatonin rhythm. Interestingly, pineal melatonin content always peaks during the dark phase of a 24-h light-dark cycle, regardless of whether the animal is dark-active (nocturnal), such as the rat, or light-active (diurnal), such as the chicken. A few other areas appear to have melatonin-synthesizing capabilities, such as the eyes or the harderian glands, but these tissues apparently do not contribute significantly to blood melatonin levels. A host of other enzymes and biochemicals within vertebrate pineals also show daily rhythms, including norepinephrine, serotonin, ribonucleic acid, cyclic nucleotide phosphodiesterase, and 5-hydroxyindole-3-acetic acid.

Regulation of rhythmicity. Pineal rhythmicity is regulated by a system with several components (Fig. 3): photic information is perceived by the eyes and is sent to the suprachiasmatic nuclei in the hypothalamus via a direct neural pathway; the suprachiasmatic nuclei seem to function as a biological clock which can be synchronized by light perceived by the eyes and which controls pineal biochemical activity via a multisynaptic neural pathway to the superior cervical ganglia; and the superior cervical ganglia in turn innervate the pineal. Consequently, destruction of the suprachiasmatic nuclei or interruption of the neural pathway from the suprachiasmatic nuclei to the pineal (such as removal of the superior cervical ganglia) abolishes pineal rhythmicity. The suprachiasmatic nuclei appear to be a central biological clock in mammals, since their destruction abolishes not only pineal rhythms, but other daily rhythms as well, including rhythms in body temperature, feeding, drinking, locomotor activity, and adrenal corticosterone output. Most significantly, a daily rhythm in electrical activity persists in the suprachiasmatic nuclei even after the nuclei have been isolated from the rest of the brain by knife cuts, showing that the suprachiasmatic nuclei are capable of autonomous self-sustaining rhythmicity.

Melatonin synthesis. The details of the control of melatonin synthesis have been well elucidated. Daily light-dark cycles, perceived by the eyes, entrain the biological clock located in the suprachiasmatic nuclei. The suprachiasmatic nuclei control the cyclic daily release of norepinephrine from the sympathetic nerve endings which terminate on the pineal. The rhythm of norepinephrine release is driven by the suprachiasmatic nuclei such that maximal norepinephrine release occurs at night. The interaction of norepinephrine with β-adrenergic receptors on pineal cells activates adenyl cyclase, which initiates the synthesis of cyclic adenosinemonophosphate (cyclic AMP) from adenosinetriphosphate (ATP). Cyclic AMP in turn stimulates the production (and activation) of the rate-limiting enzyme in melatonin synthesis, N-acetyltransferase. Consequently pineal melatonin synthesis increases during the night.

Photosensitivity in juveniles. In contrast to the adult, recent evidence suggests that the pineal of neonatal mammals may be directly photosensitive. For example, in juvenile rats a daily light-dark cycle can directly influence rhythms in pineal serotonin content and hydroxyindole-O-methyltransferase activity, even after the sympathetic innervation to the pineal has been abolished. This capacity, however, is a developmental phenomenon which is lost in the adult.

Role of pineal in reproduction. The mammalian pineal has been implicated in the control of a number of glands, including the gonads, the adrenal, and the thyroid. Most research has focused on the role of the pineal in reproduction. In many cases the pineal has an antigonadal role in that its removal promotes growth or maturation of the gonads. The most marked effects of pinealectomy are seen in strongly photoperiodic mammals such as the golden hamster. Melatonin, arginine vasotocin, and several partially purified peptides have been implicated as antigonadal substances produced by the pineal. A tripeptide, threonylserinyllysine, has recently been isolated and purified from bovine pineals, and has antigonadotropic properties.

Avian pineal. Daily rhythms in indoleamine metabolism are observed in avian pineals as well.

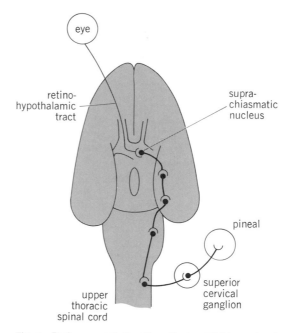

Fig. 3. Pathway mediating the effects of light on pineal rhythms in a mammal.

Unlike the mammalian pineal, however, biochemical rhythmicity in the avian pineal can continue in the absence of sympathetic innervation. The avian pineal contains in itself a biological clock or clocks which can regulate pineal indoleamine synthesis. For example, pineals removed from chickens can be maintained in organ culture, and such pineals still exhibit daily rhythms in NAT activity and melatonin release. The NAT and melatonin rhythms in cultured pineals are responsive to illumination and can persist, for a few cycles, in constant darkness. The rhythm of NAT activity persists in dispersed cells of cultured chicken pineals and remains responsive to light. Although the possibility of cellular interaction cannot be completely eliminated, this study raises the intriguing possibility that each pineal cell contains a photoreceptor, a daily (circadian) clock, and the melatonin-synthesizing machinery.

A number of studies, using primarily the house sparrow, have shown an important role for the pineal in controlling daily rhythms in birds. In the house sparrow removal of the pineal organ of birds free-running (showing their endogenous daily rhythm) in continuous darkness abolishes the daily rhythm in perch-hopping activity, whereas implantation of a pineal organ into the anterior chamber of the eyes of previously pinealectomized house sparrows reintroduces rhythmicity. Significantly, the induced rhythm emerges with the phase shown by the donor bird. Inasmuch as interruption of the nervous input and output of the house sparrow's pineal does not abolish rhythmicity, the avian pineal seems to be hormonally, rather than neurally, coupled to other components of the circadian system. In the starling pinealectomy occasionally abolishes the perching activity rhythm, but usually elicits a shortening in the period of the activity rhythm of birds free-running in continuous darkness and increases the instability of the rhythm.

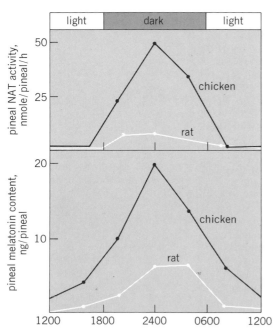

Fig. 2. Daily rhythms in pineal N-acetyltransferase (NAT) activity and melatonin content.

The current model for the role of the avian pineal in the circadian system suggests that the pineal is the site of a self-sustaining master circadian clock or pacemaker that synchronizes other self-sustaining secondary clocks located outside the pineal. In the absence of the pineal, rhythmicity is abolished if the remaining secondary oscillators become uncoupled from one another and free-run with their individual frequencies. If some coupling remains among the secondary oscillators after pinealectomy, a change in the period of the coupled system is observed. These secondary clocks may be located in the suprachiasmatic nuclei, since lesions of the suprachiasmatic nuclei abolish the rhythm of activity in house sparrows. Melatonin has been implicated as the hormonal messenger between the pineal and the rest of the circadian system, since exogenous administration of melatonin can affect the period of the avian activity rhythm if given continuously, or entrain (synchronize) the rhythm if given as a daily injection.

Studies on a role for the pineal in the endocrine system of birds are sparse and involve mainly the reproductive system. Major effects of pinealectomy have been noted on the reproductive systems of some species, such as the domestic duck or the Indian weaver bird, whereas in other species, such as the chicken, effects are minimal.

Pineal of lower vertebrates. A few studies have shown daily rhythms in pineal indoleamines in fish, turtles, and lizards. The pineal of the lower vertebrates may have a role in the animals' circadian system similar to that seen in the bird, since removal of the pineal in fish or lizards has major effects on the animals' daily activity rhythms, and melatonin can affect the activity rhythm of lizards. Pinealectomy effects on the reproductive system of fish and lizards have also been noted.

For background information *see* ENDOCRINE MECHANISMS; PERIODICITY IN ORGANISMS; PINEAL BODY (VERTEBRATE) in the McGraw-Hill Encyclopedia of Science and Technology.

[HERBERT UNDERWOOD]

Bibliography: R. J. Reiter, *The Pineal*, vol. 3, 1978; B. Rusak and I. Zucker, *Physiol. Rev.*, 59: 449–526, 1979; J. S. Takahashi and M. Menaker, *Federation Proc.*, 38:2583–2588, 1979.

Planet

During 1978–1979 a rapid flourishing of knowledge of planetary atmospheres occurred as a result of six spacecraft missions. In 1978 the four *Pioneer Venus* probes and the *Venera 11* and *Venera 12* probes entered the Venusian atmosphere; in 1979 *Voyagers 1* and *2* encountered Jupiter and its satellites, and *Pioneer 11* encountered Saturn and Titan. The sophisticated instrumentation flown on these space missions was augmented by spectroscopic observations using Earth-based telescopes and by earlier missions to Venus, Mars, and Jupiter. The range of facts now assembled concerning Venus, Mars, and Jupiter facilitates consideration of two important questions concerning their atmospheres. The first concerns the chemical cycles responsible for maintaining the presently observed atmospheric composition; the second concerns the manner in which these atmospheres have evolved over the age of the solar system.

Venus. The bulk of the atmosphere of Venus is CO_2. The surface pressure is 90 bars (9 MPa), and the surface temperature is 750 K. The detailed atmospheric composition has been measured in place by using the *Pioneer Venus* main probe and bus neutral mass spectrometers, the *Venera 11* and *12* neutral mass spectrometers, the *Venera 11* and *12* and *Pioneer Venus* gas chromatographs, and the *Venera 11* and *12* spectrophotometers. These experiments, together with earlier ground-based ultraviolet (uv) and infrared (ir) spectrometers, indicate the following volume mixing ratios for minor species: N_2 (3.5–4.5%), Ar (40–110 ppm), Ne (4–20 ppm), Kr (0.4–0.8 ppm), He (20 ppm), SO_2 (0.1 ppm at 70 km, 180 ppm at 22 km, less than 25 ppm near surface), CO (100 ppm at 65 km, 32 ppm at 52 km, 20 ppm at 22 km), H_2O (10 ppm at 65 km, 200 ppm at 48 km, 20 ppm at surface), HCl (1 ppm), HF (10 ppb), and S_2, S_3, and S_4 (50 ppb below 30 km). In addition, H_2S has been detected in large but as yet unquantified amounts below 22 km. Finally, although COS was not unambiguously detected, it could also be present in significant amounts near the surface.

There is no new spacecraft data which contradict the ground-based evidence implying that the principal component of the clouds which completely shroud the planet in the 48–70-km altitude region are composed of droplets of concentrated sulfuric acid. The *Pioneer Venus* cloud particle size spectrometer and nephelometers show a distinct layering within this cloud and a remarkable trimodal (1, 3, 8 μm in diameter) particle size distribution. Below the base of this cloud at 48 km there is a very tenuous haze surviving to the 31-km level, and nothing detectable below that. However, the instruments involved were insensitive to modest populations of particles with diameters less than 0.6 μm. The earlier *Venera 9* and *10* entry probe nephelometers showed evidence for significant particulate concentrations down to at least 20 km.

Investigations of the chemistry of the Venusian atmosphere have focused on four outstanding questions: Why is the atmosphere predominantly CO_2 when solar uv radiation continuously decomposes it to CO and O_2? What causes sulfuric acid to be formed and create the cloud layer above 50 km altitude? Why is Venus so deficient in water (the water in the Earth's oceans contains the equivalent of hundreds of bars of H_2O vapor)? Why does the atmosphere contain so much more HCl, HF, and sulfur compounds than the atmosphere of the Earth?

The answers to the first two questions appear to lie largely in the coupled sulfur and CO_2 cycle. Carbon dioxide is continuously dissociated at very high altitudes into CO and O. The CO and O diffuse downward, where the O is partly converted to O_2. Also on the way down, both the O and O_2 are partly recycled back to CO_2 in catalytic cycles involving H, OH, HO_2, ClO, and ClO_2. Despite these recycling mechanisms, a significant flux of CO and O_2 makes it down to the cloud-top level.

At the cloud-top level, the O_2 encounters H_2S, SO_2, and elemental sulfur particles which have diffused upward from the lower atmosphere. The H_2S photodissociates at wavelengths below 270 nm to yield SH radicals. The photodissociation of HCl at wavelengths less than 220 nm produces H and Cl atoms which can combine with the O_2 to form the oxidizers OH, HO_2, H_2O_2, and ClO. The SH and SO_2 can be rapidly oxidized to SO_3 and thus H_2SO_4 by these strong oxidizing agents. Also dissolution of H_2O_2 into cloud particles can lead to oxidation of any elemental sulfur in these particles to H_2SO_4. Elemental sulfur as an additional component in the visible clouds could serve to explain the observed uv light absorption by the clouds; variations in the supply of O_2 would enable longer survival of elemental sulfur in O_2-poor regions and thus provide an explanation for the remarkable spatial patterns seen in the cloud uv absorption.

Below the cloud base at 48 km, sulfuric acid evaporates, yielding SO_3 and H_2SO_4 vapor. As these vapors diffuse downward, they encounter progressively warmer temperatures, enabling thermochemical reactions to proceed. Elementary reactions (1) and (2) reverse the upper atmospheric

$$H_2SO_4 \rightarrow H_2O + SO_3 \qquad (1)$$

$$SO_3 + CO \rightarrow SO_2 + CO_2 \qquad (2)$$

oxidation of SO_2. Similar sequences of elementary reactions lead to the net reactions (3)–(5), which

$$SO_2 + 2CO \rightarrow S + 2CO_2 \qquad (3)$$

$$S + H_2 \rightarrow H_2S \qquad (4)$$

$$S + CO \rightarrow COS \qquad (5)$$

reverse the upper atmospheric reduction of CO_2 and oxidation of elemental sulfur and H_2S and also produce COS.

In addition, S_3 and S_4 in the lower atmosphere absorb near-uv and visible light which is able to penetrate through the upper cloud layer. These species may thus photodissociate, reactions (6) and (7), causing decomposition of COS to elemen-

$$S_3 + h\nu \rightarrow S_2 + S \qquad (6)$$

$$S_4 + h\nu \rightarrow S_2 + S_2 \qquad (7)$$

tal sulfur through reactions (8)–(10). The proposed

$$S + COS \rightarrow S_2 + CO \qquad (8)$$

$$S_2 + COS \rightarrow S_3 + CO \qquad (9)$$

$$S_3 + COS \rightarrow S_4 + CO \qquad (10)$$

overall cycles for CO_2 and sulfur are summarized in the illustration. However, there are many aspects of this cycle needing further elucidation and verification; the available data are suggestive but far from definitive in this respect.

Concerning the third and fourth outstanding questions about Venusian chemistry, two competing hypotheses have been advanced to explain the lack of H_2O on Venus relative to Earth. The first argument involves a study of the conditions in which Venus and Earth are expected to have accreted in the massive spinning gas and dust disk (primitive solar nebula) which was the precursor to the solar system. Venus formed closer to the primitive Sun and thus at higher temperatures than the Earth. Water was incorporated into these planets in the form of hydrous minerals, for example, tre-

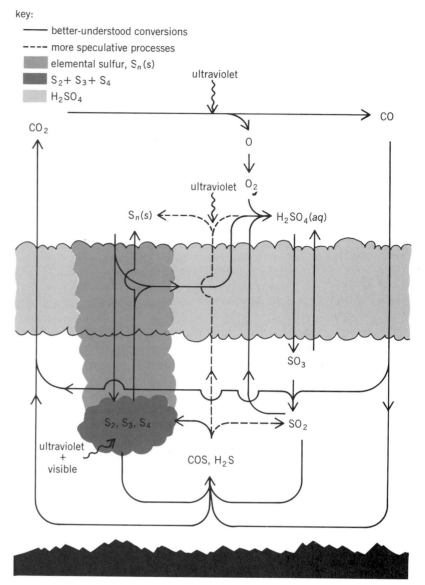

key:

— better-understood conversions

-- - more speculative processes

▨ elemental sulfur, $S_n(s)$

▨ $S_2 + S_3 + S_4$

▨ H_2SO_4

Carbon dioxide and sulfur cycles on Venus.

molite and serpentine, which are stable only at primitive solar nebula temperatures less than 500 and 350 K, respectively. Current models suggest that Venus was formed at temperatures too high to have received a large complement of these hydrous minerals relative to Earth. The competing hypothesis is that Venus began with a similar H_2O content to the Earth. However, its higher surface temperatures resulted in the H_2O residing in a predominantly water vapor atmosphere rather than in oceans. Photodissociation of H_2O, yielding H_2 and O_2, is rapid in this primitive atmosphere. The H_2 escapes to space and the O_2 is retained as CO_2 or buried in ferric minerals on the surface. The atmosphere thus ultimately reaches its present-day state highly depleted in H_2O. Both hypotheses have their strong adherents.

Finally, the presence of HCl, HF, and sulfur compounds in the atmosphere has been hypothesized to be a result of thermochemical equilibrium between the very hot surface minerals and the atmosphere. For example, the equilibrium shown in reaction (11) can enable significant quantities of

$$2NaCl(s) + CaAl_2Si_2O_8(s) + SiO_2(s) + H_2O(g) \rightleftharpoons$$
$$2NaAlSiO_4(s) + CaSiO_3(s) + 2HCl(g) \qquad (11)$$

HCl to exist in the atmosphere at 750 K.

Jupiter. The atmosphere of Jupiter is predominantly H_2 and He, with much smaller amounts of CH_4 and NH_3. In the last few years several additional and unexpected species have been discovered by Fourier transform spectroscopy from Earth-based telescopes. In particular, C_2H_6, C_2H_2, PH_3, H_2O, CO, and most recently GeH_4 have been observed. Of these newly discovered species, only H_2O was predicted to be remotely observable in a chemical-equilibrium solar-composition model of the atmosphere.

Present understanding is that the nonequilibrium species appear as a result of at least three disequilibrating processes. For C_2H_2 and C_2H_6, the sole required process appears to be the photodissociation of CH_4 by solar uv radiation. For PH_3 and GeH_4, the only feasible source appears to be rapid upward mixing from deep levels of the atmosphere where pressures exceed 100 bar (10 MPa) and temperatures exceed 800 K. At these levels PH_3 and GeH_4 are thermodynamically stable. For CO, the observed mixing ratio is also consistent with upward mixing from deep levels. The required rate of mixing agrees with that predicted from free convection driven by Jupiter's internal heat source. However, the observed CO appears to be at least partly concentrated in the upper atmosphere, and models involving bombardment of the planet with oxygen in meteoritic material and from the inner Jovian satellites have been proposed to explain an upper atmospheric CO layer.

Gaseous PH_3 is additionally interesting because of its possible role in explaining some of the colorations on the planet. In the stratosphere it photodissociates, and is expected to produce elemental red phosphorus particles. The redness of a particular region is limited by the rate at which PH_3 is mixed up into the stratosphere from below. Observations imply a rapid decrease in PH_3 concentrations with altitude in the stratosphere, in agreement with these ideas. It has also been proposed that H_2S gas and NH_4HS particles, which are both expected in a solar-composition Jovian atmosphere, are dissociated by uv light. This results in production of particulates of sulfur and ammonium and hydrogen polysulfides whose colors range from pale yellow to dark brown. However, because H_2S condenses as NH_4HS below the uppermost white ammonia crystal clouds, it is expected that these yellows and browns would be visible only in regions where these uppermost clouds are absent (for example, in regions of sinking dry air).

Voyagers 1 and *2* have augmented the earlier ground-based spectroscopic studies by confirming the presence of H_2O, C_2H_2, C_2H_6, PH_3, and GeH_4; by providing more accurate estimates of mixing ratios; and by showing the variability of species over the planet (for example, there is less C_2H_2 relative to C_2H_6 in polar regions than in equatorial regions). In addition, the imaging and temperature-sensing experiments on the Voyagers have provided important new information on the circulation of Jupiter's atmosphere. This information has considerable impact on understanding of atmospheric and cloud chemistry.

The Great Red Spot, an oval weather system with a size equivalent to the surface area of the Earth, appears to be a region of cool continuously rising and anticyclonically rotating air. This is consistent with the earlier hypothesis that red phosphorus particles produced from PH_3 are responsible for its red coloration. However, there are also smaller oval spots on Jupiter which are meteorologically similar to the Great Red Spot but which are white not red. One explanation for this apparent contradiction is that the red spot circulation goes to great depths, whereas the smaller white spot circulations are relatively shallow. The colors may also be modulated by downward mixing of C_2H_2 which could inhibit phosphorus formation.

In contrast to the Great Red Spot, the dark-brown elongated ovals in the North Equatorial Belt appear to involve warm cyclonically rotating and sinking air. Under these circumstances it is expected that the lower regions, where brown colors due to sulfur and polysulfides are predicted, would be visible. Present information concerning the Great Red Spot, the dark-brown ovals, and other yellow and brown regions of the planet are consistent with the idea that sulfur compounds, together with phosphorus, can explain the remarkable colorations of the planet. However, these consistencies are merely tantalizing. The true roles of these colored inorganic compounds and of other suggested colored organic compounds on Jupiter must await detailed measurements of cloud composition.

Atmospheric evolution. One of the most important goals of the studies of planetary atmospheres is to understand their origin and evolution. In gaining this understanding, two questions, both depending on comparative studies of planets, loom large: Is there a consistent physical picture for the retention of volatile elements and molecules during planetary formation? What general or planetary-specific evolutionary processes have led to the present atmospheres on Venus, Earth, Mars, and the major planets?

For the terrestrial planets, the new spacecraft data on Venus which have been specifically discussed can be studied together with the data on the Martian atmosphere gained from the earlier *Viking 1* and *2* missions and with the very extensive existing data on the Earth's atmosphere. It is readily seen that the absolute abundances of N_2 on Venus and Earth are almost the same. Similarly, the abundance of CO_2 on Venus is comparable to estimates of CO_2 in the form of carbonates on Earth. The Martian atmosphere is definitely deficient in N_2 and CO_2, but there are no estimates of the carbonate or nitrate content of Mars. It is also apparent that the absolute abundance of [36]Ar and the ratio [36]Ar/^{14}N decrease systematically by several orders of magnitude from Venus to Earth to Mars. In contrast, the ratios of [40]Ar to [36]Ar increase rapidly from Venus to Mars. The ratio [20]Ne/[36]Ar is very similar for all the terrestrial planets and also for meteorites, and is very different from that expected in the primitive solar-system nebula where they are the dominant Ne and Ar isotopes. Finally, as discussed in detail earlier, Venus is highly depleted in H_2O relative to the Earth. The question of the amount of water stored in the Martian crust is controversial.

The picture which best explains the origin of terrestrial atmospheres involves condensation or dissolution of volatiles into grains in the primitive solar nebula, followed by accretion of these grains to slowly form the planets. As the planets form, partial or complete melting driven by accretion energy and by decay of various radioactive isotopes (for example, ^{40}K) causes outgassing of some or all of these volatiles onto the planetary exterior. Depending on the planetary surface temperatures and the chemical composition of the outgassed volatiles, subsequent evolution of the atmosphere is possible. In this theory ^{36}Ar, which is the dominant form of Ar in the nebula, must have preferentially dissolved in grains closer to the proto-sun. For this to occur, there presumably must have been a substantial pressure gradient and a relatively small temperature gradient in the inner solar system during the time the grains were impregnated with ^{36}Ar. However, an apparently competing theory is based on fitting the observed bulk densities of the terrestrial planets, which requires a strong temperature gradient in the inner solar system. It could be that impregnation of ^{36}Ar occurred at a different time than bulk grain condensation, but for the moment this contradiction is unresolved. This contradiction forces consideration of at least two rival ideas for lack of H_2O on Venus.

Models are being developed for the evolution of the atmospheres of Venus, Earth, and Mars based on current ideas of outgassing history, atmospheric, hydrospheric, and crustal chemistry, thermal and nonthermal escape of light species from the planet, and interaction of the planet with the solar wind. For the Earth, the model must also include the dominant role of the biosphere in maintaining the present atmosphere. No one model of atmospheric evolution is widely accepted, even for the Earth; more data and more detailed study are apparently required before the many available hypotheses are either elevated to theories or discarded.

As far as Jupiter is concerned, an interesting model is emerging in which the planet accreted enough gases and solids from the solar nebula to possess a subnebula of its own. The elemental composition of this subnebula was roughly solar, but perhaps with some enhancements of elements other than H and He. The regular Jovian satellites were accreted from solid material condensed in this subnebula in the equatorial plane, in much the same way that the planets themselves accreted in the primitive solar nebula; this is an exciting prospect, for it gives another way of gaining insight into the process of planetary and atmospheric evolution.

For background information *see* JUPITER; PLANET; SPECTROSCOPY; VENUS in the McGraw-Hill Encyclopedia of Science and Technology.

[RONALD G. PRINN]

Bibliography: T. Gehrels (ed.), *Jupiter*, 1976; Pioneer Venus Mission team, *Science*, 205: 41–201, 1979; R. G. Prinn, *Geophys. Res. Lett.*, 10: 807–810, 1979; Voyager Mission team, *Science*, 206:925–996, 1979.

Plant physiology

During the last few years there has been an increasing interest by plant scientists in two environmental stress factors: salt and ultraviolet radiation. This article describes how these studies are being conducted and some of the progress to date.

SALT TOLERANCE

From studies of salt tolerance in higher plants and algae two concepts in particular have emerged: the ion selectivity of the cytoplasm and the requirement for ion compartmentation between cytoplasm and vacuole; and the osmoregulatory role of compatible organic solutes in the cytoplasm. These concepts have allowed a coherent but not comprehensive hypothesis for the biochemical basis of salt tolerance at a cellular level to be developed. The practical relevance of such work arises partly from the realization that regions of high solar irradiation but saline waters represent a major underutilized human resource. Also, the encroaching salination and alkalization of soils in many arid and semiarid regions are a serious threat to food production. One approach to these problems requires the breeding of salt-tolerant cultivars of existing species or the development of agronomically valuable halophytic crops. An understanding of the basis for salt tolerance will contribute to such development work, as well as to shedding light on saltmarsh ecology and fundamental problems of ion transport in plants.

Salt stresses. It has long been recognized that saline environments impose a number of possible stresses on plants. The high ionic strength of the bathing solution generates a high extracellular osmotic pressure which will collapse the intracellular turgor or hydrostatic pressure and ultimately dehydrate the cell's contents unless counteracted by an equivalent or near-equivalent rise in the intracellular osmotic pressure. This problem is exacerbated by high evapotranspiration rates. Saline environments contain high levels of Na^+, Cl^-, and often Mg^{++}, SO_4^{--}, borate, and other ions, which, although required as micronutrients, are toxic at high concentrations. In addition to direct toxicity, the high concentrations of the ions may interfere with either the uptake or internal metabolism of other essential nutrients, particularly K^+, $H_2PO_4^-$, and NO_3^-.

Adaptations to salinity. Many adaptive features have been recognized, including morphological, cytological, physiological, and biochemical changes, although only the last two will be discussed in this article. Several paradoxes have been revealed by the physiological studies. While the tolerance of cultivars of moderately sensitive plants is highly correlated with their ability to exclude Cl^- and Na^+ from their shoots, resistant halophytes characteristically accumulate high concentrations of these ions. Often, but not invariably, these ions are used to regulate the cell-sap osmotic pressure at levels higher than that of the external medium. Osmotic adjustment by net salt accumulation, as well as partial water loss from tissues, also occurs in many glycophytes but does not keep the growth of these plants from being seriously impaired. In laboratory cultures, the examination of the comparative salt sensitivities of enzymes and organelles from glycophytes and halophytes showed no major differences; both were almost equally sensitive to relatively low electrolyte concentrations. It was originally suggested in the 1940s that many halophytes have a high affinity for K+ compared to K+. More recently evidence has emerged that they also have capacity to selectively take up K+ from concentrated NaCl solutions, whereas in more sensitive species the ability to take up K+ is impaired.

A hypothesis has been developed in the last few years and supported by broad experimental evidence to account for these observations. Plant cells appear to have the ability, although probably differing from species to species, to selectively compartmentalize ions between their cytoplasm and vacuoles. Data from the microchemical analysis of tissues with low vacuolation, flux analysis, electrophysiological work, and x-ray microprobe analysis show that K+ is selectively accumulated in the cytoplasm, and that in some cells Na+ and Cl− are largely occluded in the vacuole. From a broader comparison of the physiological chemistry of animal and vacuolated and nonvacuolated plant cells, it was concluded that a high K+/Na+ ratio and low Cl− content are characteristic of most and possibly all eukaryotic cytoplasms, and may be related to a fundamental and conserved biochemical character such as the ionic requirements of protein synthesis. As high electrolyte levels are generally deleterious to cytoplasms, this model also requires the accumulation of nontoxic organic cytosolutes to generate high cytoplasmic osmotic pressure. Evidence for the accumulation of a number of such solutes has been obtained in the last few years in higher plants.

Compatible cytosolutes. The term compatible solute is used in the higher plant context to describe a solute which can be accumulated in the cytoplasm without impairing metabolic function.

In higher plants, as in salt-tolerant organisms other than the extreme halophilic bacteria, the putative compatible solutes fall chemically into three classes: amino acids and the N-methyl-substituted derivatives referred to by the trivial name betaines; sugar alcohols or polyols; and possibly methylsulfonium dipoles. The structures of some of the more important examples of the compounds are shown in the table.

Of the nitrogenous compounds, glycinebetaine and proline have been most widely studied in higher plants. Glycinebetaine has been found to be accumulated in a number of halophytic higher plants and its concentration to be closely correlated with the sap osmotic pressure over a range of salinities, including those where a growth stimulation by NaCl is observed. The compound does not inhibit a range of enzymes, including protein synthesis, at relatively high concentrations, and may be used as an osmoticum for mitochondrial and chloroplast isolation. Glycinebetaine partially protected some, but not all, enzymes against salt inhibition, an effect which is considered to be mediated indirectly via the solvent and not by directed binding to the enzyme. Intracellular glycinebetaine also appears to alter the tonoplast Na fluxes and increase the vacuolar content of that ion. Evidence from a number of sources shows that the

compound is largely cytosolic, but the degree of compartmentalization is not clear, nor is it known if it can alter in response to physiological conditions. Thus there is a strong case to consider glycinebetaine a cytoplasmic-compatible solute in some halophytes, but its effect on Na+ fluxes implies that it has more than a purely passive role.

There is little doubt that proline accumulation in some plants is associated with salt tolerance, as exemplified by detailed studies on *Triglochin maritima*. However, the situation regarding proline is complicated by the nearly universal observation of proline accumulation in severely water-stressed tissues, an observation whose interpretation is disputed. Nevertheless, it must be noted that proline is also generally compatible with enzyme and organelle function. Proline, in marked contrast to glycinebetaine, is highly labile and rapidly disappears after the withdrawal of stress, and thus could also act as a carbon, nitrogen, and energy store.

In addition to glycinebetaine, two other betaines have been found to be accumulated in plants under saline conditions. β-Alaninebetaine accumulation (possibly as the choline ester) has been observed in *Limonium vulgare* and *Armeria maritima*, while prolinebetaine is accumulated in a salt-tolerant cultivar of alfalfa (see table).

There have been fewer studies on polyol accumulation in higher plants, but the accumulation of sorbitol has been described in *Plantago maritima*, and this compound is well known to be compatible with enzyme and organelle activity.

Recently evidence has been produced to suggest that the methylated sulfonium compounds, particularly dimethylsulfoniopropionate, originally referred to as dimethylpropiothetin, may be involved in osmo- and ionoregulation in a few higher plants and a number of intertidal algae.

The adaptation of the cellular cytosolic osmotic pressure to hypersaline stress by the accumulation of polyols or amino acids (or their derivatives) is well characterized in bacteria, fungi, microalgae, and animals, particularly marine invertebrates. Thus the recent work on higher plants has confirmed that this is a very general biochemical adaptive mechanism. However, many detailed problems remain unresolved.

Adaptive strategies and taxonomy. It has been observed that certain organic solutes, usually the putative compatible solutes, are characteristically accumulated in particular families, for example, glycinebetaine in the Chenopodiaceae, and sorbitol in the Plantaginaceae (see table). However, there are dangers in assuming that only a single compatible solute is accumulated in a given species. Species also differ in their discrimination for and against K+, Na+, and possibly Cl−. On one hand, the shoots of halophytic Chenopodiaceae have a very high Na+/K+ ratio and low absolute K+ levels, whereas some of the tolerant Gramineae (for example, *Puccinellia*) have a relatively low Na+/K+ ratio and high absolute K+ levels, as have shoots of *Triglochin maritima*. Despite these differences, these halophytic species utilize ions to bring about the largest proportion of the osmotic adaptation. Even in such species there is evidence that their phloem Na+ and Cl− levels are low, and that phloem-fed tissues, particularly grain, still largely exclude these ions.

Chemical structure and broad taxonomic distribution of some compatible solutes in higher plants and algae

Compound	Structure	Principal taxonomic groups	Specific example
Nitrogen dipoles			
Glycinebetaine	(structure)	Chenopodiaceae, Gramineae (Hordeae, Chlorideae), some Compositae, some Solanaceae	*Atriplex, Hordeum, Spartina, Aster tripolium, Lycium*
Proline*	(structure)	Gramineae (Festuceae), Juncaginaceae, microalgae	*Puccinellia, Triglochin maritima*
Prolinebetaine	(structure)	Labiatae(?), Capparidaceae(?), a few Leguminoseae	*Medicago sativa*
β-Alaninebetaine†	(structure)	Plumbaginaceae	*Limonium vulgare*
Polyols			
Glycerol	(structure)	Microalgae	*Dunalliella*
D-Mannitol	(structure)	Microalgae, Phaeophyceae	*Platymonas*
D-Sorbitol	(structure)	Plantaginaceae	*Plantago*
D-Pinitol	(structure)	Caryophyllaceae(?)	*Honkenya*
Sugars			
Sucrose		Microalgae	*Chlorella emersonii*
Sulfonium dipoles			
Dimethylsulfoniopropionate	(structure)	Chlorophyceae, some higher plants(??)	*Ulva lactuca, Spartina*

*Also accumulated widely in water-stressed higher plants.
†Found as choline ester, and not necessarily functioning as a compatible solute.

A number of monocotyledenous species which show some enhanced salt tolerance seem to exclude Na^+ and Cl^- rather efficiently from the shoots and to accumulate sugars to provide a significant proportion of the osmotic adjustment. The free sugar levels in these plants (100 to 300 mM on a tissue-water basis) are such that they cannot be limited to the cytoplasmic compartment and must contribute to the vacuolar osmotic pressure. It is not clear as yet whether sugar can also act generally as compatible cytoplasmic solutes in plants. The evidence is somewhat conflicting; sucrose is a probable cytoplasmic osmotic in some *Chlorella* species, but in higher plants this compound is generally considered to be accumulated in the vacuoles.

Agronomic significance. Understanding of the biochemistry of salt tolerance is incomplete, and much more requires to be known about ion transport and membrane chemistry in the various species. Nevertheless, there has been a major advance in recent years leading to an appreciation of the role of compatible solutes and ion compartmentation. It is also apparent that various strategies have evolved in higher plants to combat the cytotoxicity of Na^+ and Cl^-, and no doubt other ions, while accommodating the osmotic stress. The appreciation of the strategies is important in developing agronomically valuable crops. Thus the type of strategy illustrated by some Gramineae, such as *Puccinellia* and *Diplachne fusca*, may be appropriate for fodder production where high Na^+ and Cl^- levels could be deleterious to the grazing animals. Plants able to utilize Na^+ and Cl^- as osmotica but partially exclude them from their grain could be advantageous in other ways. The strategy of salt exclusion and sugar accumulation found in some moderate halophytes could also be exploited, although it is possibly not a realistic way of exploiting the most saline soils.   [R. G. WYN JONES]

ULTRAVIOLET RADIATION

Without solar radiation, life on Earth would not exist. Despite its obvious importance in the evolution and development of living organisms, the ultraviolet (UV) portion of the electromagnetic spectrum has received relatively little attention from plant scientists until recently. This is in sharp contrast to the interest expressed in the region of visible radiation or photosynthetically active radiation (400–700 nm) and in the far-red and infrared region (700–1000 nm).

The possibility that inadvertent release of nitrogen oxides (from supersonic transports and nitrogen fertilizers through denitrification), chlorofluoromethanes (used as acrosols, foaming agents, and refrigerants), and other anthropogenic contaminants might increase UV radiation reaching the Earth's surface has accentuated interest since the early 1970s in this small but potentially damaging portion of the solar spectrum.

In this article, the link between stratospheric ozone content and solar UV flux will be described, indicating how plant scientists are attempting to study the effects of increased UV radiation, and some of the progress to date.

Stratospheric ozone and solar UV flux. Ultraviolet radiation may be divided into three wavelength regions (Fig. 1): UV-A (320–400 nm), UV-B (280–320 nm), and UV-C (100–280 nm). The UV-B region is known as the middle UV or erythemal region since it is responsible for causing sunburn of the skin (erythema). It is also involved in the development of skin cancer, and the formation of vitamin D. The UV region is also sometimes divided into the vacuum UV (100–200 nm), the far UV (200–300 nm), and the near UV (300–400 nm).

During primeval times, UV-C radiation down to 170 nm is believed to have freely impinged upon the surface of the Earth. Evolution of oxygen-yielding photosynthetic process by blue-green algae during the Precambrian Era is believed to have triggered changes in the composition of the atmosphere that led from a reduced state to an oxidized one. This in turn led to the immediate formation of ozone in the stratosphere at a height of 10 to 50 km above the Earth through photochemical dissociation of oxygen and recombination. This process involves the splitting of a normal oxygen molecule by solar UV radiation at wavelengths of 176 to 242 nm and the subsequent combination of the freed oxygen atoms with another oxygen molecule to form ozone (O_3). Ozone is decomposed in the stratosphere to molecular oxygen primarily by UV radiation in the 200- to 290-nm range and catalytically by oxides of chlorine, nitrogen, and hydrogen.

The formation and destruction of ozone in the stratosphere are highly complex, dynamic processes involving nearly a hundred different reactions, each with different rate constants. The rate of formation of stratospheric ozone is usually maintained in equilibrium with the rate of destruction. The rate of formation involves a feedback mechanism and is believed to depend almost entirely on the flux of incoming solar UV radiation and the amount of stratospheric oxygen concentration and is, therefore, independent of human influence. Any

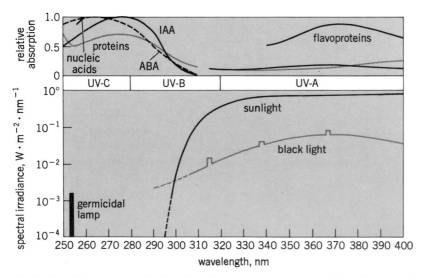

Fig. 1. Spectral irradiance at 30 cm from two common UV lamps and midday sunlight as would occur in the summer at temperate latitudes. The absorption spectra of some common chromophores in plants are represented: ABA = abscisic acid; IAA = indoleacetic acid; P_{fr} and P_r = two forms of phytochrome. Major subdivisions of the ultraviolet spectrum are indicated. (*From M. M. Caldwell, Plant life and ultraviolet radiation: Some perspective in the history of the Earth's UV climate, BioScience, 29(9): 520–525, 1979*)

factor, such as release of chlorofluoromethanes and other substances, that increases the rate of ozone destruction above that naturally occurring may be expected to shift the equilibrium toward reduced ozone levels (or concentrations).

The concentration of ozone in the stratosphere is of the order of a few parts per million (ppm). Hence the term ozone layer is somewhat erroneous. If a sample of the stratosphere is compressed to a pure gas at standard temperature and pressure (STP), its ozone content at mid-latitude would occupy a vertical column only about 0.25–0.32 cm thick (referred to as 0.25–0.32 atm-cm).

The strong link between ozone in the stratosphere and UV radiation is based on the absorption properties of ozone. At 290 nm and below, ozone and other atmospheric components absorb virtually all incoming UV radiation. At longer wavelengths in the UV, however, the absorption of ozone falls off rapidly so that it is nearly 50% at 310 nm and close to zero at 330 nm. As a result, the UV cutoff for solar radiation is at approximately 290 nm. A decrease in ozone content of the stratosphere is of concern to biologists, since it would shift this cutoff and allow shorter wavelengths of UV-B radiation to reach the Earth. It would also permit more UV penetration at the Earth's surface above this cutoff. As a result, sensitive cell constituents (such as nucleic acids, proteins, and hormones) which absorb strongly in the UV region may be adversely impacted (Fig. 1).

For decreases in stratospheric ozone of less than about 10%, there would be about a 2% increase in the level of biologically effective UV-B reaching the Earth for every 1% decrease in stratospheric ozone content. For decreases in stratospheric ozone content of more than 10%, the increase in biologically effective UV would be greater than twofold.

Approximately a 28% reduction in stratospheric ozone content (down to 0.23 atm-cm) is thought to be the upper limit of ozone reduction that might result at a steady state (in 50 to 100 years or longer) from chlorofluoromethane release. If this amount of ozone reduction were to be attained, it would result in about a 60% increase in the level of biologically effective UV radiation reaching the Earth.

Studies on UV-B enhancement. To conduct studies on the effects of increased UV-B radiation: (1) suitable sources of UV irradiance must be obtained; (2) precise instrumentation must be utilized to establish the desired level of UV-B flux and to monitor any changes with time; and (3) strict guidelines must be established in conducting and reporting the results of such studies.

Solar simulators are not readily obtainable, and when available, they are expensive. Plant physiologists, therefore, generally rely on artificial lamp sources to provide UV-B enhancement conditions. These usually consist of fluorescent sunlamps. Xenon arc lamps are also sometimes employed in broad-band and narrow-band UV studies.

To eliminate wavelengths of UV-C radiation present in most of these UV sources and to tailor the spectral cutoff in the UV-B region close to that of the solar spectrum, lamps should be used that are made of suitable glass or covered with appropriate wavelength cutoff filters (by means of collars or drapes, for example). Polycarbonate filters may be used on control (zero UV) setups to absorb all wavelengths less than about 385 nm. To serve as a UV-B control, Mylar filters may be used; these absorb essentially all UV irradiation below 315 nm, transmitting only UV-A irradiation. Filters of cellulose acetate or cellulose triacetate may be used to allow passage of UV-B irradiation; these materials transmit both UV-A and UV-B radiation down to about 288 nm. Depending upon the thickness of the cellulose acetate filter, the UV cutoff may be adjusted to simulate that projected for various levels of stratospheric ozone reduction.

A precise, automated spectroradiometer (Fig. 2)

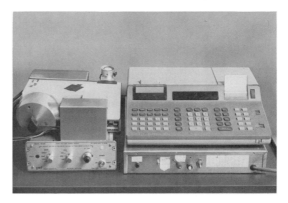

Fig. 2. Automated spectroradiometer used for making measurements of UV spectral irradiance under enhanced UV-B conditions. (*Courtesy of K. H. Norris, USDA*)

has been developed that is capable of measuring UV radiation from 250 to 400 nm with a wavelength precision of 0.1 nm (1 Å), and is a significant improvement in UV instrumentation. Since a 1-nm change in solar UV irradiance is equivalent to approximately a 16% change in ozone concentration, such wavelength precision is needed to detect small changes in UV radiation.

Significant design features of the instrument include a single or double monochromator with holographic gratings, a variable-speed motor drive to scan the entire UV region in less than 5 min, a specially designed Teflon bubble diffuser for cosine correction of incoming UV radiation, a solar blind filter or solar blind phototube, and an amplifier output digitized with a digital voltmeter interfaced with a programmable calculator.

With this instrument, rapid measurements of absolute and biologically effective UV radiation (based on preprogrammed action spectra) over the desired bandwidths are obtained. Broad-band and narrow-band portable UV radiometers have also been developed for use in UV studies.

Since the spectral characteristics of UV lamps may be influenced by various physical and environmental factors, such as lamp degradation and filter absorption changes over time, the UV source must be characterized as carefully as possible. Measurements should be taken under the lamp-filter combination used in the study. Since spectral irradiances under a fluorescent lamp bank can vary greatly from one location to another depending upon the number, spacing, and distance of the lamps, measurements with a high-quality spectroradiometer should be obtained at selected locations throughout the UV setup. Portable broadband radiometers, which can be calibrated against the spectroradiometer, may be used to estimate the level of biologically effective UV radiation at each plant location under the setup. This greatly facilitates routine monitoring.

Since biological responses to UV radiation are highly wavelength-dependent, different weighting values must be assigned to each wavelength to assess the biological effectiveness. This process is simplified and expedited by using an interfaced programmable calculator.

A comparison of UV spectral irradiance at 30 cm from two UV lamps (commonly used to study UV-C and UV-A effects) and midday sunlight as would occur at temperate latitudes in the summer is shown in Fig. 1. Although not shown, the spectral curve for fluorescent sunlamps filtered with cellulose acetate resembles that of sunlight in the UV-B region and of black light in the UV-A region.

Plant response to increased UV-B irradiance. There is a considerable variability in plant response to increased UV-B irradiation, depending upon such factors as genetic makeup, physiological and developmental status, and environmental factors.

Genetic differences in plants lead to differences in UV-B sensitivity at both the species and cultivar levels (Fig. 3). Some species and cultivars are sensitive to levels of UV radiation that presently occur, while others are extremely resistant, even to a 100% or greater increase in the level of biologically effective UV irradiation. Sensitive crops include

some cultivars of pea, soybean, cucumber, wheat, blueberry, poinsettia, and coleus. Tolerant crops include peanut, rice, cabbage, and alfalfa. Woody plants tend to be tolerant of increased UV-B irradiation. Shade plants are generally more sensitive than sun-loving plants.

In general, plants grown under field conditions are considerably more tolerant to increased UV-B irradiation than those grown under greenhouse or growth chamber conditions. Significant seasonal differences have been also found in the UV-B responses of poinsettia, cucumber, marigold, and other plants to supplemental UV radiation. Those treated in the greenhouse during fall and winter months tend to be more sensitive to UV exposure than those treated in the spring and summer. Differences in plant response to increased UV-B irradiation in the growth chamber are also observed at different levels of visible radiation, that is, photosynthetically active radiation (PAR). These findings suggest evidence for a photorepair system in high plants dependent upon the level of PAR and perhaps on other portions of the electromagnetic radiation spectrum.

Leaves of many higher plants can orient themselves to avoid full exposure to the Sun. It has been suggested that such leaf movements are of primary benefit in mitigating solar IR (thermal) excesses.

Avoidance in higher plants is largely accomplished by means of pigments such as flavonoids and colorless UV-absorbing substances contained in the epidermal cells of the leaves. Recent studies indicate that the epidermis, a single-cell layer, together with cuticular waxes and other cell wall constituents, can reduce the incident UV radiation by one or two orders of magnitude, providing at least partial protection to mesophyll cells in the leaf. In cotton and certain other plants UV-B radiation has been found to cause an increase in flavonoid pigments, while in coleus a breakdown in anthocyanin (a sugar-bearing flavonoid) has been observed.

A number of repair systems exist in bacterial cells and other organisms. Photoreactivation, an enzymatic mechanism activated by UV-A and visible radiation, is the primary system involved in repairing pyrimidine dimers, the most common lesion in DNA caused by UV-C radiation. Terminal oxidases in plants may also serve an important protective function by quenching certain free radicals produced by UV. Evidence for a photoreactivating enzyme has been obtained in higher plants. However, understanding of repair processes in higher plants is still incomplete. Recent findings by Soviet workers suggest that three types of repair systems exist in plants: (1) those which occur within the first few minutes following UV irradiation; (2) those which occur up to 6 hours afterward; and (3) long-term repair.

Characteristics of UV injury. Since plants exhibit a wide range in sensitivity and tolerance to high levels of biologically effective UV irradiation, evidence of UV injury may be overt or subtle. Signs of UV damage may be manifested at the anatomical, physiological, or biochemical level.

Visual evidence of UV-B damage to sensitive plants such as pea and cucumber includes chlo-

Fig. 3. Differential sensitivity of two cultivars of cucumber to a 40% increase in biologically effective UV irradiation. This level is equivalent to approximately a 20% decrease in stratospheric ozone reduction. (a) Ashley, a slightly sensitive cultivar; (b) Poinsett, a highly sensitive cultivar. Both plants were irradiated 6 hours a day (10:00 A.M. to 4:00 P.M.) for 18 days from the time of planting the seed in the greenhouse. (*Courtesy of D.T. Krizek*)

rosis (yellowing) of the leaves, reduction in leaf size, and abnormal leaf development (Fig. 3). Other evidence of UV-B injury in many plants includes bronzing of the upper leaf surface and cellular necrosis. Bronzing is thought to be caused by the presence of oxidized, polymerized, phenolic compounds. The extent of glazing or bronzing may serve as an indication of the extent of UV absorption and damage to the tissue.

Inhibition of photosynthesis under elevated UV-B conditions has been reported. In many cases, however, the levels of UV-B enhancement were far in excess of those projected for chlorofluoromethane reduction of stratospheric ozone. In some cases, there is evidence that plants may require a small amount of UV radiation in order to develop a normal green color.

Dark respiration rates have been increased in some studies and unaffected in others by UV irradiation. Increased levels of UV-B irradiation may close stomates in soybean and thereby reduce transpiration rates. In this crop, as in other plants studied, it appears that the response to elevated UV-B irradiation depends greatly upon concomitant factors such as the photosynthetically active radiation level and the long wavelengths available. It is difficult therefore to extrapolate from studies conducted under low photosynthetically active radiation conditions in the laboratory, growth chamber, or greenhouse to what might actually occur under projected levels of UV solar flux in the outdoor environment.

Although many crop plants may be resilient to elevated UV-B levels, there is growing evidence that a number of plant species are sensitive to present levels of solar UV radiation reaching the Earth's surface. Many of these (such as cantaloupe, muskmelon, honeydew melon, blueberry, apple, soybean, cowpea, bean, tobacco, and some of the forage legumes) exhibit various types of solar injury ranging from sunburn to sunscald.

Conclusions. Considerable progress has been made during the past several years in assessing the biological impact of possible ozone reduction in the stratosphere, especially in the areas of UV instrumentation and research methodology. Careful dose-response studies have revealed significant information of potential value to astute policymakers concerned with impact assessments and possible regulatory decisions. Despite these achievements, many questions remain. One of the most urgent needs is to uncover the secrets of the repair process and tolerance mechanisms in higher plants. Until plant physiologists and photobiologists can unravel the complexities of these processes, only speculation can be made as to how much can be extrapolated from the findings of studies conducted in the laboratory, the growth chamber, and the greenhouse of what might occur under projected solar UV fluxes in the natural environment.

For background information *see* ELECTROMAGNETIC RADIATION; EVAPOTRANSPIRATION; OZONE, ATMOSPHERIC; PHOTOSYNTHESIS; PLANT, WATER RELATIONS OF; RADIATION INJURY (BIOLOGY); SALTMARSH; ULTRAVIOLET RADIATION in the McGraw-Hill Encyclopedia of Science and Technology.   [DONALD T. KRIZEK]

Bibliography: M. M. Caldwell, The effects of solar UV-B radiation (280–315 nm) on higher plants, in A. Castellani (ed.), *Research in Photobiology*, pp. 597–607, 1977; M. M. Caldwell, *BioScience*, 29(9):520–525, 1979; T. J. Flowers, P. F. Troke, and A. R. Yeo., *Annu. Rev. Plant Physiol.*, 28:89–121, 1977; R. M. Klein, *Bot. Rev.*, 44:1–127, 1978; D. T. Krizek, Biological effects of ozone reduction, in *Fluorocarbons and the Environment*, Report of the Federal Task Force on Inadvertent Modification of the Stratosphere (IMOS), pp. 53–64, 1975; H. I. Schiff, *Stratospheric Ozone Depletion by Halocarbons: Chemistry and Transport*, 1979; J. M. Skelly et al., Air pollution and radiation stresses, in B. J. Barfield and J. F. Gerber (eds.), *Modification of the Aerial Environment of Crops*, ASAE Monogr. no. 2, pp. 114–138, 1979; R. G. Wyn Jones et al., in E. Marrè and O. Ciferri (eds.), *Regulation of Cell Membrane Activities in Plants*, 1977; R. G. Wyn Jones, C. J. Brady, and J. Speirs, in D. L. Laidman and R. G. Wyn Jones (eds.), *Recent Advances in the Biochemistry of Cereals*, 1979; R. G. Wyn Jones and R. Storey, in L. G. Paleg and D. Aspinall, *The Physiology and Biochemistry of Drought Tolerance in Plants*, 1980.

Plant reproduction

Traditional studies of the reproductive cycles of conifers made during the first half of this century are being augmented by studies of a greater variety of species, by the inclusion of a broader view of conifer reproductive biology, and by the use of reproductive biology to solve practical problems such as enhancing cone and seed production. Conifer reproduction begins with the initiation of seed cones and pollen cones followed about a year later by pollination which, in turn, is followed by a period of seed development lasting from 3 to 18 months. Many stages are involved in this long period of development, and each stage may affect the desired outcome: the production of viable seed.

Cone initiation. The initiation of a reproductive apex occurs either by the transition of a vegetative terminal apex which previously produced leaves or by the differentiation of an undetermined lateral apex which previously produced bud scales. The reproductive apices then begin to initiate bracts if a seed cone is to develop, or microsporophylls if a pollen cone is to develop. Changes that occur in the apex during the transition to a reproductive apex are similar in flowering plants and conifers regardless of whether a seed cone or pollen cone is being initiated. Chemical changes first occur in nucleic acids, and these are followed by an increase in cell divisions which results in an increase in apical size and a change in apical shape. The physiological changes which stimulate the apices to undergo this transition are not well understood in conifers.

The times and patterns of cone initiation (Fig. 1) have been determined for several conifers based on the chemical and anatomical changes that occur in the apices during cone initiation. Since all conifers are perennials, the time of cone initiation is related to the growth cycles of the vegetative shoots on which the cones are borne. Several studies have shown that cone initiation correlates with the end of the phase of bud scale initiation and rapid shoot elongation. Since the end of the rapid phase of shoot elongation has been shown to be photoperiodically mediated, the time of cone initiation is indirectly mediated by the change from increasing to decreasing day lengths. Consequently, it is not surprising that most conifers begin cone initiation during June and July (Fig. 1). This differs only in *Pinus*, where seed and pollen cones are initiated within a complex long-shoot vegetative bud; and bud primordia, which can differentiate into seed cones, are initiated late in the growing season rather than very early in the growing season, as in most other conifers.

Cone induction. Knowledge of the time of cone initiation has made it possible to induce or enhance cone production in many species by cultural and hormonal treatments. This is particularly useful in juvenile trees, which normally are not reproductive, because it can greatly shorten the time between generations in order to genetically improve the species or to produce seeds in young trees in seed orchards for reforestation.

Cultural treatments including girdling, fertilization, root pruning, and drought have proved, at times, very successful in cone induction, although in some situations they have had little or no effect or have been detrimental. Hormonal treatments began with the successful use of gibberellin A_3, which induced cones in certain members of the Cupressaceae and Taxodiaceae but not within the Pinaceae. The analysis of endogenous gibberellins in the Pinaceae, of which there are at least 50, revealed that cultural treatments which enhanced cone production also caused an increase in gibberellins that were less polar than gibberellin A_3. The less polar gibberellins, especially A_4, A_7, and A_9, applied in varying concentrations and combinations, usually with cultural treatments, at the predicted time for normal cone initiation have successfully induced cones in at least nine species within the Pinaceae. This method of cone induction is still at the experimental stage, but the rapid progress shown thus far indicates that it is likely to be a successful and economical method for cone induction and cone enhancement both for genetic studies and for seed production for reforestation.

Cone development. Although the failure of cones to be initiated is the most common cause for poor seed production, many other factors affect cone and seed development. In most species (Fig. 1) cones are initiated before winter dormancy, and cones overwinter as small seed-cone or pollen-cone buds. Dormant seed-cone buds possess bracts and ovuliferous scales which bear the primordial ovules, which in turn will develop into seeds. Dormant pollen-cone buds possess microsporangia or pollen sacs, but variation occurs between species in the stage of pollen development. In some species of *Juniperus* and *Chamaecyparis*, pollen is formed and matures before pollen cones enter winter dormancy. Pollen mother cells differentiate before dormancy but do not develop further in *Pinus*, *Picea*, and *Abies*. Pollen mother

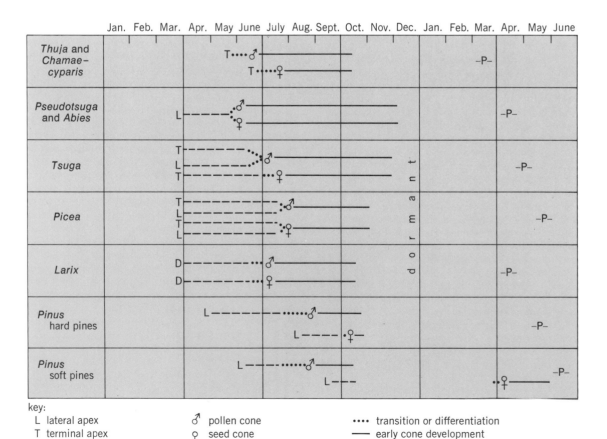

key:
L lateral apex
T terminal apex
D dwarf lateral apex
♂ pollen cone
♀ seed cone
--- bud scale initiation
···· transition or differentiation
— early cone development
–P– pollination

Fig. 1. Times and patterns of cone initiation.

cells in *Pseudotsuga, Larix, Tsuga,* and *Thuja* begin meiosis before dormancy, but meiosis becomes arrested and pollen mother cells overwinter at a diffuse diplotene or pachytene stage after chromosome duplication and pairing have occurred. Studies in Finland have shown that, even in *Pinus,* apparently dormant pollen mother cells are metabolically very active and undergo many ultrastructural changes during winter. Earlier studies in Sweden suggested that the stage at which pollen mother cells overwinter affects their resistance to frost damage, which in turn may affect pollen viability.

Pollination mechanisms. Conifer pollen is dispersed by wind, and in most species pollination occurs in the spring. Following dormancy the tips of the ovules within the seed-cone buds initiate a ring of meristematic tissue which develops into a specialized pollen-catching structure characteristic of the species. Three general types of pollination mechanisms have evolved in modern conifers, and these can be clearly observed with the scanning electron microscope (Figs. 2 and 3).

The first type of pollination involves the formation of a pollination drop. In many conifer families erect flask-shaped ovules occur, each having a narrow neck with a small opening, the micropyle (Fig. 2). The most common type found within the

Fig. 3. Stigmatic tip of *Pseudotsuga* (Douglas fir) showing pollen entangled in the stigmatic hairs and the slit-like micropyle.

and be picked up when the drop is exuded, or pollen may land directly on the drop.

The second type of pollination found in the Pinaceae occurs in *Abies* and *Tsuga,* where the tip of the ovule is very unspecialized and just forms a funnel. There is no pollination drop, and pollen simply lands in or on the rim of the funnel and the funnel slowly crimps in, eventually sealing the pollen inside the micropyle.

The third and most elaborate mechanism found in the Pinaceae occurs in *Pseudotsuga* and *Larix,* where the two arms around the micropyle elongate unequally, develop long epidermal hairs, and form a large, bilobed, hairy stigmatic surface. No pollination drop is exuded; rather, pollen becomes entangled in the stigmatic hairs (Fig. 3). After several days the outer cells of each stigmatic lobe elongate, cells adjacent to the micropyle collapse, and both stigmatic lobes with their entangled pollen are engulfed into the micropyle much as a sea anemone engulfs its prey. Engulfment takes several days and occurs whether pollen is present or absent.

Seed development. The time between pollination and seed maturation varies from 3 to 18 months in different conifers, and three patterns of development occur. Understanding these patterns is essential for a correct interpretation of the factors which cause seeds to abort early in development, enlarge normally but remain empty, or possess only rudimentary embryos. The most frequently described but least typical pattern of seed development occurs in *Pinus.* Pollination occurs in the spring; pollen tubes penetrate part way into the ovule; then, in late summer, development stops. Ovules and pollen tubes resume development after winter dormancy, fertilization occurs, and the ovule matures into a seed in the fall. In *Pinus,* pollen is necessary for further ovule and often cone development, and flattened (aborted) seed in mature cones usually results from the failure of ovules to be pollinated. In most conifers fer-

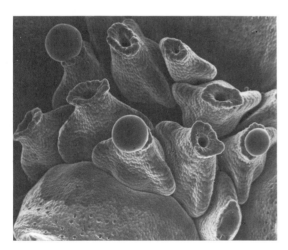

Fig. 2. *Chamaecyparis* (yellow cypress) seed cone at pollination showing the flask-shaped ovules, some with a large pollination drop.

Pinaceae differs in that the ovule tip develops as two prongs with the micropyle between. In both types an aqueous pollination drop is exuded out of each ovule through the micropyle. Pollen enters the drop, and the drop, laden with pollen, is withdrawn into the ovule which is then sealed. Pollination drops are not exuded from all ovules at one time, each ovule may exude a drop more than once, and drops appear to be exuded in the morning and withdrawn in the afternoon. The pollination drop is exuded whether pollen is present or absent, but it may be withdrawn more rapidly when pollen enters the drop and alters the surface tension. Pollen may adhere to the tip of the ovule

tilization occurs a few weeks after pollination, and ovule and seed cone development do not depend upon the presence of pollen. In this pattern, seed that externally appears normal may develop from unpollinated ovules. A third pattern occurs in only a few conifers where fertilization occurs several weeks after pollination but embryo development is not completed before winter dormancy. In this instance, seed cones overwinter with only partially developed embryos which complete their development during the second summer.

In order to interpret the many possible causes for the presence, absence, or abundance of cones and viable seed, the entire reproductive cycle of each conifer species must be known. The renewed interest in, and the intensive research on, the reproductive biology of conifers eventually will provide this needed information.

For background information *see* GIBBERELLIN; REPRODUCTION (PLANT); SEED GERMINATION in the McGraw-Hill Encyclopedia of Science and Technology.   [JOHN N. OWENS]

Bibliography: B. S. Lill and G. B. Sweet, *N. Z. J. For. Sci.*, 7:21–34, 1977; J. N. Owens and M. Molder, *Proceedings: Flowering and Seed Development in Trees—A Symposium*, Starkville, MS, pp. 25–32, 1978; J. N. Owens and M. Molder, *Can. J. Bot.*, 55:2746–2760, 1977; R. P. Pharis and C. G. Kuo, *Can. J. For. Res.*, 7:299–325, 1977.

Plant secretory structures

Laticifers are particular cell types present in only a limited number of families of angiosperms, or flowering plants. These types include the articulated laticifer, which is typically composed of superimposed cells often joined into a network of tubes within the plant, and the nonarticulated laticifer, which is a single cell that grows intrusively between other adjacent cells to form an elongated cell within the plant body (see table). Laticifers are characterized by their highly osmotic fluid content, termed latex, which contains various specialized secondary products synthesized within these cells. The high osmoticum of latex results in its exudation from the plant upon injury, and this property

of latex is employed to advantage for the procurement of several important commercial products, including natural rubber (from *Hevea*) and opiate alkaloids (from *Papaver*).

Laticifer distribution. Laticifers in flowering plants are restricted to a small number of unrelated families, where they occur in all genera or natural groups of the family. The differences in origin and development, and their unrelated distribution among families, indicate that the two types of laticifers evolved independently of each other. Both laticifer types are similar in having developed a dense protoplasmic content, often colored in appearance. The occurrence of laticifers in advanced plant families and their absence from the majority of families, including primitive angiospermous groups, indicate that laticifers represent a cell type of relatively recent evolutionary origin in angiosperms.

Latex. Latex represents the protoplasmic contents of laticifers, and contains the various organellar and membranous components present in typical cells. Components in the cytoplasm often impart a color to latex. It is commonly white (*Euphorbia*), although in some plants it is clear (*Nerium*), yellow (*Stylophorum*), or red (*Sanguinaria*). Latex cytoplasm differs from that of other cells in the abundance of vesicular components, which are presumed to be related to the accumulation of specialized secondary substances accumulated in these cells. Initial experiments on exuded latex have shown it to possess various salts, proteins, carbohydrates, and lipids present in other cells. In addition, exuded latex possesses the mechanisms to synthesize unusually high levels of specialized products characteristic of particular plants, such as rubber (*Hevea*), glycosides (*Asclepias*), alkaloids (*Papaver*), terpenoids (*Euphorbia*), and proteolytic enzymes (papain in *Papaya*). While other cells also are reported to produce specialized products, the laticifers are unique in that they synthesize these products in unusually large quantities; rubber, terpenoids, and alkaloids can constitute 50% or more of the dry weight of exuded latex. Latex from a considerable length along the cell exudes from the wound upon injury to the cell; however, only a portion of the protoplasm is lost from the cell, and the wound surface is quickly sealed. Homeostatic metabolic mechanisms within the laticifer rapidly synthesize latex components to reestablish their original levels in the latex.

The functions of latex and laticifers are not well understood. The specialized compounds in latex are typically large, complex molecules requiring the input of considerable metabolic energy for their synthesis. Such compounds are formed even when the plant is under stress, and therefore they do not appear to be metabolic reserves. There is little experimental evidence that identifies a common function for latex, such as protection against insect attack or herbivores. Future experimental studies of laticifer function very probably will show that these cells evolved to perform different functions in different plants.

Articulated laticifers. Articulated laticifers are described as either anastomosing, in which adjacent files of these cells are interconnected into a

Laticifer distribution in angiosperms

Laticifer type	Family	Genus
Articulated		
Anastomosing	Campanulaceae	*Campanula*
	Caricaceae	*Carica*
	Cichorieae	*Taraxacum*
	Euphorbiaceae	*Hevea*
	Papaveraceae	*Papaver*
Nonanastomosing	Liliaceae	*Allium*
	Musaceae	*Musa*
	Convolvulaceae	*Ipomoea*
	Papaveraceae	*Chelidonium*
	Sapotaceae	*Achras*
Nonarticulated		
Branched	Apocynaceae	*Nerium*
	Asclepiadaceae	*Asclepias*
	Euphorbiaceae	*Euphorbia*
	Moraceae	*Ficus*
Unbranched	Apocynaceae	*Vinca*
	Eucommiaceae	*Eucommia*
	Cannabaceae	*Cannabis*
	Urticaeae	*Urtica*

network, or nonanastomosing, in which the vertical cell files remain separated from each other. The anastomosing form prevails in the dicotyledons, while the nonanastomosing is present in families of both monocotyledons and dicotyledons (see table).

Laticifer initials originate as numerous files of cells in the phloem tissue of procambium or vascular cambium during plant development. In some plants these files are interconnected to form an anastomosing network within the plant, while in other plants the files of individual cell remain as separate columns in association with the conductive tissue. The protoplast of these uninucleated initials rapidly differentiates, giving rise to a characteristic vesiculated appearance, as evidenced in *Papaver*. The endoplasmic reticulum gives rise to numerous vesicles of different sizes which fill the cells. Laticifer vessel formation is initiated during cellular elongation and is characterized by the development of wall perforations. Preliminary cytochemical studies have shown the presence of cellulase in the vicinity of developing perforations, indicating an involvement of this enzyme in the wall-perforation process. Perforations develop in both end and lateral walls between contiguous elements to form a complex vessel system throughout various organs of the plant body. Nuclear division ceases at an early stage in laticifer development. Plastids, differing from typical chloroplasts in having a large inclusion within the stroma and few thylakoids, become abundant organelles within the vesiculated cytoplasm in *Papaver*. These organelles have been shown to possess enzymic activity related to alkaloid production.

Nonarticulated laticifers. This laticifer is an unusual cell type in its ontogeny and physiology. It differs from other cells in its capacity for unlimited growth potential and multinucleated protoplast. Two forms of this cell type are recognized, one of which becomes branched, whereas the second form remains unbranched during their growth (see table).

Laticifer initials arise in the cotyledonary node shortly after the initiation of the cotyledons during embryogeny. The number of such enlarging cells varies among genera and even species, although it appears constant for a species. In *Nerium oleander* 28 initials are formed, whereas in *Euphorbia marginata* only 12 arise in the embryo. No additional laticifers arise during later stages in plant development in the primary or secondary body. Following their origin, the initials develop branches which grow toward the base of the shoot and root meristems, and subsequently keep pace with meristem development in the adult axis. The branches grow intrusively along the middle lamella region between cells. The latex from these laticifers has been found to contain an abundance of pectinase, which may be secreted at the tips of the laticifer and may function to solubilize the pectinaceous middle lamella to facilitate laticifer penetration between cells. Tips of the laticifers can be observed in the vicinity of the meristems, which may contribute to the physiology of laticifer growth.

Meristems do not contribute new cells to the laticifers during growth, as they do for other tissue systems. Therefore nonarticulated laticifers are

cytologically isolated from other tissues within the axis of the plant. The original nucleus divides by mitosis, as do the progeny, to give rise to a multinucleated cell. The mitotic spindle and anaphase movements of the chromosomes appear typical during division. In late telophase the spindle apparatus disappears, leaving no evidence for cell plate formation processes, which appear to have been lost during the evolution of this cell type. Mitotic activity occurs in a successive pattern, involving numerous nuclei along the cell axis. This pattern of division suggests that a chemical stimulus is responsible for division in the laticifer, which responds differently than adjacent vertical files of cells that never exhibit a similar pattern for division.

The cytological isolation of the laticifer from the meristem may aid in explaining the accumulation of unusual structural features or secondary products in the laticifer. Thus changes of a mutagenic nature in a laticifer can give rise to features that become characteristic for a species. Quite possibly the presence of inordinately high levels of specialized products present in laticifers may be a result of this phenomenon. For example, analyses of the sterol profile of latex from numerous species of *Euphorbia* show each species to possess a specific combination of sterols characteristic for the species, very much like a fingerprint. Similarly, plastids in the nonarticulated laticifers in *Euphorbia* have evolved and specialized, in contrast to plastids in other cells, to synthesize starch grains of unusual configuration, such as rod, osteoid, lobed, and discoid shapes. A particular species possesses a particular shaped grain in the laticifer, whereas other cells possess small, more or less rounded grains.

For background information *see* SECRETORY STRUCTURES (PLANT) in the McGraw-Hill Encyclopedia of Science and Technology.

[PAUL G. MAHLBERG]

Bibliography: D. Biesboer and P. Mahlberg, *Planta*, 143:5–10, 1977; P. Mahlberg, *Amer. J. Bot.*, 62:1167–1173, 1975; C. Nessler and P. Mahlberg, *Amer. J. Bot.*, 64:541–551, 1977; C. Nessler and P. Mahlberg, *Amer. J. Bot.*, 66:266–273, 1979; K. Wilson, C. Nessler, and P. Mahlberg, *Amer. J. Bot.*, 63:1140–1145, 1976.

Pollen

A number of angiosperms (Magnoliophyta) have pollen grains in permanent tetrads. Such a tetrad structure results when the four pollen grains, which are formed by the meiotic division of each pollen mother cell, do not subsequently separate from one another. Because tetrad pollen occurs in some or most members of approximately 50 families, many of which are not closely related, it is clear that this condition has evolved independently a number of times. These families include the Begoniaceae (begonias), Droseraceae (sundews), Ericaceae (heaths), Gentianaceae (gentians), Leguminosae (legumes), Malvaceae (mallows), Nepenthaceae (pitcher plants), Nymphaeaceae (water lilies), and Onagraceae (evening primroses) among the dicotyledons (Magnoliopsida), and the Cyperaceae (sedges), Juncaceae (rushes), Orchidaceae (orchids), and Typhaceae (bulrushes) among the monocotyledons (Liliopsida). Tetrad pollen occurs within four families of the most primitive angiosperms, the order Magnoliales, of the subclass Magnoliidae of the dicots. The Annonaceae are the largest in this group, and nearly a third of the approximately 130 genera have tetrad pollen. In contrast, the Lactoridaceae comprise one almost extinct species, *Lactoris fernandeziana*, which occurs on one of the Juan Fernández islands. Only a few species of one member of the Monimiaceae, *Hedycarya*, have tetrad pollen, but all eight genera of the Winteraceae, frequently cited as possessing a greater number of primitive characters than any other flowering plant family, have tetrad pollen. Even within this single order, Magnoliales, it is generally considered that the tetrad configuration has evolved independently in each of these four families, for they are not very closely related. Recent studies on pollen development and pollination ecology in some of these families have revealed some interesting features and their significance.

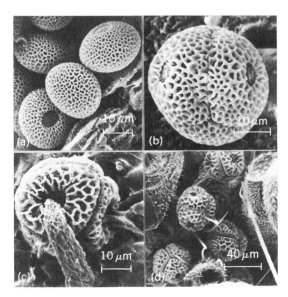

Fig. 1. Scanning electron micrographs of pollen of Winteraceae. (a) Immature solitary pollen grains of *Zygogynum pomiferum* var. *pomiferum*. (b) A pollen tetrad of *Z. baillonii*; two of the four apertures are visible. (c) A germinated pollen tetrad on the stigma of *Pseudowintera axillaris*; in this view only two (arrows) of the four pollen tubes can be clearly seen. (d) Several pollen tetrads of *P. colorata* on the head of a cranefly (tipulid).

Pollen of Winteraceae. Tetrad pollen has existed for a long time in the Winteraceae. Their fossil grains, in tetrads, date from the Cretaceous, the period in which flowering plant fossils first appeared, through to the present day, in Australia and New Zealand. F. B. Sampson recently discovered that a species of *Zygogynum*, a New Caledonian genus (which the fossil record shows was once more widespread in the Australasian region), has nontetrad pollen. A second species was subsequently found to have monad (solitary) pollen grains, too, in contrast to the other four species of *Zygogynum*.

Does this type represent the primitive condition from which tetrad pollen evolved in the Winteraceae? It seems not. The monad and tetrad pollen grains both have a single round aperture (Fig. 1a and b) through which the pollen tube emerges at germination. As with tetrad pollen, the aperture in the monads is located at what is morphologically the distal pole of each grain. This pole is opposite the proximal pole, which is located at the center of the meiotic tetrad at the time of its formation. The round aperture of monad pollen suggests a tetrad heritage, for there are no other primitive angiosperms with solitary pollen grains with such a restricted porelike aperture. This type of aperture seems to have evolved to suit the tetrad condition, so that when all four grains of a tetrad germinate, as is the usual condition in Winteraceae so far investigated (Fig. 1c), the four emerging tubes are the maximum distance from one another, thereby minimizing any chance of them becoming entangled. A single, elongated, furrowlike aperture centered at the distal pole of a boat-shaped grain, as in *Magnolia*, seems to be the most primitive type, and was the first to appear in the fossil record. It is likely that ancestors of the Winteraceae had such pollen. A further reason for considering the monad type to be derived from the tetrad is that, in the majority of its features, *Zygogynum* is one of the most specialized members of the family. Nevertheless, it seems remarkable that when this solitary form evolved, it regained full sculpturing on its morphologically inner side, identical with that of the distal region. In species with permanent tetrads, no such proximal sculpturing is visible when grains are separated from each other forcibly.

Significance of monad versus tetrad pollen. The adaptive significance of monad pollen in the two *Zygogynum* species is not known. Cohesion of pollen into tetrads can increase a plant's pollination efficiency, because four seeds can be formed from a single transfer of a pollen unit (tetrad) from a male (stamen) to a female floral organ (carpel or compound gynoecium). The acquisition of tetrad pollen may lead to the evolution of more ovules (potential seeds). J. W. Walker found a significant correlation between tetrad pollen and high ovule number in angiosperms. He found, too, within the Annonaceae, among closely related genera with either monad or tetrad pollen, that the genus with tetrad pollen had a greater number of ovules. He concluded that in *Annona* of this family, which has some species with monads and others with tetrads, the solitary grains, which were present in the more advanced species, had evolved secondarily from tetrads. Tetrad pollen is no advantage to *Annona*, insofar as there has been evolutionary reduction to a single ovule in each carpel. No such correlation between solitary pollen and low ovule number was found in *Zygogynum*. *Zygogynum pomiferum*, which possesses solitary pollen grains, has as many, or even more, ovules per carpel than species with tetrad pollen. Perhaps the monad pollen is an adaptation to particular pollination strategies in the two *Zygogynum* species, but at present no relevant information is available.

Pollen development in Winteraceae. The development of pollen tetrads is not identical in all members of the Winteraceae. In some species all four grains of a tetrad are at exactly the same stage when each pollen nucleus divides to form vegetative (tube) and generative cells (Fig. 2a). In other species the members of a tetrad are never at exactly the same stage of division (Fig. 2b). The synchronous division is attributable to the presence of gaps in the internal walls of a tetrad, similar to those shown in Fig. 3c, so that they share a common cytoplasm. The gaps are not present in species undergoing asynchronous division, but examination of their internal walls at high magnification, with the transmission electron microscope, shows irregularities in the walls, suggesting the presence of former gaps which might have been closed at an earlier stage of development. It therefore seems that the asynchronous type has evolved from the

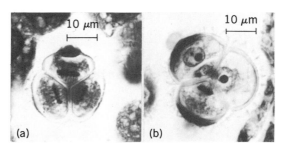

Fig. 2. Sections through three pollen grains of a tetrad (Winteraceae). (a) An unnamed species of *Belliolum* showing synchronous division, with nuclei at the metaphase stage of division. (b) *Drimys winteri* showing asynchronous division; division into vegetative cell and smaller proximal generative cell has been completed in grain at upper left; division has not begun in grain at center right; grain at lower left has a nucleus at the metaphase stage of division.

synchronous. Comparable cytomictic channels link the cytoplasm of pollen mother cells within each pollen sac at an early stage of their meiotic division in all flowering plants. These early division stages are synchronous, too. Soon after division is completed to form a small generative cell which is cut off against the proximal pole of each grain (Fig. 2b) and a large vegetative cell, the gaps in the walls of species with synchronous division are closed. Closure is by deposition of cellulosic material which forms the inner wall (intine) of pollen grains. As is true for all flowering plants, the generative cell becomes detached from the wall of the pollen grain and "floats" in the cytoplasm of the vegetative cell. Pollen is shed in the two-celled condition, a characteristic of less advanced angiosperms, and division of the generative cell to form two male gametes (sperms) does not occur until after the pollen germinates on the stigmatic region of the female part of the flower.

It is difficult to envisage any selective advantage which asynchronous division has over synchronous, or vice versa. In both types division within a tetrad is soon completed. Certainly one attribute of plants with synchronous division is that their tetrads are less readily broken by force into monads, as a result of the gaps in their walls being filled with strong cellulose microfibrils which link with the intine of the inner internal wall of each grain.

Only one type of division has been found within a single genus of Winteraceae, but many species have yet to be studied.

Pollination in primitive angiosperms. There is considerable current interest as to the pollinators of early angiosperms. Current opinion strongly favors the idea that early angiosperms were pollinated by insects rather than wind. It has been suggested that beetles (Coleoptera) were important early pollinators of flowering plants, for they were among the first insects to appear in the fossil record and had attained considerable diversity and abundance before angiosperms originated. Although flies (Diptera) appeared after beetles, they too were well developed by the time angiosperms had evolved. In the absence of relevant fossil evidence, a study of present-day primitive angiosperms may provide clues as to the nature of early pollinators.

L. B. Thien recently discovered several insect pollinators of Winteraceae, including beetles and flies, and found that flowers have different adaptations for different pollinators. The possession of tetrad pollen does not therefore restrict the Winteraceae to a particular class of pollinators. A New Guinea species of *Tasmannia* (*Drimys* sect. *Tasmannia*) was pollinated by several species of flies attracted to nectar produced by the fragrant white flowers, which remained functional on the plant for up to 12 days. All species of *Zygogynum* studied were beetle-pollinated. The comparatively large flowers of *Z. baillonii*, for example, have yellow-orange petals, a strong "burnt orange" fragrance, and a mechanism for cross-pollination. Each flower is functional for only 2 days. On the first day of flowering a whorl of four outer petals fold back at right angles to the flower stalk (pedicel). They form a landing platform for flying beetles. Two inner whorls of petals only open enough to expose the central female part of the flower. Here the beetles feed on small amounts of nectar and deposit any pollen which may have adhered from a previous visit to another flower. The two inner whorls of petals close tightly over the flower in the late afternoon of the first day. When the flower opens fully on the second day, the central female part is no longer receptive to pollen, and pollen is now shed from the stamens to be transported by the beetles to other flowers.

Scott Norton studied the pollination ecology of *Pseudowintera*, a genus confined to New Zealand. The small flowers are long-lived and lack floral movements. They have a generalized pollination system, as is the typical situation in New Zealand, where a diversity of insects visit flowers of a particular species. Pollination is by means of flies (Fig. 1d), thrips (Thysanoptera)—a group of insects which appear in the fossil record later than flies, but before the first angiosperms—and the larvae (caterpillars) of a species of moth.

There is therefore a variety of insect pollinators in the few species of Winteraceae which have been investigated. Further studies are needed before it can be established with any certainty which, if any, of these groups pollinated the ancestral plants from which the family evolved.

Pollen and pollination in Hedycarya. Although *Hedycarya arborea*, the New Zealand member of the Monimiaceae, has tetrad pollen, it has separate male and female trees and is wind-pollinated. Tetrad pollen might seem a distinct disadvantage to a wind-pollinated species, for tetrads would not travel as far as solitary grains; in fact, most dicots with tetrad pollen are insect-pollinated. However, individual pollen grains are small in *Hedycarya*, and the tetrad diameter of about 33 μm lies within the size range (20–40 μm) of pollen units of wind-pollinated plants. It has been stated that grains or tetrads larger than 40 μm fall too soon in air currents, whereas those grains smaller than about 20 μm fail to attach to the stigmatic surfaces of the carpels.

Male flowers of *H. arborea* have a shallow bowl shape with longer pedicels than female ones. They are partly drooping, and even light winds remove considerable pollen from mature flowers. The smaller female flowers are grouped in more erect clusters. The upper half of each carpel has a knobbly stigmatic surface to which tetrads readily adhere. Such relatively simple flowers, which are a considerable contrast to the sophisticated ones of more recently evolved wind-pollinated families such as the grasses (Gramineae), may well be quite

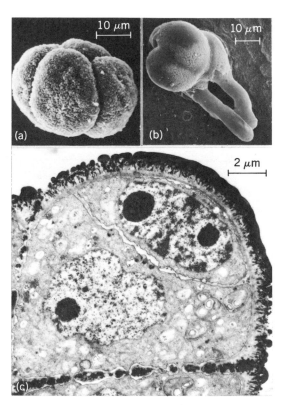

Fig. 3. *Hedycarya* pollen. Scanning electron micrographs of (a) pollen tetrad of *H. angustifolia*, and (b) germinated pollen tetrad of *H. arborea* on agar, in which three pollen tubes are visible. (c) Transmission electron micrograph of a section through part of a tetrad of *H. arborea*, showing a pollen grain in which the smaller generative cell (upper right) has recently been formed against the distal pole; one of the cytoplasmic connections between this grain and an adjacent one is shown (arrow). (From F. B. Sampson, Pollen tetrads of Hedycarya arborea J. R. et G. Forst. (Monimiaceae), Grana, 16:62–73, 1977)

similar to those of early wind-pollinated angiosperms.

Pollen tetrads of the sole Australian species, *H. angustifolia*, lack a specialized aperture (Fig. 3a), and the pollen tube presumably emerges through any part of the exposed surface of each grain. *Hedycarya arborea* has a reinforced cap of thicker outer wall material (exine) at the distal pole of each grain, which strengthens the most protuberant parts of the tetrad. Since the extent of this reinforcing varies from region to region over the plant's range, it may be a recent development. Pollen tubes cannot penetrate this distal polar region, and always emerge through the thinner, more papillose part of the external wall (Fig. 3b). Division to form generative and vegetative nuclei is synchronous within each tetrad of *H. arborea*, and as anticipated, there are large gaps in the internal walls at this stage (Fig. 3c). In contrast to the Winteraceae, the generative cell is formed against the distal pole of each grain (Fig. 3c).

There is, then, within the few families of primitive angiosperms with tetrad pollen, considerable diversity in pollen structure and pollination ecology.

For background information *see* MAGNOLIOPHYTA; PALYNOLOGY; POLLEN in the McGraw-Hill Encyclopedia of Science and Technology.

[F. B. SAMPSON]

Bibliography: F. B. Sampson, *Grana*, 14:11–15, 1974, and 16:62–73, 1977; L. B. Thien, *Biotropica*, 12:1–13, 1980; J. W. Walker, in C. B. Beck (ed.), *Origin and Early Evolution of Angiosperms*, pp. 241–299, 1976.

Polymer properties

A new class of photoresponsive polymers that change their dimensions when exposed to light has been discovered recently. A bulk piece of such material can be addressed very conveniently with light and caused to deform locally where the light impinges. The result is a raised image on the surface and a deformation throughout, to the depth of penetration of the light beam. A variety of possible applications have been proposed, including relief pictures or three-dimensional photography, raised type printing plates for letterpress (possibly to be generated by computer-guided light beams), information storage, light-triggered actuators, and actinometers.

The principle of operation of these materials is based on mechanochemistry, a branch of polymer sciences that deals with conversion of chemical energy into mechanical energy. The basic difference between these photodeformable materials and chemically deformable materials is that photochemistry is employed to cause the effect. Thus the name mechanophotochemistry has been chosen to describe the phenomenon.

Mechanochemistry. Two major classes of materials display mechanochemical effects: amorphous polyelectrolytic gels, and crystalline polymers that can be induced to melt by a chemical perturbation. In the case of amorphous polyelectrolytic gels, it was observed as early as 1950 that changes in pH cause dimensional changes. Polyacrylic acid gel fibers expand in base and contract in acid by 15% of the longitudinal size (computed for the contracted dimension). The functional groups of polyacrylic acid are neutral in acidic pH and ionized in basic pH. Consequently the chemical potential of the gel changes when the state of ionization of the functional groups is altered due to modification of pH. The liquid diluent that is present in the gel interacts with the polymer, and the quantity of this diluent present in the gel is subject to an equilibrium that balances the chemical osmotic pressure and the internal pressure repulsive forces. Thus the quantity of diluent retained by the gel is sensitive to the chemical potential present in the gel. It is not surprising therefore that when the pH is raised enough to neutralize the acid functional groups in the gel, more liquid penetrates into the gel matrix by a process similar to osmosis.

The other mechanism can be understood in terms of rubber elasticity. Normally in an elastic material the individual polymeric chains are randomly kinked. When the material is stretched, the individual chains align themselves in the direction of the stretching force and the size of their end-to-end vector increases. At one point, when these chains are stretched to their maximum, the elastic material crystallizes. During the stretching process the entropy of the system decreases steadily. When the external force is removed, the tendency of the entropy to maximize is the driving force behind the restoration forces that ultimately cause the elastic material to shrink back to the original size. Crystalline polymers contain well-oriented polymeric segments that are held by lattice forces (such as hydrogen bonds) in stretched positions. If the lattice forces are diminished, due to chemical perturbation, the tendency persists to randomize the polymeric chain configuration precisely as in the case of rubber elasticity. This corresponds to shrinkage of the material. A classic example of this phenomenon is crystalline collagen fiber that contracts in aqueous lithium bromide (LiBr) solution, due to melting, by 70% of the original size.

The photodeformable materials mentioned in this article function by the osmotic mechanism. The functional groups of the polymers have been designed to be photoionizable. Since the cause of the ionization is immaterial from the point of view of mechanochemistry, it was expected and observed that photoionization of functional groups can lead to deformations.

Design and synthesis. A general class of photoionizable materials has been identified. These are donor-haloalkane mixtures that undergo a photochemical charge-transfer reaction. This reaction leads to ionization of the donor. In many examples the resulting ions are stable. Such is the case with p-phenylenediamine, which photoionizes in the presence of carbon tetrachloride. N,N-Dimethyl-p-phenylenediamine can be easily connected to a polymer via the unsubstituted amine group. The polymers chosen to carry the photosensitive groups have acid functional groups, since acids and amines can be easily reacted to form amides. In practice the polymers were polyglutamic acid and polyacrylic acid. The condensation was effectuated by dicyclohexylcarbodiimide to yield poly[p-(N,N-dimethylamino)-N-γ-D-glutamanilide] and poly[p-(N,N-dimethylamino)-N-acrylanilide], respectively.

Mechanophotochemistry. Gels were made out of these polymeric materials by 1.5% cross linking, with N,N-dimethylformamide–carbon tetrabromide mixtures as diluents. Free-floating films made from these gels display large mechanophotochemical effects. When such films are irradiated for 4 min at a wavelength of 365 nm with a low-power light source (1680 μW/cm^2), they dilate by as much as $35 \pm 2\%$ in each dimension (the acrylanilide by 14%). It was possible to demonstrate that the photoionization reaction shown in Fig. 1

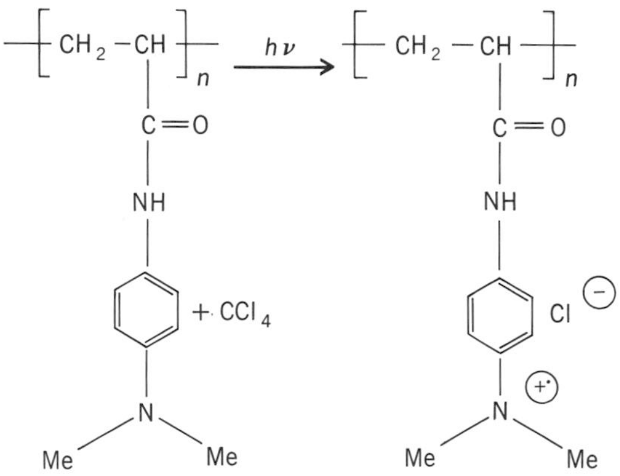

Fig. 1. Photochemical reaction that leads to mechanophotochemistry of poly[p-(N,N-dimethylamino)-N-acrylanilide].

occurred during the irradiation, and it is postulated that this photoionization is responsible for the observed effect. In separate experiments the films were anchored to solid surfaces by chemical bonding. Such films were free to deform only in the dimension normal to the surface. In these cases irradiation of selected areas of the surface produced raised images on the planar films. The films were prepared by casting carbon tetrabromide and a cross-linking agent containing polymer solutions on pretreated silicon wafers. The pretreatment consisted of chemically reacting 3-chloropropyltriethoxysilane with the wafers. This silane acted as a bridge between the wafer and the polymer film, and tied the polymer to the surface. Solventless anchored polymeric films were irradiated in the absence of N,N-dimethylformamide through masks. No deformation was observed at the end of the irradiation; however, when the irradiated films were immersed in N,N-dimethylformamide, the irradiated areas stood out above the rest of the surface. Nonirradiated areas also expand in N,N-dimethylformamide, but less than exposed areas. Figure 2 shows a raised pattern that was produced as described.

The materials described are the first two examples of a new class. If further developed, these unique materials may qualify for a number of important technological applications.

For background information *see* EQUILIBRIUM, CHEMICAL; OSMOSIS; POLYMER PROPERTIES in the McGraw-Hill Encyclopedia of Science and Technology. [ARI AVIRAM]

Bibliography: A. Aviram, *Macromolecules*, 11:

POLYMER PROPERTIES

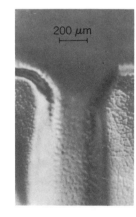

Fig. 2. Relief image produced on a film due to exposure through a mask. Exposed regions are raised relative to rest of surface. Material used was poly[p-(N,N-dimethylamino)-N-γ-D-glutamanilide].

1275–1280, 1978; P. J. Flory, *Science*, 124:53–60, 1956; L. Mandelkern, *Annu. Rev. Phys. Chem.*, 15: 421–428, 1964; G. Smets, J. Bracken, and M. Irie, *Pure Appl. Chem.*, 50:845–856, 1978; I. Z. Sternberg, A. Oplatka, and A. Katchalsky, *Nature*, 210: 568–571, 1966.

Population dynamics

Although on first inspection land plants appear to represent a vast variety of forms, life histories, and physiologies, there is now reason to believe that evolutionary specializations in plants conform to a set of basic alternatives which can be conveniently described as the primary strategies. These primary strategies may be defined as groupings of similar or analogous genetic characteristics which recur widely among plant species or populations and cause them to exhibit similarities in ecology. Recent attempts to identify the primary strategies are related to the urgent need to find a concise theoretical framework in which to deploy the wealth of detailed ecological information now available for plants.

Threats to plant existence. One approach to the identification of the primary strategies is to classify the major threats to the existence of plants. These can be summarized under three headings: stress, disturbance, and competitive exclusion.

Stress. In certain environments, such as those in arctic or arid regions or beneath forest canopies, the most potent threats to survival arise from factors such as shortages of heat, water, mineral nutrients, or light. These constraints may be an inherent characteristic of an environment, or they may be induced or intensified by activities of the vegetation, such as continuous shading or sequestration of minerals in the biomass. In contrast to the circumstances associated with competitive exclusion, this article is concerned with habitats in which severe stresses are operating more or less continuously throughout the year and affect all species present in the environment. Under these conditions there is little opportunity for characteristics of morphology or phenology to provide mechanisms of stress avoidance.

Disturbance. A second threat to the existence of plants is that associated with frequent and severe destruction of the vegetation by herbivores, pathogens, or humans (trampling, mowing, and plowing), or by phenomena such as wind damage, fire, or climatic fluctuations (frost, drought, and floods).

Where constant and severe disturbance coincides with extreme stress (for example, footpaths in mature forest, areas subject to soil erosion in arctic and alpine habitats, and overgrazed grasslands and scrub in arid regions), the result may be the total elimination of vegetation. In more productive conditions, the influence of severe and repeated disturbance is to select plants with life histories short enough to exploit the intervals between successive disturbances.

Competitive exclusion. Where resources are plentiful and there is a low incidence of disturbance, conditions encourage the development of a large, rapidly expanding biomass dominated by perennial plants with the potential for high rates of resource capture. Despite the presence under such conditions of a large reservoir of resources,

the activity of the plants during the growing season produces expanding zones of depletion, the most conspicuous of which are for light (expanding upward from the soil surface) and for water and mineral nutrients (expanding downward from the soil surface). In this type of vegetation, high rates of mortality during the growing season and low rates of reproduction are characteristic of those plants which are outgrown by their neighbors and become "trapped" in the depleted zones.

Ruderals, competitors, and stress tolerators. There is growing evidence that each of the three major threats to survival has been associated with the evolution of a distinct type of strategy. In Table 1 plants conforming to the three primary strategies have been described as ruderals, competitors, and stress tolerators, and an attempt has been made to list some of their most consistent characteristics.

A consistent feature of the ruderals is the ephemeral life cycle, a specialization clearly adapted to exploit environments intermittently favorable for rapid plant growth. A related characteristic of many ruderals is the capacity for high rates of dry-matter production, a feature which appears to facilitate rapid completion of the life cycle and maximize seed production.

The competitors are perennials, and possess in common such features as a dense canopy of leaves, the production of which coincides with an extended period when environmental conditions are conductive to high productivity. High competitive ability aboveground depends upon the development of a large mass of shoot material, which itself depends upon high rates of uptake of mineral nutrients and water. Hence, although the mechanism of competition may culminate in aboveground competition for space and light, the out-

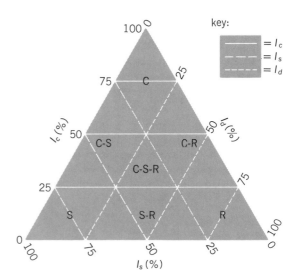

Fig. 1. Model describing the various equilibria between competition, stress, and disturbance in vegetation and the location of primary and secondary strategies. I_c represents relative importance of competition; I_s, relative importance of stress; I_d, relative importance of disturbance. (*From J. P. Grime, Evidence for the existence of three primary strategies in plants and its relevance to ecological and evolutionary theory, Amer. Natural., 111: 1169–1194, by permission of the University of Chicago Press, 1977*)

come may be influenced or even predetermined by earlier competition belowground.

The stress tolerators exhibit a range of features which are adaptations to continuously unproductive conditions. These features include morphological reductions; inherently slow rates of growth; evergreenness; long-lived organs; storage and slow turnover of carbon, mineral nutrients, and water; low phenotypic plasticity; infrequent flowering; the presence of mechanisms which allow the plant to capture resources during temporarily favorable conditions; and low palatability to unspecialized herbivores.

Triangular classification. The three strategies are, of course, extremes. The genotypes of the majority of plants appear to represent compromises associated with particular equilibria between competition, stress, and disturbance. At their respective corners of the triangle in Fig. 1, competitors, stress tolerators, and ruderals become the

Table 1. Some characteristics of competitive, stress-tolerant, and ruderal plants

Characteristics	Competitive	Stress-tolerant	Ruderal
Morphology			
Life forms	Herbs, shrubs, and trees	Lichens, bryophytes, herbs, shrubs, and trees	Herbs and bryophytes
Morphology of shoot	High, dense canopy of leaves; extensive lateral spread above and below ground	Extremely wide range of growth forms	Small stature; limited lateral spread
Leaf form	Robust, often mesomorphic	Often small or leathery, or needlelike	Various, often mesomorphic
Life history			
Longevity of established phase	Long or relatively short	Long to very long	Very short
Longevity of leaves and roots	Relatively short	Long	Short
Leaf phenology	Well-defined peaks of leaf production coinciding with period(s) of maximum potential productivity	Evergreens, with various patterns of leaf production	Short phase of leaf production in period of high potential productivity
Phenology of flowering	Flowers produced after (or, more rarely, before) periods of maximum potential productivity	No general relationship between time of flowering and season	Flowers produced early in the life history
Frequency of flowering	Established plants usually flower each year	Intermittent flowering over a long life history	High frequency of flowering
Proportion of annual production devoted to seeds	Small	Small	Large
Perennation	Dormant buds and seeds	Stress-tolerant leaves and roots	Dormant seeds
Regenerative strategies*	V, S, W, B_s	V, B_{sd}	S, W, B_s
Physiology			
Maximum potential relative growth rate	Rapid	Slow	Rapid
Response to stress	Rapid morphogenetic responses (root-shoot ratio, leaf area, root surface area) maximizing vegetative growth	Morphogenetic responses slow and small in magnitude	Rapid curtailment of vegetative growth; diversion of resources into flowering
Photosynthesis and uptake of mineral nutrients	Strongly seasonal, coinciding with long continuous period of vegetative growth	Opportunistic, often uncoupled from vegetative growth	Opportunistic, coinciding with vegetative growth
Acclimation of photosynthesis, mineral nutrition, and tissue hardiness to seasonal change in temperature, light, and moisture supply	Weakly developed	Strongly developed	Weakly developed
Storage of photosynthate and mineral nutrients	Most photosynthate and mineral nutrients are rapidly incorporated into vegetative structure, but a proportion is stored and forms the capital for expansion of growth in the following growing season	Storage systems in leaves, stems, and/or roots	Confined to seeds
Miscellaneous			
Litter	Copious, often persistent	Sparse, sometimes persistent	Sparse, not usually persistent
Palatability to unspecialized herbivores	Various	Low	Various, often high
DNA content (per cell)	Low	High or low	Very low

*V = vegetative expansion; S = seasonal regeneration in vegetation gaps; W = numerous, small, widely dispersed seeds or spores; B_s = persistent seed bank; B_{sd} = persistent seedling bank.

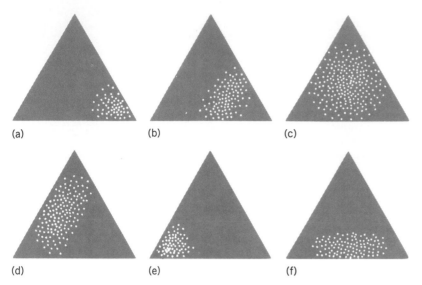

Fig. 2.   Diagrams describing the range of strategies encompassed by (a) annual herbs, (b) biennial herbs, (c) perennial herbs and ferns, (d) trees and shrubs, (e) lichens, and (f) bryophytes. For the distribution of strategies within the triangle, see Fig. 1. (*From J. P. Grime, Evidence for the existence of three primary strategies in plants and its relevance to ecological and evolutionary theory, Amer. Natural., 111:1169–1194, by permission of the University of Chicago Press, 1977*)

exclusive constituents of the vegetation, and the remaining zones of the model correspond to conditions favorable to four types of intermediate or secondary strategy consisting of: competitive ruderals (C-R), adapted to circumstances in which there is a low impact of stress and competition is restricted to a moderate intensity by disturbance; stress-tolerant ruderals (S-R), adapted to lightly disturbed, unproductive habitats; stress-tolerant competitors (C-S), adapted to relatively undisturbed conditions experiencing moderate intensities of stress; and C-S-R strategists, adapted to habitats in which the level of competition is restricted by moderate intensities of both stress and disturbance.

As shown in Fig. 2, the triangular model may be used to describe the strategic range of particular life forms and taxonomic groups. The widest range

of strategies is that displayed by perennial herbs and ferns. Annual herbs are mainly ruderals, whereas biennials are restricted to areas of the model corresponding to the competitive ruderals and stress-tolerant ruderals. Trees and shrubs include competitors, stress-tolerant competitors, and stress tolerators. Lichens are almost exclusively stress tolerators, whereas bryophytes have a distribution centered on the stress-tolerant ruderals.

Regenerative strategies. The strategies considered so far refer to established, mature plants. However, because of their relatively small size, the seedlings and vegetative offspring of plants often differ very substantially from their parents with respect to their ability to compete for resources and to tolerate or evade stress and disturbance. It is necessary therefore to devise another classification of strategies referring exclusively to the regenerative (immature) phases of the life cycles of land plants.

Especially in productive, relatively undisturbed vegetation, the most severe threats to the survival of seedlings and vegetative offspring are due to the variety of stresses which arise from close proximity to established plants. In many circumstances, successful regeneration depends upon exploitation of local conditions in which the cover of established plants and litter has been disturbed. It would appear that regenerative strategies have evolved primarily through the development of mechanisms whereby the juvenile stages in life histories tolerate or evade the effects of established plants. On this basis, five types of regenerative strategy may be classified as shown in Table 2.

Combinations of strategies. The regenerative strategies occurring most commonly in association with the primary strategies are indicated in Table 1. It is not unusual for the same plant to exhibit two or more regenerative strategies. Such multiple regeneration may be expected to enlarge the range of conditions in which regeneration can occur, and is characteristic of some of the most widely successful herbs, shrubs, and trees. Conversely, many of the most rare and precarious plant species appear to be dependent upon a single method of regeneration.

Table 2. Habitat conditions of the five regenerative strategies

Regenerative strategy	Habitat conditions to which strategy appears to be adapted
Vegetative expansion by persistent stolons, rhizomes, or suckers	Productive or unproductive habitats subject to low intensities of disturbance
Seasonal regeneration (from seed or vegetative propagules) in vegetation gaps	Habitats subjected to seasonally predictable disturbance by climate or biotic factors
Regeneration involving a bank of persistent (usually buried) seeds	Habitats subjected to temporally unpredictable disturbance
Regeneration involving numerous widely dispersed seeds or spores	Habitats subjected to spatially unpredictable disturbance
Regeneration involving a bank of persistent seedlings and immature plants	Unproductive habitats subjected to low intensities of disturbance

Vegetation dynamics. Information concerning established and regenerative strategies is providing a basis for interpretation of the sequence and rate of some successional and cyclical changes in vegetation, and for predicting vegetation responses to changes in management. Following the clearance of vegetation from a productive environment, the usual sequence of recolonization is ruderals → competitors → stress tolerators. The first step in this sequence is due to competitive exclusion, whereas the second step is the result of resource depletion arising from the development of a large plant biomass. The rate of vegetation change in this succession is initially rapid, but falls progressively with the incursion of slow-growing, long-lived plants and declining opportunities for regeneration. Succession in unproductive environments proceeds more slowly, and the role of competitors is much reduced.

In landscapes subjected to occasional and spatially unpredictable vegetation clearance, ruderal and competitive species appear to lead a fugitive existence, and the most effective regeneration strategy is that involving the production of numerous widely dispersed seeds or spores.

Where clearance occurs relatively frequently in the same place, the process of vegetation change is cyclical rather than successional. This is because the effect of frequent clearance is not merely to arrest succession but to bring into prominence those among the ruderal and competitive species which regenerate by persistent seed banks.

For background information *see* ECOLOGY, PHYSIOLOGICAL (PLANT) in the McGraw-Hill Encyclopedia of Science and Technology.   [J. P. GRIME]

Bibliography: J. P. Grime, *Nature*, 250:26–31, 1974; J. P. Grime, *Plant Strategies and Vegetation Processes*, 1979; P. J. Grubb, *Biol. Rev.*, 52:107–145, 1977; R. H. Whittaker and D. Goodman, *Amer. Natur.*, 113:185–200, 1979.

Population genetics

The study of gene frequencies at the population level. Initially the approach was almost entirely mathematical. It was assumed that the fitness of an individual is determined by genetic factors alone, being a constant assigned to each genotype. However, the real world is more complex since phenotypic fitness depends upon the genotype and the environment. Population genetics is now based upon an increasing input of laboratory and field observations under an array of environments, using organisms as diverse as the insect *Drosophila*, mosquitoes, various butterflies and moths, snails, and vertebrates such as mice. Much of this work involves the documentation and interpretation of genetic variability in natural populations.

Electrophoretic variants. In the last 20 years, high levels of genetic variability of proteins have been revealed by the technique of electrophoresis (Table 1). These levels of polymorphisms are far greater than predicted from earlier theoretical considerations. Is variation in gene frequencies at such loci neutral, being determined by chance, or is it due to the balancing forces of natural selection? The attempt to decide between these two possibilities—the "neutral" and "balance" schools—has been a major preoccupation of much experimental and theoretical work. This debate really marks the beginning of the recent phase in population genetics.

In several organisms, latitudinal gradients in polymorphic gene frequencies occur, suggesting that temperature plays an important role as a selective agent. In addition, in an organism such as *Drosophila*, considerable variation in available substrate type or concentration would be expected, perhaps in some way paralleling temperature. Obtaining definitive data is difficult, since it now appears best to regard selection as acting upon integrated metabolic phenotypes controlled by genes at several loci. However, tolerance to environmental ethanol is a measurable trait with direct ecological significance. Tolerances of *D. melanogaster* are higher inside an Australian wine cellar than outside, and they increase with proximity to the cellar (and hence ethanol) during vintage. But there is little relationship between ethanol tolerance and genotypes at the alcohol dehydrogenase (*Adh*) locus in this population. Indeed, both tolerance to environmental ethanol and alcohol dehydrogenase enzyme activity are controlled by genes

Table 1. Surveys of genic heterozygosity in a number of organisms*

Species	Number of populations	Number of loci	Proportion of polymorphic loci	Heterozygosity per locus
Human				
Homo sapiens	1	71	.28	.067
Mice				
Mus musculus musculus	4	41	.29	.091
M. m. brevirostris	1	40	.30	.110
M. m. domesticus	2	41	.20	.056
Peromyscus polionotus	7	32	.23	.057
Insects				
Drosophila pseudoobscura	10	24	.43	.128
D. persimilis	1	24	.25	.106
D. obscura	3	30	.53	.108
D. subobscura	6	31	.47	.076
D. willistoni	10	20	.81	.175
D. melanogaster	1	19	.42	.119
D. simulans	1	18	.61	.160
Horseshoe crab				
Limulus polyphemus	4	25	.25	.061

*Modified from R. C. Lewontin, *The Genetic Basis of Evolutionary Change*, Columbia University Press, New York, 1974.

in addition to those at the *Adh* locus in Australian and Californian populations. Therefore the relationship between the *Adh* locus and environmental ethanol is rather obscure, although there is some laboratory evidence for ethanol-mediated selection at the *Adh* locus. Different populations could adapt in varying ways to environmental ethanol, especially if ethanol-mediated selection acts upon metabolic phenotypes rather than upon single-locus phenotypes. The single-locus approach therefore cannot in itself explain the maintenance of genetic variability in natural populations, even though the importance of environmental selective agents is occasionally demonstrable at this level.

Environmental stresses and ecological marginality. The success of a *Drosophila* population depends upon its adaptation to annual and diurnal climatic cycles, especially in temperate regions. The annual cycle of the temperate region gives two major and largely density-independent climatic stresses: a combination of high temperature–desiccation, and low temperature. These stresses lead to the evolution of resistant races in climatically extreme habitats, so that the *Drosophila* gene pool is apparently made up of climate-associated races (Table 2).

Certain "flexible" *Drosophila* species have higher incidences of chromosomal polymorphism at the centers of their distribution than at the peripheries. One proposed explanation is that central populations are adapted to their diverse environments when there is a high level of heterogeneity resulting from a high degree of polymorphism. However, in some populations of the North American species *D. pseudoobscura*, polymorphisms vary in time (seasonally, temporally) and space (altitudinally) but not in others. In addition, some species are chromosomally rigid across diverse ecogeographical regions but not others. Electrophoretic polymorphism levels appear to vary less between central and peripheral populations in several *Drosophila* species. Indeed arguments have been presented suggesting that genic heterozygosity should be high in marginal populations because the temporal instability of such environments means that no particular genotype is favored for long periods. There are some laboratory experiments in support of this.

Two definitions of extreme habitat are being considered: peripheral in the geographic sense, and marginal in an ecological sense. An ecologically marginal population is one in which physical stresses tend to be extreme, so that resources may be unpredictable and short-lived. In these circumstances selection should favor life-history characteristics leading to rapid population growth, referred to as r-selection by ecologists. Ecological marginality should therefore be marked by a shift toward increased r-selection. Such populations tend to be extreme with low variability for climate-dependent and resource-utilization traits (Table 2). A species then becomes a potential colonizer, whereby a shift into more extreme habitats is possible if resources are available. Of extreme evolutionary interest are shifts that have been documented in the Queensland fruit fly, *Dacus tryoni*, and are apparently occurring in several *Drosophila* species.

Life-history, environmental stress, and resource utilization studies may therefore provide an understanding of marginal populations. These studies are phenotypic assessments which are directly relatable to field environments. By contrast, electrophoretic and chromosome polymorphism assays, which are effectively genotypic assessments, can provide only indirect assessments of marginal populations.

Habitat selection. Direct environmental selection may lead to differing genotypic frequencies according to the microhabitat in an ecologically heterogeneous area. Such genetic heterogeneity may be enhanced by habitat selection, or differential choice of a place to live by different genotypes. The distribution of *D. persimilis* in a small-scale Californian environment made up of various vegetational types is best explained by invoking habitat selection. In the malarial mosquito, *Anopheles arabiensis*, some genotypes are adapted to dry environments and others to more humid environments. In West African villages, dry-adapted genotypes enter relatively dry huts during the night to take blood meals, primarily by biting humans. However, genotypes adapted to more humid conditions remain outside, primarily biting animals. Only during the adult phase of the life cycle does habitat selection occur for behavioral differences of profound epidemiological significance.

Habitats are largely determined by the physical features of the environment and the resources utilized. The genetic determination of habitat selection must therefore be extremely complex. Integrated studies of habitat selection have, however,

Table 2. A comparison of two Australian Drosophila melanogaster populations

Characteristic	Population A	Population B
Location	Melbourne	Townsville
Latitude	37°S	19°S
Climate	Extreme temperate	Optimal tropical
Ecological marginality	Higher	Lower
Survival of −1°C stress	High	Low
Survival of desiccation stress	High with low variability	Low with high variability
Development times	Fast	Slow with high variability
Ethanol resource utilization	Extreme and homogeneous	Relatively heterogeneous
Overall strategy	r-strategy maximized	

been well developed in vertebrates such as the deer mouse of North America, *Peromyscus*, in which there is some evidence for behavioral phenotypes corresponding to habitats.

In the early development of population genetics, R. A. Fisher argued that polymorphism resulted from heterozygotes having a selective advantage over homozygotes. Although such overdominance has been well documented in a few cases, the fitness differentials implied are theoretically untenable at the population level as a general explanation for polymorphic variability. However, if fitnesses vary spatially across environments, conditions for the stable maintenance of polymorphisms are greatly relaxed compared with the overdominance model. In experimental terms, climatic races and habitat selection may provide a system whereby an array of genotypes are differentially distributed across environmental conditions. This conclusion is very similar to Sewall Wright's earlier theoretical arguments for the importance of population fragmentation for an understanding of the genetics of populations.

Random mating. The theoretical foundations of population genetics are based upon the assumption that random mating occurs, which is useful for mathematical modeling, but in practice rarely occurs. In humans nonrandom mating occurs for a variety of reasons, such as relative attractions based on appearance and intelligence levels. In *Drosophila* there is good evidence for rapid male mating as an important component of fitness, whereby certain genotypes are favored. Diverse mating patterns such as these provide mechanisms enhancing genetic heterogeneity. To these can be added the phenomenon of rare-male advantage, whereby a higher proportion of less frequent genotypes take part in matings than would be expected assuming random mating. An initially rare genotype will increase in frequency, associated with decreasing advantage as it becomes commoner, and so leading to an equilibrium. Hence polymorphism is maintained without associated fitness differentials due to frequency-dependent selection.

Frequency-dependent selection. Bryan Clarke argues that the major factors affecting population size are predators, parasitism, and competition, all involving frequency dependency. He contends that: (1) predators which concentrate disproportionately upon common varieties of prey overlook rare ones which therefore increase, (2) frequency-dependent selection by parasites could be important in maintaining biochemical diversity in their hosts, and (3) frequency-dependent interactions occur when members of the same species compete for a limited resource (including potential mates).

Therefore, frequency-dependent selection is almost certainly a powerful force in maintaining genetic diversity. It is probably of greater importance in optimal than marginal habitats, since in marginal habitats direct climatic selection may reduce its importance. This leads to the conclusion that future work in population genetics will depend upon detailed studies of the relationship between genes, environment, and organism at the population level, assaying the phenotype in a variety of ways, including molecular, behavioral, and ecological components. Combined with an analysis of mating systems, variation in natural populations will slowly become understood.

For background information *see* POLYMORPHISM (GENETICS); POPULATION GENETICS in the McGraw-Hill Encyclopedia of Science and Technology.

[PETER A. PARSONS]

Bibliography: R. J. Berry, *Inheritance and Natural History*, 1977; B. C. Clarke, *Proc. Roy. Soc. London*, B205:453–474, 1979; P. A. Parsons, *Evolutionary Biology*, vol. 9, 1980; E. B. Spiess, *Genes in Populations*, 1977.

Powder metallurgy

Powder metallurgy (P/M) is an automated metalworking process for forming precision metal components directly from metal powders instead of from metal that has to be melted (Fig. 1). Components made by this process have many applications, and will have greater usage in the 1980s because of increasing emphasis on energy savings, material conservation, lower labor input, and improved productivity. Recent technological trends include increased use of P/M forgings, P/M tool steel, and P/M titanium and superalloys.

Powder particles are of specific sizes and

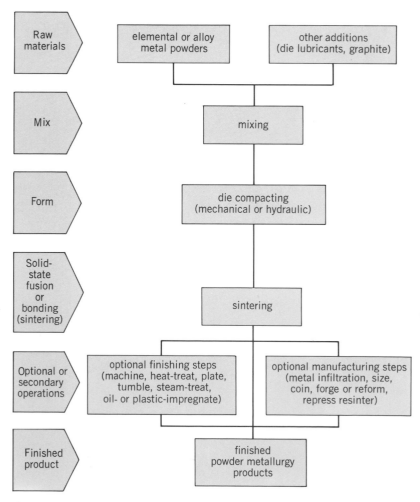

Fig. 1. Powder metallurgy process.

shapes. They are not merely ground-up chips or scraps of metal. The major methods for making metal powders are atomization, reduction of oxides, electrolysis, and chemical reduction. The single most important distinction of P/M materials is controlled density. The density can range from the porosity of filters and bearings to high-density structural components.

Materials. The most common metals available in powder form are iron, tin, nickel, copper, aluminum, and titanium, as well as refractory metals such as tungsten and molybdenum. Prealloyed powders, such as low-alloy steels, bronze, brass, and stainless steel, in which each particle is itself an alloy, are also produced.

Iron powder is the most widely used P/M material. Its major use is in the production of structural components. Low-, medium-, and high-density parts can be produced from iron powders, giving a tensile strength from 16,000 to almost 60,000 psi (110 to 413 N/mm²). By heat treatment, strengths can be increased to over 90,000 psi (620 N/mm²).

Frequently, small additions of other powders, such as carbon, copper, or nickel, are added to iron powders to improve the mechanical properties of the components produced. Adding copper to iron powder increases strength and tends to increase hardness. Low-, medium-, and high-density copper-steel parts and bearings can be produced with tensile strength ranges from 20,000 to more than 80,000 psi (138 to 552 N/mm²). Heat-treated copper-steel parts have strengths up to 100,000 psi (689 N/mm²).

Nickel in the order 2–8%, with or without copper, can be used with iron powders to obtain especially high-strength parts having improved toughness and fatigue strength. Strengths of 90,000–100,000 psi (621–690 N/mm²) can be achieved in nonheat-treated parts; heat treatment increases strength to as high as 180,000 psi (1241 N/mm²).

Fig. 2. Powder metallurgy forgings.

Applications. The major applications for P/M parts and products include automotive vehicles, business machines, jet engines, home appliances, electrical and electronic equipment, watches, cameras, and farm equipment.

The automobile industry represents the largest market for P/M parts; the average automobile may contain up to 15 lb (7 kg) of P/M parts. Typical automotive applications include shock-absorber piston and rod guide, converter turbine hub, clutch pilot bearing, transmission hub and gears, differential pinion gear, brake piston and valve spacer, rack-and-pinion steering gear, oil-pump gears, and rocker arm. P/M tungsten contact points are used in automobile distributors.

Office copier machines and postage meters are another large market, with one Xerox model containing 1200 P/M parts (75 different ones).

A P/M tungsten alloy, which is 50% heavier than lead, is used for counterweights on airplane controls and to balance helicopter rotor blades. Tungsten carbide, one of the hardest manufactured materials, is used to make forming dies and cutting tools for machining, as well as bits and drills for mining and oil well drilling.

Although the single most important use of metal powders is in the production of components and shapes, many other uses exist. Nickel powder as a catalyst accelerates chemical reactions such as hydrogenation and petroleum reforming. Iron powder is used in welding electrodes. Brass powder is used in printing inks and finishes. Aluminum powder is used in bridge paint, mining explosives, rocket fuel, and deodorants.

Technological developments. Recent P/M developments include purer base-metal powders, higher-compressibility powders, new alloys, larger compacting presses with greater pressures, and production techniques such as cold and hot forging and hot isostatic pressing.

Producing fully dense parts with improved properties through P/M hot forming or forging is a rapidly growing sector of the industry. The process begins with pressing and sintering a preform, which is then subjected to a forging treatment to bring the part up to, or close to, theoretical density.

Fatigue life of P/M hot-formed tapered roller bearings has exceeded that of wrought steels by 3.5–4 times. The introduction of ultrapure iron powders could increase this figure significantly. Also, P/M offers average material savings of 50% on bearing cup and cone production. High-temperature sintering of forged preforms will further expand the use of P/M forging, because reduction of chromium and manganese oxides in the low-dew-point atmosphere is possible at about 1232°C.

There are more than 80 alloy steel P/M forgings or hot-formed components being made on a commercial basis in the United States and around the world. Figure 2 shows examples of P/M forgings made in the United States, Japan, England, Germany, and Sweden. Compared to conventional machining operations, the process offers advantages for parts with cam contours, splines, serrations, and gear teeth and recesses. The automotive market should continue to dominate forging applications. The Hydra-Matic Division of General

Motors Corporation has made more than 16,000,000 P/M hot-formed cam stators. One GM transmission contains four hot-formed P/M parts, in addition to nine conventionally made P/M parts, having a combined weight of 8 lb (36 kg).

The Porsche 928 V-8 is the first automobile to use P/M forged-steel connecting rods. A finished rod weighs just under 2 lb (0.9 kg), and is about 9 in. (229 mm) long. Porsche is pleased with the cost and energy savings of P/M.

Making tool steels from metal powders either by cold or hot isostatic pressing or by conventional P/M compacting is entering a new growth phase. Applications include spade drills, knife blades, slotting cutters, insert blades for gear cutters, watch cases, and components for diesel engines. Tool steel materials include : M-2, T-15, M-3, (type 2) M-42, M-4, and M-2, resulfurized.

Developments are also taking place in high-temperature P/M materials such as superalloys. The term superalloy describes a group of high-strength, high-temperature nickel-base alloys made up of some of the following elements: iron, cobalt, chromium, tungsten, molybdenum, tantalum, columbium, titanium, aluminum, zirconium, vanadium, carbon, and boron. Superalloys can operate for extended periods of time at temperatures about 1200°F (650°C), and provide resistance to hot corrosion and erosion. In the past, P/M superalloy applications were limited to military aircraft, but P/M parts recently qualified for civilian engines to be used in the Boeing 707, 747, and 757 and DC-10.

Powder metallurgy will continue to grow in this particular field because of its near-net shape capability, more homogeneous microstructure, superior mechanical properties, and materials conservation—the last very important when considering the use of scarce and expensive metals such as chromium, tantalum, and cobalt. Use of P/M in place of forging of conventional materials for jet-engine components results in cost savings that range from 25 to 75%.

For background information *see* POWDER MET-ALLURGY; SINTERING in the McGraw-Hill Encyclopedia of Science and Technology.

[PETER K. JOHNSON]

Primates

Recent studies have provided new information on the most ancient phases of human ancestry and on the early development of human cultural traits.

HUMAN ORIGINS

The order Primates, including the living squirrellike "prosimians," monkeys, apes, humans, and their diverse extinct relatives, has a long history. The earliest known fossil primate, *Purgatorius* (represented by one tooth), lived alongside the last dinosaurs in Montana some 65 million years (MY) ago. Since then, many distinctive lineages have evolved and adapted to diverse modes of life, including human life. The evolutionary history of this group is thus of interest not only as a scientific exercise, but for what it might tell about the processes which formed the human species. Recent advances have been made in many areas of this history, especially in knowledge of

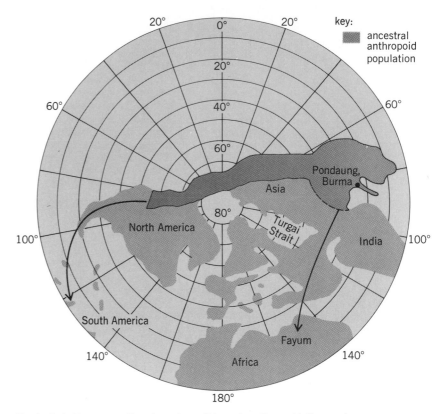

Fig. 1. Late Eocene continents and possible early anthropoid dispersal.

three groups of animals which may have been close to the origin of humanity.

Today there is a consensus (strongly opposed by some active scholars) that the living tarsier of Southeast Asia is the lower primate most closely related to the anthropoids (higher primates: monkeys, apes, and humans). Fossils widely accepted as related to tarsiers first occurred in North America and Eurasia at the start of the Eocene Epoch, 54 MY ago, diversifying and then dying out over the succeeding 30 MY. The earliest anthropoids are known from the southern continents of South America (Bolivia) and Africa (Egypt) early in the Oligocene Epoch, about 34–30 MY ago. These latter anthropoids are already members of the two geographically distinct modern anthropoid subgroups Platyrrhini and Catarrhini, from the New and Old Worlds, respectively. A still older primate known as *Pondaungia*, from the late middle Eocene (about 40 MY ago) of Burma, had sometimes been suggested as a possible anthropoid, but it was represented only by damaged fragments of teeth. New, slightly more complete finds of this animal appear to confirm its anthropoid status, in fact indicating that it was already a protocatarrhine. If so, it would appear that the circumpacific lands played a larger role in anthropoid dispersal than is usually acknowledged (Fig. 1).

Increased understanding of plate tectonics and the past positions of the continents has promoted the theory that anthropoids arose in (or near) Africa and divided into the two subgroups known today. Ancestral platyrrhines then drifted to South America on rafts of natural vegetation masses, following the past currents deduced from deep-sea fossils. But no African catarrhine fossil is a likely ancestor for platyrrhines, the two groups being morphologically distinct from their earliest occurrences. If, however, a group of protoanthropoids evolved in eastern Asia or western North American during the early Eocene from known tarsier-like ancestors, its descendants might have crossed the Bering Straits to form a circumpacific distribution. The protoplatyrrhines might then have entered South America from the north via the tectonically unknown Caribbean region, while protocatarrhines (such as *Pondaungia*) expanded west and south to cross the Tethys seaway separating Africa from Asia. Just as *Pondaungia* is most similar to some Egyptian early catarrhines, representatives of other mammalian groups have a fossil distribution which suggests that Burma-Pakistan-(Turkey?)-Egypt was a plausible route of migration and expansion of range in the later Eocene. The actual crossing of the Tethys barrier might have been simplified by an apparent world-wide(?) lowering of sea level at the end of the Eocene.

Oligocene early catarrhines. Recent expeditions to the richly fossiliferous Fayum beds of Egypt have greatly expanded knowledge of the collateral relatives of *Pondaungia*. In addition to clarifying some points of taxonomy (and, as usual, clouding others), this research has confirmed the essentially monkeylike nature of the locomotor system of all the Fayum catarrhines. What is intriguing, however, is that the monkeys which offer the closest comparisons are not those of Africa or Asia, but instead the platyrrhines of the New World. The latter thus appear to have retained many of the features of limb bone function which characterized early catarrhines, and, presumably, their common anthropoid ancestors. *Propliopithecus* (or *Aegyptopithecus*), the Fayum form most like *Pondaungia* dentally, may have already begun a trend to increased use of the forelimb for support and locomotion which culminated in the modern apes (and the ancestors of humans). Moreover, *Propliopithecus* (and other Fayum forms as well) appears to have been characterized by a pronounced degree of sexual dimorphism in both canine tooth size and lower jaw proportions, indicating significant differences in adaptation between the two sexes. Elwyn Simons and his collaborators have taken these morphological features as indicators of a polygynous, rather than monogamous, social system, by analogy with modern higher primates. *See* AEGYPTOPITHECUS AND PROPLIOPITHECUS.

Early Miocene hominoids. A gap of several million years separates the Fayum forms from the succeeding *Dryopithecus* group of early hominoid species. The Hominoidea, or humanlike (rather than monkeylike) catarrhines, are now generally considered to include a conservative family Pliopithecidae, for the Fayum "apes" and some younger relatives, and one or two families (Hominidae and perhaps Pongidae), for the Miocene to modern species; here only Hominidae is recognized, with the human line classified as hominines (the subfamily Homininae). Thus *Dryopithecus* (sometimes divided into the genera *Proconsul*, *Rangwapithecus*, and others) is the oldest recognized hominid, known in eastern Africa between about 23 and

at least 14 MY ago. Two species are also known from western Europe between 15 and 11 MY ago. As is usual in fossil primates (and other mammals), *Dryopithecus* is known especially by partial dentitions, supplemented by rare cranial parts and often unassociated limb bones (which are thus difficult to allocate to species defined from teeth).

The dentition of *Dryopithecus* is quite similar to that of modern apes, with a relatively thin coating of enamel on the molar teeth, although the incisors are not as large (compared to molars) as in living forms. The mandibular corpora of both *Dryopithecus* and modern great apes are similar in being relatively thin and deep, thus not adapted to withstanding heavy chewing stresses, and the two groups reveal similar proportions between molar tooth size and body size as estimated from limb bone size. Taken together, these features suggest an adaptation to eating foods which were neither tough nor gritty, perhaps forest fruits rather than savannah roots or tubers. This matches well with the known diet of living great apes as well as with the evidence from the fossils themselves: African *Dryopithecus* especially had a rather monkeylike limb structure, rather than any special adaptation to terrestrial life, and the paleoenvironments appear to have been most like modern montane rainforests.

Africa was an island continent for most of the middle (and early?) Cenozoic, with a fauna distinct from that of Eurasia in general, but about 18 MY ago, northeastern Africa contacted Eurasia as a result of tectonic movement (continental drift), and the faunas of the two regions began to mingle. *Dryopithecus* species only slightly different from those found in eastern Africa probably reached western Europe by expanding along the northern shore of the Mediterranean and then northward into France and Germany, a route followed by other forest-dwelling animals of this time period as well. But a second group of more advanced apes, which can be called the sivapiths, also made its first appearance about 15 MY ago.

The sivapiths, or ground apes, as they have been termed by E. L. Simons and others, are important because they show the first clear evidence of anatomical features later found only in the close relatives or ancestors of modern humans. Several of these derived characteristics imply a greater adaptation to eating tough or gritty foods, such as might be found on the ground in a forest-fringe or savannahlike environment. They include: thicker enamel on the occlusal (chewing) surface of the molars and premolars, thicker and more heavily buttressed jaws (mandibular corpus and symphysis, maxillary alveolar processes), and apparently larger teeth relative to limb bone size, all by comparison to *Dryopithecus* (and modern apes). The limbs themselves are not well known, but there is not as clear evidence for terrestriality as was once thought (and implied by the name ground apes). There are three main varieties of sivapiths, here recognized as genera: *Sivapithecus*, *Ramapithecus*, and *Gigantopithecus*. Two of these appeared about 14–15 MY ago in Turkey, East Africa, and Czechoslovakia, then continued until approximately 8 MY; *Gigantopithecus* is only known in southern Asia around 9 MY and in southern China around 1 MY ago. Although authorities differ on the allocation and interpretation of the fragmentary remains of the sivapiths, it seems that *Sivapithecus* is the most conservative form, with incisors and canines unreduced from the *Dryopithecus* condition and relatively monkeylike limb joints. The four recognized species compare in size with modern chimpanzees and gorillas. *Gigantopithecus* is somewhat larger and more robust, with gorilla-sized teeth in massive jaws. It appears more advanced morphologically, with one elbow fragment showing gorillalike adaptations, perhaps to ground life. The front teeth are all reduced, but the canines, although low-crowned and not strongly sharpened by their opponents, are thick and were used as additional food-grinding surfaces. *Ramapithecus*, the smallest and most poorly known of the sivapiths, may be the one most similar to later humans. It has especially thick and shallow mandibular corpora, its incisors and canines appear to be small, and the canine may be rotated slightly to align more with the incisors. In addition, the ultrastructure of the enamel prisms shows the so-called keyhole pattern. All of these features, in addition to those noted above for all sivapiths, are found in Pliocene *Australopithecus*, although some of them may be reduced in Pleistocene *Homo*.

New finds. Most of the sivapith species were first recognized by 1930, but only recently has their importance become clear. This new understanding is due mainly to the recovery of many new specimens, especially in the Potwar Plateau ("western Siwaliks") of Pakistan. There, D. R. Pilbeam directed a team which carefully investigated the many thousand feet of fossiliferous sediments. Rather than relying mainly on specimens brought in by local villagers, Pilbeam's crew collected intensively in small areas, so that each fossil could be carefully tied into the rock sequence, some of which can now be dated by geophysical means. The new fossils have often been more complete than those found earlier in this classic area, and in 1980 a real prize was recovered: a partial face and nearly complete lower jaw of *Sivapithecus*, now being reconstructed for study. A team from the Peking Institute of Vertebrate Paleontology and Paleoanthropology discovered contemporaneous fossils in a new site in Lu-feng County, Yunnan Province (southwestern China). So far, this locality has yielded two nearly complete mandibles, over 200 teeth, and a crushed partial skull, probably also of *Sivapithecus*. Comparable specimens, including a palate and a fine series of mandibles revealing strong sexual dimorphism, have been recovered from a single locality in northern Greece, and less complete groups of specimens have come from Turkey, Hungary, Kenya, and of course the classic Indian Siwaliks. New studies of the paleoenvironments of some of these localities have generally produced a picture of open woodland landscapes, the interface between forest and steepe/savannah, including fluctuating ecotonal microhabitats. In such areas, it would appear that the human lineage became evolutionarily distinct.

Several implications arise from this interpretation of the middle and late Miocene hominoids. First, the presence in the sivapiths of derived (advanced) features found only in humans among modern hominoids strongly suggests that they fall on the human side of the ape-human divergence, thus dating that divergence as older than 15 MY. Studies of the proteins of living hominoids have shown that humans are more similar to chimps and gorillas than any of these are to orangutans. Thus, the Asian-African ape divergence was probably even older. On the other hand, some preliminary studies suggest resemblances between orangs and *Sivapithecus* in details of enamel ultrastructure and thickness; along with their geographical proximity, this has led some authors to postulate an ancestor-descendant relationship between the two genera. This ties in closely with a more contested aspect of the protein studies, their use as a "molecular clock" to date the time of divergence of the groups studied. Without considering the theoretical controversy over such clocks, the molecular results have suggested an ape-human divergence date of less than 8 MY, far too recent in light of the fossil evidence. Some paleontologists, including Pilbeam, have suggested that the sivapiths might include the ancestors of both great apes and humans, but their derived traits shared with *Australopithecus* make this view untenable to others, including the present writer. It is, of course, possible that these features were evolved in parallel by the sivapiths and the later hominines (*Australopithecus* and *Homo*), but this is unparsimonious and thus to be rejected as the hypothesis of choice unless simpler theories are falsified. In fact, the morphology of the earliest *Australopithecus* is more similar to that of the sivapiths than to living or earlier apes, as might be expected if the two groups were genealogically related.

Many problems concerning the sivapiths remain to be solved, from their formal names to their area and time of origin. For example, the group has been recently called or placed into Hominidae, Homininae, Dryopithecinae, Dryopithecidae, Sivapithecidae, Ramapithecidae, Ponginae, and Sugrivapithecini, among other taxonomic units. Some authors recognize only two genera, combining *Ramapithecus* into *Sivapithecus*, while others accept seven or more, plus other early Miocene African genera here called *Dryopithecus*. East Africa, Arabia, Turkey, central Asia, and other areas have been suggested as the locale of their origin from a *Dryopithecus*-like stock, and numerous theories have been put forward to account for their appearance. More detailed analyses of their functional morphology, taxonomy, and habitat, involving new fossils and comparisons with close relatives, may help to solve these problems, at least temporarily, but they will doubtless raise others to keep paleoanthropologists busy with ancestors of modern humans.

[ERIC DELSON]

HUMAN CULTURE

Recent discoveries in East Africa have substantially altered speculative models for the origin and development of human culture. One view, generally accepted since the early 1960s, held that the use of stone tools, brain size and intelligence, and bipedalism evolved together through the lower Pleistocene as an adaptive complex, with increases in any one encouraging further develop-

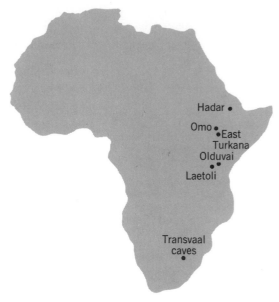

Fig. 2. Map showing the location of major fossil and archeological localities discussed in the text.

ment of the others. Discovery in Tanzania of hominid footprints dating 3.6–3.8 MY ago and analysis of early (3.2–3.9 MY old) *Australopithecus* postcranial remains from Ethiopia have shown this scenario to be incorrect. According to these findings, bipedalism evolved some 2 MY before either

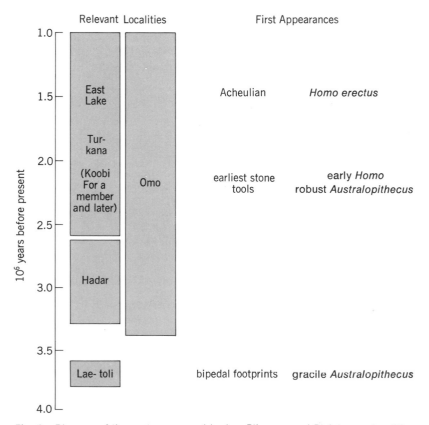

Fig. 3. Diagram of time spans covered by key Pliocene and Pleistocene localities mentioned in the text, with first appearance of hominid taxa and key features.

substantial increases in hominid brain size or the first appearance of stone tools. Intensive work in somewhat more recent fossil localities indicates consistent first appearances of robust *Australopithecus*, the earliest members of the genus *Homo*, and stone tools in the 2.0–2.2 MY range. The picture is made even more interesting by the rapid evolutionary changes, both physical and cultural, among hominids in the 2.0–1.0 MY range. These findings require that paleoanthropologists and archeologists rethink and treat as two separate issues the evolution of bipedalism and the development of primitive cultural traits.

Recent findings. Because of its geologic history, East Africa remains the most informative area in early hominid studies. Rifting and associated volcanism since late Miocene times have created both the sedimentary basins in which Plio-Pleistocene sediments, bones, and artifacts accumulated, and the intercalated deposits of tuff and lava amenable to radiometric dating. The early footprints discovered at Laetoli, near Olduvai Gorge in northern Tanzania (Fig. 2), were in fact made in freshly fallen volcanic ash. The fossil-rich limestone cave breccias of the Transvaal region in South Africa (Fig. 2) lack associated volcanics and must be dated by less precise faunal correlation.

The Laetoli footprints are small but quite similar to those of modern humans, reflecting a well-developed striding gait. Postcranial materials recovered from the Hadar region of Ethiopia (Fig. 2) indicate that the earliest hominids now known (from 3.3 to 3.8 MY ago, combining the Hadar and Laetoli materials) were fully bipedal with small brains and rather primitive dental and cranial features. All the early hominid remains from both Laetoli and Hadar can be attributed to a small (1.2–1.5 m tall), gracile form of *Australopithecus*.

Thus far, no indisputable artifact occurrences have been reported in the older Laetoli and Hadar deposits. Intensive survey in both the East Lake Turkana and Omo exposures yielded hominid fossils but no stone tools prior to about 2.2 MY ago (Fig. 3). Between 2.0 and 1.5 MY ago, stone tools, singly and in clusters, are present in the fossil record at Omo, East Lake Turkana, and Olduvai Gorge. About the same time that the first stone tools occur, there also appear remains of a robust australopithecine *(A. boisei)* and fossils now classified by most workers as the earliest members of the genus *Homo*. By 1.5–1.2 MY ago, *H. erectus* appears in the deposits at East Lake Turkana and, in more fragmentary form, in the Omo, as well as in somewhat younger deposits at Olduvai and at least one of the Transvaal caves. Robust australopithecines have been recovered from South Africa to the Omo, including Olduvai, East Lake Turkana, and other East African sites in this time range, apparently going extinct sometime between 1.5 and 1.0 MY ago. Stoneworking swiftly evolved from the early flake and chopper forms to more elaborately wrought pieces, with the Acheulian or hand-ax industry, a widespread tool-working tradition associated with *H. erectus*, appearing by 1.2–1.5 MY ago.

The coincidence of a more specialized australopithecine lineage, early *Homo*, and stone artifacts, after some 2 MY of hominid bipedality, is thought-provoking. While these "first appearances" depend on negative evidence and could be modified by future finds, many researchers are now asking what selective pressures may have been acting on hominids in this time range that favored accelerated physical cultural evolution. Of interest is the fact that several localities yield evidence of a change toward more open, somewhat drier biotopes during this time span.

Early artifacts and sites. The earliest stone tools were manufactured from a variety of raw materials; at both Olduvai and Lake Turkana, stone was worked and discarded at considerable distances from its sources. Implements are unstandardized but reflect a mastery of basic stone-working skills. They include "core-chopper" forms and flakes removed from such nodules, some of which bear secondary retouch, as well as pounders or hammerstones (Fig. 4). Their knifelike and heavy-duty cutting edges would be useful in breaking down large animals and some vegetable foods. The artifacts do not constitute weapons of the hunt, although some might have been used to prepare crude digging sticks or spears.

Aggregations of artifacts have been excavated at Olduvai, Lake Turkana, and Omo. Some are simply clusters of worked stone; others are associated with bones of either one large mammal (such as a hippopotamus) or of several animals. The sites have been found in a variety of settings: on a lake margin at Olduvai, along ephemeral streams at East Lake Turkana, and in a major river's floodplain at Omo. Taking into account possibilities of differential preservation of sites closer to water, it nonetheless appears that early hominids manufactured and discarded their implements relatively close to sources of fresh water, as is expected of a water-dependent species. Although hominid fossils and even footprints have been found in the Turkana lake margin zone, thus far no stone tools have been recovered from this environment.

Behavior and adaptation. The scanty evidence makes inferences about early hominid behavior somewhat speculative. Archeologist G. Ll. Isaac notes that modern hunter-gatherers differ from nonhuman primates in their use of a "home base" and heavy reliance on implements in foraging, in division of foraging activities along age and sex lines, with males hunting, and in food sharing among adults and from adults to immatures. Isaac contends that the association of artifacts with bones in Plio-Pleistocene sites results from regular meat consumption by early hominids. He further argues that hunting, sexual specialization of foraging in roles, and food sharing may already have evolved in these times.

Not all researchers accept Isaac's scenario. L. R. Binford has challenged the basic assumption that all bones found in deposits with stone tools are there due to hominid action. Others contend that scavenging or opportunistic killing of debilitated animals, rather than coordinated hunting, could account for the bones in the sites. Another unanswered question is whether the sites reflect a home-base strategy like that of modern hunter-gatherers, or even whether such behaviors were

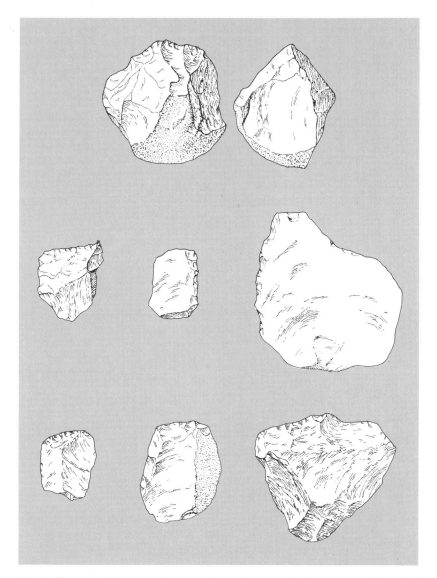

Fig. 4. Representative early stone tools from site DK, Olduvai Gorge (about 1.75 MY old): chopper and various implements on flakes (note secondary retouch on edges of pieces in the lower two rows). (From M. D. Leakey, Olduvai Gorge, vol. 3, Cambridge University Press, 1971)

daily or sporadic. Occurrence of artifacts in only some of the Lake Turkana environments that yield hominid fossils raises the possibility that early toolmaking was not so ubiquitous in hominid foraging patterns as it was to be later. Likewise, presently there is no evidence for the subsistence or home-base behavior of early hominids prior to the advent of stone tools.

Analysis of dental wear patterns and stable carbon isotope ratios in fossil bone may soon provide information on these earlier patterns of foraging and diet, as well as on changes that may have accompanied the new physical and cultural developments around 2 MY ago.

For background information see FOSSIL MAN; PRIMATES in the McGraw-Hill Encyclopedia of Science and Technology.   [DIANE P. GIFFORD]

Bibliography: L. R. Binford, *Bones: Ancient Men and Modern Myths*, 1980; N. T. Boaz, *Annu. Rev.*

Anthropol., 8:71–85, 1979; E. Delson and A. L. Rosenberger, in R. L. Ciochon and A. B. Chiarelli (eds.), *Evolutionary Biology of New World Monkeys and Continental Drift*, 1980; L. O. Greenfield, *Amer. J. Phys. Anthropol.*, 52:351–365, 1980; G. Ll. Isaac, *Sci. Amer.*, 238:90–108, 1978; R. F. Kay, E. L. Simons, and J. G. Fleagle, *Amer. J. Phys. Anthropol.*, vol. 54, 1981; M. D. Leakey and R. L. Hay, *Nature*, 278:317–323, 1979; D. R. Pilbeam, *Annu. Rev. Anthropol.*, 8:333–352, 1979; D. R. Pilbeam et al., *Postilla*, Yale Peabody Museum, 1980; F. S. Szalay and E. Delson, *Evolutionary History of the Primates*, 1979.

Proton

For many years physicists believed that the spin of the proton was not very important in violent high-energy collisions. Although spin is quite important in low-energy atomic physics, they reasoned that the tiny energy associated with spin could not much affect proton collisions with very high energy. This belief was shattered by a recent experiment studying proton-proton elastic scattering, using a polarized proton beam and a polarized proton target to control the spins. The importance of spin increased dramatically as the collisions became more violent. This result suggests that spin forces may dominate the violent hard-scattering events which directly probe the proton's constituents. It is very difficult to get such large spin forces from protons constructed of three normal quarks. Thus this discovery may change the concept of the nature of the proton's constituents. *See* QUARKS.

Elastic proton-proton scattering. In a typical high-energy physics experiment, a "beam" proton is accelerated to an energy or momentum of perhaps 12×10^9 electronvolts (12 GeV) and then fired at a "target" proton contained in liquid hydrogen. There are several different ways that these two protons can interact or scatter. A scattering event is called elastic when all the energy of the incoming beam proton is carried off by two rebounding protons. In inelastic scattering events, some of the energy goes into creating other particles such as π-mesons and K-mesons.

The interior of the proton can be probed by measuring elastic scattering events and studying how their probability or cross section, $\sigma(\theta)$, varies with the scattering angle, θ. Lord Rutherford invented this scattering experiment in 1911. A glancing collision where the protons almost miss each other gives a small-angle scattering, which reveals only the long-range nature of the protons' forces. Scattering events at large angles come from almost head-on collisions, which directly probe the inner structure of the proton. Any hard objects inside the protons would increase the number of large-angle scattering events. The cross section's behavior can be interpreted in terms of a variable called the transverse momentum transfer, $P_\perp = P \sin \theta$, which directly measures the violence of the sideways kick when a proton of momentum P scatters through an angle θ. Experiments in the 1960s found an excessively large number of violent high-$P_\perp$ scattering events; this suggested that each proton contains internal constituents, possibly quarks.

Spin scattering experiment. Each proton has an intrinsic angular momentum called spin, which probably exists because each proton is actually spinning about its own axis. This spin is a vector quantity, which has a magnitude $h/4\pi$ for protons, where h is Planck's constant. Quantum mechanics allows these spin vectors to point only in two exactly opposed directions, corresponding to clockwise and anticlockwise spin. These two directions are called up and down, and denoted by $\uparrow$ and $\downarrow$.

A spin scattering experiment is a refinement of Rutherford's classical experiment, which allows an even more detailed look at the structure of the proton. These pure spin experiments require a polarized proton beam and a polarized proton target, each of which have most of their protons' spins pointing in the same direction. The polarization of the beam and the target are each given by Eq. (1), where $N_\uparrow$ and $N_\downarrow$ are the number of pro-

$$P = \frac{N_\uparrow - N_\downarrow}{N_\uparrow + N_\downarrow} \qquad (1)$$

tons with their spins pointing up and down. In earlier experiments neither the beam protons nor the target protons were polarized, and only an average cross section could be measured. This averaging masked the behavior of the different spin cross sections, and hid important information about the protons' forces.

One reason that few earlier experiments used polarized protons is that a polarized target is a highly sophisticated and complex piece of apparatus, and only a small number exist in the world. Moreover, in 1973 the 12-GeV zero-gradient synchrotron (ZGS) at Argonne National Laboratory, near Chicago, became the first accelerator in the world able to accelerate polarized protons to high energy. The ZGS was permanently shut down in 1979, and now there are no high-energy polarized proton beams in the world. Brookhaven National Laboratory, near New York City, recently ap-

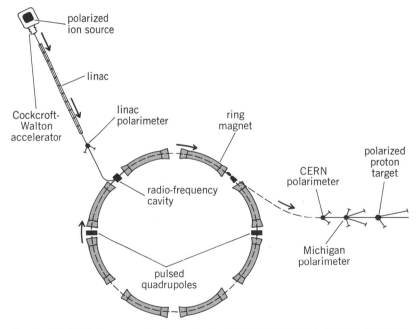

Fig. 1. The 12-GeV zero-gradient synchrotron (ZGS) showing polarized-proton-beam operation.

proved a plan to modify its 33-GeV alternating-gradient synchrotron (AGS) to allow the acceleration of polarized protons.

Polarized beam. To obtain high-energy polarized protons, one needs a polarized ion source to produce polarized protons, and an accelerator which can accelerate them to high energy without destroying their polarization. Figure 1 shows the polarized-beam operation of the ZGS, which contained three stages. First, the protons were accelerated to an energy of 750,000 eV by the Cockcroft-Walton electrostatic accelerating column. The 200-ft-long (60-m) linear accelerator (LINAC) then gave the protons an energy of 50×10^6 eV, and injected them into the main synchrotron. The 600-ft-circumference (180-m) ring of synchrotron bending magnets steered the protons in a circular orbit, allowing them to pass repeatedly through the rf accelerating cavity, which accelerated them to an energy of 12 GeV in about a million passes.

The polarized protons started as hydrogen gas injected into the polarized ion source, where several stages converted this gas into the polarized protons. The first stage dissociated the hydrogen gas molecules into hydrogen atoms, which each contain one proton and one electron. A sextupole magnet used the Stern-Gerlach effect to separate these atoms; the electron-spin-up atoms passed through the sextupole magnet, while the electron-spin-down atoms were stopped. The proton's spin was almost unaffected by these magnets because the proton's magnetic moment is 660 smaller than the electron's. The next stage contained rf radia-tion at a frequency which flipped the spins of those atom's whose proton's spin were down. Thus in the emerging beam of atoms all the proton spins were up and the electron spins were half up and half down. The final ionizer stage stripped off the electrons and emitted a beam of about 10^{11} polarized protons per pulse with a polarization of about 75%.

The ZGS was not designed to accelerate polarized protons, but the high quality of its ring magnets helped since magents can depolarize spinning protons. Because the spins were oriented vertically, any horizontal magnetic fields could cause depolarization. During the acceleration cycle, the horizontal fields normally seen by the protons tended to cancel, and the net depolarization was small. However, at certain energies a resonant condition occurred where the protons encountered identical magnetic fields on each turn around the ring. Each little kick then added coherently, causing a rapid depolarization. These depolarizing conditions were avoided by using special pulsed quadrupole magnets installed inside the ZGS. These were pulsed at exactly the correct time to jump rapidly through each resonance by slightly altering the orbit of the protons. There were 29 depolarizing resonances in the ZGS acceleration cycle, which were all successfully jumped to maintain a beam polarization of 70% up to 12 GeV. The polarization was measured by the three polarimeters shown in Fig. 1.

Polarized proton target. A polarized proton target operates by first polarizing some electrons and then transferring the polarization to the protons.

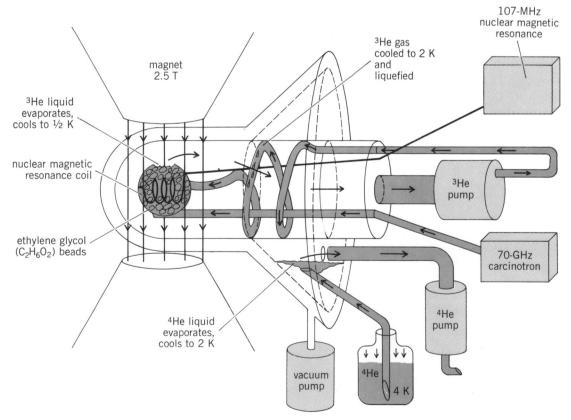

Fig. 2. Michigan PPT-V polarized proton target.

The target must be maintained in a high magnetic field, B, of 25,000 gauss and kept at a very low temperature, T, of 0.5 K. The polarized protons are in the hydrogen atoms of ethylene glycol (antifreeze), $C_2H_6O_2$, which is doped with potassium dichromate, $K_2Cr_2O_7$, and then frozen into small beads. The magnetic field polarizes the unpaired inner-shell electrons of the chromium, giving them the polarization value shown in Eq. 2. The quan-

$$P = \frac{e^{+\frac{\mu \cdot B}{kT}} - e^{-\frac{\mu \cdot B}{kT}}}{e^{+\frac{\mu \cdot B}{kT}} + e^{-\frac{\mu \cdot B}{kT}}} \qquad (2)$$

tity μ is the electron's magnetic moment, and k is Boltzmann's constant. These electrons form temporary pairs with nearby hydrogen protons. For this temperature and magnetic field, the polarization of the chromium electrons is almost 100%. Therefore in each electron-proton pair there are only two possible spin states: $(\uparrow, \uparrow)$ or electron up – proton up, and $(\uparrow, \downarrow)$ or electron up – proton down. Microwave irradiation at 70 GHz then flips both spins in the $(\uparrow, \downarrow)$ state, driving it into the $(\downarrow, \uparrow)$ state. This 70-GHz microwave irradiation does nothing to the $(\uparrow, \uparrow)$ state because the frequency is wrong. Thus the microwaves depolarize the electrons while polarizing the protons in the spin-up $(\uparrow)$ state. This "dynamic polarization" technique was developed in the late 1950s by A. Abragam and C. Jeffries, and gives a target proton polarization of 75% or more.

Figure 2 is a schematic diagram of the polarized target showing its main components. Cooling of the target beads uses two different types of helium, normal ^{4}He and the rare ^{3}He isotope, which has good low-temperature properties. Liquid ^{4}He, at 4 K, is fed into the ^{4}He cryostat from a dewar and circulated. The ^{4}He is cooled to about 2 K by the evaporation caused by pumping. ^{3}He gas is circulated in a separate cryostat, where it is turned into liquid by the 2-K cooling provided by the ^{4}He cryostat. The liquid ^{3}He moves into the target cavity, where it is cooled to 0.5 K by the evaporation caused by very large "Roots Blower" pumps. The

70-GHz microwaves produced in a carcinotron tube are carried by a waveguide to the target cavity. A coil inside the cavity measures the proton polarization by using a nuclear magnetic resonance (NMR) technique.

Spin cross sections. The polarized proton beam was scattered from the polarized proton target, and the two pure-spin elastic cross sections were measured: $\sigma_{\uparrow\uparrow}$ (beam spin parallel to target spin) and $\sigma_{\uparrow\downarrow}$ (beam spin antiparallel to target spin). When the ratio of these spin cross sections was plotted against the $P_\perp$ variable, some very surprising behavior was found (Fig. 3).

For small-$P_\perp$ glancing collisions, both spin cross sections behave in a similar way, and the ratio is close to 1. However, in violent high-$P_\perp$ collisions, the ratio becomes very large. The ratio $\sigma_{\uparrow\uparrow}:\sigma_{\uparrow\downarrow}$ dramatically changes from 1 to 4 as one passes from "soft" small-$P_\perp$ scattering to "hard" large-$P_\perp$ scattering. Thus the spin forces are most important in the violent collisions which probe most deeply into the interior structure of the proton. If these violent collisions are really caused by the direct scattering of the proton's constituents, as most physicists believe, these constituents must have enormous spin forces. It seems impossible to get such large spin forces from the conventional quark model. Thus this unexpected result forces a reexamination of the concept of the proton's constituents. The result $\sigma_{\uparrow\uparrow}:\sigma_{\uparrow\downarrow} = 4$ may indicate something about these constituents, but the message is not yet understood. Perhaps the higher-energy polarized protons at the Brookhaven AGS will make this message understandable.

For background information *see* ION SOURCES; NUCLEAR TARGETS, POLARIZED; QUARKS; SCATTERING EXPERIMENTS, NUCLEAR; SPIN (QUANTUM MECHANICS) in the McGraw-Hill Encyclopedia of Science and Technology.

[A. D. KRISCH]

Bibliography: D. G. Crabb et al., *Phys. Rev. Lett.*, 41:1257–1259, Nov. 6, 1978; A. D. Krisch, *Sci. Amer.*, 240(5):68–80, May 1979.

Quarks

The subatomic particles found in nature fall into two classes: leptons, such as the electron and neutrino, which appear pointlike and structureless; and hadrons, such as the proton, neutron, and many related unstable particles, which interact with strong forces not felt by leptons and which appear to have complex internal structure. The past decade has seen tremendous progress in the understanding of the substructure of hadrons and of the origin of the strong forces with which they interact. In particular, a new quantum number has been discovered which bears the same relation to the strong interactions as electric charge bears to electromagnetism. This new charge is known as "color" (though it bears no relation at all to familiar, visual color), and the theory of strong interactions based on color-dependent forces is known as quantum chromodynamics (QCD) in analogy to the quantum theory of electromagnetism—quantum electrodynamics (QED).

Confinement and color. For many years it has been thought that hadrons are composed of yet more fundamental particles known as quarks. All

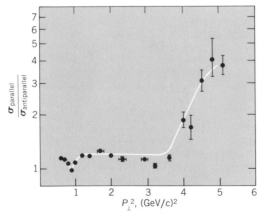

Fig. 3. The spin cross section ratio $\sigma_{\uparrow\uparrow}:\sigma_{\uparrow\downarrow}$ plotted against the square of the $P_\perp$ variable for elastic proton-proton scattering at 12 GeV.

attempts to separate quarks from hadrons have failed. Initially there was resistance to assign a fundamental significance to objects which had never been observed in isolation, but with the continued success of quark models the notion arose that quarks are indeed the fundamental constituents of hadrons and for dynamical reasons are "confined" to the interior of the hadrons they compose. The idea of confinement has no parallels in the history of physics: in all previous "atomic" systems, given a modest amount of energy it is possible to free the constituent particle from the composite.

It is now believed that there is a deep connection between the color quantum number and confinement. This connection is responsible, in part, for the indirect fashion in which the existence of color was inferred. Quarks carry the color quantum number (just as electrons and protons carry charge), but all known hadrons are combinations of quarks which are color-neutral (just as atoms are neutral combinations of charged electrons and protons). Charge was discovered rather early in the history of atomic physics because it is possible to ionize atoms—that is, to remove a charged constituent. If quarks are confined, it is not possible to "ionize" a hadron and observe color directly. The existence of color was first suspected by O. W. Greenberg in 1965. According to the quark model, baryons, which are hadrons related to the proton and neutron, are composed of three quarks. Certain baryons appear to contain three identical quarks in the same quantum state in violation of Pauli's exclusion principle. Greenberg pointed out that the exclusion principle could be saved if quarks came in three colors and if the baryon in question contained one of each color. Between 1965 and 1973 the idea languished, although it was pointed out then that if quarks came in three colors several quark model predictions were altered by factors of 3 and agreed more closely with experiment.

Asymptotic freedom. The case for color became compelling when H. D. Politzer, and D. J. Gross and F. Wilczek, realized that color-mediated forces between quarks could explain a perplexing phenomenon known as asymptotic freedom. Experiments had shown that although quarks interact strongly at distances of order the size of the proton $(10^{-15}$ m$)$, their interactions weaken considerably at shorter distances, hence the term asymptotic freedom. All known relativistic quantum theories except one predict that interactions, if they change at all, become stronger at shorter distances. Quantum chromodynamics is the exception. In quantum chromodynamics the forces between quarks are carried by eight colored, massless, spin-1 quanta known as gluons. Gluons are to color as the photon is to electric charge with one important difference: gluons themselves are colored whereas the photon is neutral. The color carried by gluons is responsible for most of the unfamiliar features of quantum chromodynamics, including asymptotic freedom.

Because the interactions between quarks become weak at short distances, it is possible to use perturbative techniques in this regime. In recent years the predictions of perturbative quantum chromodynamics have been verified in detail by numerous high-energy experiments, culminating in 1979 with the discovery in electron-positron annihilation of jets of hadrons coming from gluon decay.

Two-phase structure. Shortly after the discovery that quantum chromodynamics is asymptotically free, it was pointed out that the effects responsible for the weakening of forces at short distances appeared to make the forces between quarks grow strong at increasing distances. To separate a quark from a hadron, it is necessary to enter this regime of strong forces. It was therefore natural to speculate that color-dependent forces are responsible for quark confinement. Unfortunately, when forces grow strong, perturbative techniques break down. Until just recently there were no known reliable methods for studying relativistic quantum systems in regimes where the interactions are strong. In the past several years nonperturbative techniques have been developed by several groups and applied to quantum chromodynamics. Their results can best be understood in a language borrowed from statistical mechanics. It appears that quantum chromodynamics is a theory with two phases. In one phase, quarks and gluons move about more or less freely and interact more or less weakly like an almost ideal gas. In the other phase, a condensate of quarks and gluons forms and expels color electric fields or any particles which carry them. An analogous phenomenon occurs to magnetic fields in a superconductor: in the superconducting phase a condensate of electron bound states (Cooper pairs) forms and arranges itself so as to expel magnetic fields. In the normal, or nonsuperconducting, phase, magnetic fields may propagate more or less freely. If both phases are present, the magnetic fields are confined to the normal regions. In quantum chromodynamics it is thought, but not proved, that the condensate phase has lower energy than the almost ideal gas phase. Because of this, most of space is filled with the condensate—it appears to be the vacuum. Quarks and gluons carry the color charge. Therefore they are expelled from the vacuum and confined to "normal regions" which are recognized as hadrons. The region of space close by a quark or gluon is thought to remain in the normal phase because the strong color fields near a color charge drive out the condensate. This effect too has an analog in superconductivity. Although weak magnetic fields are expelled from a superconductor, a very strong magnetic field breaks up the Cooper pair condensate and returns the metal to the normal phase.

MIT bag model. As often happens in physics, a model with just these properties was developed several years before calculations showed that quantum chromodynamics might imply a two-phase structure for the subatomic world. This MIT bag model—named after the Massachusetts Institute of Technology where it was developed, and the flexible spatial regions where quarks are supposed to reside—was originally motivated by the paradoxical free but confined behavior of quarks already discussed. According to the model, quarks and gluons interact weakly by color-mediated forces inside of bags but are forbidden by explicit boundary conditions from leaving. The equations which govern the model are fixed almost uniquely by the requirements that they be consistent with special relativity. In the model the confined quarks generate a pressure which must be compensated by a pressure exerted by the vacuum on the surface of the hadron. The magnitude of this pressure is immense. It has been measured indirectly and is found to be roughly 10^{32} atm (10^{37} Pa). It is now believed that the bag pressure is a reflection of the different energies of the condensate and normal phases of quantum chromodynamics. This energy difference—about 3×10^{33} J/m³—measures the energy given off when a bit of hadron is returned to the vacuum.

At present none of the nonperturbative techniques which have been applied to quantum chromodynamics is sophisticated enough to allow detailed calculations of the properties of hadrons. It is necessary, therefore, to rely on quantum chromodynamics–motivated models like the MIT bag model for quantitative descriptions of subnuclear phenomena. Since the interactions inside hadrons in the model are quantum chromodynamics–like and since confinement is introduced in a manner consistent with the two-phase structure expected in quantum chromodynamics, the model calculations are thought to be trustworthy. Perhaps the best test of a model is the accuracy with which it is possible to extrapolate its predictions away from the regime where the model was originally formulated. One of the long-standing puzzles in particle physics was the occurrence in nature of only those hadrons made of three quarks (baryons) or of a quark and an antiquark (mesons). Higher configurations are allowed by the principle that hadrons should be color-neutral, but they did not seem to occur. It is as though the only nuclei were deuterium, tritium, and ³He, instead of the vast periodic table. The bag model provides a good description of mesons and baryons. In addition, it is formulated with sufficient precision to allow extrapolation to higher configurations. It predicts a large number of additional quantum states beginning with two quarks and two antiquarks, but also that these higher states should manifest themselves rather indirectly, not as the prominent scattering resonances characteristic of the quark-antiquark and three-quark system. Recently some of these subtle two-quark–two antiquark effects have been identified in low-energy scattering experiments, confirming the model's predictions and bolstering the belief that confined quantum chromodynamics can give a unified and complete description of the subnuclear world.

For background information *see* COLOR (QUANTUM MECHANICS); ELEMENTARY PARTICLE; QUANTUM FIELD THEORY; QUARKS; SUPERCONDUCTIVITY in the McGraw-Hill Encyclopedia of Science and Technology.   [ROBERT L. JAFFE]

Bibliography: G. 't Hooft, *Sci. Amer.*, 242(6): 104–138, June 1980; R. L. Jaffe, *Nature*, 268: 201–208, 1977; K. Johnson, *Sci. Amer.* 241(1): 112–121, July 1979; Y. Nambu, *Sci. Amer.*, 235(5): 48–60, November 1976.

Remote sensing

In the past few years the use of advanced laser technology in combination with remote-sensing techniques has made it possible to measure the concentration of gases and to detect the presence of particles in the atmosphere from ground mobile, airborne, and space-borne platforms. Laser-based measurement of remote gases has applications in enforcement of air-pollution regulations, testing of air-pollution plume models, study of atmospheric chemistry, identification of gases of military interest, detection of leaks in gas pipelines, and petroleum exploration. Remote-particle detection is used to locate plumes and clouds of dust and smoke and to study thermal inversion layers. Laser radar technology also has the potential to detect a far greater number of gases than has been demonstrated to date, and to identify the composition, as well as the concentration, of remote particles.

Principles of laser radar. Unlike "passive" remote-sensing systems, which collect radiation scattered or emitted by a remote source, laser radar is "active" in that it uses a laser beam to illuminate a remote target and collect scattered or reradiated energy. Wherever they are mounted, laser radar systems provide three-dimensional coverage. Called lidar (light detection and ranging), laser radar is analogous to traditional radar but uses shorter wavelengths and a laser transmitter. As explained below, several types of lasers have been combined with a variety of remote-sensing techniques, depending on the characteristics of the gases or particles under study.

Measurement of gases. The remote measurement of gases can be accomplished by using one of three techniques: fluorescence, Raman, or differential absorption lidar (DIAL). The fluorescence technique has proved to be very useful for measuring metallic atoms and ions in the ionosphere, but it is not useful for tropospheric applications because of the suppressive effect of the ambient atmosphere. The Raman technique has the advantage of simultaneous multiple-species measurements, but the extremely small scattering cross section of the Raman effect seriously limits its sensitivity. The DIAL technique currently offers the longest range and highest sensitivity for remote sensing of gases in the troposphere.

The lasers of a DIAL system transmit pulses of radiation into the atmosphere. Molecules and naturally occurring particles suspended in the atmosphere scatter the radiation, a small fraction of which is collected by the receiver system. The atmosphere thus produces a distributed scatter that provides a continuous backscattered signal. Two laser wavelengths are transmitted, one of which is absorbed by the gas to be measured and one of which is not. The signal derived from the nonabsorbed wavelength is used to calibrate the response of the system and to establish the backscatter from the atmosphere. The differential between the return signals on the absorbed and the nonabsorbed wavelengths is used as a direct measure of the concentration of the absorbing species. Since the lidar system operates with laser pulses, and the receiver is timed in radar fashion, range-resolved measurements of the concentration of gases can be obtained.

The typical DIAL system (Fig. 1) is designed to transmit wavelengths near 300 nm, which are required for remote sensing of sulfur dioxide, SO_2, and ozone. In the system one laser is tuned to the absorbed wavelength, and the other to the nonabsorbed wavelength. The two pulses are then transmitted within 100 μs of each other to eliminate scintillation in the atmosphere and variations in atmospheric backscatter. The two dye laser systems are tunable to wavelengths near 600 nm. This resulting laser radiation is then doubled in frequency by using the nonlinear crystals to yield a 300-nm beam. The Glan prism is used to combine the two beams of different polarization very efficiently. A beam expander is used to match the divergence of the laser beam with a field of view of the receiver. The telescope collects the backscattered radiation and focuses it onto a photomultiplier tube. The optical filter is used to reduce background radiation. The electronic signal from the photomultiplier tube is amplified and digitized. The digital signal can then be processed by a minicomputer or microcomputer to afford a real-time display, or it can be stored on magnetic tape for later analysis. The system shown in Fig. 1 has operated within the following parameters: a pulse energy of 5 mJ per pulse, a range resolution of 15 m, a pulse integration of 450 pulse pairs, a measurement interval of 45 s, and a receiving-telescope diameter of 20 in. (508 mm).

Typical measurements of a sulfur dioxide plume from a power plant are shown in Fig. 2. These three traces show the resulting concentration of a plume approximately 1.8 km from the lidar system. The data obtained at shorter ranges than the location of the plume indicate a negligible ambient concentration of sulfur dioxide. The variations in the concentration in this region are indicative of the noise level of the system. The three large peaks show structural changes in the shape of the sulfur dioxide plume over time. The larger random error at ranges beyond the plume is caused by absorption within the plume, which results in a decreased signal level beyond its immediate location. The measurement error in the concentration of these plumes appears to be approximately 0.2 parts per million. This measurement uncertainty can be decreased to approximately 0.02 parts per million by the use of longer-range cells and additional pulse integration. Such a sensitivity level would enable measurement of ambient concentrations in many locations and would also permit measurement of very diffused plume concentrations.

Results similar to those above have been obtained for the measurement of ambient concentrations of ozone and nitrogen dioxide, NO_2. However, sulfur dioxide, ozone, and nitrogen dioxide are the only three species of primary interest that have been detected by the ultraviolet and visible DIAL systems. One current focus of research in DIAL systems is to extend the wavelength range into the infrared in order to permit measurement of the hundreds of species that absorb in that spectral region. Infrared DIAL systems have been constructed and successfully tested for the measurement of gases such as hydrogen chloride (HCl), methane, nitrous oxide, water vapor, ethylene, sulfur dioxide, and ozone, and of temperature.

Measurement of particles. Single-wavelength visible and near-infrared lidar systems measure the backscatter that is proportional to the concentration of the backscattering particles. While these systems enjoy wide applicability and considerable success, civilian, military, and industrial needs go

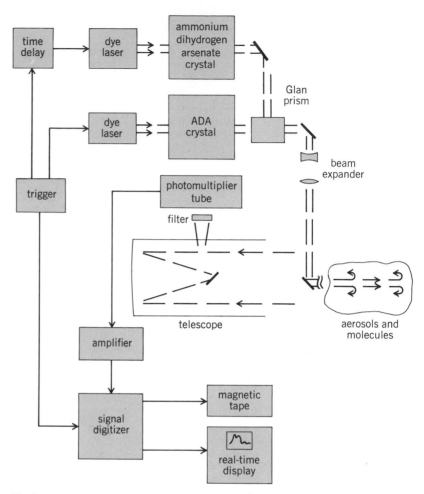

Fig. 1. Experimental DIAL system used to perform remote range-resolved measurements of sulfur dioxide, SO_2, in a plume from a power plant.

beyond simply the measurement of backscatter. Remote measurements of particle composition would find significant applications in the study of aerosol chemistry in real time, enforcing air-pollution regulations, measuring particles of military interest, providing early warning of icing threats to aircraft, and controlling industrial chemical processes.

New types of lidar systems are therefore being developed to fill this void in instrumental capabilities. For example, lidar systems based on fluorescence techniques are being used for the measurement of biologically active particles. When irradiated with ultraviolet radiation, such particles fluoresce in the visible and near-ultraviolet ranges. Measurements of this fluorescence can be used to indicate the composition of the particles.

Another technique being developed, the differential scatter (DISC) technique, uses wavelength-dependent scattering from particles in the infrared spectral region. Particles have scattering signatures in the infrared range that are uniquely related to particle composition. Figure 3 shows calculated values of scattering signatures for six different materials as a function of infrared wavelength. The values have been calculated from Mie theory for spherical particles of an assumed submicrometer particle size distribution. Each material has its own scattering signature. Even different

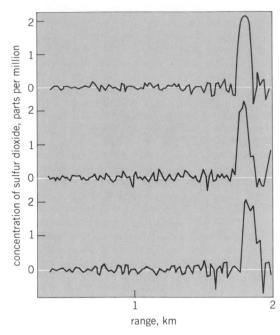

Fig. 2. Lidar-measured sulfur dioxide concentration in a plume at a distance of 1.8 km from the lidar.

concentrations of acid in aqueous solution yield different signatures.

The DISC technique relies on the measurement of backscatter of a number of carefully selected wavelengths. Wavelengths are usually selected by using a computer code that maximizes the sensitivity to the compound of interest while at the same time rejecting interference. The measured scatter-

ing signature is then correlated with a data base of signatures of known candidate materials. It is expected that this technique will be limited to measuring the composition of the major constituents of particles; minor constituents probably will contribute insufficiently to the scattering signature to permit their detection.

Recent results from laboratory measurements indicate that a two-wavelength system is sufficient to measure the composition of sulfuric acid, H_2SO_4, in water-laden particles. Additional wavelengths will be required to measure the composition of typical urban aerosols, which include sulfates and nitrates.

Results obtained by using a carbon dioxide laser with the DISC technique have shown a scattering difference between cirrus clouds, which are composed of ice, and clouds containing liquid water droplets. Ice crystals cause no threat of icing to unprotected classes of aircraft. However, supercooled liquid water freezes and causes ice buildup on the cold surfaces of aircraft. Remote sensing using such a system could permit pilots to distinguish threatening from nonthreatening cloudy regions, and thus could improve the safety of unprotected classes of aircraft.

Future developments. Continuing research, along with rapid advances in laser technology, will provide expanded capabilities of gas and particle remote-sensing systems. In addition to anticipated improvements in the capabilities of the DIAL and DISC systems, it is likely that the two types of lidar will ultimately be combined into a single remote-sensing system that analyzes both gases and particles. The DIAL and DISC lidar systems are compatible in the sense that they use much of the same hardware; they share the same lasers, telescopes, detectors, and computers. The primary differences between them are in the wavelength being transmitted, and in the software involved in the data analysis.

Ground mobile and airborne lidar systems are already used extensively in field operations. In the future, lidar systems mounted on satellites may routinely monitor the global atmosphere.

For background information *see* AIR-POLLUTION CONTROL; FLUORESCENCE; LASER; RADAR; SPECTROSCOPY in the McGraw-Hill Encyclopedia of Science and Technology.

[EDWARD R. MURRAY]

Bibliography: R. A. Baumgartner, L. D. Fletcher, and J. G. Hawley, *J. Air Pollut. Contr. Ass.*, 29: 1162–1165, November 1979; R. L. Byer and E. Murray, Remote monitoring with laser sources, pp. 406–449 in R. Perry (ed.), *Air Pollution Analysis*, 1977; W. B. Grant and R. D. Hake, *J. Appl. Phys.*, 46:3019–3023, July 1975; M. S. Shumate and R. T. Menzies, *Proceedings of the 4th Joint Conference on Sensing of Environmental Pollutants*, pp. 420–422, American Chemical Society, 1978.

Renin

The renin-angiotensin system is involved in body salt and water homeostasis and in the regulation of systemic arterial blood pressure. Moreover, the renin-angiotensin system has been implicated in the development and maintenance of several forms of experimental hypertension. There have

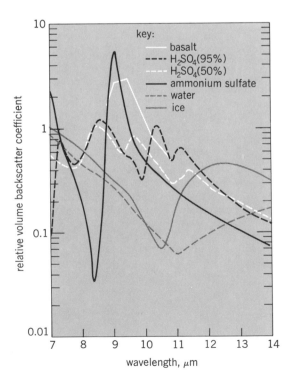

Fig. 3. Calculated relative backscatter for several atmospheric particles of different composition.

been major new developments in understanding the role of renal sympathetic nerves in controlling renin secretion. This article summarizes the present understanding of the role of sympathetic nerves in the regulation of renin secretion in light of the significant new developments.

Renin-angiotensin system. Renin is a proteolytic enzyme (its molecular weight is approximately 45,000) secreted into the systemic circulation by cells in the preglomerular afferent arterioles. Renin cleaves its substrate, angiotensinogen (an alpha-2 globulin), to release the decapeptide angiotensin I. Angiotensin-converting enzyme, a dipeptidyl-carboxypeptidase located primarily in pulmonary vessels, cleaves the carboxyl terminal dipeptide from angiotensin I to produce the octapeptide angiotensin II. Angiotensin II stimulates vascular smooth muscle to contract, and is thus a potent vasoconstrictor. It also stimulates the synthesis and release of aldosterone from the adrenal cortex. Since renin release is rate-limiting for angiotensin II production, this enzyme is ultimately involved in the control of systemic arterial pressure via vasoconstrictor effects on blood vessels, and via an effect on the renal handling of salt and water through an influence on aldosterone secretion.

Basic mechanisms of secretion. Three mechanisms control the release of renin from the kidney. The first is the vascular baroreceptor which increases renin secretion in response to a fall in renal arterial pressure. The second is the tubular macula densa mechanism which increases renin secretion in response to a decrease in distal tubular sodium chloride delivery or concentration. The third mechanism, which will be discussed in detail, is the influence of the renal sympathetic nerves. Increases in renal nerve activity increase the release of renin.

Direct neural release. Stimulation of renal nerves results in release of renin, the magnitude of which is dependent on the intensity of the stimulation. The adrenergic mechanisms which are involved in mediating these changes in renin secretion are uncertain, since either alpha or beta adrenergic receptor blockade prevents neurally mediated release of renin. This is an area of controversy in need of additional study. Interpretation of many experiments in which renal nerves have been stimulated also is clouded by the fact that the nerves were stimulated at frequencies (2–10 Hz) which altered renal vascular resistance and the distal tubular delivery of sodium chloride. Thus stimulation of renal nerves may have had direct effects on renin secretion which were independent of their effects on the tubular macula densa and vascular baroreceptor mechanisms, but these independent neural effects are difficult to separate from indirect effects mediated through nonneural mechanisms.

Reflex activation of renal nerves. There are sensory endings in the heart and great vessels which respond to changes in arterial pressure (sinoaortic arterial baroreceptors) and blood volume (cardiopulmonary receptors subserved by afferent vagal fibers). These groups of sensory endings have been shown to exert a tonic inhibitory influence on the vasomotor centers of the central

nervous system, and thus on activity which travels in renal sympathetic nerves. If these sensory receptors which respond to changes in arterial pressure and in circulating blood volume provide important feedback in the control of these variables, it would seem logical that the input from these sensory endings to the vasomotor centers should also influence the secretion of renin by the kidney. Data now support this view. When renal arterial pressure is held constant, carotid sinus hypotension (that is, reduced vasomotor center inhibition from carotid baroreceptors) results in a large increase in renin secretion. When renal perfusion pressure is allowed to increase during carotid hypotension, no increase in renin secretion is observed, presumably due to an inhibitory influence of the intrarenal vascular baroreceptor. Interruption of afferent traffic passing in vagal fibers subserving receptors in the heart and lungs also results in an increase in the secretion of renin. In contrast to the carotid baroreflex influence on renin, this effect of cardiopulmonary deafferentation is observed even when renal perfusion pressure is allowed to rise. This difference is difficult to explain on the basis of existing data, but may be the result of fundamental differences in the populations of sympathetic nerves influenced by carotid as opposed to cardiopulmonary baroreceptors. Moreover, each of these reflex influences on renin secretion (carotid or cardiopulmonary) is most apparent when the influence of the other receptor group has been eliminated or is held constant; that is, renin secretion results from vagal interruption mainly when the carotid baroreceptor input is held constant or is eliminated, and vice versa. This interaction between cardiopulmonary receptors with vagal afferents and arterial baroreceptors in the control of renin secretion probably accounts for the controversy which has existed regarding the influence of these receptor groups in the reflex control of renin secretion.

Recent experiments have demonstrated that cardiopulmonary receptors with vagal afferents may have a particularly important influence on the reflex control of renin secretion. It has been shown that during hemorrhage of 4 ml/kg, arterial pressure remained constant but renin secretion increased (Fig. 1). Under these circumstances, the tonic inhibitory influence from cardiopulmonary receptors was reduced, while the influence of arterial baroreceptors and of the intrarenal vascular baroreceptor was unchanged. Following section of the vagal nerves (and interruption of vagal afferents), similar hemorrhage failed to increase renin secretion. These data show that a modest decrease in blood volume which is not sufficient to alter the input from the arterial baroreceptors can acutely increase renin release by reducing the tonic inhibitory influence of cardiopulmonary receptors on renal nerve activity.

Renal nerves and nonneural mechanisms. Each of the three mechanisms for renin secretion (vascular receptor, macula densa receptor, and renal nervous) has been investigated repeatedly in a manner which allowed it to be studied in the absence of, or with minimal influence from, the other two mechanisms (that is, each mechanism has been examined in a relatively isolated fashion).

Recent evidence indicates that there are important interactions between renal nerves and nonneural mechanisms in the control of renin secretion. Renin responses to constriction of the suprarenal aorta have been examined in dogs in which the renal nerves were sectioned. Responses were assessed before and during stimulation of the renal nerves at a very low frequency (0.25 Hz) which did not change basal renin secretion, renal blood flow, or urinary sodium excretion. Thus this level of renal nerve activity was without direct measurable effect on renal function or renin secretion.

However, the renin secretion response to aortic constriction during concomitant renal nerve stimulation was significantly augmented as compared to aortic constriction alone (Fig. 2). This low-level renal nerve stimulation was similarly shown to augment the renin secretion response to administration of furosemide, a diuretic. If one accepts the widely held view that aortic constriction and furosemide administration have their principal influence on renin secretion through the intrarenal vascular receptor and tabular macula densa receptor, these data are consistent with the view

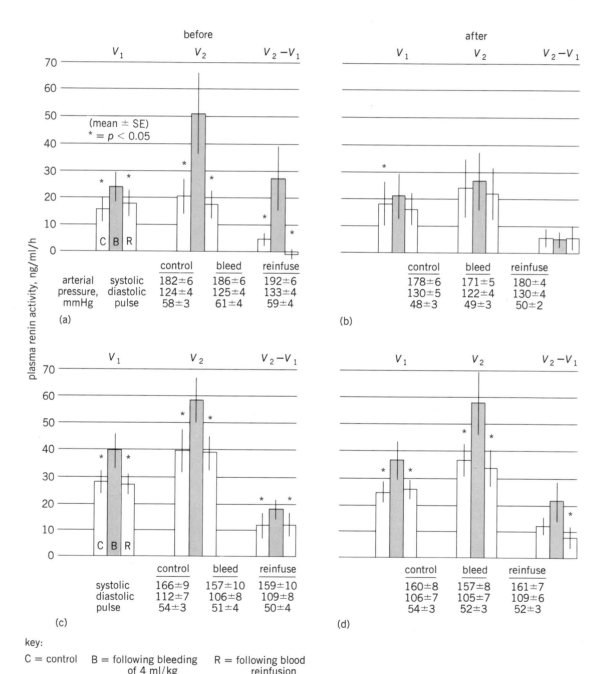

key:

C = control   B = following bleeding   R = following blood
                      of 4 ml/kg   reinfusion

Fig. 1. Plasma renin activity measured upstream (V_1) and downstream (V_2) of the renal veins. V_2 minus V_1 difference is an index of renin secretion. Responses are shown (a) before and (b) after vagotomy, and (c, d) after sham vagotomy. (From M. D. Thames, M. Jarecki, and D. E. Donald, Neural control of renin secretion in anesthetized dogs: Interaction of carotid and cardiopulmonary baroreceptors, Circ. Res., 42:237–245, 1978)

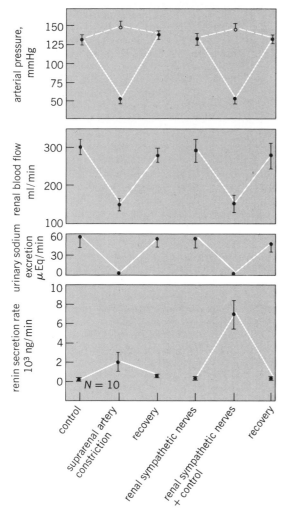

Fig. 2. Responses to constriction of the suprarenal aorta both before and during low-level stimulation (0.25 Hz) of the renal sympathetic nerves. Control and recovery values also are indicated. (*From M. D. Thames and G. F. DiBona, Renal nerves modulate the secretion of renin mediated by non-neural mechanisms, Circ. Res., 44:645–652, 1979*)

al pressure may not change, the resultant alterations in distal tubular sodium chloride may cause an increase in renin secretion mediated in part through the macula densa mechanism. The decrease in blood volume could simultaneously reduce the discharge of sensory endings in the cardiopulmonary area which are subserved by vagal afferents, thus reducing their inhibitory influence on the vasomotor center. This would tend to augment renal nerve activity, and thereby increase the renin response to sodium restriction. Similarly, changes in posture can alter cardiopulmonary blood volume, and may decrease the inhibitory influence of cardiopulmonary receptors on renal nerve activity and thus increase renin release. If the neural mechanisms influencing renin release are important, what happens to renin secretion from denervated kidneys during salt restriction? Studies in humans with renal transplantation have shown that it increases. However, the rate at which renin secretion increases is much less in denervated than in innervated kidneys. These observations indicate that renal nerves enable a much more rapid adaptation to the stress of salt restriction.

In conclusion, renal nerves play an important role in the overall control of renin secretion by causing direct neural release of renin and by modulating the release of renin mediated by nonneural mechanisms. The latter influence may well be the more important one. Cardiovascular reflexes which contribute to the control of renin secretion may have their most important influence by inducing small changes in renal nerve activity which modulate the secretion of renin mediated through nonneural mechanisms.

For background information *see* KIDNEY (VERTEBRATE); NEUROPHYSIOLOGY in the McGraw-Hill Encyclopedia of Science and Technology.

[MARC D. THAMES; JEFFREY L. OSBORN]

Bibliography: M. Jarecki, P. N. Thoren, and D. E. Donald, *Circ. Res.*, 42:614–619, 1978; G. Mancia, J. C. Romero, and J. T. Shepherd, *Circ. Res.*, 36:529–535, 1975; M. D. Thames, M. Jarecki, and D. E. Donald, *Circ. Res.*, 42:237–245, 1978; M. D. Thames and G. F. DiBona, *Circ. Res.*, 44:645–652, 1979.

Rheology

This science, which relates the deformation and flow of matter to the associated forces, has developed considerably in recent years. Formerly, each kind of material was studied by itself, so that there was little interaction between subjects such as biorheology (of biological products, such as blood), inks, lubricants, or polymer properties. Today an integrated treatment of the rheology of all materials is available, in which understanding is deepened by means of microscopic or molecular descriptions of how the observed phenomena come about.

The precise goal of rheology is to connect two quantities: stress, or force divided by area; and strain or deformation. These quantities have three important aspects. (1) They have directional properties, which distinguish an elongation from a shear, for example. Although a mathematical

that renin secretion mediated by these nonneural mechanisms is modulated by the renal nerves. Whether this modulation is occurring at one or both of these two nonneural receptors is uncertain. In addition, the role of alpha or beta adrenergic mechanisms and of intrarenal hormones such as prostaglandins in mediating this modulating influence remains unknown.

Physiological role of renal nerves. Renal nerve activity probably varies from minute to minute throughout the day. At times, increases in renal nerve activity sufficient to cause a direct neural release of renin may occur. However, under normal circumstances, changes in renal nerve activity may be very modest but still sufficient to modulate renin responses mediated by the nonneural mechanisms. For example, during modest sodium deprivation there is contraction of total extracellular, and thus of intravascular, volume. Although arteri-

description then requires a 3×3 matrix or a second-rank tensor, a single number is sufficient for simple geometries. (2) The stress and strain may depend on time. Here, too, simple situations occur: the unchanging equilibrium state; or a steady flow in which strain (typically a shear) increases at a constant rate (the shear rate, for example). (3) The magnitude or size is of importance: if doubling the strain leads to a doubling

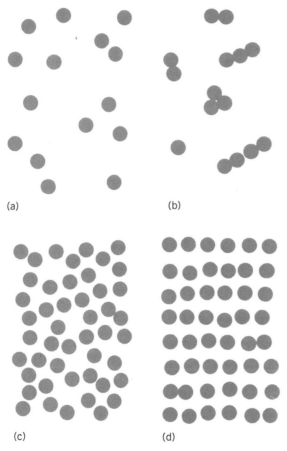

(a)   (b)

(c)   (d)

Fig. 1. Particle arrangements. (a) Dilute, random. (b) Dilute, organized into chains or aggregates. (c) Concentrated, random. (d) Concentrated, organized into layers. *(From C. E. Chaffey, Mechanisms and equations for shear thinning and thickening in dispersions, Colloid Polymer Sci., 255:691–698, 1977)*

of the stress—the geometry and time patterns being identical—then the behavior is called linear, because a graph of stress against strain is a straight line. Linear behavior is typical of small strains or strain rates, but for the large strains and rapid strain rates now common in industrial practice, nonlinear phenomena are the rule. Lately there has been important progress in the field of nonlinear rheology.

Viscoelasticity. The two limiting ideal materials are the solid and the fluid. A solid can hold a definite shape; ideally it shows elasticity and obeys Hooke's law, and the more general kinds of solid behavior are called anelasticity. Conversely, a fluid takes on the shape of its container, the ideal-

ization being the Newtonian fluid, which is characterized by its viscosity. Materials intermediate between perfectly solid and perfectly fluid are called viscoelastic. The white of an egg is an example; it can be poured from a jug, but if the falling stream is cut off below the spout, then the upstream part tends to spring back, showing some elasticity. In a purely viscous fluid, the stress disappears the moment the straining stops, whereas an elastic solid retains a constant stress as long as it is held in a strained shape. Again, a viscoelastic material shows a stress that gradually relaxes when the strain is kept constant. The time required for the stress to relax to a definite fraction ($1/e$ or 37%) of its initial value is the relaxation time.

How rheological behavior actually appears depends on the time scale of the process in which it is observed. The ratio of a material's relaxation time to the time over which behavior is observed is called the Deborah number. As the Deborah number becomes smaller, the behavior changes from solid to fluid. Asphalt acts as a brittle solid in operations that take seconds or minutes, but it flows like any fluid when months or years elapse.

Molecular effects. In the past few years, research into molecular or microscopic phenomena has explained how many kinds of viscoelastic or non-Newtonian behavior arise. Suspensions and dispersions of solid particles in a fluid have received particular attention; in addition to the factors previously discussed, one must also consider the effect of concentration, size, shape, and arrangement of the dispersed particles. Some idealized particle arrangements are shown in Fig. 1.

In general, the mechanical work needed to make a material deform or flow is used in three ways: first, it is stored reversibly, so that the material can give back the energy as mechanical work; second, it is transformed to chemical energy, in bonds or weak links between particles; and third, it is dissipated into heat and lost.

Among all the deformations and flows investigated by rheologists, one in particular has predominant importance: steady shear flow. Here the geometry is shearing, the dependence on time is a steady increase in the shear, and the magnitude is arbitrary. The three variables viscosity (defined as the ratio of stress to shear rate), shear rate, and stress can all be shown on a logarithmic plot such as Fig. 2. On it, a Newtonian fluid gives a straight line, curve 1. The diagonal lines in Fig. 2 are lines of constant stress. By straightening out the parallelograms bounded by them into squares, one can transform this graph into either a plot of stress against shear rate, or a plot of viscosity against stress.

Shear thinning. The most common non-Newtonian behavior is shear thinning (or pseudoplasticity), curve 2 in Fig. 2. It is characteristic of viscoelastic materials such as polymers. In these, elasticity arises from the tendency of the polymer segments to take on a random equilibrium arrangement due to their thermal motion. A viscous contribution results from friction as one segment slides past another. This friction also tends to drag the molecule out of its equilibrium shape. At low shear rates, however, the viscous stress is too small to do

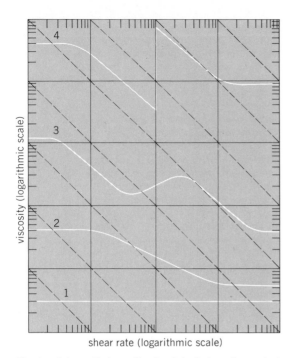

Fig. 2. Schematic logarithmic plot of viscosity against shear rate. Curve 1 is Newtonian; curve 2 is typical of polymers or dilute dispersions; increasing the concentration of a dispersion alters the curve toward 3 or even 4. (*From J. Mewis and A. J. B. Spaull. Rheology of concentrated dispersions. Advan. Colloid Interface Sci.. 6: 173–200. 1976*)

viscosity (logarithmic scale)

shear rate (logarithmic scale)

this, so that the shape of the molecule and also the system's viscosity do not change; thus the left-hand end of curve 2 in Fig. 2 is horizontal. Only at the larger shear rates can the viscous stress deform the molecule into a shape past which flow occurs more readily, lowering the viscosity. The middle part of curve 2, with a negative slope, results. Finally, at a high shear rate, the molecule cannot be deformed any more. The shear rate can then be increased indefinitely without a further drop in viscosity, as represented by the right-hand horizontal portion of curve 2. Often, though, a series of measurements will not extend over a wide enough range of shear rates to reveal a complete curve such as 2, only a portion of it actually being observed.

The equilibrium configuration of the molecule is restored when shearing stops, so that the deformation is a means of storing mechanical work recoverably. Other systems, which have different ways of storing work on a microscopic scale, likewise are shear-thinning. Energy can be stored, for example, by like electrical charges being forced near each other in colloidal systems, or by emulsion drops being distorted from the spherical shape that surface tension gives them at rest, or by stabilizing surfactant layers on dispersed particles being pressed into each other.

Chemical transformation of the mechanical work put into a system is typically associated with changes in the bonding or linking between particles. For example, energy could be used to break links initially holding particles in chains or aggregates (Fig. 1*b*), leading eventually to all the parti-

cles being unbonded (Fig. 1*a*). Since the aggregation of particles causes a greater disturbance to the flow, such a system would be shear-thinning. It would also be thixotropic (viscosity decreasing with the passage of time) when steady flow was started.

Shear thickening. The opposite process, in which link formation is promoted by shearing, the particles sticking together when they collide, is less likely; it would lead to shear thickening (dilatancy) when steady-state viscosity was plotted against shear rate, and an increase in viscosity with time (rheopexy) at the onset of flow. Actually these rare phenomena occur only in highly concentrated dispersions of particles, and another explanation is more probable.

When a dispersion is concentrated, randomly arranged particles interfere with each other greatly if flow occurs (Fig. 1*c*). Rearrangement into layers (Fig. 1*d*) allows freer flow. If the interactions between particles promote layering, and if a certain time is needed to accomplish this, rearrangement will be more perfect in the slower flows at low shear rates. The viscosity should then increase with shear rate. Outside such a shear-thickening region, the usual mechanisms for shear thinning apply. The flow behavior of these systems is actually observed to follow curve 3 in Fig. 2, though again only a part of the curve has been seen for any one such material.

Dispersions. It is now evident that one grossly oversimplifies the complex behavior of a dispersion by stating a single viscosity value, or by trying to interpret viscosity in terms of concentration and particle size alone. Indeed, measurements of the dependence of viscosity on concentration (Fig. 3) show widely divergent values, even for spherical particles. Clearly, the shear rate and chemical properties of the system need to be controlled for reproducible results. Only at the lowest concentrations, a few percent by volume, does the viscosity of the dispersion increase linearly with concentration. Further increases in concentration lead to progressively larger viscosities, so that when the volume fraction of particles is 40–50%, the dispersion has a viscosity roughly 10 times that of the suspending medium or solvent. At a volume fraction a little over 60% for uniformly sized spherical particles, the limit of random close packing is reached. The particles are now in contact through the whole system, and flow is impossible. As the concentration approaches this limit, the viscosity increases very steeply.

However, if the particles are unequal in size, very small ones can fit into gaps between larger ones, allowing flow at higher concentrations than for equal particles. Conversely, at the same concentration, a dispersion with a wide range of particle sizes has a lower viscosity than one with uniform particles. In highly uniform dispersions, arrangement into layers (Fig. 1*d*) can be quite perfect; any disruption then leads to a catastrophic change in organization. What has been observed is an abrupt shear thickening when a critical shear rate is exceeded, curve 4 in Fig. 2. Such anomalies are eliminated by broadening the distribution of particle sizes.

Particle shape also has an effect, needlelike par-

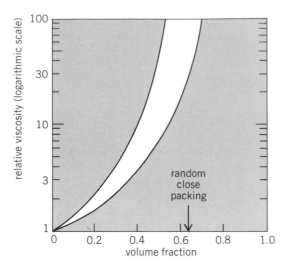

Fig. 3. Logarithm of relative viscosity (ratio of viscosity of dispersion to that of pure suspending fluid) plotted against concentration, expressed as a volume fraction (ratio of volume of all particles to the total volume of the dispersion). Data for spheres fall between the curves, the spread being due to differences in distributions of particle sizes and to chemical effects. *(From D. J. Jeffrey and A. Acrivos, The rheological properties of suspensions of rigid particles, Amer. Inst. Chem. Eng. J., 22:417–432. 1976)*

ticles raising the viscosity more than an equal volume of spheres. On the other hand, the distinction between spheres and cubes has no significant rheological importance.

The influence of the size of the particles is secondary to that of their concentration. In simple theories of the flow of dispersions, in which particles act only as a solid boundary to the region occupied by fluid, no dependence of rheological properties on particle size is predicted at all. For this simplification to be applicable, the particles must be relatively large, greater than approximately 0.1 mm in diameter. To account for the observed dependence of viscosity on particle size, one must consider active interactions between the particles by forces typically of electrostatic or colloid-chemical origin. These forces become relatively more important as size is reduced and surface area for fixed volume increases, so that viscosity usually increases as the particles are made smaller.

Forces between particles can be much altered by small amounts of suitable chemical additives. Recently, several processing aids have appeared on the market that effect a considerable reduction in the viscosity (and thus the energy used up in processing) of materials such as reinforced plastics melts and ore slurries. The detailed reason why these products are effective is an active area of current research, with a view to still further improvement. On the other hand, other additives raise the viscosity or impart thixotropy to fluid systems; they find application in coatings, inks, lubricants, cosmetics, and other products.

This article has emphasized advances in understanding of the rheology of polymers and of dispersions. The recent progress in instrumentation, elongational flow, and nonlinear viscoelastic phe-

nomena such as stress overshoot have not been discussed.

For background information *see* ANELASTICITY; BIORHEOLOGY; FLUID, NON-NEWTONIAN; FLUID FLOW, NON-NEWTONIAN; FLUIDS; INK; LUBRICANT; POLYMER PROPERTIES; RHEOLOGY; ROCK MECHANICS; VISCOSITY OF LIQUIDS in the McGraw-Hill Encyclopedia of Science and Technology.   [CHARLES E. CHAFFEY]

Bibliography: R. B. Bird, R. C. Armstrong, and O. Hassager, *Dynamics of Polymeric Liquids*, vol. 1, *Fluid Mechanics*, 1977; J. D. Ferry, *Viscoelastic Properties of Polymers*, 3d ed., 1980; R. S. Lenk, *Polymer Rheology*, 1978; W. R. Schowalter, *Mechanics of Non-Newtonian Fluids*, 1978.

Root (botany)

Plant roots can contribute significantly to the stability of steep slopes. They can anchor through the soil mass into fractures in bedrock, can cross zones of weakness to more stable soil, and can provide interlocking long fibrous binders within a weak soil mass. In deep soil, anchoring to bedrock becomes negligible, and lateral reinforcement predominates. After trees are removed, the root system begins to decay, and the soil-root fabric progressively weakens. If a forested slope is marginally stable, landslides may increase after trees are removed. As the deforested slope revegetates, however, the soil mantle becomes progressively reinforced as new roots occupy the soil.

Soil shear strength. The strength of forest soil is difficult to measure directly. Evaluating the effect of roots on soil strength increases that difficulty. In 1968 a shear box was developed to measure the contribution of small alder *(Alnus glutinosa)* roots to the strength of relatively homogeneous nursery soil in Japan. The weight of roots explained 53% of the variation in measured soil strength. The shear box was later modified to study the contribution to soil strength by roots of a mixed old-growth forest of Douglas-fir *(Pseudotsuga menziesii)*, western redcedar *(Thuja plicata)*, and western hemlock *(Tsuga heterophylla)* growing on glacial till subsoils in British Columbia, Canada. The weight of roots in the soil sample was the most significant of seven variables tested, accounting for 56% of the variation in measured soil strength. In both studies, many of the roots larger than 5 mm in diameter pulled out of the soil block rather than failed along the test shear plane.

To reduce the number of roots which pull from the soil block, the shear box was redesigned and tested in the relatively simple soil-root system of a mature shore pine *(Pinus contorta)* forest growing on coastal sands in northern California. The dry weight of the live roots less than 17 mm in diameter was the only significant variable contributing to soil shear strength among 32 soil and vegetative variables tested. The shear box tests resulted in Eq. (1), in which soil strength is in kilopascals and

$$\text{Soil strength} = 3.13 + 3.31 \text{ root biomass} \quad (1)$$

root biomass is in kilograms per cubic meter. The equation explained 79% of the variation in measured soil strength. The mean biomass of the less than 17-mm-diameter live roots was 1.77 kg/m³, which represented 64% of the total root biomass. Adding more variables did not significantly improve the regression equation.

Strength of individual roots. Roots become stronger as they become larger; the logarithm of root shear strength is closely related to the logarithm of the diameter of the root. The strength of roots also varies between species. In Canada small Douglas-fir roots are about 10% stronger than western redcedar roots. In the Soviet Union poplar (*Populus deltoides*) roots are the strongest, followed by birch (*Betula pendula*), oak (*Quercus robur*), linden (*Tilia cordata*), and spruce (*Picea abies*). Poplar roots are about 40% stronger than spruce roots. Tree roots are estimated to be 1.5–3 times stronger than the roots of grassy plants of the same diameter. In northern California the roots of brush, such as *Ceanothus velutinus* and *Sambucus callicarpa*, are about twice as strong as the roots of conifer trees, such as *Abies concolor*, *Pinus lambertiana*, and *P. ponderosa*.

Roots and soil strength. Roots increase the strength of soils. The forces involved in the failure of a section of bank along the Moscow River in the Soviet Union were evaluated. The size and number of roots protruding from the wall of the collapsed soil block were measured. The tensile strength of linden roots was determined in the laboratory. The total force required to break the soil mass reinforced by linden roots was calculated to be about 137 metric tons, of which 130 tons were required to break the roots and 7 tons to tear the sandy loam soil mass from the bank. Breaking the linden roots took 95% of the total force, although the total cross-sectional area of all the roots constituted less than half a percent of the wall area of the collapse.

It has been calculated that the root network accounted for 71% of the shear strength at saturation of glacial till soils on 35° slopes in British Columbia. It was observed in Sweden that an imposed load may be 70% greater before soil rupture in soils with a root network than in soils without roots.

Slope stability problems will likely develop as the tree root system decays after timber cutting on steep slopes where the predominant strength is contributed by the binding action of the roots. As the root system decays following deforestation, the relative reinforcement by the roots will decline (Fig. 1). Within 2 years after deforestation, about 50% of the original root reinforcement is lost and 90% is gone within 9 years.

The rate of strength loss varies according to species, root size, and the activity of decay organisms. Small roots decay most rapidly, while large decay-resistant roots may remain in the soil for decades. For example, intact roots have been found greater than 15 cm in diameter from western redcedar trees which had been cut 50 years earlier. However, redcedar roots 1 cm in diameter had lost about 50% of their tensile strength within 5 years of cutting. Douglas-fir roots decay more rapidly than redcedar roots, and the rate of decay is related to geographic location. The strength of 1-cm-diameter Douglas-fir roots decreased by about 50% within 3 years after cutting in coastal British Columbia. It was found that 50% of the Douglas-fir

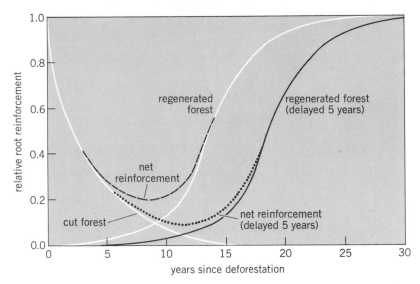

Fig. 1. Conceptual model of changes of relative root reinforcement with time after deforestation for residual roots from the cut forest and for new roots from the regenerated forest. Net reinforcement is the sum of reinforcement by the cut forest and the regenerated forest.

roots 1 cm in diameter decayed within 1½ years in the Rocky Mountains, and the same proportion was gone within 1 year in coastal Oregon. About 90% of the Rocky Mountain roots decayed in 12 years, whereas 90% of the Oregon roots were gone in less than 5 years.

As vegetation reoccupies the deforested area, new roots begin to progressively reinforce the soil. For example, in Fig. 1, about 14 years are required until the new forest provides 50% of the root reinforcement supplied by the original forest before cutting, and 23 years until the soil in the deforested area returns to the strength of that in the uncut forest. The actual rate of soil strength recovery can vary, and depends on many more environmental variables than does the rate of strength loss through decay. In severe sites the recovery of root reinforcement can be lengthy. In logged mixed conifer forests in northwestern California, calculated root reinforcement in areas logged 25 years earlier was only about 40% of that in adjacent uncut areas.

The net reinforcement of the soil by roots is the sum of the reinforcement by residual decaying roots of the cut trees and the reinforcement by new roots of the regenerating forest. In Fig. 1 the net reinforcement in a promptly regenerated forest reaches a minimum about 9 years after deforestation. Net reinforcement then is about 20% of that in the uncut forest. It becomes greater after 9 years, as the roots of the new forest continue to develop in the cut areas.

If regeneration is delayed by 5 years, decay of the residual root system of the cut forest will continue for 5 years before the new root system begins to add strength. The net soil reinforcement will then reach a minimum which is substantially lower than in areas where regeneration is prompt. The minimum net reinforcement with a 5-year delay in revegetation occurs 12 years after logging, and is only about 7% of that in the uncut forest. The

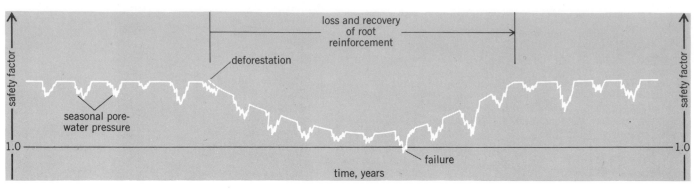

Fig. 2. Conceptual representation of the interaction between seasonal pore water pressure and the loss and recovery of root reinforcement and the effect on safety factor. Safety factor is an index of relative stability of slopes and is the ratio of available shear strength to shear stress.

Stability of slopes. Roots help stabilize steep slopes. Engineering stability analyses have been applied to slopes with and without roots. A safety factor—defined as the ratio of the available shear strength to the shear stress—provides an index of the relative stability of slopes. A slope with a safety factor of less than 1.0 cannot remain stable and must fail.

Strength of a soil or resistance to failure can be described as a modification to Coulomb's law, Eq. (2).

$$s = (c + r) + \sigma \tan \phi \qquad (2)$$

where
s = soil shear strength
c = effective soil cohesion
r = apparent cohesion due to roots
σ = effective normal stress
ϕ = effective internal friction angle

On a given slope not subjected to excavation, the two principal effects of deforestation are modifications to the root strength parameter r and the effective normal stress σ through changes in the saturated soil water regime or pore water pressures.

Consider a case where all other factors are held constant and the safety factor equals 1.0 when the relative root reinforcement is 0.15 (Fig. 1). Then, if prompt regeneration follows deforestation, the net reinforcement always remains above 0.20, and the slope would not fail. However, if regeneration is delayed 5 years, the net reinforcement would fall below 0.15 from 8 years until 16 years after deforestation, and the slope would fail 8 years after cutting.

All factors, however, do not remain constant. Pore water pressures change seasonally in response to precipitation, and are often the driving mechanism which ultimately leads to slope failure (Fig. 2). In this example, the slope would not fail because of either seasonal pore water pressures or loss of root strength alone. When both factors are considered together, the loss of root strength lowers the safety factor to a level that a moderate change in pore water pressure would result in slope failure.

For background information *see* FOREST SOIL; ROOT (BOTANY) in the McGraw-Hill Encyclopedia of Science and Technology.   [ROBERT R. ZIEMER]

Bibliography: E. R. Burroughs and B. R. Thomas, *USDA For. Serv. Res. Pap. INT-190*, 1977; C. L. O'Loughlin, *Can. J. For. Res.*, 4:107–113, 1974; T. H. Wu, W. P. McKinnell III, and D. N. Swanston, *Can. Geotech. J.*, 16:19–33, 1979; R. R. Ziemer, *Int. Ass. Hydrol. Sci. Publ. 132*, 1980.

Salamander

Terrestrial salamanders are subjected to heavy predation by snakes, birds, and mammals. This pressure has led to the evolution of a wide variety of antipredator mechanisms in salamanders, including tail autotomy, noxious and toxic skin secretions, biting, behavioral posturing, ribs protruding through the skin, and bright color patterns. Recent research has concentrated on demonstrating the effectiveness and interrelationships of these adaptations in repelling predators. The predators used in the most recent studies have been birds and shrews which are natural predators on terrestrial salamanders.

Skin secretions. The skin glands of amphibians produce secretions which are released onto the skin surface. The secretions of the mucous glands primarily function to keep the skin moist and reduce desiccation. The granular glands (also called poison glands) produce secretions which primarily serve a protective function by being noxious or toxic.

The skin secretions of some salamander species are effective in repulsing would-be predators after the salamanders are attacked. These skin secretions may be repulsive to a wide variety of predators, including insects, reptiles, birds, and mammals. There is a wide range of palatability of salamanders, ranging from the completely palatable, such as *Desmognathus ochrophaeus*, to nearly completely unpalatable, such as *Notophthalmus viridescens*. Most salamanders are somewhat unpalatable to avian and mammalian predators but can be eaten in times when food is scarce. The noxiousness could be the result of irritation of the mouth and eyes, the glutinous nature of the secretion, or illness caused by a toxin. Several toxins, including the very toxic tetrodotoxin, have been isolated from salamander skin secretions, and the presence of other unidentified toxins is indicated by recent research. At least one species of salamander, *Taricha granulosa*, is known to be toxic enough to kill humans. Despite the toxicity of this and other species of salamanders, none of them is protected from all predators. *Taricha* is eaten by the garter snake *Thamnophis sirtalis*, which is highly resistant to the salamander's toxin. Noxious or toxic skin secretions are probably the most important of all salamander antipredator adaptations because they are necessary for the evolution of most of the other specializations.

Biting. Members of the genus *Desmognathus* that have been studied lack noxious skin secretions but have been observed to repel predators by biting. The largest member of the genus, *D. quadramaculatus*, postures the body with the mouth open, exhibiting the white mouth lining when approached by the shrew *Blarina brevicauda*. As the predator comes closer, the salamander lunges toward the shrew, snapping the mouth closed with a sound clearly audible 1 meter away. If the shrew approaches closely enough or bites the salamander, the response of the salamander is to bite the shrew. The salamander normally retains its grasp on the shrew (Fig. 1); this is often effective in repelling shrews. It is unlikely that many salamanders have the dentition or jaw musculature to repel would-be predators, but being bitten by a salamander is a startling experience for a collector and presumably, for a predator. Dropping a salamander for an instant allows it a chance to escape.

Postures. Antipredator postures include any positioning of the body of a salamander which might enhance the chance of surviving contact with a predator. The aggressive threat display considered above, immobility, and active displays are all antipredator postures.

Immobility refers to a rigid maintenance of position after contact with a predator. Immobility of the prey might reduce the intensity of attack. Immobile salamanders are ignored by avian predators, while moving salamanders are attacked. Flash behavior, flipping, or protean behavior often precedes immobility in salamanders. This sudden change of shape, position, and location startles, confuses, and misdirects the predator.

While immobility might convey a selective advantage upon a prey, it is always associated with some other lines of defense in salamanders. Presumably immobility in salamanders is a form of camouflage, and as such, any alteration in shape of the salamander, such as coiling or contorting the body, will project stimuli which may not fit the search image of the predator. Particularly important may be the positioning of the limbs during immobility. If the limbs are clasped along the body, the salamander resembles a twig or straw.

Immobility is further advantageous in those salamanders which have noxious skin secretions since the predator contact with the salamander will be of less intensity, and after the predator has experienced the noxiousness, avoidance will follow. Immobility thus reduces the potential of injury to an inedible salamander.

Immobility is important in most antipredator displays and in salamanders seems to be an evolutionary precursor. The display postures of salamanders are associated with the presence and distribution of noxious skin secretions and seem to be of two basic types: rigid posturing to display ven-

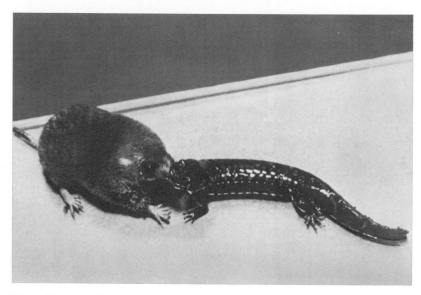

Fig. 1. *Desmognathus quadramaculatus* biting the nose of the shrew *Blarina brevicauda*. (*From E. D. Brodie, Jr., Biting and vocalization as antipredator mechanisms in terrestrial salamanders, Copeia, no. 1, pp. 127–129, 1978*)

tral aposematic coloration (Fig. 2); and orienting the body so as to position concentrations of glands toward the predator (Fig. 3).

Rigid displays exposing bright ventral coloration may be considered to be a type of immobility but are not simply immobility. Whereas immobility is usually a cryptic display, a rigid display of bright colors has the opposite function, that is, to warn visual predators. This type of posture is exhibited only by those members of the family Salamandridae with bright ventral pigmentation and relatively evenly distributed skin glands. These salamanders, which are essentially inedible to predators because of skin toxins, respond to predator attack by elevating their tail and often the forebody, thereby exposing the bright ventral surface.

Salamander species which have the granular

Fig. 2. Rigid posture of *Taricha rivularis* exposing the reddish underside of the chin and tail. (*From E. D. Brodie, Jr., Salamander antipredator postures, Copeia, no. 3, pp. 523–535, 1977*)

glands concentrated in patches exhibit postures which position these glands toward the attacking predator. The most common areas of glandular concentrations are along the dorsal ridge of the tail and in a pair of patches on the back of the head, known as the paratoid glands. The European salamanders of the genus *Salamandra* and several species of the North American genus *Ambystoma* have well-developed paratoid glands and have similar antipredator postures. When touched on the head or forelimb, the salamander stands high on the forelegs and flexes the head downward, which brings the paratoid glands into direct proximity of the stimulus. The head is turned, and the body is leaned toward the stimulus.

Most salamanders with noxious skin secretions have them concentrated along the dorsal ridge of the tail. This is true for all members of the family Ambystomatidae and most members of the family Plethodontidae. All of these salamanders use the tail to some extent in antipredator displays. Perhaps the simplest of active antipredator displays are those in which salamanders respond to predators by elevating the rear portion of the body on the hindlimbs, arching the tail, and lashing the tail forcibly toward the stimulus. After stimulation a salamander often remains in this posture and wags the tail from side to side; if the salamander is touched anywhere on the head, limbs, or body, the tail is lashed again.

The displays of some salamanders are more elaborate. Some species hold the tail vertically and undulate it slowly from side to side. The tail is lashed into the predator from this position. Other species coil the body with the head under the hindquarters while continuing to elevate and undulate the tail.

These tail displays cause the attacks of predators to be directed to the tail, which is the most dispensable part of the salamander. In salamanders the tail is often autotomous (easily broken off),

Fig. 3. Antipredator posture of *Ambystoma laterale* with the tail elevated and undulated. (*From E. D. Brodie. Jr., Salamander antipredator postures. Copeia. no. 3, pp. 523–535. 1977*)

Fig. 4. Sharp rib tips penetrating through the skin of *Pleurodeles waltl* during its antipredator posture. (*From R. T. Nowak and E. D. Brodie, Jr., Rib penetration and associated antipredator adaptations in the salamander Pleurodeles waltl (Salamandridae), Copeia. no. 3. pp. 424–429, 1978*)

and in all forms which elevate and undulate the tail there is a concentration of granular glands along the tail dorsum. This ensures that the attacking predator will first encounter the most noxious part of the salamander.

The body of the salamander is sometimes arched or the hindquarters are elevated as the tail is elevated and undulated. This raises the tail even higher and increases the apparent size of the salamander.

A unique antipredator posture has evolved in the genus *Pleurodeles*. When contacted or grasped by a predator, this salamander arches the body, causing the sharp tips of the ribs (which are longer than in other salamanders) to penetrate the skin (Fig. 4). The toxic skin secretions present on the skin surface coat the tips of the ribs as they penetrate the skin. A predator attacking this salamander in the body region will contact the sharp ribs. Should the ribs pierce the mouth of a would-be predator, it would cause intense pain and repel the predator.

The many antipredator adaptations of salamanders interact in such a way as to produce a synergistic effect in which the total protection accruing to the salamander is greater than the sum of each of the adaptations alone. Salamanders are thus not protected by single devices but by complexes of antipredator adaptations.

For background information *see* ECOLOGICAL INTERACTIONS; POISON GLAND; PROTECTIVE COLORATION in the McGraw-Hill Encyclopedia of Science and Technology.   [EDMUND D. BRODIE, JR.]

Bibliography: E. D. Brodie, Jr., *Copeia*, no. 1, pp. 127–129, 1978; E. D. Brodie, Jr., *Copeia*, no. 3, pp. 523–535, 1977; E. D. Brodie, Jr., R. T. Nowak, and W. R. Harvey, *Copeia*, no. 2, pp. 270–274, 1979; R. T. Nowak and E. D. Brodie, Jr., *Copeia*, no. 3, pp. 424–429, 1978.

Saturn

The *Pioneer 11* spacecraft has given the first close-up view of the ringed planet Saturn and its system of moons. This space mission began with the launch of *Pioneer 11* from Cape Canaveral, FL, on April 5, 1973. The spacecraft first flew by the giant

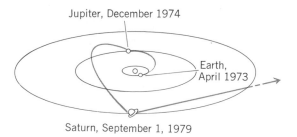

Fig. 1. Trajectory of *Pioneer 11*, showing position of Earth at launch in April 1973 and the position of Jupiter and Saturn during fly-bys of those planets in December 1974 and September 1979.

planet Jupiter on December 2, 1974, passing within 42,760 km of its cloud tops and taking the only existing pictures of its polar regions. Jupiter's massive gravitational field was used to swing *Pioneer 11* back across the solar system toward Saturn (Fig. 1). Additional maneuvers were executed in 1975 and 1976 to place the spacecraft on a suitable trajectory, with the final aiming point selected in 1977.

On September 1, 1979, the spacecraft, now designated *Pioneer Saturn*, reached Saturn after 6 years in flight. It passed through the ring plane outside the edge of Saturn's A ring and then swung in under the rings from 2000 to 10,000 km below them. At the point of closest approach, it attained a speed of 114,100 km/h (71,900 mph) and came within 21,400 km of the planet's cloud tops. While it was approaching, encountering, and leaving Saturn, the spacecraft took the first close-up pictures of the planet, showing 20–30 times more detail than the best pictures taken from Earth, and made the first close measurements of its rings and several of its satellites, including the largest satellite, the planet-sized Titan. Titan, along with Mars, has been considered by many scientists to be the most likely place to find life in the solar system.

Pioneer Saturn unraveled many mysteries. It determined that Saturn has a magnetic field and trapped-radiation belts, measured the mass of Saturn and some of its satellites, and studied the character of Saturn's interior. It confirmed the presence and determined the magnitude of an internal heat source for Saturn. Its instruments studied the temperature distribution, composition, and other properties of the clouds and atmospheres of Saturn and Titan, and took photometric and polarization measurements of Iapetus, Rhea, Dione, and Tethys. Pioneer may also have photographed a previously unknown satellite of Saturn (Fig. 2). The spacecraft measured the mass, structure, and other characteristics of Saturn's rings, and passed safely through the outer E ring, which posed a potential hazard for *Pioneer*. It also discovered two new rings. One of these rings, the F ring (Fig. 2), lies just outside the A ring, while the other, called the G ring, lies well outside the F ring.

Pioneer Saturn spacecraft. *Pioneer* carries a scientific payload of 11 operating instruments. Two other experiments, celestial mechanics and S-band occultation of Saturn, use the spacecraft radio to obtain data. *Pioneer Saturn* is a spinning spacecraft, which gives its instruments a full-circle

scan 7.8 times a minute. It uses a nuclear source for electric power because the sunlight at Jupiter and beyond is too weak for a solar-powered system.

Two booms project from the spacecraft to deploy the nuclear power source about 3 m from the sensitive spacecraft instrumentation. A third boom positions the magnetometer sensor, about 6 m from the spacecraft. Six thrusters provide velocity, attitude, and spin-rate control. A dish antenna is located along the spin axis and looks back at Earth throughout the mission, adjusting its view by changes in spacecraft attitude as the spacecraft and Earth move in their orbits around the Sun.

Tracking facilities of NASA's Deep Space Network, located at Goldstone, CA, and in Spain and Australia, supported *Pioneer Saturn* during interplanetary flight and encounter. *Pioneer*'s radio signals, traveling at the speed of light, took 85 min to reach Earth from Saturn, a round-trip time of almost 3 h, somewhat complicating ground control of

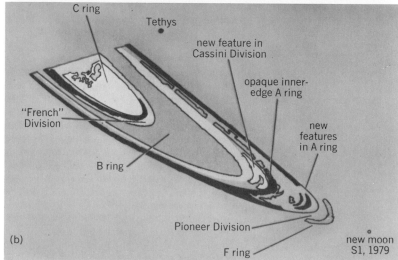

Fig. 2. Saturn's ring system shown in (*a*) photograph taken by *Pioneer 11* and (*b*) sketch identifying objects on photograph. Photograph was taken for a back-lighted configuration with the Sun at an angle of 2° below the ring plane while *Pioneer 11* was at an angle of 6° above the plane of Saturn's rings and 943,000 km (585,950 mi) from the planet. (*NASA, Ames Research Center, and the University of Arizona*)

the spacecraft. Almost 10,000 commands were sent to the spacecraft in the 2-week period before closest approach. Continued communications should be possible through at least the mid-1980s.

After the spacecraft passed Saturn, it headed out of the solar system, traveling in the direction the solar system moves with respect to the local stars in the Galaxy. *Pioneer Saturn* has a plaque attached to it which contains a message from Earth for any intelligent species that may intercept the spacecraft during its endless journey through interstellar space.

Structure of planet. *Pioneer Saturn* has already greatly expanded knowledge of Saturn and its rings and moons. It is now known that Saturn, in many ways, represents an intermediate case between Jupiter, the largest planet in the solar system, and Earth. The composition of Saturn's interior is essentially the same as Jupiter's, differing only in the size and extent of the various internal layers. Measurements for Saturn are consistent with a central core of molten heavy elements (probably mostly iron) which is the approximate size of the entire Earth, but about three times more massive. Surrounding the central core is an outer core of highly compressed, hot, liquefied volatiles such as methane, ammnonia, and water. This outer core is equivalent to approximately nine Earth masses. However, these core regions represent a very small fraction of the planet, which is composed primarily of the very lightest gases, hydrogen and helium, and is almost 100 times the mass of the Earth. Because of the high pressure in Saturn's interior, the hydrogen is transformed to its liquid metallic state. Above this metallic hydrogen shell are liquid molecular hydrogen and Saturn's gaseous atmosphere and clouds, which make up the rest of the planet.

Magnetic field. Electrical currents set up within the metallic hydrogen shell produce Saturn's magnetic field, which was measured by *Pioneer*. In spite of Saturn's large size, the magnetic field at the cloud tops is only slightly weaker than the field at the Earth's surface. Saturn is unique in that its magnetic axis is nearly aligned with its rotation axis, unlike Earth and Jupiter.

Saturn's magnetic field is also much more regular in shape than the fields of other planets. At great distances from Saturn, the magnetic field is deformed by the inward pressure of the solar wind. Near the noon meridian (close to the inbound *Pioneer* trajectory), the solar wind causes a compression of the field; in the dawn meridian (close to the outbound trajectory), the field is swept back and presumably forms a long magnetic tail. In both cases, *Pioneer Saturn* crossed the outer boundary of the magnetic field several times as the field moved in and out, responding to changing solar wind pressure.

Radiation belts. The magnetic envelope surrounding Saturn is intermediate in size and energetic-particle population between those of the Earth and Jupiter, the only two other planets known to be strongly magnetized. The three other planets investigated thus far (Venus, Mars, and Mercury) and Earth's Moon have little or no magnetism. The spacecraft found rings of particulate material and evidence for several small satellites near the rings, which strongly affect Saturn's trapped radiation. These features provide important diagnostic capabilities. A unique finding is the nearly total absence of radiation-belt particles at distances closer to the planet than the outer edge of the visible rings.

The inner region of the thick magnetic envelope of Saturn, called the magnetosphere, contains trapped high-energy electrons and protrons, with some evidence for heavier nuclei. A new discovery was made of a diffuse ring of particulate matter in the region from about 10 to 15 planetary radii (1 Saturn radius = 60,000 km) from Saturn. This ring has been tentatively designated as the G ring. The G ring clearly causes particle absorption near the equatorial region.

Inside about 10 planetary radii, the trapped radiation shows a high degree of axial symmetry around Saturn and is consistent with a centered dipole magnetic field. Saturn's rings annihilate all trapped radiation at the outer edge of the A ring, leaving a shielded region close to the planet in which the radiation intensity is extremely low. This shielding prevents the further buildup of electron intensities at lower altitudes, which otherwise would have been present and would have made Saturn a strong radio source observable from Earth.

It was found that several of Saturn's satellites absorb trapped particles from the radiation belts, producing prominent dips in the intensity. The effectiveness of absorption at the moons Tethys and Enceladus is particularly astonishing, and supports the idea that radiation belt particles are drifting inward slowly across the satellites' orbits.

A precipitous decrease in particle intensity, lasting only for about 12 s, was observed over a wide range of energies for both protons and electrons at a distance near 2.53 Saturn radii, 23 min after *Pioneer* crossed the Saturn ring plane inbound. This phenomenon has been tentatively interpreted as indicating the presence of a nearby massive body absorbing the trapped radiation. The estimated radius of this object lies in the range of 100 to 300 km, based on the effectiveness with which it absorbed the high-energy radiation. The total radiation dose received at Saturn was equivalent to only 2 min in the Jovian radiation belts because Saturn's radiation belts were so much weaker.

Imaging photopolarimeter observations. In addition to images of Saturn, the brightness, color, and polarization of the reflected light were also measured by the imaging photopolarimeter of *Pioneer Saturn*. These measurements are used to study the cloud layers of Saturn and Titan and to model the vertical structure of the atmospheres of these two bodies. In the scans that made the images, the banded structures of Saturn and of the rings were obtained in fine detail. These are essential in studying the atmosphere, rings, and satellites. A new Saturnian ring, which has been tentatively designated the F ring, was discovered in the images. It is narrower than 500 km in width, but is important because it forms an outside barrier to the bright A and B rings. The gap between the F and A rings has been designated the Pioneer Division (Fig. 2). A small satellite, which either was previously unknown or had been previously discovered from Earth but lost again, was found in the *Pioneer Saturn* images (Fig. 2). After its initial discovery, this satellite continued on its 17-h orbit around the planet and passed near *Pioneer* as the spacecraft entered the ring system. It is quite conceivable that this satellite is the same one that perturbed the radiation-belt particles.

Infrared observations. Infrared observations obtained during the Saturn fly-by revealed the temperatures in the atmospheres of Saturn and the rings and in the atmosphere of Titan. It was found that Saturn has a temperature of about 100 K (about 280°F below zero) and, according to these observations, has an internal heat source of enough strength that the planet emits approximately 2.5 times as much energy as it absorbs from the Sun. The equatorial yellowish band observable in many of the images was found to be several degrees colder than the planet at other latitudes, and is probably a zone of high clouds resembling similar zones of Jupiter. As expected, the rings were extremely cold, 65–75 K (about 330°F below zero), at the time of encounter. The temperature differences between the illuminated and unilluminated sides of the rings, and the rate of cooling as the ring particles go into Saturn's shadow, suggest that the ring particles are at least several centimeters in diameter and the rings themselves are many particle diameters thick. The very minor perturbation to *Pioneer*'s trajectory, as it passed under the visible rings, indicates that the rings probably consis of ices.

Micrometeoroids. As *Pioneer* passed through Saturn's ring system, very sensitive meteoroid detectors observed the impact of five particles on the spacecraft, particles that were about 10 μm (0.0005 in.) in diameter. Two impacts occurred while the spacecraft was above the rings, and three while the spacecraft was below the rings. No impacts were detected going through the ring plane, but the *Pioneer* instrument cannot detect individual impacts that occur less than 77 min apart. This characteristic would have prevented detection of ring particles because of the impacts detected just before both ring plane crossings. It is uncertain whether the micrometeoroids detected by *Pioneer Saturn* were stray ring particles deflected out of Saturn's ring plane or whether they were particles from interplanetary space drawn inward toward Saturn by its strong gravitational field.

Ionospheric absorption. Close to the point of closest approach to Saturn, the spacecraft's radio transmissions were affected by Saturn's ionosphere. The manner in which the radio signals were absorbed indicates that Saturn has an extensive ionosphere composed of ionized atomic hydrogen with a temperature of about 1250 K in its upper regions. This high temperature requires an extensive energy source other than the Sun. This phenomenon was also observed at Jupiter.

Ultraviolet glow. *Pioneer* measured ultraviolet glow throughout the Saturnian system. This ultraviolet glow is due to the scattering of the light from the Sun by atomic hydrogen. The observations of ultraviolet emission from an extensive cloud of hydrogen gas surrounding Saturn's visible rings are especially interesing. The rings themselves arc

presumably the source of this hydrogen. On the planet's disk, the ultraviolet observations show significant latitude variations, suggesting the possibility of an aurora near Saturn's polar regions. A similar extensive cloud of hydrogen was also seen partially surrounding Titan's orbit.

For background information *see* SATURN; SPACE PROBE in the McGraw-Hill Encyclopedia of Science and Technology.   [JOHN WOLFE]

Sedimentary rocks

Hummocky cross-stratification is the term coined by J. C. Harms and others in 1975 for an important primary sedimentary structure that is interpreted as an inner shelf to a lower shoreface feature produced by storm waves. The same structure was first recognized in 1966 in the Cretaceous Gallup Sandstone of New Mexico by C. V. Campbell, who considered it a special type of wave ripple. During the early 1970s, other workers also recognized the structure as distinctive and attributed it variously to episodic storm waves, tidal influences, or both. Hummocky stratification occurs in both progradational and transgressive sequences wherein it generally is interstratified with burrowed mudstones or shales. Examples are now known from Cretaceous strata throughout the Rocky Mountains; Cretaceous, Eocene, and Miocene of southwestern Oregon; Jurassic of western Canada; Pennsylvanian of Texas; Silurian of Nova Scotia; and Cambrian of Wisconsin. Other occurrences are being identified as more geologists have learned to recognize the nature and significance of hummocky stratification.

Characteristics. Hummocky stratification is characterized by laminae that are both concave-up (swales) and convex-up (hummocks), the latter being the most distinctive feature (Fig. 1). It occurs most typically in fine sandstone to coarse siltstone (approximately 0.30–0.05 mm), and commonly

Fig. 1. Well-exposed hummocky stratification in Upper Cretaceous strata at Cape Sebastian, OR. Note the sharp, first-order scoured base where erosion cut into underlying intensely burrowed sandstone, clearly defined undulatory laminae, well-developed convex-up hummock, and a second-order scoured boundary within the hummocky bedset. Such second-order surfaces probably represent temporary intensification of waves during a major storm event.

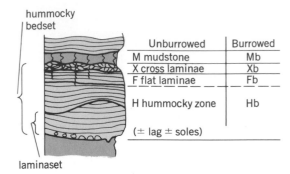

Fig. 2. Ideal or model hummocky sequence showing characteristic zones within a complete hummocky bedset unit, and fair-weather mudstone or shale separating the hummocky units. Relative degree of burrowing and relative thicknesses of the zones vary greatly. The notation scheme for zones provides a shorthand characterization of different types of hummocky units. (*From R. H. Dott and J. Bourgeois. Geol. Soc. of Amer. Abstr.* (*with programs*), *11(7):414, 1979*)

contains abundant mica and fine, carbonaceous plant detritus. It occurs in laterally persistent sedimentary units or bedsets that are from a few centimeters to at least 1 m in thickness. Individual laminasets vary from about 2 to 20 cm. Spacing of hummocks or swales is from about 50 cm to several meters. The lower bounding surface of a hummocky unit is sharp and in many cases is clearly erosional. In favorable exposures, current-formed sole marks may be present. The basal laminae tend to parallel an undulating, scoured basal surface, whereas higher laminae become progressively flatter (Fig. 2).

Conversely, a few units show flat basal laminae succeeded upward by characteristic undulating ones. The laminae typically thicken and thin laterally, which accounts for the upward-flattening tendency. Inclinations of laminae are generally less than 10° (rarely up to 15°), and the directions of inclination are essentially random, even though associated groups of symmetrical ripple marks are conspicuously oriented. In poor exposures, it may be difficult to identify hummocky stratification with confidence, for it is easily confused with either trough cross-stratification or small scour-and-fill structures. The low angles of inclination, the

presence of unique convex-up hummocks, and lack of preferred orientation help to distinguish hummocky stratification.

Depositional conditions. The genetic interpretation of Harms and others may be paraphrased as follows. Hummocky stratification is formed by strong wave action with surges of greater displacement and velocity than those required to form ordinary wave-ripple lamination. The seabed is first eroded into low hummocks and swales that lack any significant orientation. This topography is then mantled by laminae of material swept over the hummocks and swales. As the storm disturbance wanes, ordinary, smaller wave ripples may form for a brief time, and then conditions revert to slow, fair-weather deposition of fine, muddy sediments, and a burrowing fauna becomes reestablished. Harms and others implied that the hummock-and-swale topography is primarily erosional and that the laminae form by pulsating settling from the water, draping the hummocky topography (see Fig. 1, basal zone). But if the very sharply defined, parallel laminae form simply by vertical fallout from suspension, size grading would be expected. The sandstones tend to be so fine-grained and well sorted that grading is very difficult to discern. Nonetheless, a common, subtle color gradation reflects the concentration of flake-shaped mica and plant detritus at the tops of many laminae, which does indicate a settling gradation. On the other hand, the characteristic parallel, convex-up laminae prove that some hummocks represent hydraulically produced depositional bedforms rather than simple erosional features (Fig. 1, middle of the unit). Therefore, much of the deposition must have involved lateral transport as well as vertical fallout. Parallelism of the laminae suggests an analogy with the upper flow regime established by flume experiments. Under unidirectional flow much sand is moved by high velocity over a relatively smooth, flat bed; apparently under high-velocity oscillatory flow, a hummocky bed may result.

In the most typical cases of hummocky units, which are interstratified with shale that was burrowed by animals, clearly the sand had to be introduced from some other adjacent environment and then molded into hummock and swale bedforms. The most probable scenario is that an offshore muddy bottom is eroded slightly as a storm disturbance develops. Simultaneously a storm surge piles water shoreward (Fig. 3). As the winds abate and the surge retreats seaward, either en masse or as many vigorous rip currents, a great deal of fine sand is eroded from the shoreface and transported seaward. This process has been documented clearly for the 1961 Hurricane Carla in the northern Gulf of Mexico and for a river flood and storm surge in southern California. The sand is then deposited as hummocky strata in the lower shoreface zone (Fig. 3). Deposition is by a combination of vertical fallout and the to-and-fro motion of the water as each individual wave passes. Individual laminae probably represent deposition either from the pulse of a single wave or a wave train. Because storm-disturbed seas have extremely confused surface-wave patterns and multidirectional bot-

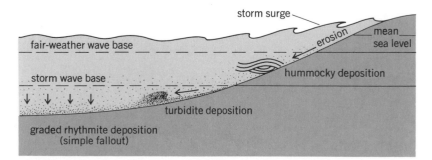

Fig. 3. Conceptual diagram showing the inferred relationship of hummocky stratification to storm-surge erosion, wave base, and possible turbidity current and suspension-fallout deposition. (*Modified from A. P. Hamblin and R. G. Walker, Can. J. Earth Sci., 16:1673–1690, 1979*)

tomwater motions, the hummock-and-swale bed-forms are not well oriented. Typically during a stormy period, winds diminish and then accelerate several times on a time scale of hours to days. Periods of waxing wave conditions within a single storm episode account for prominent second-order scoured surfaces generally observed within single hummocky bedsets (Figs. 1 and 2).

A. P. Hamblin and R. G. Walker offered further insight into possible depositional conditions for hummocky stratification based upon Jurassic strata in western Canada. There the structure occurs conformably above a sequence of turbidity-current deposits and below beach deposits. Current-formed sole marks are conspicuous on the bases of many of the hummocky sets. These marks are well oriented and conform to the directions of similar features on bases of the underlying turbidites. Therefore Hamblin and Walker argue, by analogy with deposits formed during the 1961 Hurricane Carla, that retreating storm surges carrying sand seaward not only can produce hummocky stratification in the depth interval between the fair-weather and storm-wave base, but also may generate turbidity currents that can carry some sand even farther offshore to be deposited as turbidites below the storm-wave base (Fig. 3).

Variability of stratification. R. H. Dott and J. Bourgeois noted a considerable spectrum of variability of hummocky stratification in terms of thickness of the bedsets, proportions of interstratified mudstone, degree of burrowing of the hummocky units, presence or absence of ripples at the tops of units, degree of amalgamation (blending of one bedset upon another without intervening mudstone), and proportions of flat versus typically hummocky lamination within bedsets (Fig. 4). For example, whereas burrowed shale or siltstone is the typical, slowly deposited, fair-weather sediment occurring between hummocky units, in some cases it is totally lacking (Fig. 4). One hummocky sandstone bedset lies upon and is amalgamated with another through stratigraphic intervals that may be tens of meters thick. In such cases, either storms were so frequent that mud never accumulated, or else they were so vigorous that whatever fair-weather mud did accumulate was eroded entirely by each successive storm. In such amalgamated sequences, it may be difficult to identify the true first-order bedset boundaries that delineate the deposits of separate storm episodes. Symmetrical ripples and animal burrowing at the tops of such bedsets, if present, provide important clues.

Idealized hummocky sequence. Because of the variation, Dott and Bourgeois proposed an idealized hummocky sequence analogous to the Bouma sequence for graded turbidite deposits (Fig. 2). This can serve as a descriptive standard or norm against which individual real examples should be compared. Near the extreme limit of fine-sand transport by storm-induced waves, a micro-hummocky stratification seems to be represented by thin, discontinuous lenses of very fine sand observed in some shale sequences (Fig. 4). Such lenses vary from 1 to several centimeters in maximum thickness and are typically about 10 to 100 cm long. Internally they display faint convex-up

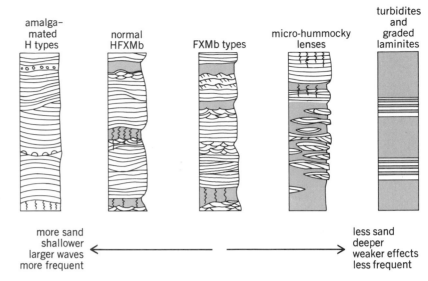

Fig. 4. Hypothetical continuum among variations of hummocky stratification and possible turbidites and graded rhythmites. Some of the factors causing these variations are listed. (From R. H. Dott and J. Bourgeois, Geol. Soc. Amer. Abstr. (with programs), 11(7):414, 1979)

and convex-down laminae, and they even contain some internal low-angle scoured boundaries. Seemingly these lenses formed by storm molding of the seabed where only a small amount of very fine sand was available. Another, perhaps still more distal structure is the graded rhythmite characterized by sets of several thin planar laminae that become successively finer and thinner upward (Fig. 4). These have been interpreted to represent a simple vertical settling of sand thrown into suspension and carried offshore by storm waves (Fig. 3).

Individual examples of hummocky stratification and related structures can be interpreted as members of a continuum in terms of such genetic factors as relative sand supply, relative proximity to shore, relative depth, and the relative magnitude and frequency of storms (Fig. 4). It is anticipated that rigorous comparison with the idealized hummocky sequence will help to stimulate and guide the future investigation of both modern and ancient shallow marine sediments and ultimately to help assess the relative importance of storms in the construction of the geologic record.

For background information see MARINE SEDIMENTS; NEARSHORE PROCESSES; SEDIMENTARY ROCKS; SHALE in the McGraw-Hill Encyclopedia of Science and Technology.

[R. H. DOTT, JR.; JOANNE BOURGEOIS]

Bibliography: C. V. Campbell, *J. Sediment. Petrol.* 36:825–828; 1966; R. H. Dott, Jr., and J. Bourgeois, *Geol. Soc. Amer. Abstr.* (with programs), 11(7):414, 1979; D. E. Drake, R. L. Kolpack, and Fischer, in D. J. P. Swift et al. (eds.), *Shelf Sediment Transport: Process and Pattern*, pp. 307–331, 1972; A. P. Hamblin and R. G. Walker, *Can. J. Earth Sci.*, 16:1673–1690, 1979; J. C. Harms et al., *Soc. Econ. Paleontol. Miner. Short Course Notes*, no. 2, 1975; M. O. Hayes, *Bur. Econ. Geol. (Tex.) Report of Investigations*, no. 61, 1967.

Sedimentation (geology)

Tidal flat environments are common features of modern coastlines. For example, nearly one-quarter of the total coastline of North and South America can be described as mud flat or mangrove flat. The largest tidal flats occur in macrotidal estuaries (which have a tidal range of more than 4 m) such as the Bay of Fundy; however, they are common in mesotidal environments as well (which have a 2–4-m tidal range). Recent studies have expanded understanding of these complex environments.

Historical perspective. Much of the early work on tidal flat sedimentation was done by the Dutch sedimentologists in the Wadden Sea of northern Europe. Their studies led to the formation of the settling-lag and scour-lag hypotheses.

The settling lag considers the amount of time it takes particles to sink to the bottom after the critical threshold velocity decreases below the settling velocity. A particle settling in a slackening current does not immediately drop straight to the bottom, but is transported landward and gradually sinks to the bottom and becomes deposited. When the tide turns and the water begins ebbing, the particle does not get transported back to its original position because it had been displaced landward of its critical threshold velocity. By a series of steps, then, the finest particles migrate landward and accumulate in the upper intertidal zone.

The scour-lag hypothesis considers the difference between the critical velocity for erosion and the lowest velocity for transportation. Because fine-grained sediment particles are cohesive, the critical velocity for erosion is much greater than the transport velocity. A given-size particle will not be removed from the bottom until after the initial transport velocity has been exceeded, and therefore, once the particle is eroded, it will not be transported as far seaward.

Finally, the Dutch sedimentologists recognized an additional factor that helped account for fine-grained sediment accumulation in the tidal flat environment—the asymmetrical shape of the tidal velocity curves. The tidal velocity asymmetry studies in particular indicated that there was a longer period of time of lower velocities when the tide was over the tidal flat which permitted the settling of fine-grained sediment.

Spatial changes in texture. The spatial variations in sediment texture over a tidal flat are fairly consistent, with a general decrease in grain size from the lower intertidal zone toward the upper. The decrease in grain size is related to physical and biological processes.

Tidal currents generally decrease in strength from the lower intertidal zone toward the upper. The lower zone is also more affected by waves. Wind waves passing from the lower to upper intertidal area lose much of their energy through frictional effects on the gently sloping bottom.

The exposure time is also a factor in understanding the spatial changes in sediment texture. The lower intertidal zone is underwater for a greater length of time, subjecting the area to a greater time percentage of tidal currents and wave action. The upper intertidal zone, which is exposed longer, has a chance to dewater, which may lead to the increased stabilization of the sediment surface. The upper intertidal zone also experiences maximum variations in water temperatures over a tidal cycle. Temperatures as high as 40°C have been observed in the very shallow water just flooding over the upper intertidal region. These high temperatures mean lower water viscosities and higher settling rates, encouraging accretion of the finer sediments in the upper intertidal zone.

Biological activity also affects the spatial changes in sediment texture. Sediment trapping by marsh grasses commonly takes place in the highest portion of the upper intertidal zone. Benthic diatoms which are common on the tidal flat secrete a binding slime that may aid in the depositional process of the upper intertidal zone. Infaunal populations, especially marine worms, often increase in numbers toward the upper intertidal area. Many of the organisms in the intertidal area are suspension feeders which create larger fecal pellets out of the finer particulates they extract from the water column. The pellets, which have hydrodynamically large settling velocities, tend to remain in place and increase mud deposition.

Temporal changes in texture. The surface sediments of the tidal flat can be separated into four main textural components. First, there is a coarse fraction greater than 1 mm in diameter that usually represents a shell hash or in some cases organic detritus derived from the adjacent saltmarshes. Second, there is a sand fraction, represented by the sediment between 1 mm and 0.062 mm. Third, there is a mud component representing the silt and clay fraction of the sediment less than 0.062 mm. Finally, there is a pellet component representing biologically produced aggregate whose size is usually, but not always, within the sand range.

The combination by weight of each one of these components will vary from season to season over the tidal flat. In the winter and early spring, the tidal flat surface is enriched in the coarse fractions of sand and shell hash. These are the stormy seasons when increased wind-wave resuspension and rainwash tend to reduce the mud fraction in the sediment closest to the surface. In northern temperate regions, ice scouring may also aid in coarsening the sediment.

At low tide, sediment will freeze to the bottom of the ice in the intertidal zone. As the tide floods, the ice will rise and currents will move across a fresh scar of recently removed sediments, while at the same time some of the sediments on the underlying block of ice will fall off and be winnowed during resettling to the bottom. The gouging of the tidal flat surface by ice will also increase the microrelief, which will interact with the tidal currents, causing increased winnowing. During this same period, the weight percentage of pellets will be at a minimum as the animals have slowed down their feeding rates in correspondence with the decrease in temperature.

By late spring and summer, the factors change and the tidal flat becomes finer-grained. Wind-wave resuspension generally lessens except during storm events, and rainwash is minimized during the drier summer months. Pellet production begins to peak, while at the same time the surface sediments undergo binding by benthic diatoms.

In the late fall months, the sediments begin coarsening as storms pick up and wind-wave resuspension and rainwash increase. The weight percent of pellets continues to increase, however, because although pellet production lessens, the breakdown by bacterial action is reduced at an even greater rate, increasing the half-life of the remaining pellets.

Changes in sediment volume. Recent marker-bed studies and leveling surveys of tidal flats have shown them to be more dynamic than originally supposed. Although the tidal flat environment is probably accretional in the long run, short-term erosion does take place. Single storm events have been documented to erode up to 8 cm of sediment.

The recent leveling surveys have shown that erosion is most pronounced in the late winter-–early spring. These are the seasons when the tidal flat is subjected to the most wind-wave resuspension as well as sediment removal by ice. Maximum deposition occurs in the summer months when mucopolysaccharide-bound pellet production is the greatest and benthic diatoms are stabilizing the surface sediment.

Conclusions. Recent studies have shown that tidal flats are a more dynamic environment than previously thought. Detailed leveling surveys indicate that tidal flats experience periods of erosion and deposition. This means that the tidal flats can serve as storage areas for particulate matter which periodically is flushed into nearby estuarine and coastal waters.

The causes of the spatial distribution of tidal flat sediments are more complex than initially envisioned by the Dutch sedimentologists. In fact, its unlikely that the scour lag or setting lag actually occurs except perhaps in tidal channels. Physical processes such as wave resuspension, rainwash, ice effects, and water temperature changes combine with biological processes, including pelletization and stabilization by benthic diatoms, and tend to blur simplistic model approaches.

Recent investigations have also shown that the sediment texture varies considerably from season to season. Again the same physical and biological processes that characterize the spatial changes will vary from month to month.

In summary, the sedimentology of tidal flats is as variable as nature itself. The complexities are just beginning to be understood as sedimentologists, zoologists, and chemists begin working jointly on a common problem.

For background information *see* ESTUARINE OCEANOGRAPHY; MARINE SEDIMENTS; NEAR-SHORE PROCESSES; SEDIMENTATION (GEOLOGY) in the McGraw-Hill Encyclopedia of Science and Technology.

[FRANZ E. ANDERSON]
Bibliography: F. E. Anderson, *Northeast. Geol.*, 1:122–132, 1979; F. E. Anderson, *J. Sed. Petrol.*, 42:602–607, 1972; R. Dolan et al., *Classification of the Coastal Environments of the World*, pt. 1, Dep. Environ. Sci., Univ. Va., Tech. Rep. 1, 1972; P. Groen, *Neth. J. Sea Res.*, 3:564–574; 1967; H. Postma, *Hydrography of the Dutch Wadden Sea*, unpub. Ph.D. diss. University of Groningen, 1954; D. R. Sasseville and F. E. Anderson, *Rev. Geogr. Mont.*, 30:87–93, 1976; L. M. J. U. Van Straaten and P. H. Kuenen, *Geol. Mijnbouw*, 19:329–354, 1957.

Seismology

The International Deployment of Accelerometers (IDA) Seismic Network is a global array of specially designed ultralong-period seismometers. The purpose of the network is to provide scientists with data for the study of the Earth's normal modes and the earth tides. The earth tides are a phenomenon of the solid Earth analogous to the ocean tides. The motion of the Earth and the Moon in their orbits changes the gravitation field of the Earth in a periodic manner. This change causes a distortion of the figure of the Earth which, among other things, causes a change in its mean radius. Just as the ocean tides manifest themselves in the semidiurnal rise and fall of the sea, so the earth tides produce a semidiurnal rise and fall of the solid Earth by some 20–30 cm. Studies of this phenomenon provide information on the internal structure of the Earth and on the shape of the deep ocean tides.

The main purpose of the IDA network is to study the Earth's normal modes. Following a large earthquake, the entire planet is set into oscillations much as a bell rings when struck. These oscillations occur at a discrete set of frequencies called the normal modes. They are the resonant frequencies of the Earth, and range from the longest-period mode of 54 min to periods of about 30 s, where the density of modes becomes so great that the oscillation frequencies are nearly a continuum. A study of these normal modes yields detailed information about the internal structure of the Earth, as well as details of the mechanism of the earthquake that generated them.

The IDA network consists of specially designed seismometers that respond to periods longer than 20 s. Usually seismometers are most sensitive to periods from 20 to 1 s, since this is the band of traditional seismology. The IDA sensors are electrical-feedback LaCoste and Romberg gravity meters, originally designed to aid in petroleum exploration. These instruments have two important features for these studies: first, they employ what is known as a zero-length spring, which is a mechanical device that allows a spring pendulum to have a period that corresponds to an exceedingly long normal spring; and second, the instrument exhibits a very low mechanical drift. This latter property is important when one is attempting to measure minute fluctuations in the acceleration of gravity—often 10^{-11} g—at periods of 1 h and longer. The data are recorded on site by using a digital cassette which is changed about every week by a local operator. The cassettes are then mailed to the IDA Data Center, in California, where compter-compatible magnetic tapes are produced for distribution by the World Data Center-A for Solid-Earth Geophysics in Boulder, CO.

The study of these global phenomena requires a global network. The illustration shows the distribution of stations on a map where the Earth's surface

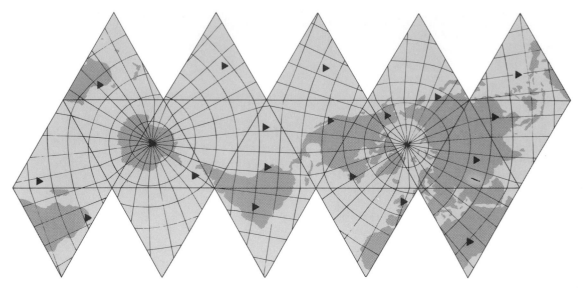

Distribution of stations in the IDA network.

has been divided into 20 equal-area triangles, an icosahedron. An attempt is made to locate one station near the center of each face.

For background information *see* EARTH, DEFORMATIONS AND VIBRATIONS IN; EARTH TIDES; SEISMOLOGY in the McGraw-Hill Encyclopedia of Science and Technology.   [JONATHAN BERGER]

Bibliography: D. Agnew et al., International Deployment of Accelerometers: A network for very long period seismology, *EOS Trans. Amer. Geophys. Union*, 57:180–188, 1976.

Sewage

A difficult problem in understanding possible environmental effects of ocean dumping of human waste has been determining where the waste goes. Recently, acoustical techniques have been shown to be of value in determining waste material transport patterns in the ocean. Furthermore, there is some evidence that acoustical techniques may be used to ascertain oceanic water column density structure, which in turn may give indications as to the longer-term dispersal of dumped waste material.

Sludge dumping. For many years barges and ships have been dumping sewage sludge from New York City into a section of the nearshore Atlantic Ocean known as the New York Bight. A particular section of the bight is designated as the sewage sludge dump area, about 4.5 km² and centered roughly 24 km away from the Rockaway–Sandy Hook transect. A basic problem for the New York sewage sludge dumping program, and indeed for ocean waste dumping in general, has been determining whether sludge in the dump zone remains there. As associated problem is assessing whether ocean-dumped sewage sludge can cause environmental damage. To answer these questions it is necessary to obtain chemical samples of the sewage sludge after it has been dumped in the ocean. It is also necessary to determine the time it takes for the subsurface plume caused by the ocean-dumped sewage sludge to disperse and how far the dispersed plume reaches.

The concentration field of the sewage sludge within the ocean is represented by the expression $C(r,t)$, where C is the concentration of the sewage sludge (for example, milligrams/liter), r is a position vector denoting a certain point within the water column, and t is time. If the concentration field can be determined, then by taking chemical samples at different depths at different times, the dilution rate of the sludge can also be determined. Until the introduction of acoustical technology in 1975, it was not possible to obtain even a partially complete picture of the subsurface concentration field. Since that time, acoustical technology improved the strategy for chemical sampling in the ocean for pollution events.

Acoustical data. Acoustical data are obtained by scattering sound from particles of dump material within the ocean. In a typical experimental situation a research ship, towing a downward-pointing acoustical transducer, passes over a sewage sludge dump with the transducer emitting pulses of sound. The pulses encounter the dump particles and scatter. The intensity of the backscattered sound signal is representative of the concentration or number of particles in the water volume.

A very important feature of the acoustical data is that the data are obtained in real time or live time. This means that as the data are gathered and observed, immediate choices can be made concerning optimum depths and locations for chemical sampling. An example of real-time acoustical data gathered on September 22, 1975, is shown in Fig. 1. In this experiment, 20kHz and 200kHz acoustical frequencies were used simultaneously. The water depth in the New York Bight sewage sludge dump zone is about 22 m. The darkened area in each of the data sections indicates the presence of sewage sludge. In the 20-kHz data, the dark band extending from about 2 to 7 m across all the data is an electronic artifact and does not indicate the presence of sludge. At the time of this experiment the oceanic water column was well mixed down to a depth of about 18 m; below 18 m a pronounced thermocline was present, extending

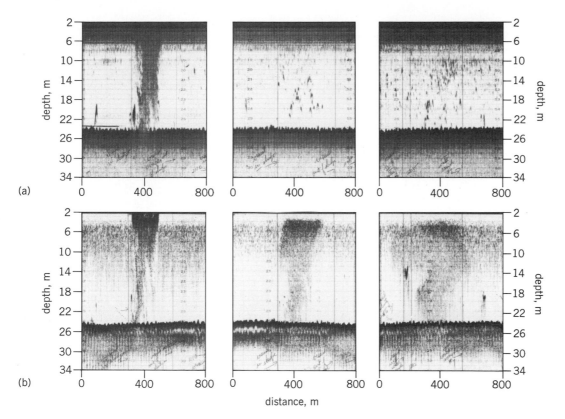

(a)

(b)

distance, m

Fig. 1. Segments of acoustic records at two frequencies: (a) 20 kHz and (b) 200 kHz. The data were gathered from three passes over the sewage sludge dump. The

first pass occurred 9 min after the sewage was dumped, the second occurred 45 min after the dump, and the third occurred 68 min after the dump.

almost (within a meter or so) to the bottom. As can be seen from Fig. 1, some of the sewage material penetrates the oceanic water column and reaches the ocean bottom within 9 min after dumping. Thirty-six min later the horizontal extent of the dumped material has increased throughout the water column. Most of the material which is more easily detected by the 20-kHz acoustical system has disappeared from the water column by this time. However, a series of 20-kHz point acoustic scatterers are present in the water column. The origin of these scatterers is unclear. Two suggestions are: they are ocean-dwelling creatures responding to the presence of the sludge; or they are flocs or large aggregates of sludge particles. Again from Fig. 1, it can be seen that at 68 min after the dump, the sewage material has dispersed still further horizontally. The 200-kHz data show that the horizontal rate of dispersion of material in the first 10 m of the water column is larger than elsewhere in the column. The subsurface cloud of sewage material, in addition to dispersal by turbulent forces, is also being distorted by vertical current shear; that is, current flow within the water columns is not uniform. This is particularly evident when comparing the location of sewage material at 14 m depth with that at 22 m depth. In addition to being dispersed and sheared, the sewage material is also being advected, that is, the gross movement of material, by current systems. The effect of advection can be seen from examination of Fig. 2 in which the backscattered acoustical intensity coming from a depth of 18 m is plotted against distance for three different time intervals

(corresponding to the three time intervals shown in Fig. 1). In addition to recording the acoustical backscattered intensity on paper for live-time presentation and use, the backscattered intensity is also recorded on magnetic tape for later analysis. It is the magnetic tape–recorded intensity that is shown in Fig. 2. Note that there is a set southward movement, that is, advection, of the sewage material at a depth of 18 m. In about 75 min the center of concentration of the material (at 18 m) has moved to the south a distance of about 500 m.

The same experiment was repeated in July 1976,

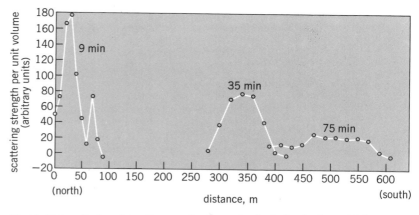

Fig. 2. Net scattering strength per unit volume due to sludge for three 45° traverses of the east-west-line sludge dump. The distance scale represents a projection of the distance along the ship track against a north-south line to show the actual width of the sludge cloud and its southerly drift.

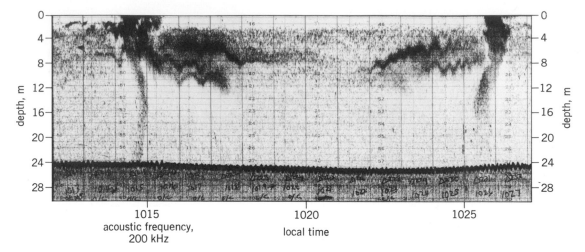

Fig. 3. Two successive passes over a spot dump. The dark reflecting layers of particulate matter are clearly indicated.

since the temperature structure of the ocean changes with the season. A change in water column temperature and density structure will profoundly influence the way in which the subsurface sewage sludge material is dispersed. In September 1975, the oceanic water column was uniformly mixed (that is, no relatively sharp increases in water density with depth were present) down to a depth of roughly 18 m, while in July 1976, the water column was well mixed only down to a depth of about 5 or 6 m. Below this depth a relatively strong density increase or stratification was present.

Figure 3 shows the real-time data obtained in July 1976, during two passes over the dump material. The ship reversed course at about 1020 local time. This time much of the sewage material remained within the upper water column between about 4 and 10 m depth. This is due to the presence of the increased water density in this depth range which seems to prevent the sewage from going deeper. Nevertheless, some of the heavier components of the sludge do penetrate the denser water and reach the ocean bottom. This too can be seen from Fig. 3. Some of the sewage material in the 4- to 10-m depth range spreads horizontally on a water density surface from the center of concentration of material. The initial horizontal movement was about 50 cm/s.

Applications. It is clear that acoustic techniques provide a powerful new tool for studying the oceanic dispersion and advection of dumped waste material. Furthermore, acoustics can be used to guide chemical sampling of sewage dumps, effectively guiding the chemist in selecting depths and locations for gathering water and sewage samples. Since 1975, acoustical techniques have been beneficial in detecting subsurface oil, ocean-dumped pharmaceutical wastes, petrochemical wastes, and dredge material. Research into further uses of acoustics is continuing at various government laboratories as well as in industry and private institutes.

For background information *see* SEWAGE; UNDERWATER SOUND in the McGraw-Hill Encyclopedia of Science and Technology.

[JOHN PRONI]

Bibliography: M. Orr and F. Hess, *J. Geophys. Res.*, 83(C12):6145–6154, 1978; J. R. Proni et al., *Science*, 193:1005–1007, Sept. 10, 1975; J. R. Proni and D. V. Hansen, *Deep Sea Res.*, in press.

Soil

One of the principal activities of soil scientists is the collection of data which are used to characterize, classify, and locate soils and to predict their performance. Overall, the main justification of the data collection effort has been to improve the use, management, and conservation of soils. This should also be the central objective of soil information systems.

Commonly stated objectives include deciding the use of land, developing management systems for soils to achieve stated objectives of use, understanding the basic productivity of soils to aid in evaluation for taxation, and maintaining or improving the productive capacity of soils. Accurate prediction of the consequences of choices in the use and management of soils, and of land, can replace costly trial and error.

Data organization. Organizing and effectively using the vast amount of existing soil data are no small challenge. Data collection is costly. If future data collection efforts are to be most useful, it is essential that greater emphasis be placed on organizing the data and on making them readily accessible.

A highly important task in the development of any land information system is the selection of primary land units or individuals around which to organize and present data. To be most useful, such units must have high interpretive value and must be closely correlated with other important attributes of land such as plant communities, water resources, climate, topography, and geology. Emphasis should be given to the kind of soil, that is the soil series, phase, or map unit, as the primary land unit for aggregating and presenting data because it represents the basic feature of land having the highest interpretive value and highest correlation with other land attributes.

Underlying the development of an information system should be an explicit statement of functions that the system must perform. Without such a statement, development of the system suffers long delays while a guiding philosophy is established, or else the system is developed and then found to be lacking.

A fully comprehensive soil information system that makes full use of automatic data processing has not yet been developed. This is not to say that excellent soil information systems do not exist. There are some in the traditional form that provide the surest guidance in developing automated systems.

The major functions of a soil information system are determination of the properties of pedons; development of the soil classification system; classification of pedons; prediction of soil performance; identification of the location of soils; and assembly of soil data for publication. Each of these major functions is discussed below. This list is arbitrary and oversimplified; it is necessary, however, to structure the functions into a few broad categories in order to identify the basic requirements of data flow.

Determination of pedon properties. An understanding of the in-place properties of soils is the base of any soil survey or other data collection project in soil science. Through these data, important relationships in soil genesis, soil behavior, and most aspects of applied soil science are elucidated.

Determining those soils for which various kinds of data are needed depends to a great extent on the amount and kind of data already available for each soil. Data from thousands of pedons are stored in the Pedon Data System of the U. S. Soil Conservation Service which also includes a subsystem to encode soil engineering test data.

The familiar soil profile description has been recorded in the field on mark-sense forms. This technique is thought to save time in the field and, where it is necessary to transcribe the coded field data in the office, there is added time saved. The Pedon Data System will also translate the coded pedon description to standard terminology. Relating laboratory results and field descriptions is generally a time-consuming task when done manually. Many comparisons and tests can be quickly performed by the computer.

Research has proved the value of computer analysis of remotely sensed data in making the field time of soil surveyors more efficient. Reductions in the number of field observations required and more accuracy in their location are valuable contributions.

Soil classification system. Many complexities were involved in the development of Soil Taxonomy, a classification system of the Soil Conservation Service. Overall, it is estimated that between 500 and 1000 worker-years were expended over a period of nearly 20 years in developing Soil Taxonomy.

The necessary soil data were scattered widely throughout the United States and the world. A great deal of effort was expended in obtaining these data. A centralized, computerized registry would have greatly reduced this effort. Certainly, if data on the properties of pedons were stored in computers, their accessibility for use in developing

a soil classification system would be greatly enhanced.

To date, there have been no comprehensive efforts to use computers in developing a soil classification system. However, it is likely that, in the future, computers will be used to test and refine existing systems; the ability to marshal huge volumes of data will be invaluable.

Pedon classification. Although not needed for the common field classing of pedons, there are some situations where the computer might be used to advantage to classify pedons. A common one is where changes are made in the definition of soil taxa. Without a computer system, the wholesale reclassification of many soil pedons is an extremely tedious, time-consuming, and error-prone operation.

Another situation is where it is desirable to classify a pedon in each of several classification systems. Examples could include placement in Soil Taxonomy, in the Canadian Soil Classification System, and in the system advanced by the U.S. Food and Agriculture Organization, or in both a pedological system and a system used in soil engineering such as the Unified or the American Association of State Highway and Transportation Officials (AASHTO) systems.

Prediction of soil performance. The main purpose of soil surveys and of most other data collection activities in soil science is to provide a guide to the use and management of land. To achieve this, it is necessary to record the performance of soils under various land uses and management systems.

A performance file is fundamental to the operation of the soil information system. Numerous automated subfiles are needed. One such subfile is the Soil-Woodland File (SIDEX) of the Soil Conservation Service, which includes measurements of the growth rates of trees on specific kinds of soils. Data from 15,000 sites have been encoded. Another is the Range Data System of the Soil Conservation Service, which includes measurements of forage production and species composition of native plant communities. Both files include complete soil descriptions, permitting analyses of the influence of soil, climate, and topography on the plant community. In addition to these two soil performance files, many others are needed, both for agricultural and nonagricultural soil uses.

Using soil performance data and proved relationships between soil properties and performance, the Soil Conservation Service has developed the computerized Soil Interpretations Record. In this file, 30 to 40 kinds of soil interpretations are included for each of the 12,000 soils series in active use in the United States. Each interpretation was developed manually through reference to all available yield and performance data. Computer programs have been developed for use in testing the accuracy and consistency of the ratings developed from this guide.

Identification of soil location. In the United States, detailed soil mapping has been completed for about 60% of the land area, or about 560,000,000 ha. Using computers to encode, manipulate, and print-out soil maps can enhance the availability of the data for numerous users. Considerable experience has been accumulated in this area over the past 10 years or so, and numerous systems are operational. The basic kinds of systems in use, grid and polygonal, differ widely in operation and cost.

The Map Information Assembly and Display System (MIADS) has been extensively used. For example, soil surveys and land use have been encoded for the entire state of Oklahoma, and work is progressing in several other states. Cell sizes range from about 4 to 250 ha.

In the United States, the Advanced Mapping System (AMS) is now operational. This system features automatic scanning and is capable of encoding soil maps for about 15,000,000 ha per year. Using an automatic plotter, the Advanced Mapping System generates maps of excellent quality for the presentation of soil interpretations.

Assembly for publication. Automated techniques are also being used in the publication of soil surveys. All interpretive tables are prepared by accessing from the Soil Interpretation Record of the Soil Conservation Service the interpretations required for each soil mapping unit in the survey. All entries in this file have already been edited; the need for typing and editing the tables is eliminated.

Computer-generated tables on climate are obtained on tape from the National Climatic Center. The table on acreage and proportionate extent is developed by the computer from the correlated legend. The entire text is keyed into a word-processing machine. Full retyping is thus not necessary after editing. The tape from the word-processing machine, containing the complete text of the soil survey, is used to drive the automated typesetter of the Government Printing Office. Typesetting that requires several weeks when done manually is thus accomplished in minutes, and the overall costs of printing and binding are reduced by one-half.

Summary. A number of effective computer subsystems have been developed over the past decade for the handling of soils data. These subsystems are now effectively saving a sizable amount of staff time in carrying out the functions that are required for the collection and effective application of soils data to planning the use and management of land.

The greatest current shortcoming is thorough identification of the logical data flow requirements of fully integrated soil information systems. The likely reason is that it is not entirely clear just what such systems ought to do. Until more thorough work is done in outlining essential functions, systems are likely to remain fragmentary.

For background information *see* LAND-USE PLANNING; SOIL in the McGraw-Hill Encyclopedia of Science and Technology.

[D. E. MC CORMACK]

Bibliography: G. Decker, J. Nielson, and J. Rogers, *Automated Data Processing for Soil Inventory*, Soil Conservation Service–Montana Agricultural Experiment Station, 1975; C. G. Johnson, *Soil Conservation Service Advanced Mapping System*, Comm. 5 Publ. Pap., FIG 15th Int. Congr. Surv., Stockholm, Sweden, 1977; D. E. McCormack, A. W. Moore, and J. Dumanski, A review of soil information systems in Canada, the United States, and Australia, *Int. Soil Sci. Soc. Symp.*, Edmonton, Alberta, Canada, 1978; Soil Conservation Service, *National Soil Handbook*, pt. II, sect. 407: Computerizing soil survey interpretations, 1976; Soil Conservation Service, *Pedon Coding System for the National Cooperative Soil Survey*, 1973.

Soil chemistry

Soils are complex chemical systems in which the dissolution and precipitation of solid phases largely control solubility relationships.

Historically soils have been viewed as systems in which adsorption onto clay surfaces, cation exchange reactions, and organic matter transformations have been considered as the major solubility control mechanisms. Consequently, most soil research in the past has been directed along these lines. Recent concern for environment quality has focused attention on soils as the ultimate depository for wastes. Since many wastes contain elevated levels of heavy metals, there is concern that their accumulation in soils may lead to long-term metal toxicity problems. Consequently, greater effort is now being made to determine specifically what happens to metals in soils and to identify the minerals that control their solubilities.

Solubility relationships. Recently W. L. Lindsay demonstrated that basic chemical principles could be applied to many aspects of soil chemistry by utilizing solid-phase/solution equilibria and free-energy relationships. For example, even though the specific minerals controlling solubility relationships in soil are not known, observed solubilities can often be expressed by means of soil-metal reactions. Examples are given in reactions (1)–(4).

$$\log K^0$$

$$\text{Fe(OH)}_3\text{(soil)} + 3\text{H}^+ \rightleftharpoons \text{Fe}^{3+} + 3\text{H}_2\text{O} \qquad 2.7 \qquad (1)$$

$$\text{Soil-Cu} + 2\text{H}^+ \rightleftharpoons \text{Cu}^{2+} + \text{soil} \qquad 2.8 \qquad (2)$$

$$\text{Soil-Zn} + 2\text{H}^+ \rightleftharpoons \text{Zn}^{2+} + \text{soil} \qquad 5.8 \qquad (3)$$

$$\text{SiO}_2\text{(soil)} + 2\text{H}_2\text{O} \rightleftharpoons \text{H}_4\text{SiO}_4^0 \qquad -3.1 \qquad (4)$$

With further research these empirical relationships can often be associated with specific minerals. For example, the solubility relationships of zinc were measured in several soils, and reaction (3) was used to express the findings. While it was not known which mineral, if any, was controlling zinc solubility, it was later shown that ZnFe_2O_4 (franklinite) is most likely the mineral controlling Zn^{2+}.

As illustrated in Fig. 1, franklinite contains iron as well as zinc; therefore, iron solubility in soils will also affect zinc solubility. When iron solubility is controlled by soil-Fe [reaction (1)], Zn^{2+} in equilibrium with franklinite is only slightly less soluble than soil-Zn (Fig. 1). Initial precipitates in soils are often amorphous, and frequently they contain other substituted ions. Such precipitates are generally more soluble than well-characterized crystalline minerals. Therefore, an amorphous solid similar in composition to franklinite may control Zn^{2+} solubility near the soil-Zn line in Fig. 1. If iron solubility drops to that maintained by $\gamma\text{-Fe}_2\text{O}_3$ (maghemite) or $\alpha\text{-FeOOH}$ (goethite), Zn^{2+} solubility will rise. On the other hand, if amorphous iron oxides temporarily control Fe^{3+} solubility, then Zn^{2+} solubility is depressed.

The average soil contains approximately 40,000 ppm of iron and only 50 ppm of zinc. Thus, zinc solubility can be affected by iron oxides. Other heavy metals such as copper and nickel also precipitate as insoluble ferrite minerals in soils. These include $CuFe_2O_4$ (cupric ferrite), $Cu_2Fe_2O_4$ (cuprous ferrite), and $NiFe_2O_4$ (nickel ferrite). Several of these solubility relationships were recently explored, and it was concluded that ferrite minerals exert important solubility controls on many metals in soils. Heretofore the low solubilities of such metals in soils was attributed to adsorption phenomena, because the simple oxides, hydroxides, carbonates, and silicates of these elements could not explain their low solubilities.

Figure 2 summarizes the solubility level of several metal ions that have been examined in soils. These levels may be temporarily elevated by additives such as wastes, fertilizers, or other materials, but in time the chemistry of the soil will readjust and return them to approximately the levels shown here.

Soils differ according to the parent materials or minerals from which they are derived. Likewise, weathering conditions, temperature, rainfall, vegetation, and drainage greatly modify the chemical environment of soils and permit different kinds of minerals to become stable. For example, high rainfall causes leaching, which removes the more soluble elements like calcium, magnesium, potassium, sodium, sulfur, and boron. Such soils usually become acid and develop certain characteristics inherent to acid soils. On the other hand, submerging soils enables reduction to take place, and those elements which change oxidation state precipitate as new minerals under reducing environments.

Models. A knowledge of solubility relationships of the various elements in soils is very useful. It enables models to be developed where chemical information can be put to work to predict what will

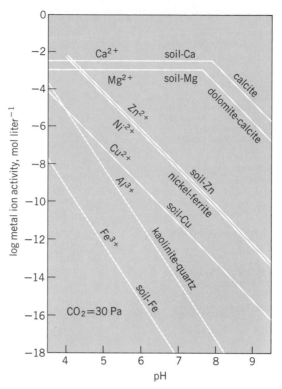

Fig. 2. Solubility of various metal ions in soils when the indicated solid phases are controlling them.

happen and what solubility relationships can be expected. These relationships include metal chelate stability, anion precipitation, complex formation, sulfide precipitation, and redox control mechanisms. Computers can be used to store vital chemical information that can be used to develop predictive models and to interpret observed solubility relationships. A great deal of research concerning solid-phase solution equilibria in soils is anticipated in the future because of the major effects that soils have on the environment.

For background information *see* SOIL CHEMISTRY in the McGraw-Hill Encyclopedia of Science and Technology.

[W. L. LINDSAY]
Bibliography: W. L. Lindsay, *Chemical Equilibria in Soils*, 1979; W. A. Norvell and W. L. Lindsay, *Soil Sci. Soc. Amer. Proc.*, 33:86–91, 1969.

Soil phosphorus

Developments in computer modeling of nutrient uptake and crop growth and in nutrient behavior in waste disposal have focused attention on needs for solution properties of nutrient ions in soils. Soil phosphorus has unique solution properties, and some recent research has been directed toward parameters of those solution properties.

Depletion. Orthophosphate, the chemical form that is absorbed by plant roots, is highly reactive with polyvalent cations (or compounds thereof) which are present in soils. These reactions tend to cause low phosphorus concentrations in solution, thereby requiring relatively rapid replenishment from solid-phase soil phosphorus in order to supply plant needs. Phosphorus may diffuse from about 0.2 to 2 mm to plant roots. These distances imply that 1–5% of the soil in the rooting zone is in-

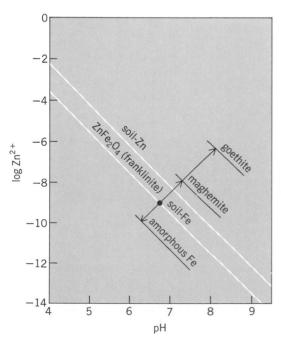

Fig. 1. Effect of various iron minerals on the solubility of zinc maintained by franklinite compared to soil-Zn.

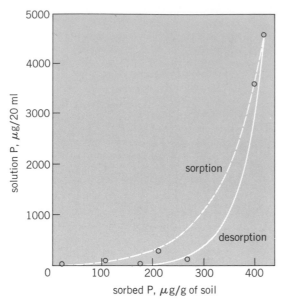

Fig. 1. Hysteresis in sorption-desorption of orthophosphate by soil.

volved in supplying phosphorus during seasonal growth. The relative stability of concentrations during phosphorus removal is indicative of buffering capabilities of soils, which is especially important in plant uptake at low solution concentrations. Buffer factors, porous matrix diffusion coefficients, and solution concentrations are soil properties used in equations describing nutrient movement to plant roots. Soils exhibit sorption and desorption buffering properties by removing phosphorus from solution during enrichment and releasing it to solution during depletion. However, sorption and desorption curves form a hysteresis loop (Fig. 1), possibly because of multiple bonding that occurs after initial sorption. Phosphorus depletion profiles were calculated in soil adjacent to plant roots by using both sorption and desorption estimates of buffer capabilities; contrasting results of the two methods are evident in Fig. 2. Calculated diffusive flux rates at the root surface (μg phosphorus/cm^2 root surface/s) using the same equations and same parameters are paired in the table with release of phosphorus to an anion exchange resin sink. Diffusive flux rates and depletion profiles based on desorption parameters correspond qualitatively with desorption to anion resin. Calculations based on sorption parameters did not correspond to phosphorus release capabilities.

Release. Two additional soil phosphorus properties are the rates (or kinetics) of release and the average concentration that is maintained during

release. Though this concept is simple, it is difficult to obtain accurate measurements of these chemical properties. Plant absorption and anion resin extraction are probably two of the more valid measurements of release rates. However, soil phosphorus absorption by plants is often not a linear function of time, and the quantity of soil contributing to the absorbed phosphorus can only be estimated. Uptake rates by plants are between 1×10^{-5} and 5×10^{-7} μg/g/s. If only 1–5% of the soil contributes the phosphorus utilized by plants, the release rates from the depleted soils ranges from 2×10^{-4} to 5×10^{-5} μg/g/s. Anion resin extraction of soil-water suspensions in several studies have indicated phosphorus release rates from 3×10^{-5} μg/g/s. These values are in qualitative agreement with the plant uptake rates just discussed. Perhaps a rule-of-thumb value would be 5×10^{-5} μg/g/s for soil adjacent to plant roots.

Concentration. Concentration of phosphorus in soil solution is of course the dominant factor in availability to plants and in environmental concerns. Critical concentrations for supplying adequate phosphorus to plants vary somewhat according to soil properties and experimental technique, but productive agricultural soils usually contain 0.04–0.06 μg of phosphorus per milliliter of soil solution. Increasing critical values from 0.02 to 0.095 μg/ml have been associated with fourfold decreases in buffer factors.

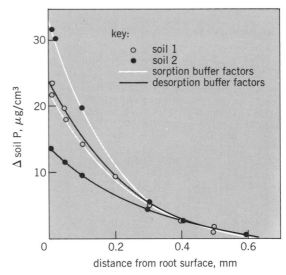

Fig. 2. Calculated depletion profile (10-day profiles) of available phosphorus as influenced by buffer-factor measurement and distance from root surface.

Conclusion. It is clear from research findings that the chemistry of most soils places restraints on solution phosphorus levels, but that field-produced crops can be produced satisfactorily within these restraints providing that the buffer capabilities, release kinetics, and root flux areas in the soil-plant system arc within normal ranges. It is also clear that phosphorus contamination of groundwater by water percolating through adequately fertilized agricultural soils normally should not occur.

Though the basics of soil phosphorus chemistry have been known for years, the composite effect of these chemical properties on rates of solution re-

Calculated phosphate diffusive flux at root surface based on sorption or desorption buffer factors of two similar soils compared to observed release of phosphorus by the soils

Soil	Calculated flux, (μg/cm^2/s) $\times 10^{-8}$		Measured release, μg/g soil
	Sorption factor	Desorption factor	
1	52	56	45
2	65	37	34

plenishment, buffering capabilities, and diffusion rates has been resolved only in recent years. Such knowledge will be useful in more efficient food production, crop growth modeling, and safe waste disposal.

For background information *see* SOIL CHEMISTRY in the McGraw-Hill Encyclopedia of Science and Technology.

[DOYLE E. PEASLEE]

Bibliography: I. Holford and G. Mattingly, *Plant Soil*, 44:337–389, 1976; P. Nye and P. Tinker, *Solute Movement in the Soil-Root System*, 1977; S. Olsen and W. Kemper, *Advan. Agron.*, 20:91–151, 1968; D. Peaslee and R. Phillips, in P. H. Dowdy (ed.), *Chemistry in the Soil Environment*, Amer. Soc. Agron. Spec. Publ., 1980.

Soil variability

Soil variability is studied, in part, because the world's growing population makes increasing demands on land and resources. The modern-day grower needs help to answer the challenges of newly imposed economic, energy, and social constraints. Not only must the land provide more and more food, but its future capacity to produce must be assured. The soil plays a central role in the environment with a remarkable capability to recycle wastes into useful or harmless substances. How water and water-carried materials move in the soil is an immense area of scientific concern because effects are not always obvious. These materials move within the soil slowly, and impacts can be months, years, or even centuries away. In fact, variability within the soil may determine whether an effect shows itself in a few months or several years.

Soils reflect the factors that led to their formation—source material, time for development, climate, topography, and vegetation. The variation is observed whether looking at a small garden area, a grower's field, or an entire river basin. Even a handful of soil shows differences in particle sizes and composition. Manifestations of soil variability include differences in color, in water intake rates, and in productivity. The variability occurs vertically and horizontally.

Soil scientists have long recognized differences in soils and have developed comprehensive schemes to classify all of the soils of the world. In the United States, over 10,000 soil series have been defined. The definitions enable the soil classifier to draw lines on a map. The scale and the determining criteria depend on the purpose. For example, less detail would be required for mapping a sparsely inhabited rangeland than a new solid-waste disposal site. Different criteria would be used for residential development than for forests. Regardless of the scale of a map, variability within the units exists.

Distributions. Soil properties are defined quantitatively in order that measurements can be meaningful. One such property is the bulk density, defined as the mass of dry solids per unit bulk volume. If other factors are the same, a high value of bulk density means more compaction and less pores (approximately 50% of the soil is water- and air-filled pores). Bulk density thus is an important factor influencing root development, seedling emergence, and aeration.

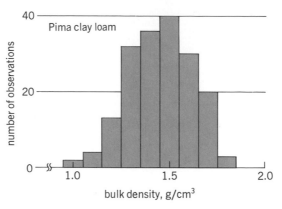

Fig. 1. Frequency histogram for bulk density of Pima clay loam.

Bulk density values were measured on 180 samples obtained at 5 depths and 36 sites over 85 hectares of Pima clay loam in Arizona. The samples were taken by digging a hole to each desired depth (30, 60, 90, 120, or 150 cm), then extracting a core with a special apparatus to minimize alteration in the structure of the removed soil. The samples were dried and weighed in the laboratory. The bulk density value was determined from the dry weight and the known volume of the core. Values ranged from 1.0 to 1.8 g/cm³, but most were closer to 1.5. Figure 1 shows the distribution of values. Each bar shows the number of samples within a range of 0.1 g/cm³, that is, the value given below the bar ± 0.05. Relatively few values exist near the extremes. There are many more in the center, the center bars each having 36–40 samples compared to only 2 or 3 on the outside. If a smooth curve were connected through the top of each bar, the result would be a bell-shaped curve. A symmetric, bell-shaped curve is characteristically a normal distribution, one of the most common types of statistical distributions describing a population. Here the population consists of the bulk density values throughout the sampling area.

The distribution of bulk density values may be described in terms of its mean and standard deviation. The arithmetic mean is approximated by dividing the total of the sample values by the number of samples. The standard deviation reflects the spread of the curve, large values correspond to wide, flat curves, and small values to narrow, peaked curves. The quotient of the standard deviation divided by the mean and expressed as a percentage is known as the coefficient of variation (CV) The CV is independent of the units and allows a comparison between types of measurements. Some meaningful soil physical properties grouped into three relative categories according to size of CVs are:

Low variation
  Bulk density.
  Porosity.
Medium variation
  Percent sand, silt, or clay.
  Water retained at given suction.
High variation
  Hydraulic conductivity.
  Apparent diffusion coefficiencies.

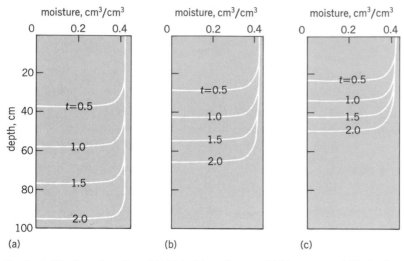

Fig. 2.   Infiltration of profiles of (a) high, (b) medium, and (c) low permeability for four given times, corresponding to the 75, 50, and 25 percentile values.

Here, porosity refers to the fraction of the total space occupied by pores. Water retained at a given suction reflects the pores smaller than a specified size. The hydraulic conductivity expresses the soil's ability to transmit water. The apparent diffusion coefficient reflects the tendency for spreading which occurs for solutes and such as they move with water through the soil. Thus, the value of the CV would be lowest for bulk density and porosity and highest for the hydraulic conductivity. The exact CV might be 0–10% for the low variation, 10–100% for the middle, and greater than 100% for the high. For a homogeneous soil the values would be less, but probably the relative positions would remain about the same. An important fact is that the hydraulic conductivity and the parameters most closely related to water and solute movement have the highest variation. Not only are they difficult to measure, but also they have a large inherent variability.

Variations in moisture intake rates are shown in Fig. 2. The surface of the soil was assumed saturated and moisture values calculated for profiles having a high, intermediate, and low permeability as based on values measured within a 150-hectare site of the Panoche soil of California. The variability of hydraulic properties was incorporated into a scaling factor with the three profiles of Fig. 2, corresponding to the 75, 50, and 25 percentile values.

The wetting fronts are labeled at $t = 0.5$, 1.0, 1.5, and 2.0 days in each case. For any given time, the water profile is deeper for the highly permeable case compared to the least permeable case. For example, after 2 days, the highly permeable profile has a wetting front at 95 cm compared to 48 cm for the low-permeability profile. Intakes for approximately 25% of the field would be even faster than the highly permeable example, 25% between the high and medium shown, 25% between the medium and low permeability, and the final 25% even slower.

For the bulk density frequency histogram, the classes were determined by the size of the bulk density value, not on the basis of where the sample came from within the field. However, values which are from adjacent sites would be expected to be

about the same, or at least more alike than values from sites far apart from each other. Graphically, the principle is shown in Fig. 3 as a correllelogram. The correllelogram gives the degree of correlation between values as a function of separation distance. For two samples close together, the separation length is small and the degree of correlation is high. For two samples farther apart, the separation distance is large and the correlation becomes negligible. Such spatial correlations have been investigated only for limited cases in soils. Applications of the general principles have been made extensively to estimate ore reserves and manage mining operations.

Ramifications. Variability is a factor that must be considered in the many different kinds of studies involving soil.

Sampling. Interpretation of the analysis of a given soil sample requires evaluation of the accuracy of the value and information about the sampled area. Often, closeness to the true mean, that is, average for the field (or area to be measured), is an adequate criterion. Generally, more samples lead to better estimates. The accuracy is also affected by the variation. If the property were exactly the same everywhere, one sample is all that would be necessary. For highly variable parameters, many samples are needed. For the parameters listed under low variation, maybe 2 or 3 samples would give as close an estimate to the overall mean as 1000 or more for the hydraulic conductivity.

Mapping. Soil classifiers have devoted major efforts to mapping soils throughout the world. Whenever lines are drawn by mappers, there is some type of differentiation between soils. Quantitative methods using sophisticated statistics have been used to discriminate between kinds of soils. Such an approach allows data synthesis and flexibility for adaptation to many special purposes. The maps drawn mechanically can be augmented by fieldworkers as desired.

Remote sensing. For the last quarter-century, air photos have aided soil surveyors and farm managers. Vast amounts of data are generated by satellites and aircraft scanning the Earth's surface, displaying differences in the soil. One of the problems involves developing methods for sifting through what may be literally millions of data points. Also, ground verification is obviously of great importance.

Resource estimation. Soil is a key natural resource. In order to estimate productivity for timber, crops, and livestock, meaningful samples have to be taken and reliability of the answers is needed. The confidence of results is strongly dependent upon variability.

Environmental protection. The soil is the eventual depository for all sorts of wastes and by-products (nearly all water and air pollutants eventually end up in the soil). Also, the soil environment is continuously being altered by anthropogenic as well as natural phenomena (erosion remains a major problem worldwide). Accurate assessments and projections are necessary. Water movement and the soil's capacity to react and render harmless a myriad of chemicals are keys to this problem. Both factors are highly variable across the landscape. Suggested solutions are subject to interpretation and need to be judged soundly.

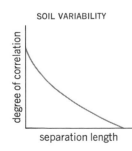

Fig. 3.   Relationship of correlation to separation length.

SPACE FLIGHT   363

Food and fiber production. Worldwide pressures accelerate demands for food. Farmers are resourceful. However, their survival requires that they change continuously with the times. A natural concern is the derivation of the greatest benefits from a given unit of input, such as a unit of water or labor or energy or fertilizer. Soil variability determines optimum combinations of field size, crop rotations, crop selection, water management, or method of weed control. For irrigated areas, the method, amount, and frequency of water application should be correlated with the spatial variation of the soil. For example, because of a few spots which need water first, a higher frequency of added water may be required.

The variation in soil parameters needs integration into a usable form for each problem situation. Small fields may be of concern in some cases and large geographical areas in others—for example, examination of sediment loading in a major river. The ideal result is a solution to the physical problem compatible with economic and social constraints.

For background information *see* SOIL in the McGraw-Hill Encyclopedia of Science and Technology.   [A. W. WARRICK]
Bibliography: M. A. Coehlo, *Spatial Variability of Water Related Soil Physical Properties*, Ph.D. dissertation, University of Arizona, 1974 (available as 75–11, 061 from Xerox University Microfilm, Ann Arbor, MI); D. R. Nielsen, J. W. Biggar, and K. T. Erh, *Hilgardia*, 42:215–259, 1973; A. W. Warrick and A. Amoozegar-Fard, *Water Resources Res.*, 15: 1116–1120, 1979; R. A. Webster, *Quantitative and Numerical Methods in Soil Classification and Survey*, 1977.

Space flight

The brisk pace of Soviet spacecraft launchings continued in 1980, in sharp contrast to American activity (only one NASA satellite was scheduled for launch in calendar year 1980). The United States continued making space shuttle preparations. Potential competition for the space shuttle as a launch service materialized upon the first flight of Ariane, an expendable vehicle developed by the European Space Agency. The agency has firm commitments from the International Telecommunications Satellite Consortium to fly at least one *Intelsat V*, numerous options for launch (including *Arabsats 1* and *2*, and two Indonesian satcoms), and expectations for a variety of European payloads which would ordinarily have been launched by NASA. Major space missions are listed in the table.

The profit and loss aspects of spacecraft launching and space utilization have grown enormously, as evidenced by the severe financial impact resulting from the December 10, 1979, loss of the *Satcom 3* spacecraft during its apogee kick motor firing. The spacecraft and its Delta vehicle launch service were insured for $50,000,000, and there was up to $27,000,000 of insurance coverage for unrealized profits by RCA American Communications, Inc. The satellite would have been used primarily by cable television organizations.

The reentry destruction of the *Pegasus 2* spacecraft assembly occurred on November 3, 1979. Launched in 1965, the vehicle served to gather micrometeoroid data for use in the design of spacecraft.

Launch requirements. For several years the majority of payloads orbited by NASA have been for a rapidly growing number of "customers" (consortia, other agencies, other countries, and private industry). Lauch projections for the 1980s are exceeding the most ambitious flight rates proposed in the past for the space shuttle. Growth in the communications satellite market alone is projected to exceed the launch capacity of NASA's anticipated fleet of four shuttle orbiters. There will, therefore, continue to be a need for expendable boosters such as the McDonnell Douglas Delta into the late 1980s. Indeed, to accommodate foreseen heavier satellites, NASA has authorized development of an improved model (Delta 3920) to increase the maximum payload placeable into a geosynchronous transfer orbit to 2750 lb (1250 kg), up 700 lb (320 kg) from the current limit.

Competition for use of the shuttle may force some customers with time requirements to launch on the more costly expendable boosters. For example, the added cost of using expendable launchers instead of the shuttle for the first two satellites of the Satellite Business Systems Company is estimated at about $34,000,000.

Space shuttle. Complexity, difficulties, and delay marked development work for 1980 on the space shuttle orbiter, *Columbia*. The major technical problem has been the laborious process of obtaining good bonding of the 30,922 thermal-protection tiles covering 70% of the orbiter's exterior for insulation against the heat of reentry. NASA management expects to solve this and other problems in time for a first launch by the end of March 1981.

At launch, the shuttle will be powered by two solid rocket motors and three main engines. The latter, mounted at the base of the orbiter, will burn liquid hydrogen and liquid oxygen, both of which will be carried in a large external tank. Each main engine generates a rated sea-level thrust of 1,670,080 newtons (375,000 lb) and is throttleable from 50 to 109% of rated power. These are the first crew-rated engines capable of throttle control during flight. The main engines will burn for about 8 min after launch and in tandem with the solid rocket motors for the first 2 min of flight.

Failures of some main engines under test led to redesigns and improvements. Space shuttle main engines, singly and in clusters, have been tested more than 620 times, and have accumulated over 85,000 s of operation, with more than 50,000 s of firing at 100% of rated power level or higher. Successful full-duration firing of a cluster of three main engines was first achieved in December 1979. Because of modifications prompted by lessons learned in the testing program, the first set of flight-rated engines was sent back to the Mississippi engine test center, where successful requalification firing occurred. These were engines 2005, 2006, and 2007, since mounted in *OV 102*, the first shuttle orbiting vehicle. They will be exercised again on the launch pad early in 1981, when the flight readiness firing is conducted.

Development of the solid rocket motors for *OV 102* was successfully completed, as was the final qualification of seven static test firings, and the motors were stacked on the mobile launch plat-

Space missions from September 1979 to September 1980

Payload name	Launch date	Payload country or organization	Purpose and comments
HEAO 3	9/20/79	United States	High Energy Astronomy Observatory to survey cosmic and gamma radiation
Cosmos 1129	9/25/79	Soviet Union	Cooperative biosatellite with France, United States, and Czechoslovakia; recovered 10/14/79
Ekran 4	10/3/79	Soviet Union	TV broadcaster over Indian Ocean
Magsat	10/30/79	United States	NASA study of Earth's magnetic field; Applications Explorer Mission (*AEM 3*)
Intercosmos 20	11/1/79	Soviet Union	Eastern Bloc data relay and ocean surface comprehensive study
Satcom C	12/6/79	RCA	Domestic communications satellite; lost contact on apogee motor firing
Soyuz T	12/16/79	Soviet Union	First flight of advanced ferry craft; docked with *Salyut 6* station
Ariane 1	2/2/80	France	First test flight of European Space Agency booster
NavStar 5	2/9/80	United States	Carries improved atomic clocks; joins constellation of spacecraft for three-dimensional global positioning and navigation system
SMM	2/14/80	United States	Solar Maximum Mission; NASA science
Ayame 2	2/22/80	Japan	Communications; apogee kick motor failed
Soyuz 35	4/9/80	Soviet Union	Carried cosmonauts L. Popov and V. Ryumin to *Salyut 6* to begin record stay
Cosmos 1174	4/18/80	Soviet Union	Fifth demonstration of a single-orbit intercept by antisatellite device
Ariane 2	5/23/80	France	Second test flight; failed in launch
Soyuz 36	5/26/80	Soviet Union	Carried a Hungarian and Soviet to *Salyut 6* visit
NOAA-B	5/29/80	United States	Weather; suffered staging failure
Soyuz T-2	6/5/80	Soviet Union	First crewed flight to *Salyut 6* of new ferry craft
Rohini 1	7/18/80	India	Experimental; orbited on India's SLV-3 launch vehicle
Soyuz 37	7/23/80	Soviet Union	Carried a Vietnamese and a Soviet to *Salyut 6* visit
GOES-4	9/9/80	United States	Geosynchronous Operational Environmental Satellite (NOAA weather)
Soyuz 38	9/18/80	Soviet Union	Carried a Cuban and a Soviet to *Salyut 6* visit

form, awaiting integration.

In January 1980 NASA completed a 2-week systems test which teamed astronauts in the *Columbia* with flight controllers in Houston, TX, in computer simulations of the shuttle's countdown, launch, orbital flight, reentry, and landing. This major accomplishment verified compatibility of flight and ground equipment.

Soviet launch activity. In the first 9 months of 1980, there were 67 Soviet launches to orbit or beyond. The calendar year 1979 produced 87 Soviet launches, involving 126 payloads grouped generally as follows: science, 15; radar calibration, 3; Earth-orbital launch platforms, 12; regular communications satellites, 10; tactical military radio store and transmit, 19; electronic ferret, 6; weather, 3; navigation, 6; Earth resources, 1; recoverable military photographic, 36; early warning, 2; ocean surveillance, 2; crewed Earth-oriented, 3; and crew-related but uncrewed Earth-oriented, 8.

Soviet crewed flight. In mid-December 1979 the Soviets flew without crew the first *Soyuz T* crew-rated advanced ferry spacecraft and docked it with the unoccupied *Salyut 6* space station in a 100-day test of the design. It was returned to Earth on March 26, 1980. Improvements in the basic Soyuz design included more efficient use of fuel, an improved Earth-touchdown system to soften landings, use of solar panels, higher telemetry data rates to the ground, and the use of pressure-fed main engines instead of turbine-driven engines. The absence of solar panels on previous Soyuz craft saved weight, but also limited electrical power. This earlier power design has forced immediate

and hazardous emergency landings by crews who missed their initial docking attempts and had no chance for another attempt.

A new world-record duration stay in orbit began on April 9, 1980, with the launch of *Soyuz 35* carrying cosmonauts Leonid Popov and Valery Ryumin to rendezvous with *Salyut 6*. The 185-day flight ended on October 11, 1980, surpassing the previous record of 175 days.

The record breakers were visited in May by *Soyuz 36* bearing the Soviet Valery Kubasov and Hungarian Bertalan Farkas, who returned to Earth 8 days later in *Soyuz 35*, leaving the fresher *Soyuz 36* still docked with Salyut. Two other visitors were launched on June 5 in the new *Soyuz T*, in its first crewed flight. The cosmonauts were the Soviets Yuri Malyshev and Vladimir Aksenov, and they returned June 9 in *Soyuz T*. Another visitation launch began on July 23, when *Soyuz 37* carried to orbit the Soviet Victor Gorbatko and Vietnamese Lt. Col. Pham Tuan. They returned in the *Soyuz 36* vehicle after 7 days at the Salyut station. On September 18, 1980, *Soyuz 38* carried the Soviet Yuriy Romandenko and Cuban Arnaldo Mendez to Salyut. They did not exchange ferries, returning in *Soyuz 38*. Uncrewed resupply missions to *Salyut 6* were also conducted by Progress spacecraft during the record duration flight.

Solar-terrestrial studies. A satellite designed to observe the peak years of activity (1980 and 1981) in the Sun's 11-year cycle was launched February 14, 1980, on a Delta 3910 into a 358-mi (576-km) orbit inclined 28.5°. Called the Solar Maximum Mission (SMM) spacecraft, it observes the range of wavelengths from orange light at 658.3 nm to high-energy gamma rays. The latter are produced in the most energetic of the sudden, violent explosions, called flares, which occur on the Sun. Along with solar x-rays and ultraviolet rays which are also not measurable on Earth, gamma rays provide the most definitive data for understanding flare mechanisms.

Seven instruments aboard SMM, especially designed to study solar flares, are providing correlated evidence, along with data from a worldwide network of ground-based solar observatories, which is forcing revisions in theories of the origin and evolution of flares. SMM has disclosed, for example, flare core temperatures of 56,000,000°C, higher than anyone had thought could occur on the Sun, where normal surface temperatures range from 4400 to 5500°C. The behavior of gas in these regions is also puzzling, for it densifies instead of expanding when heated, as a free gas would. This indicates a pressuring or confinement mechanism at work, and will perhaps provide a model for researchers who are trying to contain high-temperature, high-pressure, magnetized, ionized gas on Earth for thermonuclear fusion energy production.

Other theory-disturbing data indicated an apparent lack of acceleration of protons in the magnetic field of flares. Scientists had previously believed that every major flare accelerated protons.

Because flares appear and vanish rapidly, SMM does some of its own aiming of instruments. For example, when one x-ray sensor with a wide field of view detects a flare start, it provides automatic positional data to more narrowly focused instruments which can turn around in seconds to observe.

Earth's climate may depend critically on variations in the amount of incident sunlight. While it was long assumed that solar radiance was constant, SMM's active cavity radiometer instrument has measured day-to-day variations with unprecedented accuracy (on the order of 0.04%). Variations measured so far are not large enough to have a detectable effect on terrestrial weather. SMM is the first satellite whose instruments were placed on a novel bus called the modular multimission spacecraft, a vehicle with standardized components prepackaged into self-contained modules and mounted on a standardized frame. The scientific instrument package was designed to match the new bus's bolt patterns and electrical connections, thus locking into the necessary systems for attitude control, data handling, power supply, and communications. The entire assemblage is designed for refurbishment in orbit or retrieval from orbit by space shuttle crews. SMM is also the first spacecraft planned to transmit data through the Tracking and Data Relay Satellite System when that system is launched by the space shuttle.

The scientific value of simultaneous multipoint measurements has been further demonstrated by the International Sun Earth Explorers (*ISEE 1, 2,* and *3*). They have made possible a productive operational space plasma physics research program. An important process recently discovered in space is that in which magnetic field lines, disturbed and opened by interactions with charged particles, recombine and result in the heating of these particles. In addition, *ISEE 1,* passing through the tail

of the magnetosphere, has detected ions that, by virtue of their observed energy, must have been accelerated away from the Earth's ionosphere.

ISEE 1 and *2* follow nearly identical, highly elongated trajectories with a controllable spacing between the two, while *ISEE 3* is permanently in the solar wind at 235 earth radii upstream of the Earth in a halo orbit about the forward libration point. It was thought that placing *ISEE 3* at this location would position it outside the influence of the Earth. Yet its instruments have detected particles moving upstream from the Earth's bow shock due to acceleration out of the solar wind. *ISEE 3*'s continuing measurements of solar wind composition are conducted under most conditions of flow, and are showing new phenomena in the multicomponent plasma. In addition, *ISEE 3* detected the first burst of gamma rays to have been positively attributed to a celestial object. Triangulation by instruments on a network of satellites identified the source as a supernova remnant in the Large Magellanic Cloud.

Astrophysics. The final High Energy Astronomy Observatory, *HEAO 3,* was launched on September 20, 1979, on a scanning mission to observe high-energy radiation (from such sources as stars, quasars, pulsars, and black holes) which the atmosphere prevents studying from Earth. Preliminary results indicate that its gamma-ray spectrometer, seeking the most energetic form of electromagnetic radiation, gamma-ray line emissions, obtained evidence of the electron-positron annihilation line at 511 keV coming from the center of the Galaxy. Strong gamma-ray emission was detected from major features such as the Crab Nebula, Cygnus X-1, and Centaurus A, with over 90 gamma-ray lines so far identified.

HEAO 3's French and Danish instruments to study the isotopic composition of cosmic rays have reported on the relative abundances of iron and cobalt isotopes. A third instrument is studying the elemental composition of heavy nuclei (high-atomic-number elements) in cosmic rays. Early results indicated that those with atomic charge in the 30s probably originated in a neutron-poor environment, but this still leaves room for theories that elements with higher nuclear charge numbers could have originated in supernova explosions.

Meanwhile the focusing x-ray telescope on *HEAO 2* produced the first detection of x-rays from Jupiter. Earth is the only planet other than Jupiter from which x-rays have actually been detected. In each case the mechanism may be electrons spiraling into the atmosphere from the planet's trapped radiation belts and colliding with atoms and molecules of the atmosphere. *HEAO 2* also discovered that the coronas of certain main-sequence stars emit x-rays with greater intensity than current theory can explain. Other theories were disturbed by *HEAO 2*'s surprising discovery that hot neutron stars did not exist in the remnants of supernova explosions. *HEAO 2* also obtained the first x-ray images of bursts in globular clusters (Fig. 1). A typical 10-s x-ray burst releases more energy than the Sun does at all wavelengths in 1 week. It is believed that the burst phenomenon represents the x-ray analog of optical novae in which helium is explosively burned on the surface of a hot neutron star, resulting in a brilliant flash of x-rays.

Before *HEAO 1,* it was believed that only a very

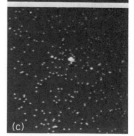

Fig. 1. X-ray photographs obtained by *HEAO 2* of the x-ray burster in the globular cluster Terzan 2, (*a*) before, (*b*) during, and (*c*) after a burst. Duration of burst was 50 s. These are the first photographs of such a burst. (NASA)

small fraction of the radiation from stars would be x-rays. However, that satellite revealed that a class of star called a cataclysmic variable included some that were strong x-ray emitters. The sensitivity of *HEAO 2* revealed a wide range of stars as relatively bright x-ray producers. *See* X-RAY ASTRONOMY.

Environmental observations. Orbiting instruments measuring constituents and processes in the upper atmosphere have changed the understanding of atmospheric photochemistry. In the 1950s it was believed that only oxygen chemistry and ultraviolet radiation were important in the formation and destruction of ozone. Since then the coupling among water, nitrogen oxides, and the chlorine family has also been demonstrated to be important. NASA's current estimates of the detrimental effect of artificial chlorofluoromethanes on the ozone layer, 14 to 18% depletion by about the year 2050, agree well with the National Academy of Science's estimate of 16%. Also, new insights have been developed on the role of bromine (important if the concentration in the upper atmosphere should increase), the role of temporary sinks (such as peroxynitric acid, HO_2NO_2, and hydrogen hypochlorite, $HOCl$), and the chemistry of the peroxy radical, HO_2.

The 7-year collection of data by the *Nimbus 4* backscatter ultraviolet (BUV) sensor has been made available by NASA to the scientific community to serve as an initial data base for global ozone climatology studies. Other data from the *Nimbus 7* scanning BUV and the total ozone monitoring spectrometer (TOMS) are being processed. They will be complemented with measurements from the *Nimbus 7* limb infrared monitor of the stratosphere (LIMS) experiment. These instruments have provided valuable daily information on temperature, ozone, water vapor, nitric acid, and nitrogen dioxide over the entire globe in the study of interactions that determine ozone distribution and variability.

To precisely validate ozone-related data collected by satellite sensors, an international rocketsonde intercomparison project was conducted by the United States, Japan, India, Canada, and Australia. The rocket-borne sensor data for ozone profile shape agreed well with that from satellite sensors, but the values differed enough to show a need for more uniform absolute calibration procedures.

Winds in the stratosphere can be measured indirectly from the temperature maps obtained by the Nimbus limb radiance inversion radiometer (LRIR). Results show a variability of winds as a function of latitude and altitude. The wind knowledge is vital to an understanding of where species in the stratosphere originate, how they move about in the atmosphere, and how they disseminate around the globe. The accuracy of satellite-derived wind maps compares favorably with those developed from rocketsonde measurements (Fig. 2), while providing better spatial and temporal coverage.

To help understand climate changes which may result from the injection of volcanic aerosols into the atmosphere, aircraft observations of volcanic ash dispersion from Oregon's Mount St. Helens eruption of May 1980 were coupled with similar measurements from the Stratospheric Aerosol and Gas Measurement Experiment (SAGE) satellite. In April 1979 SAGE measured and tracked volcanic stratospheric debris from the eruption of La Soufriere in the Caribbean. In November 1979 it observed the little noticed eruption of the Sierra Negra volcano in the Galapagos Islands. SAGE data showed that this eruption injected at least 10 times more material into the atmosphere than did La Soufriere. These data are important in assessing short-time climate phenomena. Recent sensitivity calculation indicated that, on a short-term basis (climatologically, $0.5-2$ years), major volcano eruptions have a much greater effect on global surface temperatures than do carbon dioxide increases.

Weather observations. The operational phase of the Global Weather Experiment (GWE), culminating 10 years of planning and preparation, was successfully completed in November 1979. It was the largest international meteorological experiment ever conducted, and utilized balloons, aircraft, buoys, ships, and 10 spacecraft to gather data. Early results show that in some cases weather analyses based on satellite data alone appear to be superior to those compiled from conventional sources. Significant improvements in forecasts for up to 5 days are expected to result from data including satellite observations.

The GWE observations disclosed that atmospheric circulation at mid-latitudes in the Southern Hemisphere ($40-60°S$) is more intense than formerly believed. Measurements indicated that the tropics may influence the middle and high latitudes faster and more substantially than theory suggested, implying the need for routine meteorological observations of the tropics to improve weather predictions.

In January 1980 problems with the *Tiros-N* onboard command/control system prompted the call up of *NOAA-B* as a replacement. Launched on May 29, 1980, by an Atlas-F booster, it went into uncontrolled flight during the orbital insertion phase,

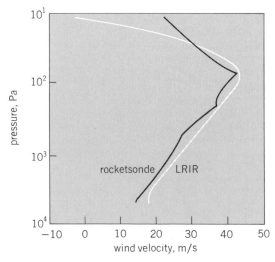

Fig. 2. Stratospheric wind velocities derived from temperature data obtained by the *Nimbus 6* limb radiance inversion radiometer (LRIR) compared with those obtained from rocketsonde measurements. Wind velocities are plotted as a function of pressure; the pressure range corresponds approximately to an altitude range of $22-65$ km. (NASA)

entered an improper orbit, and was declared inoperable. Fortunately, the on-board computer problems of *Tiros-N* stabilized, restoring its successful operation and making another replacement launch unnecessary.

Research at the National Severe Storms Forecast Center in Kansas City, MO, benefitted from satellite weather scans now made available in 6 min instead of the previous time of 45 min. Results produced an NOAA decision to make interactive graphics computers a basic element of future weather service operations.

A new Geostationary Operational Environmental Satellite (*GOES 4*) was launched from Cape Canaveral, FL, on September 9, 1980, on a Delta 3914. It will provide the first continuous operational soundings of temperature and water vapor in the atmosphere as a function of altitude by means of a new sensor called VAS, the visible-infrared spin scan radiometer. Soundings will be continuous, an important factor in the detection and observations of sudden, severe storms and hurricanes. A more complete three-dimensional analysis of weather conditions should result. This is the first United States civil spacecraft orbited under a program for orbiting combined research (the VAS sensor) and operational instruments.

The high-latitude stratospheric measurements made with the SAM II sensor on *Nimbus 7* have demonstrated for the first time that stratospheric clouds are very common in the Antarctic during winter. Sightings of such clouds in the Northern Hemisphere during winter have also been made, but even though much greater in number than previously reported, they are not nearly as frequent in occurrence as in the Antarctic. They may be a major sink for stratospheric water vapor.

Earth observations. Commercial marine operators are benefitting from wind, current, ice, and water color information derived from the satellite data sets from *Goes*, *Seasat*, and *Nimbus 7*. Quantitative wind field data have been extracted from the measurements of the defunct *Seasat*, and its altimeter data have been used in accurately charting coastal tides. For ice monitoring, space-borne microwave radiometers and altimeters have shown value for monitoring the ice edge, drift rates, and ice types, information important to ship operators and navigators for planning routes and resupply operations in polar regions.

To study Earth crustal strain and the causes of earthquakes, NASA deployed seven laser satellite tracking systems in the United States and the Pacific Basin. They became part of an expanding international network of lasers and microwave facilities which use signals from radio stars such as quasars to measure movements on the Earth's crust. Other laser systems use corner cube reflectors placed on the Moon by the United States and the Soviet Union. The tracking of laser-reflecting satellites in precisely known orbits permits pinpointing the tracking site to within 3–5 cm. The ground-based laser can pulse a beam which special reflectors on the satellite return directly to its source. Precise measurement of the beam's round-trip time provides the essential information for location. Six satellites that are currently in use have the retroreflector capability. The latest, *Lageos*, the American Laser Geodynamics Satellite, was designed to be a reference point in space.

Launched on October 30, 1979, *Magsat*, the Magnetic Field Satellite, exceeded its estimated 4.5-month lifetime by 3 months, providing useful data until a few hours before reentry. The mission yielded a complete survey of Earth's magnetic field. For the first time, vector field measurements were made in near-Earth orbit. From the data the orientation and shape of magnetized crustal features can be inferred. The quality of the data was three times better than the goal of 3-gamma (3-nanotesla) accuracy in the scalar field and 6-gamma (6-nT) accuracy in each component of the vector field. Some 20,000,000 vector measurements were made at altitudes ranging from 560 to 170 km. *Magsat* was placed in a Sun-synchronous, nearly polar orbit. Satellite measurements of the geomagnetic field began with *Sputnik 3* in May 1958, and continued with the American Pogos (Polar Orbiting Geophysical Observatories). Unlike Pogo, *Magsat* instruments were able to discern the direction as well as the magnitude of Earth's main magnetic field and to obtain a higher-resolution view of crustal anomalies. Results evidently confirm a previously detected decrease in the intensity of Earth's magnetic field, but without clarifying whether this is a normal fluctuation or a movement toward a periodic field reversal. The decrease is a rate of decline of less than 1% per decade. At this rate it would take 1200 years for the field strength to fall to zero. From there it would likely rebuild with opposite polarity. Measurements of the magnetic field, in ancient lava fields, indicate that reversal occurs about every 500,000 to 1,000,000 years. The last major reversal occurred about 700,000 years ago.

Trouble developed for the 5-year-old *Landsat 2* spacecraft on November 5, 1979, when its yaw attitude flywheel stopped functioning, probably because of insufficient bearing lubrication. Without this unit the attitude control system was unable to keep the instruments pointed toward Earth. The spacecraft was officially retired on January 22, 1980. Engineers then began trying to achieve pointing control by using the Earth's magnetic field to interact with the magnetic coil on each axis of the spacecraft. By energizing coils at selected times and attitudes, torquing forces were created. Suddenly, on May 5, the flywheel responded to an "on" command and resumed spinning. Normally the torquing action of jetting attitude-control gas must be used in combination with flywheel speed regulation for operation of the attitude control system. Because the gas had become depleted, engineers utilized their new skills in magnetic torquing while controlling the wheel speeds, allowing *Landsat 2* to return to operational service near its previous capacity.

Oil and mineral geologists are purchasing enhanced Landsat imagery in a product that General Electric's Space Division is marketing under the name Geopack. Purchasers get a geologic analysis of an Earth area plus a geometrically corrected, custom contrast-stretched and edge-enhanced color, plus black and white, Landsat image of the area. Using such techniques, one petroleum geologist predicted 11 potential oil-drilling locations. All 11 areas yielded producing wells. NASA and the minerals exploration industry are experimenting with improvements in geologic mapping based on data from the Heat Capacity Mapping Mission satellite and the Seasat synthetic aperture radar. The data have shown that thermal infrared and microwave techniques can provide crustal structure information not obtainable from visible imagery alone.

Saturn encounters. After a 6-year trip of 3,000,000,000 km, *Pioneer 11* reached Saturn on September 1, 1979. Within a month scientists had listed over 560 new discoveries. Closest approach occurred on the night side of the planet at 21,400 km (versus 42,760 km at Jupiter in 1974). *Pioneer 11* moved toward Saturn from above the tilted ring plane which is lighted by the Sun from below. This permitted the first optical measurements of a back-lighted ring structure. The spacecraft flew safely through the E or outer ring region in about an eighth of a second (at 85,000 km/h). It then skimmed from 2000 to 10,000 km below the rings, and made another shallow-angle crossing of the ring plane on ascent away from the planet and out of the solar system. *Pioneer* experienced two hits above the rings and three hits below the rings by particles of at least 10 μm in diameter. Transmission of data to Earth required 86 min at the speed of light. The 2.75-m dish antenna was able to look at Earth continuously as the spacecraft's attitude was adjusted for the Sun-orbiting movements of *Pioneer* and Earth. The spacecraft uses nuclear sources for electric power because sunlight beyond Jupiter is too weak for use of solar cells.

Saturn was found to have a magnetic field, a magnetosphere, and radiation belts. The startling discovery was made that Saturn's dipole magnetic field is aligned within 1° of the axis of rotation, compared with 10° for the other magnetic planets, Earth and Jupiter.

Voyager 1's approach to Saturn entered the observatory phase on Aug. 23, 1980. This phase of planetary encounter procedure starts when proximity provides images of quality better than those obtainable from Earth. It will continue until Saturn's size fills the view field of the narrow-angle camera (at about 16,000,000 mi or 26,000,000 km). It will be succeeded by the phase called far encounter 1, in which the entire planet will be captured by a four-image mosaic until nearness requires other mosaic techniques in far encounter 2. Following that will come near encounter, and finally, post encounter, preceding exit from the solar system. The next arrival, *Voyager 2*, should make its closest approach to Saturn in August 1981, and arrive at Uranus in January 1986. *Voyagers 1* and *2* have detected radio emissions which fluctuate with Saturn's rotation. The measurements show Saturn's rotation or "day" to be 10 h 39.9 min, some 25 min longer than that estimated from visual observation of the cloud-covered planet. *See* SATURN.

Venus observations. The hard-landing probes of the Pioneer Venus mission plunged through the Venusian atmosphere in December 1978, while using radio interferometry to track the wind-caused perturbations in their path of descent. The data show a "pattern" in which the whole atmosphere seems to be one large, westward-blowing (retrograde) wind whose upper regions circle the planet about every 4 days. Near the surface the winds blow slowly, at 1 m/s (about 2.25 mph). In the second of Venus's three distinct cloud layers, the wind speed is about 150 m/s (330 mph).

Sensing through these clouds, the radar on Pioneer Venus has produced a platewide portrait, mapping over 93% of the surface. It shows what appears to be a "one-plate planet," with no equivalents to Earth's mid-ocean ridges (where new crust is born) or to the subduction zones where old material from one tectonic plate thrusts downward under the edge of the next. Radar shows a pair of huge highland regions, continent-sized by Earth standards. Except for these and some areas considered to be below "sea level" (the mean planetary radius), Venus emerges as a vast rolling plain which varies only about 1000 m from highest to lowest point. The northern hemisphere highland, called Ishtar Terra, is as large as the contiguous United States. On it is a huge massif called Maxwell Montes which towers 11 km above "sea level" (2 km taller than Mount Everest).

Aphrodite Terra (Fig. 3) the other huge highland, centered about 5° south of the equator, is half as large as Africa, about 9700 km (6000 mi) long and 3200 km (2000 mi) wide. It consists of eastern and western mountain ranges, rising 4 and 8 km above "sea level," respectively, separated by a somewhat lower region. The huge circular region in the foreground of Fig. 3, 2900 km (1800 mi) in diameter, resembling a dome or basin structure, is a mystery to planetary geologists. There appear to be many primordial impact craters scattered over the surface, with diameters larger than 75 km.

Pioneer series. A number of aging Pioneer spacecraft are contributing to studies of solar wind dynamics and to the properties of the heliosphere at the extreme edge of the solar system where the influence of the Sun wanes into interstellar space. The 15-year-old *Pioneer 6* and the 12-year-old *Pioneer 9* are still in operation in solar orbit. In August 1980 *Pioneer 8* returned to life after failing 3 years earlier. Revival occurred on its first track at perihelion; proximity to the Sun may have caused the Sun sensor to function. *Pioneer 10* is receding from the solar system at 3.2 AU per year (1 astronomical unit is the distance from Sun to Earth 150,000,000 km) and is beyond 24 AU.

Mars observations. Relatively uneroded Mars has well-preserved flow scarps, lava channels, basaltic flood plains, wrinkle ridges, and calderas which appear in photos from the two Viking orbiters. The features infer that Mars has a history of volcanic activity so heavy that it resurfaced as much as two-thirds of the planet (92,700,000 km²).

The *Viking Orbiter 1*, on February 22, 1980, obtained exceptionally clear views of unusual meteorological features. One was a sharp, dark line curving north and east from the huge volcano Arsia Mons. Believed to be either a weather front or an atmospheric shock wave, it is unique in Martian observations. The second unusual phenomenon was four small clouds hovering just north of Lowell Crater at an altitude of 28 km (17 mi) and casting four clearly separate cloud shadows on the planet's surface.

Planetary quarantine requires that the Viking orbiters not impact Mars before the year 2020. *Viking Orbiter 2*, shut down in July 1978, will meet

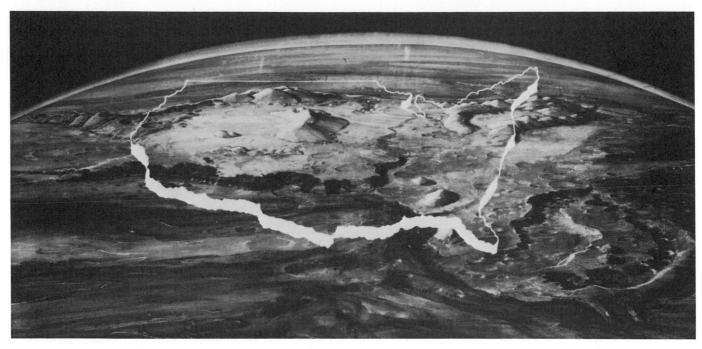

Fig. 3. Aphrodite Terra, the largest highland region on Venus, based on radar measurements by the Pioneer Venus Orbiter spacecraft. Outline of continental United States is provided for scale. (NASA)

this requirement, as will *Orbiter 1*, shut down on August 7, 1980, after it ran out of attitude control gas. In total, the Viking Orbiters sent to Earth over 52,000 pictures, functioning for years beyond their design life of 90 days in orbit. *Viking Lander 2* was shut down in June 1980. *Viking Lander 1* continues as the only remaining operational spacecraft of the original complement of two orbiters and two landers which arrived at Mars in August 1976. It is programmed so that, on interrogation, it will transmit (directly to Earth) imaging, meteorological, and engineering data. This near-autonomous operation can continue until 1994. A citizens' group called the Viking Fund has offered NASA publicly solicited money to ensure the continuation of the downlink sessions which cost $9000 each. The legality of the concept is under study.

International activity. The maiden flight of the Ariane launch vehicle occurred successfully on December 24, 1979, from the Korou site in French Guayana. This booster, primarily intended to place payloads in geostationary orbits, is expected by European Space Agency officials to capture 20–25% of the launch business market. ESA has approved construction of six Arianes. The first flight in the four-flight qualification test of this three-stage booster carried only hardware to monitor flight performance, but it also orbited its own third stage and 1602 kg of ballast. The launcher length is 47.388 m; liftoff mass is 210,269 kg; first- and second-stage fuels are nitrogen tetroxide and unsymmetrical dimethyl hydrazine (UDMH). The third stage is fueled by liquid hydrogen.

The second Ariane launch occurred on May 23, 1980, and ended in failure when all four of its first-stage engines stopped firing prematurely. Also lost were Ariane's first payloads: an amateur radio satellite called *Oscar 9* (Orbiting Satellite Carrying Amateur Radio), and the firewheel experiment from the Max Planck Institute. The latter was de-

signed to eject barium and lithium at high altitude to study visible effects of Earth's magnetic field. The Ariane launcher program will eventually be taken over by a commerical satellite organization (called Arianespace) made up of European aerospace companies.

Indicative of the vigor and dedication of the budding European space program was the decision by the science program committee of ESA to launch its own fly-by of Halley's Comet in mid-1985. The decision to fly alone on an Ariane launcher will stand unless the United States makes the venture a joint mission by supplying a Delta vehicle and the use of NASA's Deep Space Network.

The Chinese Academy of Sciences in January 1980 signed an agreement with NASA to install near Beijing a ground station purchased from the United States to read out Earth resources data from *Landsat-D*. China's annual access fee of $200,000 will help defray costs of operating the satellites. Landsat ground stations already exist in Canada, Brazil, Italy, Sweden, Japan, Australia, India, and Argentina.

On July 18, 1980, India launched its third spacecraft, the Robini Research Satellite *(RS 1)*. The two previous Indian satellites had been launched by Soviet boosters. India's SLV-3 booster is a solid propellant launcher using inertial guidance. First-stage thrust is about 95,000 lb (423 kilonewtons), and the payload capacity is about 90 lb (41 kg) in an elliptical orbit.

Canada's *Anik-B* communications satellite system is now operational. The Canadian Federal Department of Communications uses three of *Anik-B*'s 12-GHZ transponders to run pilot projects in health, conferencing, and direct television broadcasting to small, low-cost terminals.

Japan over the past 2 decades has launched 19 satellites and over 300 sounding and observation rockets. Japan intends to launch all planned spacecraft itself, and is pushing development of three new large launch vehicles while accelerating the improvement of its current launchers, the N-1 and the M-3. All of its space programs are limited to civil purposes. The *Ayame 2* satcom mission of February 1980 failed during launch, when its solid propellant apogee kick motor apparently malfunctioned. In the preceding February the *Ayame 1* mission failed when the spacecraft recontacted the upper stage of its booster following launch.

For background information *see* MANNED SPACE FLIGHT; MARS; METEOROLOGICAL SATELLITES; SATELLITES, APPLICATIONS; SATELLITES, SCIENTIFIC; SATURN; SPACE FLIGHT; SPACE PROBE; TERRAIN SENSING, REMOTE; VENUS; X-RAY ASTRONOMY in the McGraw-Hill Encyclopedia of Science and Technology.

[CHARLES BOYLE]
Bibliography: *Air Space Mag.*, 3(5):8–9, May-June 1980; *Aviat. Week Space Technol.*, issues from Sept. 10, 1979, through Sept. 15, 1980; *Defense/Space Business Daily*, issues from Sept. 5, 1979, through Sept. 2, 1980; F. R. Harnden, Jr., *Smithsonian Mag.*, 11(6):110–114, September 1980; *NASA Activities*, monthly issues from September 1979 through August 1980; 1979 aerospace highlights, *Astronaut. Aeronaut.*, 17(12):34–91, December 1979; *Sci. News*, weekly issues from Sept. 4, 1979, through Sept. 6, 1980; Statements before the Subcommittee on Space Science and Applications, Committee on Science and Technology, U.S. House of Representatives, by A. J. Calio (2/7/80), T. A. Mutch (2/20/80), S. I. Weiss (9/16/80), A. J. Calio (9/17/80), R. E. Smylie (9/17/80), and A. J. Stofan (9/18/80).

Speciation

Several models of speciation are current. As the detailed evolutionary histories of living populations are generally unknown, it is usually impossible to adequately test these competing explanations. The fossil record of the deep sea, however, offers a unique opportunity to perform such tests because in pelagic sediments continuous high (100–1000-year) resolution records of the past few million years are available. Pelagic microplankters are preserved in great abundance throughout these sedimentary sequences, providing an amazingly detailed record of floral and faunal change. Recent work with the radiolarian genus *Pterocanium* has shed new light on speciation patterns in the Radiolaria and, by implication, on other planktonic organisms as well.

Piston cores and Deep-Sea Drilling Project cores from throughout the world ocean have been sampled for the radiolarian faunas contained in them. Sediments of equivalent age (as determined by magnetic reversal stratigraphy) have been used to reconstruct the biogeographical patterns of *Pterocanium* distribution, and the various species present at each level have been identified and described. When several levels are viewed in succession, a complicated picture of speciation and phyletic evolution emerges. By using quantitative measures of character change, the patterns can be analyzed statistically as well.

Observed patterns. Seven million years ago, three different species of *Pterocanium* existed—one each in low, middle, and high latitudes. Each species is distinctly different from the others over most of its geographic range, but in regions of intense upwelling intermediate forms between species are occasionally seen. Differences are also seen within the low-latitude Indian and Pacific faunas. High abundances of the subtropical species *Pterocanium audax* and *P. trilobum* in the Indian Ocean reflect a subtropical aspect in the fauna of this region that persists to the present day. With passage of 1,000,000 years of time, the morphologies and abundances of these species change. *Pterocanium audax*, *P. trilobum*, and intermediate forms are still common in subtropical sediments. In *P. korotnevi*, a high-latitude form, changes in shell morphology are seen which prefigure later evolutionary trends. In the equatorial sediments of the Pacific, *P. trilobum* morphology becomes less common, and the population mean shifts more toward a new form—*P. sp.* "*A.*" However, at 6 Ma (mega-annum, or million years before present), only the first stages of speciation have taken place.

Early Gilbert levels (4.5 Ma) record further changes. Subtropical *P. audax* has extended its range into transitional subpolar sediments, as has *P. trilobum*. Typical *P. korotnevi* forms are still present, but many specimens are morphologically intermediate between this species and another, newly evolved one, *P. praetextum*. This species is already well established in Gilbert times in the equatorial Indian Ocean. Rare transitional forms are seen between *P. praetextum* and *P. trilobum*, which has become more common in low latitudes. "Early" forms of *P. prismatium* are seen, primarily in the equatorial Pacific. Thus, in the 1,500,000 years between the 6,000,000- and 4,500,000-year time slices, speciation has taken place, with the ancestral *P. korotnevi* giving rise to *P. praetextum*. The presence of abundant intermediate forms indicates that this speciation event is not yet complete.

Two million years later (at 2.5 Ma), the picture has finally stabilized. Although both *P. audax* and *P. trilobum* are still present, intermediate forms are not. Well-developed *P. praetextum* is found in most pelagic sediments. *Pterocanium prismatium* is distinctly demarcated from other forms and has become a common member of the equatorial fauna.

From 2.5 Ma to recent time, extinction rather than speciation occurs. *Pterocanium audax* disappears at approximately 2.0 Ma, *P. prismatium* at 1.8 Ma. The recent fauna therefore contains only four species in this genus: *P. trilobum*, *P. sp.* "*A*," *P. korotnevi*, and *P. praetextum*. Figure 1 summarizes this admittedly complicated story.

Biologic meaning of radiolarian morphology. The morphological changes described above are assumed to be due to underlying genetic changes. Radiolarian biology is insufficiently well known to establish this with any certainty. Ecophenotypic variation could conceivably produce at least part of the observed pattern. For this reason, the interpretations which follow must be treated with caution. Only further biologic work will resolve this problem.

Role of hybridization. Speciation in these radiolaria occurs in several different ways. One mechanism may involve hybridization. In these cases,

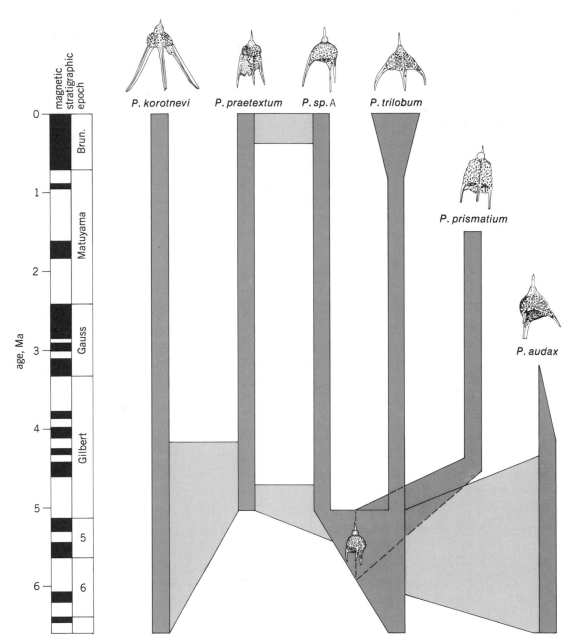

Fig. 1. Phylogeny of the species complex *Pterocanium*. Lightly screened sections shows zones of hybridization.

speciation occurs first by an extension of the range of morphology in the appropriate direction. Rare *P. korotnevi/P. praetextum* are first observed in 6,000,000-year-old high-latitude sediments. Only later do these forms become common elsewhere. Thus the second stage in speciation is an extension of range into new areas. Hybrid forms and further change occur in the new region. For example, in Gilbert-level Indian Ocean samples, well-developed *P. praetextum* are present, along with rare *P. praetextum/P. sp. "A"* transitional forms. In high-latitude sites, *sp. "A"* is not present, and neither are these particular transitional forms. Change has occurred in these latitudes, but the ancestral range is in fact not the location of the complete evolutionary transition. It is in those regions, where hybridization with other forms is occurring, that the final transition first takes place. In this particular example, *P. praetextum* is in

many respects morphologically intermediate between *P. korotnevi* and *P. sp. "A."* In areas where gene flow from both the ancestral *P. korotnevi* and from *P. sp. "A"* is occurring, relatively rapid establishment of the descendant form can take place. Establishment of true *P. praetextum* morphology takes place in the ancestral range (high latitudes) later in the Gauss (2.5 Ma). Hybridization also occurs over most of the range of *P. audax*. However, unlike *P. praetextum*, this does not result in the creation of a new species. Both *P. audax* and *P. trilobum* retain their identities as distinct species, and no successful intermediate species arise.

Nonreticulate modes of speciation. The origin of *P. sp. "A"* and *P. prismatium* is due to two other forms of speciation. *Pterocanium trilobum* gives rise to a well-differentiated population in the late Miocene of the equatorial Pacific. *Pterocanium trilobum* concurrently disappears from these sedi-

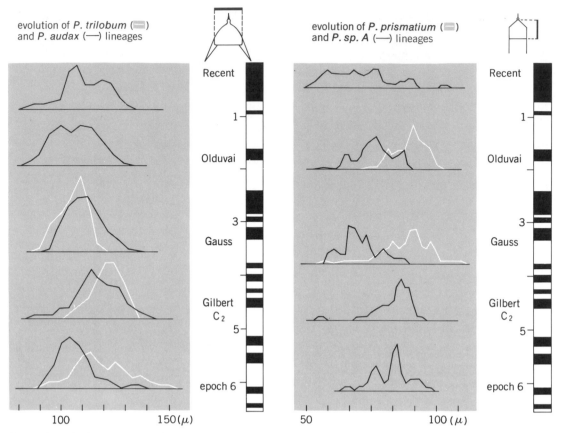

Fig. 2. Quantitative measures of phyletic change in *Pterocanium* species. Parameters measured are shown in the upper right by a heavy bar.

ments and becomes restricted to subtropical and equatorial Indian Ocean environments. With the close of the Miocene, the low-latitude population begins to differentiate. *Pterocanium sp. "A"* forms, already present within the range of observed morphologies, become more common. Eventually a distinct species appears. Speciation here is more the result of loss of intermediate forms than the result of the evolution of them. *Pterocanium prismatium* displays a different pattern. Elaborating on an extreme morphology, *P. prismatium* evolves by rapid development of a new mode and displacement of its mean away from that of the ancestral population. By 4.0 Ma, *P. prismatium* is well developed, and its evolution slows down dramatically.

Phyletic evolution. Speciation processes are but one type of evolutionary change. Change without splitting (phyletic evolution) may also occur. In the genus *Pterocanium*, speciation itself is a relatively slow process. The transition from *P. korotnevi* and *P. sp. "A"* to *P. praetextum*, and the evolution of *P. prismatium* and *P. sp. "A,"* each require about 1,000,000 years for completion. Subsequent to their emergence, these species continue to evolve phyletically. A new subspecies of *P. praetextum* gradually appears in the equatorial Pacific over a 1,000,000-year interval. After reaching a size maximum early in its history, *P. prismatium* follows a long-term trend toward smaller sizes. *Pterocanium sp. "A"* also decreases in size. In recent faunas, this trend in *P. sp. "A"* is enhanced by hybridiza-

tion with the much smaller *P. praetextum* species. *Pterocanium audax* evolves by changes in pore architecture and mean thoracic size over the entire interval studied. *Pterocanium trilobum* populations become recognizably bimodal in several characters in Pleistocene sequences. This bimodality may represent the early stages of a new speciation event. Only *P. korotnevi* seems to remain stable throughout the interval studied. Some simple measures of these changes are shown in Fig. 2. Stasis within a species is rare.

Summary. What can be said in general about speciation mechanisms in the deep sea? Speciation by large-scale hybridization may be a feature of evolution restricted to simple organisms such as radiolarians. Speciation by splitting and divergence is also seen. These splitting events are observed in successive populations from a single location. Allopatry, if it existed, was restricted to the vertical zonations often seen in deep-dwelling radiolarians. Sympatric speciation may be an important mode of evolutionary change in pelagic organisms. It is hoped that further work with deep-sea fossil records will shed more light on these processes.

For background information *see* SPECIATION in the McGraw-Hill Encyclopedia of Science and Technology.

[DAVID LAZARUS]

Bibliography: D. E. Kellog and J. D. Hays, *Paleobiology*, 1(2):150–160, 1975; D. R. Prothero and D. B. Lazarus, *Syst. Zool.*, vol. 29, no. 2, 1980.

Spin-polarized atomic vapors

Atoms in an atomic vapor, under normal conditions, are in thermal equilibrium, occupying the lowest electronic energy state (ground state). Each electron of an atom has an intrinsic angular momentum called spin. Associated with the intrinsic spin is a magnetic moment of invariable magnitude; that is, each spin acts like a tiny bar magnet. The spins of pairs of electrons in atoms couple in such a way that their magnetic moments cancel. However, some spins do not pair, and the atom must possess a resultant magnetic moment. A nonzero resultant magnetic moment must always be the case for atoms and molecules with odd numbers of electrons. A magnetic moment is also associated with orbital motion of the electron.

When an atomic vapor is subjected to an external magnetic field, the magnetic moment of each atom experiences a torque. The applied magnetic field tries to orient the moments along the field direction. This orientation is opposed by thermal agitation in the vapor which tends to create a random distribution of the direction of each magnetic moment (spin axis). When the spins are pointing in random directions, the atomic vapor is said to be unpolarized.

In 1950 A. Kastler proposed a method of orienting paramagnetic atoms in their ground state by illuminating them with circularly polarized optical resonance radiation. By this process, called optical pumping, considerable orientation of the atomic spins along a preferred direction can be achieved, and the atomic vapor is said to be spin-polarized. The degree of polarization is a measure of the orderly orientation relative to the chaotic thermal distribution. Figure 1 is a schematic representation of electron spin orientations in unpolarized and polarized atomic vapors.

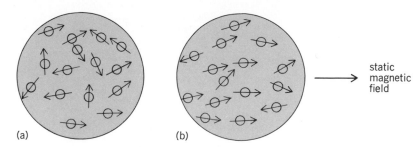

Fig. 1. Simplified schematic representation of (a) unpolarized and (b) polarized atomic vapor. The symbol denotes an atomic spin which is a quantum phenomenon.

Production. Experimentally, it is quite simple to spin-polarize an atomic vapor (Fig. 2). A small amount of the sample (for example, cesium metal) is distilled into an evacuated glass cell. Varying amounts of buffer gases such as helium or nitrogen are also added, and the cell is sealed off. It is then placed in a region of uniform magnetic field of a few gauss ($1 \text{ G} = 10^{-4} \text{ T}$). The sample is vaporized by heating the glass cell. Circularly polarized light from a resonance lamp illuminates the glass cell. Usually the light from the resonance lamp is filtered to narrow the spectral distribution and to thereby improve the pumping efficiency.

The optical pumping process is illustrated by considering a simple paramagnetic atom with a $^2S_{1/2}$ ground state and $^2P_{1/2}$ excited state (Fig. 3). Alkali atoms fall into this category. Under the influence of external magnetic field, the ground and excited states are each split into two sublevels (Zeeman effect) designated as A,B and C,D respectively, as diagrammed in Fig. 3. Under normal conditions of thermal equilibrium, when all the atoms are in the ground state, with nearly equal numbers occupying the sublevels A and B, the vapor is unpolarized. When the sample is illuminated with the

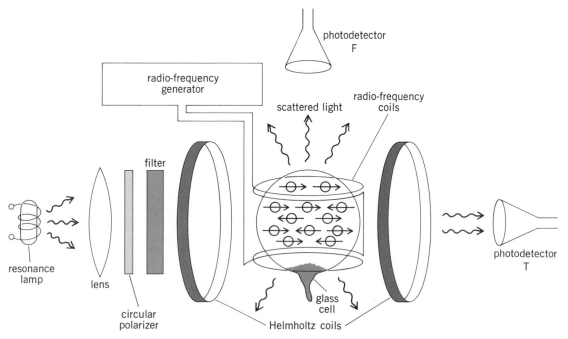

Fig. 2. Experimental arrangement used in the generation of spin-polarized atomic vapor.

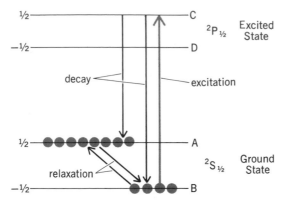

Fig. 3. Effect of the optical pumping process on a simple paramagnetic atom. The solid circles denote an atom.

resonance light, the atoms in the $^2S_{1/2}$ ground state are excited to the $^2P_{1/2}$ state. Circularly polarized resonance light is absorbed by the ground state atoms in only one of the two sublevels. For example, right circularly polarized light can excite only the atoms in sublevel B to sublevel C. The atom in the excited state decays back to the ground state in usually a hundredth of a microsecond. Although the excited atom is more likely to fall to the B level than to the A level, the pumping light can soon return it to the C level. In the absence of any relaxation mechanism, all atoms will eventually be pumped to the A sublevel, and the atomic vapor will be highly spin-polarized.

The degree of spin polarization in an atomic vapor is characterized by the equation,

$$\text{Percentage polarization} = \left[\frac{n_A - n_B}{n_A + n_B}\right] \times 100$$

where n_A and n_B are the number of atoms in ground sublevels A and B respectively.

Detection. When all atoms are pumped to one of the two ground sublevels, the vapor will become transparent to the incoming pumping light, and intensity in the photodetector T in Fig. 2 will increase. This is referred to as transmission monitoring. When complete pumping has taken place, there will no longer be any resonantly scattered light, and the intensity of the fluorescent light reaching the photodetector F will vanish. This is called fluorescence monitoring.

A large class of atoms and ions have been spin-polarized by this technique. Alkali metals cesium, rubidium, and sodium are most widely used as they pose fewer experimental problems.

Properties. In a spin-polarized atomic vapor there is a nonthermal population distribution of atoms in the ground sublevels. The natural tendency of the vapor is to restore the thermal distribution, that is, to nearly equalize the sublevel populations. The pumping light has to counteract this tendency of the spins to relax. This relaxation is described as the time it takes for the spin-polarized vapor to reach its equilibrium state of no polarization in the absence of pumping light, and is also known as relaxation in the dark. There are several ways of studying this. Recent techniques utilize probing of the atomic vapor with a weak light.

Collision of spin-polarized atoms with the internal walls of the glass cell was found to be a major source of spin relaxation. The inert buffer gases in the glass bulb lengthen the relaxation times significantly by slowing down the diffusion of spin-polarized atoms to the walls. The relaxation times can be considerably lengthened by using a paraffin or silicon coating on the walls. Spin relaxation times have been observed on the order of several seconds in oriented alkali vapors.

The relaxation of the atomic spins is considerably slowed down by the nuclear angular momentum of the atoms. The nucleus behaves like a massive flywheel with a large capacity to store angular momentum and release it slowly to the orbiting electrons. Experiments are in progress to obtain a clearer understanding of the role of nuclear spin in polarized atomic vapors.

In general, the inert buffer gases are relatively ineffective in disorienting the spins of the paramagnetic atomic vapor. Their relative effectiveness increases rapidly from helium to xenon by almost a factor of a million. Recent studies of the spin-relaxation effects of the inert buffer gases have revealed new mechanisms responsible for disorientation. In the case of rubidium with xenon (or krypton), it has been conclusively demonstrated that rubidium is disoriented by xenon by forming a relatively long-lived Rb-Xe molecule rather than by short binary collisions. Recent experiments with sodium vapor have also revealed such effects. These studies are of fundamental importance in understanding the atom-atom interaction.

The spin-polarized atomic vapor can be quickly depolarized by applying a radio-frequency magnetic field (Fig. 2). The exact frequency depends upon the strength of the uniform static field. At resonance, the light transmitted through the vapor and reaching the photodetector T decreases (fluorescent light reaching F increases). This powerful technique of magnetic resonance spectroscopy has revolutionized fundamental understanding of nature. With long spin-relaxation times, very narrow resonances have been obtained. This has led to high-precision measurements of many fundamental physical constants such as magnetic moments and hyperfine constants. Present effort is directed toward further extending the accuracy of the experimental measurements. This is crucial for testing some of the fundamental theories of physics.

In the last several years tunable dye lasers have found extensive applications in the generation and study of spin-polarized atomic vapors. With conventional resonance lamps, capable of only a few microwatts of useful light output, significant spin polarization can be achieved only in optically thin vapors (typically $10^{10} - 10^{11}$ atoms/cm^3). In contrast, a high degree of spin polarization has recently been achieved in dense vapors ($10^{14} - 10^{16}$ atoms/cm^3) by using hundreds of milliwatts of light output from dye lasers. Weak atomic interactions play a significant role in spin relaxation at high vapor density and are now being studied in many laboratories.

Another interesting property of the spin-polarized vapor is the effect of spin exchange collisions. By mixing vapors of different elements, spin polarization can be transferred from one easily polarizable specie to another that is not so easily accessible by direct optical pumping. The transfer of spin takes place through collisions. By using this powerful technique known as spin exchange spectroscopy, many atoms that cannot be readily spin-polarized have been investigated. Spin exchange collisions often broaden and shift the magnetic resonance lines, but it was recently demonstrated that in dense vapors under certain conditions, the spin exchange collisions can cause narrowing of magnetic resonance lines. In a spin-polarized atomic vapor, spin exchange collisions between like atoms usually do not contribute to any disorientation of the resultant magnetic moment; however, recent studies with dense cesium vapor optically pumped with a tunable dye laser (blue light of wavelength 459.3 nm) have demonstrated significant violation of this rule. A mixture of cesium vapor and hydrogen gas becomes chemically reactive when irradiated with blue laser light (450 nm). The reaction products have been identified by using spin-polarized cesium vapor. Efforts are also under way to use spin-polarized atomic vapors to detect small concentrations of free radicals during photolysis. Spin polarization of transient paramagnetic atoms in hydrocarbon flames at atmospheric pressure has been reported.

Uses. An impressive variety of atoms and ions have been studied using spin-polarized atomic vapors. In their excited states these vapors have revealed a wealth of new information regarding the properties of excited atoms. This has led to the refinement of many theoretical ideas and development of new ones. Spin relaxation studies in polarized atomic vapors are extensively used to study atomic interactions, such as molecular formation, in great detail. Chemists, too, are using spin-polarized atomic vapors to gain deeper insight into reaction dynamics. With the advent of laser-induced photochemistry, many new applications of spin-polarized atomic vapors are anticipated.

Spin-polarized atomic vapors are utilized as detecting elements in magnetometers. At present, magnetometers that use either spin-polarized rubidium vapor or metastable helium gas are being commercially manufactured. They have been widely employed in rockets, satellites, and space probe equipment.

Odd isotopes of inert buffer gases such as ^{3}He and ^{131}Xe (these have a nonzero nuclear spin) have had their nuclear spin polarized using spin exchange collisions in a mixture of spin-polarized alkali vapor and the inert buffer gas. Nuclear spins are hardly affected by either wall collisions or intermolecular collisions. As a result, their spin relaxation times are several orders of magnitude longer than atomic spin relaxation times. Such oriented nuclei, when produced on a large scale, will find ready applications as polarized targets in nuclear physics experiments. A leading industrial laboratory is involved in research and development of a nuclear magnetic resonance (NMR) gyro using nuclear polarized inert gases in inertial navigation systems.

For background information see MAGNETIC MOMENT; MAGNETIC RESONANCE; OPTICAL PUMPING; SPIN (QUANTUM MECHANICS); ZEEMAN EF-

FECT in the McGraw-Hill Encyclopedia of Science and Technology.   [N. D. BHASKAR]

Bibliography: A. L. Bloom, *Sci. Amer.*, October 1960; R. A. Bernheim, *Optical Pumping: An Introduction*, 1965.

Stereology

Stereology is the study of solids by using mathematical probability theory to obtain very precise and complex structural information quite simply; otherwise the information could be obtained only by painstaking serial reconstruction techniques. This multidisciplinary technique is used, for example, in materials sciences, metallurgy, petrology, mineralogy, life sciences, astronomy, and image analysis.

Samples and profiles. Subjects of experimental interest often contain many smaller internal structures (or phases) that when added together are responsible for providing the overall shape and characteristics of what is recognized as a solid object. Rocks consisting of crystals and tissues of cells are familiar examples. To study the properties of an object, it is particularly useful to quantify its internal structures in terms of their frequencies, sizes, shapes, surfaces, volumes, and lengths, or, in other words, to describe them morphometrically. However, when the structures are very small and extremely numerous—as in biology, where only a few grams of tissue may contain more than a billion cells—the problem of obtaining useful information from such an enormous population would seem formidable indeed. The solution to the problem is to collect structural information from a very small sample and then to extrapolate this information to the entire object. This small but representative sample is obtained by selecting planes randomly oriented (for example, a polished surface of a hard object, such as rock, alloy, or ceramic, or a thin slice of a soft object, such as biological tissue, food product, or plastic) and then collecting measurements of the cut structures, which are now called profiles. The only problem remaining is to know what the quantitative relationship is between the two-dimensional (planar) profile measurements and the actual three-dimensional structures in the "solid" object. Stereological equations, which are derived from geometric probability theory, define most successfully this relationship between two- and three-dimensional space. Although a thorough understanding of the mathematical theory behind these equations is often reserved for mathematicians, the equations themselves are surprisingly simple and very easy to solve. Before considering some of the equations, however, a brief example of the reasoning behind the theory might be useful for developing an understanding of how they bridge the gap between the dimensions.

Volumes of internal structures. Illustration *a* shows a cube containing a central partition that divides it into two compartments (top and bottom) of equal volume. The problem is to estimate stereologically the relative volumes of the two structural compartments in the cube (object). Since they have already been defined as being equal, the answer is 50% top and 50% bottom. The stereological solution, in contrast to the geometric one, is obtained by placing points *P*—at random—into the

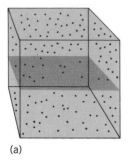 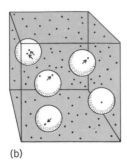 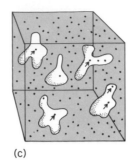

(a)   (b)   (c)

Three applications of stereology for estimating the relative volume of a structural compartment when it occurs within a larger containing or reference volume (in this case the cube): (a) cube divided into two compartments of equal volume; (b) same cube containing spheres of equal size; (c) same cube containing irregular structures. The relative volumes are found by solving Eqs. (1), (2), and (3), respectively.

cube and then scoring their location in the top, P (top), or bottom, P(bottom), compartment. Since both compartments have identical volumes, the chance of receiving a point is likewise identical, namely, 50% for the top and 50% for the bottom. The stereological equation for determining the relative volume of the top compartment would therefore be Eq. (1), where the volume of the top com-

$$\frac{V(\text{top})}{V(\text{cube})} = \frac{V(\text{top})}{V(\text{top} + \text{bottom})} = \frac{P(\text{top})}{P(\text{top} + \text{bottom})} \quad (1)$$

partment V(top) is expressed as a percent of the volume of the cube V(cube). It should be noted that stereological equations provide estimates rather than exact solutions because they represent approximating functions. For example, as the number of randomly placed points in the cube increases, the estimate for the relative volumes of the compartments comes progressively closer to the exact solution of 50% top:50% bottom.

Illustration b shows the same cube containing several spheres of equal size. The relative volume of these spheres can once again be obtained geometrically by using a radius to calculate first the volume of a single sphere, and then taking the sum of the five spheres and dividing it by the volume of the cube (the answer is 5%). A similar answer is obtained stereologically by using the random points. The proportion of the points in the two compartments depends entirely on the relative volumes of the compartments (in this case, the total volume of spheres is considered for the calculation). Therefore, when 100 random points are placed within the cube, a typical experimental result might be as given in Eq. (2). Had 1000 points

$$\frac{V(\text{spheres})}{V(\text{cube})} = \frac{P(\text{spheres})}{P(\text{cube})} = \frac{4}{100} = 4\% \quad (2)$$

been used instead, the answer would have been much closer to the true value of 5%.

Illustration c shows a problem similar to those in illustration a and b, except that the very irregular shapes of the internal structures no longer allow a direct geometric solution. Illustration c is the situation that most often occurs in nature. At this point, stereology becomes a particularly powerful method because it provides an estimate for the relative volume of the internal structures with the

same ease as described for the first two problems as in Eq. (3). In practice, random planes through

$$\frac{V(\text{structures})}{V(\text{cube})} = \frac{P(\text{structures})}{P(\text{cube})} \quad (3)$$

the object are collected, and either random or systematic sets of points (called a test system) are applied to the planes. Points overlying the cut profiles are counted as being "inside." These point-counting data are interpreted three-dimensionally by using Eq. (4), where the volume density V_V of a

$$V_{V(i)} = \frac{P_i}{P_T} \quad (4)$$

structural compartment i (i = the general case) is equal to the number of points falling on profiles of structure i divided by the total number of points, P_T, used for the counting.

Surface area of internal structures. The total surface area of the internal structures shown in illustration b and c can also be estimated stereologically, but, in addition to the points used for the volume density estimate, a second measuring probe is needed. In this case, many linear probes or "needles" are placed randomly within the cube, and the number of times they penetrate the surface of the structures divided by their total length is proportional to the surface area of the structures. This relationship has been defined stereologically by Eq. (5), where the surface density (total surface

$$S_{V_i} = \frac{2I_i}{L_T} \quad (5)$$

area/unit volume) of a structure i equals twice the number of intersections (where a linear probe enters a structure) divided by the total length of the linear probes, L_T.

A variety of stereological methods are available for estimating many different morphological characteristics using equations similar to those just described.

Biomedical applications. One of the goals of biomedical research is to characterize the complex responses of cells to a variety of experimental and pathological conditions. Stereology is particularly useful in this respect because it allows the assessment of changes in the structure of cells and organelles with considerable accuracy. Structural characteristics, such as volume, surface area, length, and frequency, are in reality the visual representations of highly ordered arrays of biological molecules, which, of course, can be assayed biochemically. By combining such structural and functional data into equations that can then be integrated into mathematical models, opportunities are being created for dissecting complex cellular responses into simpler, more interpretable events.

[ROBERT P. BOLENDER]

Bibliography: R. P. Bolender, Correlation of morphometry and stereology with biochemical analysis of cell fractions, *Int. Rev. Cytol.*, 55: 247–289, 1978; R. P. Bolender, Morphometric analysis in the assessment of the response of the liver to drugs, *Pharm. Rev.*, 30:429–443, 1979; R. T. DeHoff and F. N. Rhines, *Quantitative Microscopy*, 1968; E. R. Weibel, *Stereological Methods*, vol. 1, 1979.

Stormscope

Severe weather conditions, such as thunderstorms and tornadoes, are a serious threat to safe air travel, as well as to life and property. The Ryan Stormscope is an airborne device that generates a map of severe weather areas, and displays, in real time, weather conditions in all directions. The Stormscope was developed by Paul Allen Ryan; it is based on the association of electrical activity with severe weather.

Technological foundation. Atmospheric convective wind shear that is associated with severe weather produces a separation of positive and negative electrical charges, mostly as a result of the friction of the moving air currents. As electrical charge separation accumulates, electrical discharges occur. The electrical discharge activity primarily depends upon vertical wind shear and

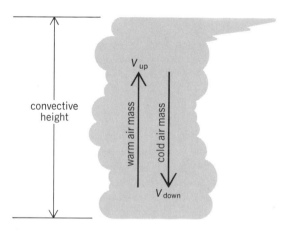

Fig. 1. Convective flow of air currents associated with thunderstorm development.

convective height, and continues to occur repetitively so long as sufficient convective shear persists. Figure 1 illustrates the convective flow of air currents associated with thunderstorm development; V_{up} represents an updraft, and V_{down} a downdraft.

It is the degree of convective shear currents that represents storm intensity, hazards to air travel, and storm danger on the surface of the Earth. As thunderstorms become more intense, convective wind shear increases, and when a severe thunderstorm spawns a tornado, convective wind shear is the greatest. The electrical discharge activity generated by atmospheric convective wind shear becomes a direct measurement of storm severity.

Radiated signals. An electrical discharge is like a powerful transmitting antenna, radiating in all directions, with the speed of light, an electrical signal that is so powerful it can be received and detected at great distances (Fig. 2). These signals can be received at the aircraft by the Stormscope system from all directions almost at the same time, thereby providing data for constructing a map of thunderstorm and severe weather activity for pilot viewing.

Stormscope system. The stormscope includes a very small flat antenna that can readily be used by

Fig. 2. Electrical signals radiated by electrical discharges and received at the aircraft.

any aircraft to receive the electrical signals radiated by storms. The Stormscope system sorts and analyzes these signals in an unusual manner. Basically the system is designed to recognize certain electrical "fingerprints," and perform special analysis upon them. It can be determined first of all whether the signal came from a storm having convective wind shear, and second, where the signal originated. A specialized computer system then accepts the sorted data for further processing. Mathematical functions are performed, arranging the electrical image in a maplike presentation for display purposes. Data are held in a digital memory, and updated as required, for continuous display. The image is conected to a display unit which constructs a map of storm-generated electrical activity. Storm activity more than 300 mi (480 km) away can be displayed.

Image. Each electrical discharge within the atmosphere is analyzed as to bearing and range relative to aircraft heading by the Stormscope system. A bright green dot is placed upon the display

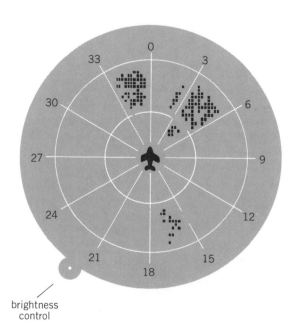

brightness control

Fig. 3. Display image on the Stormscope.

screen, showing the azimuth and range of the discharge relative to aircraft heading. Since the electrical discharges are momentary, the image is held in memory for continuous display. As repetitive discharges occur within the atmosphere, clusters of bright green dots form a maplike image of the thunderstorm's activity. Push-button panel selectors can be used to select 40-, 100-, or 200-nautical mile (74-, 185-, or 370-km) range settings. The range settings refer to the outer range circle on the display, with the inner range circle representing one-half of the range setting. On the most distant range setting, the edge of the display is calibrated for 260 naut mi (482 km) from the center of display. With the 360° continuous viewing capability of the Stormscope, the pilot can view thunderstorm activity over 212,000 mi² (728,000 km²) simultaneously.

Figure 3 illustrates a typical display image of thunderstorm convective activity. Assuming this image is generated while set on the 200-naut mi (370-km) range setting, the display indicates that there are convective shear currents at the 11:30 position from the aircraft between 100 and 200 naut mi (185 and 370 km). There is similar activity centered at the 1:30 position, extending closer to the aircraft position, and finally a smaller cluster of activity at the 5:30 position.

For background information *see* AIR NAVIGATION; ELECTROMAGNETIC RADIATION; LIGHTNING; STORM DETECTION; THUNDERSTORM in the McGraw-Hill Encyclopedia of Science and Technology.   [PAUL ALLEN RYAN]

Strip mining

Lignite overburden contains silicates, carbonates, and sulfates as major mineral groups. It also contains sulfides, which are uncommon in soils and which are major causes of acidity as they alter to produce sulfates and oxides in the weathering mine spoil. In the resulting acidic oxidizing environment, chlorite weathers to vermiculite and eventually to smectite. Lignite overburdens in Texas have mineralogy and a particle-size characteristics that are conducive to reclamation, provided loamy materials are used and proper fertilization and management practices are applied. The potential for abrupt development of extreme acidity in mine spoils which contain sulfides requires careful management practices to maintain vegetative cover.

Lignite occurs in Alabama, Arkansas, Louisiana, Mississippi, and Texas. It is being mined in Texas at several locations, and 21 mines are in operation or projected for operation by 1984. Lignite is extensive in North Dakota, Montana, South Dakota, and Wyoming, and it occurs in Alaska in several locations. Lignite and other low-rank coals occur in Canada, in the Soviet Union, and in western Europe and southward to Turkey. A few areas of low-rank coal are mapped in Australia, Argentina, and Chile.

In eastern Texas, where lignite surface mining is currently expanding, soils have low native fertility and very low permeability. Also, they are difficult to manage. Properly managed spoil can effectively support crop production, and the new soils that are forming from spoil have more desirable physical properties than the natural soils in much of the area. The properties of these new soils are largely determined by the mineral composition and the particle-size distribution because the active organic matter content is low.

Sulfide minerals. Pyrite and marcasite are the major inorganic sources of acidity in lignite mine spoil. The abundance, distribution, and properties of these iron sulfides influence acidity development during reclamation. Lignites of the Wilcox group (Eocene age) and associated strata of Texas contain pyrite as individual crystallites, massive forms, framboids, and polyframboids (Fig. 1). The surface area of framboidal pyrite is greater than that of the massive pyrite of the same particle size, and surface area is correlated with the rate of acid formation. The pyritic sulfur content of lignite and lignite plus overburden shale strata

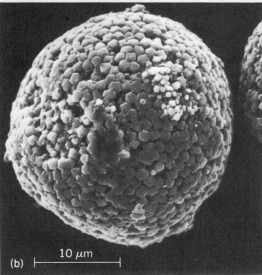

Fig. 1. Pyrite in lignite. (*a*) Massive particles may have smooth (arrow) or rough surfaces. (*b*) Framboids composed of many individual crystals. (*From C. E. Pugh, Influence of Surface Area and Morphology on the Oxidation of Pyrite from Texas, unpublished Ph. D. dissertation, Texas A&M University, p. 50. 1978*)

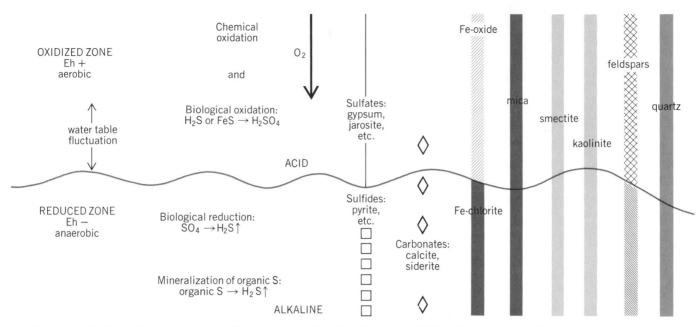

OXIDIZED ZONE
Eh +
aerobic

Chemical
oxidation

and

Biological oxidation:
H_2S or $FeS \rightarrow H_2SO_4$

O_2

water table
fluctuation

Sulfates:
gypsum,
jarosite,
etc.

Fe-oxide

feldspars

mica

quartz

smectite

kaolinite

ACID

REDUCED ZONE
Eh −
anaerobic

Biological reduction:
$SO_4 \rightarrow H_2S\uparrow$

Sulfides:
pyrite,
etc.

Fe-chlorite

Carbonates:
calcite,
siderite

Mineralization of organic S:
organic $S \rightarrow H_2S\uparrow$

ALKALINE

Fig. 2. Schematic illustration of the composition and weathering of lignite overburden in eastern Texas. (*From J. B. Dixon, L. R. Hossner, and A. L. Senkayi. Mineralogical properties of lignite overburden as they relate to mine spoil reclamation. in J. A. Kittrick, D. S. Fanning, and L. R. Hossner, eds., Acid Sulfate Weathering: Pedogeochemistry in Relation to Manipulation of Soil Materials. Soil Science Society of America, publication pending*)

in the Wilcox formation of Texas is 0.1 to 1%. Pyrite content varies greatly in a given lignite bed; usually it is highest near the top and bottom of the bed. Marcasite rarely has been identified in the lignite overburden of Texas, but it is more abundant than pyrite in the overburden of the Frechen lignite mine in West Germany.

Sulfate minerals. These are among the first reaction products of sulfide oxidation in lignite and lignite overburden. Jarosite, gypsum, barite, melanterite, szmolnokite, and rozenite have been identified in lignite overburden or spoils of Texas (see the table). Distinct yellow jarosite efflorescences often form on lignite overburden cores. Jarosite is believed to form only under very acid (pH 3–4) conditions. The low solubility of jarosite makes it a useful indicator of the weathering history of the deposit.

Gypsum occurs as macroscopic crystals associated with the lignite spoil. The amount of native sulfate sulfur in Wilcox group lignites of Texas is small (0.1 to 0.2%), and it often is present in local concentrations. The importance of gypsum is mostly related to its use as an indicator of the oxidizing conditions in the deposit. Gypsum is produced by the oxidation of pyrite in the presence of calcium, which may come from calcium carbonate. Each of the sulfate minerals can be a useful indica-

tor of the oxidation status of the lignite overburden or mine spoil.

Carbonate minerals. Calcite nodules have been observed in overburdens of the Wilcox group lignites of eastern Texas, but their supply is limited. Siderite occurs in lignite overburden of Texas, often in veins about 10 cm thick. It is transformed to a goethite-encrusted mass by weathering that produces a hard rind of goethite. Goethite-coated rocks containing siderite accumulate on eroded soil surfaces and also become a prominent feature of reclaimed spoil. Large tabular masses of siderite up to 1 m in length have been observed on the Claiborne and Wilcox group lignite overburdens of Texas. Since the weathering of iron and manganese carbonates normally involves oxidation of the cations and consumption of hydroxyl during precipitation as oxides or oxyhydroxides, they may not aid in neutralizing acidity as does calcium carbonate.

Silicate minerals. When lignite mine spoil weathers under acid oxidizing conditions, ferromagnesian (mafic) chlorite alters to smectite and releases iron and magnesium in solution. Chlorite from lignite mine spoil of Texas was transformed to vermiculite and eventually to smectite by acid oxidation in a laboratory experiment. Where chlorite is a significant component of sediments above

Sulfide, sulfate, and carbonate minerals related to lignite mine spoil reclamation

Mineral	Formula	Mineral	Formula
Marcasite	FeS_2	Calcite	$CaCO_3$
Pyrite	FeS_2	Siderite	$FeCO_3$
Barite	$BaSO_4$	Gypsum	$CaSO_4 \cdot 2H_2O$
Jarosite	$KFe_3(SO_4)_2(OH)_6$	Melanterite	$FeSO_4 \cdot 7H_2O$
Rozenite	$FeSO_4 \cdot 4H_2O$	Szomolnokite	$FeSO_4 \cdot H_2O$

lignites, it may contribute a considerable amount of smectite to soils formed in exposed spoil. Increased smectite content of spoil could lead to development of a clay pan or clayey horizon with poor physical properties. Addition of calcium carbonate during reclamation may retard smectite formation, but the effectiveness of this treatment needs investigation.

Iron-chlorite is absent from the weathered (oxidized) upper layers of Wilcox group overburdens of eastern Texas. Yet it is present in the reduced shale that overlies the lignite (Fig. 2). The oxidized zone is mottled with brown, red, and yellow colors of iron oxides that indicate precipitation of iron released from chlorite and pyrite. Although this weathering pattern has been determined at only a few locations, it appears to be a likely result of weathering in eastern Texas and other southeastern states with similar lignite overburden. The depth of weathering is extremely variable, depending on the particle-size distribution of the overburden.

Mica is present in small to moderate amounts in the clay fractions of strata overlying the Wilcox group in Texas. Mica in lignite overburden clay is a muscovite type, and it is expected to release potassium very slowly to plants. Exchangeable potassium will provide an immediate short-term supply of potassium for plants.

Vermiculite is a small component of lignite overburdens studied thus far in Texas. Vermiculite from the shale and vermiculite formed by weathering is likely to remain a small percentage of the clay fraction. Yet it may be important in these materials because it selectively adsorbs potassium.

Smectite is one of the most important minerals in the clay fractions of Wilcox group lignite overburdens in Texas. The high cation-exchange capacity, high surface area, and high shrink-swell potential make the influence of smectite greater than most other clay minerals. The role of smectite in holding plant nutrients is a positive factor for the reclamation of spoil after mining. Where the overburden is clayey and smectite is the dominant layer silicate, the potential for developing an acid soil that is extremely difficult to reclaim is a decisively negative factor. Smectitic clayey acid soils of the southeastern United States are among the most unproductive and difficult to improve.

Kaolinite is an abundant clay mineral in the clay fraction of Wilcox group lignite overburdens. Kaolinite adds to the anion-holding capacity and reduces the cohesive and adhesive properties of clay compared to smectite. Since iron oxides and humus, which also reduce cohesiveness, are deficient in many lignite overburden materials, moderate kaolinite content is a beneficial constituent during reclamation. The kaolinite in Wilcox group lignite overburdens of Texas is well crystallized and primarily coarse clay in size. Presumably it has a low cation-exchange capacity, as do other coarse kaolinite clays.

Feldspars occur mostly in the coarser fractions (sand and silt sizes) of lignite overburdens. They are less abundant in the upper weathered layers than below. Potassium and plagioclase feldspars are present. The amounts and properties of the feldspars are the subject of current investigations.

Quartz is the major component of the coarser fractions (sand and silt sizes) of lignite overburdens. It does not contribute any significant chemical activity to soils, but it is a contributor to the physical properties. Quartz particles are less cohesive than layer silicate particles (such as smectite), and they improve the friability of soils.

The minerals or organic constituents in lignite overburdens of Texas do not provide adequate nitrogen or phosphorus for revegetation. Nitrogen and phosphorus must be supplied for revegetation when these materials are reclaimed.

Summary. The mineralogy of lignite overburden is an important consideration in reclaiming mine spoils. The changing mineralogy of the spoil is an indicator of the course and rate of weathering. Changes in silicates and formation of sulfates and oxides are common features of the weathering process. Fertilization and management choices are related to the mineralogy of the spoil. For background information see COAL; SOIL CONSERVATION in the McGraw-Hill Encyclopedia of Science and Technology.

[J. B. DIXON]

Bibliography: K. Brinkmann, *N. Jb. Miner. Abh.*, 129(3):333–352, May 1977; J. B. Dixon, L. R. Hossner, and A. L. Senkayi, in J. A. Kittrick, D. S. Fanning, and L. R. Hossner (eds.), *Acid Sulfate Weathering: Pedogeochemistry in Relation to Manipulation of Soil Materials*, Soil Science Society of America, publication pending; F. M. Hons et al., in W. R. Kaiser (ed.), *Proceedings of Gulf Coast Lignite Conference: Geology, Utilization and Environmental Aspects*, Rep. Investig. no. 90, Bureau of Economic Geology, University of Texas at Austin, 1978; L. R. Hossner et al., in *Lignite Symposium*, Texas A&M University, April 1980.

Supercritical fields

Quantum electrodynamics, that is, the theory of electrons and positrons and their interaction with the radiation field, has been one of the most successful disciplines of physics. Its formal framework, which essentially dates back to the early work of P. A. M. Dirac, W. Heisenberg, and V. F. Weisskopf in the 1930s and which was completed in the late 1940s by R. P. Feynman, J. S. Schwinger, S. Tomonaga, and others, allows the calculation of atomic properties with virtually arbitrary precision. In spite of the somewhat unsatisfying divergences in the renormalization scheme, quantum electrodynamics may be regarded as a completed theory. There is, however, one phenomenon in quantum electrodynamics which has only recently been fully understood and which leads to a qualitatively new concept: the charged vacuum in strong (electrostatic) fields.

Charged vacuum. The best starting point for a discussion of this idea is the following question: what happens to the atomic electrons if the charge of the nucleus is considerably increased? As discussed below in detail, relativistic effects will become dominant and will qualitatively alter the level spectrum (leading, for example, to a very large fine-structure splitting). A first attempt to account for these effects is the Sommerfeld fine-structure formula, Eq. (1). Here, $\kappa = \pm 1, \pm 2, \ldots$;

$$E_{nj} = m_e c^2 \left[1 + \left(\frac{\alpha Z}{n - |\kappa| + (\kappa^2 - (Z\alpha)^2)^{1/2}} \right)^2 \right]^{-1/2} \quad (1)$$

$n = 1, 2, \ldots$; $\alpha = \frac{e^2}{\hbar c} \cong 1/137$ is the fine-structure constant, m_e and e are the electronic mass and charge; c is the speed of light; and $\hbar$ is Planck's constant divided by 2π. This equation describes the spectrum of electronic bound states E_{nj} in the external Coulomb potential $A_0(r) = Ze^2/r$ of a point with charge Ze. In this case the appropriate equation, the Dirac equation, can be solved analytically.

Because of the term $(\kappa^2 - (Z\alpha)^2)^{1/2}$ Eq. (1) obviously breaks down at $Z\alpha > |\kappa|$. For example, the energy levels as a function of the central charge Z for the $1s_{1/2}$ state with $E_{1s} = m_e c^2 \sqrt{1 - Z^2 \alpha^2}$, and all other states with total angular momentum $j = 1/2$ cease to exist at $Z = 1/\alpha \cong 137$ (Fig. 1). The corresponding wave functions diverge at the origin and become nonnormalizable. This, however, does not imply that the Dirac equation has no solution at high Z, as was first believed. Taking into account the finite extension of the nucleus, one can trace any energy level E_{nj} down to a binding energy of twice the electronic rest energy mc^2, if the central charge is increased as a parameter. At the corresponding charge number, which is called critical (Z_{cr}), the state reaches the negative energy continuum of the Dirac equation (the "Dirac sea") which, according to the hole-theory hypothesis, is totally occupied by electrons. If the strength of the external field is further increased, the bound state "dives" into the continuum. The overcritical state obtains a width and is spread over the continuum. The electron charge still remains localized.

The related phenomena have been investigated and analyzed very carefully. The most important aspect is that the overcritical vacuum state is charged. As already mentioned, in hole theory, the states of negative energy of the Dirac equation are occupied with electrons (Fig. 1). The Fermi surface lies at $E_F = -m_e c^2$. (In the study of weak fields in quantum electrodynamics, it is convenient to put $E_F = 0$, which is equivalent to $E_F = m_e c^2$, since there are in this case no (bound) states between $E = -m_e c^2$ and $E = 0$.) The negative energy continuum states occupied with electrons represent the model for the neutral vacuum of quantum electrodynamics. The infinite charge of that state is renormalized to zero. In field theoretical formalism the same result is obtained by symmetrizing, because of charge conjugation symmetry, between states with $E < E_F$ (occupied with electrons) and $E > E_F$ (occupied with positrons); the vacuum expectation value of the current (charge) operator is then zero: $<0|\hat{j}_\mu|0> = 0$. If now an empty atomic state dives into the negative continuum, it will be filled spontaneously with an electron from the Dirac sea with the simultaneous emission of a positron (that is, a hole) that moves to infinity. The remaining electron cloud of the supercritical atom is necessarily negatively charged. While in ordinary undercritical physics a vacuum state $|0>$ can be defined without charges or currents by choosing the Fermi surface (up to which the levels are occupied) below the lowest bound state,

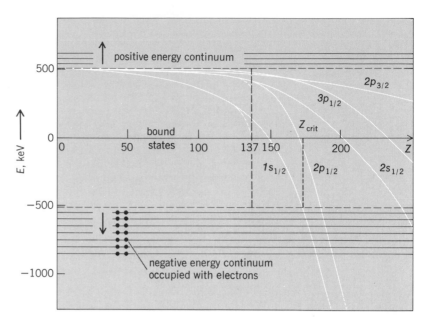

Fig. 1. Energy spectrum of electrons in very strong central electric fields as a function of the central charge Ze. The various shells—energy levels at which the electrons orbit the nucleus—are indicated by curved lines.

this is not possible in the overcritical case ($<0|j_0|0> \neq 0$). Thus the concept of the charged vacuum must be introduced. This is a fundamentally new concept leading to a new understanding of the vacuum as a physical object. As the external field strength increases, each time an electronic bound state joins the negative continuum, the vacuum undergoes new phase transitions and acquires a successively higher charge. Thus, at $Z = 173$ the $1s$ electron level (the lowest energy level; Fig. 1) dives into the negative energy continuum, and the neutral vacuum breaks down. If the central charge is further increased to $Z = 184$, the diving point of the $2p_{1/2}$ level, the vacuum becomes even more highly charged, and so on.

In the limit of an overcritical ($Z\alpha > 1$) point charge, infinitely many electron states (all the $ns_{1/2}$ and $np_{1/2}$ levels) will dive at $Z = 137$ (Fig. 1). One might assume that in this case the vacuum would become infinitely charged. It must, however, now be treated self-consistently, which is, in fact, already necessary for vacuum charges higher than $Z_{vac} \geq 2$. It has been shown that in the point-charge limit any original central charge $Z > 137$ will be shielded by the vacuum in such a way that there cannot exist any point charge with $Z > 137$. In other words, the maximal point charge is $Z_m = 137$ due to the decay of the vacuum. Expressed differently, the maximal coupling constant for point charges in quantum electrodynamics does not exceed $Z_m \cdot \alpha = (137) \cdot (1/137) = 1$. By taking the point-charge limit, an onion-type of shell structure for the vacuum charge is observed, the shells becoming closer and finally coinciding in the point limit.

Clearly, the charged vacuum is a new ground state of space and matter. The normal, undercritical, electrically neutral vacuum $|0>$, whose physical existence is reflected by the well-known vacuum polarization (displacement of electron-

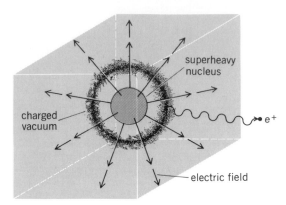

Fig. 2. Formation of charged vacuum in an overcritical field through emission of antiparticles.

positron charges) in strong fields, is no longer stable in overcritical fields. It decays by either positron (for attractive electric fields) or electron (for repulsive electric fields) emission into the new stable, but charged, vacuum.

If the vacuum is defined as a part of space free of real particles, this vacuum can be subject to certain conditions, such as penetration by fields. If these become strong enough, the particle-free vacuum can no longer exist: it must contain real particles (in the special case studied here, these particles are electrons). This is shown in Fig. 2, where the sphere in the center represents the superheavy nucleus, the source of the electric field is indicated by arrows, and the diffuse cloud represents the electrons of the charged vacuum. If this electron cloud is pumped away, new positrons, represented by e^+, will be emitted, and the cloud will reappear.

This necessitates a more general definition of

the concept "vacuum." The standard definition "region of space without free particles" obviously cannot be true for very strong external fields, and the new and better definition "energetically deepest and most stable configuration of space" seems to be more appropriate. The stability of the charged vacuum, in the case of fermions considered here, is assured by the Pauli principle. In the case of a boson vacuum, it must come from the interaction between bosons (sometimes effectively described by nonlinearities in the field equations).

Superheavy quasimolecules. The static overcritical atomic phenomena have been studied since 1968, essentially by two schools, at Frankfurt am Main and at Moscow. Historically, the first detailed investigation of supercritical quantum electrodynamics was inspired by the possible existence of superheavy nuclei. The deformed shell model of theoretical nuclear physics indicates that nuclei near the magic proton numbers 114 and 164 should have strongly enhanced lifetimes. The production of long-lived superheavy elements by fusion of ordinary nuclei would allow the extension of the exact measurements of atomic spectroscopy into a new region of electronic binding energy. The overcritical region ($Z>172$) cannot be reached. Luckily, however, it is at least possible to assemble a supercritical charge for a short period of time in the collision of very heavy ions. Such collisions are semiadiabatic with respect to the electron motion (the ratio of ion velocity to electron velocity is on the order of 1/10), and therefore the electrons will form molecular orbits during the various steps of such an encounter (Fig. 3), orbiting both nuclei together and following their motion. This leads to the concept of intermediate superheavy quasimolecules. For example, in U-U collisions near the Coulomb barrier, the lowest bound state joins the negative energy continuum for some 10^{-20} s. This should be long enough to observe the decay of the neutral vacuum by detecting emitted positrons. The internuclear distance $R(t)$ varies with time. In the limit of vanishing $\dot{R}$, corresponding to the point of closest approach for collisions at the Coulomb barrier, even the electronic structure of a quasiatom with combined central charge $Z_1 + Z_2$ can be reached (Fig. 4).

Thus, heavy-ion physics provides a means of directly testing the formation of quasimolecules. Figure 3 shows how the inner shells are emptied when electrons are excited into higher shells (labeled A) and when they are excited into the upper continuum (labeled B). The processes labeled C and D are, respectively, molecular x-ray transitions in the intermediate quasimolecule and atomic x-ray transitions in the rearranged atom. Measurements of the spectra of emitted x-rays were first carried out for the L- and M-shells and then for the most important K-shell. The quasimolecular production of δ-electrons, K-holes, and positrons suggests conclusions on the shape of the molecular correlation diagrams, that is, the dependence of the binding energy on separation (Fig. 5), and permits the investigation of details of the diving process. These investigations will be impeded somewhat by nonadiabatic and smearing-out effects as a consequence of the finite collision time which is dictated by the Coulomb repulsion

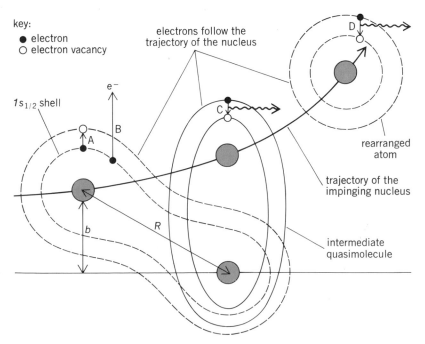

key:
● electron
○ electron vacancy

Fig. 3. Collision of two heavy ions in which a quasimolecule is formed. Distance between the nuclei is indicated by R, the impact parameter by b..

and nuclear attraction of the nuclei.

Some of the processes mentioned are schematically depicted in Fig. 5. A and B denote the Coulomb excitation or ionization which produces a hole in the $1s\sigma$ level. This hole may be set free spontaneously (C) during the diving without any energy. Due to the scattering dynamics, energy can be drawn from the nuclear motion leading to nonadiabatic filling of the hole even at distances larger than the critical distance R_{cr}. This effect has been called an induced transition and will enhance the positron production cross sections. From the shortness of collision time, which is about 2×10^{-21} s, and the corresponding large energy uncertainty, the importance of induced transitions is evident. Process D describes the induced decay of the vacuum or, differently expressed, the dynamically induced electron positron pair production. Finally, process E of Fig. 5 describes the direct production of electron-positron pairs due to the dynamical evolution of the system. It can be shown that it corresponds to the shake-off of the vacuum polarization cloud, that is, the displacement charge originating from the displacement of electrons and positrons in strong fields. It turns out that the vacuum polarization charge cannot follow the swift movement of the two ions, hence it is stripped off, which results in a strong Z-dependence of the cross section [$\sigma_{\text{shake-off}}$ is on the order of $(Z_1 + Z_2)^{19}$].

Some experiments performed so far are depicted in Fig. 6. In general they show good agreement with the theory and confirm the induced and direct (shake-off) positron production. There is not yet evidence for the spontaneous positrons stemming from the decay of the vacuum, but there is justified hope that the use of curium as a target will lead to a sufficiently increased spontaneous positron cross section due to the higher charges (higher overcriticality), so that these positrons may be identified.

It may be possible to delay the time period of overcriticality by making the superheavy system stick together longer than the collision bypass. This can be arranged by using deeply inelastic nuclear reactions in which nuclear forces may bind the system long enough (on the order of 10^{-20} s). The positron spectra should then show the tendency of building up a positron emission line whose natural width (in the case of infinite lifetime of the superheavy system) would be the decay width of the neutral vacuum into the charged one. The theoretical investigation of this idea is encouraging. *See* NUCLEAR REACTION.

Overcritical phenomena in other fields. So far, the behavior of the electron field under the influence of a strong binding potential has been discussed. If the binding energy of a state exceeds the threshold for particle creation ($2m_ec^2$), there results the spontaneous creation of real electron-positron pairs that end as a bound electron and an emitted positron. The charged-vacuum state thus produced is stabilized against further particle creation by the Pauli principle. This is a very fundamental phenomenon. Its investigation and understanding may shed new light on field theories and, in fact, may open a new and exciting area for theoretical and experimental investigation.

The concept of overcritical fields and the change

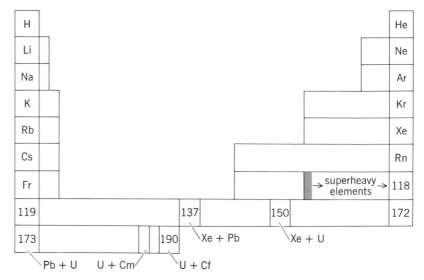

Fig. 4. Extension of the periodic table by quasiatoms and quasimolecules. A few typical superheavy quasiatoms are depicted, and the reactions through which they are obtained are indicated.

of the vacuum state is not restricted to the electron-positron fields and electromagnetic interactions. It may also occur for other fermion fields (nucleons, quarks) and be caused by strong interaction.

Overcritical boson fields. Also, for bosons strong binding in a sufficiently deep potential well can lead to the possibility of particle creation. Here, however, the exclusion principle is not in action to stabilize the vacuum. The production mechanism in overcritical boson fields can be stopped only by the mutual interaction of the created particles or by the strong interaction with the source. This effect was first treated by assuming a self-interac-

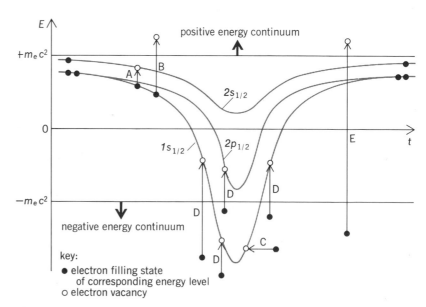

Fig. 5. Inner electron energy levels in a superheavy quasimolecule as a function of time t. At the deepest point of the $1s$ level, the colliding nuclei are at the distance of closest approach.

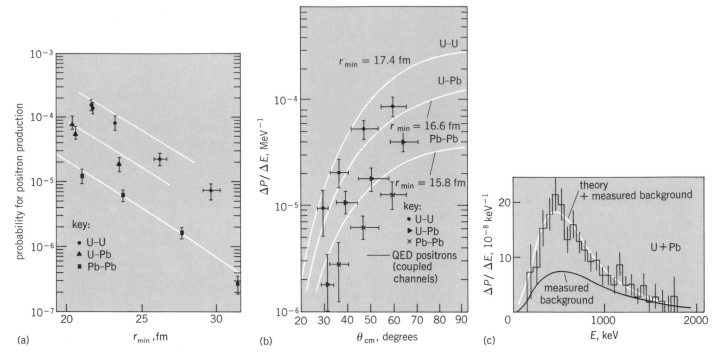

Fig. 6. Various positron production cross sections from superheavy quasimolecules. Theoretical values are indicated by curves, experimentally measured values by data points. (*a*) Excitation function of total positron cross sections as a function of the distance of closest approach r_{min}. (*b*) Positron cross sections observed in coincidence with a scattered projectile as a function of the center-of-mass scattering angle θ_{cm}. $\Delta P/\Delta E$ is the probability that positrons will be produced having energies within the unit energy interval ΔE. (*c*) Positron spectrum for U + Pb quasimolecule.

tion of the type ϕ^4, where ϕ represents the boson field. Study of the quantization of overcritical boson fields under the influence of an electromagnetic potential showed that the resulting very strong (mutual) vacuum polarization will, in fact, stabilize the vacuum. Contrary to the fermion case, the binding energy as a function of potential depth does level off (if the potential vanishes faster than $1|r$), not reaching the threshold for real particle production.

Pion condensation in dense nuclear matter. The pion spectrum in nuclear matter has been investigated. At a certain critical density (which plays the role of a critical interaction strength), the normal nuclear matter becomes unstable. A phase transition occurs leading to pion condensation, that is, a state containing collective particle-hole excitations with the quantum numbers of a pion. The phase transition to pion condensates seems to play an essential role in setting up quasihydrodynamic conditions in relativistic collisions of nuclei on nuclei, causing shock waves and thus high-density compressions in nuclear matter. This could lead to the exploration of nuclear matter at high densities and high temperatures, and as a highly isobaric gas with density isomerism. Detailed calculations will have to treat the time-dependent development of pion condensates in finite, high-temperature nuclei.

The treatment of vacuum excitation for boson fields thus requires advanced many-body techniques. The basic phenomena, however, for such processes as pion condensation and the charge of the vacuum in quantum electrodynamics are closely related.

Strong gravitational fields. A further example of strong binding is the field of a gravitational source with an event horizon (black hole). To understand the nature and development of singularities and horizons, much attention is presently being paid to the interaction of a classically described metric with quantized fields of matter. Recently the possibility of particle emission from a black hole for a time-dependent, collapsing, or static geometry has been discussed.

Guided by the example of the charged vacuum in quantum electrodynamics, the solution of the Dirac (and Klein-Gordon) equation in a Schwarzschild or Reissner-Nordstrøm (black hole with charge and angular momentum) background has been studied. A continuum of solutions for all particle energies was found, the wave functions showing an infinite number of oscillations when approaching the outer coordinate singularity. The appearance of the continuum was studied in the model of an extended gravitational source. When its radius r shrinks to the Schwarzschild radius r_s, all the discrete bound states drop to zero energy. Embedded in the newly emerged continuum are, as in the quantum electrodynamics case, bound-state resonances. Their width results from the possibility of decay into the black hole. The position of these resonances can be understood from the effective potential given by Eq. (2), where M, Q,

$$V_{\text{eff}} = \frac{eQ}{r} \pm \left(m^2 + \frac{L_z^2}{r^2}\right)^{1/2} \left(1 - \frac{2MG}{r} + \frac{Q^2G}{r^2}\right)^{1/2} \quad (2)$$

and L_z are the mass, charge, and angular momentum projection of the black hole respectively, G is the gravitational constant, and units are chosen such that $c = \hbar = 1$. The effective potential is shown in Fig. 7. Even in the case of a neutral black hole (Fig. 7*a*), the gap between the negative and positive continuum (particle and antiparticle states) is narrowed by the attractive gravitational interaction and vanishes for $r = r_s$. In the case of a charged center the charge conjugation symmetry is lifted. For a negative central charge (Fig. 7*b*), the negative continuum is varied in energy. If the Fermi energy exceeds $m_e c^2$, spontaneous electron emission will occur and the vacuum will be unstable. The same is true for opposite charges

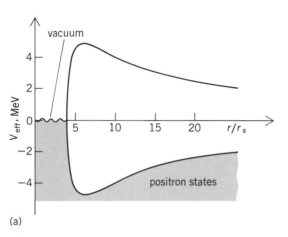

(a)

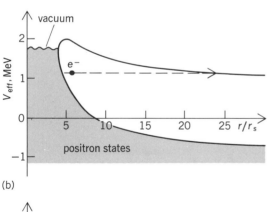

(b)

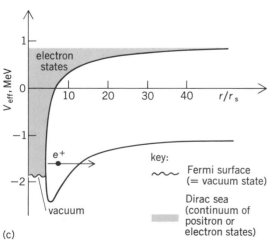

(c)

Fig. 7. Effective potentials V_{eff} of the electron in the field of a gravitational source, as a function of the radius r of the collapsing mass: (*a*) neutral star; (*b*) positively charged star; (*c*) negatively charged star.

(Fig. 7*c*). This transition to a charged electron-positron vacuum leads to a limiting stable charge for a black hole given by Eq. (3), or approximated by Eq. (4). Only one electron has to be created

$$\frac{Z_{lim}e^2}{r_s} = m_e c^2 \qquad (3)$$

$$Z_{lim} = \frac{2GMm_e}{e^2} \qquad (4)$$

since the antiparticle is swallowed by the black hole. The limiting charge to mass ratio, given by Eq. (5) reflects the double ratio of the gravitational

$$\frac{Z_{lim}e}{M} = \frac{2Gm_e}{e} \approx 4.8 \times 10^{-43} \, e/m_e \qquad (5)$$

to the electromagnetic coupling constant. Consequently, black holes, if they exist, are practically neutral as a result of vacuum decay.

For background information *see* BLACK HOLE; QUANTUM ELECTRODYNAMICS; QUASIATOM in the McGraw-Hill Encyclopedia of Science and Technology.   [WALTER GREINER]

Bibliography: H. Backe et al., *Phys. Rev. Lett.*, 40:1443–1446, 1978; J. Rafelski, L. P. Fulcher, and A. Klein, *Phys. Rep.*, 38C:227–361, 1978; J. Reinhardt and W. Greiner, *Rep. Prog. Phys.*, 40: 219–295, 1977; J. Reinhardt, B. Müller, and W. Greiner, *Progress in Particle and Nuclear Physics*, vol. 4, pp. 503–545, 1980; J. Reinhardt et al., *Progress in Particle and Nuclear Physics*, vol. 4, pp. 547–578, 1980.

Superheavy elements

An island of relatively stable atomic nuclei has been predicted to exist beyond the present periodic table by theoretical extrapolations of nuclear properties. Vigorous efforts have been made to discover these so-called superheavy elements in nature and to produce them in the laboratory. Although these efforts have not led to convincing positive evidence, there are still some observations on natural radioactivity which deserve attention. Also, reactions between heavy nuclei have not yet been fully explored as a pathway to superheavy elements in the laboratory. Whatever the final outcome, the search for superheavy elements will lead to a better understanding of the forces that terminate the periodic table at its upper end.

Nuclear and chemical properties. Heavy nuclei are energetically unstable against fission into two fragments. However, a fission barrier due to the interplay of two forces, the attractive nuclear force between nucleons and the repulsive electrostatic force between protons, prevents them from undergoing fission. In the ground state of a heavy nucleus, the attractive force dominates slightly, keeping the nucleus together. Relatively little energy is sufficient to lift the nucleus over the barrier. With increasing atomic number, the barrier decreases in height and thickness, and it should vanish around element 114. Hence, such nuclei should promptly fission whenever they are formed. Furthermore, nuclei can tunnel through the barrier in a process called spontaneous fission, characterized by a half-life which decreases with decreasing barrier height and thickness. This general trend can be overcome by the extraordinary stability

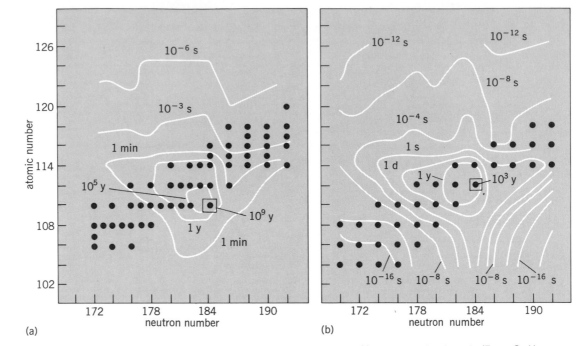

Fig. 1. Calculated overall half-lives for superheavy nuclei shown as contours of constant half-life plotted against atomic and neutron number. (*a*) Earlier treatment. (*b*) More recent treatment. (*From G. Herrmann, Superheavy-element research, Nature, 280(5723):543–549, Aug. 16, 1979*)

gained in nuclei by the completion of major nuclear shells. Such closed-shell nuclei are spherical and stiff against deformation. In the region considered here, shell-closure may produce sizable fission barriers which are now entirely due to shell effects.

Most theoretical studies agree that the next shell closures beyond the existing elements should occur at proton number 114 and neutron number 184. Hence, maximum stability against spontaneous fission is expected for the isotope of element 114 with mass number 184. In addition to spontaneous fission, other decay modes must be considered, namely, α-decay, β^--decay, and electron capture. Figure 1 shows two sets of calculated overall half-lives of superheavy nuclei. The contour lines connect nuclei with equal half-lives, and the points indicate nuclei expected to be stable against β^--decay and electron capture. In both sets the longest overall half-life is not obtained for the doubly closed-shell nucleus $^{298}_{114}\text{X}^{184}$, but for $^{294}_{110}\text{X}^{184}$ with a 10^9-year half-life (Fig. 1*a*) or for $^{296}_{112}\text{X}^{184}$ with a 10^3-year half-life (Fig. 1*b*). This shift to somewhat lower atomic numbers is caused by the α-decay mode which dominates around element 114 and beyond. Both sets of calculations result in an island of superheavy nuclei with a broad shore to the west from where the island is approached in attempts to synthesize nuclei by nuclear reactions. Isotopes with odd atomic numbers (Z) may have longer half-lives than those given in Fig. 1 for even-Z isotopes.

In addition to half-lives, the energy released in different decay modes of superheavy nuclei is of interest, since it forms the basis for specific and sensitive detection techniques. The spontaneous fission decay mode has been widely used in the search for superheavy elements. Theoretical estimates indicate that fission events with extremely

high kinetic energy and a large number of emitted neutrons should constitute a characteristic fingerprint for such nuclei.

The position of superheavy elements in the periodic table has been deduced from quantum-mechanical calculations of their ground-state electronic structure. Accordingly, element 110 is expected to be a homolog of platinum, element 112 a homolog of mercury, and element 114 a homolog of lead. From these positions and from systematic trends within the table, the chemical behavior of superheavy elements has been estimated. For example, high volatility is predicted for several of these elements. Such predictions are relevant for the selection of samples in a search for superheavy elements in nature, and for chemical separation procedures applied to both natural samples and irradiated targets.

Search in nature. Since early estimates indicated half-lives comparable to the age of the Earth for nuclides in the center of the island, many groups were encouraged to search for superheavy elements in nature. In addition to such long half-lives, formation of superheavy elements in nucleogenesis is a prerequisite for their existence in nature. They could be formed by the *r*-process occurring in supernovae, a sequence of neutron-capture and β^--decay processes in which the latter step increases the atomic number. In order to proceed to superheavy elements, the *r*-process has to pass a region of highly fissionable nuclei. Most recent studies consider the production of superheavy elements in the *r*-process to be unlikely.

Nonetheless, experimentalists have examined hundreds of natural samples, including lead from old cathedrals, manganese nodules from the ocean floor, water from hot springs, meteorites, and samples from the lunar surface. Extremely sensitive techniques were developed, permitting the detection of 10^{-12} to 10^{-14} g of superheavy elements per gram of sample in routine measurements, and of 10^{-20} g/g when large-scale chemical enrichment followed by mass separation was applied. Most of these searches produced negative results, but there are two observations which deserve further attention.

Cheleken hot springs. In hot springs from the Cheleken Peninsula in the Caspian Sea, a group at the Dubna laboratory observed a spontaneously fissioning activity after passing a huge amount of water through large ion-exchange columns. This spring water is known to be rich in volatile elements, and is thought to collect material escaping from great depths in the Earth's mantle. By further chemical treatment, a substantial enrichment of the activity was achieved, permitting a quite precise measurement of the neutron emission accompanying spontaneous-fission decay. The neutron multiplicities are lower than expected for superheavy elements, but are in agreement with those of actinide isotopes around curium ($Z = 96$). Further work is required to define the atomic number of this activity.

Carbonaceous chondrites. Strange isotopic composition of xenon isotopes in a certain class of primitive meteorites, the carbonaceous chondrites, has been traced back to spontaneous fission of a now-extinct superheavy progenitor. The evidence collected by a group at the University of Chicago comes from a correlation of the concentration of this xenon component with that of volatile elements such as thallium, bismuth, and indium, and from the enrichment of the strange fission xenon in fractions composing less than 1% of the bulk meteorite. From the experimental data, it was concluded that element 115, eka-bismuth, is the most likely progenitor, but elements 114 and 113 are other possible candidates. However, alternative explanations for these data which do not postulate extinct superheavy elements have been offered.

Searches at accelerators. Heavy-ion reactions seem to be the only practical way of producing superheavy elements in the laboratory: an attempt is made to jump from heavy to superheavy elements in one step. Recent attempts in this direction are closely related to the construction of heavy-ion accelerators at Berkeley in the United States, at Darmstadt in West Germany, and at Dubna in the Soviet Union. The first approach applied is fusion of two nuclei to form a nucleus with an atomic number in the region of maximum stability. The problem lies in the extreme neutron excess associated with superheavy nuclei which cannot be reached by any realistic fusion reaction. This can be illustrated by reference to reaction (1),

$$^{248}_{96}Cm^{152} + ^{48}_{20}Ca^{28} \rightarrow ^{296}_{116}X^{180} \qquad (1)$$

which provides the closest approach to the center of the island of any reaction at the present time. The low neutron number (180) of the compound nucleus is even further reduced by neutron evaporation. This follows because a certain kinetic energy of the projectile is required to overcome the electrostatic repulsion between the two interacting nuclei, and in the fusion process only part of this energy is consumed. Hence, the resulting compound nucleus has enough excitation energy to evaporate several neutrons, as shown in reaction

(2). In each of these subsequent neutron evapora-

$$^{296}_{116}X^{180} \rightarrow ^{292}_{116}X^{176} + 4\,^{1}_{0}n \qquad (2)$$

tion steps, prompt fission competes to an extent determined by the fission barriers of the intermediate nuclei. In an overshoot reaction such as those shown in reactions (1) and (2), a chain of fast α-transitions and electron-capture (EC) transitions follows which increases the neutron-to-proton ratio:

$$^{292}116^{176} \xrightarrow{\alpha} {}^{288}114^{174} \xrightarrow{EC} {}^{288}113^{175} \xrightarrow{EC} {}^{288}112^{176}$$

The final nuclide produced in this chain, $^{288}112$, has a calculated spontaneous fission half-life of about 1 h.

The $^{48}Ca + {}^{248}Cm$ reaction has been extensively studied by a Berkeley-Livermore collaboration and a Dubna group. Such experiments are difficult, since neither the target ^{248}Cm, with a half-life of 3.6×10^5 years, nor the projectile ^{48}Ca, with 0.19% natural abundance, are generally available. The experiments produced negative results. Figure 2 gives upper limits for the production cross sections plotted as a function of the half-life range covered with a particular technique. The Dubna group (D) applied two different chemical separation methods and inspected the samples by spontaneous-fission (SF) and α-particle counting (α). At Berkeley (BL) two different chemical procedures (CHEM) were used, and an experiment designed to detect extremely volatile elements was performed (GAS). Attempts were also made to observe short-lived species by counting reaction products collected in catcher foils (FOILS) and on a rotating-wheel

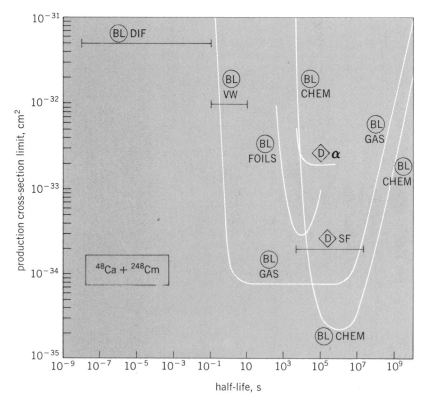

Fig. 2. Upper limits of production cross sections for superheavy elements in the reaction of ^{248}Cm with ^{48}Ca as a function of the half-life of superheavy nuclei. Different curves indicate different experimental techniques. (*From G. Herrmann, Superheavy-element research,* Nature, *280(5723):543–549, Aug. 16, 1979*)

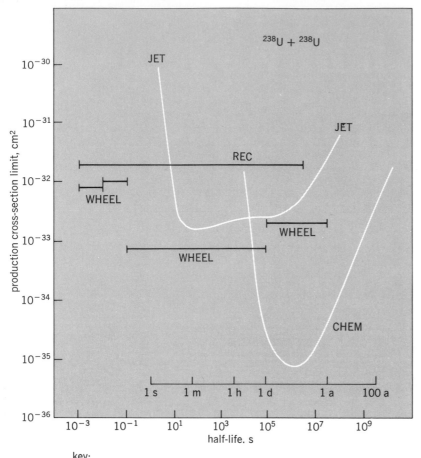

key:
JET = Marburg-Giessen group, gas-jet system
REC = Darmstadt-Heidelberg collaboration, recoil implantation
WHEEL = Darmstadt-Mainz collaboration, rotating wheel system
CHEM = Darmstadt-Mainz collaboration, chemical separations

Fig. 3.   Upper limits of production cross sections for superheavy elements in the re-action of ^{238}U with ^{238}U as a function of the half-life of superheavy nuclei. Different curves indicate different experimental techniques.

counting system (VW). Spontaneous-fission decay in flight was also searched for (DIF). The failure of these experiments can be explained either by a fusion probability smaller than expected or by competition of prompt fission that is stronger than expected. The latter would probably indicate that fission barriers are much lower than those on which the calculations shown in Fig. 1 are based. Apart from the Ca + Cm reaction, a large number of projectile-target combinations have been studied which yield compound nuclei with atomic numbers 110 to 128, but no positive evidence has been obtained for production of measurable quantities of these elements.

An alternative pathway was opened in the first studies of the interaction between colliding urani-um nuclei carried out at the Darmstadt laboratory with the observation of a large exchange of nu-cleons in so-called damped collisions. In such a collision the nuclei stick together to form a com-posite system that lives for only about 10^{-20} s. On the average the system decays back into two prod-ucts with atomic numbers and mass numbers close to those of the original nuclei, but in events of low probability a substantial number of protons and neutrons can be transferred. Hence, it might be expected that reaction (3) or one similar to it can

$$^{238}_{92}U^{146} + ^{238}_{92}U^{146} \rightarrow ^{298}_{114}X^{184} + ^{178}_{70}Yb^{108} \qquad (3)$$

occur. Complementary products with $Z \simeq 70$ have indeed been identified with cross sections of about 10^{-28} cm^2, indicating that nucleon transfer pro-ceeds into the region of superheavy elements. But what fraction of the excited atoms of element 114 formed in such a reaction would survive fission? The main drawback of damped collisions lies in the fact that, in spite of the short contact time, the kinetic energy of the projectile is largely trans-formed into internal excitation of the reaction products, thereby decreasing the stabilizing action of closed nuclear shells. However, a broad distri-bution of excitation energy results even for ex-treme nucleon transfer. Thus, one might hope that the tail of this distribution extends to low excita-tions. From the theory and systematics of damped collisions and calculated survival probabilities against fission, a production cross section of 10^{-34} cm^2 is estimated for element 114 isotopes close to neutron number 184. This corresponds, with ac-cessible beam intensities, to production rates of a few atoms per day or week, a rate just exceeding the detection limit achieved with the most sensi-tive methods.

Searches for superheavy elements in ^{238}U + ^{238}U collisions at the Unilac accelerator at Darmstadt have failed thus far. Figure 3 summarizes the up-per limits for production cross sections obtained with several techniques: chemical separations (CHEM) by a Darmstadt-Mainz collaboration, a rotating-wheel system (WHEEL) by the same group, transport of reaction products in a gas-jet by a Marburg-Giessen group (JET), and counting of reaction products implanted by recoil into solid-state detectors by a Darmstadt-Heidelberg collabo-ration (REC).

How could the production cross sections be in-creased? Application of higher bombarding en-ergies in ^{238}U + ^{238}U collisions appears to be unattractive, since the cross sections for highly fissionable actinide isotopes go through a maxi-mum at energies close to the interaction barrier. Much more attractive is the bombardment of a higher-Z target, that is, $^{248}_{96}$Cm instead of $^{238}_{92}$U, with ^{238}U ions. Theoretical estimates indicate that cross section might increase by a factor of 100. Experiments using this reaction are being per-formed at the Unilac by a Berkeley-Darmstadt-Livermore-Mainz-Oak Ridge collaboration.

For background information *see* ELEMENTS AND NUCLIDES, ORIGIN OF; SUPERTRANSURANICS in the McGraw-Hill Encyclopedia of Science and Tech-nology.

[GÜNTER HERRMANN]

Bibliography: H. Gäggeler et al., M. Schädel et al., and N. Trautmann et al., in *Annual Report 1979*, Ges. Schwerionenf. Darmstadt, pp. 60–62, 1980; G. Herrmann, *Nature*, 280(5723):543–49, 1979; E. K. Hulet et al., *Phys. Rev. Lett.*, 39:385–389, 1977; J.-V. Kratz, *Search for Surviving Acti-nides and Superheavy Nuclei in Damped Colli-sions of ^{238}U with ^{238}U and ^{248}Cm*, Ges. Schwer-ionenf. Darmstadt Rep. no. GSI 80-1, 1980; M. A. K. Lodhi (ed.), *Proc. Int. Symp. on Super-heavy Elem.*, Lubbock 1978; S. G. Nilsson et al., *Phys. Lett.*, 28B:458–461, 1969; Yu. Ts. Oganes-sian et al., *Nucl. Phys.*, A294:213–224, 1978.

Switched capacitors

A switched capacitor module consists of a capacitor with two metal oxide semiconductor (MOS) switches connected as shown in Fig 1a. These elements in the module are easily realized as an integrated circuit on a silicon chip by using MOS technology. The switched capacitor module is approximately equivalent to a resistor, as shown in Fig. 1b. The fact that resistors are relatively difficult to implement gives the switched capacitor a great advantage in integrated-circuit applications requiring resistors. Some of the advantages are that the cost is significantly reduced, the chip area needed is reduced, and precision is increased. Although the switched capacitor can be used for any analog circuit realization such as analog-to-digital or digital-to-analog converters, the most spectacular application has been to voice-frequency filtering.

Although a switch has long been used as an element in circuits and systems, it was not until the late 1970s that its potential and practicality in integrated-circuit design was realized, especially by D. A. Hodges, R. W. Brodersen, and P. R. Gray, with their students at the University of California at Berkeley.

Integrator circuits. A conventional RC integrator circuit is shown in Fig. 2a. The output voltage is given by Eq. (1), where $1/s$ indicates the opera-

$$V_{out} = \frac{1}{R_1 C_2} \frac{1}{s} V_{in} \qquad (1)$$

tion of integration (s is the differential operator), showing that integrator performance depends on

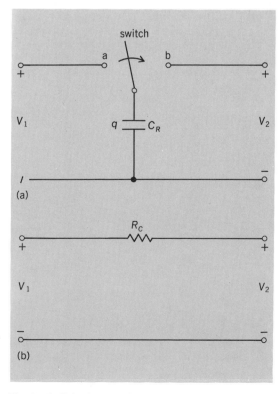

Fig. 1. Switched capacitor. (a) Basic circuit. (b) Equivalent resistive circuit.

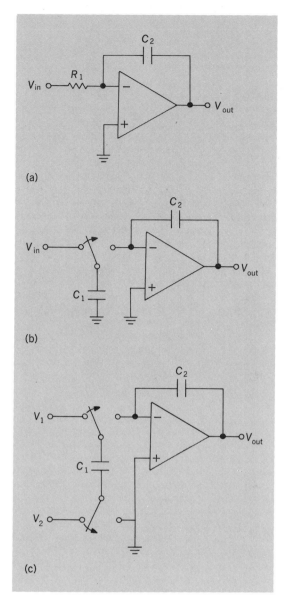

Fig. 2. Integrator circuits. (a) Conventional RC integrator circuit. (b) Single-input switched-capacitor integrator. (c) Switched-capacitor differential integrator circuit.

R_1. In MOS integrated circuits, the value of R_1 cannot be controlled better than 20% by using standard fabrication techniques. In addition, considerable chip space is required to realize resistors in the megohm range. In contrast, the switched capacitor realizations shown in Fig 2b and c depend on the ratio of capacitors which can be controlled with great accuracy. For example, the output voltage of the differential integrator shown in Fig. 2c is given by Eq. (2), where f_c is the clock fre-

$$V_{out} = \frac{C_1 f_c}{C_2 s} (V_2 - V_1) \qquad (2)$$

quency. With C_1 and C_2 in the range of 1 picofarad and f_c at 100 kHz, this circuit has an equivalent resistance of 10 MΩ, and the gain of the circuit is about 10^4. The silicon chip area required to implement the capacitors is about 0.01 mm². If a resistor is used in place of the switched capacitors, an area at least 100 times larger would be required.

Equivalent resistance. Returning to the switched-capacitor circuit of Fig. 1a, the operation may be visualized as follows. With the switch in position a, the capacitor C_R is charged to the voltage V_1. The switch is then thrown to position b, and the capacitor discharged at voltage V_2. The amount of charge transferred is then $q = C(V_2 - V_1)$. If the switch is thrown back and forth at a clock frequency f_c, the average current will be $C(V_2 - V_1)f_c$. The size of an equivalent resistor to give the same value of current is given by Eq. (3). From this

equation, it is seen that with $C = 1$ pF and $f_c = 100$ kHz, the value of 10 MΩ used previously is obtained.

$$R_c = \frac{1}{Cf_c} \tag{3}$$

The accuracy of the equivalence between the switched capacitor and the resistor depends on the relative size of the clock frequency f_c and the frequencies in the signal being processed. If the switching frequency is much larger than the signal frequencies of interest, the equivalence is excellent and the time sampling of the signal can be ignored in a first-order analysis, such that the switched capacitor is a direct replacement for a conventional resistor. If this is not the case, then sampled-data techniques in terms of a z-transform variable must be used for accuracy.

Analog operations. The switch that has been used in describing the switched capacitor is actually realized by an MOS transistor to which a pulse of voltage at the clock frequency is applied to produce the off and on conditions of the switch. This periodically operating switch is used for a number of analog operations, such as addition, subtraction, inversion, and integration. These operations are essential in the construction of analog filters, as well as in other applications of switched capacitors. These operations may be explained in terms of the circuits of Fig. 2. In Fig. 2a and b the analog operation of integration is accomplished. In addition, these circuits are of the inverting type, meaning that a sign reversal is accomplished in addition to integration. The sign reversal of a voltage can be accomplished directly by using switches and a capacitor, as seen in Fig. 2c. Assume that V_1 is grounded or $V_1 = 0$. The operation of the switches is such that the voltage applied to the MOS operational amplifier is the negative of V_2. With the switch operating from left to right, V_2 with respect to ground is reversed. With V_1 not grounded, the circuit of Fig. 2c is a differential integrator, meaning that the output voltage is a function of the voltage difference, $V_2 - V_1$. In conventional active-filter design, these analog operations are accomplished by means of additional stages incorporating operational amplifiers. In switched-capacitor design, these analog operations are implemented with switches.

Filter design. Although there are many strategies for filter design, the discussion will be restricted to the case of filters based on the passive LC ladder with resistive terminations at both ends. Extensive tables are available giving element values to achieve various forms of frequency response, such as Butterworth, Chebyshev, and Cauer (elliptic). All tables are given in terms of a normalized termination of 1 Ω, and a normalized frequency of $\omega_o = 1$ rad/s and for the low-pass case. It is standard procedure to make use of frequency transformations to realize high-pass, band-pass, band-elimination, and similar kinds of responses, and to use frequency and magnitude scaling to give practical element values. The passive ladder structure with double terminations is chosen because it has low sensitivity of changes in transmission with changes in element sizes.

Starting with the low-sensitivity, low-pass ladder

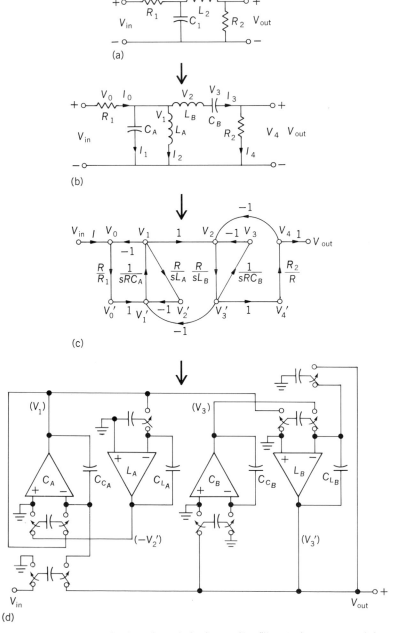

Fig. 3. Steps in the realization of a switched-capacitor filter. (a) Low-pass prototype filter. (b) Corresponding band-pass filter. (c) Signal flow graph representation of the circuit of b. (d) Final switched-capacitor band-pass filter.

structure, a frequency transformation is first accomplished. From these steps, a structural simulation is then carried out by replacing the actual filter by its signal flow graph representation. The flow graph is chosen so that most of the operations required are integration. The elements in the flow graph are then simulated by circuits like that shown in Fig. 2c.

An example of filter design is shown in Fig. 3. The ladder network shown in Fig. 3a is known as the low-pass prototype. In the usual case, $R_1 = R_2 = 1\ \Omega$. The elements C_1 and L_2 are determined from tables, depending on the form of frequency response required. Going from Fig. 3a to b accomplishes a low-pass to band-pass transformation in which all element values are determined from those in Fig. 3a, and the specification of the center frequency and bandwidth of the band-pass case. In Fig. 3c the filter of Fig. 3b is represented by its flow graph, in which the lines and arcs with arrows indicate the structure of the circuit of Fig. 3b, and the associated symbols represent the impedance or admittance. To this structural simulation of the ladder filter, an element simulation is next applied. In particular, all elements of a form such as $1/RC_4s$ are realized by using the integrator of Fig. 2c with differences of voltages accomplished by the switched capacitors. The final result, that shown in Fig. 3d, is then implemented as an integrated circuit containing only switched capacitors, ordinary capacitors, and operational amplifiers. The chip area required to realize a filter of modest order might be 100 mils (2.5 mm) on each side.

For background information see INTEGRATED CIRCUITS in the McGraw-Hill Encyclopedia of Science and Technology.

[M. E. VAN VALKENBURG]

Bibliography: G. M. Jacobs et al., Electronics, 52(4):105–112, Feb. 15, 1979; G. M. Jacobs et al., IEEE Trans. CAS, 25:1014–1020, December 1978.

Technological forecasting

Technological forecasting (TF), an offspring of military research and development, has been defined as the prediction, with a level of confidence, of a technical achievement in a given time frame with a specified level of support. The field is still dominated by specialists in the defense and aerospace companies. A wide variety of TF methods were experimented with in the 1960s, reached their apex of popularity early in the 1970s, and since then have lost some of their following. Forecasts were often popularized by news magazines, especially when characterized by optimistic scenarios on future technology and society. When these predictions failed to materialize, many forecasters turned pessimistic and saw apocalypses caused by population explosion, nuclear terrorism, or limits of growth. More rigorous and mathematical orientation has superseded the earlier flights of fancy, and the most prevalent approach today is trend extrapolation in various forms.

Technological forecasting may be broken down into six categories, each with several variants: trend extrapolation; expert opinion; systems analysis; parameter analysis; mathematical models; supplementary methods. These are described briefly below.

Trend extrapolation. In trend extrapolation the forecaster takes known trends and extends them into the future on the assumption that what has occurred in the past will continue at a predictable rate.

Linear projection. One way to project a trend is to continue it in a simple linear fashion. The slope provides the rates of change, which constitute the basic information the forecaster desires. An estimate of when an event or the need for an event will occur can then be made. Unfortunately, most trends are not linear.

Trend curves. The S-shaped growth curve is well known. A variety of curve-fitting techniques can be found in most textbooks on statistics. Forecasters using curves frequently make judgment modifications to help fit the trend to a curve.

Precursor events. This model seeks out a relationship between two sets of events, one tending to lag behind the other by an almost constant interval. As Fig. 1 shows, improved speed of military aircraft precedes by several years improved speed of transport aircraft. If this lag remains reasonably constant, forecasts of the lagging technology may be ventured.

Envelope curve forecasting. Robert Ayres developed a method of plotting the hypothetical maximum performance available for any particular functional characteristic, tangential to the individual performance trends. It has been applied to many problems, including predicting the future ratio of high-speed memory capacity (in bits) to add-time (in seconds) for digital computers. Figure 2 plots the maximum energy conversion efficiency in thermal power plants from the year 1700, with an initial curve of 1 or 2% to today's striking progression.

Diffusion studies. Any large financial investment

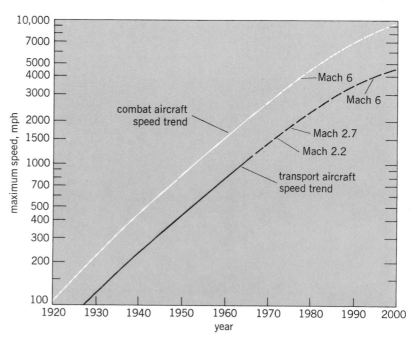

Fig. 1. Speed trends of combat versus transport aircraft, showing lead trend effect. (From J. R. Bright, ed., Technological Forecasting for Industry and Government, Prentice-Hall, 1968)

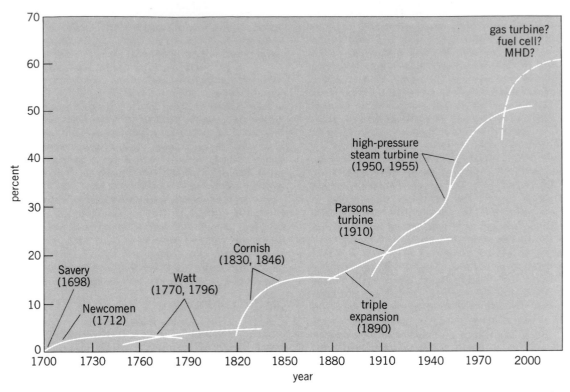

Fig. 2. Efficiency of external combustion energy conversion systems. *(From J. R. Bright, ed., Technological Forecasting for Industry and Government. Prentice-Hall. 1968)*

in existing technology slows adoption of a newer technology. This adoption rate can sometimes be forecast. Diffusion of high-speed jet aircraft into the commercial field, for example, has been found to be highly predictable in terms of capital investment and obsolescence factors.

Correlation and regression analysis. Correlation or regression analysis helps quantify the relationship between the variables. Several cause-effect relationships may at least be hypothesized from their statistical correlation and be extrapolated into the future as a forecast.

Technological changeover points. The future performance of two competing technologies may be plotted to indicate when one technology will be superior. One such analysis, for example, showed the diminishing cost per mile of underground highway construction against the increasing cost per mile of surface and above-surface highways in urban areas. This analysis suggested that it would be cheaper in the future to build highways as tunnels.

Systems analysis. An analytical approach can help suggest technological solutions for future operating problems.

Examination of present systems. If the weaknesses of existing systems can be identified, a forecaster can stipulate some of the operating parameters of future replacement systems. United Technologies Corporation, for example, approached the very-high-speed train problem in this manner. Its analysis determined that the deadweight of trains was one of the main drawbacks of this kind of proposed transportation system. It therefore concentrated on designing new light-

weight cars plus a lightweight propulsion system. Figure 3 shows an application to a critical parameter in air cushion vehicles (ACV). Performance, in terms of a factor of merit, is projected for a vehicle with a bare bottom and with a skirt and a trunk system.

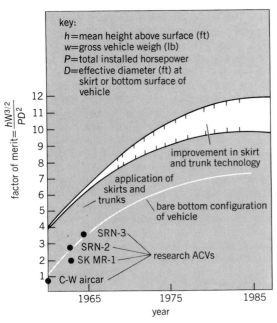

Fig. 3. Air-cushion vehicle (ACV) performance. *(From J. R. Bright, ed., Technological Forecasting for Industry and Government. Prentice-Hall. 1968)*

Hypothetical future problems. This technique is popular in military and foreign policy planning. A "what-if" problem is posed and a systems solution developed. Problems posed might range from the question of contamination in outer space to devising a policy in the event of Communist penetration of South America. Written scenarios are developed, first to describe the hypothetical problem, and then to elaborate on possible ramifications and alternate solutions.

Impact studies. Somewhat similar to the previous example, an impact study starts with a possible future technology and asks the question: If one had this, how would it affect related areas? For example, if one could provide a rental car with a coin-operated meter, and it could be picked up and dropped off anywhere within an urban area at half the cost of local taxi fare, how would it affect the urban and suburban transportation systems?

Parameter analysis. Forecasters sometimes select a single critical factor or parameter to analyze instead of looking at the entire system. There are many variations possible, but the three best known are described below.

Theoretical limits test. In this technique a phenomenon is pushed to its theoretical performance limits in an attempt to visualize its potential application. This has been done notably for the laser to forecast new application areas and markets.

Figures of merit. A. L. Floyd developed a methodology for projecting a "figure of merit" (for example, thrust-to-weight ratio for jet engines), beginning with ordinary trend extrapolation and then calculating the probability of improving performance through various levels of applied effort. Much of its accuracy depends on consistency of the related environment.

Analyzing unique properties. This technique seeks to identify entirely new market applications for an existing product by focusing on a selected unique property or function, such as a high strength-to-weight ratio. Future uses of such properties first are hypothesized, and then an analysis is made to determine the cost/performance requirements to make that product competitive in the future.

Mathematical models. Some helpful models have been developed to predict the future effects of changes. These generally provide mathematical descriptions of present systems, which may then be manipulated to gage what will happen to Y if X is changed by a certain amount.

Input-output tables. Developed in the 1950s, input-output tables usually deal either with a section of the economy or with the economy as a whole. They are based on the principle that each industry consumes a portion of its own output in direct or changed form. A change in the output of a particular industry thus will have a series of effects on the inputs and outputs of all industries. An individual industry using such a table sometimes can forecast the total effect of a change caused by a new technology.

Historical analogy. Using the lessons of history, a forecaster can develop a "genetic" model to study the impact of a new technology. A classic model of the power generation industry was based on the belief that future changeover from fossil fuel to nuclear fuel is analogous to the 19th century's electricity generation changeover from hydroelectric to fossil fuel.

Biological growth analogy. This overlooked method uses a biological analogy (for example, human population compared with that of fruit flies) to predict changes in either a biological or nonbiological system. Psychologist Mason Haire uses a mathematical biological growth analogy to forecast the organization size of a modern company.

Expert opinion. Expert opinion is especially popular for very-long-range forecasts and for social transmutations. They are easy to conduct, popular with the press, and usually wrong. The three following approaches are notable.

Visionary forecasting. This is simply personal prophecy, relying on insights, inspiration, and sometimes a few selected facts. Characterized by subjective guesswork and imagination, not scientific method, it can be highly dramatic and either utopian or despairing in the vistas offered. Tone generally reflects the optimism or angst of the decade.

Panel consensus. This technique assumes that more than one chosen prophet will come up with a better forecast than the loner. It also assumes that a qualified panel has been assembled. Consensus has indeed proved to be more valid, but only when there is prompt agreement. Agreement that comes after lengthy discussion among panelists has been found to reflect group social factors at work.

Delphi questionnaires. To minimize some of the many pitfalls in obtaining and integrating expert opinion, some TF researchers use questionnaires mailed to the panelists. In the so-called Delphi technique, four rounds of questions are generally used, each probing the respondents' answers from the previous round. However, feedback to each respondent of what the other respondents said probably influences impartiality.

Supplementary methods. The following methods elude simple categorization but are worthy of attention.

Possibility trees. These are constructs that show future product possibilities for a new technology. Usually each branch of the tree pinpoints a definite date when specific products will appear on the competitive scene or when these products will be ready for development.

Morphological analysis. This term refers to methods for identifying and counting all possible means to a given technological end at various levels of abstraction or aggregation. Closely related to possibility trees, this is an elaborate checklist for organizing a broad new-product study, and can be used as a technological grid to identify invention opportunities and markets that might otherwise be overlooked.

Technological mapping. Using network techniques, the forecaster in this method diagrams all significant approaches to a desired technological success. Then from publications, patents, and other commercial intelligence sources, he or she plots competitors' technological status in relation to each important path on his or her chart, color-coding the heavily traveled roads to show where

THIGMOMORPHOGENESIS
AND THIGMONASTY

Effects of mechanical perturbation on coiling of pea tendrils (pertubed tendrils below).

other firms dominate (and by contrast, where the remaining opportunities lie). Trends in future development can be estimated from the plotting, and strategies inferred.

Life cycle models. Particularly in the pharmaceutical field, forecasters construct models of life cycles of leading products in order to predict their longevity and dates of their likely decline. This anticipation of cyclical product erosion guides defensive research and the planned elimination or phasing out of one's own vulnerable products.

Demand assessment. This approach is based on the hypothesis that market demand—not technological state of the art—is the prime force that stimulates breakthroughs. James Brian Quinn points out that if an anticipated demand is strong enough it will generally call forth the human and physical resources necessary to attack its technological problems. Consequently, technological forecasters sometimes predict future demands for energy or food, waste disposal, communications channels, traffic control, and so forth. They also seek to determine the performance specifications that a product or system must achieve if it is to solve these future problems, and this method is often an integral part of good market forecasting.

Dynamic forecasting and algorithms. This approach aims at finding out to what extent one can influence progress with specific resources at one's disposal and where those resources should be applied. It tries to identify all the factors that will be involved in achieving certain breakthroughs, quantifies them, and then trades them off against each other to determine which solutions are best.

A transfer coefficient is assigned to each of the variables, and when all the factors are put together, along with their transfer coefficients, the result is a single expression, an algorithm. This can then be manipulated mathematically to optimize any variable; the algorithm yields an infinite number of solutions and consequences from which one may make a choice.

Technology scanning. This method requires a carefully selected panel of five outstanding experts to record on a specially designed grid its consensus on the feasibility and effectiveness of various proposed solutions to highly specific problems in areas such as pollution, transportation, biomedicine, and education. The consensus is arrived at over several months of meetings, chaired by a TF methodology expert who challenges each forecast and its data. A series of weighted indexes is arrived at similar to those in a payoff table. These indexes are then used to help set targets for research and development programs.

[SPENCER J. HAYDEN]

Bibliography: R. U. Ayres, *Technological Forecasting and Long Range Planning*, 1969; J. R. Bright, ed., *Technological Forecasting for Industry and Government: Methods and Applications*, 1968; S. J. Hayden, *Vantage Point: Year 2000—The Influence of Technological Innovation on the Future of Connecticut*, Conference Summary, Connecticut Research Commission, 1968; R. C. Lenz, *Technological Forecasting*, ASD-TR-62-414, Wright Patterson Air Force Base, 1962; J. P. Martino, Technological forecasting for the chemical process industries, *Chem. Eng.*, Dec. 27, 1971; J. B. Quinn, Technological forecasting, *Harvard Bus. Rev.*, pp. 89–106, March/April 1967.

Thigmomorphogenesis and thigmonasty

All plants that have been studied exhibit reactions to mechanical perturbations which may be tropic, nastic, or morphogenetic. Thigmotropism may be seen in roots which grow away from stones or other barriers in the soil, and in the stamens of *Portulaca grandiflora*, which quickly bend toward an insect which lands in the flower, and deposit pollen upon it. Thigmonasty is fairly widespread among flowering plants; examples may be seen in the contact coiling of the tendrils of many species (see illustration), and the insect leaf traps of *Dionea muscipula* and *Drosera* species. Thigmomorphogenesis, which consists of changes in growth and form (morphogenetic), can be seen in the stem growth of all plants so far tested. An increased stem thickening, coupled with decreased elongation, is caused by mechanical perturbation. In addition, certain species display other, more specialized responses, such as the induction of dormancy or leaf senescence, following mechanical perturbation. Thigmomorphogenesis will be discussed first.

Mechanical perturbation. A plant growing in a windy area will be more likely to withstand high winds if its stem is short and thick than if it is long and thin. That thigmomorphogenesis does in fact occur in the field has been demonstrated with both woody and herbaceous species. In kidney beans (*Phaseolus vulgaris*), a species that has been studied in some depth, the anatomical basis for thigmomorphogenesis has been elucidated. The stem thickening is due to increased lateral divisions of the inner tissues (pith and secondary xylem), coupled with increased lateral enlargement of the cells of the outer cortical tissues. Conversely, the inhibition of elongation is caused by decreased vertical divisions of the cells of the inner tissues and decreased elongation of the outer tissue.

When a bean stem is mechanically perturbed, the sensory function absorbs the mechanical energy and, within tenths of a second, transduces it to bioelectrical activity which can be measured as a propagated action potential or a drop in tissue resistance, which recovers within 3 min. During this 3-min period, there is a transient acceleration of elongation, followed by a complete stoppage of elongation for about 30 min. Elongation then resumes for 2 or 3 days, but at about half the rate preceding mechanical perturbation.

Role of ethylene. It has been known for some time that mechanical perturbation induces the evolution of the gaseous plant hormone ethylene. The way in which ethylene acts as an internal trigger of thigmomorphogenesis recently has been demonstrated in beans. Mechanical perturbation activates the production of 1-aminocyclopropane-1-carboxylic acid (ACC) from s-adenosyl methionine (SAM) in the form of a transient burst above background levels, which peaks at 30 min after mechanical perturbation. As the ACC content of the stem drops between 30 and 90 min, ethylene evolution begins. It peaks at 2 h and ceases by 4 h after mechanical perturbation. When methionine or SAM is applied to the plant, no response occurs, but when ACC or ethylene is added, they mimic the effect of mechanical perturbation in causing thickening and inhibition of elongation of the stem. Furthermore, if inhibitors of the conversion of SAM to ACC, or inhibitors of the oxidation of ACC to ethylene, are applied to the plants, they completely inhibit mechanical perturbation–induced thickening and partially block mechanical perturbation–induced inhibition of elongation. High levels of indoleacetic acid (IAA), a plant hormone which is a natural inducer of ACC production, when applied to the stem, also mimic the effect of mechanical perturbation on inhibition of elongation.

Collectively, these data indicate that endogenous ethylene directly mediates mechanical perturbation–induced thickening, although its role in mediating the inhibition of elongation is less clear. Recent work suggests that mechanical perturbation may lower the endogenous level of the plant growth hormone gibberellin, and increase the endogenous levels of IAA and abscissic acid (a plant hormone which can inhibit growth), so that it is possible that the thigmomorphogenetic syndrome is controlled by complex interactions of many plant hormones, which are perhaps triggered by ethylene.

Contact coiling of tendrils. While there is a plethora of thigmonastic plant systems, one, the contact coiling of tendrils, will be dealt with in a comparison of thigmomorphogenesis. As in the case of thigmomorphogenesis, propagated action potentials have also been observed in mechanically perturbed tendrils, and an electrical stimulus has been found to mimic mechanical perturbation in the induction of coiling. Furthermore, mechanical perturbation–induced ion efflux from cells of tendrils occurs. Following these bioelectrical events, which are necessarily related to the plasmalemma membranes of the sensory cells, there is a large increase in ethylene production. Furthermore, when applied asymmetrically in the form of the ethylene-releasing compound Ethrel, ethylene mimics the effect of mechanical perturbation and induces coiling. When the tips of tendrils of various species are dipped into solutions of IAA and other auxins, a somewhat atypical type of coiling is induced. This effect is not related to a change in the distribution or rate of basipetal auxin transport. The role of auxin in contact coiling of tendrils remains unresolved.

Thus in two such seemingly disparate mechanically perturbed reactions as stunting of stems and coiling of tendrils, there appear to be similarities in the mechanisms of action. In both systems, reception of the mechanical energy is followed by conversion to electrochemical energy, and in both systems the plant hormone ethylene acts as the internal trigger to set off the reaction. It is interesting to note that where they have been looked for, these properties also exist in other plant systems which respond in some way to mechanical perturbation. Bioelectrical action potentials occur in

the Venus's-flytrap in response to touch, and ethylene seems to be involved in the touch-stimulated stamen movements of the moss rose (*Portulaca grandiflora*).

For background information *see* PLANT HORMONES; PLANT MOVEMENTS in the McGraw-Hill Encyclopedia of Science and Technology.

[M. J. JAFFE]

Bibliography: J. Grace, *Plant Response to Wind*, 1977; M. J. Jaffe, *Bioscience*, 30(4):239–243, 1980; R. L. Satter, in W. Haupt and M. E. Feinleib (eds.), *Encyclopedia of Plant Physiology*, vol. 7, pp. 422–484, 1979.

Thyroid hormone

Although the endocrine functions of the thyroid gland have been recognized since the second half of the 19th century, significant progress in elucidating the mechanism of action of thyroid hormones at a molecular level has occurred only in the past 15 years. In this article, recent progress in this area will be reviewed, with special attention directed to the nature of the nuclear receptor and its relationship to thyroid hormone action.

Background. Although the major product of thyroidal secretion is thyroxine (3,5,3'5'-L-tetraiodothyronine, or T_4), the more active iodothyronine is triiodothyronine (3,5,3'-L-triiodothyronine, or T_3), which is largely derived from the conversion of T_4 by peripheral tissues. Although it appears probable that T_4 has some intrinsic hormonal activity, it serves largely as a prohormone.

Traditionally, the functions of thyroid hormone have been closely associated with oxygen consumption and thermogenesis. Recently a large proportion of the thyroid hormone–dependent oxygen consumption has been attributed to the maintenance of the Na-K transcellular gradient. It is postulated that thyroid hormones stimulate the formation of the enzyme Na-K ATPase, presumably through the nuclear mechanisms which will be described below. Nevertheless, it is clear that thyroid hormone action cannot be understood exclusively in terms of enhanced oxygen consumption, especially since a tissue such as brain, which appears highly responsive to thyroid hormone during development, fails to exhibit alterations in oxygen consumption in response to the deprivation or administration of thyroid hormone. Thyroid hormones are also known to stimulate the formation and degradation of carbohydrates, lipids, and proteins in what appear to be "futile cycles." In addition to a generalized effect on protein synthesis, the capacity of thyroid hormones to support and stimulate the synthesis of many specific proteins is well documented. Lastly, thyroid hormones are known to play a critical role in growth and development. In amphibian metamorphosis, a well-coordinated series of biochemical events follows the rise in circulating thyroid hormone, resulting in the transformation of the aquatic tadpole to the terrestrial frog. Thyroid hormones also play an analogous role in mammalian development by ensuring the normal ontogeny of the central nervous system and a normal rate of linear bone growth.

Initiation of hormone action. The description of specific nuclear binding sites in various rat tissues in 1972 provided a focus for reexamination of the mechanism of initiation of thyroid hormone action. The relevance of the nuclear sites to such initiation was inferred from the strong correlation between the biologic activity of thyroid hormone analogs and the binding of such substances by nuclear sites. When due account is taken of the known distributive and metabolic characteristics of these analogs, no exceptions in rank correlations between binding and activity were apparent among some 40 compounds tested. The steric configuration of the nuclear receptor (see illustration) has been inferred from these data.

Studies in the mid-1960s also indicated that administration of thyroid hormone to hypothyroid animals was followed within 6 h by increased incorporation of ^{14}C-labeled orotic acid into rapidly labeled nuclear RNA, an event which preceded protein synthesis and enhanced oxygen consumption. These findings have since been confirmed, and it has also been shown that there is a generalized increase in the labeling of poly(A)-containing RNA, which represents the bulk of cellular messenger RNA. When specific mRNA sequences coding for proteins known to be stimulated by thyroid hormone are measured in translational assays, an increase in mRNA becomes apparent even earlier, within approximately 2 h after T_3 administration. The specific messenger sequences which have been analyzed to date include those for α_{2U}-globulin, an exportable rat hepatic protein of unknown function, cytosolic malic enzyme, and pituitary growth hormone. The relatively rapid sequence of nuclear events after receptor occupation thus supports a nuclear site of initiation.

The possibility that thyroid hormone may also have extranuclear sites of initiation has been proposed by several investigators. In particular, it was demonstrated that T_3 stimulates deoxyglucose transport into rat thymocytes, a process which appears to be linked to calcium transport and which is clearly independent of new protein synthesis. The concentrations of hormone required to demonstrate the augmented transport processes under culture conditions, however, appears to be somewhat higher than the ambient concentration of hormone in body fluids. Whether the processes stimulated under the culture conditions are an accurate reflection of analogous processes in the living animal remains to be determined. Reports of specific mitochondrial sites have not been confirmed. Although there is general agreement that thyroid hormone action is initiated at least in part at the nucleus, the possibility that there are extranuclear sites of initiation remains a matter for further exploration and documentation.

Characterization of nuclear receptors. The nuclear receptor is a member of the class of nonhistone chromatin proteins which are believed to be important in the regulation of gene expression, and has an approximate molecular weight of 50,000, a sedimentation coefficient of 3.5S, and a frictional coefficient of 1.4. Up to 200-fold purification has been reported, probably still several orders of magnitude below the degree of purification required for definitive isolation of the receptor.

Studies have also shown that the specific cytosolic binding is not required for translocation of the hormone from the cytoplasm to the nucleus, as appears to be the case for steroid hormones. A kinetic analysis has shown a rapid equilibration of thyroid hormone between nuclear and cytoplasmic pools. In turn, there also appears to be a rapid interchange of T_3 between plasma T_3 and cytosolic hormone. The possibility that local generation of T_3 from intracellular T_4 establishes a significant intracellular free T_3 gradient has recently been shown in rat pituitary and brain.

Preliminary studies have suggested that although the number of receptors per milligram of DNA varies from tissue to tissue, the sites themselves appear to be similar, with the same apparent association constant, with the same binding spectrum for thyroid hormone analogs, and with identical chromatographic mobility on DEAE-Sephadex.

It has been postulated that there is a basic structural similarity between nuclear binding proteins for T_4 and T_3. A common core structure is proposed for both classes of binding sites. The T_3 site appears to be relatively labile, and heating is believed to convert the T_3 receptor to a T_4 binding site. Further evidence favors the concept that the stability of the T_3 receptor site is assured in the body by an interaction of the core receptor with histone proteins.

Number and turnover of receptors. Considerable interest has been directed to assessing the potential role of receptors as a control point in thyroid hormone action. On the basis of studies in a rat pituitary tumor cell line (GH_1), the "down regulation" of thyroid hormone nuclear receptors in the presence of thyroid hormone has been postulated. Thus, with a high concentration of thyroid hormone in the media, the level of nuclear receptors of GH_1 cells falls to approximately one-half of control values. Growth hormone production by the GH_1 cells appears to be proportional to the occupation of the depletable nuclear sites by thyroid hormone. The mechanism responsible for such deple-

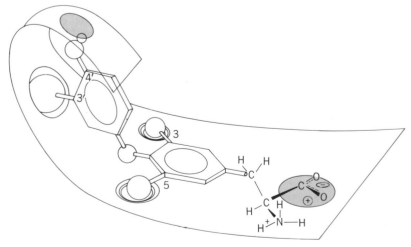

Representation of the conformation of the triiodothyronine nuclear receptor site. The requirements for binding include substituents on the 3′,3, and 5 positions of the thyronine nucleus represented as points of contact with the receptor. A phenolic grouping on the 4′ position is also required as indicated. (*From E. C. Jorgensen, Thyroid hormone structure-function relationships, in S. C. Werner and S. H. Ingbar, eds., The Thyroid, Harper and Row, 1978*)

tion is probably related to accelerated metabolism of the receptor sites. However, quantitation of nuclear receptors in hypothyroid and hyperthyroid animals has in general not shown perceptible alterations in the hepatic nuclear binding capacity. It is possible to reconcile these two observations by suggesting that in the intact animal an accelerated rate of receptor depletion is accompanied by a compensatory increase in receptor synthesis. Thus no changes in receptor content would be expected in the steady state. On the other hand, failure of GH_1 cells to demonstrate a comparable synthetic response would result in the observed receptor depletion.

In the intact animal the number of hepatic receptor sites per milligram of DNA decreases in starvation, in partial hepatectomy, and after glucagon administration. The decrease in binding capacity of hepatic nuclear sites generally does not exceed 50% of the control value. Since each of these stimuli is known to be associated with several other intracellular effects, it is unclear whether or not the alteration of the receptor sites represents the only or the principal point at which these stimuli modify thyroid hormone action. It was recently reported that there is an almost complete disappearance of nuclear receptor sites in GH_1 cells as a consequence of enhanced acetylation, triggered by the addition of butyrate to the medium, an agent known to inhibit deacetylase. The physiological significance of these striking findings, however, has not been clarified.

Postreceptor mechanisms. Independent studies have indicated that thyroid hormone receptors are associated with the linker region of chromatin which connects segments of DNA between nucleosomal regions. There appears to be a specific association between the receptor and a 5.8S chromatin particle. Studies with nuclease I and micrococcal nuclease indicate that the bulk of receptor is associated with the more extended regions of chromatin which are believed to be active in transcription.

Very little is known about the steps which bridge the gap between nuclear occupancy of T_3 and the accumulation of the specific mRNA sequences. A generalized increase in phosphorylation has been observed by a number of investigators, but in the main, these effects are observed relatively late and fail to provide information about the early molecular events. Increases in both polymerase I and II occur only after an increase in specific mRNA species. Although this may represent an inherent limitation in the sensitivity of the polymerase assay, it is also possible that polymerase activity is not rate-limiting in specific mRNA production.

Multihormonal and multifactorial control. Converging lines of evidence indicate that thyroid hormones may act in a coordinated fashion with other hormones and metabolites in the expression of certain genes. Investigations have shown that the induction of α_{2u}-globulin in the rat is contingent on the presence of dihydrotestosterone, cortisol, T_3, and growth hormone. The absence of any one of these essentially abolishes the inductive effect of the others. Cortisol, dihydrotestosterone, and T_3 operate at the level of mRNA formation, whereas growth hormone apparently exerts its role at the level of translation. A somewhat similar interaction between cortisol and T_3 in GH_1 cells has been investigated. Again, there appears to be a synergistic interaction of the two hormones in the generation of specific mRNA sequences coding for growth hormone.

More recently studies have shown that there is a synergistic interaction between a high-carbohydrate diet and T_3 administration in the induction of malic enzyme and other lipogenic enzymes in the rat, and that this interaction is manifested at the level of mRNA formation. Thus there appears to be an interaction of the T_3 nuclear signal with the signal generated as a result of the administration of the high-carbohydrate diet which results in the formation of the specific mRNA sequence coding for malic enzyme. An analysis of the kinetics of interaction of T_3 and the high-carbohydrate diet suggests a multiplicative process. Insulin is not essential in this interaction since administration of fructose, which can bypass the insulin-limited steps in hepatic glucose utilization, is fully capable of interacting with T_3 in the generation of mRNA for malic enzyme. The proximate stimulus for lipogenic enzyme induction may therefore be an intermediate of carbohydrate metabolism. In support of this inference, it was recently shown in primary hepatocyte cultures that glucose per se stimulates the induction of malic enzyme. The possibility should be considered that, in general, thyroid hormone may work by interacting in a synergistic fashion with other hormones and metabolic factors. The structural basis for such interaction, however, remains to be defined.

For background information *see* HORMONE, THYROID in the McGraw-Hill Encyclopedia of Science and Technology.

[J. H. OPPENHEIMER]

Bibliography: L. J. DeGroot, *J. Endocrinol. Invest.*, 1:79, 1978; N. L. Eberhardt et al., *Proc. Nat. Acad. Sci. USA*, 76:5005, 1979; J. H. Oppenheimer, *Science*, 203:971, 1979; H. H. Samuels et al., *J. Biol. Chem.*, 255:2499, 1980.

Transportation engineering

Within large metropolitan areas at certain "activity centers" (those without rail rapid transit), the access and traffic circulation provided by today's transportation modes (auto and bus) are often slow and tortuous. Among the types of activity centers are large airports, central business districts, shopping centers, large universities, industrial complexes, and parks. Within the last 10 years, however, the Downtown People Mover (DPM), operating independent of existing and often congested streets, has been offering an alternate mode of transportation.

DPM and SLT. DPM is not viewed as a replacement of conventional mass-transit facilities but rather as a complement of other systems in the whole transportation pattern. As an example, if a person takes a shopping trip or quick business trip into any major central business district, such as Los Angeles, Miami, or Jacksonville, the person could drive to a specially designed auto-intercept parking facility, lock the car, and walk a short distance to a People Mover station. After depositing a fare and punching a button indicating a destination, a small electric car arrives within 90 s. Its doors open and the person steps into a small, clean, modern, air-conditioned interior. The doors immediately close and the car moves away at 25–30 mph (11–13 m/s) nonstop toward the destination. The vehicle glides along its own independent guideway—sometimes elevated, sometimes at ground level, and sometimes below grade—on almost noiseless rubber tires. There is no operator. The vehicle is controlled by its own built-in computer memory. Entering the central business district, the vehicle comes to a stop at the destination, and the person steps out.

The foregoing example is not science fiction. It is happening in Morgantown, WV. A similar type of system is operating at the Dallas–Fort Worth airport. DPM systems are already in the design phase for Los Angeles and Miami, and projects are in the planning state for Houston, Baltimore, and Jacksonville.

A variation of the DPM is the Shuttle-Loop Transit (SLT), in which vehicles move along fixed guideways with few or no switches. In a simple shuttle system the vehicles move back and forth on a single guideway. They may make intermediate stops. Examples of SLT are located at the Tampa and Miami airports.

Some of the inherent features of a true Downtown People Mover in contrast to the Shuttle-Loop Transit are as follows: small, personalized vehicles giving a reliable, comfortable, clean ride; origin and destination on demand, possibly with an alternate schedule mode; close headways with absolute safety; automatic control/management with no vehicle operators; separate guideway systems capable of construction on, below, or above grade; all stations off-line except for terminals; rapid, safe switching; low air and noise pollutant, esthetically acceptable; automatic diagnostic checks on vehicles and their equipment; operation in various environments; low labor requirement.

Variations of DPM systems have been operating for several years at various amusement parks (Hershey, PA, and Disney World). The United States Department of Transportation partially funded four systems for demonstration. The vehicles ran on short shuttle-type guideways, and some did not have the basic characteristics or components of DPM. Various aspects could be shown and demonstrated, however, such as vehicles, controls, stations, and switching. Most important was that the actual ride could be experienced.

In addition to the four funded projects, several other systems were shown, all of which were basically rubber-tired vehicles running on a fixed guideway.

Morgantown PRT. In 1969 the Department of Transportaion funded the first U.S. Center City People Mover located in Morgantown, WV. The Morgantown Personal Rapid Transit (PRT) System became the first of its kind in the world. It was in the implementation of this project that fundamental experience was gained and lessons learned for future DPM systems.

Morgantown is the home of West Virginia University (WVU). The original campus is in downtown Morgantown, and two additional campuses are located just north of the downtown area. Each of the three campuses is separated from the next by about 1.5 mi (2.4 km). Prior to the PRT system, intercampus public transportation was by bus. At each class break, some 1200 students were carried between campuses by a fleet of buses which, especially when there were only 10 or 20 min between classes, was inadequate. The rugged topography at Morgantown, with differences in elevation of up to 300 ft (90m), discouraged walking or cycling. Also, the use of private autos added to the already acute parking problems and traffic congestion on the only two streets connecting the two main campuses. The application of a People Mover system was suggested as the solution to the problem and with Department of Transportation funding, a feasibility study was undertaken which concluded that a fully automated system would best suit the transportation needs of the university.

The Morgantown system was designed to have five stations, 64 cars, a maintenance-and-control center, and a twin-track guideway 3.5 mi (5.6 km) long. Rubber tires and concrete roadway were selected in preference to steel wheel/steel rail for which the 10% design grades were too steep.

Because DPM systems are generally located in densely populated urban areas, land is costly, so routes must be selected carefully. In Morgantown this was accomplished by using rights of way along existing streets, a railroad, underdeveloped land owned by the university, and unsymmetrical sections where the normal section would conflict with existing buildings. As a result, only one house and one business (a metal salvage yard) had to be relocated.

Even though the PRT project exceeded its original budget by a wide margin and had severe operating problems in the beginning, the Morgantown People Mover is beginning to prove it can be done. During the severe snowstorm of 1978 when virtually nõ other means of transportation moved, the Morgantown system operated easily since the concrete guideways were kept free of snow and ice by buried steam pipes.

Other DPM programs. In 1975 the Department of Transportation launched a new DPM program in which six cities would be selected to implement systems in their central business districts, with the Federal government paying 80% of the cost. Los Angeles and Miami had been planning such a system for many years and were among the first to apply for this new source of transportation funding.

Miami will have a 3.7-mi (6.0-km) double-tracked elevated-guideway DPM system. It will consist of a loop and a north-south leg. Fifteen stations are contemplated to serve major activity centers. Two stations are planned as modal transfer points for the system. In Los Angeles the DPM system will be a 3.0-mi (4.8-km) dual-line north-south guideway with 11 on-line stations.

In the future, Downtown People Movers will continue to be utilized as a complement to other modes of transportation in many cities throughout the world.   [EUGENE D. JONES]

Bibliography: J. P. Cunliffe, *Personal Rapid Transit* (*with Special Reference to the Morgantown*

Project), Soc. Automot. Eng. 730439, April 1973;
F. E. LoPresti, *Civil Eng.*, p. 1, November 1972.

Underwater navigation

"Navigation" has a specialized meaning when applied to United States and United Kingdom nuclear-powered ballistic-missile submarines (SSBNs). Although an SSBN navigation system generates the classical navigation outputs continuously, its primary purpose is to provide initial conditions (latitude, longitude, heading, roll, pitch, velocity, and time) to the missile at launch. Its other functions are necessary, but secondary, such as ship navigation, providing inputs to fire control for defensive weapons, sonar operations, and communications.

Navigation advances embodied in Trident SSBNs emphasize enhanced survivability of the SSBN, more efficient operation of the navigation center, and improved producibility and long-term support of the navigation equipment—all with minimal design modifications. The major advance is in the area of survivability, where navigation operations have been made more covert by introduction of an electrostatically supported gyro monitor (ESGM) to limit the error growth of the ship's inertial navigation system (SINS) and reduce dependence on external position fixes.

Development of the U.S. Navy's strategic navigation capability has, from its inception, been managed by the Systems Management unit of the Sperry Division of Sperry Corporation under the cognizance of the Navy Strategic Systems Project Office. The supplier of the inertial navigation systems (SINS, ESGM) has been Rockwell International. The basic configuration of the navigation subsystem was adopted in 1957 for the planned deployment of the Polaris SSBN in 1960. Polaris accuracy, reliability, and operability were im-

proved continually, and major advances were made for the deployment of the Poseidon SSBN in the early 1970s and of Trident in the early 1980s. The approach to navigation has remained essentially the same throughout: use of SINS for generation of the desired navigation outputs; continuous damping of the SINS erection loops with a waterspeed reference (electromagnetic log); and periodic correction of position, heading, and gyro drift. The corrections are based on external position fixes derived from loran, from celestial tracking (in early Polaris SSBNs), from the Navy Navigation Satellite System (Navsat, Transit), or from sonar terrain matching, along with operator monitoring and control and extensive data processing.

Poseidon navigation. Evolutionary changes implemented for Poseidon were primarily related to significantly improving accuracy and expanding on-board data processing. They included use of survey maps for compensation of vertical deflections due to gravity anomalies, use of a Kalman filter for optimal estimation of SINS errors from a history of position fixes, improved inertial heading determination and self-calibration, implementation of at-sea calibration of the electromagnetic log, loran improvements, data transmission improvements, and introduction of a relatively advanced central navigation computer (CP-890) to handle the new requirements for memory, speed, and word length. The star-tracking periscope used in Polaris ships was dropped, and equipment and calibration improvements also permitted removal of one of the three SINS; both of these changes were incorporated in the United Kingdom Polaris A3 SSBNs and in the United States Polaris SSBNs as opportunity permitted.

The increased capacity and programming flexibility of the CP-890 computer made it feasible to

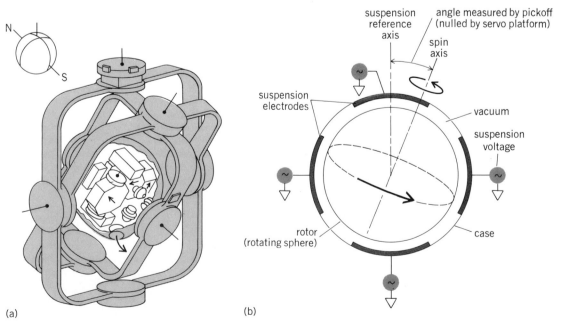

Fig. 1. Electrostatically supported gyro monitor (ESGM). (*a*) Space-stabilized, four-gimbal inertial system. Orientation of Earth and its polar axis are shown. (*b*) The gyroscope, an electrostatically supported spinning ball.

introduce additional changes in Poseidon after deployment. These included further improvements in reset and monitoring procedures and "phase-shift" loran, in which a cesium-beam atomic frequency standard was used with a new loran sensor to provide range-range operation from two stations in addition to the normal three-station hyperbolic time-difference mode.

Trident navigation. The transition to Trident included improvements in controls and displays, simplification of maintenance, and introduction of the ESGM.

ESGM for covertness. The ESGM was initially developed for the Navy in the mid-1950s and tested extensively as a laboratory device. It was tested at sea by the mid-1960s. It underwent considerable further development and at-sea testing on the USS *Compass Island*, the navigation test ship, when proposed for installation on Trident. The ESGM has now had very successful operational exposure through backfit installation on some Poseidon ships that have been converted to the Trident missile, as well as on some of the shorter-range Poseidon C-3 missile SSBNs.

The ESGM is a state-of-the-art, extremely stable inertial system. It is used in the Trident navigation subsystem to provide position data to SINS. Position data from the ESGM are processed in the computer-programmed 15-state SINS correction filter (SCF) to generate corrections to SINS position, heading, and gyro drift. Although the ESGM is basically capable of generating all of the desired navigation outputs, its role in Trident, in the interest of minimizing changes and risk associated with a new development, has been limited to SINS correction.

The ESGM is a space-stabilized, four-gimbal inertial system (Fig. 1a). The space-stabilized platform is equipped with three electromagnetic accelerometers (EMAs) and two two-degree-of-freedom electrostatically supported gyros (ESGs), oriented with their spin axes nominally polar and equatorial. The heart of the ESG is a 10-cm ball within an evacuated cavity, supported electro-statically while it is spinning at 30,000 revolutions per minute (Fig. 1b).

Because of the inherent drift stability of the ESGs, the Trident navigation subsystem will maintain accuracy many times longer than the basic Poseidon navigation subsystem between exposures for external fixes. Position fixes from external references such as Transit will still be used at very long intervals to correct (reset) both SINS and ESGM.

More efficient operation of navigation center. Navigation operations have been improved in the Trident navigation subsystem through such measures as centralizing control and diagnostic functions (and thereby reducing the number of work stations), improving displays, and simplifying equipment maintenance and calibration.

The new navigation control console (NCC) retains most of the capabilities of the Poseidon unit and features a new flat-panel display of sonar, navigation, and teleprinter data (Fig. 2). It includes remote controls for the sonar and centralized switching; provides controls and displays for functions carried over from the now-eliminated Poseidon navigation operational checkout console (a built-in calibration console and output function tester); and, in general, reduces the overall amount of navigation-center hardware.

Maintenance and reliability improvements include wide use of closed-loop cooling, equipment modifications to make use of newer components, improved accessibility to selected equipment, further increases in mean time between failures, and additional use of automatic, computer-driven diagnostic routines to isolate failures to a module or small assembly. Repair by module replacement is still emphasized; soldering or welding on modules is prohibited.

Redesign and reengineering of equipment. Several pieces of Poseidon equipment that were considered to be outmoded or to have near-future support problems have been redesigned or reengineered for Trident to increase reliability, increase producibility, and minimize support problems.

The Navsat (Transit) receiver was redesigned with newer circuits and components to be one-seventh the size of the earlier model and is now under the direct control of the central navigation computer. The redesign essentially eliminates the need for alignments and adjustments on patrol. (Approximately 100 alignments or adjustments had previously been used for proper operation.) About 50% commonality with modules of the loran receivers was also achieved. Automatic system test and fault isolation down to the single-module level provide 95% probability of correct initial diagnosis of defects. The new Navsat receiver is operated from the navigation control console, and interfaces with existing equipment that remained unchanged.

The sonar receiver/transmitter now has a solid-state design, and is equipped with automatic power-level control for the minimization of transmitted power. The computer-controlled flat-panel display in the navigation control console replaces the analog sonar recorder. The flat-panel display provides a monitoring capability in which the computer is used to superimpose the tracked bottom on the raw returns. Like the new Navsat receiver, the reengineered sonar interfaces with existing unchanged equipment and can be controlled from the navigation control console.

A cesium-beam-tube frequency standard is used for internal clocking in the central navigation computer. Precise time signals, digital time of day, and one-pulse-per-second signals, with drift rates limited to less than 0.5 ms over a year, are also transmitted to the fire control subsystem.

Additional possibilities. An intensive advanced technology program is being conducted for the whole weapon system to investigate the limits in accuracy and performance improvement that might be implemented in roughly the next 2 decades. Weapon system synthesis studies, error analyses, and error audits are being done for current and speculative configurations, including both evolutionary improvements and use of more radical techniques such as integral mid-course and terminal guidance. Some of the techniques have been reduced to practice for confidence and validation testing. Interest in the capabilities of the stellar-inertial missile, improved covertness, and increased effectiveness (through improvement of

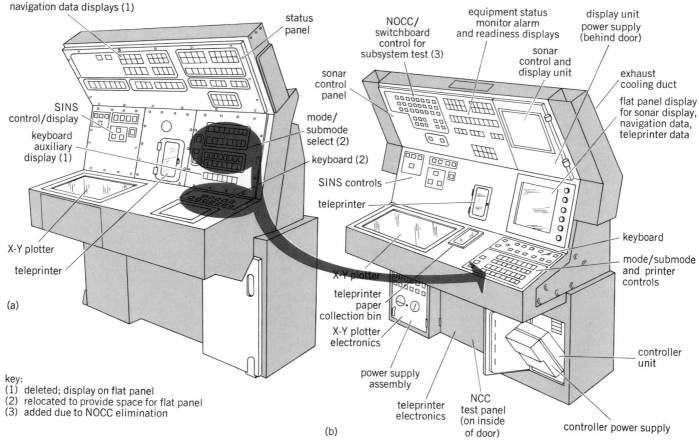

navigation data displays (1)

status panel

NOCC/ switchboard control for subsystem test (3)

equipment status monitor alarm and readiness displays

display unit power supply (behind door)

sonar control and display unit

sonar control panel

SINS control/display

keyboard auxiliary display (1)

mode/ submode select (2)

keyboard (2)

SINS controls

teleprinter

exhaust cooling duct

flat panel display for sonar display, navigation data, teleprinter data

X-Y plotter

teleprinter

keyboard

mode/submode and printer controls

(a)

X-Y plotter

teleprinter paper collection bin

X-Y plotter electronics

power supply assembly

teleprinter electronics

NCC test panel (on inside of door)

controller unit

controller power supply

key:
(1) deleted; display on flat panel
(2) relocated to provide space for flat panel
(3) added due to NOCC elimination

(b)

Fig. 2. Conversion of navigation control console (NCC) from (a) Poseidon to (b) Trident.

accuracy, possibly to the level of the land-based missile) has led the navigation subsystem industrial team and contractors, special test activities, think tanks, laboratories, and the navigation test ship (now the USNS *Vanguard*) into new research and development efforts.

The new missile design (C-4) includes a stellar sensor that can be used to observe misalignments between the true celestial sphere and one that is inferred from the navigation position and heading outputs. For the near term, the missile design places a premium on the reduction of errors that do not contribute to this misalignment, that is, verticality errors that are common to navigation and fire control (mainly those due to anomalous gravity) and ground velocity errors. Current research and development, therefore, emphasize ground velocity techniques and sensors and methods for mitigating or compensating anomalous gravity.

The near-term candidate for ground velocity improvement is the velocity-measuring sonar. Competition in testing between the Doppler concept and correlation concept was conducted in a semi-operational at-sea environment. (The Doppler sonar uses a fan of transmitted beams; the correlation sonar uses a single beam at a depression angle of 90°.) Test results were almost equally good; the correlation sonar concept was selected for further development and testing. Selection of the correlation sonar was based on many con-

siderations, including accuracy, covertness, reliability, computer complexity, producibility, and sensitivity to the SSBN special environment.

The near-term approach to compensation of anomalous gravity is to expand the current operational data bank to use compensation maps derived from satellite altimetry integrated with data from surface-ship surveys conducted at sea with a gravimeter.

A longer-term approach, which will be capable of capturing the shorter gravity wavelengths and thus providing passive velocity improvement (as well as verticality improvement), is the gravity sensors systems (GSS). The GSS is a stabilized platform containing a gravity gradiometer and a gravimeter. The gradiometer measures the spatial rate of change of the gravity vector, and the gravimeter measures its magnitude. These instruments, in conjunction with a statistical model of anomalous gravity, can be used for real-time estimation of deflections, anomalies, and geoidal heights. Feasibility of the gravimeter has already been demonstrated at sea. Other potential applications of the gravity sensors system include missile inflight gravity compensation (via launch point measurements) and at-sea precision calibration of guidance accelerometers.

Potential for improved covertness includes possible use of the gravity sensors for position fixing as well as velocity improvement. In addition, the existing ESGM can be used as an ESG navigator

(ESGN) with only minor hardware changes (although with considerable software changes). The ESGM would thus be converted to a prime inertial navigator with the long-range goal of permitting an SSBN to patrol port to port without an external fix.

For background information *see* GYROSCOPE; INERTIAL GUIDANCE SYSTEM; NAVIGATION, UNDERWATER in the McGraw-Hill Encyclopedia of Science and Techology.

[T. A. KING; H. STRELL]

Bibliography: B. McKelvie and H. Galt, Jr. (Rockwell International), *Navig. J. Inst. Navig.*, 25(3):310–322, fall 1978; M. Molny (Sperry Systems Management), in *Proceedings of the IEEE 1980 Position Location and Navigation Symposium*, December 8–10, 1980; P. J. Moonan (Sperry Systems Management), in *Proceedings of the AIAA Guidance and Control Conference*, August 11–13, 1980; H. Scoville, Jr., *Sci. Amer.*, 226(6):15–27, June 1972; S. Shinners and S. Keslowitz (Sperry Systems Management), pp. 245–250, in *Proceedings of the IEEE 1978 Position Location and Navigation Symposium*, November 6–9, 1978; Sperry Systems Management, *Trident Navigation Subsystem Training Manual*, sponsored by U.S. Navy, Publ. no. GJ-17-1037, January 1975, revised April 1978.

Vision

Present examinations of visual space perception have three different emphases: the analysis of optical stimulation; identification of the neuronal structures in the visual cortex that mediate space perception; and theoretical formulations. The following article, reflecting these three concerns, is highly selective.

Analysis of stimulation. Perception of depth is mediated by information in optical stimulation. A convenient distinction may be made between information in static stimulation and information in kinetic stimulation. The former class of variables is exemplified in the painter's creation of a perceptual world which extends into the third dimension. Analysis of static stimulation can be traced to the earliest scientific treatments of space perception. Notwithstanding, it needs little argument to establish that kinetic stimulation is at least as representative of the conditions under which the visual system has evolved. Whenever the observer moves or there is movement in the visual field, optical stimulation undergoes transformation. It may seem that changing stimulation is not favorable for the attainment of a stable perceptual world. Nevertheless, the last few years have seen a renewal of interest in the potential information in kinetic stimulation. In the view of a number of prominent theorists, kinetic stimulation is the principal support for the everyday experience of a stable perceptual world of invariant spatial relations.

The introduction of computer-generated displays has provided the technical capability for generating controlled kinetic displays. These cathode-ray-tube (CRT) displays sometimes are viewed directly and sometimes in an arrangement which eliminates the conflicting information provided by cues signaling that the display is two-dimensional. In one recent study, stimulus displays were presented consisting of computer-generated random-dot patterns that could be transformed by each movement of the observer or by the movement of the display oscilloscope to simulate the optical transformations which would be associated with various three-dimensional arrangements of surfaces. Observers manipulated stereoscopic depth to match perceived depth in the motion parallax display. Perceived depth was found to be in close agreement with the relative depth that would be present had an actual three-dimensional arrangement been present. A supplementary result of considerable interest is that the matched depth was a better fit when the optical transformations were generated by moving the observer (and also inducing appropriate transformations of the display) than by only transforming the display for the stationary observer.

Adaptive behavior requires the maintenance of invariant, stable percepts of three-dimensional objects, even in the presence of visual noise. Another study presented computer arrays of moving points which simulated three-dimensional spheres rotating around a vertical axis in depth. Observers reported the perceived direction of rotation and perceived depth of the "virtual" sphere. Noise was introduced into the displays by varying the ratio of display elements that moved consistently with the requirements of perspective geometry (signal) to those that moved randomly (noise), or by using vectors as display elements and changing the orientation of signal vectors from frame to frame in the simulation. There was a strong tendency to perceive coherent three-dimensional spheres of appropriate depth rotating in the correct direction even under adverse conditions, such as a low signal-to-noise ratio.

Extracting stable depth relations from continuously transforming stimulation is plainly an important perceptual attainment, but the perception of more complex transformations in space is also possible. One study examined the perceived nonrigid motion in depth elicited by optical transformations. Computer-generated displays were presented of an outline quadrangle in which two corners were stationary while the other pair of opposite corners moved in and out with a phase difference. Even with a minimal phase difference the quadrangle was unsymmetrical all the time, and the optical input could not be the projection of a rigid surface in motion (rotating about a fixed axis). There was very high sensitivity to the information in transforming shape. For example, with zero phase lag, 24 of the 30 reports were of an elastic surface bending in depth.

These illustrative experiments and others along similar lines are taken as evidence that there is depth information in kinetic optical stimulation and that the visual system exploits this information to generate reliable coherent veridical depth percepts.

Identification of neuronal structures. There is now ample evidence that the several portions of the brain involved in visual processing contain neurons that are selective for specific features of the optical input. For example, there are neurons which respond most vigorously to lines in a particular retinal orientation or to bars moving in a particular direction across the retina. Direct evidence of neuronal selectivity has been secured from electrophysiological recordings from single cells in animal preparations. This investigative approach is not available for studying selectivity in humans. Instead an indirect psychophysical approach called selective adaptation has been developed. Adaptation refers to the tendency of sensory systems to respond less vigorously to sustained stimulation. When this tendency can be shown to be selective for specific features of the optical input, adaptation is considered to be selective, and the nature of the selectivity is presumed to identify underlying neuronal selectivity. Investigations using the direct and the indirect approach have generated evidence of neuronal selectivity for depth features, that is, variables of optical input that impart depth information.

The work of D. Regan and his associates is singular in that it has been concerned both with single-unit recording in the cat and with psychophysical studies with human observers. In cat and monkey visual cortexes there are neurons which respond selectively to binocular disparity. Regan and colleagues have extended the studies of neuronal selectivity to kinetic stimulation. Recordings were made from area 18 of the cat visual cortex in a search for neurons tuned to the direction of motion in three dimensions. A pair of stimulus bars that oscillated independently from side to side were presented separately to the animal's left and right eyes. Adjustments of relative velocities of the two bars made it possible to vary the simulated direction of motion in depth. Considerable neuronal specificity was found; for example, some neurons responded best to relative optical motion which typically would be associated with an object approaching head on, while other neurons responded best to relative optical motion typically associated with recession. These results provide evidence of neuronal selectivity for a depth feature, symmetrical optical expansion and contraction, which has long been known to be a powerful elicitor of depth responses in human observers. In a related psychophysical study the selective adaptation procedure was used with human observers, and evidence was found of selective adaptation for specific flow patterns, such as a square generated on a CRT whose opposite sides were made to undergo oppositely directed movement. The effort to provide complementary evidence of neuronal selectivity from electrophysiological studies and psychophysical studies is likely to contribute significantly to an assimilation of the results of neurophysiological studies into theories of perception.

Theoretical formulations. Four theoretical treatments of perceived depth through motion have been advanced: (1) Perceived depth is a direct function of the information in kinetic stimulation. There is no more need to invoke mediating processes to account for the power of kinetic stimulation than there is to invoke such mechanisms in explaining the color experience elicited by monochromatic light. (2) Perceived depth is the result of the application of decoding principles applied by the observer. These decoding principles are thought to be structured in the neuronal hardware and work in a "blind-mechanical" way. (3) Depth perception is the perceptual system's preferred solution to the problem posed by the transforming stimulation regarding which environmental event has generated the optical input. Perceived depth in response to kinetic stimulation is an instance of intelligent problem solving. (4) The perceptual system's response is governed by an automatic selective rule known as the minimum principle—the tendency to select among possible perceptual representations that representation which is most economical with respect to informational load. The claim is that three-dimensional perceptual representations are more economical "interpretations" of the kinetic pattern of stimulation than two-dimensional representations. There is in progress some experimental effort to bolster these various theoretical orientations. The next few years should witness heightened interest in all aspects of the relationship between perceived depth and kinetic stimulation.

For background information *see* VISION in the McGraw-Hill Encyclopedia of Science and Technology.   [WILLIAM EPSTEIN]

Bibliography: W. Epstein, *Stability and Constancy in Visual Perception*, 1977; R. N. Haber, Visual perception, in *Annual Review of Psychology*, 1978; G. Johansson, C. von Hofsten, and G. Jansson, Event perception, in *Annual Review of Psychology*, 1980; B. Julesz, Perception, in *Annual Review of Psychology*, 1981; J. T. Petersik, *Percept. Psychophys.*, 25:326–335, 1979; D. Regan, K. Berley, and M. Cynader, *Sci. Amer.*, 241:136–151, 1979; B. Rogers and M. Graham, *Perception*, 8: 125–134, 1979; D. A. Smith and I. Rock, *J. Exper. Psychol.: Human Percept. Perform.*, 1980.

Welding

Though explosive energy has been used for a number of years to produce large-area, solid-phase metallurgical bonds, its small-area use presents problems of explosive application and detonation control. Current studies have dealt with these problems, producing a safe, reliable bonding method that is scaled to the needs of electronics manufacture. Precision explosive bonding is a welding method that employs minute amounts of special explosives to achieve small, precisely controlled, high-quality metallurgical bonds. Explosive bonding is now a promising technique for joining small electrical leads or micro assemblies. Another current application that has proved technically and economically attractive is the recladding of defective or damaged contact fingers of printed wiring boards. Some of its more favorable technical aspects stem from the absence of applied heat or chemical solutions. These factors, in addition to low equipment cost and low cost per repair, have led to wide acceptance of the process.

Capable of forming bonds between materials usually considered difficult or impossible to join, the method that has been developed contrasts sharply with more familiar joining processes. Costly bonding tools and equipment usually used in solid-phase bonding are eliminated. The need for ultraclean conditions or for precious-metal coatings to facilitate bonding is minimized by a unique contamination-removal system. The necessity for materials to flow during their bonding imposes few limitations in the case of this process. The solid-phase bonds that are produced are stronger than either of the parent materials. Because the materials can be kept well below their melting points, the

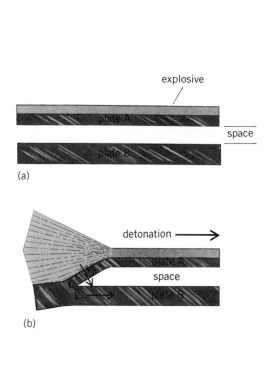

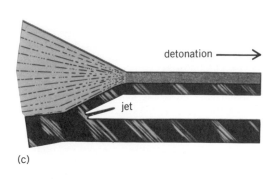

explosive

space

(a)

detonation →

jet

(c)

detonation →

space

(b)

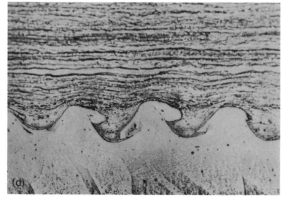

(d)

Fig. 1. Principles and conditions of explosive bonding. (a) Parallel plate arrangement. (b) Parallel plate velocity components. (c) High-velocity jet emanating from collision point. (d) Wavy or ripple interface of copper-nickel bond.

changes in material properties that are normally encountered in fusion bonding are minimized.

Bonding mechanism. The materials to be joined impose limitations upon any solid-phase bonding technique. The more familiar processes are governed by the plastic properties of the materials. However, explosive joining imposes a completely different set of parameters. The high-velocity collisions that result in intermetallic bonding are best described in terms similar to those defining kinetic energy. For example, in the formulating of reliable bond parameters for explosive joining, the bulk sonic velocity of the materials is more important than either their yield or melting points.

A typical arrangement for explosive bonding is shown in Fig. 1a. The two metal materials to be bonded are positioned parallel to each other as plates with a space between them. A layer of explosive material is applied on the top side of the upper plate. Upon initiation of the explosive reaction, a detonation wave propagates outward through the explosive from the point of initiation. The energy released by the detonating explosive accelerates the upper plate, progressively deflecting it toward the base plate to produce a collision (Fig. 1b).

The extremely high acceleration of the upper plate is a direct function of the actual detonation velocity of the explosive. The resultant pressure at the interface is very complex because it is a function of impinging velocity, material mass and density, and the surface flow of the materials. The transient pressures of the collision point produce severe surface strain and deformation, resulting in intimate contact between the metals. Experimental evidence indicates that the most favorable bond conditions are realized when "jetting" occurs between the two metal plates. This jetting phenomenon is the formation of a metallic jet between the impacting plates.

Jetting (surface disruption). Two conditions are required of this process to produce a strong bond: the velocity of the induced pressure wave in the bond material (that is, the sonic velocity) must exceed the detonation velocity of the explosive;

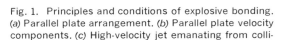

Fig. 2. Interface types as a function of bonding parameters, shown plotted for plate velocity versus mass ratio for gold-tantalum-bonds (1 ft/s = 0.3 m/s).

and the pressure generated by the explosive ahead of the collision point must exceed a critical value. During bonding, extreme pressure exceeding the dynamic elastic limit in both of the plates causes their surfaces to plastically deform and behave like nonviscous fluids for a short interval—forcing the two surface materials to spurt out between the plates (Fig. 1c). This jetting phenomenon removes surface contaminants such as oxides or absorbed gases, bringing the underlying metals into direct contact. The jetting configuration may vary from a fine metallic spray to a concentrated molten metal jet, depending on the hydrodynamics, mass, and momentum of the entire system. This configuration may be effectively altered by the use of different impinging velocities and collision angles.

Three types of metallurgical bonds can result from high-velocity collisions: a straight, solid-phase, metal-to-metal bond; a molten layer between the metals; and a wave or ripple pattern that is basically a solid-phase bond which may exhibit small molten zones. In most commercial cladding applications, the ripple interface is preferred for joint strength; this wavy interface condition is unique to explosive joining (Fig. 1d).

System dynamics. For practical purposes, the controlling factor, in terms of either energy or pressure applied at the point of collision, is a direct function of the detonation velocity. Though mass and material density affect the total system, parameter control can be achieved by velocity adjustments. To better understand the dynamics, many authorities have chosen to divide the velocity into three separate components which can be varied as a function of the geometrical arrangements (Fig. 1b). The three components described are: detonation velocity V_d, a property of the explosive material; collision point velocity V_{cp}, the velocity with which the point of collision moves in the direction of the detonation velocity; and plate velocity V_p, a component normal to the angle of inclination and a function of both the ratio of the explosive mass to the plate mass and the detonation velocity.

The parallel plate arrangement is the most straightforward system and is generally used for bonding experiments. Vector analysis, where β is the dynamic bend angle at impact (Fig. 1b), yields Eq. (1). Therefore $\sin\beta$ and V_p are expressed as

$$V_d \simeq V_{cp} \qquad (1)$$

Eqs. (2) and (3). As previously mentioned, V_p is a

$$\sin\beta = \frac{V_p}{V_{cp}} \text{ or } \frac{V_p}{V_d} \qquad (2)$$

$$V_p = V_{cp}\sin\beta \qquad (3)$$

function of the ratio of the mass of explosive c to the mass of the plate, m, being accelerated, c/m.

By using conservation of momentum, the Gurney formula, Eq. (4), defines the plate velocity.

$$V_p = V_d \frac{0.612\frac{c}{m}}{2 + \frac{c}{m}} \qquad (4)$$

Calculated results using the Gurney formula agree closely with data obtained experimentally with large cladding systems. Interface types as a function of bonding parameters are illustrated in Fig. 2.

Small-area application. Applying the explosive bonding technique to nearly microgeometries requires a method to accurately deposit a specified quantity and configuration of explosive. Because of the very small workpieces involved, the usual procedure of initiating a secondary explosive by a primary explosive detonator would be impractical; the method that has been developed uses a directly detonated primary explosive as the energy source. Commercially available as powders, primary explosives cannot be accurately patterned in this form. Hence a composition was designed for safe, easy application to small piece-parts or in patterns. The mixture is a combination of 70% by weight colloidal lead azide, PbN_6, and 30% binder consisting of 8% by weight ethyl cellulose and 92% by weight pine oil. In the wet state, as when it is applied, this composition is totally insensitive. Before the material can be converted to a combustible product, it must be dried at an elevated temperature. Drying drives off the pine oil, leaving a product which can be detonated. Normally a protective lacquer coating is sprayed over the dried lead azide surface to facilitate handling and provide a moisture barrier. A glowing tungsten filament or high-voltage spark can initiate detonation through the lacquer coating.

It is often advantageous to use an intermediate material between the explosive reaction and piece-parts to be joined. This practice is common in large metal-joining applications, and is intended to maintain the integrity of metal surfaces exposed to secondary explosives. Advantages that can be realized in small-area joining with a primary explosive are: a pattern is more easily screened on a flat buffer sheet which is then placed over a bonding array; the buffer protects surfaces from explosive by-products; and pressure disturbances can be

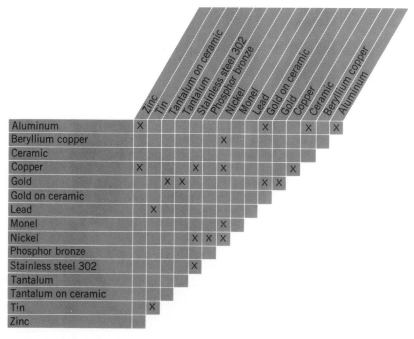

Fig. 3. Combinations of materials explosively bonded. A blank space means bonding of that combination has not been attempted; it does not mean that those combinations cannot be explosively bonded.

dampened when required. Buffer sheets are necessary when the shapes of the parts to be joined are not flat or when contamination of the materials being bonded is not acceptable. A buffer material often used in small-scale bonding is a polyimide in 1 or 2 mils ($25-50\mu$m) thickness.

The metal combinations described in Fig. 3 are typical of the systems studied. Many of the metals joined have widely different physical properties, and would be difficult, if not impossible, to join by other techniques. Pull strength is not reported as related to the specific examples; however, the cited combination exhibited bond strengths above that of the parent metals.

Conclusion. Small-area high-energy bonding has been developed to the extent that the printed wiring board repair process is a manufacturing method, and studies are now being directed to other applications. Both technical and potentially economic advantages can be projected to a variety of manufacturing processes. Some of the more interesting areas of investigation are related to wire splicing with cylindrically shaped explosives, and to the joining of metals which are normally considered incompatible.

For background information *see* WELDING AND CUTTING OF METALS in the McGraw-Hill Encyclopedia of Science and Technology.

[BEN CRANSTON]

Bibliography: M. S. Chadwick, Some aspects of explosive welding in different geometries, in *The Explosive Welding Institute: Proceedings of the Select Conference, Hove*, 1968; B. H. Cranston, D. A. Machusak, and M. E. Skinkle, *Western Elec. Eng.*, 22(4):26–35, October 1978; B. Crossland and J. D. Williams, Explosive welding, *Metals Mater. Mag.*, vol. 4, no. 7, July 1970; H. H. Holtzman and G. R. Cowan, *Bonding of Metals with Explosives*, Welding Res. Counc. Bull. no. 104, April 1965; R. F. Tylecote, *The Solid Phase Welding of Metals*, 1968.

Wigglers

Wiggler systems have been used in storage rings within the last year to increase both the intensity of synchrotron radiation available for experiments and the reaction rates in high-energy physics experiments. Multiperiod wigglers or undulators have also been used recently to make quasimonochromatic photon beams as well as to amplify existing photon beams from other sources such as in the free-electron laser. If one defines a wiggler to be any system of transverse, periodic electromagnetic fields, then recent results on photon production via charged-particle channeling in crystals also fall within this sphere. This discussion will be limited to a typical macroscopic device and how it may be used in a storage ring. The table lists some uses and characteristics of wigglers currently planned or in operation. *See* LASER.

Trajectories such as those shown in Fig. 1 can be produced with magnetic fields that are transverse to the beam direction. The planar trajectory (Fig. 1a) is produced by a series of alternating-polarity, dipole magnets, and the axial trajectory (Fig. 1b) with a bifilar, helically wound air-core magnet. The maximum angle the trajectory makes with the longitudinal axis is α. (If the lines of dots along the z axis in Fig. 1a and b are understood as lattice

sites in a monoatomic cubic crystal with plane spacing d, the planar trajectory in Fig. 1a might represent positively charged particle channeling, and the axial trajectory in Fig. 1b might represent negatively charged particle channeling. However, the latter would be greatly exaggerated because the wiggler wavelength λ_w is much larger than d.) Free-electron laser work has tended to use the helix to obtain small wiggler wavelengths λ_w, whereas applications of wigglers for production of synchrotron radiation produced in the uniform field of storage ring performance have tended to use planar wigglers. Since the radiation from these systems is intrinsically very similar, only planar wigglers will be discussed. They are easily inserted into existing synchrotrons and storage rings and also provide more capabilities in such contexts. The minimum practical wavelength achievable in either case is comparable ($\lambda_w \gtrsim 1$ cm).

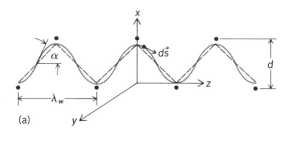

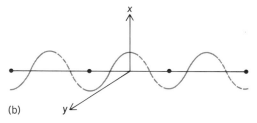

Fig. 1. Trajectory of charged particle through (a) sinusoidal planar wiggler and (b) helical wiggler.

High- and low-field wigglers. Figure 2 shows a design for an efficient high-field wiggler that produces a large range of fields with minimal perturbation to storage ring operation. From the standpoint of the radiation produced, multiperiod wigglers can have two extremes of operation that depend on the field strength and the wiggler wavelength. At "high" fields, they produce radiation with a smooth spectral distribution similar to normal synchrotron radiation produced in the uniform field of dipole magnets. However, at "low" fields, coherent emission of sychrotron radiation from all wiggles simultaneously is possible. In both regimes, the radiation falls within a cone of half-angle $\theta \sim 1/\gamma$ about the radiating particle velocity. Here, $\gamma = 1/\sqrt{1-\beta^2}$ and $\beta = v/c$, where v is the particle velocity and c is the speed of light; the energy of particles of rest mass m_0 is $\gamma\, m_0 c^2$. The synchrotron radiation seen by the observer from a wiggle or dipole element will have a time duration τ given by Eq. (1), where ρ is the bend radius in a dipole

$$\tau \sim \frac{1}{2\gamma^2}(\rho/c\gamma) \qquad (1)$$

Status of some planned or existing wiggler sources

(1) Facility-machine location	(2) E, GeV	(3) I, mA	(4) R_B, m	(5) ϵ_c^B, keV	(6) Primary usage	(7) ϵ_c^w, keV	(8) Periods (N)	(9) λ_w, cm	(10) ϵ_1, keV	(11) Gap, mm	(12) Field, kG	(13) Status, 1979
PEP (R)												
SLAC, U.S.	4–12.7	50	165.5	78.1	L/D	215	1	160	1.92	45.0	20.0	O
SSRL-SPEAR (R)												
SLAC, U.S.	1.3–4	200	12.8	11.1	SR/UR	19.2	3	34.3	0.44	44.5	18.0	O
Photon Factory (R)												
KEK, Japan	2.5	500	8.3	4.2	SR	24.9	1			10.0	60.0	P/D
SRS (R)												
Daresbury, U.K.	1–2	350	5.6	3.2	SR	13.3	1	69.0	0.055	42.0	50.0	C
NSLS (R)	0.7	1000	1.9	0.4	UR	0.13	80	2.54	0.18	12.7	4.0	C
BNL, U.S.	2.5	500	8.2	4.2		1.66			2.34			
X-Ray Ring	2.5	500	8.2	4.2	SR	20.8	1–3	14.0	0.42	35.0	50.0	D
VUV-Ring	0.7	1000	1.9	0.4	SR	0.4	1–5	16.0	0.029	50.0	12.0	D
PULS-Adone (R)												
Frascati, Italy	1.5	60	5.0	1.5	SR	2.8	3	65.4	0.033	40.0	18.5	O
VEPP-3 (R)												
Novosibirsk, U.S.S.R.	1.5–2.2	100	6.2	3.8	SR	11.3	10	9.0	0.51	8.0	35.0	O
LURE-ACO (R)												
Orsay, France	0.1–0.54	180	1.1	0.33	UR	0.078	24	4.0	0.069	22.0	4.0	C
ARUS (S)												
Yerevan, U.S.S.R.	4.5	5	24.6	8.2	SR	24.2	1			20.0	18.0	O
SIRIUS (S)												
Tomsk, U.S.S.R.	≤1.4	100	4.2	1.4	UR	0.39	10	7.0	0.27	85.0	3.0	O
Pakhra (S)												
Moscow, U.S.S.R.	≤1.2	30	4.0	1.0	UR	0.26	20	2.0	0.68	n.a.	2.7	O

(1) S = synchrotron, R = storage ring.

(2) E is the particle energy for which the wiggler is expected to operate, and not the maximum particle-beam energy.

(3) I is the beam current.

(4) R_B is the bending radius of the storage ring or synchrotron.

(5) ϵ_c^B is the critical energy of the storage ring or synchrotron.

(6) L = luminosity, D = damping rates, SR = synchrotron radiation, UR = undulator radiation.

(7) ϵ_c^w is the critical energy of the wiggler.

(9) λ_w, the wiggler wavelength, was derived from the separation between adjacent poles when specific predictions were not available.

(10) ϵ_1 is the fundamental's energy in the weak field limit computed for the highest possible particle energy, for example, $E = 18$ GeV for PEP.

(12) 1 kG = 0.1 T.

(13) P = prototype, C = construction, D = design, O = operational.

field of strength B. The associated frequency spectrum is then expected to extend up to $\omega \sim 1/\tau$. If 2α is the total angular excursion of a characteristic central trajectory through half a period or wiggle (Figs. 1 and 2), then synchrotron radiation will pass from view if the value of $\alpha\gamma$, given by Eq. (2), is

$$\alpha\gamma = \tfrac{1}{2}(q/\beta m_o c)\int B ds \simeq (q/m_o c) B_1 \lambda_w/\pi \quad (2)$$

greater than 1. Here, q is the charge of the particle, the integral is taken over one-half wavelength, $\bar{B}_1$ is the peak amplitude of a sinusoidal magnetic field, and all quantities are in SI (mska) units. The two extremes of operation are then given by inequalities (3). The undulator produces lower-

$$\alpha\gamma \gg 1 \quad \text{standard wiggler (normal synchrotron radiation)} \quad (3)$$
$$\alpha\gamma \ll 1 \quad \text{undulator or interference wiggler}$$

energy, more-directed, or more-collimated beams of radiation which are quite comparable to lasers in many characteristics. Equation (2) is independent of the parent particle energy γ and depends only on the field and wavelength for any particle of charge q and rest mass m_0. For electrons in a sinusoidal field with amplitude B_1, Eq. (2) in numerical form gives Eq. (4). For $B\lambda$ much greater than 1.0 T·cm = 10 kG·cm, one has a conventional

$$\alpha\gamma = 0.934\, B_1(\text{T})\, \lambda_w(\text{cm}) \quad (4)$$

synchrotron radiation source, and for $B\lambda$ much less than 1.0 T·cm = 10 kG·cm, one expects rather monochromatic radiation if the wiggler has a sufficient number of periods ($\Delta\lambda/\lambda \approx 0.9/N$, with N the number of periods).

Energy spectrum. In the "weak" field limit, peaks are expected at light wavelengths λ_n given by Eq. (5) for the nth harmonic, so that the critical

$$\lambda_n \simeq (\lambda_w/2\gamma^2)[1 + \tfrac{1}{2}(\alpha\gamma)^2 + (\gamma\theta)^2]/n \quad (5)$$

energy ϵ_1 of the radiation in the forward direction for $\alpha\gamma \ll 1$ is independent of field strength and given by Eq. (6), where E is the energy of the par-

$$\epsilon_1(\text{keV}) = 0.950\, E(\text{GeV})^2/\lambda_w(\text{cm}) \quad (6)$$

ticle beam. In this limit, the particle motion approximates an oscillating diode in a frame moving along the wiggler axis with the average longitudinal velocity so that only the fundamental has significant intensity. This can also be though of as Thomson scattering of virtual photons with relativistically contracted wavelength from the wiggler field. With increasing values of $\alpha\gamma$, higher harmonics increase in importance, eventually resulting in the familiar synchrotron radiation spectrum.

The "strong" field limit is of most current interest because conventional synchrotron radiation beams can be subsequently monochromatized with low background up to fairly high photon energies with intensities that are adequate for most experiments. Similarly, accelerator physics applications such as damping or beam blow-up generally require considerable radiative power loss. In this regime, the "end point" or critical energy ϵ_c of synchrotron radiation is usually defined as in Eq. (7), where $\hbar$ is Planck's constant divided by 2π.

$$\epsilon_c(\text{keV}) = \hbar\left(\frac{3}{2}\frac{c\gamma^3}{\rho}\right) = 2.218\, E(\text{GeV})^3/\rho(\text{m})$$
$$= 0.665\, B(\text{T})\, E(\text{GeV})^2 \quad (7)$$

With this definition, the photon flux varies slowly up to ϵ_c, but drops sharply thereafter with half the radiated power emitted in photons above this energy.

For a variety of technical and economic reasons, the dipole field in a storage ring is usually restricted to $B \lesssim 10-12$ kG (1.0–1.2T), whereas wiggler fields of 20 kG (2.0 T) or more are possible by using conventional iron core magnets, and fields of 50 kG (5T) or more can be obtained with superconducting magnets. Thus, standard wigglers can shift the spectrum of low-energy machines ($E < 1$ GeV) into the hard x-ray region ($\epsilon_c > 1$ keV). The signal-to-

background in the experimental areas, the operating costs, and the initial capital equipment costs will all be generally lower than for a higher-energy machine. To achieve the same maximum photon energy with an undulator would require machine energies of nearly 5 GeV. At SPEAR, the ring bending radius is 12.7 m, so that Eq. (7) implies that the ratio of the end-point energy of the wiggler ϵ_c^w to that of the ring ϵ_c^B is given by Eq. (8), or

$$\epsilon_c^w/\epsilon_c^B = 0.38\, B_1(\text{kG})/E(\text{GeV}) \quad (8)$$

nearly a 13-fold increase with a 50-kG wiggler at 1.5 GeV. This is equivalent to operating SPEAR at its upper limit of 3.5 GeV.

Irradiance and radiance. In some cases such as extended x-ray absorption fine structure (EXAFS) or topography experiments, the transverse source size is of concern, that is, the irradiance (photons/mm²/s). This can be improved by increasing the number of periods N of the wiggler, by properly selecting its location in the storage ring, or by varying its field or wavelength. In other experiments such as crystallography or photoemission, it is the radiance or brightness (photons/mm²/sr/s) that synchrotron radiation users want optimized. In this case one should choose a location in the ring lattice where the wiggler causes the least increase in beam emittance or even decreases this quantity (see below). Such decisions are based on calculation of synchrotron radiation integrals around the ring. Undulators provide high brightness and spectral brightness ($dN/dt/dA/d\Omega/d\epsilon$) with low background, but presently necessitate control of the parent beam energy according to Eq. (5).

Power. The instantaneous power P radiated by a charge (electron) is given by Eq. (9), where $\dot{p}_\perp = \gamma m_0 \dot{v}_\perp$ is the transverse acceleration, $\dot{p}_\| = \dot{E}/\beta c$, r_e

$$P = \tfrac{2}{3}(r_e c/E_0)(\dot{p}_\|^2 + \gamma^2 \dot{p}_\perp^2) \quad (9)$$

$\gamma m_0\dot{v}_\perp$ is the transverse acceleration, $\dot{p}_\| = \dot{E}/\beta c$, r_e is the classical electron radius ($q^2/4\pi E_0 m_0 c^2$), and E_0 is the particle's rest energy, $m_0 c^2$. For equal forces, there will be γ^2 more power radiated for a transverse acceleration than for a longitudinal acceleration. For circular motion in the uniform field B of a ring dipole magnet, $\dot{p}_\perp = qc\beta B$ and the power is given by Eq. (10).

$$P = \frac{8\pi\epsilon_0}{3}r_e^2 c^3 \beta^2 \gamma^2 B^2 \quad (10)$$

The energy lost per turn per charge to synchrotron radiation in a storage ring is then given by Eq. (11), where ℓ_i is the effective length of dipole i

$$U = \oint P\, ds/\beta c = \frac{8\pi\epsilon_0}{3}r_e^2 c^2 \beta\gamma^2 \oint B^2 ds \propto \gamma^4 I_2 \quad (11)$$

$$U \simeq \frac{8\pi\epsilon_0}{3}r_e^2 c^2 \beta\gamma^2 \Sigma B_i^2 \ell_i = \frac{4\pi}{3}r_e E_0 \beta^3 \gamma^4/\rho$$

having mean central field value B_i. I_2 is an important synchrotron integral proportional to the total damping rate. To maximize the synchrotron radiation power from a wiggler, B should therefore be made as large as practicable but alternating in sign so that Eqs. (12) hold.

$$\int_w B\, ds = 0$$
$$\int_w B^2 ds = \text{maximum} \quad (12)$$

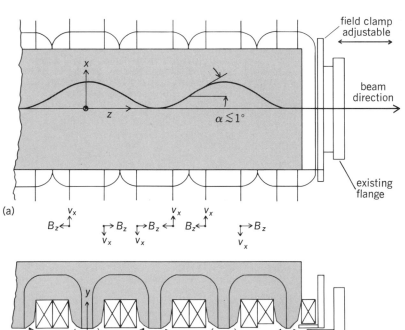

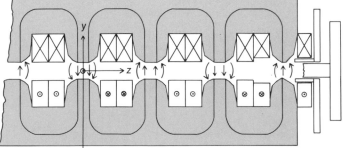

Fig. 2. Schematic layout of the three-wavelength (3λ) planar wiggler installed in SPEAR at the Stanford Linear Accelerator Center. (a) Top view. Exaggerated trajectory is shown to illustrate edge focusing for particles out of the median plane ($y \equiv 0$) of the device. (b) Side view. (From M. Berndt et al., Initial operation of SSRL wiggler in SPEAR, Proceedings of the 1979 Particle Accelerator Conference, IEEE Trans. Nucl. Sci., NS-26:3812–3815, June 1979)

Optics. With the additional constraint that the wiggler be mirror-symmetric about its midpoint (Fig. 2), there will be no deflection or displacement of the beam on passing through, as long as the fractional energy loss through the wiggler is small. This implies that the wiggler is the optical equivalent of a drift space in the dispersion plane; that is, the horizontal optics of the ring in first order are independent of wiggler excitation. Of course, the increased path length resulting from the wiggles may have to be compensated by rf changes, but this is easily done.

In the vertical direction, the wiggler exerts edge focusing for particles out of the median plane ($y \equiv 0$) of the device (Fig. 2). In other words, it is impossible to achieve a purely transverse field over the particle envelope, and the resulting small discrepancy tends to focus the particles. The longitudinal magnetic field B_z for a particle above the median plane, and the transverse horizontal velocity v_x, indicated by arrows and exaggerated trajectory in Fig. 2a, are such that the particle experiences a force toward this plane. Thus, the wiggler acts like a vertically focusing quadrupole lens, so that the beam should be made vertically small in the wiggler (that is, it should have a small vertical waist, otherwise the vertical tune of the ring may have to be adjusted).

Beam phase space. Although the ring optics experience little perturbation from the wiggler, the beam phase space may not, because of shifts in the balance between quantum excitation and damping. Local control of beam energy loss is the basis for a number of applications involving increases or decreases in transverse beam size (or emittance) or energy spread. Adding a wiggler to a ring always increases the energy loss per turn U [that is, the change in I_2 in Eq. (11) is greater than 0], so one generally expects improving particle injection rates with wiggler excitation, since injection is limited by Liouville's theorem to the order of the damping rate which is proportional to I_2. When the wiggler is at a location where $\eta = 0$ (Fig. 3), the damping rate in all degrees of freedom is increased; if $\eta \neq 0$ at the wiggler, it is possible to repartition the damping rates, that is, to increase or decrease some rates relative to what they would have been without the wiggler. The first practical application of such an effect allowed an alternating gradient sychrotron to store a beam.

Since the distribution of energies in a stored beam is a balance between damping and excitation, one only requires the relative increase in damping from the wiggler to be greater than the increase in excitation. It is easily shown that this occurs when the wiggler field B_w is less than the bend strength B of the ring, or, equivalently, when the bending radius of the wiggler ρ_w is greater than that of the ring ρ_B. That is, an undulator can generally be expected to improve the energy spread and a standard wiggler to worsen it. Similarly, placing a wiggler at $\eta = d\eta/d_{\dot{z}} = 0$, such as near the interaction region (IR) in Fig. 3, allows a significant decrease in beam emittance, whereas putting a wiggler near the symmetry point (SP) in Fig. 3 produces the opposite effect. Within limits it is also possible to maintain a constant ratio of emittance change to energy spread change as well as other effects. The table lists some uses and charac-

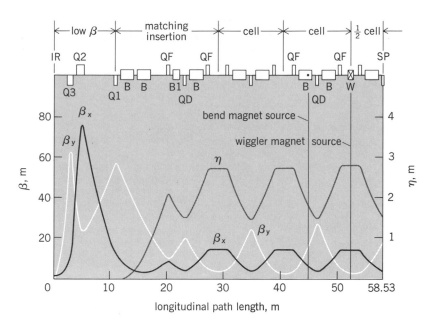

Fig. 3. Half of a SPEAR superperiod showing the machine betatron function (β_x, β_y) and the dispersion η between the interaction region (IR) used for high-energy physics and the symmetry point (SP) about which the superperiod is mirror-symmetric. The magnetic elements that are tuned to obtain a particular configuration (β_x, β_y, η) are shown at the top with the designations Q (quadrupole), B (bends), and W (wiggler). (From M. Berndt et al., Initial operation of SSRL wiggler in SPEAR, Proceedings of the 1979 Particle Accelerator Conference, IEEE Trans. Nucl. Sci., NS-26:3812–3815, June 1979)

teristics of wigglers currently planned or in use.

Since the synchrotron radiation spectrum is smooth and reasonably flat below ϵ_c, one has SR capability relevant to an enormous range of physics, chemistry, biology, and their associated technologies. Besides enhancing ongoing programs by allowing previously impractical experiments or demanding extensions of the "state of the art," new fields are anticipated which are related to time development studies, x-ray holography, pumping of gamma-ray lasers (grasers), and applications to basic research with higher-energy photons such as nuclear physics with monochromatic, polarized beams, to name just a few.

For background information *see* CHANNELING IN SOLIDS; LASER; PARTICLE ACCELERATOR; SYNCHROTRON RADIATION in the McGraw-Hill Encyclopedia of Science and Technology.

[JAMES SPENCER]

Bibliography: M. Berndt et al., *Proceedings of the 1979 Particle Accelerator Conference, IEEE Trans. Nucl. Sci.*, 26:3812–3815, 1979; D. A. G. Deacon et al., *Phys. Rev. Lett.*, 38:892–894, 1977; J. E. Spencer and H. Winick, Wiggler systems as sources of electromagnetic radiation, in S. Doniach and H. Winick (eds.), *Synchrotron Radiation Research*, 1980; R. L. Swent et al., *Phys. Rev. Lett.*, 43:1723, 1979.

X-ray astronomy

Launched on November 13, 1978, the *Einstein Observatory* is one of the largest scientific satellites ever flown. It carries the first large x-ray telescope capable of detecting x-rays from nonsolar sources, and is collecting data important to all fields of astronomy. It was developed under a scientific collaboration involving x-ray astronomers at

Fig. 1. *Einstein Observatory.*

the Harvard/Smithsonian Center for Astrophysics, Goddard Space Flight Center, Massachusetts Institute of Technology, and Columbia University.

The science of x-ray astronomy was born in 1962 with the discovery of the first cosmic x-ray source other than the Sun. This source is a dim star in the constellation Scorpius which radiates 1000 times as much energy in x-rays as in visible light. In the following years were discovered over 400 extra-solar sources embracing many different kinds of astronomical objects, all apparently embedded in a uniform diffuse x-ray background. Many of these objects are very distant and very powerful. They radiate most of their energy as soft x-rays — similar to medical x-rays but having less penetrating pow-

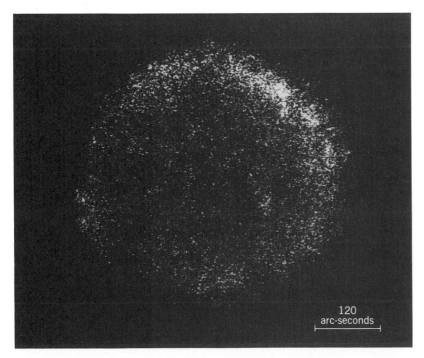

120 arc-seconds

Fig. 2. X-ray picture of the remnant of the supernova observed by Tycho Brahe in 1572. This explosion was seen as a very bright star which faded below naked-eye visibility 1 year after the explosion. The appearance of the remnant is an almost circular shell with a diameter of 8 arc-minutes, and the shell is filled with patches of strong emission, indicating a clumpiness of the ejected material or of the interstellar medium. There is no sign of emission from a neutron star at the center of the remnant. (*P. Gorenstein, Center for Astrophysics*)

er. X-rays from cosmic sources could be stopped by a few sheets of paper, and are completely absorbed in the upper layers of the Earth's atmosphere. Observations therefore must be made above the atmosphere by using instruments carried by rockets and satellites.

The *Einstein Observatory* carries a special telescope which can focus x-rays to form an image. At the focus there is one of two imaging x-ray detectors or one of two spectrometers. For the first time, it is possible to take pictures of the x-ray sky. In two respects, this is a great advance. First, the instrument is 1000 times as sensitive as most previous astronomical x-ray detectors. Very faint sources can be detected, accurately located, and usually identified with a specific optical object. Second, x-ray emission from diffuse sources can be mapped and compared with radio and optical pictures. On the average, the satellite finds 10 new sources every day. The satellite is expected to last at least 2.5 years, and so should detect and study about 10,000 sources before the end of the mission. This will be a 25-fold increase in data available and will have dramatic effects on almost every branch of astronomy.

The satellite (Fig. 1) is 23 ft (7 m) long and weighs 3 tons (2700 kg). Its panels contain solar cells which supply operating power. At the front end are a large aperture for the telescope, a square aperture for a monitor x-ray detector, and three apertures for star cameras. The star cameras are used to measure accurately the telescope pointing direction.

Stars. The Sun has an x-ray-emitting corona, an atmosphere of completely ionized gas with temperature on the order of 10^6 degrees. The x-ray luminosity of the Sun is weak on the cosmic scale (about 1 ppm of the solar radiation appears as x-rays), but because the Earth is so close, solar x-rays have a great effect on the Earth. They ionize and heat the upper atmosphere, and thus influence the Earth's climate and communications.

The existence of stellar coronae is a puzzle. The surface of a star has a temperature of only a few thousand degrees, and it is not completely understood how the star transfers energy through this relatively cool surface to maintain the outer atmosphere at a temperature of 10^6 degrees. It is also not known whether all stars have coronae, or just stars similar to the Sun. The *Einstein Observatory* has so far detected x-rays from approximately 200 stars. Contrary to previous expectations, x-rays have been seen from stars of all spectral types, luminosities, and ages.

As an example of typical results, consider the Sun's nearest neighbor — 4 light-years (4×10^{16} m) distant — a triple star system consisting of the binary star α Centauri A and B, and a companion, Proxima Centauri, the closest of the three. The satellite has detected x-rays from all three — all have a hot corona and all emit approximately the same x-ray flux as does the sun. α Cen A is a star similar in spectral type to the Sun (type G), so that this is as expected. Its companion, α Cen B, is a cooler star (type K), and the observed level of x-ray emission is more than anticipated. Prox Cen is a red (type M) dwarf star, which is 10,000 times dimmer than the other two. It is very surprising that this tiny star shines as brightly as the Sun in x-rays.

These results are not completely understood. In particular, very young stars and very old stars have been observed to have x-ray luminosities orders of magnitude higher than predicted by the previous theories.

A model favored by the satellite data is a corona in which magnetic fields and stellar rotation play the central role in heating and confinement. The satellite observations are expected to eventually yield data on hundreds of stars of all types, which will lead to a better understanding of the solar corona and to better predictability of its effects on the Earth.

Supernova remnants. On the order of once every 100 years per galaxy, a star erupts in a spectacular explosion. Light output from the star increases up to a hundred-million-fold, the star is extremely bright for a few weeks, and then fades back to obscurity in a few months. A supernova is caused by the explosion of a massive star in which the central regions collapse, releasing an enormous amount of gravitational energy. The outer layers of the star are ejected at high velocity. This material plows through the interstellar medium, and after many years the kinetic energy of the expanding debris becomes thermal energy and temperatures are commonly high enough to emit x-rays.

A supernova remnant consists of the expanding ejecta, interstellar material swept up by the ejecta, and the remains of the imploded stellar core (probably a neutron star). Supernova remnants are studied to learn the nature of the supernova explosion, the amount of energy released, the mass of the presupernova star, and the composition of the material produced in the explosion. X-rays from about 30 supernova remnants have been photographed with the *Einstein Observatory* (Fig. 2). Most of these objects appear as great shells of material expanding away from the point of the explosion, some are amorphous in nature, and only a few show x-rays from the remnant of the core. Most of the remnants show no sign of a compact object at the center. This is a puzzle, since calculations require that a supernova result in the formation of a neutron star and that the surface of the neutron star be at a temperature of millions of degrees for hundreds of years after the explosion. X-ray emission from these objects should easily be visible to the satellite's detectors, yet many young supernova remnants show no sign of a neutron star. Perhaps some supernovae produce a black hole instead, or perhaps there is no central object. The answer is at present unknown.

Extragalactic sources. The sources mentioned above are all in the Milky Way, the spiral galaxy in which the Earth is located, and which has a diameter of 10^5 light-years (10^{21} m). The nearest neighboring star systems are the Magellanic Clouds, located 2×10^5 light-years (2×10^{21} m) from the Earth. The satellite has detected about 100 x-ray sources in the Magellanic Clouds. Many of them are seen to be extended, and are thus identified as supernova remnants. There is great interest in comparing the x-ray source population in the Magellanic Clouds with that of the Milky Way. Since the distance to the Magellanic Clouds is well

known and the distance to sources detected in the Milky Way rather uncertain, the absolute luminosities of the Magellanic Cloud x-ray sources are well determined, better than those of supernova remnants in the Milky Way Galaxy. It is expected that this information will enrich knowledge of the characteristics of different classes of x-ray sources, and perhaps point out differences in source populations in the Milky Way and in the Magellanic Clouds.

M31, the great galaxy in Andromeda, at a distance of 2×10^6 light-years (2×10^{22} m), is a spiral galaxy perhaps 50% larger than the Milky Way.

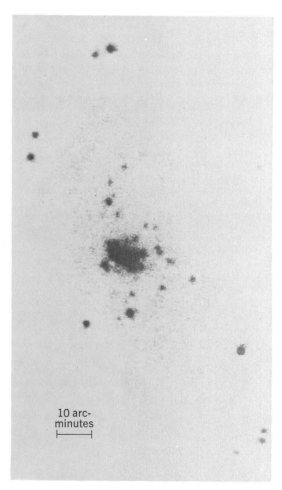

10 arc-minutes

Fig. 3. Composite x-ray picture of M31, a nearby spiral galaxy. (*L. Van Speybroeck. Center for Astrophysics*)

The satellite has detected 70 sources in M31. These sources are, of course, only the most luminous in M31, since the weaker sources, such as stars and most supernova remnants, cannot be detected at this distance. Figure 3 shows a composite x-ray picture of this galaxy. All x-rays are from individual sources. The outer sources are associated with the spiral arms, while the nucleus is seen as a group of unresolved sources. One observation showed a source at the nucleus of M31 which disappeared when observed 6 months later. A variable source at the nucleus is characteristic of many active galaxies, although the nuclei of these systems are much more luminous than that of M31. This extremely interesting observation shows the nucleus of a normal galaxy behaving in the same fashion as the nuclei of active galaxies. *See* GALAXY.

Even more distant galaxies are found not individually in space, but in clusters. Most clusters of galaxies are extended x-ray sources having x-ray luminosities comparable to the optical luminosity of all the individual galaxies. The x-rays come from tenuous hot gas filling the empty space between the galaxies, and the shape of the x-ray source maps the gravitational potential of the cluster. The satellite has measured the shape of about 100 of these diffuse x-ray sources associated with clusters of galaxies. The observations show a variety of shapes, and a study is under way which is linking shape to evolutionary sequence. It is thought that when the cluster is first formed, it is rather diffuse and irregular in shape. As time progresses, the cluster becomes more condensed, and the appearance of the x-ray source becomes smaller and more spherically symmetrical.

The satellite observations have also established quasars as a class of x-ray sources. Three quasars were known to be sources before the *Einstein Observatory*, and now x-rays have been detected from more than 100. Quasars are extremely luminous and are perhaps the most active of active galaxies. Active galaxies are characterized by violent, extremely energetic events associated with their nuclei. The farthest object detected by the satellite is a radio source known as OQ172. The red shift is 3.5, and the radiation now detected from this quasar was emitted 7×10^9 to 15×10^9 years ago. Finally, the x-ray telescope has been pointed at spots in the sky where there were no known sources of radiation. Deep surveys of these blank fields have resulted in the detection of extremely faint x-ray sources, some of which have not yet been identified. The goal is to resolve the diffuse x-ray background to see if this background can be explained as a superposition of many faint, unresolved sources. These deep surveys may result in the detection of x-ray-emitting quasars more distant than any yet known.

X-ray astronomy is no longer a science dealing with the study of a few peculiar objects. It touches upon all branches of astronomy, and to determine the nature and evolution of a given class of objects, the radio, optical, and x-ray emissions must all be taken into account. Since x-ray emission is indicative of high-temperature and high-energy processes, the discovery of new sources shows that the universe contains many strange, new objects, and that some familiar objects display energetic phenomena or perhaps a violent episode in their history.

For background information *see* GALAXY, EXTERNAL; SUN; SUPERNOVA; X-RAY ASTRONOMY; X-RAY STAR; X-RAY TELESCOPE.   [FRED SEWARD]

Bibliography: R. Giacconi, *Sci. Amer.*, 242(2): 80–101, February 1980; R. Giacconi et al., *Astrophys. J. Lett.*, 234:1–81, 1979.

McGRAW-HILL YEARBOOK OF SCIENCE AND TECHNOLOGY

List of Contributors

List of Contributors

A

Anderson, Dr. Franz E. *Department of Earth Sciences, University of New Hampshire.* SEDIMENTATION (GEOLOGY).

Appelman, Dr. Evan H. *Chemistry Division, Argonne National Laboratory.* FLUOROXYSULFATES.

Atema, Dr. Jelle. *Boston University Marine Program, Marine Biological Laboratory, Woods Hole, MA.* ECOLOGICAL INTERACTIONS (coauthored).

Aviram, Dr. Ari. *Thomas J. Watson Research Center, International Business Machines Corporation, Yorktown Heights, NY.* POLYMER PROPERTIES.

B

Banerjee, Dr. Subir K. *Department of Geology-Geophysics, University of Minnesota.* PALEOMAGNETICS.

Bank, Dr. Arthur. *College of Physicians and Surgeons, Columbia University.* HEMOGLOBIN.

Bateman, Paul C. *U.S. Geological Survey, Field Geochemistry and Petrology, Menlo Park, CA.* GRANITE.

Beck, A. C. *Farm Management, Lincoln College, University of Canterbury, Christchurch, New Zealand.* AGRICULTURAL SYSTEM (coauthored).

Berger, Dr. Jonathan. *Department of Oceanography, University of California, San Diego.* SEISMOLOGY.

Berlinguet, Dr. Louis. *Counsellor (scientific), Canadian Embassy, Paris, France.* SCIENCE AND TECHNOLOGY FOR THE DEVELOPING WORLD (feature).

Bhaskar, Dr. N. D. *Department of Physics, Columbia University.* SPIN-POLARIZED ATOMIC VAPORS.

Bolender, Dr. Robert D. *Biological Structure SM-20, University of Washington School of Medicine, Seattle.* STEREOLOGY.

Borregales, Dr. Carlos J. *La Faja Petrolifera del Orinoco, Petroleos de Venezuela, S.A., Caracas.* PETROLEUM (in part).

Bourgeois, Joanne. *Department of Geology and Geophysics, University of Wisconsin.* SEDIMENTARY ROCKS (coauthored).

Boyle, Charles, *NASA Goddard Space Flight Center, Greenbelt, MD.* SPACE FLIGHT.

Branch, Prof. David. *Department of Astronomy, University of Oklahoma.* HUBBLE CONSTANT.

Brodie, Dr. Edmund D., Jr. *Biology Department, Adelphi University, Garden City, NY.* SALAMANDER.

Bromley, Dr. David. *Hewlett Packard, Palo Alto, CA.* LIQUID HELIUM (in part).

Brown, Prof. J. *Heriot-Watt University, Riccarton, Edinburgh, Scotland.* PETROLEUM (in part).

Brown, Dr. Stephen C. *Biology Department, State University of New York at Albany.* BIOMECHANICS.

Bukowinski, Dr. Mark S. T. *Department of Geology and Geophysics, University of California, Berkeley.* HIGH-PRESSURE PHENOMENA.

C

Campbell, Dr. Richard D. *Center for Pathobiology, University of California, Irvine.* COELENTERATA.

Catsimpoolas, Dr. Nicholas. *Boston University School of Medicine.* LEUKOCYTES

Chaffey, Prof. Charles E. *Department of Chemical Engineering and Applied Chemistry, University of Toronto.* RHEOLOGY.

Cheung, Dr. Wai Yiu. *St. Jude Children's Research Hospital, Memphis, TN.* CELL PHYSIOLOGY.

Clark, Dr. Leland C., Jr. *University of Cincinnati College of Medicine.* BLOOD.

Colson, Prof. William B. *Physics Department, Rice University.* LASER (in part).

Cotton, Dr. Therese M. *Chemistry Department, Northwestern University.* BACTERIOCHLOROPHYLL.

Couch, Dr. John C. *Matson Navigation Company, San Francisco, CA.* MARINE CONTAINERS.

Cranston, Ben. *Western Electric Engineering Research Center, Princeton, NJ.* WELDING.

D

Dallas, Daniel B. *Society of Manufacturing Engineering, Dearborn, MI.* MANUFACTURING ENGINEERING.

Degenkolb, Henry J. *H. J. Degenkolb & Associates, San Francisco, CA.* BUILDINGS.

Delson, Dr. Eric. *Department of Vertebrate Paleontology, American Museum of Natural History, New York.* PRIMATES (in part).

Dent, Prof. J. B. *Lincoln College, University of Canterbury, Christchurch, New Zealand.* AGRICULTURAL SYSTEM (coauthored).

Derby, Charles D. *Boston University Marine Program, Marine Biological Laboratory, Woods Hole, MA.* ECOLOGICAL INTERACTIONS (coauthored).

Dixon, Dr. J. B. *Department of Soil and Crop Sciences, Texas A&M University.* STRIP MINING.

Dott, Dr. R. H., Jr. *Department of Geology and Geophysics, University of Wisconsin.* SEDIMENTARY ROCKS (coauthored).

Duke, Dr. Charles B. *Xerox Webster Research Center, Rochester, NY.* ORGANIC SOLIDS.

Dye, Prof. J. L. *Department of Chemistry, Michigan State University.* ALKALI METALS.

E

Epstein, Dr. Robert. *Department of Psychology and Social Relations, Harvard University.* ANIMAL COMMUNICATION.

Epstein, Dr. William. *Department of Psychology, University of Wisconsin.* VISION.

F

Fallaw, Dr. W. C. *Department of Geology, Furman University, Greenville, SC.* PALEONTOLOGY.

Fedorowski, Dr. J. *Polish Maritime Academy, Szczecin, Poland.* MARINE NAVIGATION.

Feldman, Dr. Paul D. *Physics Department, Johns Hopkins University.* COMET.

Frankel, Dr. Richard B. *National Magnet Laboratory, Massachusetts Institute of Technology.* BACTERIA.

G

Geballe, Dr. Thomas R. *Hale Observatory, California Institute of Technology.* GALAXY.

Georgi, Dr. Todd. *Department of Biology, Doane College, Crete, NE.* OSTEICHTHYES.

Gifford, Dr. Diane P. *Board of Studies in Anthropology, University of California, Santa Cruz.* PRIMATES (in part).

Gilbert, Dr. E. Robert. *Westinghouse Hanford Company, Richland, WA.* NONDESTRUCTIVE TESTING.

Golden, Prof. Robert L. *Physical Sciences Laboratory, New Mexico State Laboratory, Las Cruces.* GALACTIC ANTIPROTONS.

Green, Joseph. *Department of Oil and Gas, State of Alaska, Anchorage.* PETROLEUM (in part).

Greiner, Prof. Walter. *Institute for Theoretical Physics, University of Frankfurt, West Germany.* SUPERCRITICAL FIELDS.

Grime, Dr. J. P. *Department of Botany, University of Sheffield, England.* POPULATION DYNAMICS.

H

Hanna, Dr. Edgar E. *Laboratory of Molecular Genetics, National Institutes of Health, Bethesda, MD.* ANTIBODY (coauthored).

Harlan, Craig B. *Technical Materials, Inc., Lincoln. RI.* CLADDING.

Hay, Dr. Alastair. *Department of Chemical Pathology, University of Leeds, England.* HALOGENATED HYDROCARBON.

Hayden, Dr. Spencer J. *Spencer Hayden Company, Inc., New York.* TECHNOLOGICAL FORECASTING.

Hayes, William C. *Electrical World News, McGraw-Hill, New York.* ELECTRICAL UTILITY INDUSTRY.

Herrmann, Dr. Günter. *Institut für Kernchemie, Universität Mainz, West Germany.* SUPERHEAVY ELEMENTS.

Hinds, Prof. Edward. *Physics Department, Yale University.* PARITY (QUANTUM MECHANICS).

Hornbeck, Dr. James W. *USDA Forestry Sciences Laboratory, Northeastern Forest Experiment Station, Durham, NH.* FOREST MANAGEMENT.

J

Jaffe, Dr. M. J. *Botany Department, Ohio University.* THIGMOMORPHOGENESIS AND THIGMONASTY.

Jaffe, Prof. Robert L. *Department of Physics, Massachusetts Institute of Technology.* QUARKS.

Jain, Prof. Subodh. *Department of Agronomy and Range Science, University of California, Davis.* MEADOWFOAM.

Janzen, Dr. A. Frederick. *Photochemistry Unit, Department of Chemistry, University of Western Ontario, Canada.* PHOTOCHEMISTRY.

Johnson, Peter K. *Metal Powder Industries Federation, Princeton, NJ.* POWDER METALLURGY.

Jones, Eugene D. *PRC Harris Inc., Boston, MA.* TRANSPORTATION ENGINEERING.

K

Kay, Dr. Richard F. *Department of Anatomy, Duke University Medical Center.* AEGYPTOPITHECUS AND PROPLIOPITHECUS; APIDIUM AND PARAPITHECUS (both coauthored).

Keane, Dr. Robert W. *Department of Biological Sciences, Douglass College, Rutgers University.* MORPHOGENESIS.

King, Dr. Jonathan. *Department of Biology, Massachusetts Institute of Technology.* BIOTECHNOLOGY (feature; coauthored).

King, T. A. *Navigation Systems Section, U.S. Navy Strategic Systems Project Office.* UNDERWATER NAVIGATION (coauthored).

Kintigh, Dr. John. *Black and Veatch, Consulting Engineers, Kansas City, MO.* ELECTRIC POWER GENERATION.

Kovar, Dr. Dennis G. *Argonne National Laboratories.* NUCLEAR REACTION.

Krisch, Prof. A. D. *Physics Department, University of Michigan.* PROTON.

Krizek, Dr. Donald T. *Agricultural Research, Northeastern Region, Beltsville Agricultural Research Center, U.S. Department of Agriculture, Beltsville, MD.* PLANT PHYSIOLOGY (in part).

L

Lanford, Prof. William. *Physics Department, State University of New York at Albany.* COMPUTER (coauthored).

Lanham, Dr. J. Wayne. *McDonnell Douglas Astronautics Company, St. Louis, MO.* ELECTROPHORESIS (coauthored).

Larimer, Dr. John W. *Center for Meteorite Studies, Arizona State University.* COSMIC THERMOMETERS.

Laurance, Dr. Neal L. *Research Staff, Ford Motor Company, Dearborn, MI.* AUTOMOBILE.

Lazarus, Dr. David. *Lamont-Doherty Geological Observatory, Palisades, NY.* SPECIATION.

Ledley, Dr. Robert S. *Georgetown University Medical Center, National Biomedical Research Foundation, Washington, DC.* COMPUTERIZED TOMOGRAPHY.

Leinen, Dr. Margaret S. *Graduate School of Oceanography, University of Rhode Island, Naragansett Campus, Kingston, RI.* MARINE SEDIMENTS.

Lindsay, Dr. W. L. *Department of Agronomy and Agricultural Sciences, Colorado State University.* SOIL CHEMISTRY.

Loftus, Dr. Elizabeth F. *Department of Psychology, University of Washington, Seattle.* MEMORY (coauthored).

Long, Dr. Stephen I. *Rockwell International Electronics Research Center, Thousand Oaks, CA.* INTEGRATED CIRCUITS.

Lumpkin, Dr. Robert E. *Occidental Research Corporation, Irvine, CA.* OIL SHALE.

M

McCormack, Dr. Donald E. *Soil Survey Interpretation Staff, USDA Soil Conservation Service, Washington, DC.* SOIL.

Mahlberg, Dr. Paul G. *Department of Plant Sciences, Indiana University.* PLANT SECRETORY STRUCTURES.

Mao, Agnes L. *Geophysical Laboratory, Carnegie Institution of Washington.* HYDROGEN (coauthored).

Mao, Dr. Ho-kwang. *Geophysical Laboratory, Carnegie Institution of Washington.* HYDROGEN (coauthored).

Margon, Dr. Bruce. *Department of Astronomy, University of Washington, Seattle.* ASTRONOMY.

Melngailis, Dr. Ivars. *Lincoln Laboratories, Massachusetts Institute of Technology.* LASER (in part).

Merrett, Dr. Michael J. *Postgraduate School of Studies in Biological Sciences, University of Bradford, Yorkshire, England.* CHLOROPLAST.

Meyers, Dr. Robert A. *TRW Defense Space and Systems Group, Redondo Beach, CA.* COAL.

Midwinter, John E. *Optical Communications Systems Division, Post Office Research Centre, Ipswich, East Suffolk, England.* OPTICAL COMMUNICATIONS.

Misfeldt, Dr. Michael L. *Laboratory of Molecular Genetics, National Institutes of Health, Bethesda, MD.* ANTIBODY (coauthored).

Miyano, Dr. K. *Argonne National Laboratory.* LIQUID CRYSTAL.

Mohnen, Dr. V.A. *Atmospheric Sciences Research Center, State University of New York at Albany.* ATMOSPHERIC CHEMISTRY.

Mokhoff, Dr. Nicolas. *IEEE Spectrum, New York.* ELECTRONICS.

Murray, Dr. Edward R. *SRI International, Menlo Park, CA.* AIR-POLLUTION CONTROL.

N

Nuttli, Dr. Otto W. *Department of Earth and Atmospheric Sciences, St. Louis University.* EARTHQUAKE.

O

Olson, Dr. Gerald W. *Department of Agronomy, Cornell University.* ARCHEOLOGY.

Oppenheimer, Dr. J. H. *Department of Medicine, University of Minnesota.* THYROID HORMONE.

Orrego, Dr. Cristian. *Department of Biology, Massachusetts Institute of Technology.* BIOTECHNOLOGY (feature; coauthored).

Osborn, Dr. Jeffrey L. *University of Iowa College of Medicine.* RENIN (coauthored).

Owens, Dr. John N. *Biology Department, University of Victoria, British Columbia, Canada.* PLANT REPRODUCTION.

P

Paetkau, Dr. Verner. *Department of Biochemistry, University of Alberta, Canada.* LYMPHOKINES.

Parsons, Dr. Peter A. *Department of Genetics, La Trobe University, Bundoora, Victoria, Australia.* POPULATION GENETICS.

Parthasarathy, Dr. M. V. *Section of Botany, Genetics, and Development, Cornell University.* PHLOEM.

Patton, Dr. John S. *Department of Microbiology, University of Georgia.* FAT DIGESTION.

Peaslee, Dr. Doyle E. *Department of Agronomy, College of Agriculture, University of Kentucky.* SOIL PHOSPHORUS.

Pecoraro, Dr. V. L. *Department of Chemistry, University of California, Berkeley.* COORDINATION CHEMISTRY (coauthored).

Peterson, Dr. John B. *Laboratory for Applications of Remote Sensing, Purdue University.* AGRONOMISTS AND THE FOOD CHAIN (feature).

Pogust, Frederick B. *Eaton Corporation, AIL Division, Deer Park, NY.* MICROWAVE LANDING SYSTEM.

Postow, Dr. Elliot. *Electromagnetic Radiation, National Naval Medical Center, Bethesda, MD.* BIOELECTROMAGNETICS.

Prinn, Dr. Ronald G. *Department of Meteorology, Massachusetts Institute of Technology.* PLANET.

Proni, Dr. John R. *Atlantic Oceanographic and Meteorological Laboratories, National Oceanic and Atmospheric Administration, Miami, FL.* SEWAGE.

R

Rabinowitz, Dr. Mario. *Electrical Systems Division, Electric Power Research Institute, Palo Alto, CA.* ADVANCED ELECTRIC POWER TRANSMISSION (feature).

Raymond, Dr. K. N. *Department of Chemistry, University of California, Berkeley.* COORDINATION CHEMISTRY (coauthored).

Rayport, Dr. Stephen. *Division of Neurobiology and Behavior, Columbia University College of Physicians & Surgeons.* ANNELIDA.

Reinecke, Dr. John P. *Metabolism and Radiation Research Laboratory, North Dakota State University.* ENDOCRINE SYSTEM (INVERTEBRATE).

Richman, David W. *McDonnell Douglas Astronautics Company, St. Louis, MO.* ELECTROPHORESIS (coauthored).

Rose, Dr. James T. *McDonnell Douglas Astronautics Company, St. Louis, MO.* ELECTROPHORESIS (coauthored).

Rugg, Prof. Barry. *Division of Applied Science, New York University.* ALCOHOL.

Russell, Dr. Anthony P. *Department of Biology, University of Calgary, Alberta, Canada.* LIZARD.

Ryan, Paul Allen. *Dytronics Corporation, Columbus, OH.* STORMSCOPE.

S

Saaty, Dr. Thomas L. *University of Pittsburgh.* GRAPH THEORY.

Sampson, Dr. F. B. *Department of Botany, Victoria University of Wellington, New Zealand.* POLLEN.

Schramm, Prof. David N. *Astronomy and Astrophysics Center, University of Chicago.* COSMOLOGY.

Scott, Glen R. *Department of Psychology, University of Washington, Seattle.* MEMORY (coauthored).

Seppelt, Prof. Konrad. *Institut für anorganische Chemie, Freie Universität Berlin, West Germany.* FLUOROCARBON.

Seward, Dr. Fred. *Harvard College Observatory.* X-RAY ASTRONOMY.

Shrader, Dr. W. D. *Hermann, MO.* EROSION.

Simons, Dr. Elwyn L. *Center for the Study of Primate Biology and History, Duke University.* AEGYPTOPITHECUS AND PROPLIOPITHECUS; APIDIUM AND PARAPITHECUS (both coauthored).

Smibert, Dr. Robert. *Virginia Polytechnic Institute.* CAMPYLOBACTERIOSIS.

Smith, Dr. Ronald H. *NASA Liaison F-18 Program Office, Naval Air Test Center, Patuxent River, MD.* MILITARY AIRCRAFT.

Spencer, Dr. James. *Stanford Linear Accelerator Center, Stanford, CA.* WIGGLERS.

Strell, H. *Research Department, Sperry Corporation, Great Neck, NY.* UNDERWATER NAVIGATION (coauthored).

Swack, Dr. Norman S. *Virology Laboratory, Veterans Administration Medical Center, West Haven, CT.* ANIMAL VIRUS.

T

Tang, Dr. Man-Chung. *DRC Consultants, Inc., New York.* GIRDER.

Tapia, Dr. Richard A. *Department of Mathematical Sciences, Rice University.* NONLINEAR PROGRAMMING.

Thames, Dr. Marc D. *University of Iowa College of Medicine.* RENIN (coauthored).

Traylor, Dr. T. G. *Department of Chemistry, University of California, San Diego.* HEMOPROTEINS.

U

Uehara, Dr. Goro. *Department of Agronomy and Soil Science, College of Tropical Agriculture, University of Hawaii.* AGRICULTURAL ENGINEERING.

Underwood, Dr. Herbert. *Zoology Department, North Carolina State University.* PINEAL GLAND.

V

Van Valkenburg, Dr. Mac E. *Electrical Engineering Department, University of Illinois.* SWITCHED CAPACITORS.

Vook, Dr. Frederick L. *Sandia Laboratories, Organization 5100, Albuquerque, NM.* ANNEALING.

W

Waddington, Dr. Donald V. *Department of Agronomy, College of Agriculture, Pennsylvania State University.* FERTILIZER.

Walcott, Dr. Charles. *Neurobiology Department, State University of New York at Stony Brook.* MAGNETIC SENSE.

Warrick, Professor A. W. *Department of Soils, Water and Engineering, University of Arizona.* SOIL VARIABILITY.

Weiss, Dr. Ronald A. *McDonnell Douglas Astronautics Company, St. Louis, MO.* ELECTROPHORESIS (coauthored).

White, Dr. Samuel. *NASA Ames Research Center, Moffett Field, CA.* HELICOPTER.

White, Dr. Sidney E. *Department of Geology and Mineralogy, Ohio State University.* ALPINE LANDFORMS.

Williams, Prof. Gary A. *Department of Physics, University of California, Los Angeles.* LIQUID HELIUM (in part).

Wilson, Dr. David Sloan. *Division of Environmental Studies, University of California, Davis.* COMMUNITY.

Wolfe, Dr. John. *NASA Ames Research Center, Moffett Field, CA.* SATURN.

Woltz, Dr. Shreve S. *Agricultural Research and Education Center, University of Florida, Bradenton.* NONPARASITIC PATHOGENS OF PLANTS (feature).

Wyn-Jones, Dr. R. G. *Department of Biochemistry, Univeristy College of North Wales, Bangor.* PLANT PHYSIOLOGY (in part).

Y

Yodlowski, Dr. Marilyn L. *Rockefeller University.* PHONORECEPTION.

Z

Zebroski, Dr. Edwin L. *Nuclear Safety Analysis Center, Palo Alto, CA.* NUCLEAR REACTOR SAFETY (feature).

Zeien, Dr. Charles. *J. J. Henry Company, Inc., New York.* MERCHANT SHIP.

Ziegler, Dr. James. *Watson Research Laboratory, IBM, Yorktown Heights, NY.* COMPUTER (coauthored).

Ziemer, Dr. Robert R. *USDA Forest Service, Pacific Southwest Forest and Range Experiment Station, Arcata, CA.* ROOT (BOTANY).

McGRAW-HILL YEARBOOK OF SCIENCE AND TECHNOLOGY

Index

Index

Asterisks indicate page references to article titles.